11-13-57

G.2HC

Analytical Experimental Physics

Analytical Experimental Physics

Second Revised Edition

Michael Ference, Jr.

*Director, Scientific Laboratory, Ford Motor Company; formerly
Associate Professor of Physics
The University of Chicago*

Harvey B. Lemon

*Director, Science and Education, Museum of Science and Industry,
Chicago; Professor Emeritus of Physics
The University of Chicago*

Reginald J. Stephenson

Professor of Physics, The College of Wooster

THE UNIVERSITY OF CHICAGO PRESS

Library of Congress Catalog Number: 55-5124

THE UNIVERSITY OF CHICAGO PRESS, CHICAGO 37
Cambridge University Press, London, N.W. 1, England
The University of Toronto Press, Toronto 5, Canada

© *1943, 1946, and 1956 by The University of Chicago
Second Revised Edition Published 1956. Second Impression
1957. Composed and printed by* THE UNIVERSITY OF
CHICAGO PRESS, *Chicago, Illinois, U.S.A.*

Preface

Of the writing of textbooks of physics there seems to be no end. This will continue as long as science develops—as long as the detailed requirements of a scientific training become more and more extended—and as long as there are teachers who are interested in the efficient pursuit of their vocation. However, the reader is entitled to some brief statement as to why this particular book has been produced. This book is a revision of the original text by Lemon and Ference, first published in 1943 and revised by the same authors in 1946. For the present text a third author, Stephenson, has been included.

A very thorough revision has now been undertaken. Every chapter has been carefully evaluated in the light of the authors' experience and that of many of the users of the earlier editions. Some chapters have undergone relatively minor changes, while others have been almost completely rewritten. The type has been reset, so that the present edition, though containing almost the same number of pages as the original, is physically much smaller.

New topics have been added so that the student can have an introduction to many areas in physics where active research is currently being carried out. As examples of this, the V-2 rocket is given as an example of Newton's Third Law, the transistor as an illustration of a small portion of the research in solid state, and the Oak Ridge reactor as an introduction to the rapidly expanding field of nuclear fission. There is a brief discussion of the radioactive fall-out from the more powerful of the nuclear bombs, so that a student can appreciate some of the awful potentialities of the weapons which science has created.

While these newer developments are interesting and of importance, the fundamental aspects of physics must be known before these can be thoroughly understood. The basic concepts of physics have been discussed in such a manner that it is hoped that the student will gain an insight into physical phenomena.

However, physics is a quantitative science, and without mathematics it could never have developed to its present state. Thus mathematics must be used as a tool for the precise expression of physical phenomena. It is expected that the student will have a good working knowledge of trigonometry and elementary algebra. Relatively little calculus is used, and this is introduced very gradually and is summarized at the end of the book. The student should be taking a first course in calculus concurrently with this material, and thus it should be possible for him to get a good elementary working knowledge of the power of the calculus.

Since physics is quantitative, its results must be expressed in terms of some unit or units. Unfortunately for all of us, there is no one system of units which is universally accepted. Scientists, in general, use the centimeter-gram-second (cgs) and the meter-kilogram-second (mks) systems, while the engineers use the foot-slug-second system. In mechanics all three systems are used in this book, while in electricity and magnetism the rationalized mks system is employed. This latter is recommended by an international commission and is gradually coming into general use. It should be remembered that there is no such thing as the simplest system of units.

In order for the student to test his understanding of physical concepts, the queries at the end of the chapters have been retained, though in some cases these have been modified and added to. Many new quantitative problems have been added, with the answers to the odd-numbered ones being given. The text contains many worked examples illustrating problem-solving techniques. In all, there are over a thousand worked examples, queries, and practice problems.

This book is intended for the serious student who expects to continue in a career in the physical sciences. It is not an easy text but a challenging one. If a student's interest and enthusiasm can be aroused, he will find its study a very stimulating intellectual experience. Several sections have been set in smaller type which can be omitted as de-

v

sired. Possibly many teachers will find it profitable to spend more than the usual year on this book and thus lay a thorough foundation for later and more specialized courses.

It is a pleasure to acknowledge our indebtedness to the many students and colleagues who by their generous criticism have helped in the development of this book. In particular, we should like to thank Professor Franklin Miller, of Kenyon College, for the chapter on music, and his wife, Libuse Lukas Miller, for the abstract drawings for all the chapter headings.

Michael Ference, Jr.
H. B. Lemon
R. J. Stephenson

Table of Contents

Table of Contents

Table of Contents

Table of Contents

xiii

Table of Contents

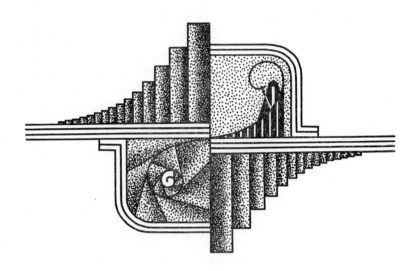

CHAPTER 1

Kinematics and Statics of a Particle

1.1. Introduction

In the study of physics, as in that of all other natural sciences, our aim is to describe, measure, and correlate phenomena in the widely diverse fields of mechanics, heat, sound, light, electricity, and atomic and nuclear physics so that we may become able to *predict* as perfectly as possible what will be the result of any given set of initial circumstances or conditions. Indeed, it might well be said that by our ability to predict unerringly the natural consequences that will follow any arbitrary initial conditions, either natural or artificial, we may judge the degree of our understanding in any field of science.

Although in physics we deal, in the main, with the simpler phenomena and processes in the world of nature, we nevertheless soon discover that many of these phenomena are extremely complicated. Therefore, we have often found it convenient to select various aspects or component parts, as it were, so that each of these may be studied by itself. From the parts we subsequently rebuild the more complex whole. In the study of these single components, each of which represents only a limited portion of the entire truth as we understand it, the student at first may acquire a totally erroneous idea that physics is a rather impractical or "theoretical" subject, dealing with many situations that cannot be realized in nature and which are at least quite imaginary, if not actually untrue. Any breadth of experience, of course, reminds us that

this particular method of study, followed in the fields of science and engineering, within the last century or so, has completely altered the life and surroundings of by far the greatest part of the populations of the world. With this preliminary comment on our method of attack, let us begin by studying motion.

1.2. Motion: Position and Time

The idea "motion" contains two primitive concepts, those of position and of time. In the present section we shall be concerned with the practical determination of position by measurement of lengths. In a following section we shall discuss units of time. These ideas will then be combined (§ 1.11) into a detailed study of motion.

If we are to specify the position of a point on a line, then we must agree on an origin, or zero point, on the line, together with a unit length or measuring rod and a sense of positive direction along the line. For locating a point in a plane we usually use the two Cartesian coordinate axes as the reference system and give the x- and y-coordinates of the point. If we are to specify uniquely the *position* of a body in space, we *may* do this by noting its distance from three mutually perpendicular planes, which we shall take as a fixed reference system.

Strictly speaking, we must take as our fundamental or primary reference system one that is fixed relative to the average motion of the stars.

We assume that the observer or experimenter who describes all the physical phenomena in this text is at rest in this frame of reference. However, calculations show that, for most phenomena that occur in the laboratory, we may take as a fixed reference system the lines of intersection of two adjacent walls and the floor of the laboratory without introducing serious error in our mode of description. In other words, we shall neglect whatever effects the rotation of the earth and its translational motion may have on our description of physical phenomena.*

The position of an object in the room, then, is found by measuring its distance from the walls and the floor in terms of some UNIT OF LENGTH. A little reflection on the part of the reader will satisfy him that the location of an object is always made by determining its position with respect to an environment which is considered to be fixed.

1.3. Measurement: Units of Length

Primitive men took a great step forward toward the creation of the sciences when they recognized that a distance could be determined in terms of some unit of length. History tells us that the first units were closely related to the trades. For example, the navigator used fathoms; the surveyor, rods; the weaver, ells; the carpenter, feet.

In modern science the unit of length universally adopted is the METER, which is defined as the distance between two marks on a platinum-iridium rod, known as the "International Prototype Meter," preserved at the International Bureau of Weights and Measures at Sèvres, France. A copy of this standard, known as "Meter No. 27," was delivered to President Benjamin Harrison on January 2, 1890, and provides the official standard of length in the United States. The meter was intended to be the ten-millionth part of the earth's quadrant, but the original measurements were not too accurate, so it is not quite that. The common subdivisions of the meter are the CENTIMETER (cm), which is $\frac{1}{100}$ of a meter, and the MILLIMETER (mm), which is $\frac{1}{1000}$ of a meter.

In 1948 the International Conference on Weights and Measures felt that the standard meter should also be given in terms of light waves

which can be accurately reproduced in any suitably equipped laboratory anywhere in the world. This conference indorsed the idea that the meter be given as 1,831,249.2 wave lengths of the green line of the mercury isotope of atomic mass number 198 whose wave length is 5460.7532×10^{-8} cm.

The BRITISH YARD is defined as 3,600/3,937.014 meter. The basis of the common commercial units of length in this country, owing to our origin as a British colony, is the U.S. YARD, slightly different from the British one, being 3,600/3,937 m. This is subdivided into 3 FEET, each of which contains 12 INCHES. From these relationships it follows that 1 yard = 0.914401829 . . . m and 1 inch = 2.54000508 . . . cm. For industrial purposes and for the majority of scientific purposes a relation between the yard and the meter has been adopted by the American Standards Association and similar organizations in other countries. The relation is 1 inch = 2.54 cm (exactly), from which it follows that 1 yard = 0.9144 m (exactly). It should be noted that the meter is the primary standard, while the yard is a derived or secondary standard.

1.4. Practical Measurement of Length

The practical measurement of length depends upon the repeated applications of foot rules or meter sticks. The accuracy with which we can measure a length depends, among other factors, upon our ability to estimate fractions of the smallest divisions of the scale. In order to facilitate such measurements, a *vernier scale* is generally attached to the measuring device.

The principle of the vernier calipers is illustrated in Fig. 1-1 (*a*). The scale, S, of the calipers is divided into millimeters (mm). The auxiliary or vernier scale, V, has 10 divisions that are equal in total length to 9 scale divisions of S. Each vernier division, then, is $\frac{1}{10}$ mm smaller than the S scale division. With the jaws of the caliper closed, the zero divisions of the V and S scales coincide. With the jaws opened $\frac{1}{10}$ mm, the first vernier division coincides with the 1-mm mark of the S scale. If the jaws are opened $\frac{5}{10}$ mm, the fifth vernier division matches an S scale division. When the jaws are opened a distance greater than 1 mm, the S scale is read up to the zero division of the vernier. To this is added the fraction of a division as read by the vernier. In Fig. 1-1 (*b*), the diameter of the ball is 1.05 cm.

---

* We may not neglect these effects in describing such large-scale phenomena as the circulation of air about the surface of the earth or the path of large projectiles (§ 6.18).

Situations often arise, however, where a direct application of the standard unit measuring rod to the distance to be measured is impracticable, if not impossible—e.g., the measurement of the height of an airplane, the width of a river, or the distance between two far-distant objects on the surface of

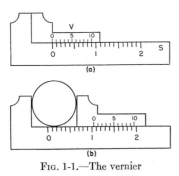

(a)

(b)

Fig. 1-1.—The vernier

the earth. In these or similar cases an indirect method of measurement must be used. For example, let us suppose we wish to measure the distance CD (Fig. 1-2) across a river. We first con-

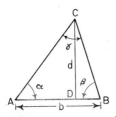

Fig. 1-2.—Indirect measurement of a length

struct, perpendicular to CD, a base line AB of length b. At positions A and B, instruments known as "transits"* are set to sight on the same object C and to make accurate measurement of angles, and the magnitudes of the angles α and β are determined. From triangle ADC we observe that $d = AC \sin \alpha$. Applying the law of sines of trigonometry to triangle ABC, we have

$$\frac{AB}{\sin \gamma} = \frac{AC}{\sin \beta}$$

or

$$\frac{b}{\sin \{ 180 - (\alpha + \beta) \}} = \frac{AC}{\sin \beta},$$

whence

$$AC = \frac{b \sin \beta}{\sin \{ 180 - (\alpha + \beta) \}}.$$

* A transit consists of a telescope attached to a graduated circular scale and can be used to measure angles in either a horizontal or a vertical plane.

Substitution for AC gives

$$d = \frac{b \sin \alpha \sin \beta}{\sin \{ 180 - (\alpha + \beta) \}} = \frac{b \sin \alpha \sin \beta}{\sin (\alpha + \beta)}.$$

Thus we have the distance across the river, d, in terms of the measured base line b and the measured angles α and β.

1.5. Units of Time

Our common and entirely subjective sense of the passage of time offers us no means of making a precise measurement of it. "Time flies" when we are actively and interestedly occupied—it crawls when we are idle and bored. Any attempt to define what we mean by "time" or by "an interval of time" in terms of subjective phenomena leads us into a maze of philosophic difficulties.

Measurement of time is based invariably upon some form of uniformly repeated or periodic motion. Of the many periodic motions that occur in nature, we conventionally adopt that of our spinning earth. Its motion as reflected in the apparent motions of the stars and of the sun make it a readily available and, as we soon discover, a very constant timekeeper.

Astronomical observations lead directly to the definition of the *sidereal day*, which is defined as the time between two successive passages of the same fixed star across the meridian of the place of observation. Similarly, a *true solar day* is defined as the interval of time between two successive passages of the sun across the meridian of the observer. Since the sun moves eastward with respect to the stars at a slightly varying rate, owing to the elliptical orbit of the earth, it follows that true solar days are longer than the sidereal day but are not all of the same duration. In scientific work and for the ordinary purposes of life the *mean solar day* is used. This is defined as the average of all the true solar days in a year. Our fundamental unit of time is the mean solar second, which is 1/86,400 part of the mean solar day. Since the number of sidereal and mean solar days in the year (tropical) are 366.2422 and 365.2422, respectively, the length of the mean solar second is thereby precisely determined by transit observations on the stars, which directly measure the sidereal day.

In the laboratory we are usually more interested in the measurement of *time intervals* than we are in the knowledge of the exact time. For this purpose we use pendulums swinging in the gravitational field of the earth (pendulum clocks), vibrations of coiled springs (chronometers and stop watches), electric motors (electric clocks) that operate synchronously on alternating currents of standardized frequency, and, recently, quartz crystals.

The International Conference on Weights and Measures in 1948 suggested that time be measured also in terms of the frequency of vibration of the ammonia molecule. This has become known as the "atomic clock" and, like the similar indorsement for the meter, can be accurately reproduced in any suitably equipped laboratory anywhere in the world.

Although the measurement of time based on the period of the earth's rotation as the fundamental unit has been, from a practical point of view, entirely satisfactory, nevertheless theoretical objections of a fundamental nature may be raised to this definition. For example, the foregoing discussion of time suggests the Newtonian concept of absolute time "that flows uniformly, regardless of the existence of any thinking being or any phenomenon or process suitable for measuring it." A. Einstein and others have shown that this concept of absolute time cannot stand critical examination. Consequently, an attempt has been made to use as a unit of time the interval required for a beam of light to travel the length of the standard meter and back again. This is based on the constancy of the speed of light.

1.6. Displacement: Vectors and Scalars

If a simple statement is made that a ship sailed 4 miles, the reader will recognize that this statement contains an element of ambiguity, for the ship could sail 4 miles either to the north, south, east, or west, or in any other direction from its starting place. However, a statement that a ship sailed 4 miles to the east of its starting place is quite explicit. Such a change in position (without any reference to the time involved) is known as a *displacement.*

In order to specify a displacement, it is, of course, necessary to give the magnitude of the displacement in terms of an appropriate unit and the direction of the displacement with reference to some standard direction. We may conveniently represent the displacement by a directed line segment, an arrow (———➤) whose length represents

the magnitude and whose direction, as indicated by the head of the arrow, is that of the displacement.

Let us pursue this train of thought a little further. If the ship, after sailing 4 miles to the east from a point A (AB, Fig. 1-3), now goes 3 miles to

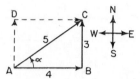

Fig. 1-3.—Addition of two displacements at right angles

the north, we find it at the point C. What is the *resultant* effect of the two displacements AB and BC? From the diagram the answer is the single displacement AC, whose magnitude we find is 5 miles (by counting the miles or using the Pythagorean theorem) and whose direction is determined by the angle α. We have added the displacement BC to the displacement AB and obtained the resultant displacement AC.

Were the ship to sail the 3 miles to the north first (AD), then 4 miles to the east (DC), it would arrive at the same position, C, with the same resultant displacement, AC. We note that the quadrilateral of Fig. 1-3 ($ABCD$) is a parallelogram, since AD is parallel to BC and BA is parallel to CD, and, further, that what we called the "resultant" is the diagonal of this parallelogram.

The discussion we have given here serves to introduce a very important and useful concept—that of the *VECTOR* quantity. Displacement is only one example of a vector quantity. A vector quantity is to be taken in sharp contrast with a *SCALAR* quantity. The latter requires only a *magnitude* for its specification. For example, no question of ambiguity arises when we state that the density of water is 1 gram per cubic centimeter (gm/cm^3) or that the volume of a tank is 60 gallons. On the other hand, such a phrase as "displacement of 1 meter" is incomplete until a direction is specified. Other common examples of vector quantities that we shall soon meet are velocity, acceleration, and force.

It is important to realize that the directed line segments of Fig. 1-3 may equally well represent vector quantities *other than* displacements.

1.7. Formal Definition of a Vector Quantity

From our discussion of displacements we are now in a position to define a vector quantity.

A vector is a directed line segment (1) which is used to represent various physical quantities having magnitude and direction and (2) which combines with similar directed line segments in accordance with the parallelogram law of (geometrical) addition so that (3) the resulting vector likewise represents the combined effects or the sum of the quantities represented by the original vectors.

The first two parts of the definition imply that only *like vector quantities* may be added. We may add velocities to velocities, displacements to displacements, but not a velocity to a displacement. The second part of the definition tells us how to combine or add vectors. The third part of the definition enables us to interpret the sum of the line segments as equivalent to the sum of the original physical quantities.

As we have seen in Fig. 1-3, addition of vectors is a geometric, rather than an arithmetic, process. If a vector 3 be added to a vector 4, the result is not necessarily a vector 7. Actually, the result may be anywhere between 1 and 7, as indicated in Fig. 1-4.

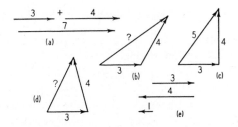

Fig. 1-4.—Addition of two vectors at various angles

In cases (*a*) and (*e*) in Fig. 1-4 the final result is given by the arithmetic sum and the difference, respectively, of the two vectors. The result in case (*c*) is given by the Pythagorean proposition that the square of the hypotenuse of a right triangle equals the sum of the squares of the other two sides. Cases (*b*) and (*d*) require further discussion.

1.8. Addition of Vectors

If two or more vectors are given, their vector sum or resultant may be determined in a variety of ways. We shall describe three useful methods: the *polygon method*, the *parallelogram method*, and the *method of components*. The numerical answer for the resultant may be obtained either *graphically*, using ruler and protractor, or *algebraically*, given sufficient data.

a) Polygon method.—In Fig. 1-5 two vectors are

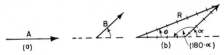

Fig. 1-5.—Addition by the vector triangle

given: *A*, a horizontal vector, and *B*, a vector making an angle α with the horizontal (Fig. 1-5 [*a*]). We add vector *B* to *A* by joining *B* to *A*, *tail to head*, as shown in Fig. 1-5 (*b*). The *vector sum*, *R*, is obtained by connecting the *tail of A* to the *head of B*. This resultant is completely specified if we know its magnitude and the direction it makes with some reference direction such as *A*, i.e., the angle θ. By means of a ruler we may measure *R* in the same units chosen for *A* and *B*; the angle θ may be measured with a protractor. This graphic solution is always possible, though its accuracy is limited by the accuracy with which the drawings can be made and measured.

An analytical method for obtaining *R*, given the magnitudes and directions of *A* and *B*, makes use of the cosine law of trigonometry. Applications of this law to Fig. 1-5 (*b*) gives us

$$R^2 = A^2 + B^2 - 2 AB \cos (180 - \alpha)$$

or

$$R^2 = A^2 + B^2 + 2 AB \cos \alpha . \qquad (1\text{-}1)$$

In equation (1-1) the student will note that α is the angle between the vector *B* and the prolongation of *A*. The angle θ may be obtained by another application of the cosine law:

$$B^2 = A^2 + R^2 - 2RA \cos \theta ,$$

$$\cos \theta = \frac{A^2 + R^2 - B^2}{2RA}. \qquad (1\text{-}2)$$

The method we have just described is capable of an obvious extension to many vectors.

The four vectors, *A, B, C, D*, shown in Fig. 1-6 (*a*) are to be added. Some one vector, such as *A*, is chosen as a reference vector; the others are added successively, tail to head (Fig. 1-6 [*b*]). The arrow from the tail of *A*, the first vector used, to the head of *D*, the last vector used, represents the vector sum or resultant, *R*. The figure is a closed polygon; hence the origin of the name *polygon*

5

method. The student will note that Fig. 1-5 is but a special case of Fig. 1-6. A numerical answer for the resultant may be found graphically, or algebraically by successive use of equations (1-1) and (1-2). By dividing the polygon into three triangles, the resultant R_1 of A and B may be computed and then added to C to obtain R_2. The resultant R_2 is added to D to give the final resultant, R.

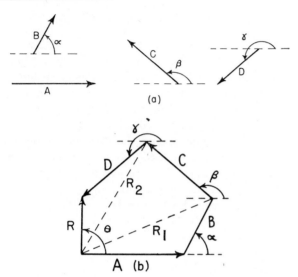

FIG. 1-6.—Polygon method for the addition of vectors

b) *Parallelogram method.*—The reader must learn that vectors can be moved anywhere, provided that their directions are not altered.*

In the parallelogram method, vectors A and B, instead of being added "tail to head," are *brought to a common origin*, and then a parallelogram is constructed, as shown in Fig. 1-7. The RESULTANT is

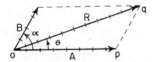

FIG. 1-7.—Parallelogram method for the addition of vectors

the diagonal of the parallelogram drawn from the common origin of A and B. Again a numerical value of R can be obtained either graphically or by employing the cosine law. Using the triangle (*opq*), the reader may show that R and θ are given by equations (1-1) and (1-2).

EXAMPLE 1: In Fig. 1–5, A is equal to three units; B, four units; and $\alpha = 60°$. Find the resultant.

* See chap. 5 for a modification of this statement.

Solution: From equation (1-1), the cosine law,

$$R^2 = 3^2 + 4^2 + 2\,(3)\,(4)\cos 60° = 37\ ,$$

$$R = \sqrt{37} = 6.08 \text{ units }.$$

The angle θ between the resultant and the vector A is obtained by a reapplication of the cosine law (eq. [1-2]):

$$\cos\theta = \frac{3^2 + 37 - 4^2}{2\,(3)\,(\sqrt{37})} = \frac{5}{\sqrt{37}}\ ,$$

$$\theta = 34°43'\ .$$

EXAMPLE 2: Add vectors of magnitude 12 and 10 making an angle of $\alpha = 49°27'30''$ with each other.

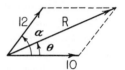

Solution:

1. $\cos 49°27'30'' = 0.65$.

From the law of cosines, equation (1-1), it follows that

2. $R^2 = 100 + 144 + 2\,(10)\,(12)\,(0.65)$,
   $R = 20$.

3. By a reapplication of the law of cosines, equation (1-2), it follows that

$$\cos\theta = \frac{400 + 100 - 144}{2 \times 20 \times 10}\ ,$$

$$\theta = 27°10'\ .$$

EXAMPLE 3: Add vectors of 10 and 12 making an angle $\alpha = 126°52'$ with each other.

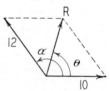

Solution:

1. $\cos 126°52' = \sin(90° - 126°52')$
   $= -\sin 36°52' = -0.6$.

2. $R = \sqrt{100 + 144 + 2\,(10)\,(12)\,(-0.6)} = 10$.

3. $\cos\theta = \dfrac{100 + 100 - 144}{2\,(10)\,(10)} = \dfrac{28}{100}$,

$$\theta = 73°43'\ .$$

It should be recognized that the resultant R and the two vectors do not coexist but that the resultant replaces or is equivalent to the two vectors.

c) *Method of components.*—The *method of com-*

ponents, in general the most useful method, especially where many vectors are involved, will be treated fully in the next section.

1.9. Method of Components

Since in Fig. 1-5 the two vectors A and B, taken together, are equivalent to the single vector, R, called their RESULTANT, we may consider the original single vector, R, as being replaced by the two equivalent vectors A and B. This process is called the *method of components*. Thus, the COMPONENTS *of a vector are those vectors which, when added together, give the original vector.*

A given vector may have *any number* of components, as shown in Fig. 1-8. In (*a*), vector A is

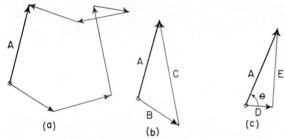

FIG. 1-8.—Components of vectors

resolved into six components. In (*b*), vector A has two components, B and C.

Of the many possibilities, certain combinations of components are quite useful. The most common and useful case is that in which the original vector is resolved into two components at right angles to each other—D and E of Fig. 1-8 (*c*). If θ is the angle between the vectors D and A, then

$$D = A \cos \theta , \qquad E = A \sin \theta . \qquad (1\text{-}3)$$

It is to be noted that equations (1-3) are simply the defining equations of the cosine and sine functions, respectively.

Any vector A may be resolved into a vector along a particular direction, OO', and one at right angles to this direction, PP', as shown in Fig. 1-9.

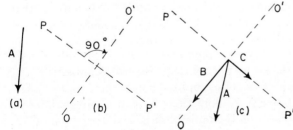

FIG. 1-9.—Components at right angles

The components of A along OO' and PP' are B and C.

To add a number of vectors, A, B, C (Fig. 1-10 [*a*]), by the method of components, we pro-

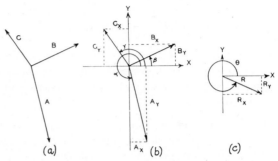

FIG. 1-10.—Addition of vectors by the method of components

ceed as follows: First superimpose upon them a rectangular system of coordinates (Fig. 1-10 [*b*]). It is immaterial how the *x*- and *y*-axes are oriented, but in many cases it is convenient to take the *x*-axis along one of the vectors. However, we choose the coordinate axes as in Fig. 1-10 (*b*). Next resolve each vector into components parallel to these axes by dropping a perpendicular from the vector head to the axis in question.

By adding all the components along the *x*-axis (A_x, B_x, etc.), we obtain ΣX, the symbol Σ meaning "sum." Adding the components along the *y*-axis gives us ΣY:

$$\left. \begin{aligned} \Sigma X = R_x = A_x + B_x + C_x , \\ \Sigma Y = R_y = A_y + B_y + C_y . \end{aligned} \right\} \qquad (1\text{-}4)$$

Now some of the components are *intrinsically negative*, e.g., C_x and A_y, since they point in the direction along their respective axes that is conventionally adopted to be negative. The resultant, R, is obtained from equation (1-4) as follows: The Pythagorean theorem gives

$$R^2 = (\Sigma X)^2 + (\Sigma Y)^2 , \qquad (1\text{-}5) *$$

while the direction θ, which is made by R measured counterclockwise from the (+)*x*-axis, is given by

$$\tan \theta = \frac{\Sigma Y}{\Sigma X}; \qquad \sin \theta = \frac{\Sigma Y}{R}. \qquad (1\text{-}6)$$

* The student will observe that the vectors A, B, C are all in one plane, are *coplanar*. Similar results may be obtained if the vectors are not in one plane. It would be necessary to introduce the third coordinate axis, z, at right angles to x and y and obtain the third components, A_z, B_z, and C_z. Then to eq. (1-4) we would add $\Sigma Z = R_z = A_z + B_z + C_z$ and replace eq. (1-5) with $R^2 = (\Sigma X)^2 + (\Sigma Y)^2 + (\Sigma Z)^2$.

If α, β, γ be the angles which the vectors A, B, C make with the $(+)x$-axis, then we see from Fig. 1-10 that

$$\left.\begin{array}{ll} A_x = A \cos \alpha , & A_y = A \sin \alpha , \\ B_x = B \cos \beta , & B_y = B \sin \beta , \\ C_x = C \cos \gamma , & C_y = C \sin \gamma . \end{array}\right\} \quad (1\text{-}7)$$

Note here that the intrinsic sign of the component, whether $(+)$ or $(-)$, *is determined by the sign of the trigonometric function*, as well as by inspection of the diagram.

EXAMPLE 4: Add the three given vectors, having magnitudes 2, 10, and 13 and angles given by $\alpha = 0$, $\sin \beta = \frac{4}{5}$, $\cos \beta = \frac{3}{5}$, $\sin \gamma = -\frac{5}{13}$, and $\cos \gamma = -\frac{12}{13}$, respectively.

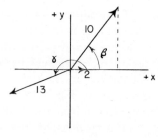

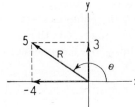

Solution:

1. $\Sigma X = 2 + 6 - 12 = -4$.

2. $\Sigma Y = 0 + 8 - 5 = 3$.

3. $R^2 = 4^2 + 3^2 = 25$,

   $R = 5$.

4. $\tan \theta = -\frac{3}{4}$; $\quad \sin \theta = \frac{3}{5}$,

   $\theta = 143°8'$.

1.10. Subtraction of Vectors

To accomplish subtraction of vectors, we must first define what we mean by a "negative vector." If we call the vector (B) positive and represent it as ($\xrightarrow{\;B\;}$), then, by definition, $(-B)$ is represented by a vector in the same line and of the same magnitude but oppositely directed ($\xleftarrow{\;-B\;}$).

Again, if it happens that a positive vector $B(+B)$ is represented as ($\xleftarrow{\;B\;}$), then $(-B)$ is ($\xrightarrow{\;-B\;}$). In other words, given a vector, the negative of the vector is represented by an arrow whose direction has been changed by 180°.

Subtraction of two vectors is treated as the addition of the negative of the vector that is to be subtracted. To illustrate, in Fig. 1-11 (a) we wish to subtract B

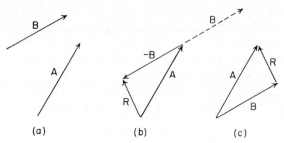

(a) (b) (c)

FIG. 1-11.—Subtraction of two vectors

from A and to call the difference R,

$$R = A + (-B) = A - B .$$

Drawing a vector equal to B but opposite in direction, we add this to A and find R (Fig. 1-11 [b]). As in any other difference, we see that the vector difference R is the vector that must be added to the subtrahend, B, to make A, the minuend (Fig. 1-11 [c]):

$$B + R = A .$$

EXAMPLE 5: Suppose we wish to subtract a vector displacement of 3 miles south from a displacement of 4 miles east. This is equivalent to adding the two vectors 3 miles north and 4 miles east. The resultant R is 5 miles in a northeasterly direction given by the angle α with the east.

1.11. Speed and Velocity

A clear understanding of the definitions and simple operations on vectors, outlined above, is necessary for graphical representations of motion. Our sense-perception provides us with a working definition of the word "motion." When we see that an object occupies different positions in space at different times, we say that "the object is in motion." For example, we observe a ball rolling along the table. We say that "the ball is in motion relative to an observer fixed with respect to the table," because we note that the ball occupies different places on the table in successive time intervals.

Motion is called "motion of translation," or

simply "translational," if the axes of a coordinate system imagined rigidly attached to the object always remain parallel to the axes of some fixed coordinate system. In Fig. 1-12, we illustrate the

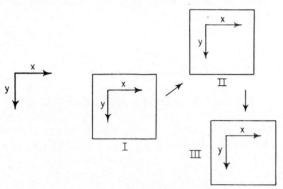

FIG. 1-12.—Translational motion

translational motion of a plane object from position *I* to *II* to *III*. Notice that this implies no rotation of the object.

By the word *speed* we shall mean distance traversed per unit time by some object or, simply, the time rate of change of distance.

An object is said to be moving with UNIFORM speed when equal distances are covered in equal intervals of time. Speed alone, however, does not suffice *completely* to describe this motion. A uniform speed along a straight line is an entirely different behavior from uniform speed in a circle. To describe uniform motion *adequately*, its *direction, as well as its speed*, must be specified. The SPEEDOMETER on a motorcar is correctly named. It indicates how rapidly you are proceeding but, giving no indication of direction, tells you nothing about where you are going.

We say that the VELOCITY of an object, particle, or point is therefore uniquely described *only if both speed and direction are specified.* Velocity is displacement in unit time and is therefore a vector divided by a scalar, that is, velocity is a vector. Hence velocity obeys the same rules as displacement and satisfies the three conditions of the definition of a vector (§ 1.7). Five miles per hour is a speed, but 5 miles per hour north is a vector.

1.12. Uniform Motion

If we represent by a graph the distance covered by a particle moving with a uniform velocity, i.e., constant speed in a straight line, this will resemble

Fig. 1-13. Arbitrary "starting points" from which to measure both distance and time are chosen. If you were s_0 miles from some chosen place when you looked at your watch, the time on which was t_0, then some later time, t, would find you s miles

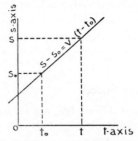

FIG. 1-13.—Uniform velocity with displacement s_0 at time t_0

away from the chosen place. By definition, then (the motion being rectilinear and the speed assumed constant), the velocity is

$$v = \frac{s - s_0}{t - t_0} \quad \text{or} \quad s - s_0 = v\,(t - t_0) \,. \quad (1\text{-}8)$$

This expression is the EQUATION OF THE GRAPH of Fig. 1-13.

Use of a stop watch in this situation would have enabled you to set the index hand at zero when you began to measure distance. Thus $t_0 = 0$ when $s = s_0$. (The reader should construct the graph for this case.)

On a measured track you might also have begun to measure distance and time simultaneously. Then $s_0 = 0$ when $t = 0$. It follows that expression (1-8) may be simplified by suitable discretion in making the measurements; thus, for constant speed in a straight line,

$$v = \frac{s}{t} \quad \text{or} \quad s = vt \,. \quad (1\text{-}9)$$

The latter relation is the equation of the graph in Fig. 1-14.

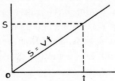

FIG. 1-14.—Uniform velocity with zero displacement at time zero.

1.13. Unit of Velocity

There is no special name that has persisted in scientific usage for the unit of velocity (The name

9

"kine" was once proposed.) In fact, very few physical units have special names, which is perhaps a good thing. Consequently, we speak of velocities in terms of the units of length and time that are involved in our measurement of the velocity, i.e., 1 mile per minute (mi/min), or its equivalents; 60 miles per hour (mi/hr); 88 feet per second (ft/sec); 2,682.24 centimeters per second (cm/sec); etc.

1.14. Average Speed

When speed is variable, we often make use of the idea of AVERAGE SPEED. By "average speed" we mean the *total* distance covered divided by the *total* time elapsed. The average value so obtained is, of course, the same as that of a uniform speed that would cover the same total distance in the same total time. In this modern age of mixed pedestrian and vehicular traffic, controlled in large part by stop-and-go lights, the distinction between one's average rate of progress between places and one's actual variable speed (at times zero) is furnished with abundant illustration. Using a capital letter to refer to total distance, average speed (v_{av}) has, for its defining equation,

$$v_{av} = \frac{S}{t}. \qquad (1\text{-}10)$$

EXAMPLE 6: From New York to Chicago is about 1,000 miles. If a train makes this run in a total elapsed time of 20 hours (hr), what is its average speed?

Solution:

$$v_{av} = \frac{1,000\,\text{mi}}{20\,\text{hr}} = 50\,\text{mi/hr}.$$

The actual train had to make some stops—to change crews, for example—hence was compelled to go much faster than 50 mi/hr at times. The problem states only that, had the train started at New York at a speed of 50 mi/hr and traveled the 1,000 miles with this constant speed, the train would have taken exactly the same length of time to reach Chicago as did the actual train going at various speeds. In other words, an average speed is equivalent to a constant speed.

1.15. Velocity as a Vector: Independence of Velocities

If an object is subject to two independent velocities, V_1 and V_2, it will move in such a direction that its velocity, V, at each instant is the vector sum of the two independent velocities. This result-

ant velocity may be obtained by means of the parallelogram law, as shown in Fig. 1-15.

As a practical illustration, consider a man who can row a boat with a velocity V_1 relative to calm water. Suppose, now, that the man rows in a

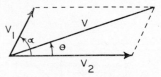

FIG. 1-15.—Resultant of two independent velocities

stream at an angle a with the direction of flow. The velocity of flow is V_2 (Fig. 1-16). All measure-

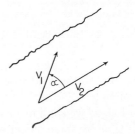

FIG. 1-16.—Man rowing across a river

ments are made by an observer on the bank of the river. The velocity of the man, V, with respect to the bank is found by taking the vector sum of V_1 and V_2 as shown in Fig. 1-15. Using equation (1-1), we obtain

$$V = \sqrt{V_1^2 + V_2^2 + 2\,V_1 V_2 \cos a}.$$

To find θ, the angle which the man's course makes with the river's bank, we have, identifying A in Fig. 1-5 (*b*) with V_2 and applying equation (1-2),

$$\cos\theta = \frac{V_2^2 + V^2 - V_1^2}{2\,V V_2}.$$

The vector V, then, is equivalent to V_1 and V_2.

Any velocity V may be resolved into two components, V_1 and V_2, or, indeed, into any other two (or more, if convenient) components, which can be combined in any order to represent the actual velocity; and, furthermore, each component may be treated for analysis independently of the others. One cannot *deduce* from any previous reasoning that such ought to be the case with respect to velocities (and displacements, accelerations, forces, etc.). The *principle of independence of velocities*, as this statement is sometimes called, enables us to analyze such motions as the flight of projectiles

(§ 1.23) and gyroscopic motion (§ 6.17) by simpler methods than would otherwise be possible. The principle of independence of velocities is merely a special case of a general principle of superposition.

An interesting experiment concerning the independence of velocities is given here. In the apparatus shown in Fig. 1-17 (*a*), a steel ball *B* hangs from a magnet. Out of view, beyond the left edge of the picture, is a device that projects a second steel ball, *A*. The apparatus is so adjusted that the ball *A* is always aimed at *B*. *At the instant* A *is projected, a contact is broken, and* B, *released from the magnet, falls freely.* The following pictures are taken with a motion-picture camera. In frame (*b*)

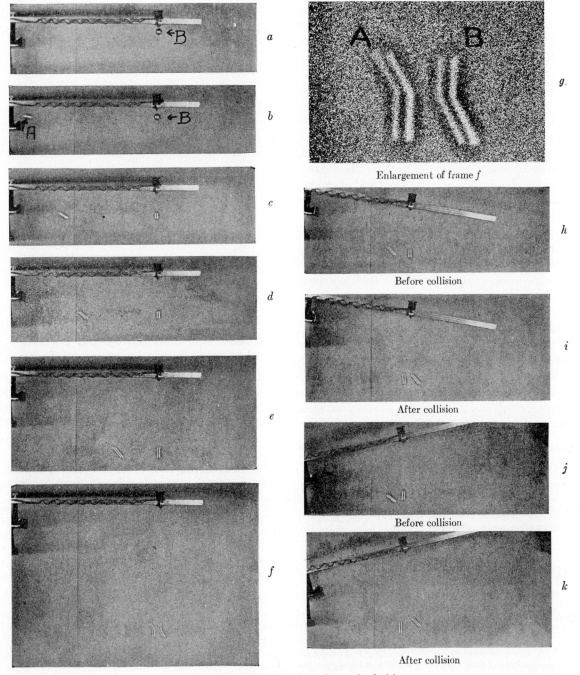

a

b

c

d

e

f

g

Enlargement of frame *f*

h

Before collision

i

After collision

j

Before collision

k

After collision

Fɪɢ. 1-17.—Experiment on independence of velocities

this has just happened—ball A is coming in from the left, and B has just started to fall. It is an experimental fact, verified by the sequence, that ball A, having been initially aimed at B, will ALWAYS HIT B. Both balls, upon being released, accelerate under the action of gravity, A falling from the projecting device and B from the magnet. Thus the vertical components of their motion are identical. The fact, astonishing to many, that the horizontal velocity imparted to A *in no way affects* its vertical velocity or the gain in the latter, cannot be derived from any previous reasoning. It is a fundamental fact of nature.

The INDEPENDENCE OF VELOCITIES, simultaneously or consecutively acquired, which enables one to treat each velocity as if the others were not present, is but a special case of the general principle of superposition, the validity of which is demonstrated by many aspects of natural phenomena. Pictures (*f*) and (*g*), on which the trail* of the two balls DURING THE IMPACT happened to be recorded and of which (*g*) is merely an enlargement of (*f*), are very revealing. It shows that, after collision, A takes up B's path and that B likewise follows A's path. When the balls are highly elastic and equal in mass (as here), it can be shown that, upon collision, they exchange velocities. Here the fact is clearly "seen" by the camera, although, of course, invisible to the eye. It is *not* necessary that ball A be projected horizontally. The stick which supports the magnet and projector may make *any angle* with the horizontal, as seen in the two pairs of frames ([h], [j]). In each the balls are shown just before ([h], [j]) and just after ([i], [k]) they have collided.

Queries

1. Would the balls collide if the stick with the magnet and projector were pointing directly upward? Downward?

2. What would be the result if the balls shown in Fig. 1-17 were made of putty?

1.16. Speed at a Point

Suppose a particle moves along the positive X-axis so that its position is x at time t and $x + \Delta x$ at a later time $t + \Delta t$, as shown in Fig. 1-18. In

* The trail drawn out by the motion of each ball during the 0.01-sec duration of each exposure is the path of highlights produced by the photoflood lamps used to illuminate the device.

this the symbol Δ, called "delta," means "a small change in," so that Δx is a small change in the x position of the particle. From § 1.14 it follows that the average speed v_{av} of the particle over the distance Δx is

$$v_{av} = \frac{\Delta x}{\Delta t}.$$

In order to find the speed at position x or at time t, we must allow the intervals to become smaller and smaller and find the limiting value of

FIG. 1-18.—Average and instantaneous speeds

this quotient. This limiting value or instantaneous speed is

$$v = \lim_{\Delta t \to 0} \frac{\Delta x}{\Delta t}$$

and is called the derivative of x with respect to t, dx/dt. Hence the speed of the particle at x and t is

$$v = \frac{dx}{dt} = \dot{x}. \qquad (1\text{-}11)\,\dagger$$

It is this speed which is indicated by a speedometer on an automobile.

1.17. Relative Velocity

We have all noticed that when we are traveling in a car and another one overtakes and passes us, it appears that the passing car has a velocity equal to the difference between the velocities of the two cars. In making this observation we are considering ourselves at rest, and the relative velocity of the passing car is then the difference between the two velocities.

Suppose an object A is moving with a velocity v_A and another object B with a velocity v_B, both

† Historically, the first men to see clearly the need of grappling with the subtleties of such questions as the meaning of the velocity at a point and other precisely similar notions were Isaac Newton and Gottfried Wilhelm Leibniz. The former used the method we have just outlined to assist him in obtaining the solutions to many problems worked out in his great *Principia*, published in 1687. Leibniz published an account of this method of approach to a continually changing quantity in 1684. Both men are accorded independently the credit for the invention of what was later to be known as the *differential calculus*. In eq. (1-11) the form to the left is expressed in the notation of Leibniz; that to the right, in the notation used by Newton. While both notations are still used, that of Leibniz is the one preferred.

velocities being measured relative to the ground (Fig. 1-19). Then the relative velocity of B with respect to A is obtained by considering A at rest, that is, by giving to B an equal but opposite velocity to that of A. The relative velocity, v, of B with

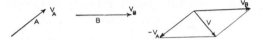

FIG. 1-19.—Relative velocity of B with respect to A

respect to A is given by the diagonal of the parallelogram having sides of v_B and $-v_A$.

EXAMPLE 7: Suppose ship A is moving northeast with a velocity of 20 mi/hr and ship B is moving north with a velocity of 15 mi/hr. The problem is to find the velocity of B relative to that of A.

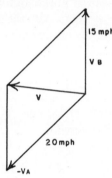

Solution: Here ship A is considered to be at rest, and hence ship B has a relative velocity v which is the resultant of two velocities, one of 20 mi/hr southwest and its own of 15 mi/hr north.

1.18. Acceleration

Motion in which the velocity changes from point to point is called "accelerated motion." For example, a motorcar speeds up from 30 to 60 mi/hr. We say that the car has been "accelerated." Technically, we shall define "acceleration" as the *time rate of change of velocity.*

The simplest case of accelerated motion occurs when the velocity of a particle moving in a straight line changes by equal amounts in equal intervals of time. The acceleration in this case is said to be "constant," and the motion "uniformly accelerated." We may write, for the acceleration,

$$a = \frac{v - v_0}{t - t_0}, \qquad (1\text{-}12)$$

where v_0 is the initial velocity, v the final velocity, t_0 the initial time, t the final time. For the case in which the acceleration is not known to be constant in magnitude, equation (1-12) defines the average acceleration over the time interval.

In general, motions in which the acceleration is constant are the exception rather than the rule. The one outstanding example of uniformly accelerated motion occurring in nature is that of a freely falling body in a vacuum near the surface of the earth. Here experiment shows that acceleration = Constant and approximately = 9.8 m/sec² = 980 cm/sec² = 32 ft/sec². Because of its widespread use in physics, this constant acceleration due to gravity is given a special letter, g. A few comments about variable accelerations are made in § 1.22.

1.19. Acceleration as a Vector

According to equation (1-12), acceleration is equal to the difference between two vectors (velocities) divided by the scalar, time, and *is itself a vector quantity* subject to the rules of addition developed in this chapter. Thus, if an object is subjected to accelerations a_1 and a_2, the resultant acceleration a of the object is obtained by adding a_1 and a_2 according to the parallelogram law (Fig. 1-20).

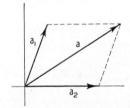

FIG. 1-20.—Parallelogram addition of accelerations

The student will observe that acceleration is *not* the time rate of change of *speed.* It is the time rate of change of *velocity* treated as a vector quantity. A particle moving with constant speed in a circle is *accelerated* in spite of the fact that the time rate of change of speed is zero. The velocity of the particle is changing continuously *since the direction of motion is changing.* Therefore, the time rate of change of velocity is not zero. Such an acceleration is called a "normal or radial acceleration" and is treated in some detail in § 1.24.

1.20. Units of Acceleration

It is evident from equation (1-12) that the units of acceleration will depend upon the unit of velocity as well as upon the unit of time during which **13** the velocity changed. This is illustrated in Example 8. In the metric system acceleration is expressed in meters per second per second, abbreviated to m/sec², or centimeters per second per second, cm/sec². Correspondingly, in the British system the unit of acceleration is usually expressed in feet per second per second or ft/sec².

EXAMPLE 8: A man on a bicycle is moving north 4 mi/hr, and 10 sec later he is moving north 12 mi/hr. What is his average acceleration during these 10 sec? (Note that if the acceleration were known to be constant, the question would have been worded simply: "What is his acceleration during the 10 sec?")

Solution: From equation (1-12),

$$a = \frac{12 \text{ mi/hr} - 4 \text{ mi/hr}}{10 \text{ sec}} = 0.8 \text{ mi/hr sec, north}.$$

EXAMPLE 9: A boat is moving north 4 mi/hr, and 10 sec later it is moving south 12 mi/hr. What is the average acceleration? Let us assume north as (+) and south as (−).

Solution:

$$a = \frac{-12 \text{ mi/hr} - 4 \text{ mi/hr}}{10 \text{ sec}} = -1.6 \text{ mi/hr sec}.$$

EXAMPLE 10: A car starts from rest and acquires a velocity of 15 m/sec in 1 min. Find the acceleration in m/sec².

Solution: The change in velocity is 15 m/sec in 60 sec, so that the acceleration is

$$\frac{15 \text{ m/sec}}{60 \text{ sec}} = 0.25 \text{ m/sec}^2.$$

1.21. Equations of Motion for a Constant Acceleration

In § 1.12 we deduced an equation for the distance s, covered by a particle in time, t, for the special case of *constant speed* in a straight line, i.e., for $a = 0$. Suppose we wish to deduce a similar relationship for the case where $a \neq 0$. On reflection, we recognize that the types of motion possible are of an infinite variety, depending essentially on how the acceleration varies with distance and time. Here we limit ourselves to but one of these innumerable types of motion and seek the equations that represent the motion of a particle subject to a *constant** acceleration. Objects sliding and rolling

* Since acceleration is a vector quantity, a *constant* acceleration implies that both the magnitude and the direction of the acceleration are constant, i.e., the motion may be along a straight line.

down inclined planes and freely falling bodies give rise to motions for which the acceleration may, to first approximations at least, be considered constant.

For any kind of motion the distance of an object from its starting point is given by the product of average velocity, $\bar{v}$, and time, t:

$$S = \bar{v}t. \qquad (1\text{-}13)$$

In the case of uniformly accelerated motion along a straight line ($a = $ Constant), the average velocity is related to the initial velocity, v_0, and the final velocity, v, by

$$\bar{v} = \frac{v_0 + v}{2}, \qquad (1\text{-}14)$$

i.e., the average velocity over any interval equals the arithmetic mean of initial and final velocities *for this particular case only.* Also the acceleration a, which is defined as the change in velocity divided by the time taken, is

$$a = \frac{v - v_0}{t} \quad \text{or} \quad v = v_0 + at. \quad (1\text{-}15)$$

Combining equations (1-13) and (1-14), we have, for the distance S covered,

$$S = \left(\frac{v_0 + v}{2}\right)t. \qquad (1\text{-}16)$$

Equations (1-15) and (1-16) are two independent relationships, each containing four of the five variables v_0, v, a, t, S. With these two equations any problem involving a constant linear acceleration can be solved. However, it is convenient sometimes to use three others which are derived from equations (1-15) and (1-16). If v is eliminated between these equations, then

$$S = v_0 t + \tfrac{1}{2}at^2. \qquad (1\text{-}17)$$

Similarly, if v_0 is eliminated, then

$$S = vt - \tfrac{1}{2}at^2; \qquad (1\text{-}18)$$

and, finally, if t is eliminated,

$$v^2 - v_0^2 = 2aS. \qquad (1\text{-}19)$$

These results can be shown graphically in the following manner. In Fig. 1-21 the velocity is plotted against the time. At zero time the velocity is v_0 and increases uniformly until at time t it is v. The acceleration is the slope of the line given by

$$\text{Slope} = a = \frac{v - v_0}{t}.$$

The total distance S traveled by the object in time t is given by the area under the curve. In a short interval of time Δt at which the average velocity is $\bar{v}$, the distance traveled is given by the

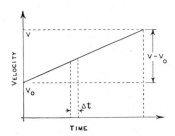

Fig. 1-21.—Velocity plotted against time for constant linear acceleration.

area $\bar{v}\Delta t$, so that the total distance moved in time t is

$$S = \frac{(v_0 + v)\, t}{2}$$

$$= v_0 t + \frac{(v - v_0)\, t}{2}$$

$$= v_0 t + \tfrac{1}{2} a t^2 \; .$$

If we now plot S against t, then a curve such as is shown in Fig. 1-22 is obtained. The total displace-

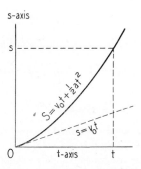

Fig. 1-22.—Distance plotted against time for accelerated motion.

ment, S, involves two contributions: $v_0 t$, a displacement due to the initial speed, and $\tfrac{1}{2} a t^2$, a displacement due to the change in speed with time. The graph of equation (1-17) is shown in Fig. 1-22.

As we have pointed out in § 1.18, the acceleration for freely falling bodies is a constant, g, equal approximately to 9.8 m/sec^2 = 980 cm/sec^2 or 32 ft/sec^2. Actually, the value of g depends upon the position of the observing station—its latitude and distance above sea-level (see § 2.18). Summa-

rizing our results for the special case where $a = g$, we get

$$
\left.
\begin{aligned}
v &= v_0 \pm g t &\quad (a) \\
S &= v_0 t \pm \tfrac{1}{2} g t^2 \; . &\quad (b) \\
v^2 &= v_0^2 \pm 2 g S \; . &\quad (c) \\
S &= \tfrac{1}{2} g t^2 &\quad (d) \\
v &= \sqrt{2 g S} \; . &\quad (e)
\end{aligned}
\right\} \quad (1\text{-}20)
$$

From equation (1-19),

For $v_0 = 0$,

and

In these equations the $(+)$ sign is used for the case of an object falling freely from a height, S (the $+$ direction of S, v_0, and g is downward); the $(-)$ sign for the case of an object thrown upward with an initial velocity v_0 (S, v_0 are positive; g negative). In this section we have discussed motion along a straight line, so that the acceleration is along the same line as the velocities. Later discussions (§ 1.24) will show that this is not always the case.

EXAMPLE 11: A body has a constant acceleration of 12 cm/sec^2 and at $t = 0$ a velocity of 20 cm/sec in the same direction. How far will it move in 4 sec?

Solution: Examine equations (1-15), (1-17), and (1-19). In this problem the final velocity is neither given nor asked for. Only in equation (1-17) is the final velocity not involved. Hence, using equation (1-17),

$$S = 20 \text{ cm/sec} \times 4 \text{ sec} + \tfrac{1}{2} \times 12 \text{ cm/sec}^2$$

$$\times \ (4 \text{ sec})^2 = 176 \text{ cm} \; .$$

EXAMPLE 12: A body has a constant acceleration. At one point in its path it has a speed of 1.0 m/sec, and 1.60 m farther along it has a speed of 0.6 m/sec. What is its acceleration?

Solution: Note that the "time" is not given. Hence equation (1-19) is indicated, since it is the one which does not include "time":

$$v^2 = v_0^2 + 2 a S \; ,$$

$$0.36 = 1.0 + 2 a \times 1.6 \; ,$$

$$3.2 a = -0.64 \; ,$$

$$a = -0.20 \text{ m/sec}^2 \; .$$

EXAMPLE 13: From a point 48 ft above the ground, a ball is thrown upward with a velocity of 64 ft/sec. How high will the ball rise, and what total time elapses before it reaches the ground?

Solution: From equation (1-20 [c]), $2gS = v_0^2 - v^2$; at the top of its path, S ft above the point whence it is thrown, $v = 0$; therefore,

$$S = \frac{v_0^2}{2 g} = \frac{(64 \text{ ft/sec})^2}{2 (32 \text{ ft/sec}^2)} = 64 \text{ ft} \; .$$

15

The time required for the rise is, from $v = v_0 - gt$,

$$t = \frac{v_0}{g} = \frac{64 \text{ ft/sec}}{32 \text{ ft/sec}^2} = 2 \text{ sec}.$$

The total height of fall is $64 + 48 = 112$ ft, so that, by reapplying $S = \frac{1}{2}gt^2$ with different meanings now for S and t, we obtain

$$t^2 = \frac{2S}{g} = \frac{2(112 \text{ ft})}{32 \text{ ft/sec}^2} = 7 \text{ sec}^2,$$

$$t = \sqrt{7} \text{ sec}.$$

Total time elapsed $= (2 + \sqrt{7})$ sec.

1.22. Instantaneous Accelerations

Just as we defined an instantaneous velocity or the velocity at a point, so we can define similar expressions for acceleration. Suppose an object is moving along the X-axis and that its velocity is v at some time t. In a short interval of time Δt between t and $t + \Delta t$ the velocity has increased by Δv and the object has moved a distance Δx. The average acceleration $\bar{a}$ during this time is

$$\bar{a} = \frac{\Delta v}{\Delta t},$$

and the instantaneous acceleration at time t or position x is the derivative of v with respect to t, or

$$a = \lim_{\Delta t \to 0} \frac{\Delta v}{\Delta t} = \frac{dv}{dt} = \dot{v}, \qquad (1\text{-}21)$$

where v is the instantaneous velocity at time t. This instantaneous velocity is given by equation (1-11) as

$$v = \frac{dx}{dt}.$$

Hence the instantaneous acceleration a may be written as

$$a = \frac{d}{dt}\left(\frac{dx}{dt}\right) = \frac{d^2x}{dt^2}. \qquad (1\text{-}22)$$

Let us now suppose that the object starts at $x = 0$ at time $t = 0$ with a velocity v_0, and proceeds with a constant acceleration a, so that, at time t_1, its velocity is v_1. During this time the object has moved a distance x_1 as represented in Fig. 1-23. At any instant t and position x the acceleration a is given by

$$a = \frac{d^2x}{dt^2} = \frac{dv}{dt}.$$

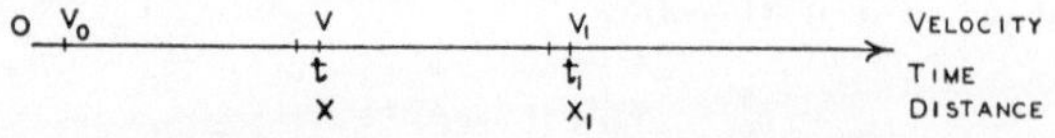

Fig. 1-23.—Instantaneous linear acceleration

To find the velocity, it is necessary to integrate the equation, giving

$$\int_0^{t_1} a\,dt = \int_{v_0}^{v_1} dv.$$

The limits of integration are at $t = 0$, $v = v_0$, and, at a time t_1, $v = v_1$. For this special case the acceleration a is a constant and does not vary with time t; hence it may be taken from under the integral sign, giving

$$a\int_0^{t_1} dt = \int_{v_0}^{v_1} dv.$$

Carrying out the integration gives

$$at_1 = v_1 - v_0$$

or

$$a = \frac{v_1 - v_0}{t_1},$$

which can be obtained from the definition of a constant linear acceleration.

At any time t the velocity is v, so the acceleration may also be written

$$a = \frac{v - v_0}{t}$$

or

$$v = v_0 + at.$$

Now the instantaneous velocity v at time t is given by equation (1-11) as

$$v = \frac{dx}{dt}.$$

Hence we may write

$$\frac{dx}{dt} = v_0 + at.$$

Integrating again between the appropriate limits gives

$$\int_0^{x_1} dx = \int_0^{t_1} v_0\,dt + \int_0^{t_1} at\,dt,$$

giving

$$x_1 = v_0 t_1 + \frac{1}{2}a t_1^2,$$

which was previously obtained in equation (1-17).

If the acceleration a is not a constant, then, in order to perform the integrations, one must have an analytic expression giving the variation of a with t. This is then placed in the equation and the integration carried out.*

* A great deal of the calculus made use of in this book is contained in the following equations found in elementary calculus texts. The derivative of x^n with respect to x is

$$\frac{d}{dx}(x^n) = nx^{n-1},$$

and the integral of $x^n dx$ is

$$\int x^n\,dx = \frac{x^{n+1}}{n+1} + C,$$

where C is the constant of integration. It is well to remember that a derivative is an instantaneous rate of change of one variable with respect to another. From this it follows that the derivative of a constant is zero, for a constant does not change.

1.23. Projectiles

Let us return now to the principle of independence of velocities and use this to analyze the motions of projectiles. We assume that all effects due to friction are either absent or negligible.

Consider, first, a particle projected horizontally from O (Fig. 1-24) with velocity V_0. This is precise-

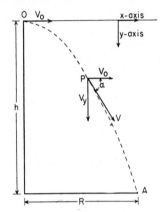

Fig. 1-24.—Bullet shot horizontally follows a parabolic path.

ly the case of ball A depicted in Fig. 1-17, to which the reader is again referred in these connections.

The point O is h m above the level of the ground. A rifle shot horizontally would meet these conditions. At the instant the bullet leaves the muzzle of the rifle, it is subject to two velocities: V_0, horizontally, and V_y, vertically. These velocities are entirely independent of each other, V_0 remaining constant in magnitude and direction,* and V_y increasing in magnitude in accordance with equation (1-20), $V_y = gt$. In time t, the distances traversed are: $x = V_0t$, horizontally, and $y = \frac{1}{2}gt^2$, vertically. If we eliminate t from these equations, we obtain

$$y = \frac{g}{2 V_0^2} x^2. \quad (1-23)$$

This is the equation of the path OPA followed by the bullet. The curve is that of a parabola. We compute the range, R, as follows: The time t_h required for the bullet to fall h m is

$$t_h = \sqrt{\frac{2h}{g}},$$

so that

$$R = V_0 t_h = V_0 \sqrt{\frac{2h}{g}}.$$

* This is justified in chap. 2, assuming no air resistance.

We may calculate the velocity V at any time t by finding V_x and V_y. But $V_x = V_0$ and, numerically, $V_y = gt$. Therefore,

$$V = \sqrt{V_0^2 + (gt)^2} = \sqrt{V_0^2 + \left(\frac{gx}{V_0}\right)^2}, \quad (1-24)$$

since $t = x/V_0$. Also, the angle a that V makes with V_0 is

$$a = \tan^{-1} \frac{V_y}{V_x} = \tan^{-1} \left(\frac{gx}{V_0^2}\right). \quad (1-25)$$

If the projectile is shot upward at an angle of θ degrees measured from the horizontal (Fig. 1-25),

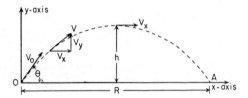

Fig. 1-25.—Bullet shot at an angle follows a parabolic path

we may resolve the velocity into horizontal and vertical velocities. At any instant of time the velocities are V_y and V_x, respectively, where

$$V_y = V_0 \sin \theta - gt, \qquad V_x = V_0 \cos \theta.$$

Again these velocities are independent of each other, V_x remaining constant, with V_y varying with time. The maximum height, h, to which the projectile will rise is given by equation (1-20 [c]), $V_y^2 = 2gh$. In this equation we must use the initial value of V_y, i.e., the value at $t = 0$, so that

$$(V_0 \sin \theta)^2 = 2gh ;$$

since at the top of the path the vertical velocity is zero,

$$h = \frac{(V_0 \sin \theta)^2}{2 g}. \quad (1-26)$$

The range, R, may be computed by calculating the time of flight, t_f, and substituting in $R = V_0 t_f \cos \theta$. The time of flight is twice that required for the projectile to rise the distance h. (Why?) The maximum range occurs for an angle of elevation of $\theta = 45°$.

EXAMPLE 14: A rifle bullet of muzzle velocity 1,280 ft/sec is shot at an angle of 30° with the horizontal. Compute the maximum rise and the range of the bullet.

Solution: From equation (1-26)

$$h = \frac{(1{,}280 \text{ ft/sec})^2 \cdot (\frac{1}{2})^2}{2 \times 32 \text{ ft/sec}^2} = 6{,}400 \text{ ft}.$$

Time of rise,

$$t = \sqrt{\frac{2h}{g}} = 20 \text{ sec}.$$

Time of flight,

$$t_f = 2 \times 20 = 40 \text{ sec}.$$

Range,

$$R = V_0 t_f \cos \theta$$

$$= (1,280 \text{ ft/sec})(40 \text{ sec})(0.866)$$

$$= 44,339 \text{ ft}.$$

1.24. Uniform Circular Motion

If a particle moves in a circle with constant speed, it is said to be in *uniform circular motion*. Although the velocity of the particle is constant in magnitude, its direction is continually changing. The velocity, therefore, treated as a vector, is changing with time; and hence the particle must be subject to an acceleration. How will we proceed to calculate the magnitude and direction of this acceleration?

As far as the direction of the acceleration, a_c, is concerned, it must be along the radius r, as shown in Fig. 1-26 (a). We may prove this by assuming

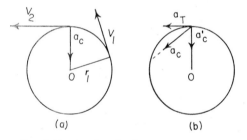

(a)   (b)

Fig. 1-26.—Acceleration for motion in a circle

that a_c is in some other direction, such as is shown in Fig. 1-26 (b). We may resolve a_c into components, a_c' along r and a_T tangent to the circle. The effect of a_T, the tangential component, is to speed up the particle. But we are dealing with motion in which the velocity *does not change in magnitude but only in direction.* Hence $a_T = 0$ and a_c must be along the radius. Being perpendicular to the path, its only effect is to change the *direction of v.* Thus it is called a *radial* or *normal acceleration.*

In order to compute the magnitude of the acceleration, consider two points, A and B, close together (Fig. 1-27 [a]). It is clear from the figure that V_1 is perpendicular to r_1, and V_2 to r_2. We wish

to measure the change in velocity, $\Delta V = V_2 - V_1$, that occurs in time Δt, for then

$$\bar{a} = \frac{\Delta V}{\Delta t}.$$

The graphical determination of ΔV is shown in Fig. 1-27 (b). Its analytical value may be obtained by observing that the vector triangle is similar to

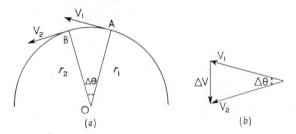

(a)   (b)

Fig. 1-27.—Analysis of motion in a circle

the physical triangle OAB; that is, treating magnitude only, we have

$$\frac{\Delta V}{V} = \frac{AB}{r}, \qquad (1\text{-}27)$$

where V is the magnitude of either V_1 or V_2.

When Δt is small, the arc AB and the chord AB are very nearly equal, so that, along the arc, $AB = V\Delta t$, and therefore

$$\frac{\Delta V}{V} = \frac{V\Delta t}{r}$$

and

$$\Delta V = \frac{V^2 \Delta t}{r},$$

so that

$$\bar{a} = \frac{\Delta V}{\Delta t} = \frac{V^2}{r}. \qquad (1\text{-}28)$$

As Δt approaches zero, the approximation we have made becomes better and better, so that, as the arc shrinks down to a point within it, we have, for the acceleration at that point,

$$a_c = \lim_{\Delta t \to 0} \frac{\Delta V}{\Delta t} = \frac{dV}{dt} = \frac{V^2}{r} \qquad \text{(exactly)}. \ (1\text{-}29)$$

The expression for the "centripetal acceleration," as a_c is called, may be put into a somewhat different form. If T is the period of motion, i.e., the time required for the body to make one complete revolution, and f is the frequency of motion, i.e., the number of revolutions for unit time, then

$$f = \frac{1}{T}. \qquad (1\text{-}30)$$

Since the speed, V, is equal to the total distance divided by the total time,

$$V = \frac{2\pi r}{T} \quad \text{or} \quad V = 2\pi f r; \quad (1\text{-}31)$$

and from equation (1-29)

$$a_c = \frac{4\pi^2 r}{T^2}, \quad (1\text{-}32)$$

or, since

$$f = \frac{1}{T},$$

$$a_c = 4\pi^2 f^2 r. \quad (1\text{-}33)$$

The angular velocity ω of the point moving in the circle is equal to the angle 2π radians in a circle divided by the time T for one complete rotation or

$$\omega = \frac{2\pi}{T} = 2\pi f \quad \text{(radians/sec)}. \quad (1\text{-}34)$$

The tangential velocity V at a distance r from the center of rotation is given by $V = \omega r$. The centripetal acceleration for motion in a circle of radius r, with constant speed V such that the period is T, frequency f, and angular velocity ω, may be written in the following manner:

$$a_c = \frac{V^2}{r}$$

$$= \frac{4\pi^2 r}{T^2}$$

$$= 4\pi^2 f^2 r$$

$$= \omega^2 r. \quad (1\text{-}35)$$

In chapter 2 we shall discuss the conditions under which this motion occurs.

Statics of a Particle

1.25. Force: Tension and Compression

In order that we may treat the important problem of simple structures in this chapter, we shall introduce a definition of the force concept particularly useful for such studies.*

By definition, whenever a coiled spring is stretched or compressed, it is said that *forces act on the spring*, one at each end. If the spring is at rest while in the state of tension or compression, the forces acting on the spring are taken to be equal in magnitude but oppositely directed, as indicated in Fig. 1-28.

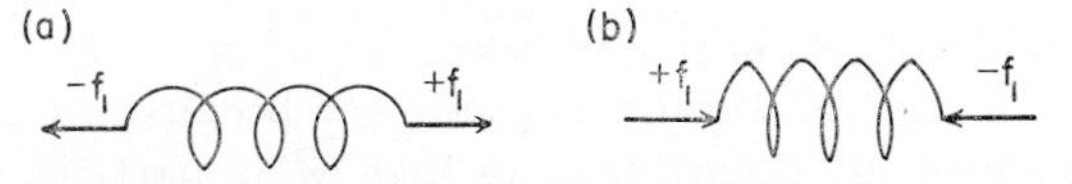

(a)   (b)

FIG. 1-28.—(a) Spring under tension; (b) spring under compression.

The magnitude of the forces acting on the spring may be measured by the amount of stretch or compression. By rigidly attaching a scale to one end of the spring and a pointer to the other end, the spring becomes the common spring balance (Fig. 1-29). We may calibrate the scales to read *pounds-force* (lbf). By definition, a pound of mass (see § 2.3), when placed in the earth's gravitational field at a standard location, has a *pound-force* ex-

erted on it. If the pound-mass is hung from one end of the coiled spring while the other end of the spring is held in the hand or attached to a rigid wall, the spring will stretch a certain amount. The scale reading corresponding to the stretch is labeled *1 pound-force*. In this way the entire scale

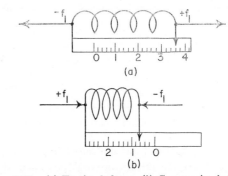

FIG. 1-29.—(a) Tension balance; (b) Compression balance

may be calibrated to read in pounds-force. The reader will observe that the scale is calibrated to read only the force at one end. Also, by definition, *the scale reading and hence the force at one end of the spring are taken as the measures of the tension in the spring.* Any agent, then, that stretches or compresses the spring gives rise to a force whose value is the scale reading of the spring balance (Fig. 1-30).

In chapter 2 (§ 2.5) we shall discuss the concept of force more generally and in greater detail.

* A dynamical definition of force following the treatment of Mach is given in chap. 2.

19

1.26. Force as a Vector

That force has direction as well as magnitude may be easily verified by the reader. A given force, as measured by the stretch of the spring, may move an object in any number of directions. For this reason we may represent forces as we did velocities, by an arrow (———→) whose length is a measure of the magnitude of the force and whose

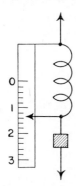

Fig. 1-30.—A spring balance

direction is that of the force. It is found experimentally that forces combine exactly as do accelerations and are therefore vector quantities subject to the rules of addition discussed in § 1.7.

A given force may be resolved into components, and an independent physical meaning given to each component.* For example, a force f is applied to a block, as shown in Fig. 1-31. What part of the

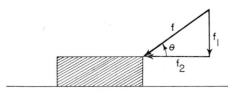

Fig. 1-31.—Components of a force

force is pressing the table and block together, and what part is effective in moving the block along the table? The force may be resolved into the components f_1 and f_2. From the diagram

$f_2 = f \cos \theta$, pushing the block along the table;

$f_1 = f \sin \theta$, pressing the block and table together.

1.27. Statics: Forces in Equilibrium

If forces acting on a particle are so oriented that the particle is at rest, the particle is said to be in

* Another example of the principle of superposition. See

equilibrium under the action of the applied forces. When this occurs it is found that the vector sum of the forces is zero, i.e.,

$$f_1 + f_2 + \ldots f_n = \Sigma f = 0, \qquad (1\text{-}36)$$

where again the symbol Σ means sum. It follows that a body cannot be in equilibrium under the action of a single nonzero force.

As a first example of these ideas, consider the case of a ball on a smooth table and subject to two horizontal forces, as indicated by the spring balances A and B (Fig. 1-32 [a]). The ball is at rest,

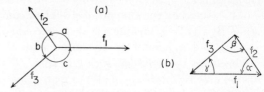

Fig. 1-32.—Equilibrium of two forces on a particle

and we note that the spring balances read the same. The force diagram for this problem is shown in Fig. 1-32 (b), where the ball is replaced by the point O, and the balances by the arrows f_1 and f_2. It follows that

$$f_1 + f_2 = 0$$

or

$$f_1 = -f_2.$$

That is, the forces are equal in magnitude and oppositely directed. It should be noted, furthermore, that a body in equilibrium under the action of two forces must itself always be in a state either of tension or of compression. We have already used this fact in our definition of force.

A case of special importance, because of its widespread occurrence, is that of a particle in equilibrium under the action of three forces: f_1, f_2, and f_3 (Fig. 1-33 [a]). We add the vectors (Fig.

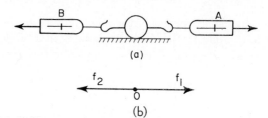

Fig. 1-33.—Equilibrium of three forces on a particle

1-33 [b]), tail to head, to obtain their sum, and find that the last vector, f_3, closes the triangle, indicating that the resultant is zero. Also, the three forces must have been acting *in the same plane.* Hence

$$f_1 + f_3 + f_2 = 0 \,.$$

Applying the law of sines to the force triangle of Fig. 1-33 (*b*), we see that

$$\frac{f_1}{\sin \beta} = \frac{f_2}{\sin \gamma} = \frac{f_3}{\sin a}. \qquad (1\text{-}37)$$

By comparing the angles of the two figures, it is evident that

$$a = 180° - a \,,$$

$$b = 180° - \beta \,,$$

$$c = 180° - \gamma \,,$$

so that equation (1-37) becomes

$$\frac{f_1}{\sin b} = \frac{f_2}{\sin c} = \frac{f_3}{\sin a}. \qquad (1\text{-}38)$$

Equation (1-38) proves to be a very useful theorem for solving many problems.

We have noted that, if a particle is in equilibrium under the action of three forces, these forces must lie in the same plane. However, if the action of four or more forces on a particle results in equilibrium, these need *not* lie in the same plane. Nevertheless, the polygon of forces is always a *closed* polygon; that is, the head of the last vector meets the tail of the first vector.

Any object may be taken as a "particle," and the preceding theorems applied, if the forces acting on the object may be considered as acting at a point. Situations in which this is not permitted are those in which rotation can take place and are treated in chapter 5.

1.28. Simple Structures

By way of illustrating the theorems of the last section, let us look at the case of a rope* stretched between two rigid unyielding walls (Fig. 1-34) and

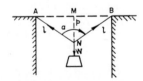

Fig. 1-34.—Equilibrium of a weight supported by three cords

supporting at its mid-point a heavy load, such as a block of weight W. The length of the rope is $2l$,

* The weight of the rope is neglected.

and the rope is depressed an amount $MN = p$. The problem is to calculate the stretching force in the two segments of the rope, NB and NA. By "stretching force" in the rope we mean the reading of calibrated spring balances (such as are shown in Fig. 1-29) placed at points A and B and connecting the rope to the walls. If the rope between N and B were cut and a spring balance inserted, it would read the same as the balance at B, the stretching force in the rope.

If we isolate the point N, we find that three forces are acting at this point: the weight, W, downward, and the forces T_1 and T_2 in the ropes. Since the point N is at rest by hypothesis, the three forces will add to zero. These forces, represented in Fig. 1-35 (*a*), are added together by the polygon

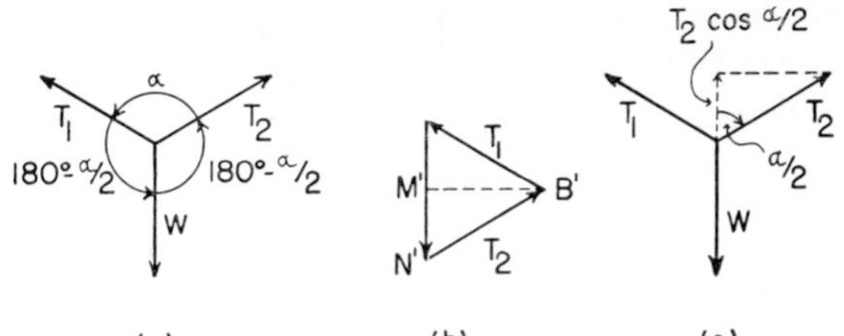

Fig. 1-35.—Three methods for analyzing equilibrium at a point.

method in Fig. 1-35 (*b*). The student will observe that, since the weight is hung from the mid-point of the rope, by symmetry the two stretching forces are the same in magnitude, $T_1 = T_2$. The force may be calculated by using four possible methods.

1. Since the forces T_1 and T_2 are assumed to be along the ropes, the angle a between T_1 and T_2 is the same as the angle between the segments of rope. Using equation (1-38), we have

$$\frac{T_1}{\sin (180° - a/2)} = \frac{W}{\sin a};$$

or

$$T_1 = T_2 = W \frac{\sin (180° - a/2)}{\sin a}.$$

2. If in the force triangle of Fig. 1-35 (*b*), which is an isosceles triangle, we construct the perpendicular $M'B'$, then $M'N' = W/2$, and the force triangle $M'N'B'$ is similar to the geometrical triangle MNB of Fig. 1-34. Therefore,

$$\frac{T_1}{W/2} = \frac{l}{p}; \qquad \text{or} \qquad T_1 = T_2 = \frac{W}{2} \frac{l}{p}. \quad (1\text{-}39)$$

3. In Fig. 1-35 (*c*), the forces T_1 and T_2 are resolved into components, parallel and perpendicu-

lar to W. Since $T_1 = T_2$, the total component parallel to W is

$$2T_1 \cos\left(\frac{a}{2}\right).$$

The perpendicular components add to zero, so that

$$2T_1 \cos\left(\frac{a}{2}\right) = W$$

or

$$T_1 = T_2 = \frac{W}{2 \cos (a/2)}. \qquad (1\text{-}40)$$

4. The force triangle of Fig. 1-35 (*b*) may be constructed accurately by ruler and protractor, and the magnitudes of T_1 and T_2 determined by measuring the number of units in T_1 relative to W. This method can always be used. We may use methods 1 and 3 if the angles between the forces are known and method 2 if the physical dimensions of the apparatus are given.

EXAMPLE 15: Let us assume that in Fig. 1-35, $a = 120°$ and $W = 100$ lbf. By equation (1-38), the law of sines,

$$\frac{T_1}{\sin (180° - 60°)} = \frac{T_2}{\sin (180° - 60°)} = \frac{100 \text{ lbf}}{\sin 120°}$$

or

$$T_1 = T_2 = 100 \text{ lbf} \frac{\sin (180° - 60°)}{\sin 120°} = 100 \text{ lbf}.$$

By equation (1-40), the method of components,

$$T_1 = T_2 = \frac{100 \text{ lbf}}{2 \cos 60°} = \frac{100 \text{ lbf}}{2 \left(\frac{1}{2}\right)} = 100 \text{ lbf}.$$

If we were given $l = 10$ ft and $p = 5$ ft, then, by equation (1-39),

$$T_1 = T_2 = \frac{100 \text{ lbf}}{2} \left(\frac{10 \text{ ft}}{5 \text{ ft}}\right) = 100 \text{ lbf}.$$

EXAMPLE 16: A weight of 10 lbf is hung from the end of the horizontal, weightless, hinged boom ab shown in Fig. 1-36. At the end of this boom is a wire, cb, attached at c to a spring balance. The wire makes an angle of 45°

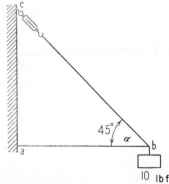

FIG. 1-36.—Equilibrium of a weightless boom

with the horizontal. Find the stretching force T of the wire and the compression C of the boom.

Solution: In solving problems of this general type the student should isolate some point at which forces are acting and should determine their directions. It is, of course, desirable to choose that point at which the magnitude and direction of some force are known. In this example we isolate the point b. The forces that are acting upon the point b are the weight of 10 lbf downward, the stretching force in the wire along the wire, and the reaction of the boom outward. These are the forces which, acting on the point b, keep it in equilibrium. The isolated forces are shown in Fig. 1-37. Applying equation (1-38),

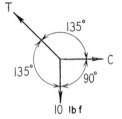

FIG. 1-37.—Equilibrium of forces at point b

we obtain

$$\frac{C}{\sin 135°} = \frac{T}{\sin 90°} = \frac{10 \text{ lbf}}{\sin 135°};$$

$$T = \frac{10 \text{ lbf}}{\sin 45°} = 10 \sqrt{2} \text{ lbf};$$

$$C = 10 \text{ lbf}.$$

If the angle a were not given but, instead, the lengths of the sides of the physical apparatus of Fig. 1-36, i.e., $ab = 3$ ft, $ac = 3$ ft, and $cb = 3\sqrt{2}$ ft, it would be convenient to construct the force triangle at the isolated point b. The force triangle, Fig. 1-38, is similar to the

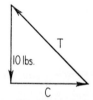

FIG. 1-38.—Force triangle for point b

physical triangle of Fig. 1-36, and hence

$$\frac{T}{10 \text{ lbf}} = \frac{cb}{ac} \qquad \text{or} \qquad T = 10 \sqrt{2} \text{ lbf};$$

$$\frac{C}{10 \text{ lbf}} = \frac{ab}{ac} \qquad \text{or} \qquad C = 10 \text{ lbf}.$$

1.29. Résumé

If you are to build your knowledge on a firm foundation, then you must know the definitions, principles, and theories and how to apply them. Test yourself by writing out from memory the definitions, principles, and laws given in the chapter. These are summarized for your convenience at the end of each chapter. When approaching a problem, do not ask what equation must be used but what physical principle must be applied, and from this it should follow what equation is applicable. Remember that the problems at the end of each chapter are on the material of the chapter, and, if you know this, you will have relatively little trouble with the problems. There is no better way of testing and improving your knowledge of physics than that of doing many problems, for the more you do, the easier you will find it to do the others.

The following is a brief listing of the definitions, principles, and theories, given in this chapter, which you should know:

Scalar and vector quantities, components of a vector, methods of addition and subtraction of vectors

Velocity, speed, and acceleration

The principle of independence of velocities

The time-distance-velocity relationships for a body moving with constant linear acceleration

Projectile motion

Uniformly accelerated circular motion

The methods of finding the resultant and equilibrant of a system of forces acting on a particle

Queries

1. Is it ever possible for an object to be accelerating without its speed changing?

2. The velocity of a train with respect to the ground is V_{TG}; the velocity of an auto with respect to the ground is V_{AG}. If V_{TA} is the velocity of the train with respect to the auto (called their "relative velocity"), can you write an expression for V_{TA} in terms of V_{TG} and V_{AG}?

3. Can you recall any motion that occurs in nature in which an object moves with a finite velocity (not zero) but whose acceleration is zero? One in which the acceleration is very nearly constant? One in which the acceleration of an object is proportional to its displacement from a starting point but oppositely directed?

4. A wire is stretched by means of two spring balances at either end, both reading T lbf. If the wire is cut at the middle and a spring balance inserted, what will it read, T or $2T$ lbf?

5. Why is it that a hammock stretched tightly between two posts is far more likely to break when a weight is at its center than one that sags considerably? Develop an equation to prove the point.

6. Is uniform circular motion (1) uniform motion, (2) uniformly accelerated motion, (3) neither?

7. Is it possible for a body to have an acceleration in a direction opposite to its motion?

8. Classify the following quantities as scalars or vectors: time, speed, velocity, volume, acceleration, force.

Practice Problems

1. In order to determine the width of a stream CD, a base line 100 yards (yd) long is established. The angles ABC and CAB, as measured by a transit, are 30° and 50°, respectively. Compute the distance CD. Ans.: 38.9 yd.

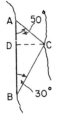

2. Two observers, O and O', a distance apart of 1,000 ft, sight an airplane. The angle between the base line and the line of sight of the airplane is 40° and 50°, respectively. What is the height of the airplane, assuming the observers and the airplane to be in a vertical plane?

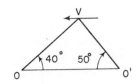

Add the vectors shown below graphically and analytically and state the magnitude and direction of the resultant.

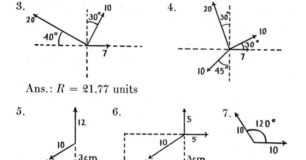

3.

4.

Ans.: $R = 21.77$ units

5.

6.

7.

Ans.: $R = 10$ units

Ans.: $R = 10$ units

8. A racing plane on a straight run passes one marker with a velocity of 200 mi/hr and 10 sec later passes the second marker, moving 300 mi/hr. Assuming constant acceleration, compute (a) acceleration between markers, (b) average velocity between markers, and (c) their distance apart.

9. A train is moving with a velocity of 90 mi/hr when the brakes are applied, reducing the speed to 30 mi/hr in 10 sec. Find (a) the acceleration of the train while being braked, assuming this to be a constant; (b) the distance the train moved while the brakes were applied; (c) how long it would take the brakes to stop the train while moving 90 mi/hr; (d) how far the train would travel before being completely stopped, the initial velocity being 90 mi/hr. Ans.: (a) −8.8 ft/sec²; (b) 880 ft; (c) 15 sec; (d) 990 ft.

10. A car is moving 20 km/hr and increases its speed uniformly to 60 km/hr in 11 sec. What is the acceleration? What is the distance traveled in this time? What is the distance traveled in the eleventh second? What is the speed at the end of the third second?

11. A car has a speed of 60 mi/hr. Find the speed of the car in ft/sec, m/sec, and km/hr, where 1 kilometer (1 km) is equal to 1,000 m. If the car is stopped in 10 sec, find the average speed in m/sec and the stopping distance in meters. Ans.: 88 ft/sec; 26.8 m/sec; 96.6 km/hr; 13.4 m/sec; 134 m.

12. A bullet is shot vertically upward with a speed of 140 m/sec. Find the height to which it rises and the time it takes to reach the ground.

13. A car can gain a speed of 15 mi/hr in 10 sec, starting from rest. If the acceleration remains constant, find (a) how long it takes the car to gain a speed of 30 mi/hr starting from rest; (b) how far from the start it moves to gain this speed; (c) how far it travels in the first second; (d) how far in the last second. Ans.: (a) 20 sec; (b) 440 ft; (c) 1.1 ft; (d) 42.9 ft.

14. If the brakes of a train permit it to lose velocity at the rate of 1.25 m/sec², how far would the train, if moving 45 km/hr, run after applying the brakes? How much time would elapse before the train stopped?

15. A ball is thrown upward from the edge of a cliff with a velocity of 100 ft/sec. (a) How high will it rise? (b) For how many seconds will it rise? (c) How many seconds before it will have fallen 50 ft below its starting point? (d) What will then be its velocity? Ans.: (a) 156 ft; (b) 3.12 sec; (c) 6.71 sec; (d) 115 ft/sec.

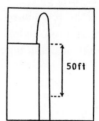

16. A ball is dropped from the Washington Monument (500 ft). How many seconds elapse before it reaches the ground? What will be its velocity at the ground? What acceleration will stop it in 2 ft? What acceleration will stop it in $\frac{1}{100}$ sec?

17. Car A is 200 m back of car B, both traveling with a constant speed of 90 km/hr. Car A accelerates and passes car B in 20 sec. Compute (a) the acceleration of car A, assuming it to be constant; (b) the distance car A traveled in 20 sec; (c) the speed of car A at the end of the 20 sec. Ans.: (a) 1 m/sec²; (b) 700 m; (c) 45 m/sec.

18. In problem 17, if the driver of car B accelerates his car uniformly at the instant car A starts to accelerate and attains a speed of 99 km/hr at the instant car A pulls in front of car B, compute (a) the acceleration of car A, assuming it to be constant; (b) the final speed of car A; and (c) the distance covered by car A in the 20 sec.

19. A particle is moving in a circle of radius 0.5 m with a constant speed of 4π m/sec. Find period, frequency, and angular velocity of motion, and centripetal acceleration calculated from each of the relationships given in equation (1-35). Ans.: $\frac{1}{4}$ sec, 4/sec, 8π rad/sec, $32\pi^2$ m/sec².

20. A particle is moving in a circle of radius $\frac{1}{2}$ ft with a constant angular velocity of 2 rad/sec. Find the period, frequency, linear velocity of the particle, and its centripetal acceleration.

21. A car goes around a circle of 50-ft radius at a speed of 20 mi/hr. What is the acceleration of the car? Direction of the acceleration? Ans.: 17.2 ft/sec².

22. A stone on the end of a string 0.5 m in length is whirled in a horizontal plane ten times per second. What are the angular velocity and the acceleration of the stone?

23. A flywheel 20 cm in radius has a period of revolution of 0.1 sec. What is the centripetal acceleration of a point on the periphery? Ans.: 7.89×10^4 cm/sec².

24. Given the vector $a = 3$ mi/hr northeast and the vector $b = 4$ mi/hr south, find the magnitude and direction of $a - b$ and $b - a$.

25. The wind is blowing 5 mi/hr from the north. A boy on a bicycle is riding east 8 mi/hr. From what direction and with what speed does the air rush past the boy? Ans.: 9.43 mi/hr.

26. A man can row a boat 3 mi/hr in still water. The water in a river is moving 2 mi/hr. How long will it take the man to row 1 mile (as viewed from the shore) up the river and back to his starting point? Show by a diagram in which direction he must row in order to move directly across the river. What will then be his speed away from shore?

27. In Fig. 1-17 the initial distance between balls A and B is 1 m. If ball A is shot out horizontally with a speed of 2 m/sec toward ball B and, at the instant ball A is shot out, ball B commences to fall, find the time in which they collide and the velocity of ball A at collision. Ans.: 0.5 sec; 5.29 m/sec.

28. A boat A is traveling north at 20 mi/hr, and boat B is moving at 15 mi/hr due west. What is the velocity of boat A relative to boat B; of boat B relative to A?

29. A force of 2 lbf and one of 4 lbf may be applied to a body so that, together, they produce the same effect as a single force of (1) 1 lbf; (2) 0 lbf; (3) 3 lbf; (4) 7 lbf; (5) 8 lbf. Which, if any, answers are correct?

30. What force must be exerted at an angle of 60° to the top of a table to push a weight along the table if the frictional force to be overcome is 10 lbf?

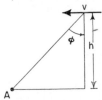

31 A bombing plane 2,000 ft from the ground and traveling 240 mi/hr releases a bomb intended for target A. What angle ϕ must the target make with the vertical? Ans.: $\phi = 63°.2$.

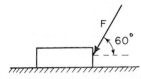

32. The barrel of an antiaircraft gun makes an angle of 70° with the horizontal when it is fired. If the muzzle velocity of the shell is 1,000 m/sec, how high will the shell rise?

33. A ball, rolling with a velocity of 300 cm/sec, leaves the top of a table, 100 cm high. Calculate the velocity (magnitude and direction) with which it strikes the floor. How far from the foot of the table does it strike? Ans.: 535 cm/sec; 135.6 cm.

34. Write the dimensions of each of the terms in equations (1-20) and (1-35).

35. An object moves along the X direction so that its distance x in meters from the origin is given by

$$x = 4 + 7t + 3t^2 ,$$

where t is the time in seconds. Find the displacement at time zero and after 2 sec. Find the velocity and acceleration at these times. Ans,: 4 m, 30 m; 7 m/sec, 19 m/sec; 6 m/sec².

36. The shell of an antiaircraft gun is shot vertically up at an airplane 6,000 ft in the air. The shell hits the plane 5 sec later. Compute (a) the speed of the shell as it left the gun and (b) the speed of the shell when it hit the airplane.

37. A weight of 100 lbf is hung from the free end of the horizontal weightless hinged boom ab. To the end of this boom is also attached a rope, cb, which makes an angle of 30° with the horizontal. Find the stretching force, T, of the rope and the compression, C, of the boom. Ans.: $T = 200$ lbf; $C = 173$ lbf.

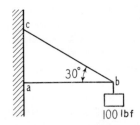

38. A weight of 40 lbf is suspended by two ropes as shown in the diagram. Compute the stretching force in each rope.

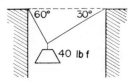

39. A train starting from rest is uniformly accelerated for 20 sec. The average velocity attained is 66 ft/sec. What is the final velocity of the train? What is the acceleration? What distance is traversed in the 20 sec? Ans.: 132 ft/sec; 6.6 ft/sec²; 1,320 ft.

40. Four vectors, A, B, C, and D, are drawn from the origin of a coordinate system such as that shown in Fig. 1-10 (b). Vector A, of magnitude 10 units, makes an angle of 30° with the positive x-axis; B, of magnitude 20 units, makes an angle of 100° with the positive x-axis; C, of magnitude 30 units, makes an angle of 180° with the x-axis; and D, of magnitude 45 units, makes an angle of 135° with the x-axis. Compute the magnitude and direction of the resultant.

41. Two vectors, V_1 and V_2, are drawn from a common origin. The vectors are each of magnitude 20 units and make an angle of 30° with each other. Find the difference $V_1 - V_2$.

42. A train has a velocity of 80 mi/hr north relative to the ground. An auto moving parallel to the train and in the same direction has a velocity of 40 mi/hr. What is the velocity of the train relative to the car? What is the velocity of the car relative to the train? If the car and train were moving in opposite directions, what would the velocity of the train be relative to the car?

43. A ball starts from rest and rolls for 8 sec with a uniform acceleration of 10 cm/sec². During the next 30 sec it moves uniformly with the acquired velocity, and finally it comes to rest after a uniform acceleration of -8 cm/sec² is applied. Calculate the total distance covered by the ball. What total time elapsed from start to finish? Ans.: 3,120 cm; 48 sec.

44. An auto traveling 60 mi/hr east crosses an intersection 2,000 ft to the north of a truck which is proceeding north at 20 mi/hr. Draw sketches showing (a) the paths of the auto and truck relative to the ground, (b) the path of the truck relative to the car, (c) the path of the car relative to the truck.

45. A ball at the end of a steel wire 0.2 m long is whirled in a circle with a constant speed of 5 m/sec. What is the period of motion? What is the frequency? What is the centripetal acceleration?

46. A projectile is shot from the floor level from an air gun in a very long tunnel 10 m in diameter. If the muzzle velocity of the projectile is 150 m/sec, what angle must the gun make with the floor of the tunnel if the projectile is to just miss the roof of the tunnel?

47. A man stands on a flat wagon holding a ball in his hand. If the wagon and man move with a constant speed of 2 ft/sec relative to the ground and the ball is dropped, what is the path of the ball as observed by the man on the wagon? As observed by a man at rest on the ground? As observed by a man on the ground moving along with the wagon? Ans.: straight line; parabola; straight line.

48. Given an object that has a constant acceleration of 2 m/sec² and at time zero has a velocity of 4 m/sec and a displacement of 1 m, find by the methods of the calculus the velocity and displacement after 5 sec.

49. If the length of the boom shown below is 10 ft, the distance ac is 10 ft, and the angle a is 30°, find the compression of the boom and the stretching force of the wire. Ans.: $C = 100$ lbf; $T = 51.9$ lbf.

* For the purposes of this chapter the trusses are assumed to be rigid and weightless and hinged at the junctions, so that forces of compression lie along the trusses.

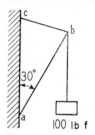

50. Given the bridge truss* shown below. If the triangle abc is an equilateral triangle, find the compression in each beam (ab) and (bc) if a weight of 50 lbf is hung from the point b.

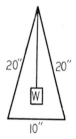

51. A light triangular frame* stands in a vertical plane and has a slant height of 20 in. A weight of 100 lbf hangs from its vertex. Find the stretching force along its horizontal base of length 10 in.

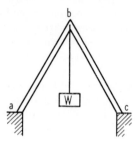

52. Find the stretching force (or compression) in each member of the pinned trusses* shown below, when loaded as shown. The sides of every triangle are in the ratio 3:4:5. *Hint:* What forces act at the points A, B, C, D, and F, respectively? Ans.: $AC = CD = 50$ lbf; $CB = 0$ lbf; $AB = BD = 30$ lbf; $CE = 60$ lbf; $DE = 0$.

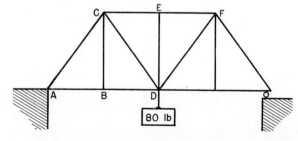

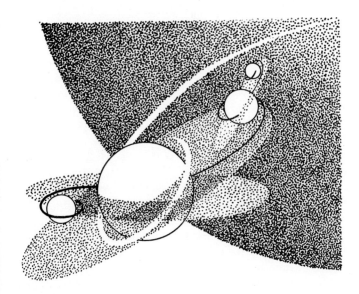

CHAPTER 2

Dynamics of a Particle: Mass—Force

2.1. The Method of Dynamics

Dynamics is the name of that branch of mechanics that treats of the motions of material objects, and not only of the motions but also of their causes and effects. The motions we encounter in ordinary experiences are usually not simple but involve many factors. For example, we see a baseball thrown. Forward velocity is imparted to it; it has rotation also; it is subject to the effects of gravity and of air resistance; it may be subjected to an impact with the bat or is slowed up and stopped finally in the catcher's glove. To analyze the motion of the ball and understand it, these various factors must be carefully separated, treated independently, and then recombined. Although we are often unable to isolate completely one single factor at a time in a complex situation, it is often possible to make this factor predominate. Then we reason from the facts observed to those that should be observed if the other factors, now already small, could be made indefinitely less and less.

The method we have briefly outlined is an important aspect of the scientific method established by the founder of modern physical science—Galileo Galilei (1564–1642), an Italian scientist of the Renaissance. Galileo made noteworthy contributions to astronomy by the application of his telescope, giving confirmatory evidence of the Copernican system. However, his most original work lay in the foundations of the experimental and mathematical science of dynamics, which we discuss in this chapter.

2.2. Concept of Mass

Our first knowledge of motion and its causes comes, perhaps, from our own attempts to set bodies in motion. The "cause" of the motion is, then, our own muscular effort, and we observe that the greater the effort, the greater the resultant motion. On the other hand, approximately the same effort appears to produce vastly differing effects, depending upon the object which is moved. A small stone, when thrown, attains a greater speed than a larger and "heavier" one, although the thrower exerts his maximum strength, let us say, in both cases. It is reasonable to suppose, then, that it is some property of the object, i.e., of matter, which determines its response to a given cause of motion. This property is called the INERTIA of the object.

The greater the tendency of an object to resist a change of velocity, the greater is its inertia. On the basis of this qualitative definition we might attempt to arrange the objects around us in the order of their inertia. For example, everyone would agree on the order in which to place a freight car, a chair, and an apple; but the definition is obviously too vague to be of much use. This property must be defined in such a way that it is amenable to precise measurement.

In order to do this, it is necessary to recognize the fundamental experimental fact that all bodies, when set in motion, tend to move with uniform velocity, i.e., in a straight line with constant speed,* with respect to an observer at rest in the primary reference system described in § 1.2. The more nearly such disturbing effects as friction, gravitational attraction, air resistance, etc., are removed, the more nearly do we observe this to be the case. In this respect, a marble on a smooth table and a freight car on a frictionless level track behave identically. Only when we attempt to *change* the velocity, is effort required, so that the inertia of bodies becomes observable *only when they are accelerated*. These considerations form the basis of our definition of inertia, or MASS, as it more often is called.

Suppose that two bodies, A and B, are mounted on frictionless wheels on a level track and connected by a spring, as shown in Fig. 2-1. When the

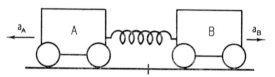

FIG. 2-1.—Cars on frictionless horizontal track

spring is compressed and released, A moves to the left and B to the right with accelerations of amount a_A and a_B, respectively, both of which can be measured. If we vary the stiffness of the spring or the amount of compression, both a_A and a_B change; but it is found *that their ratio, $-a_A/a_B$, remains constant* for all such variations as long as we retain the same bodies, A and B. It is concluded, therefore, that this constant is a property of the two objects only. We may write it as the ratio of two other constants, as follows:

$$\frac{a_A}{-a_B} = \text{Constant} = \frac{m_B}{m_A}, \qquad (2\text{-}1)$$

and call the constant m_A the mass of object A, and m_B the mass of object B.

We may use this experiment to determine when two objects have equal masses. If the ratio of the accelerations is unity, then, by definition, the two masses are equal. Having obtained a set of objects

* This empirical fact was known to Galileo but was later reformulated by Sir Isaac Newton as the First Law of Motion (cf. § 2.6).

with equal masses, we may perform a variety of experiments with the cars, using different amounts of mass in each car. For example, if we double the mass of car A, the ratio $-a_A/a_B$ is halved; if we also double the mass of car B, then the ratio becomes equal to its original value. Mass, as thus defined, is clearly a scalar quantity and possesses an additive property. *If to one body we add another, the mass of the result is the sum of the individual masses.*

If m_A, the mass of object A, is *arbitrarily* taken to be unity, then m_B is uniquely determined by the experiment. In like manner, an object C may be compared with A, and its mass number determined by measurement of the ratio of the accelerations a_C and a_A. In this way, masses could be assigned to all objects. That this definition is consistent with our earlier intuitive conception of inertia as a measure of the tendency an object possesses to resist a change in its motion can readily be seen. For, if an object X is "difficult to move" by a given muscular effort, then when the muscle is replaced by a compressed spring attached to the standard object A, this will produce a correspondingly small acceleration, a_X, so that m_X, which is given by

$$m_X = \frac{a_A}{a_X} m_A, \qquad (2\text{-}2)$$

will be large, and we say the object has a large inertia or mass. In the metric or mks (meter-kilogram-second) system the unit mass, called the "kilogram" (kg) by international convention, is adopted as being the mass of a certain block of platinum-iridium alloy, known as the "International Prototype Kilogram," which is carefully preserved, together with other standard units, at the International Bureau of Weights and Measures at Sèvres, France. In the cgs (centimeter-gram-second) system the unit mass is the gram, which is defined as one-thousandth of the International Prototype Kilogram.

2.3. Experimental Comparison of Masses

On a level track are two lightly running wagons, of unknown masses. The one on the left (Fig. 2-2) has two spring bumpers; the one on the right, a loop of cord by means of which the wagons may be held together against the compression of the bumper springs. When the string is burned, the wagons move apart with an acceleration. This acceleration lasts only while the springs are extending to their normal uncompressed length—a time interval of a small fraction of a second. If we could measure the accelerations of the two wagons, we would be in a position to determine the ratio of

Fɪɢ. 2-2.—Two wagons are tied against the thrust of the compressed springs on the bumpers.

their masses in accordance with the ideas set forth in § 2.2. Thus

$$-\frac{a_L}{a_R} = \frac{m_R}{m_L},$$

where a_L is the acceleration of the car to the left, a_R is the acceleration of the car to the right, and m_R and m_L are the masses of the two cars. However, it is quite difficult to attempt to measure these accelerations *directly*. Therefore, we make use of the fact that, whatever be their average values, $\bar{a}_L$ and $\bar{a}_R$ on each wagon, through the same short time Δt, velocities V_L and V_R are imparted to the two wagons (to the left and to the right) that are *proportional to their respective accelerations*, a_L and a_R. Hence, from kinematics,

$$V_L = \bar{a}_L \Delta t, \qquad V_R = \bar{a}_R \Delta t;$$

and, therefore,

$$\frac{\bar{a}_L}{\bar{a}_R} = \frac{V_L}{V_R}. \qquad (2\text{-}3)$$

If we neglect friction and assume that the track is level, then the wagons, after being accelerated, will drift apart with ᴜɴɪꜰᴏʀᴍ ᴍᴏᴛɪᴏɴ, i.e., with velocities V_L and V_R, respectively. These velocities may be found by measuring the *distances* covered by the wagons *in corresponding equal times*. Suppose that the mass of the wagon on the right is known, namely, 1 kg, and, by experiment, it is determined that the ratio of its velocity to that of the unknown mass M_x on the left is 2 to 1. Then from equation (2-3) it follows that the accelerations imparted to them are in the ratio of 2 to 1

and therefore the mass M_x must be twice that of the known mass or 2 kg.

In the British or fps (foot-pound-second) system the unit of mass is called the "pound." In this country, by executive order of April 15, 1893, the pound is defined as

$$1 \text{ pound (lb)} = \frac{1}{2.2046} \text{ kilogram (kg)}.$$

Practical means for comparing masses are provided by weighing, a subject to be discussed in a later section of this chapter.

2.4. Further Analysis of Mass

Because mass is a fundamental concept in physics, we feel justified in analyzing it further. We wish to learn, for example, what those specific properties of material objects are that determine the value of the ratio of accelerations in the experiment of Fig. 2-2.

Experimenting with the cars, we learn that such properties of matter as shape, color, temperature, texture, etc., do not affect the ratio. We find, further, that the motion of the cars is not changed by the state of aggregation of the material associated with them. This can easily be verified by taking one of the cars of the experiment as a standard mass and on the other placing a hollow cylinder with ice in it. We note the accelerations with the ice present, and also their accelerations when the ice is melted. We find the ratio unaltered. Similar experiments can be performed to prove that chemical reactions in closed systems will not alter the ratio and hence the mass of an object.

It is possible that you may have read that mass is "the quantity of matter in an object." That this definition is inadequate is shown in the theory of relativity, which states that the mass of a body varies with its velocity. According to this theory, if m_0 is the mass of a body at rest relative to an observer and m_v is its mass when it is moving with a velocity v relative to the observer, then

$$m_v = \frac{m_0}{\sqrt{1 - v^2/c^2}}, \qquad (2\text{-}4)$$

where c is the velocity of light in a vacuum, approximately 3×10^{10} cm/sec or 186,000 mi/sec. The velocity of light c is the limiting velocity which any object can attain. This equation has been verified many times for subatomic particles. However, for objects such as we see around us, the change in mass with velocity is negligible. Thus an airplane moving at a speed of 1,000 mi/hr has its mass increased by about 1 part in 10^{12}, a quantity much too small to be detected. The mass involved in the theory of relativity is the so-called "inertial mass." In words this theory states that it is more difficult to accelerate a body, the faster it is moving, and that it is never possible to make it move with the velocity of light.

2.5. Concept of Force

The word *force* is badly overworked in the English language. For example, we use the phrases "moral force" and "police force." However, in physics we attach to the word a technical significance. A clue to a possible definition of force is found in the fact that we intuitively associate muscular effort with the accelerated motion of objects. To illustrate, a small push will accelerate a ball (increase its velocity by a certain amount in a given interval of time). If we wish, however, to increase greatly the acceleration of this ball (i.e., cause a much greater increase in velocity in the same time interval), we must exert much greater muscular effort. *We shall define the magnitude of the resultant force which acts on an object of mass* m *as the product of the mass of the object and its acceleration:*

$$f = ma . \qquad (2\text{-}5)$$

Equation (2-5) is a relationship between the two vectors *f* and *a* so that the direction of the force is that of the acceleration. The greater the acceleration a given body possesses, the greater the force acting upon the body—a fact in complete agreement with our experience.

Equation (2-5) is not only a definition of one vector (*f*) in terms of another vector (*ma*) but is, in addition, a statement of the equality of the two vectors—*f*, which depends on the *interaction* of the *given particle* and *other particles*, and *ma*, which depends on the properties of the given particle itself. In many problems we know the nature of *f* and compute the corresponding value of acceleration.

By definition (§ 1.18), "acceleration" is the "time rate of change of velocity." Hence, if the acceleration is constant, then

$$a = \frac{v - v_0}{t}, \qquad (2\text{-}6)$$

where v_0 is the initial velocity, v the final velocity, and t the time during which the velocity changed. If we introduce the value of a of equation (2-6) into equation (2-5), we obtain

$$f = \frac{m v - m v_0}{t}, \qquad (2\text{-}7)$$

where we define the quantity mv to be the MOMENTUM of an object of mass m and velocity v. Equation (2-7) tells us that force may also be measured by the time rate of change of momentum.* Equations (2-5) and (2-7) are mathematical statements of Newton's Second Law of Motion.

Force (*f*) and momentum (*mv*) are both vector quantities subject to the rules of combination developed in chapter 1. That these quantities should be vectors checks with our intuitions. When we push or pull at anything, we always do this in a definite direction, as well as with a certain magnitude of effort. The resulting change in velocity per unit time of the object—the acceleration—we know has the direction in which we have pushed.

Quite deliberately, in our statement of the definition of "force" we said *resultant or, what is equivalent, the vector sum of all forces.* We are well aware, intuitively, that we can exert muscular force to the utmost of our physical powers and not produce motion—for example, if we were to try to push over a 20-story building or even a 2-ton truck. In such cases other forces than the one of which we are conscious are present: elastic forces of steel and concrete, great weights, large amounts of friction, and the like. Nothing moves from rest, nothing changes any velocity it may have, unless the vector sum of all existing forces adds up to something different from zero, or, in other words, unless there is an unbalanced force. Given a single force, it always produces acceleration, be the object acted upon great or small. Two simultaneous forces, however, may balance. "No acceleration" does not, of necessity, mean that there is no force—only that there is no unbalanced force, i.e., that all the forces add to zero.

A word about the choice of units. In the mks

* Strictly, eq. (2-7) should be written

$$f = \frac{d}{dt}(m v) = m \frac{d v}{dt} + v \frac{d m}{dt}. \qquad (2\text{-}8)$$

If m is constant, dm/dt is zero and

$$f = m \frac{d v}{dt} = ma .$$

system of units we choose the unit of force as that force which, acting on a mass of 1 kg, gives it an acceleration of 1 m/sec². This unit of force is called the *newton:*

$$1 \text{ newton} = 1 \text{ kg} \times 1 \text{ m/sec}^2 .$$

In general, we can say that

$$f \text{ newtons} = m \text{ kg} \times a \text{ m/sec}^2 . \qquad (2\text{-}9)$$

Similarly in the cgs system the unit of force is called the *dyne* and is that force which, acting on a mass of 1 gm, gives it an acceleration of 1 cm/sec², or

$$1 \text{ dyne} = 1 \text{ gm} \times 1 \text{ cm/sec}^2$$

and

$$f \text{ dynes} = m \text{ gm} \times a \text{ cm/sec}^2 . \qquad (2\text{-}10)$$

Since the kilogram is 1,000 gm and the meter is 100 cm, it follows that

$$1 \text{ newton} = 10^5 \text{ dynes} . \qquad (2\text{-}11)$$

Principally in physics textbooks a unit of force called the *poundal* is used. This is the force which, acting on a mass of 1 lb, gives it an acceleration of 1 ft/sec², or

$$1 \text{ poundal} = 1 \text{ lb} \times 1 \text{ ft/sec}^2 .$$

This unit, the poundal, is not used in engineering but rather the unit *pound-weight*. This we shall discuss in the section on weight.

2.6. Newton's Laws of Motion

Historically, the ideas we have been discussing were formulated as the laws of motion by the eminent English physicist and mathematician, Sir Isaac Newton (1642–1727). Newton defined "mass" as the "quantity of matter in a body as measured by the product of its density and bulk" and "force" as "any action on a body which changes, or tends to change, its state of rest, or of uniform motion in a straight line."

Newton gave the results of his and Galileo's observations and studies in the form of three laws of motion, which may be stated as follows:

I. Every body continues in its state of rest or of uniform motion in a straight line, except in so far as it is compelled by forces to change that state.

II. Time rate of change of "quantity of motion" (i.e., rate of change of momentum = msa) i

proportional to force and has the direction of the force.

III. To every action there is always an equal and contrary reaction; or the mutual actions of any two bodies are always equal and oppositely directed along the same straight line.

The First Law we have used in our formulation of the mass concept (see § 2.2).

The reader should note that "uniform motion in a straight line" means that the velocity of the particle must be constant in *magnitude* and *direction*. To change the direction of a velocity, keeping the magnitude constant, requires a force (see § 2.14).

Equation (2-5) or (2-7), we see, is the equivalent of the Second Law. We postpone discussion of the Third Law to chapter 3. A very important conclusion which we wish to draw from this discussion is that the laws of motion as formulated by Newton, or equations (2-2), (2-5), and (3-3), their equivalents, are sufficient to solve many problems in the science of dynamics.

2.7. Meaning of "Solution of a Problem in Dynamics"

By "a solution of a problem in dynamics" we mean that we should like to predict where some definite material object will be at any future time, given certain information on its initial status (i.e., initial conditions, such as its position, velocity) and the forces to which it is subjected. It should be clear that, if we know what forces act upon an object of mass m, then, assuming the forces to be constant, equation (2-5), the Second Law of Motion, tells us what the acceleration a is. Calculations of distances and velocity are then readily obtained, using the formulas of kinematics (chap. 1).

EXAMPLE 1: To illustrate this technique, let us assume that a block of mass 100 gm and velocity 100 cm/sec is moving along a smooth section of a table and suddenly comes upon a rough part which gives rise to a constant force of retardation of 200 dynes. Where will the block be 2 sec after reaching the rough part?

Solution: The initial conditions are mass of the block, velocity, and the retarding force. We wish to predict the future position of the block.

The first step in the solution of this problem is to determine the magnitude and direction of all the forces acting on the block. These evidently are the weight mg of the block downward, the equal but opposite elastic reaction of the table, and the retardation. The first two forces cancel out, leaving the retarding force of 200 dynes. From equation (2-5),

$$f = ma \ ,$$

$$200 \text{ dynes} = 100 \text{ gm} \times a \text{ cm/sec}^2 \ ,$$

$$a = 2 \text{ cm/sec}^2 \text{ in the direction of the force} \ .$$

If we take the direction of the initial velocity as a positive direction, then a is negative $= -2$ cm/sec^2. From kinematics we know that $s = v_0 t + \frac{1}{2}at^2$. Substituting the values given in the problems, we obtain for s:

$$s = (100 \text{ cm/sec} \times 2 \text{ sec}) - (\tfrac{1}{2} \times 2 \text{ cm/sec}^2 \times 4 \text{ sec}^2)$$

$$= 200 \text{ cm} - 4 \text{ cm}$$

$$= 196 \text{ cm from the start of the rough part} \ .$$

EXAMPLE 2: An automobile having a mass of 1,500 kg accelerates from rest to a speed of 13 m/sec (about 30 mi/hr) in 5 sec. Find the acceleration and the resultant force required.

Solution: The acceleration a is given by

$$a = \frac{v - v_0}{t} = \frac{13 \text{ m/sec}}{5 \text{ sec}} = 2.6 \text{ m/sec}^2 \ .$$

The resultant force acting on the automobile, from equation (2-5), is

$$f = ma = 1,500 \text{ kg} \times 2.6 \text{ m/sec}^2 = 3,900 \text{ newtons} \ .$$

2.8. Gravitational Force: Mass and Weight

We are now in a position to describe practical methods for measuring and comparing masses and forces.

A fact of common experience is that objects fall earthward. An analysis of the type of motion reveals it to be uniformly accelerated motion.* Hence we conclude from equation (2-5) that the objects while falling were subjected to unbalanced forces f_1, whose magnitudes are given by

$$f_1 = ma \ .$$

We call this very common type of force the *force due to gravity*, which originates in the mutual attraction of the object and the earth, a topic to be more fully discussed in § 2.16. Suffice it to say, however, that *all material objects are subject to gravitational forces*.

Furthermore, all objects, irrespective of their inertia (mass), fall with one and the same value of the acceleration, which is about 9.8 m/sec^2 or 980 cm/sec^2 or, in British units, 32 ft/sec^2. As we have already pointed out, this value of the acceleration

of any freely falling body is referred to so frequently that we shall use a special symbol for it, g. We call g the "acceleration due to gravity." When freed from the effects of air resistance, e g., by being placed in a vacuum, even such light objects as bits of straw or paper or feathers are also observed to fall with the acceleration g.

We wish to emphasize, however, that the statement that "all objects, irrespective of mass, fall with the same acceleration" is a result that follows readily from Newton's universal law of gravitation. A formal justification of this statement is contained in equation (2-23), together with equation (2-24) of § 2.19.

Consider two masses, m_1 and m_2, falling freely with an acceleration of g m/sec^2. The forces acting upon these objects are

$$f_1 = m_1 g \ ; \qquad f_2 = m_2 g \ . \qquad (2\text{-}12)$$

We call the gravitational force acting on m_1 *its "weight"* w_1, *and that on* m_2 *its weight* w_2. This, again, is strictly in accordance with our experience, for the expression "this lead ball is heavier than this aluminum ball" simply means that a greater force (muscular effort) is required to keep the lead ball from falling to the ground with acceleration g than is required for the aluminum ball.

* Today we have ample technology to provide high-speed, accurate timing and can measure with precision the time and distance relations in the case of a freely falling body near the surface of the ground, to show that its motion is uniformly accelerated. Galileo, it is interesting to note, reached the same conclusion most ingeniously, having available for timing only "water clocks." These measured intervals by the volume of water discharged therein.

By reducing the speed and acceleration of falling bodies by letting them roll down inclines of gentle slope, he studied and timed these less rapid motions. He assumed that the character of the motion down inclines of any slope was unchanged by the gradient and approached that of free fall as the plane became vertical.

He found his famous "odd-integer" law—to the effect that distances covered in consecutive time intervals were proportional to the odd numbers: 1, 3, 5, 7, 9, etc.

Total distances for consecutive intervals thus were proportional to 1, 1 + 3, 1 + 3 + 5, 1 + 3 + 5 + 7, etc., i.e., to the squares of the integers, 1, 4, 9, 16, 25, etc. Prove these results from eq. (1-20).

But, if distance covered is proportional to the square of time elapsed, velocity acquired is proportional to the *time* elapsed, which is true only if motion be *uniformly accelerated*.

Incidentally, Galileo thus refuted the Aristotelians, who maintained that velocity acquired was proportional to distance fallen—not to time elapsed.

Galileo's result, on the contrary, made it clear that velocity was proportional to the *square root* of *distance* covered.

Equation (2-12) becomes

$$w_1 = m_1 g \;; \qquad w_2 = m_2 g \;. \qquad (2\text{-}13)$$

By division we obtain

$$\frac{w_1}{w_2} = \frac{m_1}{m_2}. \qquad (2\text{-}14)$$

The ratio of any two masses equals the ratio of their weights. Here we have the justification for our common practice of comparing masses by "weighing." When a body of mass m_x is placed on an analytical balance, we determine the force on the mass m which keeps m_x in equilibrium. It follows that m must be attracted to the earth with the same force as that of m_x. Hence, since $w = w_x$, from equation (2-14), $m_x = m$. This technique of comparing masses depends, we see, upon the constancy of g and upon the equality of the lever arms of the analytical balances.

We wish to caution the reader against inferring from the foregoing discussion that the weight of an object is equal to its mass. We repeat that the *weight of an object* is *the force that acts on the object due to its presence in the earth's gravitational field* (see § 2.19). Weight, being a force, is dimensionally different from mass. Furthermore, weight is a vector, mass a scalar, quantity.

From the foregoing we may define that vertically downward force which the earth exerts on a 1-kg mass as 1 kg-weight (kgf). Since this mass falls with the acceleration g due to gravity, it follows that

$$1 \text{ kgf} = 1 \text{ kg} \times 9.80 \text{ m/sec}^2 = 9.80 \text{ newtons}$$

and

$$1 \text{ gmf} = 1 \text{ gm} \times 980 \text{ cm/sec}^2 = 980 \text{ dynes} \;.$$

In the metric system, using Newton's Second Law, $F = ma$, the force must be expressed in newtons (or dynes) if the mass m is in kilograms (or grams) and the acceleration a in m/sec² (or cm/sec²).

2.9. Transmissibility of a Force

If a mass is suspended from a string of negligible mass, as in Fig. 2-3, it is in equilibrium under the action of two forces, one acting downward, owing to the weight mg, and the other due to the tension T in the string, T acting upward. If the string is attached to a rigid wall, as shown in the diagram, then the string at the point P exerts a force on the

wall of value mg. We have here an illustration of the transmissibility of the force through the string to the wall. The student will note that the force in the string is equal in magnitude to the weight mg.

A spring balance is frequently used to compare forces (see § 1.25). Let us see on what principle it operates. If at some standard location, where the

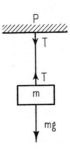

FIG. 2-3.—Transmissibility of a force

acceleration due to gravity is g, we were to restrain an object of mass m from falling to the earth by attaching it to a spring, it would be necessary for the spring, via its elastic properties, to exert an equal and opposite force of magnitude mg. By placing known masses of various magnitudes and hence known forces on the spring, we could calibrate it to read forces.

Let us attach a known mass m to the spring; it elongates, say, a distance x, as determined by a scale. Instead of calling the scale reading mg newtons, dynes, or poundals of force, we call the force m kilograms-force (m kgf), m grams-force (m gmf), or m pounds-force (m lbf), depending upon whether m is measured in kilograms, grams, or pounds. This is the basis of the Engineering system of units, which is discussed in the next paragraph.

2.10. The Engineering or Gravitational System of Units

In the so-called "absolute" system of units, mass, length, and time are taken as fundamental, and force is derived from them. On the other hand, the system of units used by engineers, the so-called "British Engineering System" of units, uses force, length, and time as the fundamental ones, and mass is derived from them. The unit of force in the Engineering system is the pound-weight, lbf, which is the force exerted on a pound-mass at a place where the acceleration due to gravity is $g = 32.174$ ft/sec². (For convenience in most calculations we shall use $g = 32$ ft/sec².)

33

The unit of mass in the Engineering system is called the *slug*, the word being associated with "sluggish," which, of course, has to do with inertia. A slug acted on by a force of 1 lbf is given an acceleration of 1 ft/sec², whereas a force of 1 lbf gives a mass of 1 lb an acceleration of 32 ft/sec². Thus the unit of mass, the slug, is equivalent to 32 lb of mass, or a slug has a weight of 32 lbf. In the Engineering system of units, Newton's Second Law is written

$$F = \frac{wa}{g}. \qquad (2\text{-}15)$$

The force F is in lbf, w is the weight of the body in lbf, and g is in the same units as a, namely, ft/sec². This is equivalent to taking the mass m of the body in slugs equal to its weight in lbf divided by the acceleration due to gravity, 32 ft/sec².

Table 2-1 sets down the various systems for comparison. Note especially the abbreviations used for mass and force units.

TABLE 2-1

COMMON SYSTEMS OF UNITS

Absolute Systems of Units	Length	Time	Mass	Force	Conversion Factors
Metric (cgs).	cm	sec	gm	1 dyne	10^{-5} newtons
British (fps).	ft	sec	lb	1 poundal	0.138 newtons
mks........	m	sec	kg	1 newton	....
British Engineering (English)..	ft	sec	slug	1 lbf	32.174 poundals

EXAMPLE 3: A force of 5 gmf is the only unbalanced force acting on a mass of 100 gm. What will be the acceleration?

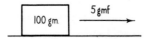

Solution: Here the weight of the block is just balanced by the upward force of the table against the block.

The unbalanced force is 5 gmf, and the mass 100 gm. In this problem we must convert the gravitational units of force into the cgs system of force. Thus

$$5 \text{ gmf} = (980 \times 5) \text{ dynes}$$

and

$$4,900 \text{ dynes} = 100 \text{ gm} \times a \text{ cm/sec}^2;$$

$$\therefore a = 59 \text{ cm/sec}^2.$$

EXAMPLE 4: Two masses, $m_1 = 20$ gm and $m_2 = 30$ gm, are attached to a string and placed over a smooth

peg, as shown in the accompanying figure. Calculate the acceleration of the masses and the tension in the string. An apparatus of this type was first used by Sir George Atwood about 1784 and is commonly called "Atwood's machine."

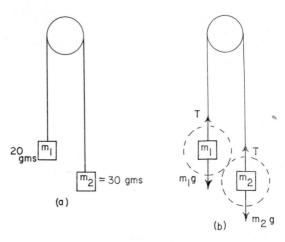

(a)

(b)

Solution: We assume in this problem that the string has no mass and hence that the tension is the same throughout the string. The first step in the solution is to isolate each mass by drawing a circle about it, as shown in (*b*), and then to ask: "What are *all* the forces acting *on the individual masses?*" On m_1 there are two forces: the tension in the cord, T, and the weight, $m_1 g$, downward. We shall choose (arbitrarily) the downward direction as positive, which means that any force, velocity, or acceleration in the downward direction is to be taken as positive and in the upward direction as negative.

The second step is to apply Newton's Second Law to each mass separately. For m_1, we have

$$m_1 g - T = m_1 a_1 \qquad (1)$$

and for m_2,

$$m_2 g - T = m_2 a_2. \qquad (2)$$

Next we obtain a relationship between the accelerations. Since the string is inextensible,

$$a_1 = -a_2. \qquad (3)$$

With three independent equations and three unknowns, we may solve for T, a_1, and a_2 as follows: Multiply equation (2) by -1 and add the resulting equation to equation (1),

$$(m_1 - m_2) g = m_1 a_1 - m_2 a_2; \qquad (4)$$

using equations (3) and (4), we get

$$a_1 = \left(\frac{m_1 - m_2}{m_1 + m_2}\right) g. \qquad (5)$$

Substituting back in equation (1) and solving for T, we see that

$$T = \frac{2\,g\,m_1\,m_2}{m_1 + m_2}. \tag{6}$$

Note that equations (5) and (6) are not general formulas but rather the algebraic solutions of equations (1) and (2) for the particular problem of this example.

Using the numerical data of the problem, we obtain

$$a_1 = \frac{20 - 30}{50}\,g = -\tfrac{1}{5}g = -196 \text{ cm/sec}^2 ,$$

$$a_2 = -a_1 = 196 \text{ cm/sec}^2$$

$$T = \frac{2 \times 980 \times 20 \times 30}{50} \text{ dynes} = 23{,}520 \text{ dynes}$$

$$= \frac{23{,}520}{980} = 24.0 \text{ gmf}.$$

The negative sign implies that m_1 goes up.

EXAMPLE 5: A car having a mass of 2 kg is accelerated along a frictionless horizontal track by means of a horizontal cord which passes over a frictionless peg to a hanging weight of 0.5 kgf, as shown in the figure. The problem is to find the acceleration of the system and the tension in the cord.

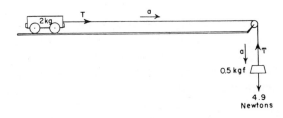

Solution: If the cord is inextensible, the acceleration a of the car is the same as that of the hanging weight. Again, if the cord is perfectly flexible and has no mass and there are no frictional forces, then the tension T in the cord is the same throughout.

Consider, first, the 2-kg car on the horizontal track. If the tension T is expressed in newtons and the acceleration in m/sec², then

$$T = 2a .$$

The 0.5-kg weight, which has a mass of 0.5 kg, is acted on by two forces—0.5×9.8 newtons downward and T newtons upward. Since the acceleration is downward, the resultant force is downward, that is, 0.5×9.8 newtons is greater than the tension T in newtons. The equation of motion for the hanging weight is

$$0.5 \times 9.8 - T = 0.5a .$$

Adding the two equations gives

$$4.9 = 2.5a ,$$

$$a = 1.96 \text{ m/sec}^2 .$$

The tension T is

$$T = 2a = 3.92 \text{ newtons}$$

$$= \frac{3.92}{9.8} = 0.4 \text{ kgf} .$$

Actually, one could find the acceleration a by noticing that the resultant force is that due to the hanging weight 0.5×9.8 newtons and the total mass moved is the sum of the two masses, namely, 2.5 kg. This is a very convenient method for calculating the acceleration of a system of connected bodies. One looks for the resultant accelerating force F on the system and the total mass M moved by this force. The acceleration is then given by Newton's Second Law, $a = F/M$.

2.11. Dimensions and Dimensional Analysis

As stated previously for the mks and cgs systems, sometimes called the "absolute systems," mass, length, and time are taken as fundamental, and the dimensions of other units are derived from these three. It is usual to use the symbols M, L, T in brackets to denote the dimensions of the quantities mass, length, and time. Thus the dimensions of velocity are [L/T] or [LT⁻¹]; of acceleration, [L/T²] or [LT⁻²]. The units used must have these dimensions. The units for velocity are m/sec, cm/sec, ft/sec, mph, etc.; and for acceleration they are m/sec², cm/sec², ft/sec², mph/sec, etc.

In any physical equation, or equation between physical quantities, *the dimensions on each side of the equation must be the same.* This is the foundation of dimensional analysis, which is used extensively in some branches of physics and in engineering. As an example consider equation (1-17),

$$S = v_0 t + \tfrac{1}{2}a t^2 .$$

The dimensions of the various terms are

$$[L] , \qquad [LT^{-1}T] , \qquad [LT^{-2}T^2] ,$$

so that each term has the dimensions of length as it must if the equation is to be of use in physics. Notice that any multiplying factor, such as the $\tfrac{1}{2}$ in $\tfrac{1}{2}at^2$, has no dimensions and makes no appearance in the dimensions of this term.

The dimensions of force in the absolute system are derived from those of mass, length, and time **35** by the use of Newton's Second Law. From this law, $F = ma$, the dimensions of force are those of mass *times* acceleration, or dimensions of force are $[MLT^{-2}]$.

In the Engineering system of units, force, length, and time are fundamental, and mass is derived. The dimensions of mass from $F = ma$ are those of force divided by acceleration, or dimensions of mass in the Engineering system are

$$\left[\frac{F}{LT^{-2}}\right] \quad \text{or} \quad [FL^{-1}T^2] .$$

When writing Newton's Second Law as $F = wa/g$, we see that the mass is w/g or force divided by the acceleration due to gravity.

2.12. Friction

More than once in our discussion of the principle of inertia (Newton's First Law) and in the definition of "force" (Newton's Second Law) we have had to make reservations whenever we said that it was a fact of common experience that moving objects, left to themselves, would continue in motion indefinitely. Common experience, we were quite aware, is *entirely to the contrary*. Every common object set in motion invariably stops—and rather quickly, too.

The fact that an object projected along a horizontal table eventually comes to rest means that it encounters RESISTANCE to its motion. This resistance, since it produces acceleration in the negative sense, is measured by a force. We call this the force of FRICTION.

Such a force is also brought into play when we attempt to start a body moving from its state of rest. Consider, for example, an iron block at rest on a horizontal table (Fig. 2-4). We attach a spring balance to it to measure the pull with which we set it in motion.

We discover that a very considerable force is exerted *before the block begins to move*. Once it has started, the maintenance of this same force produces accelerated motion. By reducing the force, once motion has begun, we find that it is possible to keep it in uniform motion without acceleration, but never by a zero force.

The force necessary to start the motion is called the force of STATIC FRICTION. The lesser force needed just to keep it moving uniformly is called the force of KINETIC FRICTION.

It is found experimentally, moreover, that *for two given surfaces* which are dry and not lubricated (e.g., that of the iron block on the wooden table) *the ratio of the tangential force needed to overcome the friction to the normal force which holds the two surfaces in contact* (here the weight of the block) *is a constant, independent of area or of the velocity with which the surfaces move*, over rather wide limits. This ratio is called μ, the COEFFICIENT OF FRICTION.

In the case of kinetic friction, as different portions of the surfaces come into contact, there are rather wide variations in the frictional force, often amounting to 10 or 20 per cent. The coefficient of static friction is larger; in addition, the value obtained by starting the object from rest a number of times is more consistent. For small velocities of sliding we may consider the coefficient of friction to be a constant.

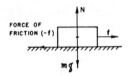

FORCE OF FRICTION (-f)

FIG. 2-4.—The force of friction

In the problems that follow we shall use only this force of sliding friction and refer to it simply as the "force of friction."

If f is the force of friction, and N the normal force, then the law of friction stated above may be written mathematically as

$$\frac{f}{N} = \mu , \qquad (2\text{-}16)$$

where μ is the constant called the "coefficient of sliding friction."

For the case of the block on the horizontal table (Fig. 2-4), if m is the mass of the block, then mg is numerically equal to the normal force N, and equation (2-16) may be written as

$$\frac{f}{mg} = \mu .$$

Suppose, for example, that the block of iron rests on a board, originally horizontal, and that the board then is tilted until a limiting angle θ_r is reached, beyond which the block will begin to slide down the board (Fig. 2-5). At this angle the component of the weight of the object along the board is just equal in amount to that necessary to overcome the force of friction. From the diagram the force down the plane is $mg \sin \theta_r$, while the normal force N is equal to $mg \cos \theta_r$. By equation (2-16) we have

$$\mu = \frac{mg \sin \theta_r}{mg \cos \theta_r}$$

or

$$\mu = \tan \theta_r . \tag{2-17}$$

The limiting value of θ_r for which equation (2-17) is true, is called the *angle of repose.* Measurement of the tangent of this angle will give the coefficient of friction of the two solids.

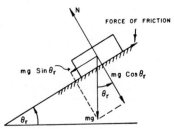

Fig. 2-5.—Limiting value of friction on an inclined plane

Table 2-2 gives the values of the coefficient of friction for various common combinations of surfaces that slide one upon the other.

TABLE 2-2

APPROXIMATE VALUES OF THE COEFFICIENT
OF FRICTION

Surfaces	μ
Dry masonry on dry masonry..	0.60–0.70
Iron or wood on stone.........	.30– .70
Wood upon wood (dry)........	.30– .50
Metal on metal (wet)..........	.25– .35
Wood upon wood (soapy)......	.20
Steel on agate (dry)...........	.20
Metal on metal (dry)..........	.15–0.20
Lubricated polished surfaces....	0.05*

* The presence of a liquid lubricant entirely alters the character of the friction, and in these cases a coefficient μ should not be defined as here. Instead, there is involved a coefficient of friction within the body of the lubricant itself. This constant is called the "coefficient of viscosity" of the lubricant.

2.13. Rolling Friction

At least passing mention should be made of cases in which circular objects, such as wheels or spheres, are used to reduce friction and thus to facilitate the motion of heavy objects along surfaces of various kinds. Friction is greatly diminished by these devices but never entirely overcome. The mechanism of rolling friction is different from that of sliding friction. If the surfaces could be made perfect-

ly hard, there would be no surface of contact, but rather a point where a sphere rests or a line when a cylinder rolls.

Actually, when a cylinder or disk rolls along a surface, deformations of the surfaces take place, as shown in Fig. 2-6 (a). The production of these de-

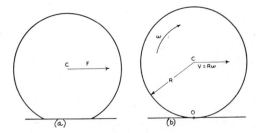

Fig. 2-6.—a, Flattening of contact between wheel and road. b, Conditions for ideal rolling of a disk on a plane.

formations is a factor causing rolling friction. From the figure it is evident that the magnitude of the force necessary to overcome a given force of friction will depend upon the point of application of the force. Under similar conditions, i.e., with f parallel to the surface, it is found that the coefficient of rolling friction may be as small as one-tenth that of sliding friction. For example, steel wheels on steel rails have a value of μ of about 0.02.

Let us suppose that the wheel shown in Fig. 2-6 (b) is rolling along a horizontal track with a speed v. If this is to roll and *not* slide, then the point or line of contact at O must be instantaneously at rest. Consider what would happen if this were not the case. Such situations arise when the wheels of an automobile spin in sand or on ice. For no spinning, the point of contact must be at rest, and the instantaneous speed of the center is $V = R\omega$.

We do not wish to create the impression that friction is always an annoyance in mechanics and serves no useful function. Power may be transmitted from one machine to another by friction belts. We are able to walk, automobiles are able to move, etc., only because of the presence of friction between objects (see § 6.14).

Finally, we should mention that frictional forces exist when objects move within fluids, liquids, or gases. These will be discussed later at a more appropriate place.

EXAMPLE 6: The problem of Example 3 ("A force of 5 gmf is the only unbalanced force acting on a mass of 100 gm. What will be the acceleration?") was idealized from actual experimental conditions, since it was assumed that there was no friction between the block and the table. Assume that the coefficient of friction between the block and the table is 0.02. Then there is a retarding force of (100) (0.02) = 2 gmf.

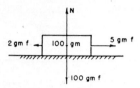

Then 5 gmf forward and 2 gmf retarding, since friction always opposes motion, leaves 3 gmf as the unbalanced applied force.

$$f = 3 \times 980 = 2{,}940 \text{ dynes} ,$$

$$2{,}940 \text{ dynes} = 100 \text{ gm} \times a ,$$

$$a = 29.4 \text{ cm/sec}^2 .$$

EXAMPLE 7: Suppose a mass of 0.1 kg hangs on the end of a cord and that the tension in the cord is 0.105 kgf. This means that, if a spring balance were inserted in the cord, the balance would read 0.105 kgf; or, if you had the upper end of the cord in your hand, you would need to pull upward with a force of 0.105 kgf in order to fulfil the conditions of the problem. Since the weight is 0.1 kgf downward, the unbalanced applied force is 0.105 − 0.100 = 0.005 kgf upward; i.e., you must be accelerating the mass upward if a tension great-

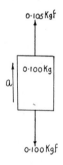

er than the weight is to be maintained in the string. To find the acceleration, the accelerating force f is

$$f = 0.005 \times 9.80 = 0.049 \text{ newtons} ,$$

and the acceleration a is

$$0.049 \text{ newtons} = 0.10 \text{ kg} \times a ,$$

$$a = 0.49 \text{ m/sec}^2 \text{ upward} .$$

To generalize this problem, let T be the tension in newtons and m be the mass in kilograms. Since f and a are vectors and m is a scalar, the direction of f, i.e., the unbalanced force, is the same as the direction of the acceleration. If T is greater than mg, the unbalanced force, and hence the acceleration, are upward:

$$T - mg = ma .$$

If T is less than mg, the unbalanced force, and hence the acceleration, are downward, and

$$mg - T = ma .$$

EXAMPLE 8: A force of 50 lbf is applied to a 100-lb mass placed on a frictionless inclined plane. What is the acceleration of the 100-lb mass?

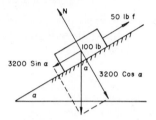

Solution: Weight = 100 × 32 = 3,200 poundals downward. Of this force, only one component, 3,200 sin a is along the plane; the other component, 3,200 cos a, is perpendicular to the plane and hence balanced by the force of the plane against the mass. Then the net unbalanced force along the plane (which is the only direction in which motion can occur) is 50 × 32 poundals up the plane less 3,200 sin a poundals down the plane:

$$(50 \times 32) - (3{,}200 \sin a) = 100a .$$

Note that here the direction of a, up or down the incline, depends entirely upon the slope a, i.e., upon whether the component of the weight along the plane is less than or greater than the given tension in the cord. For a = 30°, there are no unbalanced forces: the system is in equilibrium. For a > 30°, the acceleration is down the plane; for a < 30°, up. See problem 31.

The engineer would calculate this problem by using $F = wa/g$. The force down the plane is 100 sin a lbf; and the acceleration up the plane, if a < 30°, is

$$(50 - 100 \sin a) \text{ lbf} = \frac{100a}{32} ,$$

since the mass of the body on the plane is 100/32 slugs. As can readily be seen, the two systems using poundals or lbf and pounds-mass or slugs are equivalent. Since the engineers exclusively use slugs and lbf and not pounds-mass and poundals, we shall not use the unit of the poundal any further.

EXAMPLE 9: If, in the above example, the coefficient of friction is μ, the force of friction is $\mu \cdot 100 \cos a$ lbf and is directed down the plane if the actual motion is up the plane. Then the total unbalanced force along the plane would be

$$f = (50 - 100 \sin a - \mu\, 100 \cos a) \text{ lbf}$$

and the acceleration becomes

$$\frac{100a}{32} = 50 - 100 \sin a - \mu\, 100 \cos a\,,$$

where a is given in ft/sec^2.

In all these examples it makes no difference, as far as the resulting motion is concerned, what causes the force in the cord. The origin of the forces is immaterial. This fact may be used in determining the acceleration when several masses are connected. See problem 32.

EXAMPLE 10: A 150-lb man is in an elevator which is accelerating upward at the rate of 10 ft/sec^2. What force does he exert on the floor of the elevator?

Solution: If the elevator were stationary, the floor would exert a force of 150 lbf upward. In addition to this force, the floor must accelerate the man upward at the rate of 10 ft/sec^2. The additional force is calculated from

$$f = \frac{wa}{g}$$

$$= \tfrac{150}{32} \times 10 = 46.9 \text{ lbf}.$$

The total force F exerted on the floor of the elevator by the man is

$$F = 150 + 46.9 = 196.9 \text{ lbf}.$$

EXAMPLE 11: A stream of water impinges horizontally against a wall at the rate of 2.4 kg/min. If the velocity of the water is 9.80 m/sec, calculate the average force against the wall, assuming the water to be stopped completely.

Solution: We are essentially given the rate at which the momentum of the water is changed. Using equation (2-7), we obtain

$$f = \frac{m v - m v_0}{t}$$

$$= \frac{2.400 \text{ kg} \times 9.80 \text{ m/sec}}{60 \text{ sec}}$$

$$= 0.392 \text{ newtons or } 39,200 \text{ dynes}$$

$$\frac{0.392}{9.8} = 0.04 \text{ kgf or } 40 \text{ gmf},$$

since 1 kgf = 9.8 newtons and 1 gmf = 980 dynes.

2.14. Centripetal Forces

In chapter 1 we treated the topic of uniform circular motion kinematically. In this section we shall treat the problem from the point of view of dynamics. This solution was first given by Christian Huygens (1629–95), son of a Dutch poet and diplomat. Huygens continued the work of Galileo and was the first to solve problems in dynamics which involved several bodies. Huygens is also remembered for his valuable contributions to optics.

If a stone is attached to a string of length r and is moving uniformly in a circle, then the motion it describes can be explained on the assumption that a constant force (in magnitude) deflects the stone from the rectilinear path it would follow (because of its inertia) if no string were present. This is in accordance with Newton's First Law of Motion.

Since the value of the acceleration is, by § 1.24,

$$a_c = \frac{v^2}{r} \quad \text{or} \quad a_c = \frac{(2\pi r/T)^2}{r}$$

$$= \frac{4\pi^2 r}{T^2} \quad \text{or} \quad a_c = 4\pi^2 f^2 r\,,$$

where v is the linear speed, T the period of motion, f the frequency, and r the radius (Fig. 2-7), the

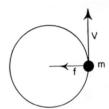

FIG. 2-7.—The centripetal force

inward radial force acting on the stone of mass m must be

$$f = m a_c = \frac{m v^2}{r} \qquad (2\text{-}18)$$

or

$$f = \frac{m 4\pi^2 r}{T^2}. \qquad (2\text{-}19)$$

Since angular velocity, ω, is defined (§ 1.24) as the angular distance, θ, measured in radians swept out per second, if an object rotates once in T sec, then

$$\omega = \frac{2\pi \text{ radians}}{T \text{ sec}},$$

and equation (2-19) may be written as

$$f = m \omega^2 r\,. \qquad (2\text{-}20)$$

The tension in the string supplies this force; by means of it, the stone is continuously deflected from its rectilinear path and made to move toward the center of the circle. If we isolate the mass by drawing an imaginary circle about it and then ask, "What are all the forces acting *on the stone?*" we find that there is *one and only one force*—that supplied by the string and acting toward the center. This radial or central force acting on the stone toward O is called the *centripetal force*. Its magnitude is given by equation (2-18), (2-19), or (2-20).

It is interesting to point out the following observation: In Fig. 2-7 we see that at any instant the linear momentum of the ball is mv pointing in the direction of the tangent to the circle. As the ball moves along the circle, *the magnitude of the momentum does not change, but its direction is continuously changing.* Treated as a vector, therefore, we have here a case in which the momentum is changing with time, and hence, according to Newton's Second Law (eq. [2-7]), a force must be present. This force is the centripetal force.*

EXAMPLE 12: A 3-lb stone on the end of a string 5 ft long is whirled in a horizontal plane with a speed of 100 ft/sec. The string breaks. Calculate the tension which broke the string. With what linear speed did the stone move just after the string broke? In what direction did the stone move just after the string broke? (In this example the force of gravity is not required.)

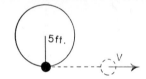

Solution:

1. $$a_c = \frac{v^2}{r} = \frac{100 \times 100}{5} = 2,000 \text{ ft/sec}^2,$$

$$f = \frac{w}{g} a = \frac{3}{32} \times 2,000 = 187.5 \text{ lbf}.$$

2. Stone left with a speed of 100 ft/sec.
3. Stone left at a tangent to the circle.

2.15. Further Examples of Centripetal Forces

a) A conical pendulum consists of a mass m attached to a string, the mass revolving about a vertical axis (PQ, Fig. 2-8) with uniform circular motion. The mass m is accelerated toward the center, O, of its circular path. To be consistent with the laws of motion, a centripetal force must be present, acting on m toward O. From the diagram (Fig. 2-8) two forces act on the mass: its weight, mg, downward, and the tension in the string along the string. These two forces, when added vectorially, must give rise to the necessary centripetal force. The reader should be able to show that the force is equal to $mg \tan \theta$.

---

* We reserve the words *"centrifugal* force" for the special phenomena of § 6.18.

b) The motion of the moon about the earth is another example of motion with centripetal force. The moon is accelerated toward the earth; the gravitational attraction of the earth on the moon is the centripetal force.

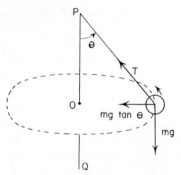

FIG. 2-8.—The conical pendulum

c) Another interesting example of centripetal forces is found in the unusual properties of rapidly revolving belts and chains. The chain and belt, flexible when at rest, become dynamically stable when revolving and exhibit properties akin to elastic solids. It is for this reason that precautions are taken to prevent rapidly moving belts from slipping off their drums.

d) A practical application of these forces is found in the so-called "centrifugal clothes-drier." Adhesive forces cause the drops of water to cling to the garments. As the garment is gradually spun around faster and faster, the adhesive force supplies the necessary centripetal force to keep the water in contact with the clothes. However, as a critical speed is reached, the adhesive force is not large enough to maintain the centripetal acceleration. As a result, the water flies off tangentially.

e) A variation of the apparatus discussed in paragraph *d* is found in the "ultra-centrifuges" designed by Svedberg in Sweden and by Beams in America for determining the molecular weights of giant protein molecules. The centrifuges are made to operate in excess of 100,000 revolutions per minute. At these speeds the centripetal force acting on any mass exceeds 750,000 times its weight.

There are a host of other practical mechanical devices that utilize centripetal forces: engine-governors, spark-timing adjustments in modern automobiles, cream-separators, and centrifuges of all kinds. The correctness of relations (2-18) and (2-19) may be tested experimentally in these many ways.

Gravitational Phenomena

2.16. Newton's Law of Gravitation

In § 2.8 we pointed out that the weight of an object is a force acting on the object by virtue of the earth's gravitational attraction. Now let us examine this statement in somewhat greater detail.

In 1687, Newton published in his *Principia* the result of an idea that had been germinating in his mind since 1666. This was the first recognition of the importance of gravitational forces and their effects in regions far beyond the immediate vicinity of the earth's surface. In addition, he gave a precise quantitative description for the evaluation of the magnitude of such forces for any bodies at any distances apart. His generalization regarding the force that pulls the apple earthward on its being loosened from the bough as but a special *local* exhibition of attractive forces that govern the motions of sun, planets, moons, and stars was, indeed, epoch-making. So important is this matter that we will devote a paragraph or two to his calculations and arguments.

From the best astronomical data of his time Newton made calculations substantially equivalent to the following, in which, however, we have used the modern values of the essential constants correct to four significant figures only:

T = period of moon's revolution about the earth

   = 27.32 days,

R = radius of moon's orbit (center of earth to center of moon)

   = 238,800 miles,

r = mean radius of earth = 3,959 miles.

On the assumption that the gravitational attraction of the earth for all bodies near or far acts as if the mass of the earth were all concentrated at its center, Newton suspected that it was this attraction, doubtless somewhat attenuated by distance, that supplied the centripetal force that held the moon in its orbit, assumed circular, about the earth. Writing, then, from our equation (2-19), the value of the centripetal force needed to hold 1 gm of the moon's mass in its orbit is

$$f = \frac{m\,4\pi^2 R}{T^2}$$

$$= \frac{1\,\text{gm} \times 4\pi^2 \times (238,800 \times 5,280 \times 12 \times 2.540)\,\text{cm}}{(27.32 \times 24 \times 60 \times 60)^2\,\text{sec}^2}$$

$$= 0.2722\ \text{dynes}\ .$$

This represents an attenuation of the force from that of gravity at the earth's surface by the ratio

$$\frac{0.2722}{983.2} = 2.763 \times 10^{-4} = \frac{1}{3,619}.$$

Now the ratio of the distances involved, moon to earth's center, compared to earth's surface to earth's center, is

$$\frac{R}{r} = \frac{238,800}{3,959} = 60.31,$$

the *reciprocal of which* SQUARED is

$$\frac{1}{(60.31)^2} = \frac{1}{3,637}.$$

The agreement between these two numbers to 1 part in 200 is, indeed, striking. That it is not more exact is due to the unwarranted approximations we made of (*a*) a spherical, instead of an oblate spheroidal, earth and (*b*) a circular, instead of an elliptical, orbit for the moon about it.

Today the vast accumulation of detailed observation of the motions of everything periodic in the solar system is in accordance with the following INVERSE SQUARE law for gravitational forces first stated by Newton:

Every particle in the universe attracts every other particle with a force that is proportional to the product of the masses and inversely proportional to the square of the distance between the particles.

If the two masses m_1 and m_2 are separated by a distance r, the gravitational force between them is given by[*]

$$f = \gamma\,\frac{m_1 m_2}{r^2}. \tag{2-21}$$

This inverse square law of equation (2-21) holds only for point masses. However, as we have noted previously, for spheres of uniform density it is correct to consider the mass, however great, concentrated *at the center*.

* The quantity γ is a constant whose magnitude depends on the units used. The units and the method of measurement are given in the following sections.

2.17. Kepler's Laws and Gravitation

Newton's universal law of gravitation (eq. [2-21]), together with his formulation of the Laws of Motion (§ 2.6), laid the foundations for modern astronomy. Newton showed, for example, that the three empirical laws of planetary motion deduced by Johannes Kepler (1571–1630) from the observational data of Tycho Brahe (1546–1601) could be given a rational explanation. These famous laws of Kepler are:

I. The orbit of each planet is an ellipse with the sun at one focus.

II. The radius vector drawn from the sun to a planet sweeps out equal areas in equal times.

III. The squares of the periods of revolution are proportional to the cubes of the semimajor axes of the elliptical orbits.

We may without serious error assume that the ellipse can be approximated sufficiently well by a circle, the semiaxes becoming the radii of the circle. The sun is then at the center of the circle, and the planets revolve about it. By the law of gravitation the sun exerts a force on the planet along the radius vector, R, and of value given by equation (2-21), i.e.,

$$f = \frac{\gamma\, mM}{R^2},$$

where M is the mass of the sun and m that of the planet. The resulting centripetal acceleration is given by equation (1-32), i.e., by

$$a_c = \frac{4\pi^2 R}{T^2},$$

where R is the radius of the circle and T the period of revolution. By Newton's Second Law of Motion,

$$f = ma, \qquad \frac{\gamma\, mM}{R^2} = \frac{m\,4\pi^2 R}{T^2},$$

and hence T^2 is proportional to R^3. This is the Third Law of Kepler. The result may be proved exactly by using elliptical orbits. The Second Law follows from an analogous argument.

Another consequence of the law of gravitational attraction is the production of *tides*. It is well known that at any place on the shore of an ocean the water level rises to a maximum and then falls to a minimum once in about every 12 hours and 25 minutes. This effect is due primarily to the gravitational attraction of the moon and sun on the water of the earth and to the rotation of the earth. In addition to these tides of the oceans, it has also been found that the earth itself yields to these tide-producing forces to a maximum extent of about 9 inches. On the basis of this result, geophysicists have decided that the earth must be as rigid as a ball of steel.

As a final example it is interesting to point out that, because of the character of the law of gravitational attraction, an object initially at the surface of the earth may escape to outer space if projected upward with a high enough initial velocity. If air friction is neglected, calculations indicate that the initial upward velocity is given by $v_0 = \sqrt{2g_0 R}$, where g_0 is the value of g at the surface of the earth and R is the radius of the earth. The escape velocity is about 7 mi/sec. This type of calculation has been applied to the problem of determining whether or not gases of the earth's atmosphere can escape into outer space. There is evidence to indicate that the temperature of the air at high levels (about 300 mi) may reach such values as to enable such gases as hydrogen and helium to escape.

2.18. Cavendish Experiment

The determination of the constant γ of equation (2-21) was first made by Henry Cavendish in 1798, using a plan described by Newton and first attempted by the Rev. John Mitchell nearly thirty years earlier. The experiment can now be performed in any laboratory.

If we suspend from a long, fine quartz fiber, as in Fig. 2-9, two small balls of equal mass, say of

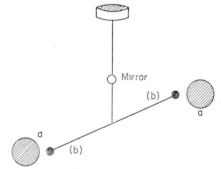

Fig. 2-9.—The Cavendish experiment

1 gm each, carefully protecting this little system from air currents, it will finally come to rest in a position in which the fiber is entirely untwisted. Now, if two massive lead balls of about 3,000-gm mass are brought into position (*a*), one on either side of the suspended system and each quite close to the small ball, the purely gravitational attraction will cause the suspended system to rotate slightly until the slight twist so produced in the fiber balances this attractive tendency and the suspended system is again at rest. Next, the positions of the two lead balls are reversed into locations (*b*). The gravitational attraction now produces a counterrotation. A spot of light reflected

from a tiny mirror on the suspended system enables this twist to be accurately measured.

Knowing the elastic constants of the fiber, by methods outlined in chapter 7, we may calculate f for equation (2-21). The factor γ, as recently measured by P. R. Heyl, is 6.670×10^{-8} dynes-cm^2/gm^2. We may write equation (2-21) as

$$f = \frac{6.670 \times 10^{-8} m_1 m_2}{r^2} \text{ dynes} . \quad (2\text{-}22)$$

The force between two 1-kg masses placed 1 m apart is (from eq. [2-21]) 6.67×10^{-6} dynes or 6.67×10^{-11} newtons. Thus the gravitational constant γ in the mks system is 6.67×10^{-11} newtons-m^2/kg^2.

The masses m_1 and m_2 present in the law of gravitation as given by equation (2-22) are sometimes designated as the *gravitational masses* of the objects, to distinguish them from their *inertial masses* as measured, for example, by the experiment in § 2.2. It is not clear from any previous arguments that gravitational and inertial masses are equivalent. It is a postulate of physical science, supported, to be sure, by experiment, that gravitational mass and inertial mass are equivalent. This postulate was used by Einstein (1879–1955) in his formulation of the General Theory of Relativity.

2.19. Meaning of Weight

If we have a mass of m kg near the surface of the earth, the force of attraction between it and the earth, of mass M kg, is

$$f = \frac{6.670 \times 10^{-11} mM}{r_1^2} \equiv \text{weight} , \quad (2\text{-}23)$$

where r_1 is the distance in meters between the center of the earth and the center of the mass, m. It is this force, f, that we had previously defined as the *weight*, w, of the object whose mass is m kg. By dividing (2-23) by m, we obtain the acceleration, a, that the mass would possess if free to move,

$$\frac{f}{m} = a = \frac{6.670 \times 10^{-11} M}{r_1^2} \text{ m/sec}^2. \quad (2\text{-}24)$$

We see from equation (2-24) that the acceleration is the same for all objects, irrespective of mass, since the mass factor, m, of the object divides out. This acceleration we have called the "acceleration due to gravity," g, that is,

$$g = \frac{6.670 \times 10^{-11} M}{r_1^2} \text{ m/sec}^2. \quad (2\text{-}25)$$

We see now why g must be the same for all objects, irrespective of size, shape, or mass. Further, we also see why g does not vary greatly for distances near the surface of the earth, for r_1 is approximately equal to 6.37×10^6 m, and a small variation in it would not appreciably affect g.

Owing to the rotation of the earth (§ 6.18) and to the fact that the earth is not spherical in shape, the value of g must also vary with geographical latitude. At sea-level and any latitude, ϕ, the following empirical equation gives g_ϕ:

$$g_\phi = (9.832 - 0.052 \cos^2 \phi) \text{ m/sec}^2 . \quad (2\text{-}26)$$

At sea-level and latitude 45°, $g = 9.806$ m/sec^2.

The standard value of the acceleration due to gravity, g_s, adopted by the International Committee on Weights and Measures is $g_s = 980.665$ cm/sec^2, or 32.174 ft/sec^2, and is the value that should be used in data that involve this constant.

The value of g (m/sec^2) at various heights h (meters) above sea-level is given by

$$g_h = g_\phi - 3.086 \times 10^{-6} h . \quad (2\text{-}27)$$

From equation (2-23) the direction of f, and therefore of g, is along the radius joining the center of m and M, and hence the direction would change from point to point along the surface of the earth. However, for points sufficiently close together, say within a city block, the directions of g are sensibly parallel. If we also confine ourselves close to the surface of the earth, so that r_1 does not vary appreciably in magnitude, then g, treated as a vector, is a constant. Within such a circumscribed region, we have what is called a *uniform gravitational field*.

2.20. Résumé

The following is a brief listing of the definitions, principles, and theories, given in this chapter, which you should know:

The concept of inertia and how the mass is measured
Newton's first two laws of motion
The distinction between mass and weight
The definitions of the newton, dyne, lbf and slug, kgf, and gmf
Newton's law of gravitation
Centripetal force
The definition of the coefficient of friction

Queries

1. Two identical closed bottles are placed on an analytical balance. A live fly is placed in one bottle. Is it possible to obtain the true weight of the fly?

2. The equation of motion (2-5) is a perfectly general one, applicable to all types of motion, provided that f and a are instantaneous values. In the illustrations given in the text, however, only constant forces were used (for mathematical simplicity). Can you cite examples from nature in which forces vary in some regular manner?

3. The following question is sometimes asked: If an object of mass m falls to the ground from a height h, with what force does it strike the ground? Has the problem an answer? Explain.

4. In § 2.14 we emphasized that the ball at the end of the string was subject to one force only, that toward the center. According to Newton's Second Law, the object should therefore be continually accelerated toward the center. Does it follow that the motion must be toward the center? Explain.

5. How far does the moon fall toward the earth in 1 sec?* If this is true, how is it that the moon never appears to get any nearer to the earth?

6. A mass M is supported by a cord A, and a similar piece of cord B is attached to the lower side of the mass. With a sudden jerk on the lower cord, it is found that cord B always breaks, while for a steady pull cord A always breaks. Explain this phenomenon.

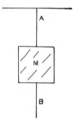

7. State the acceleration in the following problems in which the given resultant force acts on the total mass given:

	Force	Mass
a)	1 newton	1 kg
b)	9.80 newtons	1 kg
c)	1 kgf	1 kg
d)	10 newtons	2 kg
e)	10 kgf	2 kg
f)	1 dyne	1 gm
g)	980 dynes	1 gm
h)	1 gmf	1 gm
i)	100 dynes	20 gm
j)	100 gmf	20 gm
k)	1 lbf	1 slug
l)	1 lbf	1/32 slug or 1 lb
m)	10 lbf	10 lb

8. State the force in the following problems:

	Mass	Acceleration
a)	1 kg	1 m/sec²
b)	1 kg	9.80 m/sec²
c)	10 kg	200 m/sec²
d)	1 gm	1 cm/sec²
e)	1 gm	980 cm/sec²
f)	100 gm	9.80 cm/sec²
g)	1 slug	1 ft/sec²
h)	1 slug	32 ft/sec²
i)	32 slugs	1 ft/sec²
j)	1 lb	1 ft/sec²
k)	1 lb	32 ft/sec²

9. State the mass in the following problems:

	Force	Acceleration
a)	1 newton	1 m/sec²
b)	9.80 newtons	1 m/sec²
c)	1 kgf	1 m/sec²
d)	1 dyne	1 cm/sec²
e)	980 dynes	1 cm/sec²
f)	1 gmf	1 cm/sec²
g)	1 lbf	1 ft/sec²
h)	1 lbf	32 ft/sec²

Practice Problems

In the following problems the student can always get a check (except for actual numerical values) on the correctness of the equation he sets up for the solution by remembering that one can *equate* only the *same kinds of quantities*. A force can be equal only to another force, etc. DIMENSIONAL ANALYSIS may be used as a guide sometimes, in case a detailed law is not known, to find the form or function by which one quantity is related to a series of others (see § 2.11).

In solving these problems we recommend following the general pattern of Examples 1 and 3.

1. A mass of 50 gm is acted upon by a force of 10 gmf. What is the acceleration of the 50-gm mass? Ans.: 196 cm/sec².

2. What constant force in dynes and newtons acting

* This was, as a matter of fact, what Newton computed, and not the moon's acceleration, as we did in § 2.16.

for 10 sec will increase the velocity of an object of mass 100 gm from 10 m/sec to 100 m/sec?

3. What constant force acting for 5 sec will decrease the momentum of an object from 10,000 gm-cm/sec to 2,000 gm-cm/sec? Ans.: $-1,600$ dynes.

4. Two masses of 100 and 50 gm each on a horizontal frictionless table are connected with a string of negligible mass. If a force of 50 gmf acts upon the system, as shown in the accompanying figure, find the acceleration of the system and the tension in the cord between the 50- and 100-gm masses.

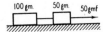

5. Two masses of 10 and 5 kg are placed on a horizontal table and connected by a string of negligible mass, as shown in problem 4. If a horizontal force of 1.5 kgf acts on the system, find the acceleration of the system and the tension in the cord between the 10- and 5-kg masses if (a) there is no friction; (b) the coefficient of friction between the table and the blocks is 0.01. Ans.: (a) 0.98 m/sec², 9.8 newtons = 1 kgf; (b) 0.882 m/sec², 8.92 newtons = 0.91 kgf.

6. Two masses of 10 and 5 lb are placed on a horizontal table and connected by a string of negligible mass, as shown in problem 4. If a horizontal force of 3 lbf acts on the system, find the acceleration of the system and the tension in the cord between the 10- and 5-lb masses if (a) there is no friction; (b) the coefficient of friction between the table and the blocks is 0.1.

7. What is the acceleration of a 1,000-gm mass placed on a horizontal table and subject to a force of 50 gmf acting 30° with the horizontal? The coefficient of friction of block and table is (a) 0.1; (b) 0.01. Ans.: (a) 0; (b) 32.4 cm/sec².

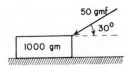

8. A 600-lb shell leaves the barrel of a gun with a muzzle velocity of 1,500 ft/sec. If the barrel is 50 ft long, calculate the force in lbf acting on the shell, assuming the force to be constant over the length of the barrel.

9. A machine gun fires 20 bullets per second into a target. If the mass of each bullet is 10 gm and the velocity is 75,000 cm/sec, calculate the average force f in kilograms-force (kgf) with which the target is pushed back. Ans.: 15.3 kgf.

10. A man weighing 150 lbf steps into an elevator and is accelerated upward with an acceleration of 2 ft/sec². What force does the man exert on the floor of the elevator? Draw a vector diagram showing all the forces acting on the man.

11. A block of mass m gm is placed on an inclined plane. When the angle of the plane is 37°, the block starts to move down the plane with uniform velocity. Calculate the coefficient of friction between the block and plane. Ans.: $\mu = 0.75$.

12. A weightless string is hung over a frictionless pulley (Atwood's machine). To the end of the string are attached masses of 20 and 10 gm, respectively. Calculate the tension in the string and the acceleration of the masses.

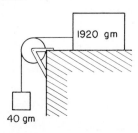

13. If there is no friction, compute the acceleration of these masses and the tension in the cord, using the mks system of units. Ans.: $a = 0.20$ m/sec², $T = 0.384$ newtons.

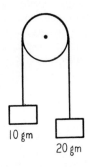

14. In problem 13, if the coefficient of friction is $\frac{1}{100}$, compute the acceleration and the tension in the cord.

15. In problem 13, if the coefficient of friction is $\frac{1}{10}$, compute the acceleration and the tension in the cord. Ans.: $a = 0$; $T = 0.04\ g$ newtons.

16. A car having a speed of 60 mi/hr can be stopped by the brakes in a distance of 200 ft. Find the time of stopping and the value of the braking force if the weight of the car is 3,000 lb. If this same car is stopped from 30 mi/hr with the same braking force, find the stopping distance and stopping time.

17. A block slides down an inclined plane of angle of inclination of 30°, with a coefficient of friction of 0.2. Find the acceleration of the block down the plane. Ans.: 0.33 g.

18. If there is no friction, compute the acceleration of these masses and the tension in the cord.

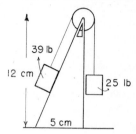

19. In problem 18, if the coefficient of friction is $\frac{1}{10}$, compute the acceleration and tension in the cord. Ans.: $a = 4.75$ ft/sec^2; $T = 28.7$ lbf.

20. In problem 18, what coefficient of friction would cause the blocks to move with no acceleration, once they were given a velocity?

21. If there is no friction, compute the acceleration and the tensions in the cords. Ans.: $a = 30$ cm/sec^2; $T_1 = 437,000$ dynes; $T_2 = 407,000$ dynes.

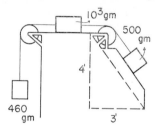

22. A block is given an initial velocity of 50 ft/sec up the plane shown in the accompanying figure. How far up the plane will it move (no friction)? How many seconds before it reaches the highest point? What will be its velocity when it gets back to the bottom of the plane? Ans.: 83 ft; 3.33 sec; 50 ft/sec.

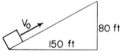

23. In problem 22, if the coefficient of friction is 0.2, how far up the plane will it move? How many seconds before it is again at the bottom of the plane? What will be its velocity when it gets back to the bottom of the plane? Ans.: 60 ft; 5 99 sec; 34 ft/sec.

24. The centripetal force on a ball of mass 50 gm moving in a circle of 10-cm radius is 1 kgf. What is its period of motion? Its angular velocity? Ans.: 0.142 sec; 44.3 rad/sec.

25. A car goes around a curve of 121-ft radius at a speed of 30 mi/hr. What must be the coefficient of friction between the wheels and the level pavement in order that the car may not skid? Ans.: 0 5.

26. The coefficient of rolling friction between a loaded boxcar and a level track is 0.02. If a car weighing 50 tons-force and moving 1 ft/sec is to be stopped in 6 in, how much force must be applied? How much force must be applied to keep the car moving 1 ft/sec?

27. Calculate the acceleration of the moon toward the earth if the moon, of mass 8×10^{19} tons, is at a distance of 238,000 miles from the earth and has a period of rotation of 28 days. With what centripetal force in tons-force must the earth pull on the moon?

28. A 50-gm body with an initial velocity of 20 cm/sec is acted on for 3 sec by a force of 2 gmf acting in the direction of its initial velocity. Find the acceleration and the final velocity.

29. Find the tension in the cord and the acceleration of each block. (Assume that the cord slides without friction under and over each pulley and that the weight of the pulleys is negligible.) Ans.: $T = 40/7$ lbf; $a_8 = 2g/7$ ft/sec^2 down; $a_{10} = g/7$ ft/sec^2 up.

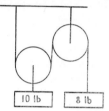

30. Calculate the force of attraction of two lead spheres 5 cm in radius and placed so that their centers are 12 cm apart. Take the density of lead to be 11 gm/cm^3.

31. Using equation (2-22), calculate the mass of the earth, given that $R = 6.38 \times 10^8$ cm. *Hint:* Let the force of attraction between the earth and a 1-gm mass be 980 dynes. Ans.: 5.96×10^{27} gm.

32. Calculate a value for the mean density of the earth.

33. In Example 8 of § 2.13 calculate the value of the acceleration when (1) $a = 45°$; (2) $a = 30°$; (3) $a = 15°$. Ans.: $0.21g$ down; 0; $0.24g$ up.

34. In Example 9 of § 2.13 calculate the value of the acceleration for $\mu = \frac{1}{100}$; $a = 30°$; $a = 45°$; $a = 15°$.

35. Solve problems 1, 2, 3, 7, 24, 28, and 30, using the mks system of units. For example, in problem 3 the momenta considered are $\frac{1}{10}$ kg-m/sec and $\frac{1}{50}$ kg-m/sec. Then, since $F = mv_1 - mv_2/t$, we see that

$$F = \frac{\frac{1}{50} \text{ kg-m/sec} - \frac{1}{10} \text{ kg-m/sec}}{5 \text{ sec}} = -\frac{4}{250} \text{ newtons}.$$

36. Find the acceleration of the cars and the tensions in the cords attached to the 0.5-kg car when (a) a constant horizontal force of 0.1 kg pulls on them; (b) a hanging weight of 0.1 kg is attached as shown in the diagram (assume the system is frictionless).

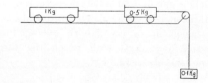

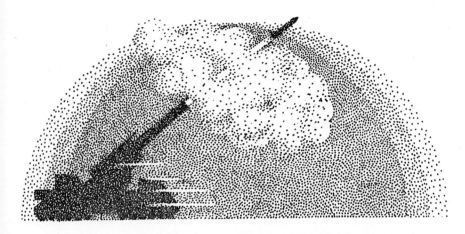

CHAPTER 3

Conservation of Momentum

3.1. Third Law of Motion*

The experiment we have already discussed in chapter 2 (Fig. 2-1) will serve to introduce Newton's Third Law of Motion. We recall that we had two small cars, freely mounted on wheels so that they might roll with very little friction on a horizontal track. The cars were then tied together against the pressure of two coiled springs mounted on bumpers between them. The string was then burned, the springs being released thereby, and the cars moved apart. Observation shows that *BOTH* CARS MOVE, AND MOVE IN OPPOSITE DIRECTIONS FROM THE POSITION THAT THEY OCCUPIED BEFORE THE STRING WAS BURNED. This is the most obvious aspect of the Third Law.

However, in addition, a definite quantitative relation is also involved, as we have already noted,

$$\frac{m_A}{m_B} = \frac{-a_B}{a_A}, \qquad (3-1)$$

where m_A and m_B are the masses of the two cars and a_A and a_B are their respective accelerations. The negative sign indicates that the cars go in opposite directions.

Also, since we have defined force as the product of m and a, i.e.,

$$f = ma, \qquad (3-2)$$

it follows from equations (3-1) and (3-2) that

$$m_A a_A = -m_B a_B$$

* The student interested in the historical development of the ideas of mechanics will find great interest in the perusal of one of the classics on this subject, *The Science of Mechanics*, by Ernst Mach, English translation by McCormack (Chicago: Open Court Pub. Co., 1902).

or

$$f_A = -f_B, \qquad (3-3)$$

i.e., the force acting on car A is exactly equal but opposite to the force acting on car B. Equation (3-3) is the analytic statement of the Third Law of Motion.

Newton's statement of the Third Law of Motion is:

To every action (force) there is always an equal and contrary reaction (force); or the mutual actions of two bodies are always equal and oppositely directed along the same straight line.

The Newtonian statement of the Third Law and the analytic representation of it, equation (3-3), contain two important ideas:

First, forces in nature always occur in pairs of equal and opposite forces; that is to say, if an object A is accelerated, then somewhere in the universe a second object (perhaps not easily discernible) is also accelerated in such a way that equation (3-1) is satisfied.

Second, *these forces always act upon different objects.*

Before going on to the analytical consequences of equation (3-3), let us first cite a number of examples of the Third Law taken from common experience.

3.2. Illustrations of the Third Law

If we fire a gun or even a heavy-caliber revolver, we are, of course, conscious of the recoil, i.e., of the force imparted to our shoulder or our hands through the barrel of the weapon as the expansion of the powder hurls the bullet swiftly forward. Fortunately, few of us know from personal experience how powerful and sledge-hammer-like is the blow which the stopping of this selfsame bullet imparts when our persons are subjected to the reacting force that brings it to rest.

Let us look at an even more simple situation, one that is static rather than dynamic—that of a book lying at rest on a table. Here, as in all such problems of static equilibrium, we must be careful not to confuse the forces that *act on the book* and produce static equilibrium with the action and reaction of Newton's Third Law.

A force (*action*) *acting on the book* is its weight, W, downward, due to the gravitational attraction of the earth on the book. The *reaction* to this force is the gravitational attraction of the book on the earth. We have here identified a pair of forces, as required by the Third Law. *If no other forces were present, then Newton's Second Law, $f = ma$, would* require that the earth and book should accelerate toward each other.

To prevent this mutual acceleration, we interpose a third body, the table. The book exerts a force (action) on the table, owing to its weight. The table, under the action of the weight, bends a little and *reacts* on the book, exerting an elastic force of amount F_T *on the book*. In accordance with the Third Law, we have again identified a pair of forces acting, respectively, on table and book.

Let us now apply Newton's Second Law, $f = ma$, to the book. *Two forces act on the book, F_T and W.* Since the book is in static equilibrium, $F_T + W = 0$, or $F_T = -W$. The table and earth are also in equilibrium in this problem. As an exercise the reader may apply $F = ma$ to each and determine the forces that maintain the table and earth in equilibrium.

The illustration we have given is a typical application of the laws to the domain of statics. Now suppose we transfer this problem to the domain of dynamics, where there are *unbalanced* forces present. What occurs when we push (accelerate) the book along the table? Since the book is accelerated, a resultant force must act on the book. According to the Third Law, an equal but opposite force must act upon the experimenter. And such is the case. Since, however, the experimenter is attached to the earth through frictional forces, the force acting on the experimenter acts then on the earth, giving it an acceleration opposite to that of the book and of

a magnitude determined by equation (3-1). However, the mass of the earth is so great in comparison with the book that its acceleration is vanishingly small. Hence the observation that the book moves but the experimenter appears at rest.

As a variation, now let us assume that both the book and the experimenter are on a horizontal frictionless plane (ice is a good approximation). Here we observe that when the experimenter pushes the book it goes forward but that the experimenter recoils, the accelerations in each case being given by equation (3-1). The points we wish to emphasize are: (1) when a body is in static equilibrium, two or more forces that add to zero must be acting on this one object; (2) the *pair* of forces that comes into play by virtue of the Third Law act on different bodies; and (3) the acceleration produced in each object equals the resultant force *acting on it*, divided by the mass of the object.

3.3. Impulse: Conservation of Momentum

In nature many motions take place quite suddenly; that is to say, the forces acting are rapidly changing in magnitude and direction. In fact, the entire period during which acceleration takes place may be compressed into a small fraction of a second. This was the case in the experiment in Fig. 2-2, where the springs relaxed and imparted acceleration in less time than the camera could record. For another example consider a golfer who tees off. The time during which the driver is in contact with the golf ball is small indeed. Furthermore, the force acting on the ball varies rapidly in magnitude. We may represent what occurs during the "tee-off" by plotting the magnitude of the force acting on the ball as ordinate and the time the force acts as

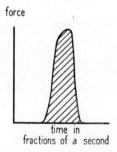

force

time in
fractions of a second

FIG. 3-1.—A variable force acting for a short time

abscissa. Figure 3-1 might be a typical representation.

According to equation (2-7), the average force

on the ball and the duration of the force are given by

$$\bar{f} = \frac{m v - m v_0}{t}, \qquad (3\text{-}4)$$

where v is the velocity of the ball after being hit, and v_0 (usually zero) is the ball's velocity before being hit.

In the golf-ball problem both f and t are extremely difficult to measure individually. But the product, ft, is quite a determinate quantity, easily measured. Furthermore, in problems of this sort we are generally interested in the *total effect of the force*, not in the value of the force or the way in which it may vary. This total effect is, by equation (3-4), the product ft equal to the total change in momentum imparted.

For this reason it is convenient to introduce a new quantity called the IMPULSE (or sometimes IMPULSIVE FORCE), which, *for a uniform force*, is defined as

$$\text{Impulse } (I) \equiv ft . \qquad (3\text{-}5)$$

This is measured in newton-seconds, dyne-seconds, or lbf-sec, depending upon the units chosen. Impulse is a vector quantity. A graph illustrating a constant force acting for a known time t is shown in Fig. 3-2. Note that the area under the graph is equal to $ft \equiv$ Impulse.

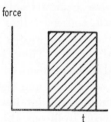

force

FIG. 3-2.—A constant force acting for a short time

Likewise, the area under the curve of Fig. 3-1 is the impulse given the golf ball. Both are measured by taking the difference of the final and initial momenta.*

In this case, where the force f depends upon the time, the impulse I is given by

$$I = \int_{t_0}^{t_1} f\, dt . \qquad (3\text{-}6)$$

---

* The idea of momentum as the central and controlling factor in cases of impact was recognized by the English physicist John Wallis and presented to the Royal Society in 1671.

Let us apply what we have just been discussing about impulse to the motion of two objects colliding with each other. Suppose that there is a collision between two balls of masses m_1 and m_2 and velocities u_1 and u_2 (Fig. 3-3). After collision, their

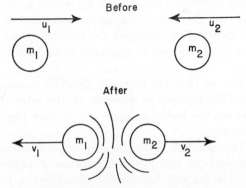

Before

After

FIG. 3-3.—Collision of two balls

velocities will presumably be different; so we shall call them v_1 and v_2. Let the time during which they are in contact be t. From the Third Law of Motion the force f_1 acting on m_1 due to m_2 is equal and opposite to f_2 acting on m_2 due to m_1 at each instant of time. Since the two balls are in contact each with the other for the same time interval, the impulse $f_1 t$ is equal and opposite to the impulse $f_2 t$, i.e.,

$$f_1 t = - f_2 t ;$$

and, since

$$\left.\begin{array}{l} f_1 t = m_1 \left(v_1 - u_1 \right) , \\ f_2 t = m_2 \left(v_2 - u_2 \right) , \end{array}\right\} \qquad (3\text{-}7)$$

regardless of the manner in which the forces between the balls may vary, it follows that

$$m_1 \left(v_1 - u_1 \right) = - m_2 \left(v_2 - u_2 \right) \qquad (3\text{-}8)$$

or

$$m_1 u_1 + m_2 u_2 = m_1 v_1 + m_2 v_2 . \qquad (3\text{-}9)$$

These are vector quantities subject to the rules of addition given in chapter 1. Their proper algebraic signs must be used. Now the left side of this equation is the sum of the momenta of the two balls before collision. The right side is the sum of the momenta after collision. Evidently, the total momentum of the system taken as a whole was unchanged by the collision.

We have here a special case of the general principle of the CONSERVATION OF MOMENTUM:

The vector sum of all the momenta for any system of bodies moving in any way and interacting in any way remains constant, provided that there are no "external" forces acting upon the system.

The proviso that no external forces act upon the system is introduced to guard against a situation such as the following. Suppose that, as the two balls of our example collide, one of the balls is given a push by some third body. This third body would contribute the external force, and we should find that the vector sum of the momenta of the two balls before and after the collision is not the same. Can friction be regarded as an "external" force?

We see that the principle depends in no way upon the manner or duration of the interaction between the bodies or upon any of their physical properties.

This principle of the conservation of momentum represents *no new fact;* but, because of its precise formulation and because the quantities in it are easily measured, the principle is by far the more useful formulation of the Third Law.

EXAMPLE 1: A United States Army rifle weighs 8.69 lbf and fires a 150-grain bullet with a muzzle velocity of 2,700 ft/sec. What is the velocity of recoil of the gun?

Solution: The momentum of the gun and bullet just before firing is zero. Hence, from the principle of the conservation of momentum, the total momentum just after firing must also be zero. However, the bullet has a forward momentum; hence the gun must have an equal backward momentum:

$$150 \text{ grains} = \frac{150}{7,000} = 0.0214 \text{ lbf},$$

$$\frac{-(8.69)(\text{velocity of gun})}{32}$$

$$= \frac{(0.0214)(2,700)}{32} \text{ slugs-ft/sec},$$

$$\text{Velocity of gun} = -6.65 \text{ ft/sec}.$$

The negative sign indicates that the velocity of the gun is in the direction opposite to that of the bullet.

EXAMPLE 2: What average force is necessary to stop, in 5 sec, a 3,000-lb car moving 88 ft/sec?

$$[F \text{ (lbf)}] \cdot [5 \text{ sec}] = \frac{[3,000 \text{ lbf}]}{32} \cdot [0 \text{ ft/sec}]$$

$$-\frac{[3,000 \text{ lbf}]}{32} \cdot [88 \text{ ft/sec}]$$

$$5F = (-3,000)\frac{88}{32}$$

$$F = -1,650 \text{ lbf}.$$

The answer is negative, since it is a retarding force.

3.4. Some Simple Examples of the Conservation of Linear Momentum

The law of conservation of momentum requires a few simple experiments to bring home its full meaning and general applicability.

A man stands at one end of a low flat wagon (both at rest) on a frictionless plane. He starts to run off the wagon, taking with him a certain amount of momentum (to the left). From the conservation principle, since the man plus the wagon had initially no momentum, then, after the man leaves the wagon, the vector sum of the momenta of the wagon plus the man must still be zero. The only way this can occur is for the wagon to move to the right with the same momentum as that possessed by the man. This is checked by measurement. See, however, query 1.

A more difficult case to visualize is involved when a ball is thrown—let us suppose directly upward—from the earth. There is no escape from the conclusion that, if this is to be done, the whole earth must be given an equal and opposite momentum. So vast is the earth's mass in comparison with that of the ball that the velocity imparted to it is correspondingly negligible, defying any human attempt to measure it.

As the ball is ascending, after having left our hand, it is, of course, being attracted by gravitation back to the earth; and, according to the universal Third Law, the earth likewise is being attracted back up toward the ball. Thus the ball, in time, loses its upward momentum away from the earth, its velocity becomes zero, and the earth likewise loses its finite but insignificant downward velocity, owing to the recoil from our effort of throwing. Subsequently, the ball, falling down, acquires momentum toward the earth; and the earth, still subject to the ball's attraction, now acquires momentum up toward the ball. The ball strikes the ground. Its momentum, still exactly equal and opposite to that acquired by the earth, is now exactly neutralized by the latter; and the original status quo before the ball was thrown is restored.

The modern rocket serves as an interesting and important application of Newton's Third Law of Motion and the principle of conservation of momentum. A rocket moves forward because of the reactive thrust F of the jet of hot gases ejected from its tail (Fig. 3-4). In other words, the gases carry momentum backward, and there is an equal amount of momentum forward given to the rocket. Since a rocket carries its own fuel and oxidizing agent with it, the rocket develops a thrust which is independent of the atmosphere or medium through which it moves. As a matter of fact, the atmosphere is a detriment, for it presents resistance or drag forces, so that a rocket moves faster in a vacuum than it would in the air.

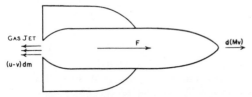

FIG. 3-4.—Diagrammatic sketch of a rocket

The German V-2 shown in Fig. 3-5 represented a major advance in the art of rocket design. The V-2 carried its own fuel and oxidizer, ethyl alcohol and liquid oxygen, and ejected the burned gases from its combustion chamber with a speed of about 6,000 ft/sec. During propulsion a sea-level thrust of about 60,000 lbf was developed, giving the rocket a maximum speed of 5,575 ft/sec or about 3,800 mi/hr. Other data for this rocket are given in Table 3-1.

As an illustration of the momentum principle we shall calculate the velocity v_b of a rocket when all the fuel and oxidizing agent have been used up. We shall assume an ideal case of a rocket in horizontal flight in a vacuum.

FIG. 3-5.—A V-2 rocket take-off

TABLE 3-1

PERFORMANCE CHARACTERISTIC OF THE V-2 ROCKET

Length	47 ft	Duration of flight	360 sec (approx.)
Diameter	5 ft	Rate of fuel consumption	275 lb/sec
Take-off weight	28,000 lb	Exhaust velocity of gases	6,000 ft/sec
Empty weight	9,000 lb	Maximum height	116 mi
Fuel weight	19,000 lb	Maximum range	220 mi
Thrust (at sea-level)	60,000 lbf	Acceleration	1–6 g
Duration of thrust	68 sec	Maximum speed	5,575 ft/sec

Let us suppose that at some time t the rocket and un-burned fuel have a mass M and a velocity v relative to the earth. The mass M represents the sum of the masses of the empty rocket, M_0, and the fuel, m, so that $M = M_0 + m$. Let the velocity of the exhaust gases at this time t be u relative to the rocket, so that, relative to the earth, their velocity would be $(u - v)$ backward.

Suppose, now, that in a small interval of time dt a mass dm of a gas is ejected from the rocket. As a result, the total mass of the rocket is decreased by this amount, and the velocity is increased by a small amount dv. This results in an increase in the momentum of the rocket by an amount $d(Mv)$. An equal amount of momentum change must be given to the ejected gas, of amount relative to the earth, of

$$dm(u - v).$$

This momentum change is in the direction opposite to that of the rocket (Fig. 3-4), so that, by the principle of conservation of momentum, for this mass dm of gas we have

$$d(Mv) - dm(u - v) = 0. \qquad (3\text{-}10)$$

Now, by differentiating the product (Mv), in which both factors M and v change, we have

$$d(Mv) = v\,dM + M\,dv. \qquad (3\text{-}11)$$

The change in mass of the rocket and fuel dM represents a decrease in mass, whereas dm of the gases produced represents an equal increase. Hence

$$dM = -dm.$$

By substitution in equations (3-10) and (3-11), we have

$$v\,dM + M\,dv + dM(u - v) = 0$$

or

$$u\,dM = -M\,dv. \qquad (3\text{-}12)$$

Now the velocity u of the ejected gases relative to the rocket is considered constant and is approximately so during the time in which the fuel is being burned up. Hence from equation (3-12) we have

$$dv = -u\,\frac{dM}{M}.$$

The integral of dx/x is $\log_e x$ or $\ln x$ with the appropriate limits of integration. In our example the velocity v is zero at the beginning when $M = M_0 + m_0$, the mass of the fully loaded rocket, and $v = v_b$, the velocity when the fuel is completely used up, or $M = M_0$. Thus

$$\int_0^{v_b} dv = -u \int_{M_0+m_0}^{M_0} \frac{dM}{M},$$

$$v_b = -u\,\ln\left(\frac{M_0}{M_0+m_0}\right) = u\,\ln\left(\frac{M_0+m_0}{M_0}\right). \qquad (3\text{-}13)$$

Let us substitute the values given for the V-2 rocket;

namely, $M_0 + m_0 = 28{,}000$ lb; $M_0 = 9{,}000$ lb; $u = 6{,}000$ ft/sec. Then

$$v_b = 6{,}000\,\ln\left(\frac{28{,}000}{9{,}000}\right)$$

$$= 6{,}000\,\ln(3.11)$$

$$= 6{,}000 \times 1.135$$

$$= 6{,}810\ \text{ft/sec}.$$

The maximum speed of 5,575 ft/sec, given in Table 3-1 is for a rocket in vertical ascent. What are the retarding forces that could account for this difference?

3.5. Cases of Elastic and Inelastic Impact

Another especially interesting application of these principles is found in the "elastic" and "inelastic" impact of two objects, where we wish to calculate the velocities of the two objects after impact, given their initial velocities.

The conservation law as given by equation (3-9) is not sufficient, in general, to solve this problem. A second condition that the colliding objects must satisfy was first discovered by Newton. He observed that the *ratio of the relative velocity after impact to the relative velocity before impact was a constant;* that is to say, he found experimentally that

$$\epsilon = \frac{v_2 - v_1}{u_1 - u_2}, \qquad (3\text{-}14)$$

where $u_1 - u_2$ is the relative velocity before impact and $v_2 - v_1$ that after impact but with the sign changed, and where ϵ is a constant whose value depends upon the substances in contact and not upon the velocities of the colliding bodies. This constant is called the *coefficient of restitution* of the two substances. For glass on glass, ϵ is about 0.93; for lead on lead, 0.20; for iron on lead, 0.12. We say that the bodies are perfectly elastic and the collision between them elastic if $\epsilon = 1$, i.e., the relative velocities before and after are the same except for sign; perfectly inelastic if $\epsilon = 0$; imperfectly elastic if $0 < \epsilon < 1$. In nature the extreme case, $\epsilon = 1$, probably never occurs, but rather the intermediate one. However, for the sake of mathematical simplicity, we shall treat the limiting cases.

Let two balls of mass m_1 and m_2 and initial velocities u_1 and u_2 collide elastically so that their velocities after collision are v_1 and v_2, as shown in Fig. 3-6. We want to calculate their final velocities.

From the law of conservation of momentum,

Momentum before impact

$$= \text{Momentum after impact} ,$$

$$m_1 u_1 + m_2 u_2 = m_1 v_1 + m_2 v_2 . \qquad (3\text{-}15)$$

If we assume an elastic impact, i.e., $\epsilon = 1$, then the relative velocity of approach $u_1 - u_2$ of the

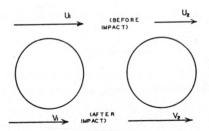

FIG. 3-6.—Elastic impact of two balls

two balls must equal their relative velocity of recession, $v_2 - v_1$, or

$$u_1 - u_2 = v_2 - v_1 . \qquad (3\text{-}16)$$

There are two equations with two unknowns, namely, v_1 and v_2. Solving these equations gives

$$\left. \begin{aligned} v_1 &= \frac{m_1 u_1 + m_2 (2 u_2 - u_1)}{m_1 + m_2} , \\ v_2 &= \frac{m_2 u_2 + m_1 (2 u_1 - u_2)}{m_1 + m_2} . \end{aligned} \right\} \qquad (3\text{-}17)$$

Suppose, now, that the balls are of equal mass and approach each other with the same velocity, so that

$$m_1 = m_2 \quad \text{and} \quad u_2 = - u_1 ; \quad (3\text{-}18)$$

then, by substitution in equation (3-17), we have

$$v_1 = - u_1 \quad \text{and} \quad v_2 = - u_2 , \quad (3\text{-}19)$$

or the balls rebound with their original speeds.

If the masses are each equal to m but the initial velocities u_1 and u_2 are different, then, from equation (3-17),

$$\left. \begin{aligned} v_1 &= \frac{m u_1 + m (2 u_2 - u_1)}{2 m} = u_2 , \\ v_2 &= \frac{m u_2 + m (2 u_1 - u_2)}{2 m} = u_1 , \end{aligned} \right\} \qquad (3\text{-}20)$$

or the balls exchange velocities.

This is illustrated in the photographs using twelve balls of equal mass suspended by strings of equal length. A single ball on the right is raised in Fig. 3-7 (a) and then let go; in (b) the balls are in-

stantaneously together, and after another short interval the ball on the left is raised (Fig. 3-7 [c]) to a height equal to that from which the original ball fell. In this experiment $u_2 = 0$, so that $v_1 = 0$ and $v_2 = u_1$, i.e., each of the twelve balls exchanges velocities on impact and then comes to rest, with the exception of the right one, which is raised in height.

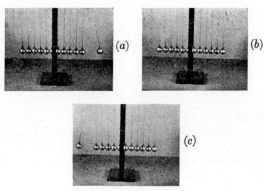

FIG. 3-7.—Elastic impact of two balls of equal mass with $u_2 = 0$, $v_1 = 0$, and $v_2 = u_1$. a, One ball raised; b, impact with exchange of velocities; c, ball on left raised, $v_2 = u_1$.

Suppose, now, some putty or wax is placed between two balls of equal mass so that $\epsilon = 0$, or the impact is inelastic. After impact the balls stick together and move off with a common velocity v, which, from the law of conservation of momentum, equation (3-15), is

$$v = \frac{m u_1 + m u_2}{2 m} . \qquad (3\text{-}21)$$

Whenever a collision is not central, then equation (3-21) must be solved by treating the terms as vectors. This is illustrated in Example 3.

EXAMPLE 3: A 100-gm ball of putty has a velocity of 20 cm/sec east. Into it is thrown a 25-gm marble with a northerly velocity of 50 cm/sec. In what direction and with what speed will the putty containing the marble move?

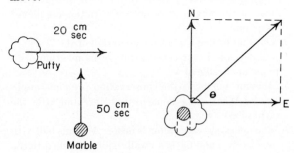

Initial momentum east $= 100 \times 20$

$\qquad = 2,000$ dynes-sec ;

Initial momentum north$= 25 \times 50$

$\qquad = 1,250$ dynes-sec .

We recall that momenta are vector quantities and are added and resolved accordingly.

The total momentum before impact is

$$\sqrt{2,000^2 + 1,250^2} = 2,358.4 \text{ dynes-sec} .$$

Hence

$$\tan \theta = \frac{1,250}{2,000}, \qquad \theta = 32°.$$

$$125 \, v = 2,358.4 ,$$

$$v = 18.86 \text{ cm/sec} .$$

Or we can work with each component of momentum separately to find velocity components from which the resultant velocity can then be calculated, thus:

The combined mass of $100 + 25 = 125$ gm must have an easterly momentum of

$$v_E \times 125 = 2,000 \text{ dynes-sec}$$

and

$$v_E = 16 \text{ cm/sec east} .$$

It must have a northerly momentum of

$$v_N \times 125 = 1,250 \text{ dynes-sec}$$

and

$$v_N = 10 \text{ cm/sec north} .$$

Then

$$v = \sqrt{16^2 + 10^2} = 18.86 \text{ cm/sec} ,$$

125×18.86

$\qquad = 2,358$ dynes-sec, the total momentum ,

$\tan \theta = \frac{10}{16}$,

$\theta = 32°$.

EXAMPLE 4: A case-hardened steel ball falls from a height h on a case-hardened steel slab and rebounds to a height h_1. The problem is to find the coefficient of restitution, ϵ.

The velocity of approach of ball and slab u, i.e., the velocity with which the ball strikes the slab, is, from equation (1-20),

$$u = \sqrt{2 g h} ,$$

and the velocity v with which the ball recedes from the slab is

$$v = \sqrt{2 g h_1} .$$

Since, by definition, ϵ is the velocity of recession divided by the velocity of approach, it follows that

$$\epsilon = \frac{v}{u} = \sqrt{\frac{h_1}{h}}.$$

In an actual experiment $h = 40.5$ cm and $h_1 = 37.0$, so that

$$\epsilon = \sqrt{\frac{37.0}{40.5}} = 0.957 .$$

On the second impact the ball started from the height of 37.0 cm and rebounded to a height of 34.5 cm, so that

$$\epsilon = \sqrt{\frac{34.5}{37.0}} = 0.965 .$$

3.6. Résumé

The following is a brief listing of the definitions, principles, and theories, given in this chapter, which you should know:

Newton's Third Law of Motion
The principle of conservation of momentum
Impulse and momentum change
Impacts and coefficient of restitution

Queries

1. In the illustrative example of the man running off the wagon (§ 3.4), would the momentum of the wagon to the right be exactly equal to the momentum of the man if the wagon were not on a frictionless plane? Explain.
2. On what principle does the operation of the "rocket" car depend? Is the conservation-of-momentum principle applicable here?
3. A bomb bursts; is the conservation principle applicable? What is the momentum before and after the explosion?
4. The following observation is made: A tennis ball with a velocity u approaches a wall rigidly attached to the ground and hence with a momentum mu toward the wall. Assuming the collision between the wall and ball to be elastic, with what velocity does the ball recede from the wall? Is the conservation principle satisfied here? Explain.
5. A man throws a ball against a wall; the ball rebounds elastically and is subsequently caught by the man. By using the conservation of momentum and taking the earth into account, analyze the momentum changes that take place.
6. The classical problem of the horse and cart is usually stated in this way. If the cart pulls on the horse with the same force that the horse pulls the cart, then how

do the horse and cart move forward? Explain.

7. A shell is falling vertically toward the earth, and in case *a* it explodes on contact with the ground and in case *b* some distance, say 200 ft, above the ground. In which case would you expect the shell to do most damage to any buildings? What is the momentum of the shell fragments after the collision in relation to that of the shell just before collision?

8. How could you get off a perfectly smooth pond of ice, and how could you initially have got to rest on the ice if it were perfectly smooth?

9. For the rocket, show that the rocket velocity is equal to the exhaust velocity when the ratio M/M_0 is e or approximately 2.8 and the rocket velocity is twice the exhaust velocity when the ratio M/M_0 is e^2 or approximately 8.

Practice Problems

1. A man whose weight is 160 lbf is standing on a wagon of mass 50 lb and holds a 10-lbf weight. Calculate the horizontal velocity which he must impart to the weight in order that he and the wagon may attain a velocity of 0.6 mi/hr. Ans.: 12.6 mi/hr.

2. An auto of weight 2.0 tons-force going at a velocity of 60 mi/hr collides at an intersection with a truck of mass 10 tons going 30 mi/hr on a street perpendicular to the street on which the auto is moving. If the cars lock together, find the resultant velocity (speed and direction) of the two after impact.

3. A shotgun of weight 8 lb fires a bullet of weight 2 oz with a speed of 1,600 ft/sec. Find the impulse given to the gun and to the bullet and the velocity of recoil of the gun. Ans.: 6.25 slugs-ft/sec; 25 ft/sec.

4. A block of wood of mass 2 kg rests on a horizontal table having a coefficient of friction of 0.1. If a 10-gm bullet is shot horizontally into the block of wood and becomes imbedded in it, find the velocity with which the bullet and block start to move, and the distance the block moves before coming to rest.

5. An apparatus consists of one straight frictionless plane between M and N, horizontal between N and P, and a rough portion, PQ, on the same level with NP.

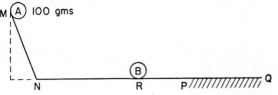

A ball of mass 100 gm rolls along MN and attains, at N, a velocity of 60 cm/sec in 4 sec in the direction NR. It moves along NP and collides elastically with ball B, whose mass is also 100 gm. Ball B comes to rest at Q, the distance PQ being 300 cm. Find (1) the momentum of ball A along NR; (2) after collision, the velocity of ball B; (3) the time required for B to travel from P to Q; (4) the velocity of A and B after collision if B is only 50 gm. Ans.: (1) 6,000 dynes-sec; (2) 60 cm/sec; (3) 10 sec; (4) 20 cm/sec→, 80 cm/sec →.

6. Two steel balls of masses in the ratio of 3:1 are mounted as pendulums in contact at the lower part of their arcs and are allowed to fall and collide after being separated equal distances on opposite sides. If the velocities of approach of the two balls are V and $-V$, calculate their velocities of recession. What will they be after the second impact? Assume $\epsilon = 1$.

7. Three bullets of mass 10 gm each and speed 700 m/sec strike simultaneously and remain lodged in a target of mass 1 kg. If the paths of the bullets are in the same plane and make an angle of 30° with each other, calculate the momentum of the target and imbedded bullets (magnitude and direction). See the accompanying figure. Ans.: 19.1 kg-m/sec.

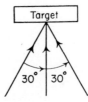

8. A ball weighing 3 lbf is thrown into a 5-lb mass of putty, giving it a velocity of 20 ft/sec. Find the original velocity of the ball.

9. A 50-gm body with an initial velocity of 20 cm/sec is acted on for 3 sec by a force of 2 gmf acting in the direction of its initial velocity. Find the final velocity, the acceleration, the distance traveled during the 3 sec, the final momentum, the change in momentum, and the impulse of the force. Ans.: 137.6 cm/sec; 39.2 cm/sec²; 236.4 cm; 6,880 dynes-sec; 5,880 dynes-sec; 5,880 dynes-sec.

10. A steel ball weighing 2 lbf and moving 50 ft/sec hits a 10-lb steel ball at rest and rebounds with a velocity of 20 ft/sec. Find the velocity given to the 10-lb ball and the coefficient of restitution.

11. A machine gun fires 400 bullets per minute. If each bullet has a mass of 1 ounce and a muzzle velocity of 200 ft/sec, find the average reaction of the gun against its support. Ans.: 2.6 lbf.

12. A ball is dropped from a height of 1 m and rebounds

49 cm from a steel plate. Find the coefficient of restitution between the ball and the plate. To what height would it go on the second rebound?

13. A shell of mass 650 lb leaves the barrel of a gun of mass 20 tons with a velocity of 1,650 ft/sec. What impulse was given to the shell by the powder? With what velocity did the gun recoil? Ans.: 3.35×10^4 lbf-sec; 26.8 ft/sec.

14. Solve problem 9, using the mks system of units.

15. A 2-kg gun fires a bullet having a mass of 0.005 kg with a speed of 500 m/sec into a block of wood having a mass of 2 kg. Find (a) the velocity of recoil of the gun, (b) the velocity of the block of wood if the bullet becomes imbedded in it. If the bullet goes down the barrel in 0.01 sec, find (c) the force on the bullet and the force on the gun. The bullet is stopped in the block of wood, after going 0.10 m into it, by the resistance force of the wood. Find (d) the retardation of the bullet in the block of wood, (e) the force exerted by the wood on the bullet. Ans.: (a) 1.25 m/sec; (b) 1.247 m/sec; (c) 250 newtons; (d) 1,250,000 m/sec^2; (e) 6,250 newtons.

16. If the rocket in § 3.4 has an empty weight of 4,000 kgf, a fuel weight of 8,000 kgf, and the speed of the exhaust gases is 2,000 m/sec, find the maximum horizontal speed of this rocket in a vacuum. If all the fuel were burned in 100 sec, what would be the average thrust during this time?

CHAPTER 4

Work and Energy

4.1. New Concepts of Work and Energy*

As was pointed out in chapters 2 and 3, all problems in mechanics can be solved (at least theoretically, in the sense discussed in § 2.7) by employing the Second and Third Laws of Motion. However, many mechanical situations arise in which the direct application of the laws would involve complicated calculations of forces and accelerations in order to obtain a relatively simple result. For example, a pendulum bob of mass m is raised a distance h and released. What is its velocity as it passes its position of equilibrium? If we were to apply $f = ma$, we should be faced with the problem of a force varying continuously and hence an acceleration that is varying. Many of the simple equations of kinematics deduced in chapter 1 would no longer be applicable. The solution of this problem by Newton's Second Law requires the use of more advanced mathematics than is suitable for these pages.

* The confusion among some of the ablest scientific men late in the seventeenth century regarding the concept of energy and the fact that it, like momentum, had a conservation principle was very great. Details of these famous controversies are admirably set out by L. W. Taylor in *Physics, the Pioneer Science* (Boston: Houghton Mifflin Co., 1941), pp. 217 ff.

However, by introducing new concepts, those of work and energy (potential and kinetic), we are able to solve this problem with ease, as we shall show in § 4.9. A more profound reason for introducing the energy concept is that energy, like momentum, is found to satisfy a conservation principle. Let us see how this comes about.

4.2. Definition of Work

If an object is moved through a distance s by the action of a force f, we say that *work* has been done on the object by this force (Fig. 4-1). We de-

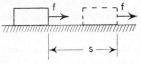

FIG. 4-1.—Work fs when the force and displacement are in the same direction.

fine the amount of work done, W, *as equal to the product of the displacement and the component of the force in the direction of the displacement*. If, as in Fig. 4-1, the force and displacement are along the same straight line, then

$$W = fs . \tag{4-1}$$

More generally, if the force and the displacement are not in the same straight line, as in Fig. 4-2,

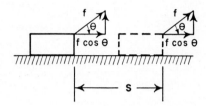

FIG. 4-2.—Work for a horizontal displacement. s is $fs \cos \theta$

then the work done for a horizontal displacement s is

$$W = f s \cos \theta . \qquad (4\text{-}2)$$

The reader will recognize from our definition of work, equation (4-2), that the two concepts, force and displacement, are here inseparably connected. If either is zero, no work is done. Furthermore, if the line of action of the force is at right angles to the displacement, $\theta = 90°$, $\cos \theta = 0$, and *no work* is done. An illustration of a situation in which force and displacement are at right angles is that of an object at the end of a string in uniform circular motion. The centripetal force here involved does no work. Another example is that of moving an object through a horizontal displacement. The work against the vertical force of gravity is zero, since again $\theta = 90°$.

Although work is the product of two vectors, force and displacement, it is a scalar quantity. It is partly the fact that work has magnitude but no direction which makes it such an important concept.

The units of work are naturally expressed in terms of units of force and units of distance. In the mks system, when f is in newtons and s in meters, the unit of work is the newton-meter, to which the name "joule" has been given. Similarly, in the cgs system the unit of work is the dyne-centimeter, to which the name "erg" has been given:

$$1 \text{ joule} = 1 \text{ newton-meter} ,$$

$$1 \text{ erg} = 1 \text{ dyne-cm} .$$

Since 1 newton = 10^5 dynes and 1 m = 10^2 cm, then it follows that

$$1 \text{ joule} = 10^7 \text{ ergs} .$$

In the British Engineering system of units the unit of work is the foot-pound-force or ft-lbf. No separate name has been given to this unit. From the magnitudes of the quantities it may be shown that, approximately,

$$1 \text{ ft-lbf} = 1.36 \text{ joules} .$$

One of the great advantages of the mks system is its use in electrical measurements, and it is largely because of this that the mks system is coming into universal use. The dimensions of work in terms of the dimensions of mass, length, and time are

$$[ML^2T^{-2}]$$

and, in terms of the dimensions of force, length, and time used in engineering,

$$[FL] .$$

In general, when an object is moved by the action of some force through a distance s, forces that oppose the motion may come into play. For example, friction may be present, in which case work is done against the *force of friction*. One may raise an object a distance s above the table against the *force of gravity;* thus one may do work against gravity. A spring may be stretched and work be done against the *elastic forces* of the spring. If one exerts a force on an object to which no other external forces are applied, the object accelerates. Work is done in *producing this acceleration* against the inherent inertia of the object. Again, forces due to friction and to gravity may be present together, in which case the applied force would do work against friction, against gravity, and against inertia if the object were accelerating. In what follows we shall treat separately work against friction, work against gravity, and work done to accelerate an object. Work done against elastic forces will be treated in chapter 7.

4.3. Work Diagrams

If the magnitude of the force applied to an object is plotted as the ordinate and the distance through which the force acts as the abscissa, the resulting graph is called the *work diagram*, because the area under the curve represents work done by the force. For example, if the force is constant, f_0, and acts through a distance s_0 (Fig. 4-3), the work done is

$f_0 s_0$, which is also the area of the rectangle of Fig. 4-3.

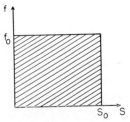

FIG. 4-3.—Work $f_0 s_0$ for a constant force

Suppose the force varies with the distance, as shown in Fig. 4-4, and that at some distance s the

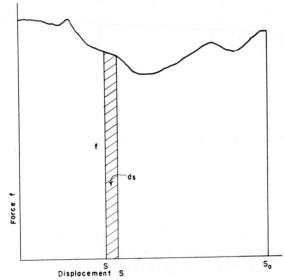

FIG. 4-4.—Work diagram for a varying force

force is f. Over a small distance ds the force is sensibly constant, and the work done in the small displacement is fds. This is represented by the small shaded area, approximately a rectangle. The total work done between O and S_0 is the sum of all such shaded areas or the area under the curve. This area is given by the summation of fds, i.e., the integral of fds. Thus the work done by the varying force f in the displacement O to S_0 is

$$W = \sum_0^{S_0} fds = \int_0^{S_0} fds. \qquad (4\text{-}3)$$

If the variation of f with s is known as an analytical expression, then the integration may be carried out; otherwise one may have to determine it by measuring the area, say, with a planimeter. In the diagram it is assumed that the force f is in the direction of the displacement at every point along the path.

4.4. Work against Friction

When we attempt to pull a heavy block horizontally along a rough table, we recognize that a resisting force due to friction is present. The resisting force increases and opposes our pull until the latter equals μmg. If our pull exceeds this amount, the body moves, and work is done against this resisting force as the block is moved through any distance. This work against the frictional force is equal to the product of the force and the distance, if the two are collinear. We note that the force of our effort necessary to keep the block in *uniform motion* must, by this very fact, exactly balance the resistance force, i.e., is equal to the latter but oppositely directed.

Work done against friction is usually a total loss mechanically, i.e., is unrecoverable mechanically.* The net result of any friction is that the surfaces become heated.

EXAMPLE 1: How much work is necessary to pull a 10-lb sled 20 ft along the ground if the coefficient of friction is 0.4 and if the pull is applied horizontally? Assume zero acceleration; i.e., assume that the sled was in motion and that its velocity did not change.

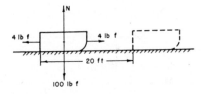

Solution: The applied force must be equal to the force of friction F:

$$F = (0.4)(10\ \text{lbf}) = 4\ \text{lbf}\ ;$$

the work done by the applied force is, then,

$$W = (4\ \text{lbf})(20\ \text{ft}) = 80\ \text{ft-lbf}\ .$$

EXAMPLE 2: In Example 1 suppose the rope makes an angle of 30° with the horizontal. How much work is done?

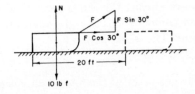

* See chap. 16 on heat engines for an elaboration to this statement.

Solution: Let F be the tension in the rope measured in pounds-force. The force of friction acting equals μ times the total force acting normal to the surface. This normal force is made up of two parts: the weight, 10 lbf downward, and the upward component of the tension, $F \sin 30°$. The horizontal force, $F \cos \theta$, must be equal to the force of friction, i.e.,

$$F \cos 30° = 0.4 (10 - F \sin 30°)$$

or

$$F = 3.75 \text{ lbf} .$$

The work done is

$$W = (F \cos 30°)(20) = (3.75)(0.866)(20)$$

$$= 64.95 \text{ ft-lbf} .$$

4.5. Work against Gravity

If an object of mass m is raised a vertical distance h against the force of gravity, then the work W done against gravity is

$$W = mgh . \qquad (4-4)$$

If we drag the object up a frictionless plane (Fig. 4-5), we see that we are doing work along a

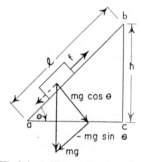

FIG. 4-5.—Work in pulling body up a frictionless plane

line of motion which is not collinear with the direction of the force of gravity. The latter is vertical. Let us resolve the vertical force (mg) into components parallel and perpendicular to the plane. These components are

$mg \sin \theta$, along the plane ;

$mg \cos \theta$, perpendicular to the plane .

As we drag the weight along the plane a distance l, we do work against the component $mg \sin \theta$, since the other component is at right angles to the displacement. The work done is

$$W = (mg \sin \theta)(l) . \qquad (4-5)$$

From the geometry of the figure, $\sin \theta$ is given by

$$\sin \theta = \frac{h}{l} ,$$

so that the work done is

$$W = (mg)\left(\frac{h}{l}\right)(l) = mgh . \qquad (4-6)$$

Thus we see that in dragging a weight up a frictionless plane, no matter what the distance may be along the plane, we do the same amount of work as we would have done, had we lifted it vertically. In Fig. 4-5 no work is done in moving along ac, since mg is at right angles to ac; work is done along bc. If, therefore, the points a and b were connected by any path whatsoever—Fig. 4-6, for example—the

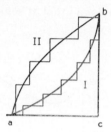

FIG. 4-6.—Work against gravity is independent of the path

work done in moving a weight along either I or II is simply mgh, where h is the vertical height between a and b. This is so because paths I and II may be broken up into paths parallel and perpendicular to the vertical, i.e., to mg.

In moving the mass m from a to b, an amount of work mgh is done against gravity by some external agent. Now, if the mass m is moved from b to a, then an amount of work mgh is done by the force of gravity, so that the total work done by or against gravity in going around a closed path, such as $abca$, is zero. For this reason the force of gravity is called a *conservative force*. The electrostatic force between two electric charges is also a conservative force, but from the discussion of the force of friction it follows that this is not a conservative force.

If m is measured in kilograms, h in meters, and g in meters per second per second, then the work W is in units of

$$\frac{\text{kg} \times \text{m}^2}{\text{sec}^2} \equiv \text{joules} ,$$

or in the cgs system, using grams, centimeters, and centimeters per second per second, the work W is in units of

$$\frac{\text{gm} \times \text{cm}^2}{\text{sec}^2} \equiv \text{ergs} .$$

For the Engineering system the force is in pounds-weight, lbf, and the vertical height in feet, so that the work is in units of ft-lbf.

4.6. Work against Inertia, To Accelerate an Object

Let us assume that we have an object such as an automobile of mass m on a horizontal, frictionless surface. The mass is subject to a resultant external force f, parallel to the surface, and acts through a distance s (Fig. 4-7). The amount of

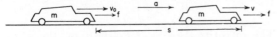

Fig. 4-7.—Work against inertia

work done on this object by the force is

$$W = fs . \qquad (4-7)$$

We wish to know what will happen to the object when work is done on it. From the Second Law of Motion, equation (2-5), we know the object will accelerate

$$f = ma .$$

Also, from kinematics, equation (1-19), we know that the object will have its velocity changed in accordance with the following equation:

$$2\,a\,s = v^2 - v_0^2 ,$$

where v_0 is the initial and v the final velocity. Combining these equations, we obtain

$$\left. \begin{array}{l} 2\,a\,s\,f = ma\,(v^2 - v_0^2) \\[4pt] f\,s = \tfrac{1}{2}\,m\,v^2 - \tfrac{1}{2}\,m\,v_0^2 . \end{array} \right\} \qquad (4\text{-}8)$$

or

And, by equation (4-7), the work done is

$$W = \tfrac{1}{2}\,m\,v^2 - \tfrac{1}{2}\,m\,v_0^2 . \qquad (4\text{-}9)$$

We see that the effect of a constant force acting through a distance is measured by taking the difference of the quantities $(\tfrac{1}{2}mv^2)$. This difference is a *measure of the work done by the force against inertia only.*

It is interesting to contrast the results of this section with that of § 3.3. There we showed that the effect of a force acting for a *given time,* t (the impulse $= ft$), also resulted in a change in velocity; but in this case the effect of the force was measured by taking the difference of the final and initial values of the momenta, equation (3-7),

$$ft = m\,v - m\,v_0 .$$

The impulse, ft, is a vector quantity; the work, fs, is a scalar.

A word about units to be used in equation (4-9). It follows that the units of $mv^2/2$ must be the same as those used for fs. In the mks system the units of $mv^2/2$ are

$$\frac{\text{kg} \times \text{m}^2}{\text{sec}^2} \equiv \text{joules} .$$

For the cgs system the units of $mv^2/2$ are

$$\frac{\text{gm} \times \text{cm}^2}{\text{sec}^2} \equiv \text{ergs} .$$

For the Engineering system equation (4-9) is written

$$W = fs = \frac{1}{2}\frac{w\,v^2}{g} - \frac{1}{2}\frac{w\,v_0^2}{g} , \qquad (4\text{-}10)$$

where w is the weight of the object moved whose mass m is w/g. The units of $wv^2/2g$ are

$$\frac{\text{lbf} \times \text{ft}^2}{\text{sec}^2} \times \frac{\text{sec}^2}{\text{ft}} = \text{ft-lbf} ,$$

since the units of v^2 are ft²/sec² and of g are ft/sec².

EXAMPLE 3: A mass of 50 gm with an initial velocity of 10 cm/sec has its velocity changed to 20 cm/sec. How much work was done on the object?

Solution: From equation (4-9),

$$W = \tfrac{1}{2}\,m\,v^2 - \tfrac{1}{2}\,m\,v_0^2$$

$$= \tfrac{1}{2}\,(50\ \text{gm})\,(20\ \text{cm/sec})^2$$

$$- \tfrac{1}{2}\,(50\ \text{gm})\,(10\ \text{cm/sec})^2$$

$$= 10{,}000\ \text{gm-cm}^2/\text{sec}^2 - 2{,}500\ \text{gm-cm}^2/\text{sec}^2$$

$$= 7{,}500\ \text{ergs} .$$

4.7. Concept of Energy: Potential and Kinetic Energy

If a weight, mg, is lifted a distance h above the plane PP' (Fig. 4-8) and then attached to one end

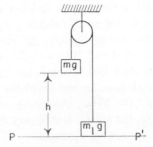

Fig. 4-8.—Potential energy of mg is mgh

of a string which passes over a frictionless pulley to another almost equal weight, m_1g, at rest on the plane, the lifted weight, by virtue of its position above the plane, *possesses a capacity for doing work.* That this is true is obvious from the diagram, for, in descending to the plane, the weight mg thereby raises the weight m_1g a distance h above the plane. The work done is m_1gh. This simple experiment serves to introduce the important concept of *energy.*

By "energy" we shall always mean the *capacity for doing work.* In general, an object or a system of objects possesses energy if the object or system of objects is capable of doing *work*, in the sense that we have defined the term (§ 4.2). The lifted weight, therefore, possesses energy, since it can raise a weight, thereby doing work. We shall call the energy of the lifted weight *gravitational potential energy* and measure quantitatively the potential energy by the amount of work required to lift the weight, mg, the distance h:

Gravitational potential energy (PE) $= m\,g\,h$. (4-11)

Potential energy, therefore, is measured in the same units as is work and is a scalar quantity. In equation (4-11) we tacitly assume that the potential energy of the weight when on the plane PP' is zero. However, with respect to a plane below PP', the weight would have potential energy. Gravitational potential energy, therefore, is always measured with respect to a reference plane, the choice of which is arbitrary. It may be the top of a table or the floor or the surface of the earth. From our discussion of gravitational phenomena in § 2.16 we know that the work done in lifting a weight is work done against the gravitational force of attraction between the mass and the earth and that, when the mass falls earthward, the forces of the field do work on the mass. For this reason we may think of the gravitational potential energy as residing in the gravitational field.

A stretched spring and a twisted rod possess energy which we may call "elastic potential energy." Here the energy resides in the stretched spring and twisted rod. The amount of potential energy in the stretched spring is measured by the amount of work done *against the elastic forces of the spring* (see § 7.10). When the spring returns to its original shape, the elastic forces may do work, as, for example, by raising a weight.

Potential energy is always associated with an object (*a*) if work is done on the object against forces to change the position of the object and (*b*) if this work can be fully *recovered* by allowing the object to return to its original position. Because of these two conditions, the potential energy is qualitatively called "energy of position."

Very similar considerations enable us to say that an object in motion possesses *kinetic energy.* In § 4.6 we found that an accelerating force, f, acting on an object of mass, m, through a distance, s, does an amount of work, W, equal to

$$W = \tfrac{1}{2}\,m\,v^2 - \tfrac{1}{2}\,m\,v_0^2 .$$

The quantity $\tfrac{1}{2}mv^2$ is, by definition, the kinetic energy that the moving object possesses. That an object in motion possesses a capacity for doing work, and hence has energy, can readily be seen from Fig. 4-9. To the object m in motion we at-

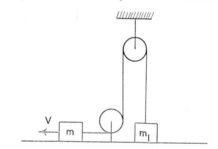

FIG. 4-9.—**Kinetic energy** of mass m raises weight m_1g

tach a string that passes over a pulley to a second mass, m_1. The weight, m_1g, is raised through some height, h. Hence m does work, finally coming to rest. From equation (4-9) it follows that the units of kinetic energy and those of work are the same.

We have seen that work against gravitational forces or gravity gives rise to gravitational potential energy; work against elastic forces to elastic potential energy; and work against inertia to kinetic energy. What has happened to work done against friction? We remarked that in doing work against friction the surfaces in contact became heated. This work against friction has gone into *heat energy.* From the point of view of mechanics, heat energy is wasted energy, whereas potential and kinetic energy can be recovered. The sum of the kinetic and potential energies of an object or system of objects is called the *total mechanical energy.*

Suppose a car of mass m and velocity v is stopped in a distance s by a braking force f. The kinetic energy of the car is removed through the work done against the braking force, and this energy goes into heat. From equation (4-9) we have

$$f s = \tfrac{1}{2} m v_0^2 \,,$$

since the final velocity is zero. For a given car, the mass m and presumably the braking force f are constant, so that the stopping distance s is proportional to v^2,

$$s \propto v^2 \,.$$

This is a very important consideration in motoring. It means that it takes four times the distance to stop when going at 60 mi/hr as when going at 30 mi/hr. You should be able to show that for this case the time for stopping is twice, not four times.

4.8. Principle of the Conservation of Energy

Of fundamental importance is the observational fact that, whenever work is done on an object, the work manifests itself either as potential energy (gravitational or elastic), kinetic energy, or heat energy, or a combination of all three.[*] If we were to measure, by methods to be described later, the amount of heat energy present, we would find (1) that the sum of the total mechanical energy and heat energy was, at each instant, constant and, further, (2) that the mechanical energy was being transformed gradually into heat energy. The first statement, known as the "law of conservation of energy," is perhaps the greatest generalization in natural science. In all the experiments to which the law was critically subjected, it was found that energy could not be created or destroyed—it could only be transformed from one kind to another. The conservation law in its most general formulation takes into account forms of energy other than those discussed so far (see chapter on heat). In 1905 Einstein showed in his theory of relativity that matter itself was a form of energy. The energy equivalence E of a mass m is given by the Einstein relationship as

$$E = m c^2 \,, \qquad (4\text{-}12)$$

where c is the speed of light or electromagnetic radiation and is about 3×10^8 m/sec, or 3×10^{10} cm/sec. One kilogram of matter has an energy equivalent of 9×10^{16} joules, and 1 gm of matter

[*] In mechanics, as a rule, we are not interested in other possible forms of energy, such as chemical energy, electrical energy, nuclear energy, etc.

an energy equivalent of 9×10^{20} ergs. This relationship is of fundamental importance in nuclear physics. The so-called "atomic bomb" is a vivid and destructive use of the energy produced by annihilating matter. The energy produced eventually goes into heat energy.

For the important cases in which there is no friction and other dissipative forces, so that the only forms of energy are kinetic and potential, the law of conservation of energy is restricted to two forms of mechanical energy. In other words, in a system in which there are no frictional forces, there is conservation of mechanical energy. That is, for any motion in the system the sum of the kinetic and potential energies is constant.

4.9. The Pendulum as an Example of the Conservation of Mechanical Energy

A very simple illustration of the transformation of energy back and forth from the potential to the kinetic form is given by the swinging of a pendulum (Fig. 4-10). The pendulum bob, being higher

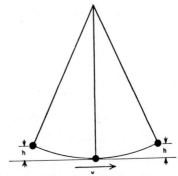

Fig. 4-10.—Conservation of mechanical energy with a pendulum.

above the surface of the ground at the extremities of its arc of swing than it is at the lowest portion of the arc, has the least potential energy in the latter position. On the other hand, at the bottom of its arc it has acquired its greatest velocity; therefore, here it has the largest amount of kinetic energy. It has zero kinetic energy, on the other hand, when it is at rest at each extremity of its swing. The conversion of potential energy at the extreme end of the swing into kinetic energy at the middle of the swing and back into potential energy at the end of its opposite excursion is entirely obvious.

Since the kinetic energy at the mid-point comes

from loss of potential energy at the extremity of the swing, we may equate

$$PE = KE ,$$

$$mgh = \tfrac{1}{2}mv^2 ,$$

or

$$v = \sqrt{2gh} .$$

This is the solution of the problem proposed at the beginning of this chapter. Friction at the point of support of the pendulum and air friction or the friction of any medium through which it swings continually dissipate the store of energy originally put into the pendulum when it was lifted by some outside agent away from the central, equilibrium position. Ultimately, therefore, the pendulum thus set in motion comes to rest; the original store of mechanical energy has been converted into heat.

EXAMPLE 4: To pull the sled referred to in Example 1 up a hill making an angle of 30° with the horizontal and where the rope is parallel to the hill slope will require how much work?

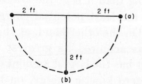

Solution: The force of friction is $0.4 \times 8.66 = 3.464$ lbf. The necessary force in the rope is, then,

$$3.464 \text{ lbf} + 5 \text{ lbf} = 8.464 \text{ lbf} ,$$

$$W = (8.464 \text{ lbf})(20 \text{ ft}) = 169.28 \text{ ft-lbf} .$$

EXAMPLE 5: A pendulum consists of a 5-lb ball hanging on a 2-ft rope. If it is held so that the rope is horizontal and then released, what velocity will the ball have at the bottom of its path?

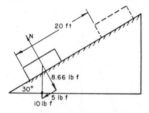

Solution: At (a) the ball has a potential energy of $2 \times 5 = 10$ ft-lbf, since that much work was necessary to lift it to the position (a) relative to the reference point (b).

Hence in position (b) the energy of the ball is all kinetic and of amount 10 ft-lbf in accordance with the law of conservation of mechanical energy. We are, of

course, neglecting any frictional effects. This KE may be written as $wv^2/2g$, where v is the velocity at (b). Thus

$$10 \text{ (ft-lbf)} = \frac{1}{2} \frac{5 \text{ (lbf) } v^2 \text{ (ft}^2/\text{sec}^2)}{32 \text{ (ft/sec}^2)} .$$

Hence

$$v^2 = 4 \times 32 \text{ (ft}^2/\text{sec}^2) ,$$

$$v = 11.3 \text{ ft/sec} .$$

EXAMPLE 6: If all the energy of a United States Army rifle bullet (mass 0.0214 lb, velocity 2,700 ft/sec) could be used to lift a man weighing 150 lbf, how high would it lift him?

Solution:

$$KE \text{ of bullet} = PE \text{ of man} ,$$

$$\frac{1}{2} \frac{0.0214}{32} \frac{(2,700)^2 \text{ lbf ft}^2/\text{sec}^2}{\text{ft/sec}^2} = 150h \text{ ft-lbf} ,$$

where h is the vertical distance in feet which the man is raised. Solving this equation gives

$$h = 16.2 \text{ ft} .$$

EXAMPLE 7: A 0.1-kg mass is pulled 1 m up a rough inclined plane by a force of 0.075 kgf parallel to the plane. If the plane makes an angle of 30° with the horizontal and the coefficient of friction is 0.2, compute (a) the total work done; (b) the work done against friction; (c) the work done against gravity; (d) the kinetic energy imparted to the mass; (e) the velocity of the mass.

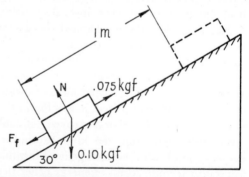

Solution: The total work done is

$$Fs = (0.075 \times 9.8 \text{ newtons})(m) = 0.735 \text{ joule} .$$

Force of friction:

$$\mu mg \cos 30° = (0.2 \times 0.1 \times 9.8 \times 0.866)$$

$$= 0.170 \text{ newton} .$$

Work against friction:

$$1 \times 0.170 = 0.170 \text{ joule} .$$

Component of the weight along plane:

$$mg \sin \theta = 0.1 \times 9.8 \times \tfrac{1}{2} = 0.49 \text{ newton} .$$

Work against gravity:

$$0.49 \times 1 = 0.49 \text{ joule} .$$

The velocity at the top of the plane can be calculated from the resultant accelerating force and the distance of 1 m, through which it acts. The resultant accelerating force up the plane is

$$0.075 \times 9.8 - 0.1 \times 9.8 \times 0.5 - 0.2 \times 0.1 \times 9.8$$

$$\times 0.866 = 0.735 - 0.49 - 0.17$$

$$= 0.0753 \text{ newton} .$$

The work done by this force is 0.0753×1 joule, and this appears as kinetic energy. Hence the velocity of the 0.1-kg mass at the top of the plane is given by

$$\tfrac{1}{2} \times 0.1 \times v^2 = 0.0753 \text{ joule} ,$$

$$v^2 = \frac{0.0753}{0.05} = 1.51 \ (\text{m/sec})^2 ,$$

$$v = 1.23 \text{ m/sec} .$$

Adding work against friction =	0.17
Work against gravity =	0.49
Work against inertia =	0.075
Total energy =	0.735 joule

4.10. Machines: Mechanical Advantage

Any device by which the amount or direction of a force is changed for the sake of gaining some particular end (such as lifting a large weight) we shall call a "machine." We have machines of such design that, by the application of a small force, a very large weight may be lifted—at the expense, however, of the speed with which the weight is moved. In some mechanisms, on the other hand, speed is wanted, and then the machine exerts a smaller force than the one originally applied. These statements are all in accord with the conservation-of-energy principle, *for the amount of work done by the machine must be equal to the work done on the machine*, assuming no friction. This last statement is sometimes called the "work principle."

We define the mechanical advantage (MA) of a machine as the ratio of the *force exerted by the machine* to the *force applied to the machine*.

An example of an elementary machine is the simple pulley in Fig. 4-11 (*a*). What is the mechanical advantage of this pulley? Since we have one continuous rope and since the system (rope and weight) is in equilibrium (not accelerated) under the action of the applied force F, it follows that the tension in the rope must be the same throughout its length and hence equal to W, the weight:

$$F = W .$$

The mechanical advantage, then, is

$$\text{MA} \equiv \frac{\text{Force exerted by machine}}{\text{Force exerted } on \text{ machine}} = \frac{W}{W} = 1 . \quad (4\text{-}13)$$

If the force F acts through a distance s, then, from geometry and the principle of work, the weight must be raised through the same distance ($x = s$):

$$\left. \begin{array}{c} F s = (W) x \\ x = s . \end{array} \right\} \quad (4\text{-}14)$$

or

In the pulley arrangement of Fig. 4-11 (*b*) the mechanical advantage is still unity. The only effect

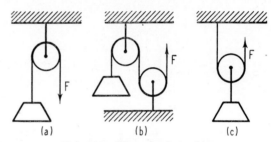

FIG. 4-11.—Three simple machines

of introducing the second pulley is to change the direction of the force F. However, in Fig. 4-11 (*c*) the mechanical advantage is 2, for here there are two forces keeping the weight in equilibrium. What are they?

The student will observe that the mechanical advantage of the pulley system of Fig. 4-12 is 8.

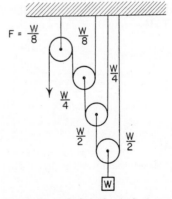

FIG. 4-12.—A theoretical mechanical advantage of 8

Another arrangement of pulleys is that known as the "block and tackle" (Fig. 4-13). Again, from geometrical considerations, the mechanical advantage is 3, since the weight on the block, W, is supported by three strands of rope. The force F applied to the extended end of but one of them, in **65** the absence of friction, results in a uniform tension of this amount (F) throughout the entire length of the rope:

$$3F = W . \qquad (4\text{-}15)$$

From the principle of work, if the force F moves through a distance s and the weight through a distance x, then

$$F s = W x . \qquad (4\text{-}16)$$

Combining equations (4-15) and (4-16), we obtain

$$s = 3x . \qquad (4\text{-}17)$$

The weight goes up only one-third the distance that the force F does.

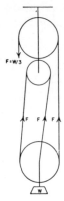

Fig. 4-13.—A theoretical mechanical advantage of 3

An inclined plane may be considered to have a mechanical advantage equal to the ratio of the weight of an object pulled up the plane to the force parallel to the plane required to pull the object.

In Fig. 4-14 the weight, W, acts vertically down-

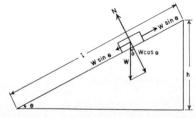

Fig. 4-14.—Mechanical advantage of a frictionless plane is l/h.

ward. We resolve the weight into two components —one parallel to the plane, $W \sin \theta$, and the other normal to the plane, $W \cos \theta$. Assuming no friction, the force necessary to hold the block at rest must be equal to $W \sin \theta$. The mechanical advantage, therefore, is

$$MA = \frac{W}{W \sin \theta} = \frac{1}{\sin \theta}$$

or

$$MA = \frac{l}{h},$$

where h is the vertical height of the plane and l is the length of the inclined plane.

Another simple machine is the screw, the jackscrew shown in Fig. 4-15 being a typical example.

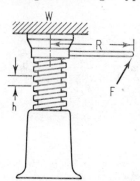

Fig. 4-15.—Mechanical advantage of a screw

Let the pitch or the distance the jack moves upward for one complete turn of the screw be h. A force is applied at a distance R from the center of the screw. Applying the principle of work, we see that for one complete revolution the work done on the machine is

$$F 2\pi R , \qquad (4\text{-}18)$$

assuming the force F to be tangent to the circle swept out by the arm. The work done by the machine is

$$W h . \qquad (4\text{-}19)$$

From the work principle it follows that

$$F 2\pi R = W h \qquad (4\text{-}20)$$

or

$$\frac{W}{F} = \frac{2\pi R}{h} \equiv MA .$$

Another simple machine, the lever, is discussed in chapter 5.

4.11. Efficiency

The efficiency, ϵ, of any machine is defined as the ratio of the output of work to the input,

$$\epsilon = \frac{\text{Work out}}{\text{Work in}},$$

or, in percentage,

$$\epsilon = \frac{\text{Work out}}{\text{Work in}} \times 100 . \qquad (4\text{-}21)$$

Only in the case of an ideal frictionless machine is the efficiency $\epsilon = 100$ per cent; in all actual machines ϵ is always less.

For example, actual experience shows that, to lift the weight W of Fig. 4-13 at a uniform rate (to avoid increasing its kinetic energy, and hence doing work against inertia), a force F' greater than F must be applied. Friction of the pulleys themselves and friction due to the bending of the rope over them are considerable. The actual mechanical advantage, as contrasted with the theoretical one, is, then, not quite 3; but $W/F' < 3$.

Experimentally, therefore,

$$F' \times 3\,s > W\,s\,,$$

and hence the efficiency,

$$\epsilon = \frac{W\,s}{3F'\,s} < 1 \equiv 100 \text{ per cent} . \qquad (4\text{-}22)$$

The student may show that the efficiency of a machine is also the ratio of the actual mechanical advantage to the theoretical mechanical advantage.

EXAMPLE 8: If in the block and tackle of Fig. 4-13 a force of 35 lbf is required to raise a weight of 90 lbf, calculate the actual mechanical advantage, the theoretical mechanical advantage, and the efficiency of the machine.

Solution: The theoretical mechanical advantage is given by equation (4-17) as 3, but

$$\text{The actual mechanical advantage} = \frac{90 \text{ lbf}}{35 \text{ lbf}} = \frac{18}{7} .$$

If the 35 lbf acts through a distance of 1 ft, the 90 lbf is raised through $\frac{1}{3}$ foot. The work input is, then, 35 ft-lbf; and the output, 30 ft-lbf:

$$\epsilon = \frac{30}{35} = \frac{6}{7}, \quad \text{or} \quad 85.7 \text{ per cent} ;$$

also,

$$\epsilon = \frac{\text{Actual MA}}{\text{Theoretical MA}} = \frac{18}{(7)(3)} = \frac{6}{7},$$

$$\text{or} \quad 85.7 \text{ per cent} .$$

4.12. Power

The term *power*, frequently used in connection with engines, motors, and all kinds of machines, refers to the time rate of doing work. In connections with the transmission of energy of all types, power refers to the rate at which energy is trans-

mitted. If E is the work done in time t, then th average power is, by definition,

$$\text{Power } (P) = \frac{E}{t}. \qquad (4\text{-}23)$$

In the mks system the unit of power is the joule/sec, which is called a "watt":

$$1 \text{ watt} = 1 \text{ joule/sec.}$$

For the cgs system the unit of power is the erg/sec, but this is seldom used in practice, and we do not find a single name used for this.

In the British Engineering system the unit of power is the ft-lbf/sec. When machines came to replace horses, a unit of power called the horsepower, hp, was introduced. By definition,

$$1 \text{ hp} = 550 \text{ ft-lbf of work per second}$$

$$= 33,000 \text{ ft-lbf of work per minute}$$

$$= 746 \text{ watts} .$$

The more common units of work and power often encountered are given in Table 4-1.

TABLE 4-1

COMMON UNITS OF WORK AND POWER

1 joule $= 10^7$ ergs	work
1 watt $= 1$ joule/sec	power
1 horsepower (hp) $= 746$ watts*	power
$= 7.46 \times 10^9$ ergs/sec	power
1 hp $= 550$ ft-lbf/sec	power
$= 33,000$ ft-lbf/min	power
1 kilowatt-hour (kwh)	
$= 1,000$ joules/sec $\times 3,600$ sec	work
$= 3.6 \times 10^{13}$ ergs	work
1 hp-hr $= 7.46 \times 10^9$ ergs/sec	
$\times 3,600$ sec	work
$= 2.6 \times 10^{13}$ ergs	work

* This unit, the American horsepower, now used by engineers in this country, is defined by the Bureau of Standards to be exactly equivalent to 746 watts. In other countries the horsepower is equivalent to 33,000 ft-lbf/min, equal to 745.70 watts.

Power also possesses a simple relation to the force that is performing the work and the velocity of the point of application of the force. Since $E = fs$,

$$P = \frac{fs}{t} = fv ; \qquad (4\text{-}24)$$

i.e., power is the scalar product of the force and the speed of its point of application.

This expression fv for power may be interpreted as the constant power delivered by a constant force f and moving an object with a constant velocity v. Otherwise, if the power is varying, it may be regarded as the expression for the instantaneous power.

Most machines deliver their power through rotating shafts, through pulleys, or through wheels of some sort. The measurement of the power of a machine and also its efficiency is a very important practical problem. One of the simplest mechanical means of measuring the power of a driven shaft or pulley goes by the name of the "Prony brake." The simplest type of Prony brake is illustrated in Fig. 4-16.

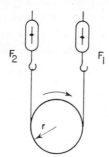

FIG. 4-16.—Measurement of the power of a machine

The mechanism whose efficiency is required is arranged to rotate the drum (shown as a circle in Fig. 4-16), against which a brake lining under tension is rubbing. By means of the spring balances, the forces F_1 and F_2 in kilograms-force acting on the drum can be recorded. Their difference, $F_1 - F_2$, is the frictional force against which the surface of the drum is being driven. If n is the number of revolutions made by the drum in 1 sec and r the radius of the drum, then the power P_0, in watts dissipated as heat in the drum and lining, using the mks system of units is,

$$P_0 = (F_1 - F_2) \, g 2\pi r n \, \text{(watts)}. \quad (4\text{-}25)$$

If the power input of the machine is P_i and the output P_o, then the efficiency is

$$E = \frac{P_o}{P_i} \times 100 \text{ per cent}.$$

In equation (4-25), $(F_1 - F_2)g$ is the resultant force in newtons, acting tangentially to the drum; $2\pi r$, the distance through which the force acts per revolution; and n, the number of revolutions per second.

EXAMPLE 9: What horsepower is necessary to draw a 10-ton load along a level road at 15 mi/hr if the coefficient of friction is $\frac{1}{10}$? Neglect air friction.

Solution: Since 60 mi/hr = 88 ft/sec,

$$15 \text{ mi/hr} = \frac{88}{4} \text{ ft/sec}.$$

The force of friction is

$$\tfrac{1}{10} (10) (2,000) = 2,000 \text{ lbf}.$$

Let P be the necessary horsepower. Then

$$550 \, P = (2,000 \text{ lbf}) (\tfrac{88}{4} \text{ ft/sec}).$$

Note that in almost all these horsepower problems both sides of the equation may be divided by 11. Making this division simplifies the arithmetic:

$$50 \, P = (2,000) (\tfrac{8}{4}),$$

$$P = 80 \text{ horsepower}.$$

EXAMPLE 10: How many miles per hour can a 480-hp engine pull a 150-ton train up a 2 per cent slope if the force of friction is 3,000 lbf?

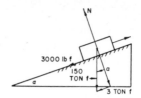

Solution: By definition of a 2 per cent slope,

$$\sin \alpha = \tfrac{2}{100}.$$

If the train moves up the incline at x mi/hr, then its speed in ft/sec is $x88/60$. By conservation of energy,

Work done/sec by engine = Work done/sec against

friction + work done/sec against gravity

$$480 \times 550 \text{ ft-lbf/sec} = 3,000 \times \tfrac{88}{60} \text{ ft-lbf/sec}$$

$$+ 0.02 \times 150 \times 2,000 \times \tfrac{88}{60} \text{ ft-lbf/sec},$$

$$480 \times 50 = (6,000 + 3,000) (\tfrac{8}{60}) \, x,$$

$$48 \times 5 = (6 + 3) (\tfrac{8}{6}) \, x,$$

$$x = 20 \text{ mi/hr}.$$

4.13. The Ballistic Pendulum

An application of the principle of conservation of momentum, supported by the principle of conservation of energy, makes it very simple to determine the speed of swiftly moving objects, such as bullets, if their size (diameter) or caliber is not too great. The bullet is fired into a box of sand, or a block suspended to swing to and fro as a pendulum. Actual photographs of these experiments are reproduced in Fig. 4-19.

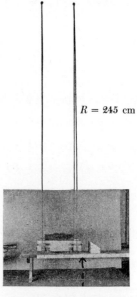

FIG. 4-17.—A bullet fired into a block of wood

In the sketch of Fig. 4-17, m_1 and m_2 are the masses of bullet and block, respectively. The initial velocity of the bullet is u_1; that of the block is zero. The final velocity of the block, i.e., bullet and block (inelastic impact), is v. According to the conservation of momentum (§ 3.3),

$$m_1 u_1 = (m_1 + m_2) v \qquad (4\text{-}26)$$

or

$$u_1 = \left(\frac{m_1 + m_2}{m_1}\right) v . \qquad (4\text{-}27)$$

We have the unknown velocity of the bullet in terms of the masses of bullet and block and of the velocity of swing of the pendulum.

The pendulum's instantaneous velocity of swing after impact is difficult to measure directly. It can be obtained indirectly in terms of the vertical distance through which the bob is raised. If h is this distance, then, from the law of conservation of energy, as applied in § 4.9, it follows that, on equating the potential energy of the bob at its extreme swing to its initial kinetic energy,

$$v = \sqrt{2 g h} . \qquad (4\text{-}28)$$

We may obtain h in terms of R, the length of the pendulum string, and of d, the horizontal distance that the bob moves. It follows from triangle OPQ (Fig. 4-18) that

or
$$\left.\begin{array}{r} (R - h)^2 + d^2 = R^2 \\ d^2 + h^2 = 2Rh . \end{array}\right\} \qquad (4\text{-}29)$$

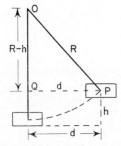

FIG. 4-18.—Relationship of displacement d and height h

Since R is made large, d is large compared to the vertical rise h; and hence h^2 may be made negligible in comparison with d^2. With this approximation,

$$h = \frac{d^2}{2R} \qquad (4\text{-}30)$$

and

$$v = d\sqrt{\frac{g}{R}}, \qquad (4\text{-}31)$$

and the initial velocity, u_1, of the bullet becomes

$$u_1 = \left(\frac{m_1 + m_2}{m_1}\right) d\sqrt{\frac{g}{R}}. \qquad (4\text{-}32)$$

An actual experimental setup for determining the velocity of a bullet is shown in Fig. 4-19. A bul-

$R = 245$ cm

31.5 cm   (a)

42 cm   (b)

FIG. 4-19.—A bullet fired into a block of wood gives a displacement, d, from (a) to (b) of 10.5 cm, or 0.105 m.

let of mass 0.00478 kg is fired into a block of wood, mass 6.30 kg, suspended as a pendulum of length 2.45 m. As a result, the block is moved a distance **69** of $0.420 - 0.315 = 0.105$ m; $g = 9.80$ m/sec^2. Hence

$$u_1 = \frac{6.30478 \text{ kg} \times 0.105 \text{ m}}{0.00478 \text{ kg}} \sqrt{\frac{9.80 \text{ m/sec}^2}{2.45 \text{ m}}}$$

$$= 277 \text{ m/sec.}$$

The bullet was found to have penetrated a distance of 0.082 m into the block of wood, that is, in the distance of 0.082 m most of the kinetic energy of the bullet was used up in doing work against the frictional force of the wood on the bullet. If $\bar{F}$ is the average value of this frictional force, then $0.082 \, \bar{F}$ is the work done by this frictional force, and this reduces the kinetic energy of the bullet. The initial velocity of the bullet is 277 m/sec, and, after becoming imbedded in the block, the bullet and block have a velocity given in equation (4-31). This common velocity of block and bullet is

$$v = d\sqrt{\frac{g}{R}} = 0.105 \text{ m} \sqrt{\frac{9.80 \text{ m}}{2.45 \text{ m/sec}^2}} = 0.21 \text{ m/sec.}$$

Thus the

Kinetic energy before impact $= \frac{1}{2} \times 4.78 \times 10^{-3}$ kg

$$\times \, (2.77 \times 10^2 \text{ m/sec})^2 = 1.83 \times 10^2 \text{ joules}.$$

Kinetic energy after impact $= \frac{1}{2} \times 6.30478$ kg

$$\times \, (0.21 \text{ m/sec})^2 = 1.39 \times 10^{-1} \text{joules},$$

and

$$\bar{F} s = \tfrac{1}{2} m v^2 - \tfrac{1}{2} m v_0^2,$$

$$\bar{F} \times 0.082 \text{ m} = 183 - 0.139 \text{ kg-m}^2/\text{sec}^2,$$

$$\bar{F} = \frac{182.861}{0.082} \frac{\text{kg-m}^2/\text{sec}^2}{\text{m}}$$

$$= 2.23 \times 10^3 \text{ newtons}$$

$$= 2.27 \times 10^2 \text{ kgf}.$$

4.14. An Interesting Experiment on Impact

The masses of the two balls shown in A and B of Fig. 4-20 are in the ratio of 1 to 3. To the extent that the elasticity is perfect, energy must be conserved, as well as momentum, during impact. This means that the total momentum before and after impact must be the same and, in addition, that the relative velocity of recession must equal the relative velocity of approach of the two balls.

As indicated along the margins of Fig. 4-20 in A, if the small mass with velocity V, taken positive to the right, strikes the large one, at rest, they must recoil with velocities $-V/2$ and $V/2$, respectively, in order that the relative velocity of recession may be

$$V = \left[\frac{V}{2} - \left(-\frac{V}{2} \right) \right]$$

and the total momentum remain mV. No other velocities will satisfy BOTH energy and momentum criteria.

In the series of figures in B the large mass is dropped onto the small one, initially at rest. Can you find the velocities of each after impact and show that now the small mass, m, comes to rest on alternate impacts?

4.15. Summary of Units and Dimensions

In Table 4-2 notice that the unit of newton-sec for impulse or momentum is the same as kg-m/sec and that dyne-sec is equivalent to gm-cm/sec. The acceleration due to gravity is approximately $g = 9.80$ m/sec^2 = 32 ft/sec^2. 1 newton = 10^5 dynes; 1 joule = 10^7 ergs. The units given in Table 4-2 form a consistent set for use in Newton's laws. For the mks or cgs one uses

$$F = ma,$$

i.e.,

$$F \text{ (newtons)} = m \text{ (kg)} \, a \text{ (m/sec}^2),$$

$$F \text{ (dynes)} = m \text{ (gm)} \, a \text{ (cm/sec}^2).$$

For the Engineering system we write

$$F = \frac{wa}{g}$$

$$F \text{ (lbf)} = \frac{w \text{ (lbf)} \, a \text{ (ft/sec}^2)}{g \text{ (ft/sec}^2)}.$$

The units for KE (or PE or work) are

$$F s = \tfrac{1}{2} m \, (v^2 - v_0^2),$$

$$(F s) \text{ joules} = \tfrac{1}{2} m \text{ (kg)} \, (v^2 - v_0^2) \text{ m}^2/\text{sec}^2,$$

$$(F s) \text{ ergs} = \tfrac{1}{2} m \text{ (gm)} \, (v^2 - v_0^2) \text{ cm}^2/\text{sec}^2.$$

For the Engineering system,

$$F s = \frac{1}{2} \frac{w}{g} (v^2 - v_0^2),$$

$$F s \text{ (ft-lbf)} = \frac{1}{2} \frac{w \text{ (lbf)}}{g \text{ (ft/sec}^2)} (v^2 - v_0^2) \text{ (ft}^2/\text{sec}^2).$$

Positive direction→

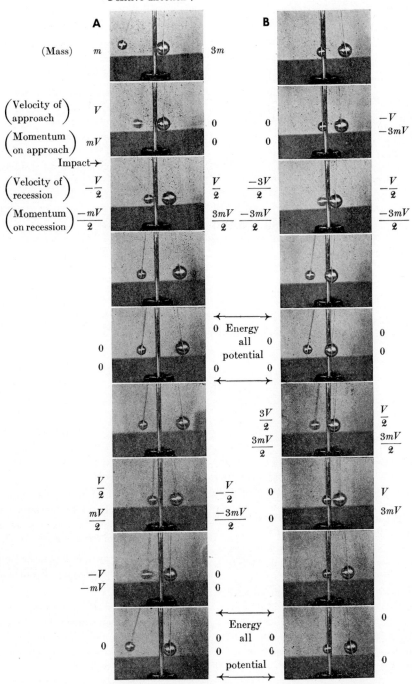

A   **B**

(Mass) m $3m$

$\begin{pmatrix}\text{Velocity of}\\\text{approach}\end{pmatrix}$ V 0 0 $-V$

$\begin{pmatrix}\text{Momentum}\\\text{on approach}\end{pmatrix}$ mV 0 0 $-3mV$

Impact→

$\begin{pmatrix}\text{Velocity of}\\\text{recession}\end{pmatrix}$ $-\dfrac{V}{2}$ $\dfrac{V}{2}$ $\dfrac{-3V}{2}$ $-\dfrac{V}{2}$

$\begin{pmatrix}\text{Momentum}\\\text{on recession}\end{pmatrix}$ $-\dfrac{mV}{2}$ $\dfrac{3mV}{2}$ $\dfrac{-3mV}{2}$ $\dfrac{-3mV}{2}$

$\longleftrightarrow$
0   Energy   0
all
0 potential 0

0 0 0
$\longleftrightarrow$

$\dfrac{3V}{2}$ $\dfrac{V}{2}$

$\dfrac{3mV}{2}$ $\dfrac{3mV}{2}$

$\dfrac{V}{2}$ $-\dfrac{V}{2}$ 0 V

$\dfrac{mV}{2}$ $\dfrac{-3mV}{2}$ 0 $3mV$

$-V$ 0 0
$-mV$ 0

$\longleftrightarrow$
Energy 0
0   all
0 0 0
potential
$\longleftrightarrow$

Fig. 4-20.—An elastic collision of two balls; masses m and $3m$

Since power is work/time, it follows that work or energy may be expressed in units of power *times* time. This is used extensively in electricity. For example, an energy of 1 kilowatt-hour (kwh) is equal to

$$1 \text{ kwh} = 1,000 \text{ watts} \times 3,600 \text{ sec}$$
$$= 3.6 \times 10^6 \text{ watt-sec}$$
$$= 3.6 \times 10^6 \text{ joules}.$$

Similarly, for an amount of energy of 1 horsepower-hour (hp-hr),

$$1 \text{ hp-hr} = 550 \text{ ft-lbf/sec} \times 3,600 \text{ sec}$$
$$= 19.8 \times 10^5 \text{ ft-lbf}$$

$$= 746 \text{ watts} \times 3,600 \text{ sec}$$
$$= 2.69 \times 10^6 \text{ joules}.$$

4.16. Résumé

The following is a brief listing of the definitions, principles, and theories, given in this chapter, which you should know:

The definition of work and of energy
Kinetic and potential energies
Work against friction and heat
Conservation of mechanical energy
Conservation of all forms of energy
Machines—their mechanical advantage and efficiency
Power

TABLE 4-2

SYSTEMS OF UNITS AND THEIR DIMENSIONS

System of Units	Acceleration	Mass	Force	Momentum or Impulse	Work or KE or PE	Power
mks...........	m/sec² $[LT^{-2}]$	kg $[M]$	newton $[MLT^{-2}]$	kg-m/sec or newton-sec $[MLT^{-1}]$	joule or newton-m $[ML^2T^{-2}]$	watt $[ML^2T^{-3}]$
cgs...........	cm/sec² $[LT^{-2}]$	gm $[M]$	dyne $[MLT^{-2}]$	gm-cm/sec or dyne-sec $[MLT^{-1}]$	erg or dyne-cm $[ML^2T^{-2}]$	erg/sec $[ML^2T^{-3}]$
Engineering.....	ft/sec² $[LT^{-2}]$	slug $[FL^{-1}T^2]$	lbf $[F]$	lbf-sec $[FT]$	ft-lbf $[LF]$	ft-lbf/sec $[LFT^{-1}]$

Queries

1. A box, entirely inclosed except for two ropes, A and B, contains a secret mechanical device. The two ropes are connected to this device. We find, on experimenting with the inclosed machine by means of the two ropes, that a weight of 10 lbf attached to A will sustain a weight of 30 lbf attached to B. If friction is neglected, how many different kinds of machines may be within the box and satisfy this experiment? Design two.

2. To take friction into account, we perform an additional experiment. We find that if 20 ft-lbf of work are done on the machine via rope A, then the machine is able to do only 18 ft-lbf of work via rope B. What are the mechanical advantage and efficiency of the machine? Design two machines.

3. If f makes an angle θ with v, how must equation (4-24) be modified?

4. In the experimental determination of the velocity of a bullet given in Fig. 4-19, how much momentum was lost at the impact? How much mechanical energy was lost? Account for any losses you find.

5. If the speed of an automobile is doubled, then by what ratio are the stopping distance and the time of stopping increased?

6. An automobile, starting from rest, can reach a speed v in a distance s and time t. For the same automobile operating under the same conditions, find the distance and time in which it reaches a speed of $2v$, $3v$, $4v$, in terms of s and t.

Practice Problems

1. A man of mass 170 lb climbs a mountain 15,000 ft high in 24 hr. What amount of work was done? At what average rate did he do work? If the commercial value of energy is 5 cents a kilowatt-hour, what is the commercial value of his day's work? Ans.: 2.55×10^6 ft-lbf; 4.8 cents.

2. A man whose weight is 190 lbf runs up a flight of stairs 20 ft high in 3.5 sec. What power did he expend?

3. A 100-lb body has an initial kinetic energy of 5,000 ft-lbf. After 20 sec it has a kinetic energy of 200,000 ft-lbf. Find the initial velocity, the acceleration, the force acting, and the distance traveled during the 20 sec. Ans.: 56.5 ft/sec; 15.1 ft/sec^2; 47 lbf; 4,150 ft.

4. A mass of 5 lb falls 20 ft to the ground and sinks in 1 in. With what average force is it brought to rest?

5. A mass of 50 lb is pulled 10 ft up a rough plane ($\mu = 0.1$), inclined 25° to the horizontal, by a force of 100 lbf acting parallel to the plane. Find (*a*) the work done by the 100 lbf; (*b*) the work done against gravity; (*c*) the work done against friction; and (*d*) the increase in kinetic energy of the body. Ans.: (*a*) 1,000 ft-lbf; (*b*) 210 ft-lbf; (*c*) 45.4 ft-lbf; (*d*) 744.6 ft-lbf.

6. The drum of the Prony brake of Fig. 4-16 is attached to an electric motor. The radius of the drum is 4.5 cm; the scale readings are $F_1 = 1,590$ gmf, $F_2 = 150$ gmf; and the number of revolutions per minute is 2,090. If the power input is 500 watts, calculate the efficiency of the motor.

7. A jackscrew has a pitch of 0.35 in. How much force must be applied at the end of the arm, 20 in from the axis, in order to lift 3 tons? Neglect friction.

8. What is the efficiency of a motor that delivers 20 hp for 2 hr and consumes 6.92×10^8 ft-lbf of energy?

9. A gun weighs 4 kgf. The bullet weighs 20 gmf and leaves the gun with a velocity of 400 m/sec. Find the velocity of recoil of the gun, the kinetic energy of the gun, and the kinetic energy of the bullet. Ans.: 2 m/sec; 8×10^7 ergs; 16×10^9 ergs.

10. A ball weighing 1 gmf is thrown into a 9-gm ball of putty, giving it a velocity of 5 cm/sec. The ball penetrates 7 cm. Compute the initial velocity of the ball and the force necessary to push the ball into the putty.

11. A block of mass M slides down a frictionless inclined plane whose vertical height is h. Compute the velocity of the block at the foot of the plane, using the conservation-of-energy principle.

12. A bullet of mass 3.3 gm is shot into a ballistic pendulum whose bob is of mass 2,000 gm. If the length of the pendulum string is 245 cm and the horizontal displacement of the bob is 18.5 cm, calculate the velocity of the bullet and the mechanical energy lost during the impact.

13. In problem 12 the bullet penetrates the block to a depth of 8 cm. What average force of resistance acted upon the bullet?

14. How many miles per hour can a 500-hp engine pull a 175-ton train up a 2 per cent slope if the force of friction is 3,500 lbf? ("Slope of 2 per cent" means

that the incline makes an angle $\sin^{-1} \frac{2}{100}$ with the horizontal.) Ans.: 17.8 mi/hr.

15. What is the coefficient of friction if a 2-hp team can just pull a 5-ton load 4 mi/hr along a level road?

16. Design a simple block and tackle by which a force of 20 lbf can just lift a weight of 50 lbf, and compute the efficiency of your system. (More than one correct answer possible [friction must be assumed].)

17. The two pulley wheels shown below are rigidly fastened together. Let r be the radius of the smaller one, and R the radius of the larger. From the principle of work compute the theoretical mechanical advantage of this "differential pulley."

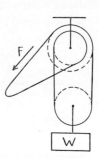

18. A certain engine can just pull a 300-ton train 20 mi/hr along a level track. If it can just pull this train 15 mi/hr up a $\frac{1}{4}$ per cent grade, find the horsepower of the engine and the force of friction, assuming each to remain constant. (See Ex. 10.) Ans.: 240 hp; 4,500 lbf.

19. How much energy is necessary to run a 5,000-watt machine for $\frac{1}{2}$ hr?

20. Two joules of work are expended against gravity by a force acting 30° from the vertical. How high does the force raise 1 kg of mass? What is the increase in potential energy of the mass? If $\frac{1}{30}$ watt of power is available, how long will it take to raise the mass? Ans.: 20.4 cm; 2 joules; 60 sec.

21. A 5-lb stone on the end of a 2-ft rope is whirled in a vertical plane. At the top of its path, the speed of the stone is 10 ft/sec. What is the tension in the rope when the stone is at the top of its path? At the bottom of its path?

22. How high above the top of the loop should the car of a "loop-the-loop" apparatus start if it is desired that the resultant force of the car against the track at the top of the loop be the same as the weight of the car? Ans.: $2h$ equals the radius of the loop.

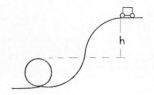

23. A weight of 1,000 lbf is hung from a cable 20 ft long. Compute the horizontal force F necessary to hold the cable at an angle of 30° with the vertical, and also the tension in the cable. If the weight is released, compute the increase in tension in the cable as it passes the vertical over its value when hanging vertically at rest. Ans.: 577.4 lbf; 1,155 lbf; 268 lbf.

24. The wedge is another example of a simple machine. As shown in the diagram, an application of a force F parallel to l causes the wedge to exert a force W at right angles to l. Using the principle of work, calculate the mechanical advantage of this machine. *Hint:* While F moves through the distance l, W acts through the distance h. In what respect does the wedge differ from the inclined plane?

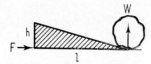

25. Show that the mechanical advantage of a machine can also be defined as

$$\text{MA} = \frac{\text{Speed of the operator}}{\text{Speed of the machine}}.$$

Hint: Use the simple pulley of Fig. 4-11 (a).

26. A freight car weighs $2\frac{1}{2}$ U.S. tons-force. If the coefficient of rolling friction of its wheels on the track is 0.03, how many joules of work will be required to drag it horizontally 1 km? Ans.: 6.78×10^5 joules.

27. Solve problems 9 and 12, using the mks system of units.

28. A ball of mass 0.1 kg and moving with a speed of 0.05 m/sec collides head-on with a stationary ball of mass 0.2 kg. Show that there is no loss in energy if the coefficient of restitution is unity. Find the loss in energy for coefficients of restitution of 0.2 and zero.

29. Show that 1 ft-lbf = 1.356 joules.

30. In the collision of the two balls having masses in the ratio of 1 to 3 (Fig. 4-20), find the velocities after collision if the coefficient of restitution is 0.6. Assume that the ball of smaller mass has a velocity of v at impact. Find the loss in energy at the first impact. Is this the same at each impact?

31. An automobile traveling at 60 mi/hr can be stopped by the brakes in a distance of 200 ft. In what distance can this same automobile be stopped when going at 30 mi/hr? At 20 mi/hr? What is the stopping time when going at 60 mi/hr, 30 mi/hr, and 20 mi/hr? Ans: 50 ft, 22.2 ft; 4.55, 2.77, 1.52 sec.

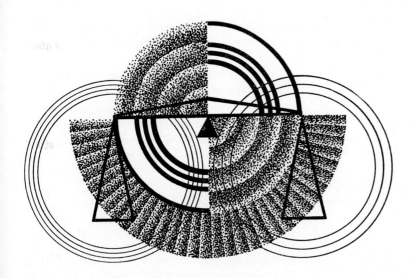

CHAPTER 5

Rigid Bodies in Equilibrium: Center of Mass

5.1. Rigid Bodies

The laws of motion discussed in the last four chapters were developed for mass particles, i.e., mathematical points that were endowed with the property of mass. Although this concept is artificial, nevertheless it proved quite useful, since, as we have seen, many natural phenomena occur that may be effectively described and analyzed by means of a mass-point interpretation. This prevented us from discussing any effects due to rotation of the object, for, by definition, a mass point has no spatial dimensions and cannot rotate. In order to generalize our discussion, however, to include objects that possess extension, we must now learn what modifications have to be made in the laws of motion so that they will be literally applicable to real objects.

In this and the following chapter we shall confine our attention to *rigid bodies*.

We define a rigid body as an extended object whose separate portions all maintain fixed distances from one another when subjected to external forces.

No physical material that we know possesses perfect rigidity in this sense; but most solids, especially crystalline ones, approximate it closely. We shall consider, for example, a meter stick to be a rigid body. Objects whose separate parts *do not* maintain fixed distances from one another when subjected to forces are known as *deformable bodies*, a special class of which will be treated in chapter 7.

5.2. Equilibrium of a Rigid Body

According to equation (1-36), the necessary and sufficient condition that a particle or mass point at rest shall remain at rest is that the vector sum of all the forces acting on the particle shall be zero. Transferring the idea to an extended rigid body, we first inquire: If the vector sum of all the forces acting on a *rigid object* is zero, will the object necessarily remain at rest if originally at rest? The answer to this question is found in the following simple experiment. To the ends of a uniform meter stick we apply equal and opposite forces, F and $-F$, as shown in Fig. 5-1. Here the vector sum

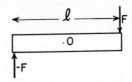

Fig. 5-1.—Translational but not rotational equilibrium

of the forces is obviously zero. However, we observe that, although no translational motion of the stick ensues, it does, nevertheless, move. It begins to rotate about the mid-point, O, as an axis. Only when the two forces are equal and opposite and *in the same straight line* does the stick remain at rest.

Satisfying the condition of equation (1-36) will, therefore, assure us only that the object will not move translationally. Now, what additional condition must be satisfied in order that the rigid body shall not rotate? Suppose we pivot the meter stick of Fig. 5-1 at any point such as O; we then obviously will have no translational motion, since the pivot can supply, by its reaction on the stick, a force of such magnitude as to satisfy equation (1-36).

If to the pivoted meter stick we apply a force F_1 at a distance l_1 from O (Fig. 5-2), the stick will be-

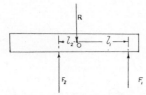

FIG. 5-2.—For equilibrium $F_1 l_1 = F_2 l_2$

gin to rotate counterclockwise about the pivot as an axis. We may prevent this rotation by applying another force, F_2, a distance l_2 from O. We see that F_2 tends to produce clockwise rotation. It is found experimentally that we may prevent rotation by applying either a force greater than F_1 at a distance from O less than l_1 or a force less than F_1 at a distance greater than l_1 or, finally, a force equal to F_1 at a distance equal to l_1. In other words, *what is important is that the product* $F_2 l_2$ *shall be equal to* $F_1 l_1$. We call such a product the *moment of a force about an axis* or a *torque*.

By definition, the *moment of a force about a point or axis is the product of the force and the perpendicular distance from the point to the line of action of the force*. This latter distance is sometimes called a "lever arm." A moment of a force may turn, or tend to turn, the object clockwise about the point or axis. In this case the force has a clockwise moment about the point. Similarly, a force can have a counterclockwise moment about a point or axis.

The "moment of a force," or "torque," is a vector quantity whose direction is taken parallel to the axis about which it tends to produce rotation (see § 6.2 for a more thorough treatment of vector representation of torques).

Our simple experiment has given us the second condition that must be satisfied if a rigid body is to be in equilibrium, i.e., to have neither translational nor rotational acceleration. *If an object is to be in rotational equilibrium, the sum of the clockwise mo-*

ments of force about some axis must be equal to the sum of the counterclockwise moments of force about the same axis.

For the case of forces that lie in a given plane, we may take the counterclockwise moments of force to be $(+)$ and the clockwise as $(-)$ and then state the second condition for equilibrium as follows. *The algebraic sum of the moments of force about any axis must be zero.* In the case of Fig. 5-2 this condition becomes

$$F_1 l_1 - F_2 l_2 = 0 \ . \qquad (5\text{-}1)$$

For translational equilibrium we have

$$F_1 + F_2 - R = 0 \ , \qquad (5\text{-}2)$$

where R is the reaction of the pivot on the stick.

If F_1 makes an angle θ with the stick (Fig. 5-3), condition (5-1) becomes

$$F_1 l_1 \sin \ \theta = F_2 l_2 \ . \qquad (5\text{-}3)$$

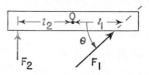

FIG. 5-3.—Moment of F_1 about O is $F_1 l_1 \sin \theta$

The reader will recognize that $l_1 \sin \theta$ is the *perpendicular* distance from O to the line of action of F_1, i.e., the lever arm.

If we represent torque, or moment of force, by the letter L and the distance from the pivot point to the point of application of the force by the letter l, then we may write

$$L = Fl \sin \ \theta \ . \qquad (5\text{-}4)$$

Note that the torque is always zero if the line of action of the force goes through the pivot point.

Although torque, L, has the same dimensions as work (force *times* distance), it is, nevertheless, a quantity distinct from work. In the definition of work, either F and s were in the *same direction*, or the component of F in the s direction ($F \cos \theta$) was used; in the definition of torque, F and the lever arm are at right angles to each other, and, if any other l is specified, the perpendicular component ($l \sin \theta$) is taken. Also, work is a scalar, torque a vector, quantity.* If F is measured in newtons or

---

* In more advanced analysis these two different kinds of products of vectors—one, like work, a scalar, and the other, like torque, a vector—are distinguished by the use of two different signs of multiplication: (·) and (×), respectively. These

dynes, l in meters or centimeters, then L is measured in meter-newtons or centimeter-dynes. If F is in kgf or gmf and l in meters or centimeters, then L is in m-kgf or cm-gmf. Also, with F in pounds-force, lbf, and l in feet, L is in foot-pounds-force, ft-lbf.

We may summarize the results of this section by stating that if a rigid body is in equilibrium, i.e., there is no translational or rotational acceleration, then

I. **The vector sum of all the forces acting on the object must be equal to zero, and**

II. **The vector sum of all the torques or force moments about any axis must be zero.**

In symbols these conditions can be written

$$\left.\begin{aligned} &\text{I. } \Sigma F = 0 \text{ (no translational acceleration)}, \\ &\text{(or, in terms of the components in a plane,} \\ &\quad \Sigma F_x = 0 \text{ , } \Sigma F_y = 0) \text{ .} \\ &\text{II. } \Sigma L = 0 \text{ (no rotational acceleration)} \text{ .} \end{aligned}\right\} \quad (5\text{-}5)$$

5.3. Equilibrium Conditions Hold for Any Axis

We have just seen that the meter stick of Fig. 5-2 is in equilibrium if

$$F_1 + F_2 - R = 0$$

and

$$F_1 l_1 - F_2 l_2 = 0 \text{ ,}$$

where the force moments were calculated about the pivot point O. A little consideration shows that, if the moments about a particular axis are zero, then the moments calculated with respect to *any other parallel axis must also be zero*. This may be easily shown analytically by reference to Fig. 5-4. The three forces F_1, F_2, and R satisfy the con-

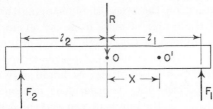

Fig. 5-4.—Moments about O and O' are zero

symbols, when so used, contain implicitly the appropriate trigonometric functions, i.e.,

$$W = f \cdot s = fs \cos (f, s) = fs \cos \theta \text{ (scalar)},$$

$$L = f \times l = fl \sin (f, l) = fl \sin \theta \text{ (vector)} \text{ .}$$

ditions of equations (5-5), the axis being through O. Let us choose another axis O', x cm from O, and calculate moments about O'. We obtain

$$\Sigma L = F_1 (l_1 - x) + Rx - F_2 (l_2 + x) \text{ .}$$

Rearranging terms, we have

$$\Sigma L = (F_1 l_1 - F_2 l_2) - (F_2 + F_1 - R) \, x$$

Since, for translational equilibrium, $(F_2 + F_1 - R)$ must be zero, and, by equation (5-1), $(F_1 l_1 - F_2 l_2)$ is zero, it follows that

$$\Sigma L = 0 \text{ ,}$$

and the theorem is proved.

5.4. Coplanar Forces on a Rigid Body

Let us consider the more general problem of a number of coplanar forces acting on a body; in particular, consider the rectangular board of Fig. 5-5 subject to four forces and in equilibrium. As

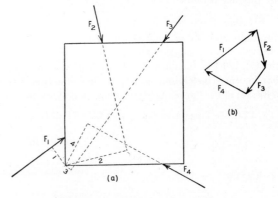

Fig. 5-5.—Coplanar forces in equilibrium

far as translational equilibrium is concerned, the vector sum of the forces must be zero, as shown by the closed polygon of Fig. 5-5 (*b*); that is,

$$\Sigma F = F_1 + F_2 + F_3 + F_4 = 0 \text{ .}$$

Also, as far as condition II for rotational equilibrium is concerned, the sum of the force moments about any axis must be zero. Let us choose the lower left-hand corner as an axis. Then

$$\Sigma L = -F_1 l_1 - F_2 l_2 - F_3 l_3 + F_4 l_4 = 0 \text{ ,}$$

where l_1, l_2, l_3, and l_4 are the lever arms of the

77

forces. (In Fig. 5-5 the respective lever arms bear the subscript numbers only, i.e., l_1 is indicated by 1, etc.)

5.5. Couples

The special case of oppositely directed parallel forces shown in Fig. 5-1 merits attention. Although there is translational equilibrium, there is no rotational equilibrium, since there is an unbalanced torque acting on the stick. The two parallel forces produce rotation in the same direction. Such a pair of equal and opposite forces not acting on the same straight line is called a *couple*. The *"moment of the couple"* (or the torque) is Fl, where F is the value of either force and l is the perpendicular distance between them. In order to produce rotational equilibrium, another couple equal and opposite to Fl must be applied. No single force can produce equilibrium, since it will, of necessity, impair the translational equilibrium.

EXAMPLE 1: A rigid board has forces acting on it as indicated by the accompanying illustration. How can we produce equilibrium with a single force?

Solution: From condition I, we need a force of 3 lbf to the right and one of 4 lbf down. These two forces are equivalent to a single force of 5 lbf, as shown in figure (*b*). Let x be the normal distance from the lower left-hand corner (the chosen axis) to the line of action of the 5 lbf, (*c*) of the figure. Applying condition II, we have

$$4(0) - 7(20) - 5(x) + 10(18) = 0,$$

$$x = 8 \text{ in}.$$

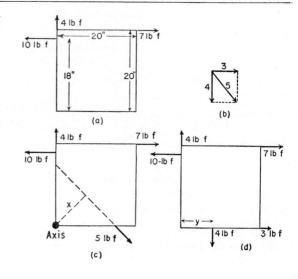

A different solution is as follows: Arbitrarily place the 3 lbf so that it has no moment about the chosen axis (fig. [*d*]), and place the 4 lbf at an unknown distance y. Then

$$10(18) + 4(0) - 7(20) + 3(0) - 4y = 0,$$

$$y = 10 \text{ in}.$$

The sum of 3 lbf to the right and 4 lbf down at a point 10 in to the right of the axis is the same as the 5 lbf in the previous solution.

Note that for a *rigid board* the vectors represent the lines of action of the forces. The forces themselves may be considered as acting *at any point along the line of the vector.*

Center of Mass (CM)

5.6. Formal Definition of Center of Mass

In the discussion of the rotation of rigid bodies there is a particular point in the body called the *center of mass* which plays a very important role. Let us consider a meter stick lying on a horizontal frictionless table. Since the meter stick is uniform, its center of mass is at its geometrical center, the 50-cm mark. If a horizontal force is applied to the stick, then, in general, it will commence to rotate as well as to move with translational motion. If, however, the force is applied at the center of mass, the stick will not rotate but will move with pure translational motion. Now suppose a mass of putty is stuck on the stick, thus changing the position of the center of mass from the 50-cm mark. A force applied at the 50-cm mark will now produce both translational and rotational motion in the stick. A new point can be found at which a force produces only translational motion. This point would be the new center of mass of stick and putty. While this might be considered a method for determining the center of mass, it would be relatively cumbersome. A much simpler and more accurate method for determining the position of the center of mass is given below.

Consider two point masses, m_1 and m_2, at distances x_1 and x_2 from O (Fig. 5-6). We locate a point, P, called the *center of mass* (CM) of the two masses as follows:

$$\bar{x} = \frac{m_1 x_1 + m_2 x_2}{m_1 + m_2}, \qquad (5\text{-}6)$$

where $\bar{x}$ is the distance of the center of mass, P, from the chosen origin O. If we have n masses, m_1, $m_2, \ldots, m_n$, along a straight line, then the dis-

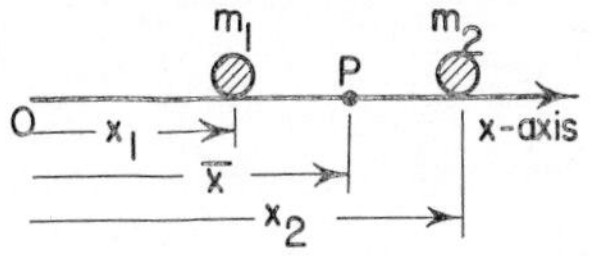

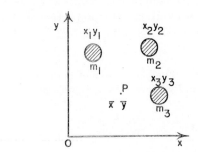

FIG. 5-6.—Center of mass of two particles

tance of the center of mass of these masses from some origin is, by definition,

$$\bar{x} = \frac{m_1 x_1 + m_2 x_2 + \ldots + m_n x_n}{m_1 + m_2 + \ldots + m_n} = \frac{\sum\limits_{i=1}^{n} m_i x_i}{\sum\limits_{i=1}^{n} m_i}, \qquad (5\text{-}7)$$

where again $x_1, x_2, \ldots, x_n$ are the distances of the masses from the origin from which $\bar{x}$ is measured. The symbol Σ means "take the sum of"; the subscript i denotes a typical mass.

It should be noted that the center of mass, P, located by the defining equation (5-6) or (5-7) is independent of the particular origin chosen to locate the masses. We may prove this important theorem as follows: In Fig. 5-6 let us choose as an origin a point x' cm to the left of O. We wish to show that the distance $\bar{x}'$ of the center of mass from O' is given by

$$\bar{x}' = \bar{x} + x' .$$

By definition,

$$\bar{x}' = \frac{m_1 (x_1 + x') + m_2 (x_2 + x')}{m_1 + m_2},$$

$$\bar{x}' = \left(\frac{m_1 x_1 + m_2 x_2}{m_1 + m_2} \right) + \left(\frac{m_1 x' + m_2 x'}{m_1 + m_2} \right).$$

The first parenthesis is simply $\bar{x}$, while in the second the masses cancel out, leaving x', the distance of O' from O. Therefore,

$$\bar{x}' = \bar{x} + x' .$$

If we have three masses not in the same straight line but in one plane, as shown in Fig. 5-7, then we

may locate the center of mass, P, of these masses as follows:

$$\left. \begin{aligned} \bar{x} &= \frac{m_1 x_1 + m_2 x_2 + m_3 x_3}{m_1 + m_2 + m_3}, \\[6pt] \bar{y} &= \frac{m_1 y_1 + m_2 y_2 + m_3 y_3}{m_1 + m_2 + m_3}, \end{aligned} \right\} \qquad (5\text{-}8)$$

where $x_1 y_1$, $x_2 y_2$, etc., are the positions of the masses from some arbitrary origin and $\bar{x}, \bar{y}$ is the

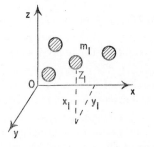

FIG. 5-7.—Center of mass of three particles

position of the center of mass from the same origin. For the case of n masses lying in a plane, we have

$$\bar{x} = \frac{\Sigma m_i x_i}{\Sigma m_i}; \qquad \bar{y} = \frac{\Sigma m_i y_i}{\Sigma m_i}. \qquad (5\text{-}9)$$

For the general case of n point masses distributed in space (Fig. 5-8), we may locate the center of mass by means of the following defining equations:

$$\bar{x} = \frac{\Sigma m_i x_i}{\Sigma m_i}; \qquad \bar{y} = \frac{\Sigma m_i y_i}{\Sigma m_i}; \qquad \bar{z} = \frac{\Sigma m_i z_i}{\Sigma m_i}. \qquad (5\text{-}10)$$

Equations (5-6), (5-7), (5-8), and (5-9) are but special cases of equation (5-10).

FIG. 5-8.—Center of mass of particles in space

Let us emphasize that the position of P is independent of the origin used to locate it. *The center of mass depends only upon the value of the mass points and the distribution of the mass points among themselves*, that is to say, for a given mass distribu-

79

tion, equation (5-10) uniquely locates a point which we have called the "center of mass."

The reader will observe that in equation (5-10) we multiply *each mass by the first power of the distance* from some origin. For this reason the center of mass is sometimes called the "first moment of mass." If in equation (5-10) we were to multiply each mass by the square of the distance, we would obtain the "second moment of mass." In fact, we do just this in chapter 6 (eq. [6-23]) when we define the radius of gyration. The properties of the second mass moment are quite different from those of the first.

EXAMPLE 2: Find the center of mass of the three masses shown. Using equation (5-9), we obtain

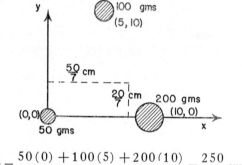

$$\bar{x} = \frac{50(0) + 100(5) + 200(10)}{50 + 100 + 200} = \frac{250}{35} \text{ cm},$$

$$\bar{v} = \frac{50(0) + 100(10) + 200(0)}{50 + 100 + 200} = \frac{100}{35} \text{ cm}.$$

5.7. Center of Mass of Extended Objects

Since an extended object, such as a meter stick, may be thought of as a system of particles very closely packed, an extended object also possesses a center of mass. Unfortunately, we cannot use equation (5-10) directly to locate this point, since we cannot count all the particles that make up the stick.

The location of this point is most easily obtained by the use of the calculus. It can be shown, however, that if a homogeneous object possesses a point, line, or plane of symmetry, the center of mass will be at the point, on the plane, or on the line of symmetry. For example, in the case of a sphere or cube, its center of mass is at the geometrical center; for a right circular cylinder, its center of mass is halfway up its axis of symmetry. For the case of irregularly shaped objects or objects in which the density varies in some unknown

manner, we must resort to an experimental method for locating the center of mass, as illustrated in § 5.8.

We shall now prove by the methods of elementary calculus that the center of mass of a uniform stick is at its geometrical center. Consider a thin uniform rod of length l and mass M (Fig. 5-9),

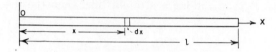

FIG. 5-9.—Center of mass of a uniform rod

placed along the X-coordinate axis. From equation (5-9) the position of the center of mass for a series of discrete masses is

$$\bar{x} = \frac{\sum_i m_i x_i}{\sum m_i}.$$

In this case the masses form a continuous object. Let us consider an element of mass of the stick of very small length Δx at point x from the origin O. Since the stick is uniform, its mass per unit length is M/l, and the mass of the element Δx is $M\Delta x/l$. The position $\bar{x}$ of the center of mass from O is given by

$$\bar{x} = \frac{1}{M} \sum_0^l \frac{Mx\Delta x}{l}.$$

Now this summation can be most easily carried out with the calculus. In order to do this, we make Δx the infinitesimally small quantity dx and go from a summation to an integral. Thus

$$\bar{x} = \frac{1}{M} \int_0^l \frac{M}{l} x\, dx.$$

Since M and l are constants for the stick, we may take these outside the integral sign, giving

$$\bar{x} = \frac{1}{l} \int_0^l x\, dx$$

$$= \frac{1}{l} \left[\frac{x^2}{2}\right]_0^l = \frac{l}{2}.$$

The center of mass is located at a point $l/2$ from the origin O or at the mid-point of the stick. From this result it can be readily shown that the center of mass of a uniform rectangle is at its geometrical center.

EXAMPLE 3: Locate the center of mass of the two plane objects of different densities shown in the accompanying figure.

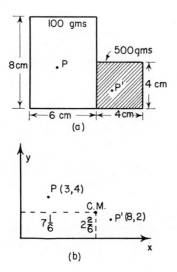

(a)

(b)

Solution: From symmetry the center of mass of the 100-gm object is at P, and that of the 500-gm object at P'. In figure (b) we replace the extended objects by the equivalent point masses. Applying equation (5-9), we obtain

$$\bar{x} = \frac{100(3) + 500(8)}{100 + 500} = \frac{4,300}{600} = 7\frac{1}{6} \text{ cm},$$

$$\bar{y} = \frac{100(4) + 500(2)}{100 + 500} = \frac{1,400}{600} = 2\frac{2}{6} \text{ cm}.$$

The center of mass of the object is, then, $7\frac{1}{6}$ cm to the right of the lower left-hand corner and $2\frac{2}{6}$ cm up.

5.8. Properties of the Center of Mass

Let us next look at some of the more important properties of the center of mass. If a set of mass particles is acted upon by forces, the center of mass of these mass points will move as if all the masses were concentrated at this point and the forces applied at the point.

Again, if a force is applied at the center of mass of a rigid object, the object will move with translational acceleration only—*there is no tendency for the body to rotate*. We may generalize these statements as follows: *If an arbitrary set of forces acts on a rigid body, the center of mass of the body will move as if all the mass and all the forces were concentrated at the center of mass.*

As an illustration of these ideas, consider a shell fired from a gun, the shell exploding in mid-air. As the shell leaves the muzzle of the gun, its center of mass describes a parabolic path (all friction is neglected). When, in mid-air, the shell explodes, *the center of mass of the fragments of the shell continues the original parabolic path of the shell.*

Furthermore, *any freely rotating body rotates about an axis that passes through the center of mass of the body.* For example, if we were to throw a long stick through the air so that it rotated as it fell, the center of mass of the stick would follow a parabolic path, while the stick itself rotated about an axis through the center of mass.

It should be noted that no proof has been given of these two properties of the center of mass of an object. Such proofs are usually found in more advanced texts, in which the calculus can be freely used. However, an experimental verification may be helpful. In the following pictures a meter stick weighing 160 gmf is supported as shown in Fig. 5-10. For such a system $\Sigma F_y = 220 + 190 -$

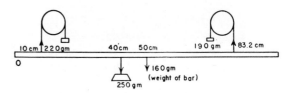

FIG. 5-10.—A uniform rod in translational and rotational equilibrium.

$250 - 160 = 0$; and, for the moments about the 40-cm mark,

$$\Sigma L = 190 \times 43.2 - 10 \times 160 - 220 \times 30$$

$$= 8,208 - 1,600 - 6,600$$

$$= 0 \text{ (approximately to the nearest millimeter setting of the 190-gm mass)}.$$

The center of mass of the 250-gm and 160-gm masses is at the 43.9-cm mark. If 20 gm are added at the 43.9-cm mark, then there is a resultant downward force of 20 gm at the center of mass, but the resultant moment about the center of mass in zero. In this case the meter stick moves downward with translational, but no rotational, motion (Fig. 5-11). Suppose, now, that the 20-gm is removed and the 250-gm mass moved to the 42-cm mark. The center of mass is now at the 45.1-cm mark (shown by the ink dot). In this case $\Sigma F = 0$, so that there is no translational motion, **81** but the sum of the moments ΣL is not zero, so that there is rotational motion, as shown in Fig. 5-12.

Finally, the weight at the 42-cm mark is increased from 250 to 270 gm. For this case $\Sigma F \neq 0$ and $\Sigma L \neq 0$, so that there is both translational and rotational motion, as shown in Fig. 5-13.

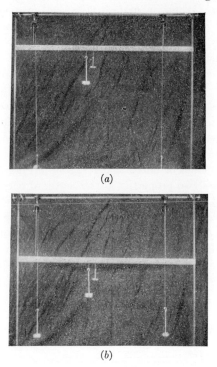

(a)

(b)

FIG. 5-11.—Motion is purely translational

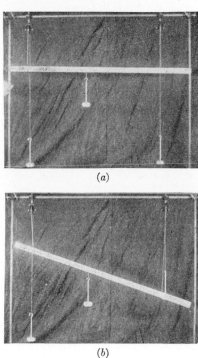

(a)

(b)

FIG. 5-13.—Motion is both translational and rotational

5.9. Center of Weight

Consider an object, such as a meter stick, in the earth's gravitational field. We know from experience that it is possible to apply a single upward force to the stick so that it will remain in equilibrium. Obviously, the magnitude of the force must be the weight of the stick, $W = Mg$, where M is the mass of the stick. We wish to locate the point of application of this single upward force. For the purpose of this calculation we shall *assume that the earth's gravitational field is uniform* (§ 2.18). Also, to facilitate calculations, we introduce the x and y reference system, as shown in Fig. 5-14. We may

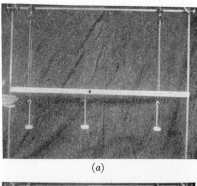

(a)

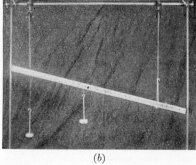

(b)

FIG. 5-12.—Motion is purely rotational about an axis through the center of mass.

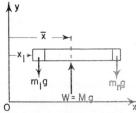

FIG. 5-14.—Center of weight or center of gravity

consider the stick as divided into a set of mass particles, $m_1, m_2, \ldots, m_n$, each being at a distance $x_1, x_2, \ldots, x_n$ from the y-axis. The single upward force is equal to

$$W = m_1 g + \ldots + m_n g = M g . \quad (5\text{-}11)$$

To insure rotational equilibrium, we must have the force moments about an axis through O, say, add vectorially to zero. Let $\bar{x}$ be the distance from the y-axis at which the weight must act. Then

$$M g \bar{x} = m_1 g x_1 + \ldots + m_n g x_n \quad (5\text{-}12)$$

or

$$M g \bar{x} = g \Sigma m_i x_i . \quad (5\text{-}13)$$

Dividing by Mg, we obtain

$$\bar{x} = \frac{\Sigma m_i x_i}{M} . \quad (5\text{-}14)$$

Thus the distance $\bar{x}$, which locates the point at which the single upward force must be applied, is also the x-coordinate of the center of mass of the stick. In a similar way we may locate a y- and a z-coordinate of the point of application of the single upward force. *As far as the weight of the stick is concerned, it may be thought of as acting through the center of mass, ignoring the distribution of mass and weight of the object.* For this reason we call the point of application of the single force the *center of weight* of the object. If the stick of Fig. 5-15 were

oriented into various positions, and for each orientation the line of action of the single upward force imagined drawn within the stick, then the lines of action of the single forces for the different orientations would intersect in a point: the center of weight, or center of mass, of the meter stick.

This *coincidence of center of weight and center of mass* came about because of the assumption that the earth's gravitational field was *uniform* (see § 2.18). If we consider the earth's field to be non-

uniform, then (1) we may still locate a point at which an upward force may be applied to maintain equilibrium, but the line of action of this upward force will not, in general, go through the center of mass of the object; (2) the lines of action of the upward forces for various orientations of the object *will not intersect at a point*, that is to say, a center of weight independent of orientation is nonexistent. To fix this idea firmly in the reader's mind, consider a uniform stick a mile long, oriented in the earth's gravitational field, as shown in Fig. 5-15. Since we are assuming a nonuniform field, the value of g at a point near m_1 is less than that near m_n. Therefore, the point at which the single upward force, $W = \Sigma m_i g_i$, must be applied is O, some distance below the center of mass. The student will see that the position of O depends upon the orientation of the stick.*

Since in practically all the problems that will concern us in mechanics the objects will be near the surface of the earth, where g is nearly constant, the center of mass and the center of weight or center of gravity are to be taken as the same point. We make use of this coincidence to locate experimentally the center of mass of irregular objects. For example, let us locate the center of mass of a thin irregular sheet as shown in Fig. 5-16 (a). We

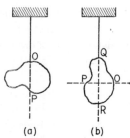

FIG. 5-16.—Experimental determination of the center of weight or center of gravity.

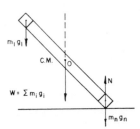

FIG. 5-15.—A rod in a nonuniform gravitational field

suspend the sheet by a thread from some point O on its edge. The center of weight will lie directly under the point of support, because then the various force moments due to the distribution of weight add to zero. Let OP be the line containing

* In many texts the phrase "center of gravity" is used where we have used "center of weight." "Center of gravity" is defined as a center of gravitational attraction, that is, the single point at which all the mass of a body can be concentrated *without changing the gravitational field of the original object*. The distinction between center of weight and center of gravity is largely academic, and we shall adopt the common procedure of using the terms interchangeably.

the center of weight. The sheet is next suspended from another point, Q. Again the center of weight must lie along the vertical QR. Since we are assuming a uniform field, the lines of action OP and QR intersect at the center of weight and hence the center of mass.

5.10. Various Types of Equilibrium

We have explained the conditions under which a rigid body may be in static equilibrium, i.e., possess neither translational nor rotational acceleration. We should like to investigate what happens to the state of equilibrium of a rigid body when disturbed momentarily in the earth's gravitational field. The analysis is divided into three parts.

A heavy cube rests on a table. Let us apply a force F to the cube, tending to rotate it about a pivot point, such as A in Fig. 5-17. We note that

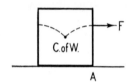

Fig. 5-17.—Body in stable equilibrium

the center of weight of the cube is raised, work is done on the cube, and hence its potential energy is increased. If the force is removed, the cube converts its increased potential energy to kinetic energy and falls back to its original position, where it is said to be in *stable equilibrium*. It is seen from the diagram that any force which tends to change the center of weight of the cube will first raise the center of weight. The dotted lines of Fig. 5-17 indicate the path described by the center of weight upon tilting the cube. In general, we may say that if the application of any force tends to raise the center of weight of a body, then the body is in stable equilibrium.

Let the cube of Fig. 5-17 rotate until it rests on its edge, as in Fig. 5-18. The cube is in equilibrium,

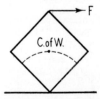

Fig. 5-18.—Body in unstable equilibrium

for it satisfies the conditions for equilibrium of a rigid body: all the gravitational torques cancel out, and the forces on the cube add to zero. However, if the cube is now subjected to any force (other than a vertical one along the line through the point of support and the center of weight), the center of weight is lowered, and the cube falls. Its potential energy is decreased. We have here a case of unstable equilibrium. In general, if the center of weight of a body is lowered when subjected to any horizontal force, the body is in unstable equilibrium. The body will revert to a position of stable equilibrium when disturbed.

The third case of equilibrium we wish to consider is illustrated by a ball resting on a table (Fig. 5-19). If the ball is subjected to any force, we see

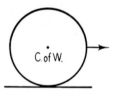

Fig. 5-19.—Body in neutral equilibrium

that the center of weight is neither lowered nor raised. Hence the potential energy of the ball is not altered. This is the case of *neutral equilibrium*.

When the ball of Fig. 5-20 is placed at A or E,

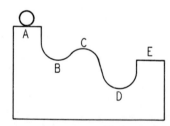

Fig. 5-20.—Positions of neutral, stable, and unstable equilibrium.

it is in neutral equilibrium; when at B or D, stable equilibrium; and when at C, unstable equilibrium.

These three cases of equilibrium of rigid bodies may also be treated from the point of view of the relationship of the center of weight of an object to a fixed *center of suspension* (CS). When an object is suspended from its center of weight, it is in neutral equilibrium. If the center of suspension is above the center of weight, the object is in stable equilibrium; when below, in unstable equilibrium.

The reader should recognize at once which of

the three cases of equilibrium are illustrated in Fig. 5-21.

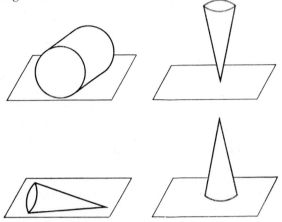

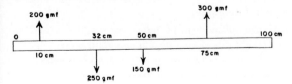

Fig. 5-21.—Bodies in neutral, stable, and unstable equilibrium.

5.11. Statics of a Rigid Body

In applying the conditions for equilibrium of a rigid body as given by equation (5-5), we treat the weight of the body as a force acting through the center of weight. In practical applications of equation (5-5) it is convenient to remember that we may resolve the forces acting on the object into components parallel to two mutually perpendicular axes. To satisfy condition I, the forces parallel to each axis must add to zero. As far as condition II is concerned, any convenient axis may be chosen for torque calculations. These points are illustrated in the following examples.

EXAMPLE 4: Given a uniform meter stick weighing 150 gmf with forces applied as indicated in the figure herewith. What *one* additional force must be applied, and at what point, for equilibrium?

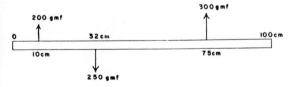

Solution: Since the stick is uniform and weighs 150 gmf, we know that, as far as force moments are concerned, we may consider its weight as concentrated at the 50-cm mark. Then we have

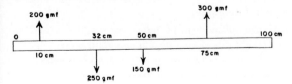

Condition I indicates that the sum of the forces must be zero. Let F be the force to be added; then

$$200 + 300 - 250 - 150 - F = 0 ,$$

$$F = 100 \text{ gmf (downward)} .$$

We do not yet know where to add this force, but we may assume that it is at an unknown distance x from the zero end of the stick.

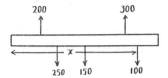

Condition II indicates that the sum of the force moments about every axis must be zero. Let us compute the sum of the moments about the zero end and, by equating this sum to zero, determine the value of x necessary for equilibrium:

$$(200 \times 10) - (250 \times 32) - (150 \times 50)$$
$$+ (300 \times 75) - (100x) = 0 ,$$
$$x = 90 \text{ cm} .$$

If rotational equilibrium were produced in some other way, as, for example, by putting 200 gmf at 45 cm, there would be, indeed, no tendency for rotation; but there would be an unbalanced downward force, and the stick would have a downward acceleration, i.e., would not be in equilibrium.

If the 100-gmf were placed at some point other than at the 90-cm mark, the stick as a whole would not have a linear acceleration; but it would tend to rotate, i.e., would not be in equilibrium. Note that there are *two* unknown quantities and *two* conditions to be satisfied.

Since condition II requires the moments about every axis to be zero, we could just as well have chosen any other point as an axis for our computations. Let us try the 50-cm mark:

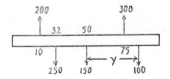

$$(-200 \times 40) + (250 \times 18) + (150 \times 0)$$
$$+ (300 \times 25) - (100y) = 0 ,$$
$$y = 40 \text{ cm} .$$

This locates the 100-gmf at the same point as the first solution. Note that the force moment of the 150-gmf was zero. Hence, if there is a force which cannot conveniently be determined by taking moments about that

85

force, its magnitude is not needed. The reader should try other points as axes until he is convinced that any point may be used. Some, of course, require a little more arithmetic for the solution than do others.

EXAMPLE 5: A ladder 60 ft long and weighing 300 lbf is carried by two men. The center of mass of the ladder is one-third of its length from one end. What force does each man exert?

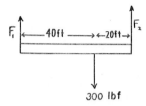

300 lbf

Solution: This problem may likewise be solved in many ways. One solution is to take moments about one end—say the left-hand end. Condition II then gives

$$(F_1 \times 0) - (300 \times 40) + (F_2 \times 60) = 0 ,$$

$$F_2 = 200 \text{ lbf} .$$

Then $F_1 + F_2 = 300$ gives $F_1 = 100$ lbf.

Note here that the forces F_1 and F_2 have a ratio inverse to that of their corresponding lever arms about the center of mass, i.e.,

$$\frac{F_1}{F_2} = \frac{20}{40} .$$

EXAMPLE 6: A 60-ft ladder rests against a smooth wall at a point 48 ft above the ground. The ladder weighs 300 lbf, and its center of mass is one-third the way up. A 150-lb man is halfway up the ladder. Find the force with which the ladder presses against the wall.

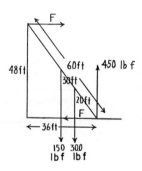

Solution: Since the forces on the wall are not known a priori, we can work with the forces on the ladder only and then use Newton's Third Law to determine the forces which the ladder exerts on the wall and on the ground. Note that a *smooth* wall can support a force only normal to the wall. Why?

Take moments about the bottom of the ladder. The distance from the bottom of the ladder to the line of action of the 300-lbf measured perpendicular to that force is 12 ft (by similar triangles). Condition II then gives

$$(150 \times 18) + (300 \times 12) - (F \times 48) = 0 ,$$

$$F = 131.25 \text{ lbf} .$$

The horizontal force of friction must also be 131.25 lbf.

In other words, if we imagine the floor and wall removed, the ladder will remain fixed in space, provided that a force of 131.25 lbf is applied at the top of the ladder and in the direction of F, a force of 131.25 lbf at the bottom of the ladder and parallel to the ground, and also a force of 450 lbf acting upward at the bottom of the ladder.

EXAMPLE 7: If the coefficient of friction between the ground and the ladder were 0.35, how far up the ladder could the man go before it started to slip?

Solution: Let x = fraction of the total length of the ladder, measured from the ground, that the man may go before the ladder slips. Applying condition II, we have

$$(x \times 36 \times 150) + (300 \times 12) - (48F) = 0$$

or

$$F = \frac{5,400x}{48} + \frac{3,600}{48} \text{ lbf} .$$

But $F = (0.35 \times 450) = 157.5$ lbf is the maximum force of friction. Hence

$$\frac{5,400x}{48} + \frac{3,600}{48} = 157.5 ,$$

$$5,400x + 3,600 = 7,560 ,$$

$$5,400x = 3,960$$

$$x = \tfrac{11}{15} .$$

Therefore, eleven-fifteenths of 60 = 44 ft is the distance the man may climb before the ladder starts to slip.

5.12. Résumé

The following is a brief listing of the definitions, principles, and theories, given in this chapter, which you should know:

Definition of moment of force
The conditions of equilibrium of a rigid body
The definition and location of the center of mass
Properties of the center of mass

Queries

1. Figure 5-2 may be taken to represent a lever of the first class, since F_2 (the opposing force) and F_1 (the applied force) are on opposite sides of the pivot point or *fulcrum*, O. What is the mechanical advantage of such a machine?

2. In Fig. 5-2, if the pivot were placed at the left end of the bar and the force F_2 reversed, so that both forces were on the same (right) side of the pivot but oppositely directed, we would have an example of a lever of the second class. What would be the mechanical advantage? If, on the contrary, in Fig. 5-2 we reverse the direction of F_1, leaving F_2 as drawn, thus making

F_1 the load, we have an example of a lever of the third class.

3. What can be said about the center of mass of loaded dice?

4. A spool of thread is lying on a table. It is possible, by pulling the thread properly, to move the spool toward or away from the experimenter. Explain by means of a diagram how this is possible.

5. In the case of a gymnast who hurls himself into the air, turning his body in somersaults, what path does his center of mass describe? While in the air, can he alter the path of his center of mass?

Practice Problems

1. A uniform meter stick weighing 200 gmf has forces applied to it as indicated. What *one* additional force must be applied, and at what point, for equilibrium? Ans.: 200 gmf applied at the 10-cm mark.

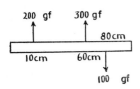

2. Work problem 1, using a different axis than you used in the first solution.

3. A man and a boy carry a 200-lbf load hung on a 10-ft pole between them. Where must the load be hung so that the man will exert three times as much force as the boy? Ans.: $2\frac{1}{2}$ ft from the man.

4. The given forces act on a rectangle of negligible weight. What one additional force must be applied, and where, for equilibrium?

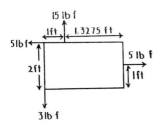

5. A 100-lb gate, 6 ft long and 3 ft high, is hung from a single hinge at the bottom corner and a horizontal string at the top corner. Find the total force on the

hinge and on the string. Ans.: Tension in string is equal to 100 lbf, that in the hinge 141 lbf, at 45°.

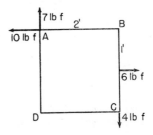

6. On a 2-ft square of negligible weight the following forces are applied. What *one* force must be applied, and where, for equilibrium?

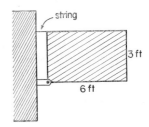

7. A 100-lb roller, 1 ft in radius, is lying against a curbing 6 in high. How much force, parallel to the ground and acting on the axle, is necessary to cause the roller to climb over the curbing? Ans.: 173 lbf.

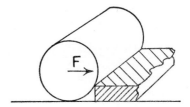

8. A 100-lb block is placed on a level floor, as indicated. What horizontal force acting on the top of the block is necessary to upset it? (Assume that the block does not slide along the floor.)

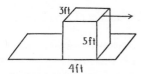

9. A 100-lb ladder leans against a *smooth* wall. The base of the ladder is 6 ft from the base of the wall, and the top of the ladder is 30 ft above the ground. The center of mass of the ladder is one-third the way up the ladder. With what force does the ladder push against the wall? What is the minimum coefficient of friction between the ladder and the ground in order that the ladder just does not slip? (*Note:* "Smooth" wall implies that there is no force of friction between the wall and the ladder.) Ans.: $6\frac{2}{3}$ lbf; $\mu = 0.0667$.

10. If a 150-lb man climbs halfway up the ladder in problem 9, what must the coefficient of friction be so that the ladder does not slide? If the man climbs two-thirds the way up, what must the coefficient of friction be?

11. A thin, uniform meter stick, weighing 120 gmf, has a 100-gm ball of negligible radius attached at the 100-cm mark. Where is the center of mass of the system? Ans.: At the 72.8-cm mark.

12. Find the coordinates of the center of mass of the three masses shown in the accompanying illustration.

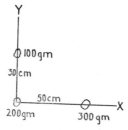

13. A thin, uniform meter stick, weighing 120 gmf, has a sphere 10 cm in radius and weighing 100 gmf attached at its 100-cm end and a ball of negligible radius and weighing 50 gmf attached at its 0-cm end. Where is the center of mass of the system? Ans.: 63 cm from the 0-cm end.

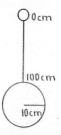

14. Three cubes, 10 cm on an edge, are cemented together as shown herewith. If the masses of the blocks are 100, 300, and 500 gm, locate the center of mass of the blocks, i.e., compute $\bar{x}$, $\bar{y}$, and $\bar{z}$.

15. A uniform boom weighing 20 kgf is hinged to a vertical wall as shown. The other end of the boom is attached to a cord so as to form a right triangle. If a 10-kg mass is hung from the end of the boom, find (*a*) the tension in the cord and (*b*) the magnitude and direction of the force exerted by the hinge on the boom. Ans.: (*a*) 12 kgf; (*b*) 24.75 kgf at 67°0 with the horizontal.

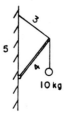

16. Suppose the cord in the figure in problem 15 is lengthened so that the boom is horizontal and the cord makes an angle of 45° with the horizontal. Find (*a*) the tension in the cord and (*b*) the magnitude and direction of the force at the hinge

17. A uniform piece of metal is cut in the form of a T, as shown in the diagram. Find the position of the center of gravity. Ans.: 0.58 m from the upper edge.

18. A uniform piece of metal is cut in the form shown. Find the center of gravity of the plate.

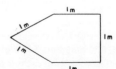

19. A uniform plate of metal of diameter d has a piece cut away of diameter $d/2$, as shown in the diagram. Find the distance of the center of mass from the center of the larger circle. Ans.: $d/12$.

20. The upper end of a uniform ladder of length 13 ft rests against a vertical wall having a coefficient of friction of 0.1. The foot of the ladder is 5 ft out from the foot of the wall, and the top of the ladder is 12 ft from the ground. If the ladder weighs 50 lb, find the minimum coefficient of friction at the ground for equilibrium.

A man weighing 200 lb goes 12 ft up the ladder. Find the new minimum coefficient of friction at the ground for equilibrium.

21. Solve problems 1 and 12, using the mks system of units.

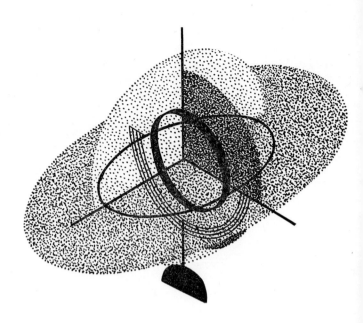

CHAPTER 6

Rotational Motion

6.1. Rotation

A rigid object moves with pure translational motion *if every particle of the object has the same velocity and acceleration.* Pure rotational motion, on the other hand, may be described as motion in which *one line of particles through the body remains fixed while all other particles move in circles of varying radii with the line as a center.* This line is the axis of the motion. A grindstone or a flywheel rotates thus about its axis.

It can be shown that any general motion of an object can be resolved into a translational motion of the center of mass and a rotation of the object about an axis through the center of mass. The reader may verify this statement for himself. For the greater part of this chapter we shall be chiefly concerned with rotations about a fixed axis.

6.2. Definition of Angular Quantities

In the discussion of translatory motion in chapter 1, we talked about "linear displacement," "velocity," "acceleration," etc. For rotational motion we shall introduce an analogous set of quantities: angular displacement, angular velocity, angular acceleration, etc.

Consider a disk of radius r that is rotating about an axis through O (Fig. 6-1). The angular dis-

placement of the disk is found by measuring the angle θ that a radius, OB, makes with some fixed reference line, such as PQ. The angle θ must be measured in *radians* (rad) rather than in degrees if one relates an arc to a radius. A radian is the

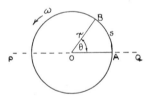

Fig. 6-1.—Rotation of a disk about a fixed axis

angle at the center of a circle subtended by an arc equal to the radius. Thus, in Fig. 6-1,

$$\theta = \frac{s}{r} \text{ rad} \qquad \text{or} \qquad s = r\theta. \qquad (6\text{-}1)$$

The radian, being a ratio of two lengths, is a pure number. From the definition of a radian it follows that in a complete circle there are 2π rad; that is,

2π rad are equivalent to $360°$,

π rad is equivalent to $180°$,

and

1 rad is equivalent to $57°.3$ (approximately) .

The time rate of change of angular displacement is, by definition, the angular velocity and is often represented by the Greek letter omega (ω). Mathematically,

$$\omega = \frac{d\theta}{dt} \text{ (rad/sec)} . \qquad (6\text{-}2)$$

The average angular velocity is given by

$$\bar{\omega} = \frac{\theta}{t} , \qquad (6\text{-}3)$$

where θ is the angle through which the object rotated during t seconds. If the object rotates through equal angles in equal intervals of time, the motion is one of *uniform rotation* or *constant angular velocity*. Note that the units of angular velocity are radians per second.

Finally, *the time rate of change of angular velocity is defined as the angular acceleration* and is denoted by the letter alpha (α). Mathematically,

$$\alpha = \frac{d\omega}{dt} \text{ (rad/sec}^2) . \qquad (6\text{-}4)$$

If the angular velocity changes by equal amounts in equal intervals of time, we speak of the motion as one of *uniform angular acceleration*, and α has a constant value. If ω_0 is the initial angular velocity, ω the final velocity, and t the time in seconds, then

$$\alpha = \frac{\omega - \omega_0}{t} \text{ (rad/sec}^2) . \qquad (6\text{-}5)$$

If the acceleration is not constant, then equation (6-5) defines the *average* angular acceleration for the time t. The units of angular acceleration are radians per second per second (rad/sec²). It will be noted that these definitions of angular quantities are set up in a manner precisely similar to that used for the corresponding linear quantities in chapter 1.

A note about the graphical representation of angular quantities is necessary. Since angular velocity, angular acceleration, and torque are vector quantities, in what direction are we to draw an arrow to represent this fact? We recall that in connection with linear quantities, such as velocity, an arrow had the following significance: (1) its length was a measure of the magnitude of the velocity; (2) the direction of the arrow indicated the direction in which the effect took place; and (3) the arrow, when combined with a similar quantity in accordance with the parallelogram law, gave rise

to a resultant. Now the arrows representing angular quantities should possess just these properties and no others. Since angular quantities were defined with respect to *axes of rotation*, we draw the arrow representing an *angular* quantity along the axis of rotation. The arrow will thus possess the desired properties. In Fig. 6-2 we represent a uni-

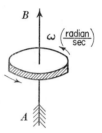

Fig. 6-2.—Vector representing angular velocity

formly rotating object by the arrow AB, drawn along the axis. The length of the arrow is proportional to the angular velocity. By convention, the arrow is pointed in a direction such that, if the thumb of the *right hand* is laid along it, the fingers curl around the thumb in the direction of rotation of the motion. This direction is such that, viewed along the direction of the arrow (i.e., from tail to head), the rotation is clockwise, or that in which a right-handed screw must be turned if it is to progress in the direction of the arrow.

6.3. Kinematics of Rotation

Following the methods used in chapter 1, we may write down a set of equations that will describe the motion of an object rotating about some axis.

Consider an object, such as a wheel, rotating about a fixed axis. Suppose the wheel has a constant angular acceleration of α rad/sec² and has an initial angular velocity of ω_0 rad/sec and after a time of t seconds the angular velocity is ω rad/sec. During the time t the wheel has turned through θ radians.

By definition, the constant angular acceleration α is the change in angular velocity divided by the time taken; then

$$\alpha = \frac{\omega - \omega_0}{t} \quad \text{or} \quad \omega = \omega_0 + \alpha t .$$

The average angular velocity $\bar{\omega}$ during this   **91** time t is $(\omega_0 + \omega)/2$, so that the angle θ through which the wheel turns in the t seconds is

$$\theta = \bar{\omega} t = \left(\frac{\omega_0 + \omega}{2} \right) t . \qquad (6\text{-}6)$$

From these two equations you should derive three others. They are

$$\left. \begin{array}{c} \theta = \omega_0 t + \frac{1}{2} a t^2 , \\[2mm] \theta = \omega t - \frac{1}{2} a t^2 , \end{array} \right\} \qquad (6\text{-}7)$$

$$2 a \theta = \omega^2 - \omega_0^2 . \qquad (6\text{-}8)$$

Derive these equations yourself, using the calculus, similar to the method given in § 1.22.

The analogous equations for a constant linear acceleration are equations (1-14)–(1-19).

6.4. Relation between Linear and Angular Quantities

For a rotating body there exist simple relations between its angular displacement, velocity, and acceleration and the corresponding *magnitude* of the linear displacement, speed, and acceleration of each and every particle of the body. In these relations the distance of the particle from the axis is an essential element.

In Fig. 6-3 let P be any point in the body at a distance r from the axis of rotation (through O).

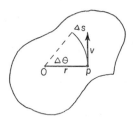

FIG. 6-3.—Rotation about an axis through O

Let us assume that in a small time interval, Δt sec, the radius r containing the point P describes the angle of $\Delta \theta$ rad. The point P describes the arc of a circle, Δs. By definition of average angular velocity,

$$\bar{\omega} = \frac{\Delta \theta}{\Delta t} .$$

Also, from geometry, $\Delta \theta = \Delta s / r$, where $\Delta \theta$ is measured in radians. Therefore,

$$\bar{\omega} = \frac{\Delta s}{\Delta t} \frac{1}{r} .$$

As we allow the time Δt to approach zero, $\Delta s / \Delta t$ approaches the linear speed, v, of the particle; i.e.,

$$\omega = \frac{v}{r} \qquad \text{or} \qquad v = \omega r . \qquad (6\text{-}9)$$

The direction of v is always at right angles to the radius. Similarly, the linear acceleration of the particle at P is given by

$$a = a r , \qquad (6\text{-}10)$$

where a is directed, at each instant of time, at right angles to r.

The three relations

$$s = r \theta , \qquad v = \omega r , \qquad a = a r , \quad (6\text{-}11)$$

are of the *utmost importance* in the study of the motions of rigid bodies. Notice that θ is measured in radians, ω in radians per second, and a in radians per second per second.

It is interesting to note that, since the velocity of the particle is constantly changing in *direction*, the particle is also subject *to a centripetal acceleration along the line joining the axis to the mass.* This centripetal acceleration, a_c, is related to r and to the angular velocity as follows:

$$a_c = r \omega^2 \qquad (6\text{-}12)$$

or

$$a_c = \frac{v^2}{r} \qquad (6\text{-}13)$$

(see eq. [1-29]). If a is zero, then the *tangential* acceleration, as given by equation (6-10), is also zero; the centripetal acceleration, a_c, however, will be zero only if either $r = 0$ (i.e., for a point on the axis) or $\omega = 0$ (i.e., when there is *no rotation*).

Although the relations given by equations (6-11) were deduced for the case of an object rotating with respect to a fixed axis, they are, nevertheless, of general validity. For example, if the axis of rotation changes from moment to moment, equations (6-11) may be used if the quantities a, ω, a, and r are measured with respect to the instantaneous axis. See § 6.14 for an application of this theorem.

EXAMPLE 1: A wheel starts from rest with an angular acceleration of 10 rad/sec². How many revolutions will it make in 30 sec, and what angular velocity will it acquire?

Solution: From equation (6-7), with $\omega_0 = 0$,

$$\theta = \tfrac{1}{2} a t^2 = \tfrac{1}{2} (10 \ \text{rad}/\text{sec}^2) (30 \ \text{sec})^2 = 4,500 \ \text{rad} .$$

Since there are 2π rad in 1 revolution, it follows that

$$N = \frac{4,500}{2\pi} \text{ revolutions in } 30 \text{ sec.}$$

Also, from equation (6-5), $\omega = at = (10 \text{ rad/sec}^2)$ (30 sec) = 300 rad/sec, the angular velocity at the end of 30 sec.

EXAMPLE 2: If the radius of the wheel of Example 1 is 10 cm, calculate (1) the tangential velocity of a particle on the rim after 30 sec, (2) the tangential acceleration of the particle, and (3) the centripetal acceleration of a particle on the rim at the end of 30 sec.

Solution:

(1)　$v = r\omega = (10 \text{ cm})(300 \text{ rad/sec})$
　　　$= 3,000 \text{ cm/sec tangential velocity}$,

(2)　$a = ar = (10 \text{ rad/sec}^2)(10 \text{ cm})$
　　　$= 100 \text{ cm/sec}^2 \text{ tangential acceleration}$,

(3)　$a_c = \dfrac{v^2}{r} = r\omega^2 = (3,000 \text{ cm/sec})^2 \left(\dfrac{1}{10 \text{ cm}}\right)$
　　　$= 9 \times 10^5 \text{ cm/sec}^2 \text{ centripetal acceleration}$
　　　directed toward the axle.

6.5. Rotational Inertia

The principle of inertia, to which we devoted much time in chapter 2 when dealing with translational motion, is of fundamental generality. It applies as well to rotational motion, to all sorts of combinations of translation with rotation—indeed, to any kind of motion.

In translational motion, as we have seen, a given mass always receives the same acceleration when subjected to the action of the same force, independent of where the force is applied. As a matter of fact, the force is *defined* by the mass-acceleration product. How large the angular acceleration of a given body under the action of a certain force will be depends upon *where the force is applied.* If the force is applied so that the line of action of the force goes through the axis of rotation, no rotational acceleration results. The farther out from the axis the force is applied, the greater becomes the acceleration which one and the same force will produce. If we make quantitative tests, we find that the acceleration produced about a given axis is proportional to the perpendicular distance of the line of action of the force from that axis—is proportional, in other words, to

the *force moment,* or *torque.* Mathematically, we may represent this as

$$L \sim a .\qquad (6\text{-}14)$$

In translational motion any object has but one value for its inertia, which we call its "mass." This neglects any relativistic change in mass with velocity. In rotational motion the same force moment may give one and the same body a wide variety of different angular accelerations, depending upon how the axis of rotation is placed with respect to the body. If we grasp a meter stick and twist or wiggle it in a horizontal plane about its axis of greatest length (Fig. 6-4 [a]), it resists very

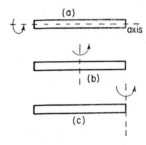

FIG. 6-4.—Rotational inertia about three axes

slightly. If we grasp it at its mid-point and accelerate it about an axis perpendicular to its length (Fig. 6-4 [b]), a greater torque is required. If, finally, we grasp it at one end and accelerate it in a horizontal plane about an axis at right angles to its length (Fig. 6-4 [c]), we find that the largest torque is required in this position for a given angular acceleration. Thus the rotational inertia of an object depends not only upon the mass but upon the *distribution of the mass with respect to the axis of rotation.* We find experimentally that for a given axis of rotation the rotational inertia is a constant. We call the rotational inertia by a name that suggests this: *moment of inertia.* Next we must proceed to discover exactly how the moment of inertia depends upon the distribution of mass with respect to an axis.

Consider a uniform stick of length r pivoted at one end O on a horizontal frictionless table, as shown in Fig. 6-5. Suppose a horizontal force F acts perpendicularly to the length of the rod at the far end, a distance r from the pivot O. This force exerts a torque $L = Fr$ on the stick about the pivot at O and produces an acceleration. In order to analyze the effect of this torque, consider the stick divided into small elements of mass m_1, m_2, ..., m_i, ..., which are at distances of r_1, r_0, ..., r_i, ..., from O.

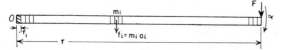

FIG. 6-5.—Horizontal stick pivoted at O and acted on by horizontal force F.

At any instant of time each particle possesses a linear acceleration. For example, to accelerate m_1 by an amount a_1 requires a force f_1:

$$f_1 = m_1 a_1 .$$

The moment of this force about this axis O is

$$f_1 r_1 = m_1 a_1 r_1 .$$

Similarly, for the other particles:

$$f_2 r_2 = m_2 a_2 r_2$$
$$\cdot \qquad \cdot$$
$$\cdot \qquad \cdot$$
$$\cdot \qquad \cdot$$
$$f_i r_i = m_i a_i r_i .$$

If we add these many equations, we obtain

$$\Sigma f_i r_i = \Sigma m_i a_i r_i . \qquad (6\text{-}15)$$

The applied torque, L, is equal to the sum of the individual torques, $L = \Sigma f_i r_i$. Also, since the stick is rigid, all the mass points turn with the same angular velocity and acceleration. Let a be the angular acceleration of the stick that results when a torque, L, is applied. Then, by equation (6-10),

$$a_i = a r_i .$$

Substituting in equation (6-15), we obtain

$$L = \Sigma a m_i r_i^2$$

or

$$L = a \Sigma m_i r_i^2 . \qquad (6\text{-}16)$$

The quantity $\Sigma m_i r_i^2$, which depends upon the *distribution of mass with respect to the axis of rotation*, is, by definition, the *moment of inertia:**

$$I \equiv \Sigma m_i r_i^2 , \qquad (6\text{-}17)$$

and so

$$L = I a , \qquad (6\text{-}18)$$

where L, I, and a are to be measured along the same axis. Stated in words: *The moment of inertia of an object about some axis is equal to the sum of the products of the elements of mass that constitute the*

* See also last paragraph of § 5.6.

object by the square of the distance of each mass element from the given axis of rotation.

The units of I depend upon the units chosen for m and r. In the mks system, I is in kilogram-meters squared; in the cgs system I is in gram-centimeters squared; in the British Engineering system, I is in slug-feet squared.

6.6. Moment of Inertia of Solids

It should be clear that the definition of moment of inertia given by equation (6-17) is applicable to any kind of point-mass distributions. A solid may be thought of as the sum of particles closely packed, to which the definition may be readily extended by replacing the summation by integrals, i.e., $I = \int r^2 dm$. For the case of regular solids, such as the disk, sphere, cone, parallelepiped, etc., such calculations have been carried out. The reader especially interested in the details may find them in any good textbook on the calculus. Our purpose will be served by merely tabulating the results for some common solids in Table 6-1. In this table note how important the position of the axis of rotation is for the value of the moment of inertia. Also note that, in general, *any dimension of an object that is parallel to the axis of rotation never appears in the expression for* I *about that axis.* Formulas for the moment of inertia of the less common solids can be found in any mechanical engineering handbook.

For the case of a thin ring or hoop whose thickness can be disregarded in comparison to its diameter, we may calculate directly the moment of inertia about a central axis, O, perpendicular to its plane (Fig. 6-6). We use equation (6-17). In this

FIG. 6-6.—Moment of inertia of a ring about an axis through the center.

solid every element of mass, m_i, is assumed to be at the same distance, r_i, from the axis; therefore, all the r_i's are the same and equal to R, the radius of the hoop:

$$I = \Sigma m_i r_i^2 = R^2 \Sigma m_i$$

or

$$I = MR^2 \, ,$$

where M is the total mass.

As another example of how the moments of inertia about an axis can be calculated, let us consider that of a

TABLE 6-1

MOMENTS OF INERTIA

Diagram of System	Solid and Axis	I
(1)	Cylinder about the axis of figure	$\dfrac{MR^2}{2}$
(2)	Cylinder about a central diameter	$\dfrac{MR^2}{4} + \dfrac{Ml^2}{12}$
(3)	Solid sphere about any diameter	$\tfrac{2}{5}MR^2$
(4)	Block about a central axis	$M\left\{\dfrac{a^2+b^2}{12}\right\}$
(5)	Thin bar, about axis at center $\perp$ to length	$\dfrac{Ml^2}{12}$
(6)	Thin bar, about axis at one end	$\dfrac{Ml^2}{3}$
(7)	Hoop about a diameter	$\dfrac{MR^2}{2}$
(8)	Hoop about a tangent line	$\tfrac{3}{2}MR^2$
(9)	Ring about an axis through its center $\perp$ to the plane of the ring	$\dfrac{M}{2}(R_1^2+R_2^2)$

solid rod about an axis through one end, the result of which is given in (6) of Table 6-1. First, we shall calculate by an algebraic method, then by the integral calculus.

Consider the rod (Fig. 6-7) divided arbitrarily up into 100 equal parts so that each element of mass is $M/100$, where M is the total mass of the rod, of length l. The distance of the center of the first element of mass from the

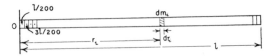

FIG. 6-7.—Moment of inertia of a rod about one end

end O is $l/200$; of the second it is $(l/100 + l/200)$; of the third it is $(2l/100 + l/200)$; . . . ; and of the last element of mass it is $(99l/100 + l/200)$. These distances can be written

$$\frac{l}{200}, \quad \frac{3l}{200}, \quad \frac{5l}{200}, \cdots, \quad \frac{199\,l}{200}.$$

The moment of inertia about the axis O is

$$I = \sum_1^{100} m_i l_i^2 = \frac{M}{100}\left[\left(\frac{l}{200}\right)^2+\left(\frac{3l}{200}\right)^2+\left(\frac{5l}{200}\right)^2 \right. $$
$$\left. +\ldots+\left(\frac{199l}{200}\right)^2\right]$$
$$= \frac{Ml^2}{100\times200^2}[1^2+3^2+5^2+\ldots+199^2].$$

The sum of the squares of the odd integers from 1^2 to $(2n-1)^2$ is given by

$$\frac{n}{3}(2n+1)(2n-1) = 1^2+3^2+5^2+\ldots+(2n-1)^2.$$

In our problem $(2n-1) = 199$, so that $n = 100$, and the sum

$$I = \sum_1^{100} m_i l_i^2 = \frac{Ml^2\times100\times201\times199}{3\times100\times200^2}$$
$$= \frac{Ml^2}{3} \text{ (approximately)}.$$

This approximation would be closer if we had divided the rod into 1,000 or even a greater number of parts. For the calculus the rod is divided into an infinite number of parts, each of mass dm, so that the summation is replaced by an integral, and the moment of inertia about O is

$$I = \int_0^l r^2 dm, \qquad (6\text{-}19)$$

where r is the distance of the mass dm from O. Now the mass dm is $(M/l)dr$, since M/l is the mass per unit length of the rod and an element of length dr has a mass Mdr/l. Hence

$$I = \int \frac{M}{l} r^2 dr = \frac{M}{l} \left[\frac{r^3}{3}\right]_0^l$$

$$= \frac{Ml^2}{3}.$$

In so far as irregular solids are concerned, although direct calculations are impossible, we may still obtain a value of the moment of inertia about a given axis experimentally. According to equation (6-18),

$$I = \frac{L}{a}.$$

To use this equation it is necessary to mount the object on a given axis, apply a known torque, L, and observe the resultant angular acceleration, a.

6.7. Theorem of Parallel Axes

Many, but not all, of the moments of inertia tabulated in Table 6-1 are about axes passing through the centers of mass of the solids involved. It is often convenient to know the relation between the moment of inertia about some axis and that parallel to the one that passes through the center of mass but at some distance, x, from it.

If I_{cm} is the moment of inertia about an axis containing the center of mass of a body whose total mass is M, then the moment of inertia about some other axis, at a distance x from the first axis and parallel to it, is given by

$$I_x = I_{cm} + Mx^2. \qquad (6\text{-}20)$$

Thus, in the cases of the hoop about a diameter (which contains the center of mass) and about a line tangent to its periphery, as listed in the seventh and eighth items of Table 6-1,

$$I_x = \underset{\substack{\text{about} \\ \text{tangent} \\ \text{line}}}{\tfrac{1}{2}MR^2} + \underset{\substack{\text{about} \\ \text{diameter}}}{MR^2} = \underset{\substack{\text{since } x \\ \text{here} \\ \text{equals } R, \\ \text{radius of hoop}}}{\tfrac{3}{2}MR^2}. \qquad (6\text{-}21)$$

Similarly, knowing the value for the moment of inertia of a disk about a line perpendicular to its plane and coincident with its axis of figure, equation (6-20) enables us to find I about a parallel axis tangent to its periphery. Equation (6-20) is known as "STEINER's theorem," or the theorem of parallel axes.

In order to prove the theorem of parallel axes for a rod, let us consider a uniform rod whose moment of inertia about the center of mass is I_{cm} and about a parallel axis distant x from it is I_x. What we wish to prove is that

$$I_x = I_{cm} + Mx^2$$

where M is the mass of the rod.

In order to do this, consider coordinate axes along the rod having their origins at the center of mass and at the intersection of the parallel axis distant x from the center of mass. From Fig. 6-8

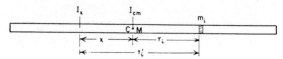

Fig. 6-8.—Moment of inertia about two parallel axes

it follows that the coordinates of an element of mass m are related by

$$r'_i = r_i + x.$$

Notice that r'_i and r_i are variables which change with the position of m_i, while x is a constant in any particular problem. Now, by definition,

$$I_x = \Sigma m_i r_i'^2, \qquad I_{cm} = \Sigma m_i r_i^2,$$

where the summations are taken over the whole length of the rod. Let us now substitute for r'_i in the expression for I_x, giving

$$I_x = \Sigma m_i (r_i + x)^2$$

$$= \Sigma m_i r_i^2 + \Sigma m_i x^2 + \Sigma 2 m_i r_i x$$

$$= I_{cm} + Mx^2 + 2x \Sigma m_i r_i,$$

since $\Sigma m_i = M$, the total mass of the rod.

The last summation on the right is zero, since, by definition of the center of mass, $\Sigma m_i r_i$ for an origin through the center of mass is zero. Relative to the center of mass there are as many positive values of $m_i r_i$ as there are negative ones, so that, for the rod as a whole, the sum of $m_i r_i$ is zero. Thus we have the relationship between the moment of inertia about the center of mass and about a parallel axis distant x from the center of mass of a body having a total mass M:

$$I_x = I_{cm} + Mx^2.$$

Notice that the minimum moment of inertia for a body is about an axis through its center of mass.

EXAMPLE 3: To find the moment of inertia of a rectangular plate about an axis through one corner and the perpendicular to the plane of the plate.

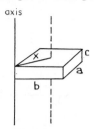

Solution: From Table 6-1,

$$I_{cm} = \frac{M(a^2 + b^2)}{12};$$

also from the figure,

$$x = \sqrt{\left(\frac{a}{2}\right)^2 + \left(\frac{b}{2}\right)^2},$$

so that

$$I = M\left(\frac{a^2 + b^2}{12}\right) + M\left[\left(\frac{a}{2}\right)^2 + \left(\frac{b}{2}\right)^2\right]$$

or

$$I = M\left[\frac{a^2}{12} + \frac{b^2}{12} + \frac{a^2}{4} + \frac{b^2}{4}\right] = M\left(\frac{a^2 + b^2}{3}\right).$$

EXAMPLE 4: To find the moment of inertia of a thin rectangular plate about an axis x cm from the center and parallel to an edge.

Solution: From entry (4) of Table 6-1 and with $a \rightarrow 0$,

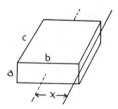

$$I_{cm} = M\frac{b^2}{12},$$

$$I = M\frac{b^2}{12} + Mx^2.$$

If $x = b/2$, i.e., axis through a side, then

$$I = M\left[\frac{b^2}{12} + \frac{b^2}{4}\right] = \frac{Mb^2}{3}.$$

EXAMPLE 5: To find the moment of inertia of a ball hanging on a thin rod, with the axis in the plane of the

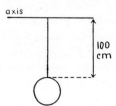

paper. Given: length of thin rod, 100 cm; radius of rod, 1 mm; mass of thin rod, 10 gm; radius of ball, 10 cm; mass of ball, 200 gm.

Solution: From Table 6-1, we obtain

I (rod about its own center)

$$= 10\left(\frac{0.1^2}{4} + \frac{100^2}{12}\right) \text{gm-cm}^2,$$

I (rod about the given axis)

$$= 10\left(\frac{0.01}{4} + \frac{10,000}{12}\right) + 10 \times 50$$

$$\text{gm-cm}^2 \text{ (theorem of parallel axis)},$$

I (sphere about its own center)

$$= \tfrac{2}{5} \times 200 \times 10^2 \text{ gm-cm}^2,$$

I (sphere about the given axis)

$$= \tfrac{2}{5} \times 200 \times 100 + 200 \times (110)^2 \text{ gm-cm}^2,$$

Total $I = 2,461,333.358$ gm-cm^2.

Note: If one neglects the mass of the rod and the radius of the ball (i.e., mass of the rod $\rightarrow$ 0, radius of ball $\rightarrow$ 0), the only remaining term is 2,420,000 gm-cm^2. This is an error of about 1.6 per cent. This sort of approximation will be used when we study simple pendulums.

6.8. Radius of Gyration

We have seen that the idea of center of mass or inertia in translational motion defines a point at which all the inertia may be considered as concentrated, so that, when forces are applied at this point, the object accelerates without any rotation. There is a corresponding idea in rotational motion, that of a RADIUS OF GYRATION, which describes a similar hypothetical concentration of rotational inertia.

The radius of gyration (k) of any body capable of rotation about some given axis is the distance out from that axis where all the inertia or mass of the body would have to be concentrated for it to have the same MOMENT OF INERTIA as that the body actually possesses about that axis.

From this definition of k and that of I,

$$I = Mk^2. \qquad (6\text{-}22)$$

For example, take the case of a disk capable of rotations about its axis (Fig. 6-9). Its radius being

R, its moment of inertia about this axis is, as we have seen above,

$$I_{\text{disk}} = \tfrac{1}{2}MR^2 .$$

But, since

$$I = Mk^2 ,$$

we see that, for a disk,

$$k_{\text{disk}} = \frac{R}{\sqrt{2}} = 0.707R .$$

In other words, if we replaced the disk by a point having the total mass of the disk and located

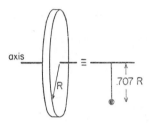

axis

.707 R

R

Fɪɢ. 6-9.—Radius of gyration of a disk equal to $0.707R$

a distance equal to 0.707 of the radius from the axis, this mass point in rotation about this axis would have the same moment of inertia as that of the disk itself.

We can obtain another expression for k^2 by making use of the defining equation for I and k^2 (eqs. [6-17], [6-22]). Thus for a body subdivided into elements of which a typical one is of mass m_i and lies at a distance r_i from some given axis of rota-

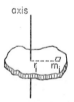

axis

r_i m_i

Fɪɢ. 6-10.—Radius of gyration of irregular-shaped body

tion (Fig. 6-10), the radius of gyration of this body about that axis will be given by the quotient

$$k^2 = \frac{\Sigma m_i r_i^2}{\Sigma m_i}. \qquad (6\text{-}23)$$

The student will note that the defining equation for k (eq. [6-23]) is similar to that used for the center of mass (§ 5.6) except that here we have obtained the second mass moment. It will be observed that, whereas the center of mass is independent of any origin but is dependent solely

on the distribution of mass within the object, *the radius of gyration depends directly upon the axis chosen for its calculation.*

In the notation of the calculus,

$$k^2 = \frac{\int r^2 dm}{\int dm} .$$

6.9. Newton's Laws of Motion for Rotation

Newton's three Laws of Motion for mass points as given in chapter 2 can be readily extended to rotating objects. A useful formulation of these laws follows:

Law I: A body rotating with constant angular velocity about a given axis will continue to rotate with the same angular velocity unless acted upon by an unbalanced torque.

The reader should note that *constant angular velocity* means constant *in magnitude and direction*. If the object is speeded up or retarded, a torque must be applied; likewise, if we attempt to change the direction of the axis and hence ω, a torque must be applied. The gyroscopic compass is a practical application of the First Law.

Law II: If an unbalanced torque is applied to an object, the angular acceleration produced is inversely proportional to the moment of inertia of the body and directly proportional to the magnitude of the unbalanced torque.

The resultant angular acceleration is parallel to the axis of the applied torque, and the moment of inertia is to be calculated with respect to the same axis. In symbols,

$$L = I\alpha .$$

If the torque varies in magnitude and direction, the Second Law is assumed to hold at each instant of time.

Law III: If object A exerts a torque on object B along a given axis, then object B exerts an equal and opposite torque on A along the same axis.

As in the case of forces, torques always appear in pairs *and act on different objects*. The "floating-power" installation of motors in the modern automobile illustrates this. The engine blocks are seen to recoil opposite to the rotation of the flywheel whenever the motor is accelerated and in the same direction whenever it is shut down·

Anyone who has ever driven a small boat with a high-speed power plant is familiar with the heeling of the boat to one side, opposite to the rotation of the propeller shaft, when the throttle valve is opened.

If an ordinary watch or alarm clock, while running, is suspended from a fiber so that the plane of its oscillating balance wheel is horizontal and the outside case observed closely, it will be found to be vibrating with the same period as the balance wheel, owing to the reaction of the latter through the coil spring that connects the wheel to the case.

6.10. Angular Acceleration Due to a Constant Torque

If an object is constrained to rotate about a given axis and a torque is applied parallel to this axis, we may compute the constant angular acceleration acquired by using equation (6-18). We may also compute the final angular velocity and displacement after any time t by using equations (6-6), (6-7), and (6-8). Examples 6 and 7 illustrate this type of problem.

EXAMPLE 6.: Given a disk of mass 100 gm and radius 10 cm. A *force* of 5 gmf acts along the circumference. Find the angular acceleration.

Solution:

$$I = \tfrac{1}{2} m r^2 = \frac{100 \times 10^2}{2} = 5,000 \text{ gm-cm}^2 ,$$

$$L = 5 \times 980 \times 10 = 49,000 \text{ cm-dynes};$$

and, since

$$L = I a ,$$

$$49,000 = 5,000a ,$$

$$a = 9.8 \text{ rad/sec}^2 .$$

If the torque acts for 10 sec, compute the final angular velocity and displacement, assuming the disk to start from rest.

By equation (6-5),

$$\omega = \omega_0 + a t \cdot \quad \omega = (9.8 \text{ rad/sec}^2)(10 \text{ sec})$$
$$= 98 \text{ rad/sec};$$

also

$$\theta = \tfrac{1}{2} a t^2 ; \quad \theta = \tfrac{1}{2}(9.8 \text{ rad/sec}^2)(10 \text{ sec})^2$$
$$= 490 \text{ rad} .$$

EXAMPLE 7: Given a cylinder of mass 1 kg and 0.10 m radius which is free to rotate about its axis. Find the angular acceleration of the cylinder and the linear acceleration of the 0.1-kg hanging mass and the tension in the cord.

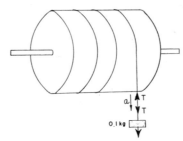

Solution: From equation (6-18) for the rotation of the cylinder,

$$L = I a .$$

The resultant torque L, assuming no friction at the bearings, is

$$L = 0.10 \times T \text{ m-newtons} ,$$

where T is the tension in the cord in newtons. The moment of inertia of the cylinder, about the axis shown in the figure, is the same as that for a disk, namely,

$$I = \frac{m r^2}{2} = \frac{0.01}{2} \text{ kg-m}^2 .$$

If a is the angular acceleration of the cylinder about its axis in rad/sec², then the linear acceleration a of the 0.1-kg mass is

$$a = 0.1 a \text{ m/sec}^2 .$$

The equation of motion of the hanging mass, by Newton's Second Law, is

$$0.1 \times 9.8 - T = 0.1 a ,$$

or

$$T = 0.98 - 0.1 a \text{ newtons} .$$

By substitution in $L = Ia$,

$$0.10(0.98 - 0.1a) = \frac{0.01 a}{0.2}$$

or

$$9.8 - a = 5a ,$$

$$a = 1.63 \text{ m/sec}^2 ,$$

$$T = 0.98 - 0.163 = 0.817 \text{ newtons} = 0.0834 \text{ kg}$$

$$a = \frac{1.63}{0.1} = 16.3 \text{ rad/sec}^2 .$$

6.11. Work Done by a Torque: Energy of Rotating Objects

Energy relations with respect to rotating bodies follow closely parallel to those we have discussed in connection with translation motion. In order to compute the work done by a torque in rotating an object through an angle θ, we shall refer to Fig. 6-11. Let us assume that the constant force F acts

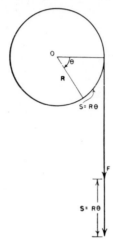

Fig. 6-11.—Work done by a torque and change in rotational kinetic energy.

through the distance s, as shown. Since F is always tangent to s, the work done by F is

$$W = F s .$$

From geometry the distance s is related to the radius, r, and angle through which it rotates, θ, by the following equation:

$$s = r \theta .$$

Therefore,

$$W = F r \theta .$$

But Fr is precisely the *torque* acting on the stick. Therefore,

$$W = L \theta .$$

It should be emphasized that the axis of angular displacement, θ, must be parallel to the axis of the torque, L.

(For the case of a rigid body spinning about an axis pivoted at a point rather than completely fixed as above, it is possible for the axis of the torque, L, to make an angle β with the axis of the angular displacement, θ. For this case, $W = L\theta \cos \beta$; and if L and θ are at right angles, no work is done. Cf. § 6.16.)

Work may be done against frictional torques; against elastic torques, such as the winding of a watch, resulting in stored potential energy; or against rotational inertia, that is, by imparting angular acceleration to a rigid body. In this case it is rotational kinetic energy that is acquired. This kinetic energy is calculated by using Fig. 6-11.

If a constant torque L acts on the disk through an angle θ, then the acceleration produced, a, is

$$L = I a .$$

Also, by equation (6-8), for constant a,

$$2 a \theta = \omega^2 - \omega_0^2 .$$

Multiplying these two equations yields

$$2 a \theta L = I a (\omega^2 - \omega_0^2)$$

or

$$L \theta = \tfrac{1}{2} I \omega^2 - \tfrac{1}{2} I \omega_0^2 . \qquad (6\text{-}24)$$

The quantity $\tfrac{1}{2} I \omega^2$ is *the kinetic energy of rotation.* Thus the work done by the torque L in turning the disk through an angle θ is equal to the increase in the rotational kinetic energy. Note that I must be calculated about the same axis as that about which ω is measured. The similarity of equation (6-24) to $\tfrac{1}{2} m v^2$ is obvious.

The dimensions of torque in terms of [M], [L], and [T] are $[ML^2T^{-2}]$ and of rotational kinetic energy are also $[ML^2T^{-2}]$. In terms of the British Engineering dimensions [F], [L], and [T] the units are both [FL]. The dimensions of moment of inertia are either $[ML^2]$ or $[FT^2L]$ and of ω^2 are $[T^{-2}]$.

The units used for the torque and rotational kinetic energy depend on the system employed. In the mks system the torque L is measured in newton-m and I in kg-m². For the cgs system L is measured in dyne-cm and I in gm-cm². The British Engineering system has units of lbf-ft for L and slug-ft² for I. In all three systems θ is measured in radians and ω in radians/sec. The units in which the energy $L\theta$ is measured are the same as for $\tfrac{1}{2} I \omega^2$, namely, joules, ergs, and ft-lbf, respectively.

The power developed by an object subjected to a torque L and rotating with angular velocity ω is

$$P = L \omega . \qquad (6\text{-}25)$$

Note the similarity of this expression to equation (4-24).

EXAMPLE 8: In Example 6 calculate the work done by the torque.

Since the torque of $L = 49,000$ cm-dynes acts through an angle of 490 rad (as computed in Ex. 6), work done is

$$W = L\theta = (49,000 \text{ cm-dynes})(490 \text{ rad})$$
$$= 2,401 \times 10^4 \text{ ergs} .$$

We may also calculate the work done by the torque by computing the difference in final and initial kinetic energy; i.e.,

$$\text{KE} = \tfrac{1}{2} I\omega^2 = \tfrac{1}{2} (5,000 \text{ gm-cm}^2)(98 \text{ rad/sec})^2$$
$$= 2,401 \times 10^4 \text{ ergs} .$$

EXAMPLE 9: Suppose the constant force F in Fig. 6-11 is 5 lbf and the radius of disk 2 ft and its mass 20 lb or 20/32 slugs. If the disk starts from rest, $\omega_0 = 0$, and the force acts for 10 seconds, the problem is to find the angular acceleration a, the angular velocity ω at the end of the 10 seconds, the angle θ the disk is turned through, the work done by the torque, and the change in rotational KE.

Solution:

$$L = 5 \text{ lbf} \times 2 \text{ ft} = 10 \text{ ft-lbf} ,$$

$$I = \frac{Mr^2}{2} = \tfrac{1}{2} \times \tfrac{20}{32} \text{ slugs} \times 2^2 \text{ ft}^2 = 1.25 \text{ slug-ft}^2 ,$$

$$a = \frac{L}{I} = \frac{10 \text{ ft-lbf}}{1.25 \text{ slugs-ft}^2} = 8 \text{ rad/sec}^2 ,$$

$$\omega = at = 80 \text{ rad/sec} ,$$

$$\theta = \tfrac{1}{2} at^2 = \tfrac{1}{2} \times 8 \text{ rad/sec}^2 \times 100 \text{ sec}^2$$
$$= 400 \text{ rad} ,$$

$$L\theta = 10 \text{ ft-lbf} \times 400 \text{ rad} = 4,000 \text{ ft-lbf} ,$$

$$\tfrac{1}{2} I\omega^2 = \tfrac{1}{2} \times 1.25 \text{ slugs-ft}^2 \times 80^2 \text{ rad}^2/\text{sec}^2$$
$$= 4,000 \text{ ft-lbf} .$$

Thus the work done by the torque is equal to the change in rotational KE, as it should be.

6.12. Angular Momentum

In chapter 2 we introduced the idea of linear momentum, mv, a concept that proved useful in the analysis of a certain class of problems. Similarly, in the rotation of rigid bodies it proves useful to introduce an analogous quantity, which we call *angular momentum*. From equation (6-18),

$$L = Ia = \frac{I(\omega - \omega_0)}{t}, \qquad (6\text{-}26)$$

where ω and ω_0 are the final and initial angular velocities, respectively, and t is the time in which the angular velocity changes.

We define the quantity $I\omega$ as the *angular momentum*, or the *moment of momentum*, of the rotating body. It is a vector quantity whose direction is that of the angular velocity. If $\bar{L}$ is the average torque acting on an object for a time t, then equation (6-26) becomes

$$\bar{L}t = I\omega - I_0\omega_0 , \qquad (6\text{-}27)$$

where $\bar{L}t$ is called the *impulse of the torque* and is measured by the difference between the final and the initial angular momenta. If we were to strike a pivoted meter stick a sharp blow with a hammer, say, it would be difficult to know the value of the applied torque and the time that the torque acted. However, the product, $\bar{L}t$, is known immediately by equation (6-27).

It should be emphasized (1) that equation (6-26) (or [6-27]) is not the most general form of the angular momentum equation, since it holds only when dealing with a fixed axis, and (2) that angular momentum and linear momentum are quite different and independent physical quantities.

6.13. Conservation of Angular Momentum

Let us assume that an object is rotating about a fixed axis and that there is no torque acting. Then, according to equation (6-27),

$$I\omega = I_0\omega_0 = \text{Constant} ; \qquad (6\text{-}28)$$

that is, the angular momentum about a given axis of a rotating body remains constant. This is known as the *principle of the conservation of angular momentum*. Stated in more general terms, it reads:

If there are no external torques acting, the total angular momentum of an object or system of objects about any axis remains constant.

According to the principle, the product $I\omega$ must remain constant. This does not imply that I and ω, separately, must be constant. With rotating objects it is possible to alter the mass distribution about an axis and thus alter the moment of inertia. Whenever this happens, however, the angular velocity of the body must alter in such a way that the product, *angular momentum*, remains the same.

The most common example of this is that of the person who sits on a piano stool or stands on a rotating platform with outstretched arms and is given a spin. By pulling his arms in close to his body he decreases his moment of inertia and increases his rate of spin correspondingly. Another example is that of a ballet dancer who pirouettes on her toes. In order to stop herself, she gracefully extends her arms and one leg, thereby increasing her moment of inertia roughly by a factor of 7 and decreasing her angular velocity to one-seventh. The technique of radically changing one's moment of inertia to gain angular velocity in accordance with the conservation principle has been profitably used by acrobats.

In the examples given thus far we have stressed the constancy of the magnitude aspect of angular momentum. We wish now to illustrate what happens when we try to change the direction of the angular momentum, i.e., turn the axis of rotation.

A student stands on a platform that may revolve only about a vertical axis and holds in his hand a bicycle wheel with a vertical axis. Let us assume that the student and the platform are initially at rest but that the wheel is spinning about its axis with an angular velocity ω'.

The initial total angular momentum, $I'\omega'$, is that due to the rotating wheel, where I' is the moment of inertia of the wheel about the axis of spin. We represent this by an arrow in Fig. 6-12 (a). The student now turns the axis of the

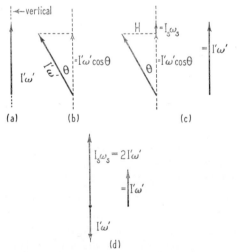

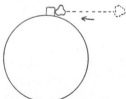

Fig. 6-12.—Vectors representing angular momentum of student and wheel.

wheel through an angle θ from the vertical; i.e., he applies a torque about a horizontal axis. The amount of angular momentum about the vertical is now only $I'\omega' \cos\theta$ (Fig. 6-12 [b]). But, according to the conservation principle, the total vertical momentum must be $I'\omega'$, since no torques were applied along the vertical. Accordingly, the student + platform begins to rotate about the vertical with an angular velocity ω_s. The extra angular momentum, $I_s\omega_s$, when added vectorially to $I'\omega' \cos\theta$, equals the original amount (Fig. 6-12 [c]). I_s is the moment of inertia of the wheel + student + platform calculated with the vertical as an axis.

If the student were to turn the axis of the wheel through 180°, the student + platform would speed up until their angular momentum $I_s\omega_s$ equals $2I'\omega'$, for then the total angular momentum of the system, student + platform + wheel, equals the original amount $I'\omega'$ (Fig. 6-12 [d]),

An important practical application of the conservation principle emphasizing the direction aspect of the angular momentum is found in gyroscopes, a topic we treat in a later section.

For convenience of reference we summarize in Table 6-2 the important angular formulas and

TABLE 6-2

EQUATIONS FOR LINEAR AND ANGULAR MOTION

$$(s = \theta r;\ v = \omega r;\ a = \alpha r)$$

Linear Motion	Angular Motion
$s = v_0 t \pm \frac{1}{2}at^2$	$\theta = \omega_0 t \pm \frac{1}{2}\alpha t^2$
$v = v_0 \pm at$	$\omega = \omega_0 \pm \alpha t$
$2as = v^2 - v_0^2$	$2\alpha\theta = \omega^2 - \omega_0^2$
$F = ma$	$L = I\alpha$
$Ft = mv - mv_0$	$Lt = I\omega - I\omega_0$
$Fs = \frac{1}{2}mv^2 - \frac{1}{2}mv_0^2$	$L\theta = \frac{1}{2}I\omega^2 - \frac{1}{2}I\omega_0^2$

their linear analogues. In Table 6-3 are listed the various units of measure for angular quantities.

EXAMPLE 10: Angular momentum, as we have remarked, is sometimes called "moment of momentum." If a point mass (m) goes around an axis with a linear speed v at a distance r from the axis, the moment of inertia of the particle is mr^2, and its angular velocity is v/r. Then its angular momentum, $mr^2 \times v/r = (mv)r$, can be considered as the linear momentum multiplied by the lever arm.

A disk of mass of 0.10 kg and radius 0.10 m has a little projection (negligible mass) on its circumference. A 0.020-kg bit of putty with a velocity of 5.0 m/sec strikes this projection and sticks fast. Find the angular velocity given to the disk.

$$I \text{ (disk and putty)} = \frac{0.10 \text{ kg} \times (0.10)^2 \text{ m}^2}{2}$$

$$+ 0.020 \text{ kg} \times (0.10)^2 \text{ m}^2$$

$$= (5 \times 10^{-4} + 2 \times 10^{-4}) \text{ kg-m}^2 \,.$$

The moment of momentum of the putty is equal to its momentum × lever arm; i.e., 5.0 m/sec × 0.020 kg × 0.10 m = 10^{-2} kg-m²/sec. Since there is no applied torque from outside the disk-putty system, the angular momentum of the system must remain constant. Then,

$$7 \times 10^{-4} \text{ kg-m}^2 \times \omega = 10^{-2} \text{ kg-m}^2/\text{sec} \,,$$

$$\omega = \tfrac{100}{7} = 14.29 \text{ rad/sec} \,.$$

If we experiment quantitatively, we find, as we would expect, that, although the angular velocity, ω, increases greatly as the masses are pulled toward the center, the angular momentum, $I\omega$, is unchanged; i.e.,

$$I_C \omega_C = I_D \omega_D \,,$$

where I_C is the moment of inertia of the rotating system when the masses are at the greatest distance from the axis and ω_C is the angular velocity, and where I_D is the moment of inertia of the rotating system when the masses are nearest the vertical axis of rotation and ω_D is the angular velocity. The angular velocities ω_C and ω_D are found to be approximately $\omega_C = 5.25$ rad/sec and $\omega_D = 15.7$ rad/sec.

The data necessary to calculate I_C and I_D are shown in Fig. 6-13. It is clear that the moment of inertia of the

TABLE 6-3

UNITS FOR LINEAR AND ANGULAR MOTION

System of Units	Mass (m)	Force (F)	Torque or Moment of Force (L)	Angle (θ)	Angular Velocity (ω)	Angular Acceleration (a)	Moment of Inertia (I)	Work or Energy
mks..........	kg	newton	newton-m	rad	rad/sec	rad/sec²	kg-m²	joule
cgs...........	gm	dyne	dyne-cm	rad	rad/sec	rad/sec²	gm-cm²	erg
British Engineering..	slug	lbf	lbf-ft	rad	rad/sec	rad/sec²	slug-ft²	ft-lbf
British (not commonly used)..	lb	poundal	poundal-ft	rad	rad/sec	rad/sec²	lb-ft²	ft-poundal

EXAMPLE 11: A quantitative measurement on conservation of momentum. Two masses can slide freely on a horizontal rod which rotates about a vertical axis. Pins prevent the masses from flying off when the rod is whirled. As shown in Fig. 6-13, two cords are attached

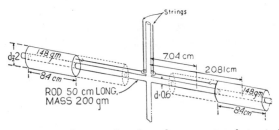

FIG. 6-13.—Conservation of angular momentum but not of angular kinetic energy.

to the masses and pass through central pulleys and then vertically along the axis of rotation. By exerting an upward force, F, on the cords, the masses may be pulled in toward the center while the device is in rapid rotation. No torque is applied to the masses while they are pulled inward. (Why?)

rotating system is that of the fixed rod, I_r, plus that of the movable masses, $2I_{cyl}$. From Table 6-1 we see that

$$I_r = m_r \left(\frac{r^2}{4} + \frac{l^2}{12} \right)$$

$$= 200 \left(\frac{0.3^2}{4} + \frac{50^2}{12} \right) = 41,700 \text{ gm-cm}^2 \,.$$

The moment of inertia of a hollow cylinder of inside radius r_1, of outside radius r_2, and of length l about an axis through its center of mass and normal to the axis of the cylinder is

$$I_{cyl} = m \left(\frac{r_1^2 + r_2^2}{4} + \frac{l^2}{12} \right) \,.$$

In order to determine the moment of inertia of the cylinder about a parallel axis h cm from its center of mass, we must add mh^2 to the preceding expression. With $h = 20.81$ cm, we obtain

$$I_{cyl} = 148 \left(\frac{0.3^2 + 1.0^2}{4} + \frac{8.4^2}{12} \right)$$

$$+ 148 \times 20.81^2 = 65,000 \text{ gm-cm}^2 \,.$$

The total moment of inertia I_C of the rotating system is

$I_C = I_r + 2I_{cyl}$

$\qquad = 41,700 + 2(65,000) = 172,000 \text{ gm-cm}^2$.

With $h = 7.04$ cm,

$I_D = 41,700 + 2(8,240) = 58,200 \text{ gm-cm}^2$.

The angular momentum $I_C\omega_C$ is, then,

$I_C\omega_C = (172,000)(5.25)$

$\qquad\qquad\qquad = 903,000 \text{ dynes-cm-sec}$.

Also,

$I_D\omega_D = (58,200)(15.7)$

$\qquad\qquad\qquad = 914,000 \text{ dynes-cm-sec}$.

The agreement is well within the errors of measurement, since the measurements may be in error by as much as 5 per cent.

Although the angular momentum of the system remained unchanged while the masses were pulled toward the center, the energy of the system was greatly increased, since work was done by the force F. We may compute this work as follows:

$KE_C = \frac{1}{2}I_C\omega_C^2 = \frac{1}{2}(172,000)(5.25)$

$\qquad\qquad\qquad = 2.37 \times 10^6 \text{ ergs}$,

$KE_D = \frac{1}{2}I_D\omega_D^2 = \frac{1}{2}(58,200)(15.7)^2$

$\qquad\qquad\qquad = 7.17 \times 10^6 \text{ ergs}$.

The increase of rotational kinetic energy is then equal to 4.80×10^6 ergs.

Since the masses were moved about 13.8 cm, the *average force* exerted on the masses is

$$\frac{4.80 \times 10^6}{13.8 \times 980} = 360 \text{ gmf} .$$

This average force is the centripetal force, f_C, accelerating the whirling masses. Since $f_C = m\omega^2 r$, it is clear that the centripetal force is by no means constant.

Query: Can the reader show that the centripetal force in this experiment varies inversely as the cube of the radius?

6.14. Simultaneous Rotation and Translation

The problems of rotation discussed thus far were concerned with rotation about fixed axes. We wish now to study those problems in which translational and rotational motions occur simultaneously. As an example of this motion, consider a solid cylinder of radius r rolling, *without slipping,*

along a horizontal plane (Fig. 6-14). Since we are assuming rigid bodies, the cylinder is in contact with the plane *along a line;* this line is the *instantaneous axis of rotation of the cylinder.* If the cylinder is rolling without slipping on a plane, then the point of contact of cylinder and plane must instantaneously be at rest. To show that this must be so, consider what would happen if the slipping was the only thing occurring, as when automobile wheels spin on ice or in sand. There is then no forward motion, for, in order to have any forward motion, the axle at O must move forward and this can take place only if the point of contact is at rest some of the time. Suppose the cylinder has a constant angular velocity ω about its axis. Then if there is no slipping, the angular velocity about an instantaneous axis of the line of contact of cylinder and plane is also ω (Fig. 6-14). During

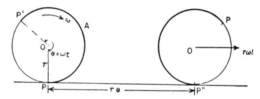

Fig. 6-14.—A cylinder rolling without slipping on a horizontal plane.

a time t the wheel turns through some angle $\theta = \omega t$ corresponding to the arc PAP' on the circumference of the cylinder. If there is no slipping, then the point of contact moves forward a distance $PP'' = \text{arc}(PAP') = r\theta = r\omega t$. For this to be so the axis of the cylinder must be moving forward with a velocity $v = r\omega$, so that in time t it moves forward a distance $vt = r\omega t$. Thus, for no slipping, the angular velocity about the instantaneous axis through the line of contact of cylinder and plane is ω, which is equal to the angular velocity ω about the axis.

Similarly, if the cylinder has an angular acceleration, α rad/sec², about its axis and the cylinder is rolling without slipping on a plane, then the angular acceleration about the instantaneous axis through the line of contact of cylinder and plane is also α rad/sec². We shall make use of these results in the analysis of cylinders and spheres rolling down an inclined plane. In Fig. 6-15 the axis is through P perpendicular to the plane of the paper. The problem here is to compute the linear acceleration of the center of mass, O, and also the linear velocity of the center of mass at the foot of the incline. We shall solve this problem using three somewhat different methods.

Method A.—Since there is no slipping, we may compute torques and angular accelerations about the instantaneous axis through P. The various forces acting *on the cylinder* are: mg, the force of gravity downward; F, the force of friction along the plane; and the reaction of the plane on the

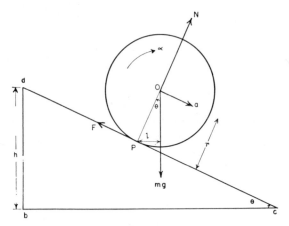

Fig. 6-15.—Cylinder rolling down an inclined plane

cylinder. With an axis through P, the only torque acting on the cylinder is that due to mg; i.e.,

$$L = mgl ,\qquad (6\text{-}29)$$

where $l = r \sin \theta$ is the lever arm. The moment of inertia of the cylinder calculated with respect to the axis through P is

$$I_p = \tfrac{1}{2} mr^2 + mr^2 = \tfrac{3}{2} mr^2 . \qquad (6\text{-}30)$$

At each instant, $L = Ia$, so that

$$a = \frac{mgl}{\tfrac{3}{2} mr^2} = \frac{mgr \sin \theta}{\tfrac{3}{2} mr^2} = \frac{2}{3} \frac{g}{r} \sin \theta \ \text{rad/sec}^2 . \ (6\text{-}31)$$

Note that, for a given slope of the plane, the magnitude of the angular acceleration about the instantaneous axis is constant. Since $a = ar$, it follows that

$$a = \tfrac{2}{3} g \sin \theta \qquad (6\text{-}32)$$

or

$$a = \tfrac{2}{3} g \frac{h}{s}, \qquad (6\text{-}33)$$

since $\sin \theta = h/s$. Equation (6-33) gives the value of the linear acceleration of the center of mass.

The final velocity attained by the center of mass of the cylinder in rolling down the plane is

$$v^2 = 2as$$

or

$$v^2 = 2 \left(\tfrac{2}{3} g \frac{h}{s} \right) s = \tfrac{4}{3} gh ,$$

$$v = \sqrt{\tfrac{4}{3} gh} . \qquad (6\text{-}34)$$

Method B.—We may treat the problem by considering independently the translational motion of the center of mass and a rotation about the center of mass. Newton's Second Law for translation of the center of mass is

$$mg \sin \theta - F = ma , \qquad (6\text{-}35)$$

where $mg \sin \theta$ is the component of the weight along the plane and F is the force of friction along the plane.

For rotation about the center of mass,

$$Fr = Ia ,$$

since all the other forces have zero lever arms with respect to the chosen axis. Also, $I = \tfrac{1}{2}mr^2$ and $a = ar$. Hence, $Fr = \tfrac{1}{2}mr^2 a/r$, so that

$$F = \frac{ma}{2} ,$$

and equation (6-35) becomes

$$a = \tfrac{2}{3} g \sin \theta ,$$

the same value as that given by equation (6-32).

Method C.—It is instructive to apply the energy equation to this problem. The potential energy of the cylinder at the top of the plane is mgh, since the center of mass is raised a distance h. At the foot of the plane all this energy is converted into kinetic energy of rotation and translation; i.e.,

$$mgh = \tfrac{1}{2} mv^2 + \tfrac{1}{2} I\omega^2 .$$

Here, again, $\tfrac{1}{2}mv^2$ represents the kinetic energy of translation of the center of mass, and $\tfrac{1}{2}I\omega^2$ the kinetic energy of rotation about the center of mass. Since $I = \tfrac{1}{2}mr^2$ and $v = r\omega$ (no slipping), it follows that

$$mgh = \tfrac{1}{2} mv^2 + \tfrac{1}{2} (\tfrac{1}{2} mr^2) \frac{v^2}{r^2} ,$$

$$mgh = \tfrac{3}{4} mv^2 ,$$

or

$$v = \sqrt{\tfrac{4}{3} gh} .$$

The problem of rolling without slipping brings into sharp focus an important function of friction. Since we have assumed *rigid objects* in contact, so that the cylinder and plane were in contact along **105** a line, *friction was not a dissipative force but rather a converter of translational energy into rotational.*

In Table 6-4 we compare the velocity of the center of mass of a hollow cylinder and sphere with the solid cylinder. Note that neither the mass nor the radius of the rolling object enters the answer.

TABLE 6-4

VELOCITIES OF ROLLING BODIES

Sphere	Solid Cylinder	Hollow Cylinder
$mgh = \frac{1}{2}mv^2 + \frac{1}{2}\cdot\frac{2}{5}mr^2\omega^2$	$mgh = \frac{1}{2}mv^2 + \frac{1}{2}\cdot\frac{1}{2}mr^2\omega^2$	$mgh = \frac{1}{2}mv^2 + \frac{1}{2}mr^2\omega^2$
$mgh = \frac{1}{2}mv^2 + \frac{1}{5}mv^2$	$mgh = \frac{1}{2}mv^2 + \frac{1}{4}mv^2$	$mgh = \frac{1}{2}mv^2 + \frac{1}{2}mv^2$
$mgh = \frac{7}{10}mv^2$	$mgh = \frac{3}{4}mv^2$	$mgh = mv^2$
$v = \sqrt{\frac{10}{7}gh}$	$v = \sqrt{\frac{4}{3}gh}$	$v = \sqrt{gh}$
$v = 1.195\sqrt{gh}$	$v = 1.155\sqrt{gh}$	$v = \sqrt{gh}$

A freely falling body would attain velocity $\sqrt{2gh} = 1.41\sqrt{gh} = v'$.
Hence

$v = 0.845v'$	$v = 0.817v'$	$v = 0.707v'$

From this it follows that if a solid sphere, a solid cylinder, and a hollow cylinder are started together from rest at the top of an inclined plane, then, assuming no slipping, the sphere will get to the bottom of the plane first, i.e., win the race, while the hollow cylinder will be the loser.

EXAMPLE 12: A string wrapped around a cylinder passes horizontally over a frictionless pulley, and a mass is hung on the end of the string. The cylinder rolls without slipping on a horizontal plate. Determine the angular acceleration of the cylinder and the linear acceleration of the 20-gm mass. The mass of the cylinder is 100 gm; its radius, 10 cm.

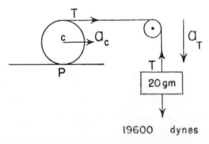

Since the necessary force of friction is unknown, take moments about P, and the value of the force of friction will not enter the equations. Let T be the tension in the string measured in dynes.

By equation (6-18),

$$T \text{ dynes} \times 20 \text{ cm} = I_p a_p \text{ gm-cm}^2 \times \text{cm/sec}^2 . \quad (1)$$

But

$$I_p = I_c + mr^2 = \frac{100 \times 10^2}{2} + 100 \times 10^2$$

$$= 15,000 \text{ gm-cm}^2 .$$

Equation (1) becomes

$$T = \frac{15,000 \text{ gm-cm}^2}{20 \text{ cm}} \times a_p \text{ cm/sec}^2 \quad (2)$$

$$= 750 a_p \text{ gm-cm/sec}^2 .$$

Applying $F = ma$ to the 20-gm mass, we obtain

$$(19,600 - T) \text{ dynes} = 20a_T \text{ gm} \times \text{cm/sec}^2 . \quad (3)$$

By equation (6-10), $a_T = a_p(2r) = 20a_p$. Equation (3) becomes

$$19,600 - T = 20 \times 20a_p . \quad (4)$$

Solving equations (2) and (4) simultaneously, we obtain, for a_p,

$$a_p = 17.04 \text{ rad/sec}^2 ,$$

so that $T = 750 \times 17.04 \text{ gm-cm/sec}^2 = 12,780$ dynes, the tension in the string; also, $a_T = 17.04 \text{ rad/sec}^2 \times 20 \text{ cm} = 340.8 \text{ cm/sec}^2$, the acceleration of a point on the periphery and that of the 20-gm mass. Also,

$$a_c = 17.04 \text{ rad/sec}^2 \times 10 \text{ cm} = 170.4 \text{ cm/sec}^2 ,$$

the acceleration of the center of mass.

We next determine the force of friction necessary for the cylinder just to roll without slipping: From

$$f = ma ,$$

where f is the sum of the tension T and the force of friction,

$$(12,780 + \text{force of friction}) \text{ dynes}$$

$$= 100 \text{ gm} \times 170.4 \text{ cm/sec}^2 .$$

The force of friction equals 4,260 dynes *forward*—forward, since the force of friction has the same sign as the forward acceleration. We may check our result by taking the moment of forces about the center. It must be remembered that the angular displacement, the angular velocity, and (in this case) the angular acceleration about all axes must be the same for a rigid body.

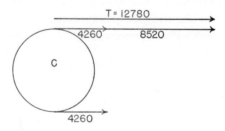

The moment of the force of friction about c is just balanced by the moment of 4,260 dynes of the tension, leaving the moment of 8,520 dynes of the tension unbalanced. Then

$$8520 \times 10 = 5,000a , \qquad a = 17.04 \text{ rad/sec}^2 .$$

6.15. Impulsive Blows and Center of Percussion

We must all be familiar with the fact that it is possible to sting the hands severely when striking a baseball with a bat. On the other hand, it is possible to grasp the bat at such a position that no sting is felt. The so-called "sting" is the result of an impulse on the hands resulting from the impulse delivered to the bat by the ball. The position of the hands for no sting depends on where the ball strikes the bat and not on the mass of the bat or the magnitude of the impulsive blow by the ball. In order to analyze this situation, let us consider a uniform thin rod which is freely suspended by a horizontal axis through O (Fig. 6-16). Suppose a

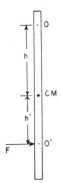

Fig. 6-16.—Center of percussion

horizontal force F is applied to the rod at O' for a short time Δt such that there is an impulsive horizontal blow of amount $F\Delta t$. If there is *no* horizontal blow on the axis through O, then the point O' is called the *center of percussion*, and we shall prove that the distances of O and O' from the center of mass h and h', respectively, are related by

$$hh' = k_{cm}^2 ,$$

where k_{cm}^2 is the square of the radius of gyration of the rod about an axis through the center of mass. Suppose the impulse $F\Delta t$ causes the rod to start rotating about the axis through O with an instantaneous angular velocity ω. This produces an instantaneous linear velocity of the center of mass of $h\omega$. Since $F\Delta t$ is the only horizontal impulse acting on the rod, it follows from § 5.8 that the linear momentum of the center of mass is equal to the impulse of

$$F\Delta t = M\omega h , \qquad (6\text{-}36)$$

where M is the mass of the rod.

Now take moments about the fixed axis through O; this gives

$$F(h+h') = I_0 a = \frac{I_0\omega}{\Delta t} , \qquad (6\text{-}37)$$

where I_0 is the moment of inertia of the rod about a horizontal axis through O and $a = \omega/\Delta t$ is the instantaneous angular acceleration of the rod produced by the blow.

From the theorem of parallel axes and the definition of the radius of gyration it follows that

$$I_0 = I_{cm} + Mh^2 = Mk_{cm}^2 + Mh^2 . \qquad (6\text{-}38)$$

By substitution in equations (6-36), (6-37), and (6-38) we have

$$M\omega h(h+h') = I_0\omega = M(k_{cm}^2 + h^2)\omega \qquad (6\text{-}39)$$

or

$$hh' = k_{cm}^2 . \qquad (6\text{-}40)$$

The two distances h' and h locate two points called the "center of percussion" (at O') and the "instantaneous axis of rotation" (at O).

If we consider the rod placed on a horizontal frictionless table and a horizontal impulsive blow applied at O', as in Fig. 6-16, then the rod will instantaneously rotate about an axis through O, since in the analysis the point O was located where no impulsive force acted. In other words, it would not matter whether there was a vertical pin attached to the table passing through O or not. No force would act on the pin, so that it need not be there. It is for this reason that the axis through O is called the "instantaneous axis of rotation." The reader will note that for every position of the center of percussion there is a corresponding instantaneous axis of rotation that is momentarily at rest. Where is the axis of rotation when the center of percussion is at the center of mass?

We can now appreciate that if the baseball bat is held at the position of the instantaneous axis of rotation for a given position of the center of percussion where the ball strikes, no impulsive blow will be felt by the hands. If at any other position, then a sting is felt.

6.16. The Gyroscope

The gyroscope shown in Fig. 6-17 consists of a wheel loaded at the rim and free to rotate about the axis aa'. The frame that holds the axle aa' is itself free to turn about bb', an axis perpendicular **107** to *aa'*. Finally, a second frame to which *bb'* is attached is free to rotate about the vertical, *cc'*. In the gyroscope shown, the three mutually perpendicular axes intersect at the center of mass of the rotating wheel and frame. Some of the simpler properties of such a gyroscope will now be discussed.

If we push downward on the axle *aa'* at the point *a'*, we find that, with the wheel at rest, the frame, together with the wheel, turns quite easily about the axis *bb'*. However, if the wheel is given a very large *angular velocity* about the axis *aa'*,

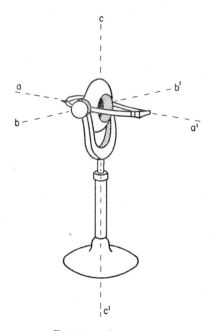

Fig. 6-17.—A gyroscope

we find, on pushing downward at *a'*, that the spinning wheel exhibits a remarkable resistance to the applied force. The sensation is similar to that received on pushing, successively, two spheres identical in appearance but with the first a hollow aluminum shell and the second a solid lead sphere. The pedestal of the gyroscope, in fact, may be turned about at random without changing the direction in space of the axis *aa'* of the spinning wheel. On turning the pedestal, no torques are transmitted to the wheel through the frames (friction is assumed negligible). Hence, by the First Law of Motion, as given in § 6.9: *A body supported at its center of mass and spinning freely about an axis tends to keep the direction of its axis of spin fixed in space unless acted upon by a torque.*

This remarkable stability of the axle of the spinning wheel can be readily used to prove the rotation of the earth. Let us suppose the spin is maintained by an electric motor and that the axis *aa'* of the gyroscope is set pointing to some object in the room. After a few hours the axis no longer points to the object, the reason being that the axle has maintained its direction in space; but the room, because of the rotation of the earth, has changed its orientation. This property of gyroscopes is embodied in the construction of the modern gyrocompass.

6.17. Precession of the Gyroscope

We learn of another important property of the gyroscope by means of the following experiment, elaborated from the observation in the preceding paragraph. With the wheel spinning with a large angular velocity and the axis *aa'* pointing horizontally, a weight is hung at the end *a'*; i.e., a torque is applied to the spin axis, tending to rotate it about *bb'*. Instead of tilting downward, as it obviously would if the wheel were not spinning, the gyroscope starts to rotate about the vertical axis *cc'*; that is to say, the gyroscope begins to *"precess" about the axis* *cc'*. At the instant the torque is applied, the spin axis is observed to wobble somewhat about the axis *bb'* as the precession starts about *cc'*. A detailed study of this motion is beyond the scope of this text. After ignoring the slight initial wobble (known as "nutation"), a steady state is soon reached in which the weight hung on the end of the axle of the wheel describes a horizontal circle* as it precesses. It is quite within our means to deduce a relation between the torque applied and the resulting *uniform precessional motion about the axis* *cc'*.

For greater clarity let us replace the apparatus of Fig. 6-17 by that shown in Fig. 6-18. The wheel is rotating as indicated, with a spin velocity of ω rad/sec. If M is the mass of the wheel and axle, then the torque, L, applied to the gyroscope is Mgl, where l is the distance from the pivot point to the center of mass of wheel and axle. As can be seen from the diagram, the torque Mgl is exerted about the axis *bb'*. Therefore, we have drawn the torque vector Mgl along this axis in the appropriate direction (see § 6.2). The wheel is observed

---

* In reality, this circle is not quite re-entrant but is a slowly descending spiral curve, due to friction in the bearing *cc'*.

to precess uniformly about cc' with angular velocity ω'.

Let I be the moment of inertia of the wheel about the spin axis, aa'. Then the angular momentum about aa' is $I\omega$ at the instant shown in the diagram. We represent this angular momentum by the vector PQ of the three-dimensional vector

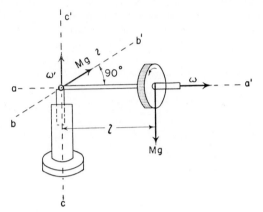

FIG. 6-18.—A gyroscope with an unbalanced torque

diagram of Fig. 6-19. In time Δt we observe that the axle of the wheel moves (precesses) through the very small angle $\Delta\theta$ in a horizontal plane. The angular momentum is now represented by the vector PR, whose length is still $I\omega$. The change in angular momentum ($\Delta I\omega$) in time Δt is, then,

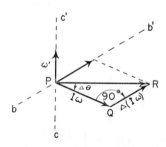

FIG. 6-19.—Angular momentum vectors

$(PR - PQ) = QR$. Note that QR is parallel to the torque axis, bb' (change in momentum has the direction of the torque producing it).

From the diagram

$$|QR| = |PQ|\,\Delta\theta , \qquad (6\text{-}41)$$

where | | means "the magnitude of." The time rate of change of angular momentum is

$$\frac{|QR|}{\Delta t} = |PQ|\frac{\Delta\theta}{\Delta t} . \qquad (6\text{-}42)$$

But the $\lim_{\Delta t \to 0} (\Delta\theta/\Delta t)$ is, by definition, the angular rate with which the axle aa' is turning about cc', i.e., the precessional velocity, ω'. Since, numerically, PQ is equal to $I\omega$, *we have, for the time rate of change of angular momentum about the axis* bb',

$$|PQ|\,\omega' = I\omega\omega' . \qquad (6\text{-}43)$$

Therefore, by Newton's Second Law (§ 6.9), *the torque needed to maintain the existing precessional velocity* is

$$L = I\omega\omega' . \qquad (6\text{-}44)$$

The direction of L is along bb'. In our case, $L = Mgl$, so that

$$Mgl = I\omega\omega' . \qquad (6\text{-}45)$$

The direction of the precessional velocity is such that the vector ω moves toward the positive direction of the torque axis L.

In this analysis we have assumed that the angular momentum of precession is small compared to the angular momentum of spin, so that the total angular momentum of the gyroscope is $I\omega$ along the spin axis. If this is not the case, then the analysis becomes much too complicated to be considered at this level.

Precessional motion is by no means confined to laboratory apparatus, as shown in Figs. 6-17 and 6-18, but occurs in many everyday experiences. We list a few common examples of it.

1. Because the earth is not a sphere but rather an oblate spheroid, the gravitational attraction of the moon (and, to a lesser extent, the sun) exerts a torque about an axis at right angles to the axis of rotation of the earth. (The reader should draw an appropriate diagram.) This causes the axis of the earth to precess slowly (since L is small and $I\omega$ very large). The pole of the earth describes a complete circle in 25,800 years.

2. Another interesting case of precession is found in the automobile, the flywheel acting as a gyroscope. When the car turns around a curve, a torque forces the axis of the flywheel to turn in the direction of the curve. As a result, the precessional velocity tends to tilt the axle upward or downward. If the wheel is rotating counterclockwise as seen from the rear of the car, the front of the car will tend to fall when the car makes a left-hand turn. Racing cars go around the tracks making left-hand turns in order to keep their front wheels on the ground. This same phenomenon occurs in an airplane when it makes a turn. Precession forces **109**

the nose of the ship either downward or upward, depending upon the direction of the turn.

3. We observe the motion of a wheel or a coin as it rolls away from us. As the wheel starts to tip over, say to the right, a force comes into play to the right end of the axle as we view the wheel. This force, together with the reaction of the ground on the wheel, constitutes a torque whose axis is along the ground in the direction of motion. This torque causes the wheel to precess about a vertical axis, so that the wheel turns to the right. What would be the gyroscopic effect of the *four wheels* of a car as the car makes a turn?

6.18. Effects of the Rotation of the Earth

Up to the present we have made use of Newton's laws in coordinate systems attached to the earth, whereas they were stated in § 1.2 as being applicable only in a nonrotating coordinate system such as could be imagined attached to the fixed stars. The question then arises of how much these laws shall be modified when they are used with a rotating coordinate such as we have on a rotating earth. The angular velocity of rotation, ω, of the earth is small, since it takes 86,400 mean solar seconds to turn through one complete revolution, so that

$$\omega = \frac{2\pi}{86,400} = 7.28 \times 10^{-5} \text{ rad/sec}.$$

It must be clear that, in the limited range of phenomena we have discussed, the effects of the rotation of the earth must be negligibly small, otherwise we should not have spent so much time in a fruitless effort. However, there are some large-scale phenomena in which the effects are not negligibly small, and these we shall now discuss. To do this we shall resort to two laboratory experiments.

The first of these concerns the *centrifugal force* which arises in rotating coordinate systems. Let us consider an observer at some place P (Fig. 6-20), latitude ϕ on the earth, of which O is the center, and $OE' = OP = R$ is the radius of the earth. Because of the earth's angular velocity ω about its axis NS, the observer at P has an angular velocity about the axis OP of $\omega \sin \phi$, the component of ω along OP. For middle latitudes of about 45° the angular velocity of rotation about the vertical axis, OPZ, is of the order of 5×10^{-5} rad/sec. This small angular velocity produces very small effects on our experiments; so, in order to amplify these, we shall con-

sider an experiment on a large horizontal table rotating about a vertical axis with a velocity of 5 rad/sec or 100,000 times as fast as that of the earth.

A student stands on the rotating table and performs his experiments just as he does in the laboratory. We shall require that the student be scrupulously honest and interpret the results of his experiments in terms of Newton's laws and forget about his normal experiences on the earth. His first experiment consists in placing an object of mass m at some distance r from him on the rotating table. He finds the object moves radially out away from him. Thus he says there is a force acting on the mass. In order to measure this, he attaches a spring balance to the mass and finds that the force registered

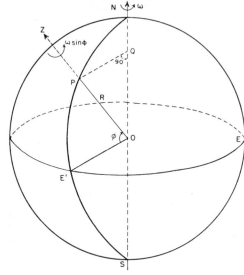

Fig. 6-20.—Observer on the earth at P latitude ϕ

depends on the angular velocity of rotation of the table ω_T, the distance the mass is from the axis of rotation r, and the mass m of the object. By varying these parameters one at a time, he finds that the inward force which is necessary to keep the ball *at rest* is given by $mr\omega_T^2$. To the student on the rotating table there is a *real outward force*, $mr\omega_T^2$, which he calls a *centrifugal force*, exerted on the mass m.

In some of the amusement parks in this country you may find, or used to find, a large circular horizontal platform which can be turned about a vertical axis just as the one in this experiment does. People sat on this platform and then it was rotated, and the problem was to stay on it. The centrifugal force pushed them off. If you have ever been on such a platform, the feeling you have is that you are on an inclined plane, and the larger the angular velocity, the steeper the plane. This is because the only experience you have had of such a force on you is that on an inclined plane, as when sliding down a hill.

A second student who is at rest relative to the rotating table describes his observations of this experiment somewhat differently. The object when free moves tangentially off the board, and, in order to keep it at rest, the observer on the rotating table must exert an inward or centripetal force of $mr\omega_T^2$. This simple experiment illustrates the importance of noting the frame of reference or coordinate system in which an observation is made. In our laboratory we are on a rotating coordinate system, and this centrifugal force exists, but it is very small, for on a 1-gm mass at latitude 45° ($r = R \cos\phi = PQ$) the force is about 10^{-3} dynes, which is of the order of 10^{-6} gmf.

If the earth were a smooth sphere, then an object placed at P (Fig. 6-21) would move toward the equator,

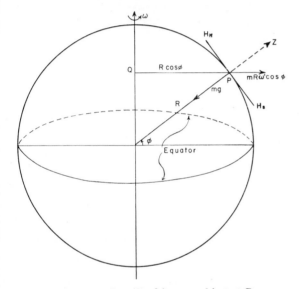

FIG. 6-21.—Centrifugal force on object at P

since there is a component of the centrifugal force along $H_N H_S$, the horizontal plane through P. In the figure, H_N is the north N point of the horizon circle and H_S the south S point. The earth, however, is not spherical but is slightly flattened at the poles, because of deformations resulting from the centrifugal forces on each particle of the earth. The shape of the earth is such that a particle at rest on its surface will remain in equilibrium under the combined action of centrifugal and gravitational forces. For the oblate earth the component of the centrifugal force along $H_N H_S$ (Fig. 6-21) is compensated by a component of the weight mg along $H_N H_S$. The component of the centrifugal force along the zenith PZ is absorbed in the measurement of g. For all experiments on the earth in which weight is involved, centrifugal forces are automatically compensated.

The second force which occurs because of our being on a rotating coordinate system is known as the _Coriolis force_, called after the discoverer of this force. Again we make use of our horizontal rotating table with a student at the center. In this case he fires a bullet horizontally with a velocity v into a block of wood AB (Fig. 6-22), placed near the periphery of the table. Although the gun is aimed at point A, it strikes the block at point B. To the student on the rotating table the bullet is deflected to the right, while to another student standing alongside he says the bullet moved in a straight line and the deflection AB is due to the rotation of the table during the flight. Again both observations are correct, but the

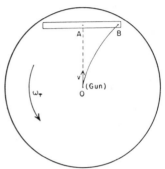

FIG. 6-22.—Bullet fired from center of rotating table into a block of wood.

rotating observer is sure that the deflection is due to a force at right angles to the direction of flight of the bullet. By a series of experiments he establishes that the force at right angles to v depends on the mass m of the bullet, its velocity v, and the angular velocity of rotation ω_T of the table. Mathematical analysis shows the Coriolis force is given by

$$F = 2 m v \omega_T$$

and is at right angles to v at all times.

Now consider an observer on the earth at P (Fig. 6-20); the angular velocity of rotation about his zenith or his vertical axis OPZ is $\omega \sin\phi$. If a bullet of mass m is shot horizontally in his horizon plane at right angles to his zenith PZ, he finds the bullet is deflected from a straight path with a force of $2mv\omega \sin\phi$. In the Northern Hemisphere this force is to the observer's right, i.e., if the bullet is shot horizontally toward the south, it is deflected toward the west, while in the Southern Hemisphere the deflection appears reversed in direction. Since the Coriolis force is perpendicular to v at all times, it does no work but merely changes the direction of motion. For a mass of 1 gm at latitude 45° and a horizontal velocity of 1,000 cm/sec and ω about 5×10^{-5} rad/sec, the Coriolis force is about one-tenth of a dyne, or of the order of 10^{-4} gmf.

For ordinary motions of automobiles, etc., on the earth the Coriolis force is negligibly small, but there are many cases of relatively large-scale phenomena in which it must be taken into account. There is appreciable force on, and deflection of long-range, high-velocity missiles; the erosion of one bank more than an- **111** other of fast-winding rivers is mainly due to this force. Winds do not move directly from high-pressure regions to low-pressure ones, but are markedly deflected by the Coriolis force. Another rather striking example of the Coriolis force is in the motion of ice floes in the Arctic. The ice does not entirely flow in the direction of the wind stress on the floe, but also has a component of velocity at right angles to this stress. Unless stated to the contrary, we shall neglect both the centrifugal and the Coriolis forces due to the earth's rotation, in the remainder of the book.

6.19. Résumé

The following is a brief listing of the definitions principles, and theories, given in this chapter, which you should know:

Definition of angle in radians, of angular velocity in rad/sec, of angular acceleration in rad/sec²
Relationship of angle, angular velocity, and time for an object moving with constant angular acceleration
Relationship of linear and angular velocities
Rotational inertia or moment of inertia and radius of gyration
The theorem of parallel axes for moments of inertia
Newton's laws of motion applied to rotation of rigid bodies
Work done and energy of rotating bodies
Angular momentum and conservation of angular momentum
Simultaneous rotation and translation of rolling bodies
Impulses and center of percussion
The gyroscope and its precessional velocity
Some effects due to the rotation of the earth

Queries

1. What principle do we employ when we change the beat of the common metronome?
2. Two massive disks are connected by a short rod of small diameter and, resting on the rod, roll, without slipping, down a narrow inclined plane. At the bottom of the incline the disks come in contact with a horizontal table top, and it is observed that the translational speed is greatly increased at this instant. Explain.
3. A man turns with an angular velocity, ω, on a rotating table, holding two equal masses at arm's length. What change takes place in his angular velocity if, without moving his arms, he drops the two weights? Reconcile your answer with the principle of conservation of momentum.
4. What is the effect on a single-engine pursuit ship of a rapid turn? On a double-engined ship?
5. Give a simple explanation for the fact that the moment of inertia of a body about an axis through its center of mass is a minimum.
6. Explain briefly in words why the angles in angular velocity and angular acceleration in the equations in the text must be given in radians and not in degrees.
7. Has any new physical principle been introduced in this chapter? Explain. Give the new physical concepts introduced in this chapter.
8. What are the dimensions of a radius of gyration, of a moment of inertia, in the mks and British Engineering systems of units?

Practice Problems

1. After the power in a motor is turned off, the flywheel, which is rotating at the rate of 240 revolutions per second (rev/sec), makes 350 revolutions (rev) before coming to rest. How long does this take? Ans.: 2.917 sec.
2. A certain wheel turns through an angle of 500 rad in 10 sec, starting from rest. Calculate the average angular velocity. If the acceleration were uniform, calculate its value and also the final angular velocity after 10 sec.
3. A wheel starts from rest with an angular acceleration of 3 rad/sec². How many revolutions will it make in 20 sec, and what angular velocity will it acquire? Ans.: $300/\pi$ rev; 60 rad/sec.
4. If the wheel in problem 3 is 1 m in radius, calculate (a) the tangential acceleration; (b) the tangential velocity of a particle on the rim after 10 sec; (c) the centripetal acceleration of the particle after 10 sec.
5. Three point masses, each of 100 gm, form the vertices of an equilateral triangle of side 10 cm. Calculate (a) the moment of inertia of the masses about an axis perpendicular to the plane and through one of the masses; (b) the moment of inertia about an axis through the center of mass and perpendicular to the plane; (c) the radius of gyration for cases (a) and (b); (d) the moment of inertia about an axis joining two of the masses. Ans.: (a) 2×10^4 gm-

cm²; (b) 10⁴ gm-cm²; (c) 8.18 cm, 5.77 cm; (d) 7,500 gm-cm².

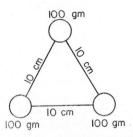

100 gm

10 cm 10 cm

10 cm

100 gm 100 gm

6. Two thin, uniform yardsticks, each of 2 lb, are fastened together, as indicated in the accompanying illustration, and an axis is passed through the vertical one 6 in from the end and normal to the plane of the paper. Compute the moment of inertia about this axis.

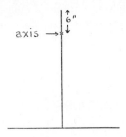

axis →

7. The same situation as in problem 6 except that the axis is passed parallel to the lower stick. Compute the moment of inertia about this axis. What is the radius of gyration about this axis? Ans.: 0.5 slugs-ft²; 2.0 ft.

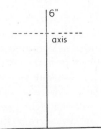

axis

8. This plate of uniform thickness has a mass of 400 gm. Compute its moment of inertia about an axis normal to the plane of the plate through the point indicated.

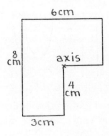

6cm

8 cm

axis

4 cm

3cm

9. A thin, uniform meter stick is mounted on an axis through the 10-cm mark. The mass of the stick is 120 gm. At the 100-cm mark a little lead ball of mass 50 gm and negligible radius is fastened. Compute the moment of inertia about the given axis. Ans.: 6.97 × 10⁵ gm-cm².

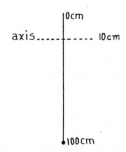

0cm

axis........|....... 10cm

100cm

10. A uniform disk of 0.3 m radius and mass 1 kg is pivoted about its axis. What torque would generate in 15 sec an angular velocity of 20 rad/sec, starting from rest?

11. A torque of 0.5 newton-m is applied to an object of mass 1 kg whose radius of gyration with respect to the axis of the torque is 0.4 m. Calculate the resulting angular acceleration and the angular velocity attained in 1 min. Ans.: 3.125 rad/sec²; 187.5 rad/sec.

12. A disk of 100 lb and radius of 2 ft is pivoted at its center. A string is wrapped around the disk, and a 14-lb weight hung from the string. Find the angular acceleration of the disk and the linear acceleration of the 14-lb weight. What will be the linear velocity and the angular velocity 10 sec after the start?

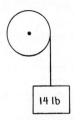

14 lb

13. A disk of 100 lb and radius of 4 ft is mounted as indicated in the accompanying figure. A string is wrapped around the disk and pulled with a constant force of 14 lbf. Find the angular acceleration of the disk and the linear acceleration of the string. Ans.: 2.24 rad/sec²; 8.96 ft/sec².

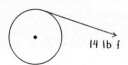

14 lb f

14. A wheel of 0.2 m radius has an initial angular velocity of 20 rad/sec and a constant angular acceleration of 8 rad/sec². What will be the angular velocity at the end of 5 sec? What will be the centripetal acceleration of a point on the circumference at the end of 5 sec? Through what angle will the wheel turn in the first 5 sec? How far (linear) will a point on the circumference have moved during the first 5 sec?

15. A wheel has a moment of inertia of 1,000 slugs-ft². At a certain instant it has an angular velocity of 10 rad/sec. After the wheel has turned through 100 rad, its velocity is 30 rad/sec. Compute the angular acceleration and the torque acting on the wheel, assuming both to be constant. Ans.: 4 rad/sec²; 4,000 ft-lbf.

16. A wheel of 20 cm radius rotates against a fixed belt with a constant angular velocity of 30 rad/sec. The tension on one end of the belt is 1,000 gmf, and on the other is 200 gmf. How many watts are necessary to keep the wheel in this motion?

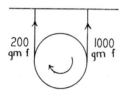

17. A frictional torque of 1 m-kgf is applied to a disk of radius 100 cm and mass 100 kg revolving at the rate of 175 rad/sec. How long will it take the disk to stop? Through what angle in radians does it turn after the brake is applied? How much work is expended? Ans.: 884 sec; 7.81 × 10⁴; 7.65 × 10⁵ joules.

18. A torque of 0.016 newton-m is applied to an object, turning it through an angle of 400 rad. If the mass of the rotating object is 1 kg and its radius of gyration 0.2 m, calculate the increase in kinetic energy of the object and its angular velocity. Initial angular velocity is 0.

19. A truck weighing 10 tons exclusive of wheels is moving at the rate of 22 ft/sec. Each of the four wheels of radius 1.5 ft has a weight of 200 lb and a radius of gyration about the center of 1 ft. Calculate the total kinetic energy and the percentage of the total due to rotation of the wheels. Ans.: 1.57 × 10⁵ ft-lbf; 3.43.

20. A uniform stick, 8 ft long and of weight 20 lb, is suspended freely from a fixed point at one end. If an impulse of 40 lbf-sec is applied at the other end, compute the angular momentum and the angular velocity imparted to the stick. Ans.: 320 ft-lbf-sec; 24 rad/sec.

21. A figure skater starts to rotate with arms outstretched with an angular velocity of 1 rad/sec and then immediately brings in his arms. Calculate the resulting angular velocity, given that the initial radius of gyration of the skater is 0.2 m and the final radius of gyration 0.05 m. Ans.: 16 rad/sec.

22. A solid sphere of radius 0.06 m and mass 0.5 kg rolls down a rough inclined plane of inclination 30°. What is the linear acceleration of the center of mass? How far will it roll in 3 sec, starting from rest? How much kinetic energy will it gain in 3 sec?

23. A uniform bar, 4 ft in length and of weight 10 lb, lies on a smooth table and is pivoted at one end. A 1-lb piece of putty moving at right angles to the length of the bar with a velocity of 10 ft/sec strikes the bar 3 ft from the pivoted end. Calculate the resulting angular velocity of the bar and attached putty. Ans.: 0.482 rad/sec.

24. A cylinder 100 cm long and of radius 6 cm is pivoted at one end and is free to oscillate about a diameter, as shown herewith. Calculate the center of percussion of the cylinder. Ans.: 16.84 cm below the center of mass.

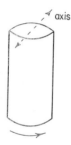

25. A gyroscope with a disk of 100-gm mass and a radius of 5 cm is mounted as indicated. The center of mass of the disk is 10 cm from the point of support. If the disk is spinning 100 rad/sec, with what angular velocity will the axis precess? In the position shown in the figure, will the right-hand end of the axis move toward or away from the observer? Ans.: 7.84 rad/sec.

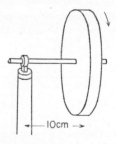

26. Given a pivoted meter stick in unstable equilibrium (see accompanying figure). Find the angular

velocity of the stick as it passes its position of stable equilibrium.

27. Given the pulley as shown in the diagram herewith. The mass of the pulley is 100 gm; its radius, 10 cm. If a rough string of negligible mass is hung over it with masses of 10 and 50 gm at each end, respectively, find the linear acceleration of the 10-gm mass. *Hint:* If the string does not slip, then the tension cannot be the same throughout the string, and the pulley must rotate. Ans.: $a = 356$ cm/sec².

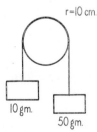

28. In problem 20 compute the impulse transmitted to the axis.

29. A four-wheeled truck has a total mass of 1,400 gm; the mass and radius of each solid disk wheel are 100 gm and 5 cm, respectively. If a force of 400 gmf acts on the truck, find the *linear* acceleration of the truck, assuming no slipping. Ans.: $g/4$.

30. A cord is wrapped around a solid cylinder. The radius of the cylinder is 10 cm, and its mass 1 kg. If the cylinder is allowed to fall vertically, as shown in the accompanying diagram, compute (a) the time required to fall 10 ft from rest; (b) the tension in the cord.

31. What is the centrifugal force acting on a 1-gm mass at the equator? Use 6.4×10^8 cm as the earth's radius. Ans.: 3.4 dynes.

32. A ball of mass 500 gm is thrown northward in latitude 45° N. with a velocity of 5 m/sec. What Coriolis force acts on the ball?

33. Solve problems 8, 24, and 25, using the mks system of units.

34. A wheel has a moment of inertia of 10^{-3} kg-m². Find (a) the torque necessary to stop the wheel in 20 sec when it is turning with an angular velocity of 20 rev/sec, and (b) the work done in stopping the wheel.

35. A gyroscope has an angular velocity of 200 rev/sec and a moment of inertia of 0.001 kg-m². It has a torque of $\pi \times 10^{-3}$ newtons-m applied to it about an axis perpendicular to the axis of spin. Find the angular velocity of precession. Ans.: 0.0025 rad/sec.

36. A meter stick weighing 150 gm lies on a horizontal frictionless table and is struck by a blow at right angles to the stick having an impulse of 1,500 dynes-sec at a distance of 20 cm from the center of mass. Find (a) the distance from the center of mass about which the stick begins to turn; (b) the linear velocity of the center of mass; and (c) the angular velocity of rotation.

37. The spool shown in the diagram consists of a central solid cylinder of radius 0.05 m and mass 2 kg, which is centrally attached to two solid disks, each of radii 0.15 m and of mass 1 kg. A cord is wound around the inner cylinder and is vertically con-

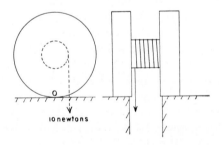

nected to a constant force of 10 newtons. The spool moves along the table with a constant linear acceleration, while the cord is kept vertical by passing through a slot in the table. Find (a) the moment of inertia of the spool about the line of contact at O; (b) the linear acceleration of the center of mass, assuming no slipping; (c) the horizontal force exerted on the spool; and (d) the vertical force exerted on the spool. Ans.: (a) 0.115 kg-m²; (b) 0.652 m/sec²; (c) 2.61 newtons; (d) 49.2 newtons.

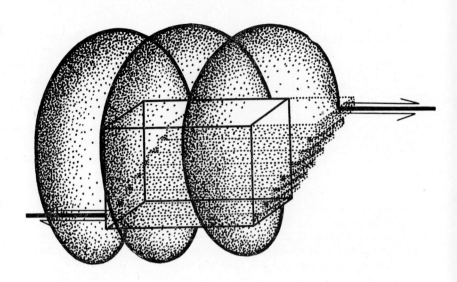

CHAPTER 7

Elasticity

7.1. Real Objects Are Not Perfectly Rigid

Following our method of development outlined in chapter 1, we have so far been dealing, for simplicity, with idealized perfectly rigid or non-deformable solids. Such solids, by hypothesis, are those in which the distances between their ultimate molecular or atomic particles are fixed and unchanging and in which the relative positions of these particles are likewise unalterable by the application of any amount of force. Now such conditions, of course, are entirely hypothetical and quite contrary to the facts of nature.

All solids are compressible—in none are the relative distances between their molecules or ultimate molecular groups unalterable. By the application of large forces the volume of any solid may be slightly reduced. Furthermore, all solids are deformable in shape. By bending, twisting, or stretching we can alter, to a slight degree at least, the relative positions of molecular groups within the solid. We can slightly change the shape of the specimen.

7.2. Elasticity

Whenever, in any solid object, by the application of a certain amount of pressure (force per unit area) or force or torque from the outside, we change either the volume or the shape, we observe that, after the removal of the impressed forces, the body "springs back," as we say, to its original

shape or volume. Only an idealized object ever springs back completely. Such an object, if it could be found, would be said to possess PERFECT ELASTICITY, even though it obviously is deformable—does not possess perfect rigidity. No solid materials of this character are known. Nevertheless, in what immediately follows we shall assume perfect elasticity in comparing the relative deformability of various solids under the action of outside forces that are impressed in various ways upon them.

When we apply forces to a solid object, changing its volume or its shape or both, we call forth internal reacting forces in the object due to its elastic qualities—forces that react against and resist the applied forces. As the deforming forces are increased, the deformation increases; and the forces resisting further deformation likewise increase. Consequently, for all values of the applied forces the yielding of the material instantly supplies an opposing force. If the impressed force decreases, the material springs back in part, and the opposing force is lessened. We have at all times, therefore, a condition of equilibrium involved—there are no unbalanced forces, no accelerations. We are dealing with statics—not with dynamics.

7.3. Hooke's Law

Robert Hooke (1635–1703) discovered, as early as 1676, the fundamental law existing between force and the resulting distortion of an elastic body. He summarized the results of his experiments in the form of a law, "Ut tensio sic vis," which, when freely translated, means that "a change in form is proportional to the deforming force."

Many years later Thomas Young (1733–1829) gave Hooke's law a more precise formulation by introducing definite physical concepts to be associated with "a change in form" and "a deforming force."

The resultant forces called forth in the deformation of any solid by outside agents are popularly called "stresses." However, the word "stress" as used in the study of elasticity has a technical significance. If a cube is subjected to a uniform compression, the forces acting on each face are distributed uniformly over that area and may not be considered as acting at a single point. If the force F is normal to the area A, then we say that the cube is subject to a *pressure whose value is* F/A. On the other hand, if the forces are parallel to the surfaces, the cube is subject to a shear whose value is also given by F/A. Such forces, distributed over areas, are called "surface forces," to be contrasted, for example, with the weight of an object that acts throughout its volume (called a "body force"). We shall always have in mind surface forces when discussing stresses. By definition,

Stress ≡

$$\frac{\text{Force that balances the applied force*}}{\text{Area over which the force acts}} = \frac{F}{A}. \quad (7\text{-}1)$$

When a stress is called forth in a solid by the application of external forces, some physical change is produced. For example, the cube under a pressure will change its volume; a wire will stretch; a rod, when twisted, will change its shape. These relative distortions are called *strains* and may be one of three kinds: (1) change in size of a body which retains the same shape, (2) change in shape of a body which retains the same volume, and (3) change in length. Hooke's law may now be stated as follows:

$$\frac{\text{Stress}}{\text{Strain}} = \text{Constant} \equiv \textit{Modulus of elasticity},$$

a term introduced by Thomas Young.

---

* Since in statics the internal force is always equal and opposite to the applied force, we can and, in fact, do use the applied force in eq. (7-1).

The constant is called the *bulk modulus* (K) if the strain corresponds to 1; a *shear* or *rigidity modulus* (n) if the strain is 2; and *Young's modulus* (Y) if the strain is a stretch or compression which corresponds to 3. We shall discuss the three moduli in the order named.

It should be pointed out that, although the modulus for a given material is a constant, for crystalline materials—in particular, in single crystals—the modulus of elasticity may vary greatly in different directions through the crystal.

In this text we shall limit the discussion to homogeneous isotropic substances. These are substances which are uniform throughout and have no preferred directions, that is, the various elastic moduli do not vary with any particular direction in the material.

7.4. Bulk or Volume Modulus

Let us apply a constant force F over the faces of a cube, as shown in Fig. 7-1. The cube then is in

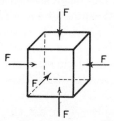

Fig. 7-1.—A uniform pressure applied to a cube

a state of compression, the pressure on each surface being F/A, where A is the area of a face. This constant pressure will produce a change in volume of the solid. The stress is, clearly,

$$\text{Stress} = \frac{F}{A}.$$

Some question arises, however, as to what to call the accompanying strain. If we call the strain the "change in volume ΔV," then the modulus obtained by taking the ratio of stress to strain would vary, depending upon the size of the original cube. It is important so to choose the strain that the modulus will truly measure a quality of the solid material itself and depend not at all upon the characteristics of any particular specimen of this material. For this reason we choose the strain as the specific change in volume; i.e.,

$$\text{Strain} = \frac{-\Delta V}{V};$$

whence, from Hooke's law,

$$\frac{\text{Stress}}{\text{Strain}} = \frac{F/A}{-\Delta V/V} = K, \text{ the bulk modulus} \quad (7\text{-}2)$$

or

$$K = \frac{\Delta P}{-\Delta V/V} = -\frac{\Delta P V}{\Delta V},$$

where the minus sign indicates that the volume decreases upon application of the stress. We use the symbol ΔP, instead of P, for the pressure, since all objects with which we experiment are under some pressure already, and we supply simply a change of pressure. Since the ratio $\Delta V/V$ is a pure number, K is a pressure; and hence its units are those used in measuring pressure. In the metric system these are commonly in kg/mm², dynes/cm², and newtons/m². In the British Engineering system the units are lbf/ft² or, more commonly, lbf/in², abbreviated to psi. The value of K is constant over only a limited range of pressures, namely, those in which Hooke's law is valid.

The type of deformation that involves only volume change applies quite as well to liquids as to solids. Indeed, the order of magnitude of the bulk moduli of solids and liquids is the same. This is only the technical way of stating the obvious fact that both solids and liquids are highly incompressible—require great pressure changes to produce even minute changes in volume.

Handbook tables often list the *compressibility* of solids and liquids instead of the bulk modulus. The compressibility, β, equals $1/K$; hence

$$\beta = \frac{-\Delta V}{V \Delta P}. \quad (7\text{-}3)$$

This number represents a *specific* change in volume (change in volume per unit volume), per unit pressure.

If the bulk modulus is given in units of dynes/cm², then the compressibility in cm²/dynes is the inverse. For example, the bulk modulus of aluminum is about 7×10^{11} dynes/cm², so that its compressibility is about 1.4×10^{-12} cm²/dyne. Since in engineering the common unit of pressure is lbf/in², it is convenient to have the conversion factor from dynes/cm² to lbf/in². This may be calculated to be 1 dyne/cm² = 1.45×10^{-5} lbf/in², so that the bulk modulus of aluminum is about 10×10^6 psi. Compressibilities are generally given in terms of contraction per unit volume per atmosphere. Now a pressure of 1 dyne/cm² is equal to 9.87×10^{-7} atm pressure where 1 atm = 14.7 lbf/in².

Thus for aluminum the bulk modulus can be expressed as 7×10^{11} dynes/cm² or as

$$7 \times 10^{11} \times 9.87 \times 10^{-7} = 6.9 \times 10^5 \text{ atm},$$

and the compressibility as

$$\frac{1}{6.9 \times 10^5} = 1.4_5 \times 10^{-6}/\text{atm};$$

so that, for each atmosphere increase in pressure, the volume decrease per unit volume is $1.4_5 \times 10^{-6}$.

Bulk-moduli determinations involve rather complex and rugged apparatus—apparatus that is not suitable for the elementary laboratory. We shall not go into any descriptive details of such experiments, therefore, but merely give herewith a table (Table 7-1) of a few of the more important

TABLE 7-1

BULK MODULI AND COMPRESSIBILITIES

Material	K	β
Tungsten...	370 $\times 10^{10}$ dynes/cm²	0.27×10^{-12} cm²/dyne
Platinum...	260	0.38
Iron ⎱	167	0.60
Gold ⎰		
Silver......	101	0.99
Glass......	45	2.23
Quartz.....	27	3.7
Calcium....	17.5	5.7
Ice........	8.3	12
Cesium.....	1.6	61
Water......	2.04	....
Mercury....	25.0	....
Ether......	0.5	....
Glycerine...	4.5	....
Kerosene...	1.6	....

coefficients and the units in which they are expressed.

EXAMPLE 1: How much pressure is required to reduce the volume of an iron specimen by 1 per cent?

Solution: If V is the volume of the specimen and ΔV the change in volume resulting from a pressure p, then

$$\frac{-\Delta V}{V} = \frac{1}{100}.$$

By equation (7-2) we see that

$$K = \frac{-\Delta p}{\Delta V/V}$$

or

$$\Delta p = \frac{-\Delta V}{V} K = \tfrac{1}{100}(167 \times 10^{10} \text{ dynes/cm}^2)$$

or

$$\Delta p = 167 \times 10^8 \text{ dynes/cm}^2 \text{ or approximately}$$

$$16,600 \text{ atm}.$$

7.5. Rigidity or Shear Modulus

The second kind of strain, called a *shearing strain* or *shear*, results whenever two adjoining layers of a solid slide relative to each other and in a direction parallel to their surfaces of contact. With this type of strain, no volume changes of the solid occur. The two blades of a pair of scissors, for example, shear the material between them by moving portions of it in opposite directions. Shear is also present in the twisting of a straight rod of circular cross-section, where successive circular sections move relative to one another (see § 7.6). A shearing strain may be produced by applying equal and opposite forces to the diagonally opposite edges of a specimen (Fig. 7-2), at the same

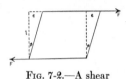

Fig. 7-2.—A shear

time preventing any rotation. A force F, then, may be thought of as distributed over the surface of the specimen. If we conceive of the specimen as made up of a set of parallel planes of molecules, each plane being attached to its neighbor by cohesive forces, then the shear results in a displacement of adjacent planes parallel to the applied forces.

From the diagram we see that an edge of the specimen whose original length was l is displaced a distance ϵ. The angular displacement of the edge is ϕ radians. We take as a measure of the strain the ratio ϵ/l. Since, in practice, ϵ is quite small, we may replace ϵ/l by the angle ϕ in radians. Then

$$\text{Strain} = \phi \,.$$

The shearing stress is the tangential force per unit area; i.e.,

$$\text{Stress} = \frac{F}{A} \,.$$

For small strains we find experimentally that Hooke's law holds:

$$\frac{\text{Stress}}{\text{Strain}} = n, \text{ the rigidity modulus} \quad (7\text{-}4)$$

or

$$n = \frac{F/A}{\phi} \,. \quad (7\text{-}5)$$

The units of n are those of a stress and, as with the

bulk modulus, may be expressed in dynes/cm² or lbf/in².

This *elastic* property of shear is not present in all materials. Only solid objects can sustain permanent shearing strains. Fluids, for example, can be distinguished from solids by the absence of *shearing strain elasticity*. This does not imply, of course, that a fluid can offer no resistance to a shearing stress.

From Fig. 7-2 it is seen that in the shearing of the cube one diagonal is lengthened while the other is shortened. Thus a shear can also be regarded as two mutually perpendicular longitudinal strains.

7.6. Torsion of a Rod

As another example of a shearing strain we shall consider in some detail the twisting of a cylindrical rod of radius R cm and length l. We shall assume that one end of the rod is securely clamped, while at the other end a torque is applied, twisting the upper circular section through an angle θ and the other sections through angles proportional to their distances from the clamped end (Fig. 7-3 [a]).

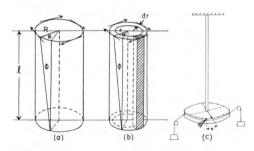

Fig. 7-3.—A shear in a cylindrical rod

The problem is to determine the relations between the angle of twist θ and the applied torque, the rigidity modulus and the dimensions of the rod.

For this purpose we must first obtain a suitable expression for the shearing strain (θ is not the strain, as will become evident). We may regard the solid rod as made up of a series of hollow cylinders or tubes, one of which, with radius $r < R$ and thickness dr, is drawn in Fig. 7-3 (b). We next consider the thin hollow cylinder divided into small segments approximating rectangular parallelepipeds by having a series of imaginary planes intersect along the axis of the cylinder. The shaded portion of Fig. 7-3 (b) represent a typical segment *prior* to being twisted. Note the similarity between Figs. 7-2 and 7-3.

The shearing strain is therefore ϕ, which is related to the angle of twist by

$$\phi = \frac{r\,\theta}{l}, \qquad (7\text{-}6)$$

$r\theta$ representing the distance that the top segment was moved relative to the bottom and l the length of the rod.

Consider, next, the force acting over the plane area of the cylindrical ring of radius r and thickness dr (Fig. 7-3 [b]). This force, distributed over the area $2\pi r dr$, gave rise to the strain computed above. From the nature of the applied twist, the resulting distributed force is everywhere normal to the radius r, and therefore the torque associated with this force, computed along the axis of the cylinder, is given by

$$dL = (2\pi r\,dr)\,(\text{stress})\,(r). \qquad (7\text{-}7)$$

The stress, on the other hand, is simply related to the strain and rigidity modulus by Hooke's law,

$$\text{Stress} = n\phi = \frac{n r\,\theta}{l}, \qquad (7\text{-}8)$$

so that equation (7-7) becomes

$$dL = \frac{2\pi n\,\theta\,r^3\,dr}{l}. \qquad (7\text{-}9)$$

To obtain the total torque of the solid rod, we add the contribution of each concentric cylindrical tube by integrating equation (7-9) from O to R,

$$L = \int^{R} \frac{2\pi n\,\theta\,r^3\,dr}{l} = \frac{\pi\,\theta nR^4}{2l}$$

or

$$\theta = \frac{2Ll}{\pi nR^4}\ \text{rad}. \qquad (7\text{-}10)$$

Equation (7-10) gives the angle of twist in radians in terms of the applied torque, L cm-dynes; the length of the rod, l cm; the radius of the rod, R cm; and the rigidity modulus, n dynes/cm². Equation (7-10) may be applied to the problem of the torsion balance and to the determination of the twist in a drive shaft that delivers power at a given angular velocity (cf. eq. [6-25] and Ex. 2).

Equation (7-10) may be written as

$$L = L_0\theta, \qquad (7\text{-}11)$$

where

$$L_0 = \frac{L}{\theta} = \left(\frac{n\pi R^4}{2l}\right)$$

is the moment of torsion of the rod. By definition, the moment of torsion of a rod is the ratio of the applied torque to the angular displacement produced. Its units are centimeter-dynes per radian. The moment of torsion is a constant for a *particular* specimen, whereas the rigidity modulus is a constant independent of the size of

the specimen. Equation (7-11) may be taken as a special formulation of Hooke's law for angular displacements. We shall use this equation in connection with vibratory motion in chapter 8.

TABLE 7-2

MODULUS OF RIGIDITY, n

Material	Dynes/Cm²	Material	Dynes/Cm²
Aluminum...	3.4×10^{11}	Platinum...	6.6×10^{11}
Brass.......	3.7×10^{11}	Glass......	2.4×10^{11}
Cast iron....	$5\text{-}8\times10^{11}$	Quartz.....	2.6×10^{11}
Steel........	8.3×10^{11}	Rocks......	$1\text{-}3\times10^{11}$

In Table 7-2 are a few values of the rigidity modulus for materials often used in these connections.

EXAMPLE 2: The solid steel drive shaft of an engine is 2.5 m long and 8 cm in diameter. While revolving at the rate of 200 rad/sec, the shaft delivers 20 hp. Compute the twist produced in this shaft.

Solution: As indicated by equation (6-25), the power delivered is equal to the product of the applied torque and angular velocity:

$$20\times746\times10^7\ \text{cm-dynes/sec} = L\ \text{cm-dynes}$$
$$\times\,200\ \text{rad/sec};$$
$$L = 746\times10^6\ \text{cm-dynes}\cdot$$
$$\theta = \frac{2Ll}{\pi nR^4} = \frac{2\times746\times10^6\times250}{\pi\times8.3\times10^{11}\times(4)^4}\ \text{rad}$$
$$= 0.00056\ \text{rad};$$
$$\theta = 0°.0321.$$

7.7. Young's Modulus or the Stretch Modulus

We consider next the stretching of a wire or solid cylinder by a force F, as shown in Fig. 7-4.

FIG. 7-4.—The stretching of a rod

Here the stress is the force F per unit area of the cross-section; i.e.,

$$\text{Stress} = \frac{F}{A}. \qquad (7\text{-}12)$$

If l_0 is the original length of the wire or cylinder and, as a result of the tensile stress, the length becomes l, then the deformation is $l - l_0$ and the longitudinal strain is

$$\text{Strain} = \frac{l - l_0}{l_0} = \frac{\Delta l}{l_0}. \qquad (7\text{-}13)$$

Young's modulus Y for the substance is defined as

$$Y = \frac{\text{Tensile stress}}{\text{Longitudinal strain}} = \frac{F/A}{\Delta l/l_0} = \frac{F}{A}\frac{l_0}{\Delta l}. \qquad (7\text{-}14)$$

The units of Y are dynes per centimeter squared. Typical values of Young's modulus are given in Table 7-3.

TABLE 7-3
YOUNG'S MODULUS, Y

Material	Dynes/Cm²	Material	Dynes/Cm²
Aluminum......	6.9×10^{11}	Silver.........	7.5×10^{11}
Brass..........	9.2×10^{11}	India rubber...	0.05×10^{11}
Phosphor bronze	12×10^{11}	Glass.........	5.8×10^{11}
Copper.........	10×10^{11}	Quartz........	5.6×10^{11}
Iron..........	20×10^{11}	Ivory.........	9×10^{11}
Steel..........	22×10^{11}	Oak wood.....	1.4×10^{11}

Rewriting equation (7-14), we obtain

$$F = \frac{Y A \Delta l}{l} \qquad (7\text{-}15)$$

or

$$F = f_0 \Delta l, \qquad (7\text{-}16)$$

where $f_0 = YA/l$ is a constant characteristic of a particular specimen of material. The constant is sometimes called the "force constant" and is numerically equal to the force required to produce unit elongation. The units of f_0 are dynes per centimeter. Equation (7-16) may be considered as another special formulation of Hooke's law. The equation brings out the linearity between applied force and the corresponding stretch.

The nearest approach to perfect elasticity in which Hooke's law holds for very wide ranges of stress in obtained when a wire, such as steel or phosphor bronze, is wound in helical form upon a spindle, making a spiral or helical "spring." On such a spring the ratio between applied force and displacement (extension or compression) produced is accurately constant over very wide limits. The deformation of the wire itself, when in this form, is beyond our present interests or scope for detailed discussion, but it consists primarily of a twisting strain of the wire.

A practical use of Young's modulus comes in the bending of beams. It is beyond the scope of this text to analyze this phenomenon, but it may be seen that if a beam is bent, one surface is extended while the other is contracted. The amount of this extension or contraction depends on Young's modulus for the material of the beam as well as the geometrical shape of the beam.

Aside from their theoretical importance in giving the physicists some idea of the forces acting between molecules in solids and liquids, the elastic moduli are extensively used by engineers in the design and construction of machinery, buildings, bridges, etc. For example, the maximum amount of power that can be transmitted along a shaft depends directly upon the shear modulus of the rod.

The stretch modulus, or Young's modulus, is used more frequently than the other two. An example will illustrate its typical use.

EXAMPLE 3: A steel rod 1 m long is to support a structure whose weight is 5,000 kgf. If the maximum allowable change in length of the rod is 2 mm, what must the cross-sectional area of the rod be?

Solution: From equation (7-14),

$$Y = \frac{F/A}{\Delta l/l} \qquad \text{or} \qquad A = \frac{Fl}{Y \Delta l}.$$

Substituting the assumed values for F, l, Y, and Δl in the appropriate units, we get

$$A = \frac{(5,000,000 \times 980) \text{ dynes} \times 100 \text{ cm}}{22 \times 10^{11} \text{ dynes/cm}^2 \times 0.2 \text{ cm}} = 11.1 \text{ cm}^2.$$

EXAMPLE 4: A steel bar 6 m long, with cross-sectional area of 25 square centimeters (cm²), stretches 2 mm under an axial load. Compute the strain, stress, and load.

Solution: With $l = 600$ cm and $\Delta l = 0.2$ cm, the strain is

$$\text{Strain} = \frac{\Delta l}{l} = \frac{0.2}{600} = \tfrac{1}{3} \times 10^{-3}.$$

With Young's modulus, $Y = 22 \times 10^{11}$ dynes/cm², the stress is

$$\text{Stress} = Y \times \text{strain}$$
$$= (22 \times 10^{11})(\tfrac{1}{3} \times 10^{-3}) = 7.3$$
$$\times 10^8 \text{ dynes/cm}^2. \qquad \textbf{121}$$

The load is

$$F = \text{Stress} \times \text{area} = (7.3 \times 10^8)(25) = 182.5$$
$$\times 10^8 \text{ dynes}.$$

EXAMPLE 5: A plank of oak wood is freely supported at its two ends and is loaded at the mid-point by a weight of 50 kgf. Calculate the depression at the mid-point due to loading if the plank is 4 m long, 30 cm wide, and 3 cm thick.

It is beyond the scope of this text to work out formulas for the bending of beams under various conditions of loading and suspension. Suffice it to say, however, that in the case of a beam the amount of bending will depend upon the shape of the cross-section and on the manner in which the ends are fastened. In the problem we have proposed, it is clear that the upper surface of the plank is compressed and the lower surface stretched by the load, while within the plank there is a surface called the *neutral surface* that is neither stretched nor compressed. In advanced treatises on elasticity it is shown that the amount of depression D of the center of a rectangular beam loaded at the mid-point is given by

$$D = \frac{Fl^3}{4\,Y\,b\,d^3},$$

where l is the distance between the freely supported ends, b the breadth, d the vertical thickness, and Y, Young's modulus for the material used. For our problem,

$$D = \frac{(50 \times 980 \times 10^3)(4 \times 10^2)^3}{(4)(1.4 \times 10^{11})(30)(3)^3} = 7.1 \text{ cm}.$$

7.8. Poisson's Ratio and the Relationships between the Various Elastic Moduli

In the discussion of the extension of a wire by a stress we have assumed that the cross-sectional area of the wire remains constant. This is not strictly correct, for a tensile stress which produces an elongation in one direction also produces a contraction in a direction at right angles. Suppose a bar of length l_0 and square cross-section of breadth b_0 is subjected to the tensile stress produced by the forces F on each end (Fig. 7-5). As a result, there is a longitudinal strain of

$$\frac{l - l_0}{l_0} = \frac{\Delta l}{l_0}$$

and a transverse strain of

$$\frac{b - b_0}{b_0} = \frac{-\Delta b}{b_0}.$$

(Explain the presence of the negative sign before Δb.)

For homogeneous isotropic substances the transverse strain $\Delta b/b$ is independent of the direction chosen at right angles to the longitudinal strain. For such substances the ratio of the transverse strain to the longitudinal strain is a con-

FIG. 7-5.—Longitudinal extension and lateral contraction

stant called "Poisson's ratio." This ratio, usually denoted by σ, is given by

$$\sigma = \frac{-\Delta b/b_0}{\Delta l/l_0}. \tag{7-17}$$

For most common substances Poisson's ratio is of the order of 0.3. The change in area of a wire subjected to a tensile stress (Fig. 7-5) is negligibly small, as may be shown by working problem 15 at the end of this chapter.

For homogeneous isotropic substances it may be shown that the various elastic moduli and Poisson's ratio are interrelated. The theory, which is usually found in more advanced texts, gives the relationship between Young's modulus Y, the shear modulus n, and the bulk modulus K as

$$Y = \frac{9nK}{3K + n}; \tag{7-18}$$

and, for these moduli and Poisson's ratio,

$$K = \frac{Y}{3(1 - 2\sigma)}; \tag{7-19}$$

and, by eliminating Y between equations (7-18) and (7-19), we have

$$\sigma = \frac{3K - 2n}{6K + 2n}. \tag{7-20}$$

It is well to remember that these are theoretical equations and that for real substances the relationships given in these equations do not exactly hold.

7.9. General Stress-Strain Diagram

In Fig. 7-6 we have drawn a typical stress-strain curve for a specimen of steel. The specimen is gradually subjected to increasing loads until rupture takes place. The corresponding strains (compression or stretch) are recorded.

The stress-strain relation is linear from O to some point P, called the "proportional" or "P" limit. It is defined as that stress at which Hooke's law is observed to fail. For manganese steel the "P" point is roughly 120,000 pounds per square inch (lbf/in² or psi). The "elastic limit" of the specimen is the point E, almost indistinguishable

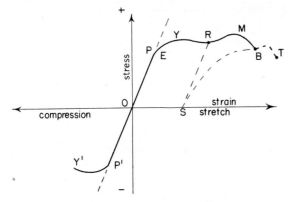

Fig. 7-6.—Typical stress-strain diagram

from "P." Since no material, when deformed, ever *exactly* returns to its original shape when the load is moved, the elastic limit, to be of practical importance, must be defined in terms of some rather marked and definite degree of failure on the part of the specimen to regain its former shape. This is frequently taken as the stress at which a specimen being stretched or compressed is permanently deformed by 10^{-5} of its length, i.e., by one-thousandth of 1 per cent. These two points, "P" and "elastic" limit, do not, in general, coincide.

After passing the proportional point, the curve shows a pronounced "kink" and is followed by a large increase in strain for constant or slowly increasing stress. This point, Y on the curve, is called the *yield point*. The material is said to "flow" at the yield point. The point M is the maximum stress to which the specimen is subjected.

It is found experimentally that ductile materials (like steel or copper, for example, but not cast iron or concrete) exhibit local weaknesses that give rise to local constrictions. This occurs just beyond the point M. The constriction quickly increases, and

at the point B the rod ruptures. The maximum stress is called the "rupture" or "breaking stress." The curve $OP'Y'$ represents the stress-strain relationship for compression.

If at some point R the stress is removed, we find that the strain is not zero; but there results, instead, a *permanent set* of magnitude OS. A completely new stress-strain curve is followed by the specimen SBT shown in Fig. 7-6.

The region in which we are principally interested is that in which Hooke's law is applicable, that is, that below the proportional limit P or elastic limit E. For such a region the ratio of stress to strain is a constant for a given substance.

7.10. Work Done on and by Elastic Forces

We have shown that the relation between applied force F and the resulting elongation Δl or S of a wire or spring is given by equation (7-16) as

$$F = f_0 S .$$

This is an equation of a straight line, which we plot in Fig. 7-7.

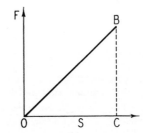

Fig. 7-7.—Work done in stretching a wire

From our definition of work in chapter 4 we see that the graph of Fig. 7-7 can also be interpreted as a work diagram. The work done by the applied force F in elongating the wire a distance S is the area under the graph OBC, from geometry, one-half the base ($OC = S$) times the height ($BC = F$); i.e.,

$$W = \tfrac{1}{2}FS , \qquad (7\text{-}21)$$

which states that the work done by a force F in stretching an elastic wire a distance S is equal to one-half the product of the force by the distance change. If we wish to eliminate the force F from equation (7-21) and introduce, instead, the force constant of the wire, we may do so by equation (7-16):

$$W = \tfrac{1}{2} f_0 S^2 . \qquad (7\text{-}22)$$

In terms of the calculus, suppose a force F is required to produce elongation of s, then, from equation (7-16), we have

$$F = f_0 s \; ;$$

and the work done in producing an infinitesimal elongation ds is

$$dW = F \, ds = f_0 s \, ds \; .$$

Notice that this assumes that the force F is constant in the infinitesimal elongation of ds. Thus the total work done in producing a finite elongation of S is

$$W = \int_0^S f_0 s \, ds = \tfrac{1}{2} f_0 S^2 \; .$$

In all actual cases Hooke's law is never exactly obeyed. A slightly larger amount of work is done in deforming the material than is returned by it when it relaxes.

7.11. Résumé

The following is a brief listing of the definitions, principles, and theories, given in this chapter, which you should know:

Definitions of stress, strain, and modulus of elasticity
Hooke's law and its limitations—elastic limit
The bulk, shear, and Young's moduli of elasticity
Poisson's ratio
Work done by elastic forces

Queries

1. If a wire is stretched so that Hooke's law is no longer applicable, does it necessarily follow that the wire has been stretched beyond its elastic limit? Explain.
2. The work done in stretching a wire is given by equation (7-22). What is the work done in twisting a rod through an angle θ by an applied torque L?
3. Assuming the conservation-of-energy principle to be applicable, what is the maximum angular velocity that the rod of query 2 will attain when released?
4. Show by dimensional analysis that equation (7-19) is dimensionally correct.

Practice Problems

1. How much force is necessary to stretch a steel wire 2 mm² in cross-sectional area and 2 m long a distance of 0.5 mm? How much work is done? Ans.: 11×10^6 dynes; 2.75×10^5 ergs.
2. A steel rod is rigidly clamped at one end. At the other a disk 10 cm in radius is attached. A bearing at the disk end supports the rod. The rod is 1 m long and 5 mm in diameter. What mass must be hung from the circumference of the disk to twist the rod 90°?

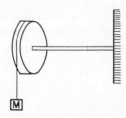

3. What is the force constant, f_0, of a brass rod 2 m long and 4 mm in diameter? Ans.: 5.78×10^8 dynes/cm.
4. What is the rigidity modulus of a rod 2 m long and 4 mm in diameter if the moment of torsion, L_0, of the rod is 5×10^6 cm-dynes/rad?

5. To the opposite faces of a cube of rubber 10 cm on an edge, parallel and opposite forces of 1.5 kgf each are applied. If the relative displacement of the faces is 2 mm, calculate the strain, the stress, and the shear modulus. Ans.: 0.02; 1.47×10^4 dynes/cm²; 7.35×10^5 dynes/cm².
6. Two balls each of mass 1 kg are attached to the ends of a steel rod 1 m long and 2 cm² in cross-sectional area. If the system is spun about a vertical axis through the center of mass and with an angular velocity of 360 rad/sec, calculate the increase in length of the rod.

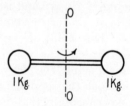

7. In the high-pressure apparatus of Professor Bridgman, pressures as high as 20,000 atmospheres (atm) may be obtained. What change of volume will a cube of quartz, 1 cm on a side, undergo if subjected

to this hydrostatic pressure? (One atm equals approximately 10^6 dynes/cm².) Ans.: 0.074 cm³.

8. A phosphor bronze rod, 2 ft long and a cross-sectional area 14 in², stretches $\frac{1}{10}$ in under an axial load. What are the stress, strain, and load?

9. A 100-gm mass stretches a spring $\frac{1}{10}$ cm. What is the force constant? If the spring is stretched 8 cm, how much potential energy is stored in the spring? Ans.: $f_0 = 9.8 \times 10^5$ dynes/cm; PE $= 3.14 \times 10^7$ ergs.

10. Calculate the twist produced in a solid steel shaft if the power transmitted is 50 hp when the shaft is revolving at 100 rev/sec. Take the length of the shaft to be 3 m and the diameter 7 cm.

11. If we assume the pressure at a point 700 miles in the interior of the earth to be roughly 3,000,000 lbf/in², what would the final volume be of a 1-m cube of quartz taken from the interior and brought to the surface, assuming Hooke's law to hold? Ans.: 1.77 m³.

12. In the design of a steel drive shaft that must deliver maximum power of 30 hp at an angular speed of 250 rad/sec, it is desired that the angle of twist should not exceed 3°. If the shaft must fit a hole 5 cm in diameter, what must the length of the drive shaft be?

13. Prove that equation (7-10) and the equation given in Example 5 are dimensionally correct.

14. Compute the bulk modulus of tungsten and mercury in the mks system of units of newtons/m².

15. A rod of copper has a length of 100 cm and a square cross-section of breadth 1 mm. It is rigidly held at its upper end, and a load of 10 kg is applied to the lower end. Find (a) the increase in length of the rod and (b) the decrease in breadth if Poisson's ratio for this substance is 0.3. Ans.: (a) 0.098 cm; (b) 2.94×10^{-5} cm.

16. A block of iron having a volume of 2 in³ is subjected to a pressure of 1,500 lb/in². Find the resultant decrease in volume.

17. From equation (7-20) find Poisson's ratio for iron and calculate Young's modulus for iron from its bulk modulus and its shear modulus.

18. Find the bulk modulus of water in lb/in² from Table 7-1, and from this find the weight-density of water at a pressure of 4,000 lb/in² if the weight-density of water at normal atmospheric pressure is taken as 64 lb/ft³.

19. If the elastic limit of brass is 60,000 lbf/in², find the maximum load which a rod of brass, 1 ft in length and of radius 0.1 in, can support without exceeding this limit. Find the resulting elongation of the rod, its decrease in radius, and its change in volume, assuming a value for Poisson's ratio of 0.3. Ans.: 1,885 lbf; 0.054 in; 1.35×10^{-4} in; 6.8×10^{-4} in³.

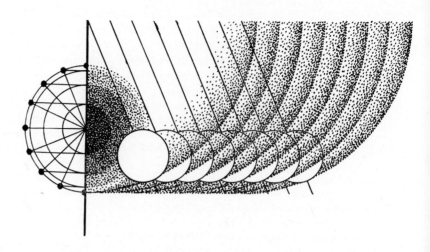

CHAPTER 8

Simple Harmonic Motion

8.1. Review of Kinematics

Each of the types of motion we have previously studied has been defined in terms of the acceleration and the path of the particle. First, the simplest motion has an acceleration of zero, a straight-line path, and a constant velocity:

$$s = vt .$$

Second, we studied motion in a straight line with a constant acceleration parallel to the initial velocity (if any):

$$v = v_0 + at ,$$
$$s = v_0 t + \tfrac{1}{2} a t^2$$
$$v^2 - v_0^2 = 2 a s .$$

Third, on the subject of projectiles we studied motion resulting from a constant acceleration *not* parallel to the initial velocity.

Fourth, when the acceleration is constant in magnitude but not in direction and is constantly normal to the velocity of the particle, the resulting motion was found to be uniform motion in a circular path:

$$a_c = \frac{v^2}{r} = \frac{4\pi^2 r}{T^2} = \omega v = 4\pi^2 n^2 r = \omega^2 r .$$

8.2. Motion with Acceleration Not Constant

We note that all the motions listed here have accelerations constant in magnitude. Consequent-
ly, we might group all other motions together as "motion with the acceleration not constant." However, a varying acceleration can change in such a variety of ways that equations covering all sorts of variable accelerations would be too general to be useful.

If we write $a = f(s, t)$, this means that the acceleration depends upon position and upon the time; but, until we know the nature of that dependence, such an expression is not very useful. A more practicable way to proceed would be to examine the sorts of motion that would result from a dependence of the acceleration either upon position or upon time, independently. Then, since we are interested in physics rather than in mathematics, we must further inquire whether any such motion actually occurs in nature. We can, of course, define and study motions which never occur in nature, but we would not consider such studies particularly "useful" to physics.

8.3. Simple Harmonic Motion

A commonly defined and practically very important type of motion has its acceleration given as a function of position and is called *simple harmonic motion*.

Linear simple harmonic motion is motion in a straight line with the acceleration proportional to the distance from a fixed point in that line and directed toward that fixed point.

This may be expressed analytically as

$$a = - kx$$

where x is the distance from a fixed point and k is a positive constant of proportionality or the acceleration per unit distance.

In order to make this type of motion more familiar, we shall consider the case of a mass vibrating on the end of a vertical spring (Fig. 8-1). If the mass is pulled down to position B a distance OB below the equilibrium position O, then, by

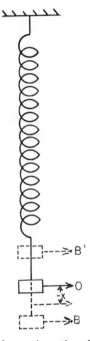

Fig. 8-1.—Simple harmonic motion of a spring

Hooke's law, there is an upward force exerted on the mass which is proportional to the displacement OB. As a result, the mass m is accelerated upward. At the equilibrium position O, the restoring force and acceleration are zero, but the velocity and kinetic energy of the mass m have their maximum values. If there is no friction, as is assumed in simple harmonic motion, then the mass moves upward to a position B', where $OB' = OB$ is called the maximum displacement or *amplitude* of the motion.

Suppose, now, that the mass is at some intermediate distance, $-x$, below the equilibrium position, then at this instantaneous position the mass has both kinetic and potential energy and an upward restoring force acting on it of

$$F_x = -f_0 x ,$$

where f_0 is the force constant of the spring or the force exerted by the spring for a unit displacement. The negative sign appears because the force is

upward when the displacement is downward. The upward positive acceleration a for the negative displacement, $-x$, is, by Newton's Second Law,

$$a = \frac{F_x}{m} = \frac{-f_0 x}{m} = -kx , \qquad (8\text{-}1)$$

where k is the constant f_0/m which depends only on the particular spring and the mass moved.

The linear simple harmonic motion defined by equation (8-1) is an example of a periodic motion, for which we must now define some terms. The *amplitude* of the vibrations is the value of the maximum displacement $OB = OB'$ in Fig. 8-1. The *period* T of the vibrations is the time for one complete vibration or the time for the mass m to move from, say, B to B' and back to B. The number of complete vibrations per second is called the *frequency* n of the motion. From these definitions it follows that the period must be the inverse of the frequency, or

$$T = \frac{1}{n} \qquad \text{or} \qquad n = \frac{1}{T}. \qquad (8\text{-}2)$$

If the period T is $\frac{1}{10}$ sec, then there are 10 complete vibrations per second, or the frequency n is 10 per second.

We now wish to determine the period of the simple harmonic motion in terms of k or f_0 and m, as well as to find the displacement at any time t or displacement x. This really involves the solution of a differential equation, and, in order to avoid this at the present, we shall resort to an analytical trick of considering the motion, projected on a diameter, of a particle moving with constant speed in a circle. This projected motion is simple harmonic.

8.4. Analysis of Simple Harmonic Motion

Consider a particle moving with constant speed v in a circle of radius r (Fig. 8-2). If T is the time for the particle to go once around the circle, then T is the period of the motion and $v = 2\pi r/T = 2\pi rn$, where n is the frequency of the motion. At any point P in its path, the particle is subjected to a centripetal acceleration along PO, variously expressed in § 1.24 as

$$a_c = \frac{v^2}{r} = \omega^2 r = 4\pi^2 n^2 r ,$$

where ω is the constant angular velocity of the particle moving in a circle and is equal to v/r or to $2\pi n$ or $2\pi/T$.

Let us now project this circular motion along a horizontal diameter BB' whose center O is the origin of the X-coordinate axis. Displacements to the right of O are considered to be positive, while those to the left of O are negative. If x is the coordinate of the point P, then from Fig. 8-2 it is seen that

$$x = r \cos \theta .$$

The acceleration of the particle at P is $4\pi^2 n^2 r$ along PO, and the projection of this acceleration along

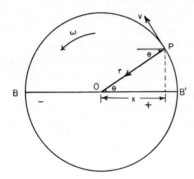

FIG. 8-2.—The projection of uniform circular motion on a diameter gives simple harmonic motion.

the diameter gives the acceleration a_x along the X-axis as

$$a_x = -4\pi^2 n^2 r \cos \theta = -4\pi^2 n^2 x = -\omega^2 x . \quad (8\text{-}3)$$

The negative sign appears because the acceleration is inward or negative for an outward or positive displacement x. Thus the projected acceleration is proportional to the displacement and in the opposite direction to the displacement. Hence, by definition, the projected motion of the particle P along the diameter BB' is simple harmonic motion. From equation (8-3) it follows that the frequency of the simple harmonic motion is given by

$$n = \frac{1}{2\pi} \sqrt{\frac{-a_x}{x}}. \quad (8\text{-}4)$$

Comparing equations (8-1) and (8-3), we see that the ratio of the acceleration $-a$ to the displacement x is f_0/m or k, so that the frequency of the simple harmonic motion given by equation (8-1) is

$$n = \frac{1}{2\pi} \sqrt{\frac{f_0}{m}} = \frac{1}{2\pi} \sqrt{k} , \quad (8\text{-}5)$$

and the period T of this motion is

$$T = \frac{1}{n} = 2\pi \sqrt{\frac{m}{f_0}} = 2\pi \sqrt{\frac{1}{k}}. \quad (8\text{-}6)$$

EXAMPLE 1: A 200-gm weight hung on the end of a vertical spring extends it a distance of 5 cm. This weight is removed and replaced by a 500-gm weight, which is then pulled down below its equilibrium position and released. Find the frequency of the resulting simple harmonic motion.

Solution: The force constant f_0 is $(200 \times 980)/5$ dynes/cm, and the mass moved is 500 gm. (It should be noted that, for consistent dimensions in eqs. [8-5] and [8-6], f_0 must be in dynes per centimeter or newtons per meter if m is in grams or kilograms.) From equation (8-5) the frequency n of the motion is

$$n = \frac{1}{2\pi} \sqrt{\frac{200 \times 980}{5 \times 500}} = 1.41 \text{ vibrations per second} .$$

EXAMPLE 2: A 6-lb block of wood is floating on water, and it is found that a downward push of 0.5 lbf on the top of the block submerges it a further distance of 1.0 in. (This further distance of submergence is found to be proportional to the downward force up to the point where the block is completely submerged.) When the block is pushed down and released, it vibrates up and down with simple harmonic motion. Find the period of this motion.

Solution: In this case f_0 is (0.5×12)lbf/ft, and the vibrating mass is 6/32 slugs. From equation (8-6) the period T is given by

$$T = 2\pi \sqrt{\frac{m}{f_0}}$$

$$= 2\pi \sqrt{\frac{6 \text{ lbf}}{32 \text{ ft/sec}^2 \times 6 \text{ lbf/ft}}}$$

$$= 1.11 \text{ sec} .$$

From Fig. 8-2 let us determine the displacement, velocity, and acceleration at any time t and position x. When a particle is executing simple harmonic motion, its velocity is zero when the displacement is a maximum, and we shall assume that the particle is started out at time zero from its position of maximum positive displacement, position B' in Figs. 8-1 and 8-2. At a time t later, the particle in Fig. 8-2 is at P or has turned through the phase angle θ, where

$$\theta = \omega t \quad \text{or} \quad \omega = \frac{\theta}{t}.$$

The displacement x along the diameter BB' is

$$x = r \cos \theta = r \cos \omega t . \quad (8\text{-}7)$$

At time t equal to zero, the displacement x is a maximum equal to the amplitude r, the radius of the reference circle. The velocity v_x of the particle along the diameter BB' at time t is the projection of the tangential velocity v along a horizontal line and is given by the following expressions:

$$
\begin{aligned}
v_x &= -v\cos(90-\theta) = -v\sin\theta \\
&= \frac{-v\sqrt{r^2-x^2}}{r} = -\omega\sqrt{r^2-x^2} \\
&\qquad = -2\pi n\sqrt{r^2-x^2}, \\
&= -v\sin\omega t,
\end{aligned}
\right\} \quad (8\text{-}8)
$$

since $v = 2\pi rn$, $\theta = \omega t$, and $\omega = 2\pi n$. Since at time zero the displacement x is equal to r, it follows that the velocity at time zero is zero and has a maximum value when the displacement x is zero.

The acceleration of the particle along BB' is given by equation (8-3) as $a = -\omega^2 x$, and from equation (8-7) this may be written as

$$a = -\omega^2 r \cos \omega t. \qquad (8\text{-}9)$$

These results are summarized below for a particle undergoing simple harmonic motion with amplitude r such that its displacement x at time t is given by

$$x = r\cos\theta = r\cos\omega t.$$

The velocity v_x is

$$v_x = -v\sin\omega t = -2\pi n\sqrt{r^2-x^2}.$$

The acceleration $a = -\omega^2 r\cos\omega t = -\omega^2 x$. The displacement, velocity, and acceleration at any time t are shown in Fig. 8-3.

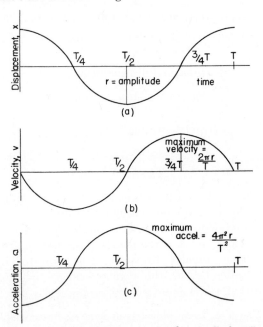

(a)

(b)

(c)

Fig. 8-3.—Graphical representation of (a) the displacement, (b) the velocity, and (c) the acceleration of a particle undergoing simple harmonic motion.

8.5. Characteristics of Simple Harmonic Motion

Because of its widespread use in physics, it is well worth our time to emphasize the important characteristics of simple harmonic motion.

1. When the displacement is at a positive maximum, the acceleration is a negative maximum and points toward the origin, while the velocity is zero.

2. When the displacement is zero, the acceleration is zero, and the velocity is a maximum equal to $2\pi rn$.

3. The period of motion is independent of the amplitude of motion.

4. The total energy of a particle executing simple harmonic motion is proportional to the square of the amplitude of motion. (See § 8.9 for a fuller discussion.)

8.6. Graphical Representation of Simple Harmonic Motion

A convenient way of representing the dependence of a, v, and x on t is to plot these quantities as ordinates against time as an abscissa, as shown in Fig. 8-3. The curve of Fig. 8-3 (a) represents $x = r\cos\omega t$; the amplitude of the motion is indicated. Fig. 8-3 (b) represents the variation of velocity with time ($v = -2\pi rn\sin\omega t$), while Fig. 8-3 (c) represents the variation of acceleration with time ($a = -4\pi^2 n^2 r\cos\omega t$).

We wish particularly to emphasize, first, that linear simple harmonic motion is a periodic motion along a straight line and, as such, is difficult to represent by means of a diagram. Second, the displacement curve, Fig. 8-3, *is not the path of motion of any particle*. The curves shown in the figure are merely graphical representations of equations (8-7)–(8-10) *and, as such, reveal the same information about simple harmonic motion as do the equations*. For example, viewing the three curves together, we observe that when the displacement is a maximum, the acceleration is also a maximum, but oppositely directed, and the velocity is zero. The maximum values that a, x, and v may attain are shown on the curves; their intermediate values may be obtained by actual measurement.

8.7. The Analysis of Simple Harmonic Motion by the Calculus

Simple harmonic motion has been defined as that motion in which the acceleration is propor- **129** tional to the displacement and in the opposite direction, as given in equations (8-1). Since we are assuming that the motion is along the X-axis, the velocity is given by dx/dt, and the acceleration by d^2x/dt^2, so that the equation for simple harmonic motion can be written as

$$\frac{d^2x}{dt^2} = \frac{-f_0 x}{m} = -kx. \qquad (8\text{-}10)$$

This is a differential equation involving the displacement x and the time t. From various lines of evidence we can reasonably assume that the displacement is a sine or cosine function of the time Let us therefore assume a solution of the type

$$x = A \cos(\omega t + a), \qquad (8\text{-}11)$$

where A and a are constants to be determined from the initial conditions of the problem. In the previous analysis of this problem we assumed that initially, or at $t = 0$, the body started from rest with a maximum displacement of r.

If at $t = 0$, $x = r$, then from equation (8-11) we have $r = A \cos a$. Also at $t = 0$ the velocity $dx/dt = 0$. Now, by differentiation of equation (8-11), we have

$$\frac{dx}{dt} = -A\omega \sin(\omega t + a).$$

Hence it follows that

$$0 = -A\omega \sin a,$$

or $\sin a = 0$ and the angle $a = 0$. Thus for our particular problem, the solution which we assume is

$$x = r \cos \omega t, \qquad (8\text{-}12)$$

so that the velocity or rate of change of distance with time is

$$\frac{dx}{dt} = -r\omega \sin \omega t, \qquad (8\text{-}13)$$

and the acceleration, which is the rate of change of velocity with time, is

$$\frac{d}{dt}\left(\frac{dx}{dt}\right) = \frac{d^2x}{dt^2} = -r\omega^2 \cos \omega t = -\omega^2 x. \qquad (8\text{-}14)$$

Thus we see that $x = r \cos \omega t$ is a solution of the differential equation (8-10) if

$$\omega^2 = \frac{f_0}{m} = k \quad \text{or} \quad \omega = \sqrt{\frac{f_0}{m}} = \sqrt{k}.$$

Since the angular velocity ω is equal to $2\pi/T$ or to $2\pi n$, where T is the period and n is the frequency of the motion, it follows that

$$n = \frac{1}{2\pi}\sqrt{\frac{f_0}{m}} = \frac{1}{2\pi}\sqrt{k},$$

as given in equation (8-5), and the velocity at any time t and displacement x is obtained from equations (8-12) and (8-13) as

$$v = \frac{dx}{dt} = -r\omega\sqrt{1 - \cos^2 \omega t} = -2\pi n\sqrt{r^2 - x^2},$$

which is given in equation (8-8).

8.8. Angular Simple Harmonic Motion

The periodic motion of the balance wheel of a watch is an example of *angular simple harmonic motion*. Here the object oscillates periodically through some angle of ϕ radians.

Angular simple harmonic motion is rotation about an axis with an angular acceleration which is proportional to the angular displacement from a given position but oppositely directed. Analytically,

$$a = -b\phi. \qquad (8\text{-}15)$$

If we look along the axis of oscillation and observe a clockwise displacement of ϕ radians, the angular acceleration would be such as to produce a counterclockwise rotation. We may obtain the equations giving the value of the acceleration, displacement, and velocity from equations (8-7)–(8-9) by substituting from Table 6-2. Thus

$$\left.\begin{aligned}
\phi &= \phi_0 \cos \omega t \ (\text{angular displacement}), \\
\Omega &= -\frac{2\pi\phi_0}{T} \sin \omega t \ (\text{angular velocity}), \\
a &= -\frac{4\pi^2}{T^2}\phi_0 \cos \omega t \ (\text{angular acceleration}), \\
T &= 2\pi\sqrt{-\frac{\phi}{a}} = 2\pi\sqrt{\frac{1}{b}} \ (\text{period}),
\end{aligned}\right\} (8\text{-}16)$$

where ϕ is the angular displacement; Ω is the angular velocity of the oscillating object and not to be confused with $\omega = 2\pi/T$; a is the angular acceleration; and T is the period of oscillation. The maximum angular displacement is indicated by ϕ_0. Notice that Ω, the angular velocity, is $d\phi/dt$ and that a, the angular acceleration, is $d^2\phi/dt^2$.

EXAMPLE 3: An object is undergoing linear simple harmonic motion with a period of 2π sec and an amplitude of 10 cm. Compute the position, velocity, and acceleration 10 sec after the motion has started.

Solution: From equation (8-7)

$$x = r \cos \omega t ,$$

$$x = 10 \cos \left(\frac{2\pi}{2\pi} \times 10 \right)$$

$$= 10 \cos (10 \text{ rad}) = 10 \cos 573°$$

$$= -10 \cos 33° = -8.39 \text{ cm} ;$$

$$v = -\frac{2\pi r}{T} \sin \frac{2\pi t}{T} = -r\omega \sin \omega t$$

$$= 10 \sin 33° = 5.45 \text{ cm/sec} ;$$

$$a = -\frac{4\pi^2 r}{T^2} \cos \omega t = -\omega^2 r \cos \omega t$$

$$= 10 \cos 213°$$

$$= 8.39 \text{ cm/sec}^2 .$$

Notice we have assumed that the object started from its maximum displacement.

8.9. Energy Relations in Simple Harmonic Motion

The spiral spring, as a typical example of simple harmonic motion, offers an excellent opportunity to study the energy changes that take place in a system undergoing simple harmonic motion.

At any intermediate position, x, of the stretched spring, its potential energy is given by equation (7-22) as

$$PE = \tfrac{1}{2} f_0 x^2 .$$

From equation (8-6) we have

$$T = 2\pi \sqrt{\frac{m}{f_0}} \quad \text{or} \quad f_0 = \frac{4\pi^2 m}{T^2}, \quad (8\text{-}17)$$

so that the potential energy may be written as

$$PE = \frac{1}{2} \left[\frac{4\pi^2 m}{T^2} \right] x^2 = \frac{m}{2} \omega^2 x^2, \quad (8\text{-}18)$$

since $\omega = 2\pi / T$.

From equation (8-7) we have $x = r \cos \omega t$; hence we may write

$$PE = \frac{m}{2} \omega^2 r^2 \cos^2 \omega t, \quad (8\text{-}19)$$

where r is the amplitude of the motion.

The kinetic energy is $mv_x^2/2$, and at any time t the velocity v_x is, by equation (8-8),

$$v_x = -r\omega \sin \omega t ,$$

so that the kinetic energy at any time t is

$$\left. \begin{aligned} KE &= \tfrac{1}{2} m r^2 \omega^2 \sin^2 \omega t \\ &= \tfrac{1}{2} m\omega^2 (r^2 - x^2) . \end{aligned} \right\} \quad (8\text{-}20)$$

The total energy of the system at any time is the sum of the potential and kinetic energies at that time, so that

$$\left. \begin{aligned} \text{Total energy} = PE + KE &= \tfrac{1}{2} m r^2 \omega^2 \cos^2 \omega t \\ &+ \tfrac{1}{2} m r^2 \omega^2 \sin^2 \omega t \\ = \tfrac{1}{2} m r^2 \omega^2 &= \frac{2\pi^2 m r^2}{T^2} . \end{aligned} \right\} (8\text{-}21)$$

Thus the total energy is a constant. At positions of maximum amplitude, the energy is all potential, while its energy is all kinetic when passing the equilibrium position. For other positions the energy is partly potential and partly kinetic, but the sum of these is a constant throughout the motion. It should be noticed that we have assumed that there are no frictional or dissipative forces present, so that the motion, once started, will continue indefinitely with the same amplitude. This is, of course, an idealized situation and not the case for any practical example. If frictional forces are present which are proportional to the velocity—a good assumption for small velocities—then the problem can be analyzed by the calculus. This shows that if the frictional force is large relative to the restoring force, then, following an initial displacement, the object returns slowly to its equilibrium position. This overdamped motion occurs when an oil dash-pot is used to damp the motion, and it is also used in galvanometers and ordinary electrical meters to prevent the pointer from oscillating. In this latter case the damping or resistance force is due to electromagnetic action. On the other hand, if the damping is small, then the object oscillates about the equilibrium position with decreasing amplitudes and comes to rest after several oscillations. The dividing line between this damped oscillatory motion and the overdamped motion is the so-called "critically damped motion," in which the displaced object comes to the equilibrium position in a minimum time (Fig. 8-4).

When frictional forces are present, work is done against these, and the original kinetic and potential energies are gradually dissipated into heat energy. This is what occurs in any practical situation, for we are aware that vibrating springs or **131** pendulums soon come to rest unless some extraneous source of energy is added to the system.

EXAMPLE 4: In Example 1 the maximum displacement is 4 cm. Calculate the kinetic energy of the mass as it passes its equilibrium position and the potential and kinetic energies for a displacement of 2 cm.

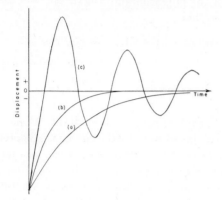

FIG. 8-4.—Damped harmonic motion: (*a*) overdamped motion; (*b*) critically damped motion; (*c*) damped oscillatory motion.

Solution: At the equilibrium position the oscillating object has its maximum velocity, which is given by equation (8-12) as $r\omega$, since the sin ωt has a maximum value of unity. In the problem of the simple harmonic

motion of a spring the value of $k = \omega^2 = f_0/m$. In the example $m = 500$ gm and $f_0 = 40$ gmf/cm $= 40 \times 980$ dynes/cm. Hence

$$\omega^2 = \frac{f_0}{m} = \frac{40 \text{ gm/cm} \times 980 \text{ cm/sec}^2}{500 \text{ gm}}$$

$$= 78.4 \text{ rad}^2/\text{sec}^2 ,$$

and the maximum velocity is

$$v_{\max} = r\omega = 4 \times 8.86 = 35.44 \text{ cm/sec}$$

The maximum kinetic energy is

$$\tfrac{1}{2} m v_{\max}^2 = \tfrac{1}{2} \times 500 \times (35.44)^2 = 314,000 \text{ ergs} .$$

The potential energy for a displacement x is given by $f_0 x^2/2$ and, since in this example $f_0 = 39,200$ dynes/cm and $x = 2$ cm, the PE $= 78,400$ dynes $\times$ cm or ergs.

From equation (8-8) the velocity v is equal to $-\omega\sqrt{r^2 - x^2}$, and for $r = 4$ cm, $x = 2$ cm, and $\omega = 8.86$ rad/sec the velocity at a displacement of 2 cm is $8.86 \times \sqrt{12} = 30.7$ cm/sec.

The corresponding kinetic energy is $500 \times (30.7)^2/2 = 235,600$ ergs. The sum of the PE and the KE at the displacement of 2 cm is $(78,400 + 235,600) = 314,000$ ergs. This is the same value as the maximum kinetic energy, which it should be, since we have assumed no frictional forces so that the energy is constant throughout the motion.

Physical Examples of Simple Harmonic Motion

In the examples that follow, the reader should be on the alert for the assumptions made in order that the periodic motion of the physical apparatus may be considered to be simple harmonic motion.

8.10. The Spiral Spring Oscillator

In § 8.3 we showed that if a mass m hanging on the end of a spring is displaced a distance x, then there is a force exerted on the mass m of $-f_0 x$, where f_0 is the force constant of the spring. From equation (8-1) we have, for the acceleration,

$$a = -\frac{f_0}{m} x ;$$

since f_0 is constant if the elastic limit of the spring is not exceeded and m is a constant, it follows that the motion of the mass m is simple harmonic, with a period of

$$T = 2\pi\sqrt{\frac{m}{f_0}} .$$

It is important to note that neither the displacement x nor the amplitude of the motion enters into this equation.

8.11. Period of a Simple Pendulum

By definition, a simple pendulum consists of a point mass at the end of an inextensible string of negligible mass. This is, of course, an ideal which can never be perfectly realized, but experimentally a small heavy ball hanging from a light string is a very good approximation to the ideal.

Let the pendulum be pulled to one side, making a small angle, θ, with the vertical, and then released (Fig. 8-5). What kind of motion does the ball or pendulum bob undergo?

The forces acting on the bob are mg downward and the tension of the string, T. The resultant force is tangent to the arc of the circle, x. This resultant force, f, is equal to $mg \sin \theta$. The acceleration at this instant is, by Newton's Second Law,

$$- m g \sin \theta = m a$$

or

$$a = - g \sin \theta . \qquad (8\text{-}22)$$

Let us suppose that θ is a small angle, less than $10°$. Then we may replace $\sin \theta$ by θ measured in radians. Also, $l\theta = x$, where l is the length of the string, so that equation (8-22) becomes

$$a = - g \frac{x}{l} . \qquad (8\text{-}23)$$

For these small angles the arc and chord are approximately equal. By comparing equations (8-23) and (8-1) we see that such a simple pendulum

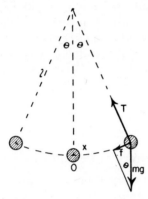

Fig. 8-5.—The simple pendulum

satisfies the condition for simple harmonic motion. By equation (8-5) the period is, with $k = g/l$.

$$T = 2\pi \sqrt{\frac{l}{g}} . \qquad (8\text{-}24)$$

From equation (8-24) we see that the *period is independent of the mass of the bob and the amplitude of the swing (provided that θ is small)*. The velocity and displacement at any time, t, are given by equations (8-8) and (8-7).

If θ is large, we may no longer tolerate the equality of chord and arc; the motion then is not simple harmonic motion, and equation (8-24) is not true. Rigorous calculation for the period of the simple pendulum gives

$$T = 2\pi \sqrt{\frac{l}{g}} \Big\{ 1 + \frac{\sin^2 (\theta_0/2)}{4} \\ + 9 \frac{\sin^4 (\theta_0/2)}{64} + \ldots \Big\} . \qquad (8\text{-}25)$$

Here θ_0 is the amplitude of motion. The reader may verify that for $\theta = 7°$ the percentage error made in using equation (8-24) is about 0.1 per cent.

8.12. Period of a Physical Pendulum

A rigid object oscillating about a fixed horizontal axis under the action of its own weight is called a "physical" or "compound" pendulum. In Fig. 8-6 the horizontal axis passes through P. The

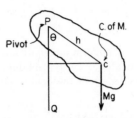

Fig. 8-6.—A physical pendulum

center of mass, c, is at a distance h from P. If the pendulum is pulled to one side and released, it will oscillate about an axis through P and perpendicular to PQ. The weight, acting through the center of mass, together with the reaction of the support, constitutes a torque of magnitude

$$L = - Mgh \sin \theta . \qquad (8\text{-}26)$$

Here $h \sin \theta$ is the lever arm. The negative sign is introduced, since, at the instant shown in the figure, θ is measured counterclockwise and L clockwise. From equation (6-18), $L = Ia$, it follows that this torque will give rise to an instantaneous acceleration about P of value

$$a = - \frac{Mgh \sin \theta}{I} , \qquad (8\text{-}27)$$

where I is the moment of inertia of the body about the axis of rotation. If θ is small, less than $10°$, then $\sin \theta \to \theta$ in radians, and we have

$$a = - \frac{Mgh}{I} \theta . \qquad (8\text{-}28)$$

Comparing equations (8-28) and (8-15), we see that the motion is angular simple harmonic motion. The period is given by equation (8-16):

$$T = 2\pi \sqrt{\frac{I}{Mgh}} . \qquad (8\text{-}29)$$

In equation (8-29) M is the mass of the object, h the distance from P to the center of mass, and I the moment of inertia of the object about the axis of rotation. The period is independent of the amplitude and the magnitude of the mass (the reader should prove this).

A simple pendulum is said to be *equivalent* to a physical pendulum when each has the same period. **133**

Thus if l_e is the length of the equivalent simple pendulum, then

$$2\pi\sqrt{\frac{l_e}{g}} = 2\pi\sqrt{\frac{I}{Mgh}} \qquad (8\text{-}30)$$

or

$$l_e = \frac{I}{Mh}. \qquad (8\text{-}31)$$

EXAMPLE 5: A uniform meter stick suspended at the 10-cm mark oscillates as a pendulum. Calculate the period and the length of the equivalent simple pendulum.

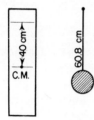

Solution: For the meter stick, $h = 40$ cm. Also, $I = I_{\text{cm}} + Mh^2$. Hence

$$I = \tfrac{1}{12}M(100)^2 + M(40)^2 = \frac{M}{12}(29{,}200) \text{ gm-cm}^2 \ ,$$

and

$$T = 2\pi\sqrt{\frac{\cancel{M}29{,}200}{\cancel{M}g(40)(12)}} = 1.57 \text{ sec} \ ,$$

$$l_e = \frac{\cancel{M}29{,}200}{\cancel{M}(40)(12)} = 60.8 \text{ cm} \ .$$

8.13. Period of a Torsion Pendulum

A torsion pendulum consists of an object attached to the end of an elastic rod so that, when the object is rotated about the line of the rod as an axis, the rigidity of the rod tends to rotate the object back toward its position of rest (Fig. 8-7).

FIG. 8-7.—The torsion pendulum

The magnitude of the restoring torque is, by equation (7-11),

$$L = -L_0\theta \ ,$$

134 where L_0 is the moment of torsion measured in

torque per radian. This torque gives the object an angular acceleration of amount

$$a = -\frac{L_0\theta}{I}. \qquad (8\text{-}32)$$

Equation (8-32) is of the form of equation (8-1), and the ensuing motion is angular simple harmonic motion. The period then is

$$T = 2\pi\sqrt{\frac{I}{L_0}}, \qquad (8\text{-}33)$$

where I is the moment of inertia of the oscillating system about the axis of oscillation. Equation (8-33) has been used to determine the moment of torsion of very fine quartz fibers.

If the torsion pendulum consists of a solid cylindrical rod with a heavy concentric disk attached to its lower end, then we can calculate the moment of inertia, and from equation (7-10) it is possible to find the rigidity modulus n of the rod.

8.14. Summary of Pendulums

There are other types of "pendulums" besides these four examples. In each case a formula for the period is obtained in a manner analogous to the method used here. The oscillations of a magnetic compass needle and the vibrations of a ballistic galvanometer will be studied at an appropriate place.

There is such a similarity in these derivations that it may help to repeat them in parallel columns (see Table 8-1).

It is to be noted that for purposes of simplicity we have completely discounted friction, assuming it nonexistent. This is never true. Only the effects of *very small* friction may be neglected.

8.15. Reversibility of the Physical Pendulum: Center of Oscillation

The equation for the period of a physical pendulum, equation (8-29), may be written in a somewhat different form by employing Steiner's theorem of parallel axes (§ 6.7) and the definition of the radius of gyration (§ 6.8):

$$I = I_{\text{cm}} + Mh^2 = Mk^2 + Mh^2, \qquad (8\text{-}34)$$

where M is the mass of the pendulum, k the radius of gyration with respect to an axis through the center of mass, and h the distance of the center of mass from the point or center of suspension. On replacing I of equation (8-29) by its equivalent value as given by equation (8-34), we obtain for the period

$$T = 2\pi\sqrt{\frac{k^2 + h^2}{gh}}. \qquad (8\text{-}35)$$

With k a fixed number for a particular pendulum, we see from equation (8-35) that the period will vary as we vary h. When $h = 0$, i.e., the object is pivoted at the center of mass, the period is in-

finitely great. We have in Fig. 8-8 a graph showing the dependence of the period (*ordinates*) on the distance of the center of suspension from the center of mass (*abscissas*).*

We observe (1) that the graph is symmetrical about a line through the center of mass, since it does not matter in which half we choose the center

* This graph was obtained experimentally by drilling a number of holes in a bar and then allowing the bar to swing from a fixed peg through any of the holes. Such a bar is shown in Fig. 8-8.

TABLE 8-1

LINEAR AND ANGULAR SIMPLE HARMONIC MOTION EQUATIONS

LINEAR SHM DEFINED BY $a = -kx$		ANGULAR SHM DEFINED BY $a = -b\phi$	
Spiral Spring	Simple Pendulum	Torsion Pendulum	Physical Pendulum

Spiral Spring

$$f = -f_0 x,$$
$$f = ma;$$
$$\therefore ma = -f_0 x,$$
$$a = \frac{-f_0 x}{m}.$$

If the displacement is sufficiently small that Hooke's law is valid, f_0/m is constant, and the motion is SHM.

The general equation is

$$T = 2\pi\sqrt{\frac{-x}{a}}.$$

For this special case,

$$T = 2\pi\sqrt{\frac{x}{f_0 x/m}},$$

$$T = 2\pi\sqrt{\frac{m}{f_0}}.$$

Simple Pendulum

$$f = -mg \sin \theta$$
$$= -mg\frac{x}{l},$$
$$f = ma;$$
$$\therefore ma = -mg\frac{x}{l},$$
$$a = \frac{-g}{l}x.$$

If the arc is sufficiently small that x may be considered as the displacement from a point in the line and the motion considered as straight-line motion, then the motion is SHM.

The general equation is

$$T = 2\pi\sqrt{\frac{-x}{a}}.$$

For this special case,

$$T = 2\pi\sqrt{\frac{-x}{(-g/l)x}},$$

$$T = 2\pi\sqrt{\frac{l}{g}}.$$

(It can readily be seen that the simple pendulum can be analyzed by the same method as that used for the compound pendulum.)

Torsion Pendulum

$$L = -L_0\theta,$$
$$L = I a,$$
$$I a = -L_0\theta,$$
$$a = \frac{-L_0\theta}{I}.$$

If θ is sufficiently small that Hooke's law is valid, L_0/I is constant, and the motion is angular SHM.

The general equation is

$$T = 2\pi\sqrt{\frac{-\theta}{a}}.$$

For this special case

$$T = 2\pi\sqrt{\frac{\theta}{L_0\theta/I}},$$

$$T = 2\pi\sqrt{\frac{I}{L_0}}.$$

Physical Pendulum

$$L = -mgh \sin \theta,$$
$$L = I a,$$
$$I a = -mgh \sin \theta,$$
$$a = -\frac{mgh}{I} \sin \theta.$$

If θ is sufficiently small, $\sin \theta = \theta$ and

$$a = -\frac{mgh}{I}\theta.$$

Since mgh/I is constant, the motion is angular SHM.

The general equation is

$$T = 2\pi\sqrt{\frac{-\theta}{a}}.$$

For this special case,

$$T = 2\pi\sqrt{\frac{\theta}{mgh\,\theta/I}},$$

$$T = 2\pi\sqrt{\frac{I}{mgh}}.$$

of suspension, and (2) that the stick has a minimum period of vibration. It can be shown that this occurs when $h = k$. For a given center of suspension we may locate *three other points* at which we may suspend the stick and obtain the *same* period of vibration. In Fig. 8-8, a, b, c, and d are just such points. A pair of unsymmetrical points, e.g., a and c, are of special interest. Let point a be the

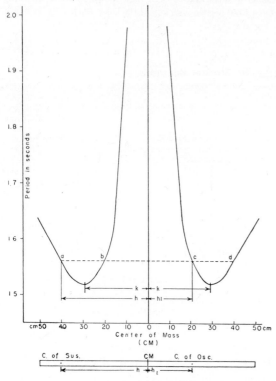

Fig. 8-8.—Period of a compound pendulum for various centers of suspension and corresponding centers of oscillation.

center of suspension at a distance h from the center of mass, and let c be the unsymmetrical point which is called the *center of oscillation*. The center of oscillation is at a distance h_1 below the center of mass. Applying equation (8-35) to these two points, we have

$$T^2 g h = 4\pi^2 (k^2 + h^2),$$

$$T^2 g h_1 = 4\pi^2 (k^2 + h_1^2),$$

which, on subtracting, gives

$$T^2 g (h - h_1) = 4\pi^2 (h^2 - h_1^2);$$

and, on dividing by $h - h_1$, since $h \neq h_1$, we obtain

$$T^2 g = 4\pi^2 (h + h_1)$$

or

$$T = 2\pi \sqrt{\frac{h + h_1}{g}}. \qquad (8\text{-}36)$$

This is the period of an equivalent simple pendulum whose length is $h + h_1$. Comparing equations (8-36) and (8-35), we see that

$$h + h_1 = \frac{k^2 + h^2}{h}$$

or

$$h_1 = \frac{k^2}{h}. \qquad (8\text{-}37)$$

Equation (8-37) gives the distance of the center of oscillation in terms of the fixed radius of gyration and the distance of the center of suspension from the center of mass. The reader will observe that equations (8-37) and (6-40) are the same; i.e., the center of oscillation and center of percussion are the same points.

EXAMPLE 6: A simple pendulum has a period of 2 sec. If this pendulum were made 1 mm longer, how many seconds would it lose in a day? Use g as 980 cm/sec².

Solution: This problem involves a peculiar example in calculation. The answer, correct to five significant figures, is 43.4958 sec.

If one were to work the problem in the most direct method, the solution would doubtless be something like this:

$$T_1 = 2\pi \sqrt{\frac{l}{g}}; \qquad l = \frac{T_1^2 g}{4\pi^2} = 99.29 \text{ cm},$$

$$T_2 = 2\pi \sqrt{\frac{99.29 + 0.1}{980}} = 2\pi \sqrt{0.1014}$$

$$= 2\pi \times 0.3184 = 2 \times 3.1416 \times 0.3184$$

$$= 2.00057 \text{ sec},$$

$$T_2 - T_1 = 2.00057 - 2.0 = 0.00057 \text{ sec}.$$

There being 86,400 sec in a day, the correct pendulum would make 43,200 complete swings. The long pendulum loses 0.00057 sec in each swing and hence loses $0.00057 \times 43{,}200 = 24.624$ sec.

This answer is evidently quite unsatisfactory. Not enough significant figures were carried in the calculation. Using more figures, one gets

$$l = 99.29475998,$$

$$T_2 = 2\pi \sqrt{0.1014232243} = 2\pi \times 0.3184701$$

$$T_2 = 2.001007 \text{ sec},$$

$$(T_2 - T_1) \, 43{,}200 = 43.50 \text{ sec};$$

but to do this required a great deal of arithmetic. Cannot a shorter method be found which will give an acceptable answer? We observe that

$$T_1 = 2\pi\sqrt{\frac{l}{g}}; \quad T_2 = 2\pi\sqrt{\frac{l+0.1}{g}} = 2\pi\sqrt{\frac{l}{g}}\sqrt{1+\frac{0.1}{l}},$$

$$T_2 = T_1\sqrt{1+\frac{0.1}{l}}.$$

From the binomial theorem of algebra,

$$(1+h)^n = 1 + nh + \frac{n(n-1)}{2}h^2$$

$$+ \frac{n(n-1)(n-2)}{2\times3}h^3 + \cdots .$$

When h is very small compared to unity, the terms of this series after the second are quite negligible:

$$(1+h)^n = 1 + nh .$$

Thus

$$\sqrt{1+\frac{0.1}{l}} = \left(1+\frac{0.1}{l}\right)^{1/2} = 1 + \frac{0.1}{2l}$$

and

$$T_2 - T_1 = T_1\left(1+\frac{0.1}{2l}\right) - T_1 = T_1\frac{0.1}{2l} = \frac{0.2}{2l} ;$$

$$\text{Time loss} = 43,200\,(T_2 - T_1) = 43,200\left(\frac{0.2}{2l}\right)$$

$$= \frac{43,200\times0.2}{2\times99.29} = 43.51 \text{ sec} .$$

This type of calculation is frequently necessary when the difference of nearly equal quantities is involved.

8.16. Lissajous Figures

In physics there are many examples of a particle undergoing two linear simple harmonic motions at right angles. The resultant path traced out by the particle is called a "Lissajous figure," after the man who first investigated it. Let us compound two simple harmonic motions of the same frequency and amplitude, which are acting at right angles. The resultant path or the Lissajous figure depends very much on the relative phase of the two simple harmonic motions. If the two motions start in phase, as shown in Fig. 8-9, then both have their maximum displacement at the same instant, at *1'* along the horizontal and *1''* along the vertical.

Initially, then, in the separate motions having the same amplitudes, the displacements at *1'* and *1''* give the resultant displacement at *1*. A

quarter of a period later, the displacements are zero for both motions, and the resultant has zero displacement at position *2*. In a further quarter-period the separate motions are at *3'* and *3''*, giving the displacement at position *3*. Following this geometrical analysis further, we see that the resultant motion is a straight line at 45° with the coordinate axes. This result can be readily ob-

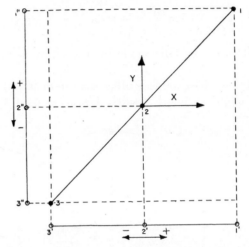

Fig. 8-9.—Compounding of two simple harmonic motions at right angles with 0° phase angle.

tained by analytical means. The motion along the *X*-axis from equation (8-7) is given by

$$x = r \cos \omega t .$$

Notice at time $t = 0$, $x = r$. Similarly for the motion in the *Y* direction we have $y = r \cos \omega t$. Hence the resultant figure, obtained by eliminating t from these equations, has the equation

$$x = y ,$$

which is the equation of the line at 45° in Fig. 8–9.

If the relative phase angle between the two motions is −90°, then, as shown in Fig. 8-10, the resultant motion is a circle. The equations of the two simple harmonic motions are

$$x = r \cos \omega t$$

$$v = r \cos (\omega t - 90)$$

$$= r \sin \omega t .$$

Eliminating t gives

$$x^2 + y^2 = r^2 ,$$

which is the equation of a circle. In this case the resultant motion is counterclockwise.

137

Had we assumed a relative phase angle of $+90$, so that

$$y = r \cos(\omega t + 90)$$

$$= -r \sin \omega t,$$

then the motion would have been clockwise. A quarter of a period $T/4$ after the start, the value of the displacement y would be

$$y = -r \sin \frac{\omega T}{4} = -r \sin \frac{2\pi}{T} \frac{T}{4} = -r,$$

that is, the particle would be in position marked *4″*.

If the phase angle is other than $0°$ or $90°$, then an elliptical figure results. Lissajous used two

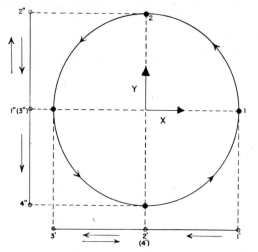

Fig. 8-10.—Compounding of two simple harmonic motions at right angles with phase angle of $90°$.

coupled pendulums or other mechanical means for producing the two simple harmonic motions. A much more convenient method today for producing the Lissajous figures is by means of an oscilloscope (see § 29.7). In this, electrons having an extremely small inertia (9×10^{-31} kg) are acted on by electric fields at right angles. By applying oscillating potentials—for example, the 60-cycle alternating voltages from the electric mains—with suitable phase differences, the electrons can be made to trace on a fluorescent screen the various patterns mentioned previously. The compounding of simple harmonic motions is of importance in the interpretation of polarized-light phenomena.

If the frequencies of the two simple harmonic motions are not equal, then quite complicated

patterns can result. Some of these are shown in Fig. 8-11. These can be traced by the geometrical means given earlier.

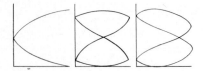

Fig. 8-11.—Lissajous figures with frequency ratio 2 to 1 and various phase angles.

8.17. Forced Harmonic Oscillations: Resonance

We have mentioned that any simple harmonic motion eventually ceases, because of friction or other dissipative forces. If the vibrations are to

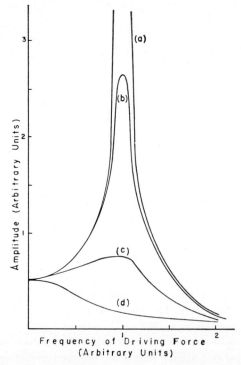

Fig. 8-12.—Amplitude of vibrations plotted against frequency of periodic driving force for (*a*) no friction, (*b*) friction small, (*c*) friction large, and (*d*) friction so large no resonance can occur.

continue without any decrease in amplitude or energy, then some periodic force must be applied to the system. The amplitude of the vibrations depends, among other things, on the frequency of the driving force. An example of this is a child's swing, in which the periodic push is adjusted to the period of vibration of the swing. Soldiers marching in step across a long bridge may cause the bridge to be set into a dangerous state of oscillation, so it is a rule to have them break step under such conditions. All clocks and watches have some mechanism for delivering a periodic driving force applied to them for maintaining the motion.

The frequency of the periodic driving force which produces maximum amplitude is called the "resonant frequency." For no friction or a very small amount, this frequency is the natural frequency of oscillation of the system. As the friction force increases relative to the restoring force, the frequency for resonance becomes progressively smaller than that of the natural frequency of the system without any damping. If the frictional force is sufficiently large, no resonance can occur at any frequency.

When a periodic driving force is applied to a system capable of vibrating, there is first set up a transient motion which dies away, leaving a steady-state motion in which the frequency of the vibrat-ing system is that of the periodic driving force. As shown in Fig. 8-12, the steady-state amplitude varies with the applied frequency, reaching a maximum at the resonant frequency.

8.18. Résumé

The following is a brief listing of the definitions, principles, and theories, given in this chapter, which you should know:

The definition of simple harmonic motion (SHM)

Definitions of amplitude, period, and frequency in SHM

Phase relationships of acceleration, velocity, and displacement is SHM

Conditions for angular SHM

The kinetic and potential energies of a body executing SHM

The theory of the motion of the simple, compound, and torsion pendulums

The compounding of SHM as in Lissajous figures

Resonance in forced vibrations

Queries

1. Why was the weight *mg* ignored in the discussion of § 8.9?

2. Strictly, one-third the mass of the spring should be added to *m* of equation (8-17). Explain in qualitative terms why this must be so.

3. What assumptions were made in deriving equation (8-33)? Should the angular displacement be small, as in § 8.12?

4. Suggest other examples of forced oscillations and resonance not given in § 8.17.

5. Can the analysis of a simple pendulum be carried out in terms of angular simple harmonic motion so as to obtain equation (8-24)?

Practice Problems

1. A spring vibrates 300 times in 3 min. What is the period? The frequency? If the amplitude is $\frac{1}{10}$ cm, what is the maximum acceleration and velocity? Ans.: 0.6 sec; 11 cm/sec^2; 1.05 cm/sec.

2. An object of weight 10 lb is moving with simple harmonic motion of a frequency of 1 vibration per second and an amplitude of 2 ft. Find the restoring force when the object is at the end of its swing.

3. An 8-lb weight vibrates with simple harmonic motion of a period of 4 sec and an amplitude of 10 in. Calculate the kinetic energy when the displacement is 6 in. Ans.: 0.137 ft-lbf.

4. An object vibrates with simple harmonic motion of a period of $\frac{1}{10}$ sec and an amplitude of 2 cm from the position of maximum displacement. Find the displacement, velocity, and acceleration 3 sec after starting from rest. If the mass of the object is 1 gm, compute the maximum force to which it is subject.

5. A disk 10 cm in radius oscillates with angular simple harmonic motion of a period of π sec and an amplitude of $\frac{1}{10}$ rad. Compute the angular displacement, velocity, and acceleration $\pi/8$ sec after starting from rest from its position of maximum displacement. If a mass of 1 gm is attached to the periphery of the disk, compute the maximum centripetal force on the mass. Ans.: 0.0707 rad; -0.1414 rad/sec; -0.2828 rad/sec^2; 0.4 dynes.

6. A spiral spring with a mass of 10 gm attached to it has a period of π sec when vibrating in a vertical path. (*a*) Find the force constant of the spring. (*b*) If the mass has 45,000 ergs of kinetic energy as it passes the mid-point of its path, what must be the amplitude of its motion? (*c*) What is the acceleration of the mass when its displacement is 4 cm?

7. A mass of 22 kg is attached to a steel wire 10 m long and 0.01 cm^2 in cross-section. If the mass is set into

vertical vibration of small amplitude, compute its period. If the object is a sphere 10 cm in radius, compute the period when set oscillating about a vertical axis. Ans.: $T = 0.1985$ sec; $T = 51.2$ sec.

8. Compute the period of the two yardsticks of problem 6 of chapter 6 as a pendulum swinging about the axis given.

9. In problem 8, if the mass of each yardstick were changed to X pounds, would the period of the pendulums be changed?

10. A disk of 200-gm mass and 10-cm radius, when hanging from a certain wire, has a period of 4 sec. If small masses of 50 gm each are placed at the opposite ends of a diameter of the disk, what will be the new period of oscillation?

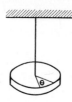

11. Compute the period of the ball and wire of Example 5 of chapter 6 (§ 6.7), swinging as a physical pendulum. Compute the length of an equivalent simple pendulum. Ans.: $T = 2.10$ sec; $l = 109.4$ cm.

12. A shelf moves vertically with a simple harmonic motion of amplitude 10 cm. What is the shortest period T that the shelf may possess in order that an object may remain in contact with it?

13. The formula $T = 2\pi\sqrt{I/mgh}$ is frequently used to determine the moment of inertia of an object. If m, g, and h were known exactly, but the determination of T were wrong by 1 per cent, what would be the error in the computation of I? See Example 6. Ans.: 2 per cent.

14. If the period of a certain pendulum on the earth is 1 sec, its period on the moon, where g is one-tenth as large, would be how long?

15. A penny is placed at the end of a rough plank which is vibrating horizontally with simple harmonic motion of an amplitude of 10 cm and a period of 4 sec. Compute the coefficient of friction between the penny and the plank if the penny is just on the verge of slipping. Ans.: $\mu = 0.025$.

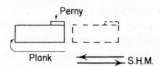

16. A simple pendulum has a period of 2.4 sec at a point where $g = 981.0$ cm/sec^2. What is the value of g at another point on the surface of the earth where the period is 2.41 sec?

17. The motion of the tides in a given harbor is simple harmonic motion. Given that the high water of 36 ft is reached at 1:00 P.M., and the low water of 28 ft at 6 hr, 10 min later. A ship which draws 30 ft of water has to enter the harbor. Find (a) the amplitude of the harmonic motion; (b) the period of the harmonic motion; (c) the displacement from the equilibrium for the required ship depth; and (d) the latest time in the afternoon the ship can enter the harbor. Ans.: (a) 4 ft; (b) 12 hr, 20 min; (c) -2 ft; (d) 6:08 P.M.

18. A solid cylinder, 20 cm in radius, is pivoted 19 cm from the center of mass and oscillates as a physical pendulum. The axis of suspension is parallel to the axis of the cylinder. Find the center of oscillation of the disk.

19. In problem 18 at what distance from the center of mass must the cylinder be pivoted in order that its period shall be a minimum? Ans.: 14.14 cm.

20. Two masses each of value 1 mg are connected by a fine steel wire 2 cm long and of mass 4 mg. The system is suspended from a fine quartz fiber of negligible mass, as shown in the diagram. What is the moment of torsion of the fiber if the period of vibration is observed to be 12 sec? What torque is required to twist the fiber through an angle of 5°?

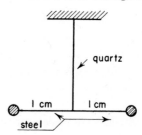

21. Solve problems 5 and 7, using the mks system of units.

22. Find the dimensions of $\sqrt{m/f_0}$ in equation (8-17).

23. By the geometrical method given in § 8.16 show that two simple harmonic vibrations of the same amplitude and frequency at right angles to each other with a phase difference of 90° give rise to a circle, and with a phase difference of 45° give rise to an ellipse.

24. If the frequencies of the simple harmonic vibrations at right angles are in the ratio of 2 to 1 and the vibrations have the same amplitude and both have zero phase difference, find by geometrical means the resulting Lissajous figure.

25. Show that the dimensions of $\sqrt{1/k}$ are time. Check the equality of dimensions in equations (8-35) and (8-36).

26. A simple harmonic motion is given as $a = -kx$. If the mass being oscillated is m, write the expressions for the kinetic and potential energies at any time and show that the total energy is a constant.

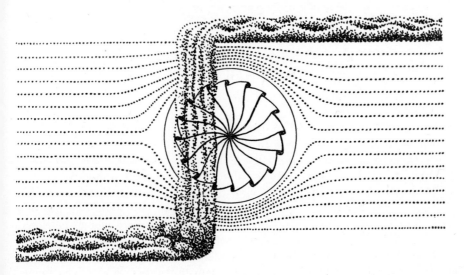

CHAPTER 9

Mechanics of Fluids

9.1. Definition of an Ideal Fluid

Of the three physical states in which matter occurs, the solid state is by far the most convenient to study and the simplest in its gross mechanical properties. The strong cohesive forces that exist between molecules when they are relatively close together in the solid state and when the temperature is sufficiently low that their thermal kinetic energies are insufficient to overcome these forces result in that property we call "rigidity." Solid objects retain their shape and form. If distorted by external stress, a slight elastic yield results. Indeed, they may also flow very slightly if this stress is long continued, but this effect is so small as to be virtually microscopic. Immediately after the stress is removed, they resume their original form to a high degree of approximation. Most exact and refined measurements are able to show that here, likewise, a minute part of the recovery does not occur at once and that, indeed, the original form of the specimen is never quite restored. These slight details are matters that we have pursued further in other connections (compare, e.g., chap. 7). For the present purpose we shall assume that the solid state is characterized by complete rigidity, i.e., by its ability to transmit shearing stresses (see also chap. 7).

The other two states of matter—the liquid and the gaseous—possess little, if any, rigidity. Together they may be classified as "fluids." *We define an ideal fluid as a substance which is incapable of transmitting shearing stresses.* The form of any given mass of a fluid is quite indefinable; it conforms to the shape of the containing vessel.

Fluids may be further subdivided into gases and liquids. A gas, like the atmosphere, for example, is a fluid that has no free surface. It can occupy any volume by expanding or being compressed until it just fills it. A liquid, like water, is a fluid which, if not completely confined, presents a free horizontal (level) surface when at rest. It possesses a definite volume at a definite temperature. Liquids in general are but slightly compressible, i.e., they possess large bulk moduli, comparable to those of solids. Gases also resist compression but characteristically expand indefinitely when the pressure is released. In the discussion that follows we shall be chiefly concerned with liquids, the properties of gases being deferred to a later chapter.

It is true that no real fluid can exactly meet the conditions of an ideal fluid, since all fluids exert some shearing stress. We shall, however, neglect this factor for the present but shall return to it when we discuss viscosity of fluids.

141

Hydrostatics

9.2. Fluids at Rest

In the case where complete absence of motion throughout the entire body of a fluid is assumed, the forces at every point within it must balance to a zero resultant. In this very special case a precise science exists, one which is so simple that some of its fundamental principles were understood by some of the early scientists—indeed, one important principle is associated with the name of the Greek philosopher Archimedes (287–212 B.C.), who may be considered the founder of HYDRO-STATICS.

Since the total forces acting throughout the body of, or along the surface bounding any portion of, a homogeneous fluid depend upon the extent of the surfaces or the magnitude of the volume of the fluid under consideration, we can immediately simplify our discussion by confining our attention to unit surfaces and unit volumes. As a result, in the case of fluids we are much less likely to refer to mass than to DENSITY, i.e., mass per unit volume,

$$\rho = \frac{m}{V}. \quad * \qquad (9\text{-}1)$$

Likewise, in many of our discussions on this topic we shall speak not of "total forces acting on the fluid" but rather in terms of *pressures*. This term was first used in chapter 7, § 7.3. We shall, however, now *define* "pressure" in connection with fluids.

Consider a small area A within a fluid; the fluid on one side of the area exerts a force F on the fluid on the other side, the force being transmitted across the area A. The *average pressure*, p, on this area A, is, by definition,

$$p = \frac{F}{A}. \qquad (9\text{-}2)$$

The pressure in a fluid may vary from place to place; therefore, let us define *pressure at a point within a fluid*. Equation (9-2) defines the average pressure on the surface A. If we allow A to become smaller and smaller, the total force transmitted across A will likewise decrease. The ratio of F to A will always be finite and give the average pressure over the area. The limit of this ratio, as we allow A to approach a point, is, by definition, the pressure

at this point. The pressure at a point may then be expressed as

$$p = \frac{dF}{dA}. \qquad (9\text{-}3)$$

The student will note the similarity of this definition to that of velocity at a point (§ 1.16). Pressure, like temperature, is a scalar and possesses a definite value at every point in a liquid.

It is clear from equation (9-2) that pressure has dimensions of force per unit area. We measure pressure in newtons per square meter (newtons/m²), dynes per square centimeter (dynes/cm²), in *bars*, in pounds-force per square inch (lbf/in² or psi), in kilograms-force per square centimeter (kgf/cm²), in atmospheres, in millimeters (or inches) of mercury, etc. The following data are useful:

$$1 \text{ atm} = 14.7 \text{ lbf/in}^2 = 1.0336 \text{ kgf/cm}^2$$
$$= 1.013 \times 10^6 \text{ dynes/cm}^2$$
$$1 \text{ bar} = 10^6 \text{ dynes/cm}^2 .$$

A logical consequence of the definition of an ideal fluid at rest is that *the force* **F** *acting across any area* **A** *acts perpendicularly to* **A**.

If this were not so, then there would be a component of the force parallel to the area, and this would cause a motion of the fluid in this direction,

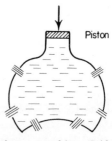

FIG. 9-1.—The force exerted by a fluid on any area is perpendicular to the area.

which is contrary to the assumption that the fluid is at rest. Thus we see that the force exerted by a fluid on any element of area must be perpendicular to that area, and, by Newton's Third Law, there is an equal but opposite force on the fluid.

This is experimentally verified by the apparatus shown in Fig. 9-1. Water is found to spout from the openings perpendicularly to the wall of the container.

* For techniques used to determine density see § 9.12.

9.3. Variation of Pressure with Depth in a Liquid

We shall now prove that the pressure in a fluid increases with the depth of fluid. In order to do this, consider an imaginary vertical cylinder of area A, shown dotted in Fig. 9-2. The pressure at the upper surface is p_1, so that the downward force exerted on the top of the cylinder is p_1A. The pressure p_1 may be due to the overlying fluid, but

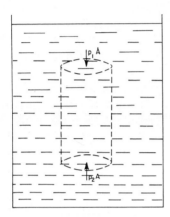

FIG. 9-2.—Pressure increases with depth in a fluid

what produces it is of no consequence here. On the lower horizontal surface of the cylinder is a pressure p_2 and an upward force of p_2A. The remaining vertical force is the weight of the fluid in the cylinder. From the density of the fluid and the volume of the cylinder the weight of the fluid on the cylinder can be obtained. Suppose the density or mass per unit volume ρ of the fluid is constant. This is really assuming that the fluid is incompressible. The volume of the cylinder is Ay, where y is the height of the cylinder measured from the top of the cylinder down. Thus the mass of fluid in the cylinder is $Ay\rho$, and the weight of this fluid is $Ay\rho g$, when g is the acceleration due to gravity. For the equilibrium of this cylinder in a vertical direction we must have $\Sigma F_y = 0$ or

$$A y \rho g = p_2 A - p_1 A$$

or

$$p_2 - p_1 = g \rho y . \tag{9-4}$$

The horizontal forces on the sides of the cylinder are also in equilibrium and, of course, make no contribution to the vertical forces. Thus from equation (9-4) we see that the pressure increases in a downward direction in a fluid in a manner proportional to the depth y.

In the cgs system of units ρ is measured in gm/cm³, g in cm/sec², and y in cm, so that the pressure p is in gm/cm-sec² or dynes/cm². Similarly, in the mks system ρ is in kg/m³, g in m/sec², and y in meters, so that p is in kg/m-sec² or newtons/m². For the British Engineering system ρ is in slugs/ft³, g in ft/sec², or is more commonly expressed as ρg, which is the weight-density in lbf/ft³. In engineering work, pressure is frequently expressed in lbf/in², written as psi.

9.4. Pressure Changes in a Compressible Fluid

Let us consider a small, imaginary vertical cylinder in a compressible fluid. If y is measured in an upward vertical direction, then, from equation (9-4), for a cylinder of height dy the increase in pressure with height is

$$dp = - g \rho d y , \tag{9-5}$$

where the minus sign indicates that the pressure p decreases with increase in height dy. For this case the density ρ is not a constant but varies with the temperature and pressure of the fluid. Assuming a perfect gas as the fluid, then, according to the general gas law given in chapter 13, the density ρ for a gas of molecular weight M, at pressure p and absolute temperature T, is

$$\rho = \frac{M}{R} \frac{p}{T},$$

where R is a universal constant for all perfect gases. Thus from equation (9-5) we have

$$\frac{dp}{p} = \frac{- gM d y}{RT} . \tag{9-6}$$

Let us now consider the variation of pressure with height in the earth's atmosphere. We still cannot integrate equation (9-6) unless we know how the temperature of the air in the atmosphere varies with height. This, in general, varies day by day, so that, in order to carry out the integration to give the variation of the pressure with height, we must know this variation. The simplest reasonable assumption, which rarely exists over any extensive region of the atmosphere, is that the temperature is a constant. With this assumption, the integration of equation (9-6) gives

$$\log_e p \equiv \ln p = \frac{- gM y}{RT} + C ,$$

where C is a constant of integration found from the boundary conditions. If $p = p_0$ at $y = 0$, that is, the **143** pressure at the surface of earth where $y = 0$, is some value p_0, then $C = \ln p_0$ and

$$\ln p - \ln p_0 = \ln \frac{p}{p_0} = \frac{-gMy}{RT}.$$

Thus, from the definition of a logarithm, it follows that

$$p = p_0 e^{-gMy/RT}. \qquad (9\text{-}7)$$

For $g = 9.80$ m/sec^2, $T = 273°$ K $(0°$ C$)$, $M = 29 \times 10^{-3}$ kg/mole approximately for air, and $R = 8.31$ joules/$°$ K, it follows that

$$p = p_0 e^{-y/8},$$

where y is measured in kilometers. With the assumption of a constant temperature, it follows from equation (9-7) that the pressure decreases exponentially with vertical height in the atmosphere, such that at a height of 8 km the pressure is $1/e$ of its initial value. Thus the height of an isothermal atmosphere is infinite. Actually, the temperature of the atmosphere decreases with height up to about 15 km, then increases. For details of this problem see chapter 13.

9.5. Forces and Moments on a Dam: Center of Pressure

In many problems of hydraulic engineering it is important to compute the total force on a vertical or inclined surface which is supporting a large body of water. Consider the dam shown in Fig. 9-3, with the water at the top of the spillway of

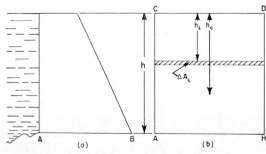

Fig. 9-3.—A dam supporting a large body of water; (a) an end view, (b) a front view.

total depth h. The pressure due to the water on the side of the dam increases uniformly from the top, at which it is zero, to the value of ρgh at the bottom. If the area of the dam wall is $A = hb$, and the average pressure at the center is $\rho gh/2$, then the total horizontal force on this wall is

$$F = \frac{A \rho g h}{2}, \qquad (9\text{-}8)$$

where ρ is the density of the water. Suppose the height h of the wall is 40 ft and its breadth b is 150 ft, and the weight-density of water ρg is 62.4 lbf/ft^3, then the total force on the dam wall is

$$F = 40 \times 150 \times 62.4 \times \frac{40 \text{ ft} \times \text{ft-lbf} \times \text{ft}}{2 \text{ ft}^3}$$

$$= 748,800,000 \text{ lbf}$$

$$= 374,400 \text{ tons weight}.$$

In problems concerned with the stability of dams it is necessary to know the point of application of the resultant force F. This point of application, sometimes called the "center of pressure," is found by the use of the theorem of moments (§ 6.5, eq. [6-15]), namely, that the sum of the moments about a given axis is equal to the moment of the resultant force about the same axis.

Let us consider the moments of all the forces about the top of the dam or the line CD (Fig. 9-3). On a strip of area ΔA_i at depth h_i there is a pressure due to the water of ρgh_i and a force perpendicular to it of $\rho gh_i\Delta A_i$. The moment of this force about the axis CD is

$$\Delta L_i = h_i \Delta F_i = \rho g h_i^2 \Delta A_i.$$

The area of the strip ΔA_i is $b\Delta h_i$, where b is the breadth of the dam. The total moment L of all the forces acting perpendicular to the vertical wall of the dam is the summation over the whole wall:

$$\left. \begin{aligned} L &= \Sigma g \rho b h_i^2 \Delta h_i = \int_0^h g \rho b h_i^2 dh_i \\ &= g \rho \frac{b h^3}{3}. \end{aligned} \right\} \qquad (9\text{-}9)$$

The area A of the wall is bh, so the total moment of the forces of the water on the wall about CD is

$$L = \frac{g \rho A h^2}{3}. \qquad (9\text{-}10)$$

This total moment of force is equal to the product of the resultant force and the distance h_c of this force from CD. From equation (9-8) the resultant force is

$$F = \frac{A g \rho h}{2},$$

and its moment is Fh_c. Hence

$$\frac{A g \rho h h_c}{2} = L = g \rho A \frac{h^2}{3}$$

and

$$h_c = \tfrac{2}{3} h. \qquad (9\text{-}11)$$

Thus the resultant force acts at a distance of $2h/3$ from the top of the dam or $h/3$ from the bottom.

In the example the height of the dam was 40 ft, so the distance of the center of pressure from the top is $80/3 = 26.7$ ft. In order to find the moment of force tending to topple the dam over, we take moments about the line AH at the bottom. The resultant force of 748,800,000 lbf acts at a distance of $40/3$ from the bottom, so the turning moment is

$$748,800,000 \times \tfrac{40}{3} = 9.984 \times 10^9 \text{ ft-lbf}.$$

To balance this turning moment, which would act along a line through B, in Fig. 9-3 (a), there would be the opposite moment of the weight of the dam acting at its center of weight about the line through B. Usually dams are exceedingly heavy structures, so there is relatively little danger of their being pushed over by the force due to the water.

9.6. Pascal's Law

The law or principle known by the name of Pascal (1623–62) was stated earlier, in 1585, by Benedette and probably also independently by Stevin in 1586. The latter two were familiar with such ideas as the lever arm and the impossibility of perpetual motion, respectively; consequently, the basic principle of work may have formed a background for this famous principle or law.

Pascal's law states the fact that *pressure impressed at any place on a confined liquid is trans-*

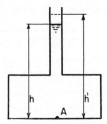

Fig. 9-4.—Pascal's law for transmission of pressure

mitted without loss to all other portions of it. The law, however, does not imply that pressures are equal at every point throughout the confined liquid but rather that the pressure at each point, whatever its initial value may be, *is increased by the same amount* when an external pressure is applied. This law follows from the principles already developed, as we shall see from an examination of Fig. 9-4.

A confined vessel with an open tube whose area

is, say, 1 cm² is filled to some height h above the base of the tank. Neglecting atmospheric pressure, the pressure on the base at A is $h\rho g$ by equation (9-4). If an additional pressure is now produced by adding some liquid in the tube, so that the liquid is h' above A, then the pressure at A becomes $\rho h' g$. Also, the pressure is the same along the horizontal. Hence adjacent points on the base must be subjected to the same increase in pressure; otherwise the liquid at this level would flow from the region of increased pressure to that of lower pressure, in contradiction to our hypothesis of a nonmoving liquid.

If an atmospheric pressure of B dynes/cm² acts on the free surface of a liquid, then, by Pascal's law, each point within the liquid has its pressure increased by B over that due to the liquid's weight. Equation (9-4) then becomes

$$p = h\rho g + B, \qquad (9\text{-}12)$$

giving the pressure at a depth h due to a liquid of density ρ in the earth's gravitational field and the free surface exposed to atmospheric pressure B.

Pascal's law of the transmission of pressures has wide practical applications—to the hydraulic press for compressing materials into bales, to the hydraulic brakes on modern automobiles, and to hydraulic elevators, to name only a few examples. We shall briefly describe the hydraulic press with the aid of Fig. 9-5. A small force f is exerted on

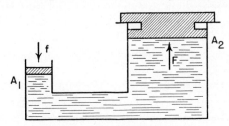

Fig. 9-5.—The hydraulic press

a piston of area A_1; this is equivalent to a pressure $p = f/A_1$, which is transmitted in accordance with Pascal's law to the piston of area A_2. The force F on A_2 is such that $F/A_2 = p$. The mechanical advantage of the machine is obviously

$$MA \equiv \frac{F}{f} = \frac{A_2}{A_1}. \qquad (9\text{-}13)$$

9.7. A Hydrostatic Paradox

According to equation (9-4), the pressure on the base of a vessel depends upon the height of the **145** liquid above the base, whatever may be the shape of the vessel. Consider four vessels with the same area of base, as in Fig. 9-6. In the case of the cylinder (*a*) the truth of the foregoing statement is obvious. However, in vessels shaped like those in (*b*), (*c*), and (*d*) the statement is not at all obvious. Vessel (*b*) may contain a volume of water many times greater than that of (*a*), yet the pressure on the base is the same; in (*c*) the volume may be less than in (*a*). We may explain away the paradox as follows:

In the case of (*b*) the force on a very small area at any point, such as *S*, is perpendicular to the

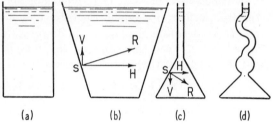

(a)   (b)   (c)   (d)

Fig. 9-6.—The pressure at the base is the same in all the vessels.

sides of the vessel. The vessel, because of its rigidity, produces a reaction, *R*, on the liquid, equal and opposite to the applied pressure. We resolve *R* into two components—a horizontal one, *H*, and a vertical one, *V*. The horizontal component can be neglected; the vertical one acts upward to support the water above the tilted side. Hence, the force on the base is less than the total weight of the water by an amount equal to the upward thrusts, and the force on the base is equal to that on the base of the vessel in (*a*).

In the case of vessel (*c*), the component *V*, due to the reaction of the sides on the water, acts downward. The base is then subjected to a force greater than that caused by the weight of the water alone. A vessel such as (*d*) can be thought of as a combination of (*a*), (*b*), and (*c*).

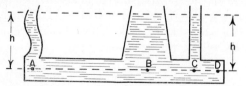

Fig. 9-7.—Pressure along a horizontal line is constant

It follows from an extension of these results, together with Pascal's law, that the pressure at A, B, C, and D of a group of connecting vessels,

as shown in Fig. 9-7, must be the same and must be dependent on the height *h*. Such a surface of equal pressure we call a "level surface."

9.8. The Barometer

In the mercury BAROMETER we have another application of Pascal's law. An open tube is inserted into an open dish of mercury, and the air is then pumped out of the upper portion of the tube, which is subsequently sealed off. If the tube is long enough, it will be found that, as the air is pumped out, the mercury will rise to a height about 76 cm vertically above the level of the mer-

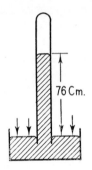

76 Cm.

Fig. 9-8.—The mercury barometer

cury in the dish—but no farther, no matter how much longer the tube may be. It is, of course, the pressure of the atmosphere on the mercury surface in the open dish which, communicated through the body of the liquid, forces it up in the tube as the pressure of the air upon it there is removed. There remains above the surface of mercury in the tube after the air is removed a slight pressure due to the mercury vapor, which at ordinary temperatures is so small that it can be neglected.

Since the normal pressure of the atmosphere at sea-level supports a column about 76 cm long in a tube of cross-section *A* and since the density of mercury at room temperature is about 13.6 gm/cm^3, we have a weight of mercury of $76 \times 13.6 \times A = 1,033.6A$ gmf supported. This is equivalent to a pressure (force/cm^2) of $1,033.6 \times 980$, a trifle over a million (1,013,000) dynes/cm^2, balancing the same pressure of the atmosphere. To take accurate readings on a well-made barometer, the mercury surface in the dish must first be carefully adjusted to the zero of the scale, with the scale hanging vertical. A vernier is usually provided for setting an index at the level of the top of the mercury in the column. The temperature of the mercury must be noted in order that the proper value of mercury density may be used.

If a barometer with water instead of mercury should be used, the tube would have to be about 13.6 times longer than the 76 cm, i.e., 10.336 m, or about 34 ft. Water-vapor pressure at room temperature amounts to several inches of water pressure and would also have to be corrected for. Such a barometer is, of course, entirely too unwieldy for practical use but was the type first made by Torricelli, the inventor of this instrument.

9.9. Buoyancy and the Principle of Archimedes

Considerations which we have previously discussed in this chapter lead us directly to the deduction that when an object is immersed in a fluid, the object is buoyed up by a force equal to the weight of fluid displaced by the object. In order to prove this statement, known as "Archimedes' principle," we shall consider a uniform cylinder of cross-sectional area A and height h placed in a fluid of density ρ (Fig. 9-9). The pressure p on the

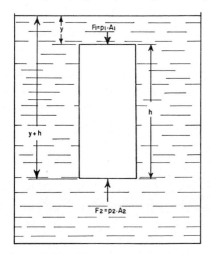

FIG. 9-9.—Archimedes' principle

upper surface, given by equation (9-4), is $g\rho y$, and on the lower surface it is $g\rho(y+h)$. Thus the resultant upward force on the cylinder is

$$F_2 - F_1 = g\rho A\,(y+h) - g\rho Ay$$
$$= g\rho Ah\,. \qquad (9\text{-}14)$$

From the figure it follows that Ah is the volume of the cylinder, $Ah\rho$ is the mass of the fluid displaced

by the cylinder, and $Ah\rho g$ the weight of fluid displaced. While this proof of Archimedes' principle is given for a right circular cylinder, it can be extended to an object of any shape. Thus any object immersed in a fluid will suffer an apparent loss of weight equal to the weight of fluid displaced by the object, a result known to Archimedes two thousand years ago. It follows from this that a solid object whose density is greater than that of a liquid will sink in the liquid, while one whose density is less will float in the liquid.

9.10. Determination of Volume and Density from Loss of Weight

While the mass of an object can ordinarily be determined quite accurately by weighing the object with a beam balance, the determination of the volume of most objects offers some difficulty. A convenient way of determining volumes depends upon the apparent loss of weight when the object is immersed in a suitable liquid.

Let the solid whose volume V we wish to determine be of density ρ greater than the density of the liquid, ρ_l, in which it is immersed. The object is weighed in air and then in the liquid. The apparent loss of weight is given by

$$\left.\begin{aligned} w_a - w_l &= V\rho_l g\,, \\[2mm] V &= \frac{w_a - w_l}{\rho_l g}\,, \end{aligned}\right\} \qquad (9\text{-}15)$$

where w_a is the weight in air, w_l the weight when submerged, V the volume of liquid displaced, and ρ_l the density of the liquid. Also, since

$$w_a = V\rho g\,,$$

it follows, upon division, that

$$\rho = \rho_l \left(\frac{w_a}{w_a - w_l}\right), \qquad (9\text{-}16)$$

which gives the density of the block in terms of the density of the liquid, which we assume to be accurately known, and the easily determined weights, w_a, and w_l. We may also use equation (9-16) to determine the density of the liquid in terms of the density of the solid.

9.11. Objects That Float

When an object at rest floats on a liquid, *it is in equilibrium under the action of two forces: its weight* **147** *and the buoyant force which is equal to the weight of liquid displaced*, in accordance with the principle of Archimedes.

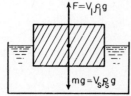

FIG. 9-10.—Forces on a floating object

Let V_s be the volume of the solid and ρ_s its density; V_l the volume of the displaced liquid and ρ_l its density. For equilibrium to take place:

$$\text{Weight} = \text{Buoyant force},$$

$$V_s \rho_s g = V_l \rho_l g ,$$

or

$$\frac{V_l}{V_s} = \frac{\rho_s}{\rho_l}. \qquad (9\text{-}17)$$

Thus, for a floating object, the ratio of the submerged volume to its total volume measures the relative density of the solid and liquid.

9.12. Density and Specific Gravity

We have already defined *density* as mass per unit volume. In the metric system the units of density are kilograms per cubic meter (kg/m³) or grams per cubic centimeter (gm/cm³). Engineers frequently define density as weight per unit volume measured in pounds-force per cubic

TABLE 9-1

APPROXIMATE DENSITIES OF SOLIDS AND LIQUIDS

Solids	Density (Gm/Cm³)	Liquids	Density (Gm/Cm³)
Aluminum	2.7	Alcohol (ethyl)	0.81 (0° C)
Brass	8.4	Benzene	0.90 (0° C)
Copper	8.9	Chloroform	1.53 (0° C)
Cork	0.24	Glycerine	1.26 (0° C)
Glass (Crown)	2.6	Mercury	13.60 (0° C)
Gold	19.3	Oils (lubricating)	0.91 (20° C)
Ice (0° C)	0.917	Turpentine	0.87 (15° C)
Iron	7.8	Water, pure	1.00 (4° C)
Lead	11.4	Water, sea	1.03 (15° C)
Uranium	(18.7)		
Wood (pine)	0.5		

foot (lbf/ft³). Water has a density of very close to 1 gm/cm³, actually 0.999973 gm/cm³ at 3°.98 C. The weight-density of water is 62.4 lbf/ft³, its mass-density being 62.4/32 slugs/ft³. (See Table 9-1 for densities of various solids and liquids.)

The *specific gravity* of a substance is defined as the ratio of the density of that substance to the density of some other substance taken as a standard. Water, with a density of 1 gm/cm³, or 62.4 lb/ft³, is the common standard chosen. For gases, either air or oxygen is used as a standard.

Density in the cgs units and specific gravity have the same numerical values. The density in the mks or in English units is not equal numerically to the specific gravity. Specific gravity is not a dimensional quantity and hence is the same in all systems of units.

9.13. Density and Specific-Gravity Measurements

The density of regular solids can usually be measured directly by determining the mass and the volume of the solid. For irregular solids equation (9-16) may be used if a liquid in which the solid is not soluble can be found.

HYDROMETERS are used for rough comparative determinations of densities of different liquids. They are cylindrical glass tubes, hermetically sealed and weighted at the bottom end so that they float vertically. The amount of submerged volume may be read off from the emergent portion of the stem. Better still, if the stem at the liquid surface is calibrated so that it is possible to read the ratio of the surmerged volume in any liquid to the submerged volume when the device is floating in water and, further, if this scale is marked so that it reads 1 when floating in pure water at 15° C, then, by general agreement, the hydrometer will read specific gravities (eq. [9-17]).

A special type of balance, known as the WESTPHAL BALANCE (Fig. 9-11), is in quite common use for the precise determination of liquid densities by weighing. The weight used on it for submergence in the liquid is that of a glass thermometer, which thus conveniently indicates the temperature of the liquid under observation. Furthermore, the balance is provided with a set of weights decimally related to one another and with ten divisions on its balance arm. By having the largest weight, when placed in the 10 division mark on the lever arm, just equal to the loss of weight of the submerged bob when the latter is in pure water, the instrument can be made to read densities directly.

A convenient way of determining the specific gravity of liquids with some precision is by use of a specific-gravity bottle known as a PYCNOMETER. It consists essentially of a small, thin-walled bottle with a glass stopper, ground to fit the neck of the bottle. A fine capillary hole runs through the stopper, so that excess liquid may run out as the stopper is placed in position. In this way a definite and reproducible volume of liquid can be obtained. If w_w is the weight of water at a given temperature and w_l the weight of the liquid whose specific gravity is required, then, by definition,

$$\text{Specific gravity} = \frac{w_l}{w_w}. \qquad (9\text{-}18)$$

place, there is equilibrium between the water replacing the ship and that outside. The buoyant force acts at the center of gravity of this water replacing the ship. Call this point B_1, the center of buoyancy (Fig. 9-12 [a]). The point C is the *center of gravity of the object*. When the points C and B_1 lie in the same vertical line OO' (Fig. 9-12), the object is in equilibrium. Let us assume that the object tips as in Fig. 9-12 (b). We see that the center of buoyancy in this position, B_2, has changed, while the center of gravity C is fixed. Through the point B_2 draw a vertical line PP'. The point where

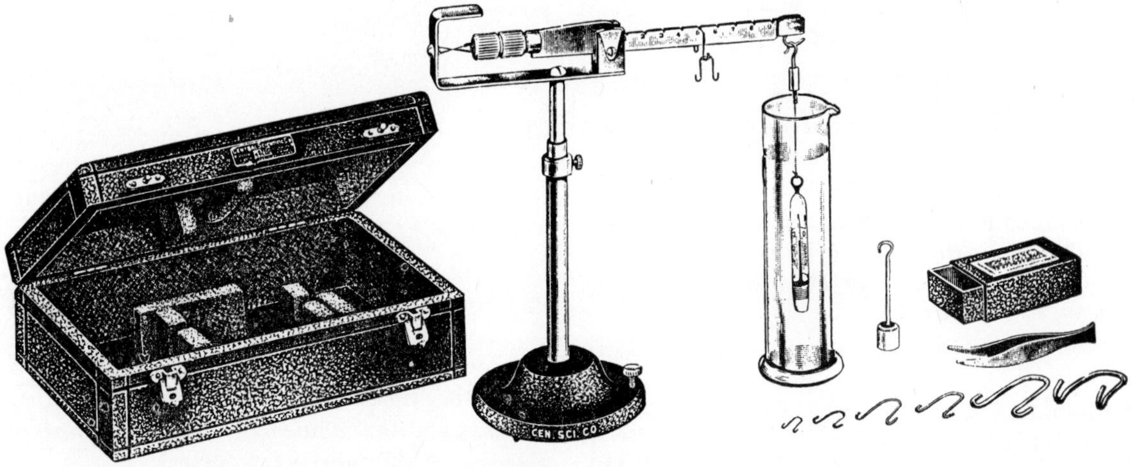

FIG. 9-11.—The Westphal balance for specific gravities (Courtesy Central Scientific Company, Chicago, Illinois)

EXAMPLE 1: An iceberg floats with 10.87 per cent of its volume above sea water, the density of which is 1.03 gm/cm³. Find the density of the ice.

Solution: For a floating body the total weight of the body is just equal to the weight of the displaced liquid. Let V be the total volume of ice. Then $[(100 - 10.87)V]/100$ is the volume of ice under water and is the volume of water displaced. Therefore,

$$1.03 \left(\frac{100 - 10.87}{100} \right) V = \rho_{ice} V,$$

$$\rho_{ice} = 0.918 \text{ gm/cm}^3.$$

9.14. Stability of Floating Bodies

We have already discussed the condition for flotation of bodies (eq. [9-17]). In this section we should like to find the conditions that determine whether a floating object is in stable equilibrium. The application to ships is obvious. The buoyant force acting on a floating body acts through the *center of gravity of the displaced liquid*. For, if we imagine the ship removed and the water back in

it intersects the center line of the object OO' is called the *metacenter*, M. From the diagram it follows that, for stable equilibrium, the metacenter M must never come below C, the center of gravity

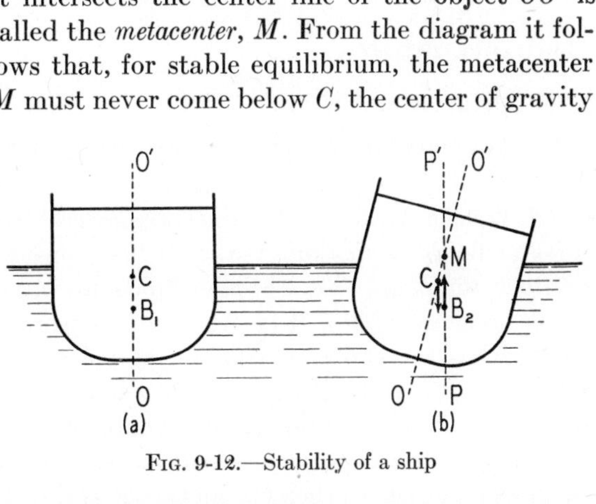

FIG. 9-12.—Stability of a ship

of the floating object; otherwise the couple acting on the object due to its weight and the buoyant force will not bring it back into equilibrium. The reader will observe that if M were below C, the couple would tend to overturn the body. In the design of ships, therefore, arrangements are made **149** to have the center of gravity of the ship as low as possible. To achieve this, some ships, when not carrying cargo, are loaded with ballast.

9.15. Work Done against a Constant-Pressure Head

Let us consider the section of a pump sketched in Fig. 9-13. The cylinder of the pump is of length

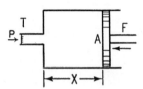

FIG. 9-13.—Work against a constant-pressure head

x, and the area of the piston is A. We assume that the pipe T is connected to some device, such as a water tower, that gives rise to a constant hydrostatic pressure P. The piston pushes the incompressible liquid from the cylinder into the pipe mains against the pressure P. How much work is done?

From Pascal's principle we know that the force on each unit area of the piston is P. Hence the total force F that must be exerted to force the liquid out is

$$F = PA .\qquad (9\text{-}19)$$

As the piston moves a distance x, the work done is

$$W = Fx .\qquad (9\text{-}20)$$

The volume of liquid ejected is

$$V = Ax .$$

By combining equations (9-19) and (9-20) we obtain

$$W = PV ;\qquad (9\text{-}21)$$

that is, the amount of work done in moving a volume of liquid against a constant pressure P is equal to the product of the volume by the pressure.

If this volume is ejected in t seconds, the horsepower developed by the pump is

$$\text{Horsepower} = \frac{PV}{550t},\qquad (9\text{-}22)$$

where P is measured in pounds-force per square foot, V in cubic feet. If the pressure is not a constant, then the work done in changing the volume from V_1 to V_2 is given by

$$W = \int_{V_1}^{V_2} P\,dV .\qquad (9\text{-}23)$$

In order to carry out this integration, we must know the functional relationship between the pressure and the volume (see chap. 13 for an application of this equation to gases).

Hydrodynamics

9.16. "Stationary" Flow

We have already noted that any detailed description of fluids in motion is somewhat complicated. We can, however, study certain general features of fluids in motion. We shall first consider *steady* or *stationary flow*. Consider fluids flowing in a pipe, as shown in Fig. 9-14. At point a, a particle of fluid has a certain velocity, v_1; at point b its velocity is v_2; while at c its velocity is v_3. If, as time goes on, the velocity of whatever fluid particle happens to be at a is still v_1, that at b is v_2, and that at c is v_3, then the motion is said to be *steady* or *stationary*. The path followed by a fluid particle in steady flow as it traverses the length of the pipe is called a *streamline*.

In Fig. 9-14 (a) we have drawn only one of the many possible streamlines. This is the path abc. In Fig. 9-14 (b) are drawn some streamlines of

liquid flowing in a constricted tube. Note the crowding-together of the streamlines in the constriction. Streamlines are characterized by the property that the tangent at any point on a streamline gives the direction of flow of the liquid.

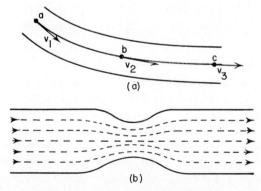

FIG. 9-14.—Stationary or streamline flow

Further, particles of fluids may not flow from one streamline to another. As suggested by Fig. 9-14, in regions where the streamlines are crowded together, the velocity of flow is increased.

A bundle of similar streamlines is called a *tube of flow*. In the steady flow of liquids in pipes we shall consider the liquid as passing in a tube of flow bounded by the pipe. It is assumed that all particles passing through such a stream tube will have the same velocity as they pass a given section normal to the direction of flow, that is, there is no friction between the pipe and the liquid.

9.17. Continuity Principle

A useful principle of hydrodynamics is revealed by considering the steady flow of liquid in a pipe (or a tube of flow) of varying cross-sectional area (Fig. 9-15). The velocity of a liquid as it passes

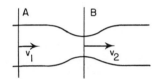

FIG. 9-15.—The continuity principle

normal to the plane A is v_1 and at B is v_2; the cross-sectional area at A is a_1, that at B is a_2. Since the liquid is assumed incompressible, there can be no accumulation of mass between A and B. It follows that the mass of liquid that passes A must be the same as that which passes B in a given time interval, provided that there are no "sources or sinks" between A and B. If ρ is the density of the liquid and t the time, then

$$v_1 a_1 \rho t = v_2 a_2 \rho t . \qquad (9\text{-}24)$$

Equation (9-24) expresses the *continuity principle.** If we divide equation (9-24) by ρt, we obtain the useful result

$$v_1 a_1 = v_2 a_2 . \qquad (9\text{-}25)$$

It follows directly from equation (9-25) that at B, where a_2 is less than a_1, the velocity of the liquid v_2 is greater than that of v_1 (refer to Fig. 9-14).

We may obtain another interesting result by applying the Second Law of Motion to the flow of liquid between A and B. A particle of liquid at A with velocity v_1 has its velocity increased at B to v_2.

* The word "continuity" is not too well chosen but is established by usage. What is "continuous" here is, of course, the same mass across all cross-sections of the tube.

Therefore, the liquid was accelerated in going from A to B. The acceleration came about because of difference in pressure acting on the particle of liquid flowing from A to B. The higher pressure, obviously, must be at A. We conclude, therefore, that in the steady flow of liquid the pressure is greatest where the velocity is least.

9.18. Bernoulli's Theorem

Let us go a step further and apply the principle of conservation of energy to the flow of a perfect incompressible liquid of density ρ in a tube of flow between the two planes A and B (Fig. 9-16). Let

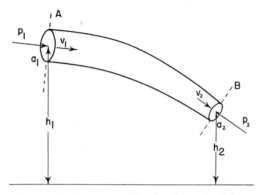

FIG. 9-16.—Energy considerations in a tube of flow

p_1 be the pressure, a_1 the cross-sectional area, and v_1 the velocity of the liquid particles at A, while p_2, a_2, and v_2 are the corresponding values at B. We may describe the motion as the entrance of Δm gm of liquid through A and the simultaneous exit of the same Δm gm through B. The forces acting on the mass during this motion are the forces of gravity and the forces $p_1 a_1$ and $p_2 a_2$ at each end of the tube of flow. The work done by the gravitational field is simply the difference in potential energy of the mass; i.e.,

$$\text{Work done by gravity} = \Delta m\, g\,\{\, h_1 - h_2\,\}, \quad (9\text{-}26)$$

where h_1 and h_2 are the differences in height of the tube from some appropriate plane.

The work done by the forces $p_1 a_1$ and $p_2 a_2$ during a time Δt is equal to

$$p_1 a_1 v_1 \Delta t - p_2 a_2 v_2 \Delta t , \qquad (9\text{-}27)$$

where $a_1 v_1 \Delta t$ is the volume of mass Δm. The net effect of the total work is to change the kinetic energy of the mass particles. We may write this change in kinetic energy as

$$\text{Change in kinetic energy} = \tfrac{1}{2}\Delta m\,(\,v_2^2 - v_1^2\,). \quad (9\text{-}28)$$

151

By the energy principle, then,

$$\Delta m\, g\,(h_1 - h_2) + p_1 a_1 v_1 \Delta t - p_2 a_2 v_2 \Delta t$$
$$= \tfrac{1}{2} \Delta m\,(v_2^2 - v_1^2). \quad (9\text{-}29)$$

Since, however, $a_1 v_1 \Delta t = a_2 v_2 \Delta t = (\Delta m / \rho)$, where ρ is the density, we may rewrite equation (9-29) in the following form, after arranging terms:

$$p_1 + \rho g h_1 + \tfrac{1}{2} \rho v_1^2 = p_2 + \rho g h_2 + \tfrac{1}{2} \rho v_2^2. \quad (9\text{-}30)$$

Equation (9-30) is known as *Bernoulli's principle* (Daniel Bernoulli, 1700–1782).

If we divide equation (9-30) by ρg, we obtain a form of Bernoulli's equation widely used by engineers:

$$\frac{p}{\rho g} + \frac{v^2}{2g} + h = \text{Constant}. \quad (9\text{-}31)$$

The *constant* is not the same for all streamlines. The dimension of each term of equation (9-31) is that of a length. The quantity $p/\rho g$ is the *pressure head* (Fig. 9-22); $v^2/2g$ the *velocity head* (Fig. 9-22); h the *potential* or *elevation head*. For the flow of liquid along a horizontal tube, Bernoulli's equation may be written for two points along a streamline in the tube as

$$\frac{p_1}{\rho g} + \frac{v_1^2}{2g} = \frac{p_2}{\rho g} + \frac{v_2^2}{2g}. \quad (9\text{-}32)$$

We wish to emphasize that Bernoulli's principle is strictly applicable only to streamline flow as defined in § 9.16.

In applying equation (9-31), we must be careful of the units used. For the mks system p is measured in newtons/m², ρ in kg/m³, g in m/sec², v in m/sec, and h in m. Similarly for the cgs system, p is in dynes/cm², ρ in gm/cm³, g in cm/sec², v in cm/sec, and h in cm. For the Engineering system the weight-density ρg is given in lbf/ft³, v in ft/sec, g in ft/sec², p in lbf/ft², and h in ft. Notice that in each system every term has the dimensions of length.

9.19. Application of Bernoulli's Theorem

a) Distribution of pressure along a rough tube.—We shall assume that the liquid that passes along the horizontal tube of uniform bore loses energy as heat, due to friction with the tube. Let τ be the energy per unit mass per unit length lost by the liquid, and l the distance

along the tube from A to B (Fig. 9-17). To the energy equation (9-29) we must add $\tau l \Delta m$, i.e.,

$$\Delta m\, g\,(h_1 - h_2) + p_1 a_1 v_1 \Delta t - p_2 a_2 v_2 \Delta t$$
$$= \tfrac{1}{2} \Delta m\,(v_2^2 - v_1^2) + \tau l \Delta m$$

or, since

$$a_1 v_1 \Delta t = a_2 v_2 \Delta t = \left(\frac{\Delta m}{\rho} \right),$$

we obtain

$$\frac{p_1}{\rho} + \frac{v_1^2}{2} + h_1 g - \frac{p_2}{\rho} - \frac{v_2^2}{2} - h_2 g = \tau l. \quad (9\text{-}33)$$

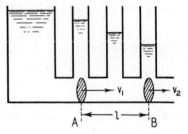

Fig. 9-17.—Energy loss in a tube of constant diameter

By the continuity principle, since the cross-sectional areas are the same,

$$v_1 = v_2,$$

so that equation (9-33) becomes, with $h_1 = h_2$,

$$\frac{p_1 - p_2}{l} = \tau \rho; \quad (9\text{-}34)$$

that is to say, a *pressure gradient* is set up along the tube proportional to the energy loss.

b) Pressure changes in a perfect liquid at a constriction.—We next consider the pressure changes that take place when a perfect liquid flows into a constricted tube (Fig. 9-18), where the tube is assumed horizontal. We are again assuming that there are no energy losses.

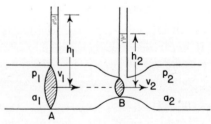

Fig. 9-18.—Reduction of pressure at a constriction of a tube.

At A the pressure is p_1, the velocity is v_1, and the area is a_1; at B these quantities are p_2, v_2, and a_2. Let a_2 be less than a_1. Applying Bernoulli's theorem, we get

$$\frac{p_1}{\rho} + \frac{v_1^2}{2} = \frac{p_2}{\rho} + \frac{v_2^2}{2}; \quad (9\text{-}35)$$

also, by the continuity equation for an incompressible liquid,

$$a_1 v_1 = a_2 v_2 . \qquad (9\text{-}36)$$

Solving these two equations for $p_1 - p_2$, we obtain

$$p_1 - p_2 = \tfrac{1}{2} \rho v_1^2 \left\{ \left(\frac{a_1}{a_2} \right)^2 - 1 \right\} . \qquad (9\text{-}37)$$

Since $a_1 > a_2$, then $v_2 > v_1$, and, by equation (9-35), $p_1 > p_2$; that is to say, in a constricted tube the velocity is increased and the pressure decreased (see § 9-17). This effect and other related ones are experimentally depicted in Fig. 9-19.

← Air flow

FIG. 9-19.—Photograph showing reduction in pressure at the central constriction and also reduction in pressure due to energy loss in tube.

We may use equation (9-37) to obtain the rate of flow of liquid in cm³/sec. The volume of liquid that passes A per second is $Q = v_1 a_1$, so that, by equation (9-37),

$$Q = a_1 \sqrt{ \frac{2 (p_1 - p_2)}{[(a_1/a_2)^2 - 1] \rho} } . \qquad (9\text{-}38)$$

The water meter whose working diagram is shown on Fig. 9-18 is known as a "Venturi flow meter." If p is in newtons/m², ρ in kg/m³, a_1 in m², then Q is in m³/sec.

c) *Flow of liquid from a small orifice.*—As shown in Fig. 9-20, liquid flows out of a tank with constant eleva-

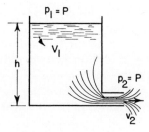

FIG. 9-20.—Water flowing out of a small orifice

tion head h; the external pressure on the two surfaces, at the top of the tank and on the emergent jet, is the same—that of the atmosphere. By Bernoulli's theorem,

$$\frac{p}{\rho} + g h + \frac{v_1^2}{2} = \frac{p}{\rho} + \frac{v_2^2}{2}$$

or

$$\frac{v_1^2}{2} + g h = \frac{v_2^2}{2} .$$

Also, $a_1 v_1 = a_2 v_2$, where a_1 and a_2 are the cross-sectional areas of the tank and orifice, respectively. If we assume that $a_1 \gg a_2$, then we may neglect v_1 in comparison with v_2 and get

$$v_2 = \sqrt{2 g h} , \qquad (9\text{-}39)$$

a relation first discovered by Evangelista Torricelli in 1641.

The actual rate of discharge depends upon the shape of the orifice. It is found experimentally that the rate of discharge (cross-sectional area times velocity) for a sharp-edged orifice is about 0.61 times the theoretical rate, a phenomenon known as the *vena contracta*, (Fig. 9-21).

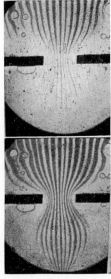

FIG. 9-21.—The vena contracta shown by ink jets some distance from the orifice.

d) *Measurement of fluid velocities with the Pitot tube.*— A very useful apparatus for measuring the velocity of a fluid is the Pitot-static tube. In Fig. 9-22, liquid is assumed to flow uniformly from left to right. This is indicated by the equally spaced streamlines. Along a streamline such as ABC we shall place an open-end manometer, called a *static tube* since it measures the hydrostatic pressure at B, p_B. The height of the stationary liquid in the tube is $p_B/\rho g$. At C we shall place the

153

bent tube known as the *Pitot tube*. The fluid comes to rest at C, and therefore the pressure here is not only hydrostatic pressure but also that contributed by the moving fluid. The total pressure is p_c. The height of the stationary liquid in the bent-tube manometer then is $p_c/\rho g$. The pressure and velocity at B and C are related

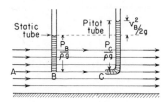

Static tube → | Pitot tube | $\frac{v_B^2}{2g}$

$\frac{p_B}{\rho g}$ $\frac{p_c}{\rho g}$

A   B   C

Fig. 9-22.—The Pitot tube

through Bernoulli's equation (9-32). Along the stream-line ABC we obtain

$$\frac{p_B}{\rho g} + \frac{v_B^2}{2g} = \frac{p_c}{\rho g} + 0,$$

since the velocity at C is zero. Rearranging terms, we get

$$\frac{p_c}{\rho g} - \frac{p_B}{\rho g} = \frac{v_B^2}{2g},$$

that is, the velocity head $v_B^2/2g$ is the difference between the static and the Pitot readings, as shown in Fig. 9-22. We may solve this equation for the free-stream velocity v_B as follows:

$$v_B = \sqrt{\frac{(p_c - p_B)\,2}{\rho}}. \qquad (9\text{-}40)$$

In actual practice this equation is modified by introducing a suitable multiplying coefficient. This modification is due to the fact that the Pitot-static tubes disturb the fluid motion somewhat. This Pitot-static combination, however, is widely used on aircraft as air-speed indicators.

EXAMPLE 2: Water flows in a tapered pipe at the rate of 50 ft³/min from a cross-section, A, of area 10 in² to a cross-section, B, of area 4 in², which is 5 ft higher than A. If the pressure at A is 1,500 lbf/ft², what is the pressure at B?

Solution: In this problem it is convenient to measure all quantities in the same units, i.e., p in lbf/ft², v in ft/sec, density in lbf/ft³, etc.

Since the rate of flow is equal to the product of area and velocity, it follows that

$$v_A = \frac{(50)(144)}{(60)(10)} = 12 \text{ ft/sec}\,; v_B = \frac{(50)(144)}{(60)(4)}$$

$$= 30 \text{ ft/sec}$$

also, we may take $h_A = 0$ and $h_B = 5$ ft; $p_A = 1,500$ lbf/ft²; and $\rho = 62.4$ lbf/ft³. Applying Bernoulli's principle (eq. [9-31]), we obtain

$$\frac{1,500}{62.4} + \frac{(12)^2}{2g} + 0 = \frac{p_B}{62.4} + \frac{(30)^2}{2g} + 5,$$

$$p_B = 460.9 \text{ lbf/ft}^2.$$

9.20. Other Examples of Bernoulli's Effects

The pressure-velocity relation of Bernoulli's theorem as deduced for a liquid in § 9.18 permits a very useful general qualitative statement for all fluids (liquid and gases) in stationary flow:

In regions where the velocity of flow is increased, the hydrostatic pressure of the fluid is less than in the surrounding fluid.

Many examples of Bernoulli's theorem are related to common experiences, some of them illustrated in Fig. 9-23. A ball can be supported on a

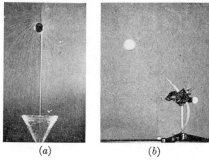

(a)   (b)

Fig. 9-23.—Light ball can be balanced in (a) a jet of water, (b) a jet of air.

swiftly moving jet either of water or of air. In the vicinity of the jet, where there is swift motion of the fluid medium, a slight reduction of pressure below the normal atmospheric pressure results. Any light object in the vicinity of the jet is thus pushed into it by this excess pressure from outside. The upward momentum of the jet is, to some extent, imparted to the object in it, which is thus supported against gravity.

The curved flight of a pitched ball requires that the ball be rotated when thrown. The combination of air flow due to the forward motion and of that due to the rotation results in different speeds of air passing the ball on opposite sides (Fig. 9-24 [a]). The ball drifts toward the region of lower pressure, i.e., curves in that direction toward which its "nose" of forward face is moving.

At side p_2 the velocity of the air is approximately the sum of that due to the forward motion of the ball and that dragged with the ball in its spinning motion, whereas at side p_1 these two velocities are in opposition. There is then a larger air velocity on side p_2 than on side p_1, resulting in an air pressure p_2 smaller than p_1 and a sidewise force toward side p_2.

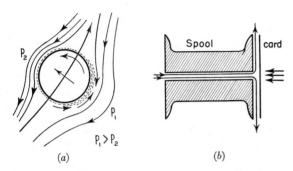

Fig. 9-24.—The curving of a spinning ball

Figure 9-25 (*a*) shows two hanging tennis balls which exhibit an apparent "attractive" force for each other when a jet of air is blown between them. Between the balls the velocity of the air is very

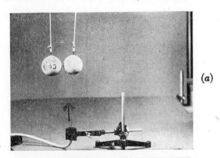

(*a*)

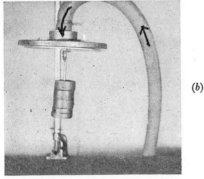

(*b*)

Fig. 9-25.—(*a*), "Attraction" produced by jet of air blown between two hanging tennis balls; (*b*), weights attached to a plate are held up by forcing a jet of air between the two plates.

large, and the pressure is therefore reduced, resulting in a sidewise force pushing the balls together. A more practical case of this sidewise force

occurs when two ships are moving parallel and close together. Unless some caution is used, they may collide. When a relatively light car passes a truck on a highway with both going at high speed, there is a motion of the car toward the truck. Suppose air is blown through a small hole in a spool, as shown in Fig. 9-24 (*b*). If a card is placed over the hole, then there is a force holding the card against the spool. When blowing through the hole against the card, the air emerging from the hole and deflected by the card into a radial flow away from the hole in all directions beneath the card is under less pressure because of its velocity than is the outside air. Thus the harder one blows, the more forcefully does the atmosphere press the card inward toward the hole in the spool. A large weight can be held up by a sufficiently large jet of air (Fig. 9.25 [*b*]).

A common laboratory device is the water aspirator pump used for partial evacuation of air from apparatus. It also operates in conformity to these principles. A jet of running water from a contracted pipe at *a*, Fig. 9-26, is directed into an

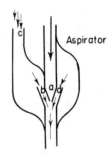

Fig. 9-26.—The water aspirator

expanded opening of a funnel at *b* and passes out into the drain. Around this jet is a region of reduced pressure, which draws in the air from all sides and thus evacuates any apparatus connected to *c*.

The forces on airplane wings or airfoils can be explained by Bernoulli's principle. When a plane moves in horizontal or slightly upward flight, there is an increase in velocity of the air above the wing relative to that below. There is thus a reduction in pressure above the wing relative to that below, producing an upward force, called "lift." As the angle which the wing makes with the line of flight —the angle of attack—increases, the streamline breaks up and turbulence sets in, resulting in a decrease in the lift force, resulting in stalling.

9.21. Viscosity: Internal Friction in a Fluid

We have so far been considered what we have called a "perfect fluid," i.e., a fluid that cannot transmit shearing stresses and on which no work is done in changing its shape. As we have already pointed out, no fluid is a perfect fluid—in fact, such liquids as honey, glycerin, and oil tar are far from perfect. They are called "viscous liquids" and possess the property known as "viscosity"—a manifestation of the internal friction of fluids.

In order to understand the nature of viscosity or internal friction of fluids, we shall begin by considering the following simple example. Suppose that we have two plane solid surfaces, one above the other, and that between them is a continuous fluid, such as castor oil or glycerin (Fig. 9-27). If we apply a constant force, f, to one of the

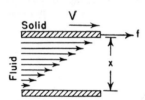

FIG. 9-27.—Viscosity of fluids

surfaces, experiment shows that it will accelerate only for a moment and will promptly acquire a constant velocity, v, moving with a uniform rate across the other surface. Furthermore, at the surface of each solid there will be fluid adhering to it and with no motion at this point with respect to the solid. If we double the force, the velocity of one plane past the other will be doubled. This limiting velocity is proportional to the force, and vice versa. If the thickness, x, of the fluid between the plates is increased, the application of a given force results in a correspondingly greater velocity between the plates. If the area, A, of the plates is increased, a corresponding diminution of speed will be observed, other conditions being the same as before. Thus velocity is proportional to f and also to $1/A$:

$$v \propto (f)(x)\left(\frac{1}{A}\right);$$

and we can write an equation for f:

$$f = \eta A \frac{v}{x}, \qquad (9\text{-}41)$$

where the constant, η, depends only upon the fluid and its temperature. Equation (9-41) *defines* the

COEFFICIENT OF VISCOSITY of the fluid. We note that this coefficient may be interpreted as that tangential force necessary to maintain a unit-velocity difference between two planes each of 1 cm² in area and 1 cm apart in the fluid; i.e.,

$$\eta = \frac{f}{A}\frac{x}{v},$$

when A, x, and v are each equal to 1. In the setting-up of this definition we have chosen a case in which the application of forces to fluids produces differences of velocity between thin adjacent sheets of layers (laminae) within the fluid. Such flow is called, for this reason, LAMINAR FLOW and is characterized by the proportionality between force and velocity that we have noted previously.

If we divide equation (9-41) by the area A, the ratio $f{:}A$ is, according to § 7.5, a shearing stress. If we call this shearing stress τ, then equation (9-41) can be written as

$$\tau = \eta \frac{v}{x}. \qquad (9\text{-}42)$$

The quantity v/x represents the space rate at which the velocity changes in the x direction. A more accurate and general representation of v/x would be dv/dx, so that the foregoing equation becomes

$$\tau = \eta \frac{dv}{dx}. \qquad (9\text{-}43)$$

Equation (9-43) is known as "Newton's law of viscosity." For many fluids equation (9-43) is correct. However, such fluids as blood and colloidal suspensions in general are exceptions.

The cgs unit of viscosity is the dyne-second per square centimeter (dyne-sec/cm²), called the "poise," and is that of a liquid in which 1 dyne is required to maintain the unit-velocity difference tangentially between two 1-cm² surfaces, 1 cm apart:

Poise = dyne-sec/cm² = [gm] [cm] $^{-1}$ [sec] $^{-1}$.

9.22. The Nature of Viscosity

Viscosity in a liquid is to be attributed to the effect of the cohesive forces between the molecules, relatively close together in a liquid. In a gas, however, in which viscosity is also an obvious and by no means negligible property, the molecules are too far apart for their cohesion to be effective, and another physical interpretation must be sought.

This is found in the phenomenon of momentum transfer in a gas both ways between two regions having a relative velocity to each other. Since one layer of gas moves past another with a relative velocity v, it is evident that molecules continually cross the boundary from one layer to the next. Hence the molecules of the faster-moving layer, on crossing the boundary, transfer a larger amount of momentum to the lower layer than do molecules that cross into the upper layer. Because of this, there is a tendency for the upper layer to slow down and the lower layer to speed up. Thus we see that the internal friction or viscosity of gases is essentially a *momentum-transfer process*.

A complete mathematical theory of such a process would predict that the coefficient of viscosity should increase with temperature. In the case of gases this is accurately established, and the explanation given previously can be considered correct. However, in the case of liquids, the coefficient of viscosity decreases with rise in temperature. This fact is in qualitative agreement with one's expectations when the viscous forces originate chiefly in molecular attraction, as they do in liquids.

9.23. Methods of Measurement of the Coefficient of Viscosity

If we constrain the motion of fluid through a cylindrical tube so that its velocity of flow is quite small (either by having a very fine capillary tube if the fluid is gaseous or a very inviscid liquid or by permitting ourselves the use of a larger, shorter tube for a liquid like castor oil or glycerin), laminar flow occurs, and we can measure the coefficient of viscosity of the fluid in terms of the amount of it that will pass in a given time through the tube in question. The dimensions of the tube are also involved in the result and, naturally, the pressure difference by means of which the fluid is forced through the tube. The analysis using equation (9-43) is too lengthy to give in detail here. The relation is given in the following:

$$Q = \frac{\pi R^4}{8\eta L}(p_1 - p_2).\qquad(9\text{-}44)$$

This equation is known as POISEUILLE'S EQUATION, in honor of the man who first derived it. In it Q is the quantity of fluid that flows out per second (cm³/sec), R is the radius of the tube (cm),

L is the length of the tube (cm), $p_1 - p_2$ is the difference in pressure in the fluid at the two ends of the tube (dynes/cm²), and η is the coefficient of viscosity which can be found by solving the equation for it in terms of the other quantities, all of which are measurable (poise).

We shall show that both sides of equation (9-44) have the same dimensions. The dimensions of the individual terms are:

$$Q\text{:}[L^3T^{-1}];\ R^4\text{:}[L^4],\ p\text{:}[MLT^{-2}L^{-2}]\text{ or }[ML^{-1}T^{-2}].$$

For the dimensions of η we use equation (9-42) or

$$\eta = \frac{\tau x}{v},$$

so that η has the dimensions of

$$[MLT^{-2}L^{-2}]\,[L]\,[L^{-1}T]\qquad\text{or}\qquad[ML^{-1}T^{-1}].$$

Thus the right-hand side of equation (9-44) has the dimensions of

$$[L^4]\,[ML^{-1}T^{-2}]\,[M^{-1}LT]\,[L^{-1}]\qquad\text{or}\qquad[L^3T^{-1}],$$

which is the same as that of the left-hand side for Q.

Another method for determining the coefficient of viscosity η relates to the terminal velocity, v, with which a sphere of radius R falls through a fluid. There is a force F exerted by the fluid on the sphere, and this force depends on the radius R, the coefficient of viscosity η, and the value of the terminal velocity v if the latter is small. The only manner in which a combination of $R\eta v$ can have the dimensions of force is that of a simple product, or

$$F = k\,(R\,\eta\,v),$$

where k is a dimensionless constant.

The student should check for himself that the dimensions of the product $R\eta v$ are that of a force. By physical analysis of this problem Stokes found the constant to be 6π. Thus the resistance force exerted on a sphere falling with constant velocity in a medium having a coefficient of viscosity η is

$$F = 6\pi\,\eta\,vR\ .\qquad(9\text{-}45)$$

Since this force F is exactly balanced by the weight of the sphere, we have

$$F = \tfrac{4}{3}\pi R^3\,(\sigma - \rho)\,g\ ,\qquad(9\text{-}46)$$

where σ is the density of the sphere and ρ that of the fluid. Notice that $4\pi R^3\rho g$ is the buoyant force on the sphere. Thus

$$\tfrac{4}{3}\pi R^3\,(\sigma - \rho)\,g = 6\pi R\,\eta\,v$$

or

$$\eta = \tfrac{2}{9}\frac{g}{v}R^2\,(\sigma - \rho)\ .\qquad(9\text{-}47)$$

157

Equation (9-45) or (9-47) is often called "Stokes's law." By measuring the terminal velocity of the sphere, the coefficient of viscosity of the fluid can be obtained from equation (9-47).

It is interesting to note that if raindrops falling from a cloud of several thousand feet were to fall without encountering any resistance forces, then each time a rainstorm began, we should have to run to cover lest we be killed by the impact. That this does not happen is due to the fact that they reach a terminal velocity in the air. This can be calculated from Stokes's law and increases with the size of the raindrops. At a diameter of about 1 mm the terminal velocity is about 4.3 m/sec, while for those of 2-mm diameter it is about 5.8 m/sec. Table 9-2 shows the values of the coefficient of viscosity of a few common substances.

TABLE 9-2

VISCOSITY COEFFICIENTS AT 20° C
UNLESS OTHERWISE STATED

Liquids		Gases ($\eta \times 10^6$)	
Sulphur (187°)....	560.00 poises	Air at 20°.2.......	181.2
Castor oil........	10.71 poises	Air at 99°........	220.3
Glycerin.........	8.30	Air at 302°.......	299.3
Water at 0°......	0.0179	He at 15°........	196.9
Water at 100°....	0.0028	H_2 at 15°........	88.9
Ether...........	0.0124	H_2O at 100°......	132.0
Mercury.........	0.0154	A at 15°.........	220.8
Liquid air (−192°)	0.0017	CO_2 at 15°.......	145
		Ether at 16°......	73.0
		Benzene at 19°....	79.0

9.24. Turbulent Flow

One of the most obvious facts about the resistance which solid objects encounter when they are dragged or driven through any fluid (gas or liquid) at any but the smallest velocities is that the resistance experienced is proportional *not* to the first power of the velocity but approximately to the *SQUARE of the velocity*. The laws of viscous friction, the formulation of which is predicated on very low velocities and on flow that is laminar in type, can be of no use to us in these connections.

It was Osborne Reynolds who first investigated the cause of these two different laws of fluid friction. He discovered that the change from the first-power law to the second-power law was not gradual but astonishingly sudden and that it occurred in any given fluid and for any given apparatus *always at the same critical velocity*. By a beautiful experiment he showed that this change in the law of resistance occurred simultaneously with a change

in the character of the flow in the apparatus from the laminar type to TURBULENT FLOW. His experiment (Fig. 9-28) consisted simply of feeding a thin thread of highly colored liquid into the center of a tube through which the same liquid, uncolored, was flowing. He could alter the velocity of flow over quite a considerable range, at each speed measuring the resistance of the tube upon the liquid within it.

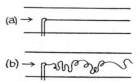

FIG. 9-28.—A diagram of an apparatus for illustrating turbulent flow.

At low speed the colored thread of liquid remained unbroken and continuous throughout the length of the tube (Fig. 9-28 [*a*]). When a certain critical speed was reached, the colored thread was violently agitated and its continuity destroyed by eddies, thus revealing turbulence of flow (Fig. 9-28 [*b*]). It was at exactly this speed that the law of resistance was found to change in the exponent of *v*.

The cause of the relatively high values of the resistance of objects not streamlined in character and moving through air is due more to the reduction of pressure on their leeward (away-from-the-wind) surfaces than to the impingement of the air blast on their windward faces.

9.25. Résumé

The following is a brief listing of the definitions, principles, and theories, given in this chapter, which you should know:

Definition of pressure

Variation of pressure with depth in an incompressible fluid

The center of pressure and the stability of dams

Transmission of pressure and Pascal's law

Archimedes' principle and the measurement of density and specific gravity

Work done in moving a volume of fluid against a pressure

Principle of conservation of fluids

Conservation of energy in fluids—Bernoulli's principle

The definition and method of measurement of coefficient of viscosity

Conditions for turbulent motion

Queries

1. A carefully ground solid cylinder of glass is placed at the bottom of a tank and then mercury is carefully poured into the tank covering the glass cylinder (see the diagram herewith). In spite of the fact that the cylinder displaces a weight of mercury far in excess of its own weight, the cylinder remains in place. Explain this paradox.

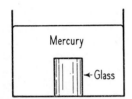

2. A tank of water is placed on one side of a beam scale, and weights are added to bring the scale into balance. A brass cylinder held by a string is lowered into the tank. Is the balance disturbed? Explain. This is known as the "inverse problem of Archimedes."

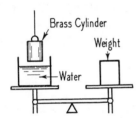

3. The following is known as "Pascal's paradox." The base of a piston is attached by a string to one end of a beam balance. The cylinder inclosing the piston is attached to some rigid support independent of the balance, as shown in the accompanying diagram. If the cylinder is filled with water to a height h, a weight W must be placed on the other end of the beam to balance. If, however, the liquid is frozen and freed from the sides of the cylinder, a weight W_1 much less than W will keep the beam in balance. Explain with the aid of Pascal's principle.

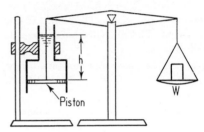

4. An object floats on the surface of mercury exposed to the atmosphere. If the air above the mercury is removed, will the object remain at its former depth, sink, or rise somewhat?

5. Explain why a large inflated balloon has a definite ceiling or height to which it will rise, while a submarine, if it starts to sink and makes no changes, will go to the bottom of the ocean.

Practice Problems

1. A block of wood of rectangular section and 10 cm deep floats in water. If its density is 0.7 gm/cm³, how far below the surface is its lower face? Ans.: 7 cm.

2. If the area of the block in problem 1 is 100 cm², what weight placed on the upper surface will sink the block to a depth of 9 cm?

3. A U-tube has water placed in the lower part, and then 10 cm of oil of density 0.8 gm/cm³ is put into one arm. How much difference will there be in the levels of the water in the two arms? Ans.: 8 cm.

4. A cylinder and piston with a cross-section of 1,000 cm² are arranged with a side tube of cross-section 2 cm², as indicated. If the cylinder is filled with water and a 75-kg man steps on the piston, how high in meters will the water be forced in the side tube?

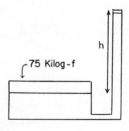

5. A lead cylinder weighs 115 gm in air and has an apparent weight of 105 gm in water and 107 gm in oil. Calculate the density of the lead and oil. Ans.: 11.5 gm/cm³; 0.8 gm/cm³.

6. A sinker weighing 38 gmf is fastened to a cork weighing 10 gm, and the two together are in equi- **159**

librium when immersed in water. Find the density of the sinker if that of the cork is 0.25 gm/cm³.

7. A balloon of volume 1,500 m³ is filled with hydrogen. If the mass of the balloon and basket is 250 kg, calculate the force tending to raise the balloon from the ground, given the densities of air and hydrogen as 1.29 and 0.09 gm per liter, respectively. Ans.: 1,550 kgf.

8. A Venturi meter has a pipe diameter of 10 in and a throat diameter of 5 in. If the water pressures in the pipe and throat are 8 and 6 lbf/in², respectively, calculate the rate of flow of the liquid in cubic feet per second.

9. A water motor uses ¼ ft³ of water per second, delivered at a pressure of 20 lbf/in². If the motor is 50 per cent efficient, what horsepower does it develop? Ans.: 0.655 hp.

10. Water is maintained at a constant height of 10 m in a tank. Calculate the velocity of efflux. Ans.: 1,400 cm/sec.

11. A horizontal capillary tube 25 cm long and with a bore of 1 mm in diameter is arranged as shown in the diagram herewith. If glycerin is poured into the thistle tube to a height of 50 cm above the capillary tube, calculate the time required for 1 cm³ of glycerin to flow out of the end of the tube after a steady state has been reached. The specific gravity of glycerin is 1.26. Ans.: 1.37×10^3 sec.

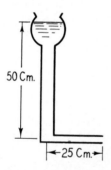

50 Cm.

←25 Cm.→

12. A sphere of radius 10 cm is placed on a sensitive analytical balance and found to weigh 100 gm. Because of buoyancy in the air, this is only the apparent weight. Taking the density of the weights to be 8.5 gm/cm³ and that of the air to be 0.00129 gm/cm³, calculate the true weight of the sphere. What percentage of error is made if the buoyancy is neglected?

13. The water mains of a city are laid over a hill 100 ft above the level of the pumping station. If the pressure at the station is 50 lbf/in², calculate the pressure at the top of the hill, neglecting all friction losses. Ans.: 6.67 lbf/in².

14. Calculate the terminal velocity of a spherical raindrop, radius 0.2 mm, using Stokes's law. Take the density of air at 20° C to be 0.00129 gm/cm³.

15. A pipe in which water is flowing at the rate of 200 ft³/min gradually changes from a cross-sectional area of 4 in² at A to 14 in² at B, a point 18 ft higher than A. Calculate the pressure at B if that at A is 1,500 lbf/ft². Ans.: 13,000 lbf/ft².

16. Water is maintained at a height of 10 m in a tank. Calculate the radius of a circular opening at ground level that will eject liquid at the rate of 10 cm³/sec.

17. Vertical walls form the sides and ends of a swimming pool 100 ft long, the bottom being a single inclined plane. The water is 4 ft deep at one end and 12 ft deep at the other. Find the hydrostatic force on one side of the pool. Ans.: 2×10^5 lbf.

18. Find the center of pressure on a vertical rectangle 12 ft by 8 ft high when the upper end is in the surface of the water, and also when the upper end is 4 ft below the surface. Find moments of the hydrostatic forces with respect to the lower end in the two cases.

19. Manometers of the type shown in Fig. 9-17 are placed 9 in apart in a horizontal tube of constant diameter through which water is flowing at a constant rate. If the heights of the water manometers are 9 and 6 in, respectively, find the energy loss per unit mass per unit length of the water in the tube. Ans.: 10.67 ft-lbf/slugs-ft.

20. What would be the difference in pressure between a static and a Pitot tube immersed in a water stream of velocity 100 cm/sec? In an air stream of velocity 300 km/hr? Density of air 0.00129 gm/cm³.

21. A Venturi meter consists of a pipe having a main cross-section of 0.5 ft² and a constriction of 0.25 ft², with a pressure difference between these two of 10 lbf/in². If water is flowing through the pipe of weight-density of 62.1 lbf/ft³, find the number of cubic feet flowing through the pipe per second. Ans.: 35.2 ft³/sec.

22. A fire hose connected with the city mains can throw water to a height of 50 feet. If frictional effects can be neglected, find the pressure of the water in the mains.

23. Find the ratio of the radii of two spheres, one of steel of density 8 gm/cm³ and the other having a density of 6 gm/cm³, if they both fall with the same velocity in oil whose density is 0.9 gm/cm³ and viscosity is 2.5 poises. Ans.: 1.18.

24. Find the radius of a water droplet which falls in air of density 0.00125 gm/cm³ and viscosity 2×10^{-4} poises with a terminal velocity of 10³ cm/sec.

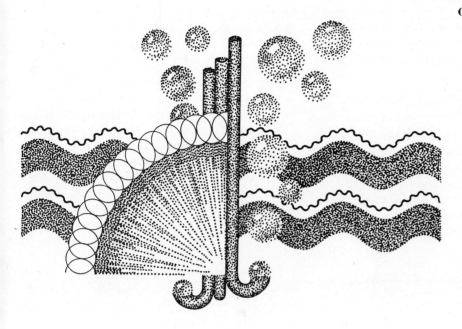

CHAPTER 10

Surface Tension

10.1. Molecular Forces

There is one important class of mechanical phenomena in connection with liquids which exhibits gross effects by no means insignificant in magnitude but which nevertheless has its underlying cause in the *fine-scale structure* of the liquid itself. We refer to those interesting properties that are associated with the free surface of any liquid or the interface between any two nonmiscible liquids. At first glance these properties might seem to suggest that there is a kind of contractile surface membrane or skin that envelops all liquids. For example, a waxed needle may float on the surface of water, even though its density far exceeds that of the liquid. Many insects are able to walk on the surface of liquids, in apparent contradiction to Archimedes' principle.

The surface properties of liquids, as we shall see, result from the existence of intense intermolecular forces—forces of cohesion. These forces do not follow an inverse-square law and are not gravitational in origin. Recent investigations in atomic and molecular structure suggest that these forces are due to the interaction of the electrical fields that surround atoms. The forces between molecules vary as inverse powers of the distance. Depending upon the types of molecules and types of

forces, these powers may range from the second to the seventh! In other words, the intermolecular forces are of extremely short range, and the details of calculating their magnitude are extremely complicated. At a distance of about 10^{-8} cm between nitrogen atoms, the intermolecular forces are of the order of 10^{-6} dynes, while the gravitational attractive forces are very much less, of the order of 10^{-45} dynes.

Many simple experiments with which we are acquainted suggest the existence of these short-range intermolecular forces. A plate of glass may be brought very close to the surface of a liquid without being especially attracted toward it; but if the glass touches the liquid, the liquid is found to cling to the plate with surprising tenacity. Again, if two accurately scraped gauge blocks (Johansson blocks) are pressed into contact, surprisingly large forces are necessary to separate them. A drop of mercury, instead of spreading out into an infinitely thin layer because of gravitational forces, forms, instead, a drop nearly spherical in shape. Soldering and welding are processes dependent upon the existence of intermolecular forces.

These intermolecular forces are directly responsible for many of the capillary phenomena: **161** the rise of liquids in fine-bore tubes; the complete wetting of a towel when one end is dropped into water; the rise of oil in wicks and of water in soil.

10.2. Contraction of a Liquid Surface: Molecular Pressure

Let us now look more closely at the situation in which a molecule finds itself when it has drifted (by virtue of its thermal agitation and its mobility) into the surface region of a liquid.

When in the interior of the liquid, this molecule was pulled radially in all directions by the forces of attraction of neighbors on all sides (Fig. 10-1 [a]).

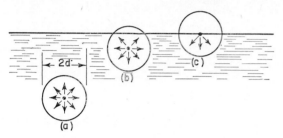

Fig. 10-1.—Forces on the molecules of a liquid

Suppose this attraction was conspicuously effective only through a distance d, called the "molecular range," whose value is roughly 10^{-7} cm.

Just as soon as any molecule wanders to within a distance less than d from the surface, the forces on it become unbalanced, since a hemisphere of this radius below the molecule in question is filled with other molecules attracting it; but the hemisphere above the molecule now extends above the liquid surface, and the upper portion of it is not filled with molecules to the same extent as is the rest of it (Fig. 10-1 [b]).

As the molecule in question approaches close to the boundary, the unbalanced forces on it increase to a maximum when it lies at the boundary (Fig. 10-1 [c]). This surface layer of thickness d is sometimes called the "active layer." Thus there is a strong tendency for the molecule to be pulled back out of the surface into the liquid and for *the surface of the liquid to contract spontaneously.*

Because of this tendency to contract, the interior of a liquid is considered to be subject to an enormous pressure, which for water has been estimated to be roughly 10,000 atm. There is, however, no direct hydrostatic way to measure this molecular pressure.

10.3. Free Energy of the Surface: Definition of Surface Tension

The fundamental property that we deduced in the previous section—that a liquid surface contracts spontaneously—implies that work must be done in increasing the free surface of the liquid. The origin of this work in terms of molecules is this: When the surface is increased, molecules from the interior of the liquid must be brought to the surface, and hence work must be done against the inward attractive forces. The work done in extending the surface a given amount is a definite quantity, since molecules have a definite size and there will always be a definite number of them in a unit area.

The work per unit area expended in extending the surface of a liquid is, by definition, the "surface tension" (T) *of the liquid.* The temperature of the surface must remain constant for a constant T, for reasons that will become clear later. The units of T are joules per square meter (joules/m²) which is equivalent to newtons per meter (newtons/m) and to ergs per square centimeter (ergs/cm²), which is equivalent to dynes/cm.

This definition of surface tension gives rise to a straightforward method for its measurement. Suppose we insert a wire frame, as in Fig. 10-2, beneath the surface of a liquid. Upon withdrawing it

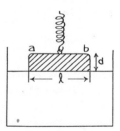

Fig. 10-2.—A method for measuring surface tension

carefully, we will notice that the surface has now been spread out over the frame, and we have literally pulled out of the original surface of the liquid an additional surface of area $2ld$, where l is the length of the frame and d is the distance the frame is out of the water. The factor 2 takes care of the fact that there are two sides to the film. By our definition, the work done in extending the liquid surface must be

$$W = 2Tld . \qquad (10\text{-}1)$$

If F is the force required to pull the film out, then, since it acts through the distance d, the work done is

$$Fd = 2Tld \qquad (10\text{-}2)$$

or

$$T = \frac{F}{2l}. \qquad (10\text{-}3)$$

One of the remarkable results we obtain from this experiment is that the force F is constant, independent of the area of film on the frame. This is different from what we would expect if the surface were an elastic membrane. Of course, the forces acting on the surface of a liquid are in no way related to elasticity.

Experimentally, the force F can be obtained by a spring attached to a movable vertical arm having an index which gives a measure of the tension in the spring (Fig. 10-3). A clean wire frame of, say, 10 cm in length

FIG. 10-3.—An apparatus for measuring surface tension

is placed in a vessel of clean water. First the wire frame is placed in the water with the vertical legs almost, but not quite, submerged. The index on the spring reads 37.4 cm. The vessel of water is then raised until the entire frame is submerged, after which the vessel is slowly lowered. A water film forms on the wire frame. The maximum tension which the film can support is given by the reading of the index on the spring of 39.5 cm. By a separate experiment, the force constant of the spring was found to be 0.715 gmf/cm. Thus the force exerted by the film is

$$(39.5 - 37.4) \times 0.715 = 1.501 \text{ gmf}$$

$$= 1,470 \text{ dynes}.$$

Since there are two sides to the film, it follows that the surface tension T of the water in this experiment is

$$2Tl = 1,470$$

$$T = \frac{1,470}{2 \times 10} = 73.5 \text{ dynes/cm}$$

$$= 0.0735 \text{ newton/m}.$$

Equation (10-3) also brings out an interesting and useful mathematical result. The surface tension T whose units are ergs per square centimeter is equivalent to a force per unit length. Because of this mathematical equivalence, it has become customary in the solution of many problems to substitute for work per unit area "force per unit length." If, as in Fig. 10-4, a line 1 cm long is drawn in the

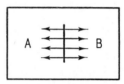

FIG. 10-4.—Tension in a liquid surface

surface of a liquid, surface A on one side of the line pulls surface B on the other side of the line with a force of T dynes. The dimensions of surface tension T are $[MLT^{-2}L^{-1}]$ or $[MT^{-2}]$.

10.4. Factors Affecting Surface Tension

The definition we have given for surface tension is, strictly speaking, that for a liquid in equilibrium with its vapor. If two nonmiscible liquids are in contact, the interfacial surface tension is less than the surface tension of the liquid with the higher tension. This can easily be seen when we realize that less work is required to bring a molecule to the interface than to the surface of liquid surrounded by its vapor because of the intense forces acting inward toward the second liquid. We may consider a contaminated surface as the interface of two liquids. Hence, in the measurement of surface tension of pure liquids, it is essential that the surface be kept scrupulously clean.

Measurements on surface tension show that, with rare exceptions, the surface tension decreases with rising temperature. It is for this reason that we defined "surface tension" as "work per unit area at constant temperature." Table 10-1 gives **163** the variation of surface tension with temperature for water and alcohol, as well as the surface tension of various liquids.

TABLE 10-1

VALUES OF SURFACE TENSION
(Dynes per Centimeter)

Temperature (° C)	Water	Alcohol
0.........	75.6	24.0
25.........	71.9	21.8
50.........	67.9	19.8
100.........	58.8	....

Mercury—air...........	513	
Mercury under water.....	392	
Turpentine—air.........	28.8	
Turpentine—water.......	11.5	
Glycerin................	63	
Ether..................	16	
Soap solution...........	25	(approx.)

10.5. Angle of Contact: Adhesive Forces

When liquids are contained in solid vessels, not only are the forces of the liquid upon itself (the cohesive forces) important, but, where the liquid is adjacent to the solid wall, phenomena occur that indicate the effect of the attraction of molecules of the solid for those of the liquid. These forces are sometimes loosely spoken of as "forces of adhesion."

Take the case of pure water in a clean glass dish (Fig. 10-5). At the edge of the water surface in con-

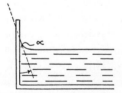

FIG. 10-5.—Angle of contact for clean water and clean glass is 0°.

tact with the glass, it will be observed that the surface of the water turns up and lies flat on the glass. The water "wets" the glass. Evidently, the attraction of glass for water is greater than that of water for itself, since the mere presence of a glass wall will pull out a thin film of water into close contact with the glass.

The case of mercury in glass is quite different (Fig. 10-6). At the edge of a pool of mercury in a

glass container the mercury surface is depressed, and there is a tendency for it to draw down away from the glass.

We define the ANGLE OF CONTACT between a liquid and its solid container as the angle a (see Figs. 10-5 and 10-6). This angle is measured, as indicated, from the vertical section of wall below the liquid around to the position of the tangent to the liquid surface at its point of contact with the

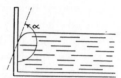

FIG. 10-6.—Angle of contact for a liquid such as mercury

wall. For a pair of clean surfaces the angle of contact is a definite quantity. A few values are listed in Table 10-2.

TABLE 10-2

ANGLES OF CONTACT

Pure water—clean glass........	$a =$ 0°
Impure water—ordinary glass...	$a \sim$ 25°
Pure mercury—clean glass......	$a =$ 148°
Mercury after exposure to air...	$a <$ 140°
Turpentine—glass.............	$a =$ 17°
Kerosene—glass...............	$a =$ 26°

10.6. Pressure at a Curved Film

Let us consider a spherical soap bubble of radius R (Fig. 10-7). Let the pressure inside the

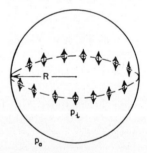

FIG. 10-7.—Pressure difference across a soap bubble

bubble be p_i and that outside the atmospheric pressure, p_a. There is then an excess pressure of $p_i - p_a = \Delta p$ across the soap film. Consider the forces acting across a plane passing through the center. The upper and lower halves are held together by the surface-tension force of $2T \times 2\pi R$, since there are two surfaces—an inner and an outer. The opposing force is that due to the pressure difference, or Δp, acting across the dimetral plane of area πR^2. Hence, for equilibrium,

$$\Delta p \, \pi R^2 = 4 \pi R T$$

or

$$\Delta p = \frac{4T}{R}. \qquad (10\text{-}4)$$

A more general relationship was given by Laplace (1749–1827), who showed that, for a stable film having principal radii of curvature of R_1 and R_2 at any point where the excess pressure is Δp, the surface tension T of the film is given by

$$\Delta p = 2T\left(\frac{1}{R_1}+\frac{1}{R_2}\right). \qquad (10\text{-}5)$$

(a)

(b)

Fig. 10-8.—(a), Two soap bubbles of approximately equal size initially; (b), the smaller bubble then blows up the larger one.

The factor 2 arises because the film has a double surface. For the case of a spherical soap bubble in which $R_1 = R_2 = R$, the excess pressure is $4T/R$, as given in equation (10-4).

The rather surprising result that the larger the radius of the bubble, the smaller the pressure dif-

ference can be verified experimentally by blowing a bubble on a tube connected to a manometer.

Another rather striking experiment is to blow two soap bubbles* on the ends of tubes connected by a manometer, as shown in Fig. 10-8 (a). Originally they are of approximately the same size, but, since it is not possible to have them of exactly the same size, we observe that the smaller bubble, with the greater pressure, decreases, while the larger bubble, with the smaller pressure, increases. The final result is shown in Fig. 10-8 (b).

10.7. Rise (or Fall) of Liquids in Capillary Tubes

As we shall see, some liquids rise in a capillary tube, while others fall below the free surface of the liquid. This rise or fall depends on the angle of contact. First, let us consider the case of a liquid such as water, which rises in a capillary tube, from the point of view of pressure differences. Let the radius of the capillary tube at the point where the surface exists be r and the radius of the concave surface, assumed spherical, be R. Then from Fig. 10-9 (b) at the concave surface it is seen that

$$r = R \cos a, \qquad (10\text{-}6)$$

where a is the angle of contact.

Just underneath the liquid surface the pressure is less than atmospheric by an amount Δp, given by equation (10-4) as

$$p = \frac{2T}{R}, \qquad (10\text{-}7)$$

since for this case there is only one surface.

Now it is this decrease in pressure which enables the liquid to rise to a height h, so that, if the liquid has a density ρ, this pressure difference is

$$p = h g \rho. \qquad (10\text{-}8)$$

Hence

$$\frac{2T}{R} = h g \rho,$$

or, in terms of the radius r of the capillary tube,

$$T = \frac{h g \rho r}{2 \cos a}. \qquad (10\text{-}9)$$

It should be noted that the tube need not have a uniform bore; all that is necessary is to deter-

* It is very difficult to produce films of pure water, since they are quite fragile. By adding soap to the water, films of considerable stability result, in spite of a marked reduction in surface tension.

mine the radius r of the capillary tube at the place where the film is formed.

For a liquid like mercury, in which the angle of contact is greater than 90°, the cosine of the angle is negative, and equation (10-9) gives a negative result. This signifies that mercury is depressed in a capillary tube below the level of the free mercury surface.

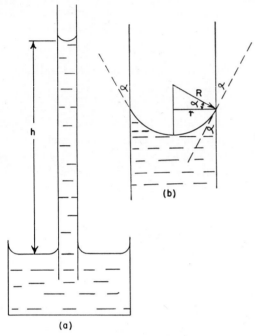

(b)

Fig. 10-9.—(*a*), Rise of liquid to height h in capillary tube; (*b*), curved surface with angle of contact α.

Another method of deriving equation (10-9) is that of considering the upward force of surface tension acting around the perimeter of the capillary tube supporting the weight of the column of liquid. If the capillary tube has a constant radius and the liquid of density ρ is raised to a height h, then the weight of liquid in the tube is $\pi r^2 h \rho g$. The upward force due to surface tension is $2\pi r T \cos \alpha$. Thus

$$2\pi r T \cos \alpha = \pi r^2 h \rho g \qquad (10\text{-}10)$$

or

$$T = \frac{h \rho g r}{2 \cos \alpha}.$$

10.8. Spreading of Liquids on Solid or Liquid Surfaces

If a drop of liquid is placed on the surface of another liquid, as shown in Fig. 10-10, it may

spread out into a film, like oil on the surface of water, or it may remain as a drop, as in the case of mercury on the surface of glass. Let us see what conditions determine whether the liquid will spread or not. The liquid in the form of a drop possesses gravitational energy with respect to the level of the second liquid. There is a tendency, therefore, for the liquid to spread so that its gravitational energy is a minimum. On the other hand, the surface energy of the drop is a minimum when in the form of a sphere; as it spreads, work must be done to increase the area. Direct computation of the energy changes due to spread-

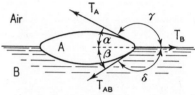

Fig. 10-10.—Surface-tension forces acting on a liquid drop placed on the surface of another liquid.

ing is complicated, since three surface tensions must be taken into account: that of the first liquid, T_A; the second liquid, T_B; and the interfacial surface tension, T_{AB}. In order to determine the conditions for spreading in terms of these surface tensions, it proves convenient to use surface tension as a force per unit length. In the diagram of Fig. 10-10 the three surface tensions are shown acting along the periphery of the drop A.

At the edge of a drop which does not spread, three fluid surfaces meet: air, liquid A, and liquid B. Let the angles between the tensions be as shown in the figure. If the drop is in equilibrium, then, from equation (1-37).

$$\frac{T_A}{\sin \theta} = \frac{T_B}{\sin (\alpha + \beta)} = \frac{T_{AB}}{\sin \gamma}. \qquad (10\text{-}11)$$

For liquids such that T_B is greater than T_A, it is possible that

$$T_B \geqq T_A + T_{AB}, \qquad (10\text{-}12)$$

and equilibrium is impossible. The liquid A will spread on B.

Water has a high surface tension and therefore does not spread on many organic liquids; organic liquids (e.g., oils), on the other hand, generally spread on water. Practically all liquids will spread on clean mercury, but mercury is so hard to keep clean that one finds water forming drops on the surface instead of spreading. In this case the mercury probably has a thin layer of oil or grease on it. A very sensitive test for the cleanliness of a mercury surface is the spreading of water on it. In Fig. 10-11 several sizes of mercury drops are

Fig. 10-11.—Drops of mercury on a clean glass plate

shown on a glass plate. This clearly indicates that, as the drops become smaller and gravitational energy is lessened, the force of surface tension becomes predominant, and the drops become more nearly spherical.

10.9. Estimate of Molecular Sizes; Monomolecular Films; Some Surface-Tension Phenomena

When a drop of liquid is present on a surface, the limit to the amount of spreading that will take place is reached when a film one molecule thick is attained. If the area of this monomolecular layer and the amount of oil required to attain it are known, an estimate of the size of a molecule can be made. Although an estimate of the size of a molecule by measuring the thickness of monomolecular films was given by Rayleigh as early as 1890, recent improvements in the technique of making manomolecular films have resulted in giving exceedingly important information about the size of complex organic molecules—their cross-sectional area and length. It has been found that the cross-sectional area of a molecule of palmitic acid is 20.5×10^{-16} cm^2.

The modern detergents, which are large-sized molecules, change the value of the surface tension rather strikingly. Placing a suitable detergent in water in which a duck is swimming causes the duck to flounder around and eventually to sink. Normally, the duck's feathers are not wetted by the water, but the detergent allows wetting to take place.

A fine-meshed sieve can be made to hold water by pouring it in very carefully (Fig. 10-12 [a]).* A piece of paper placed on the bottom of the sieve helps in doing this. The paper is later removed.

If the sieve is made oily by dipping it in hot paraffin and shaking the excess off, then the experiment is more likely to be successful. The water can be made to drip out of the sieve by drawing one's finger across (Fig. 10-12 [b]) and upsetting the previously formed water film. Any camper who has spent a rainy day in the usual canvas tent knows that he must not rub the inside of the tent with head or finger, otherwise water starts to drip inside. The place where the drip

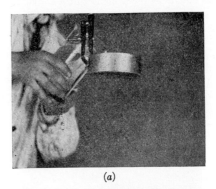

(a)

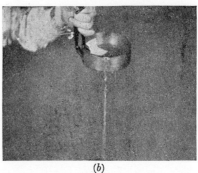

(b)

Fig. 10-12.—(a), Water being poured into a sieve and, (b), the water running out after the water film has been broken by drawing a finger across the film.

occurs can be changed to close to the ground by purposely extending the portion in which the water film is broken.

10.10. Résumé

The following is a brief listing of the definitions, principles, and theories, given in this chapter, which you should know:

Concept and definition of surface tension
Angle of contact of fluids and solids
Measurement of surface tension by various methods

* In Roman literature there may be found the story of a Vestal Virgin who, accused of violation of her vows, was condemned to death unless she could transport a certain quantity of water in a sieve. This she succeeded in doing.

Queries

1. It was stated in the text that the interior of liquids is subjected to enormous pressures, greatly in excess of the hydrostatic pressures discussed in chapter 9. If an egg is submerged in water, why is it not crushed?
2. Liquid is drawn up into the tubes (a) and (b) shown in the accompanying figure and then allowed to seek its own level. The capillary tube on the top of (b) is of the same bore as (a). Will the liquids in the two tubes remain at the same height h above the level of the tank? Explain.

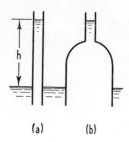

(a)   (b)

3. The following perpetual-motion machine has been proposed. A fine capillary tube is immersed in a liquid. The liquid would normally rise a distance h

in accordance with equation (10-10). By having a capillary whose length is less than h immersed in the liquid, the liquid was supposed to overflow and operate a machine. What is wrong with the proposal?

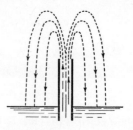

4. Explain the following phenomena: Two objects are wet by a liquid and are sufficiently near each other that the liquid surface between them is curved and, as a result, they are forced very close together. If the liquid does not wet either object, they are again forced together. But if the liquid wets one but not the other, the objects are forced apart.
5. If a liquid had an angle of contact of 90°, would the liquid be raised or lowered in a capillary tube? What would be the shape of the liquid surface in the capillary tube?

Practice Problems

1. How high will pure water at 25° C rise in a clean glass tube with a diameter of 0.1 mm? Ans.: 29.4 cm.
2. How much will mercury be depressed in the tube of problem 1?
3. By means of two wire hoops, each of 1-cm diameter, a soap bubble is produced which has cylindrical sides and spherical ends. Compute the radius of curvature of the spherical ends. Ans.: 1 cm.

4. How far below the surface of water must a bubble 0.2 mm in diameter be in order that the total pressure

inside will be twice the atmospheric pressure at the surface of the water? (1 atm = 10^6 dynes/cm².)

5. A light stick, 10 cm long, floats in water. Ether is poured on one side of the stick. What force tends to move the stick and in what direction? Ans.: 540 dynes away from the ether.
6. What types of surfaces are possible for a liquid film exposed on all sides to the open atmosphere? (I.e., p of eq. [10-5] is zero.)
7. If two rubber balloons of different radii were blown and then connected as in Fig. 10-8, would they remain in equilibrium?
8. What is the surface tension of water at 0° C in the mks system of units?
9. Derive the expression for the depression of a liquid in a capillary tube which has an angle of contact a greater than 90°. Use the pressure-difference method given in § 10.7.

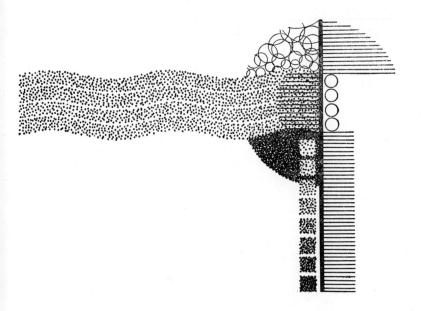

CHAPTER 11

Temperature and Its Measurements

11.1. Introduction

In the previous chapters we discussed the subject of mechanics, the central aim of which was to describe the motion of bodies in space and to interpret their behavior in terms of the fundamental laws of motion. We idealized our analyses to the extent that we avoided introducing any thermal effects that must accompany most mechanical phenomena that occur in nature. Although there are a few cases where this is not true—for example, the motion of a body *in vacuo*—thermal effects occur with, and are intimately related to, mechanical effects.

It is our purpose in the next few chapters to focus our attention on thermal phenomena. For this end it is necessary to introduce two new concepts—the *heat* and *temperature* of an object. We will show that the heat content of an object can be interpreted as the mechanical energy of the random motions of the atoms and molecules of the object and can therefore be measured in terms of mechanical-energy units. Temperature, on the other hand, is quite distinct from heat and must be introduced as a new fundamental physical quantity. Temperature plays a special role in our description and interpretation of thermal phenomena.

11.2. Meaning of Temperature; Temperature Equilibrium

Let us begin by introducing the concept of temperature through the following simple experiment. Three beakers containing the same quantity of water are kept over a fire for varying lengths of time and are then removed. In order to describe the sensations produced upon placing our fingers in the three beakers, we shall invent some words and say that the water in the first beaker is "warm," that in the second "hot," and that in the third "very hot."

We have in this experiment an illustration of what we call our "temperature sense." This qualitative description of the temperature of the object, however, is not especially useful in physics. In the first place, our temperature sense is not reliable, for we find that if the right hand is placed in hot water and the left in cold water and then both are placed in lukewarm water, the lukewarm water will feel hot to the left hand and cold to the right hand. Furthermore, psychologists have shown that the relation between sensation and stimulus is not a linear one. Also, the range of our temperature sense is quite limited. We wish, obviously, to avoid such ambiguities in a quantitative measurement of temperature. We want to be able to give a *temperature number* to each object—large numbers for the "very hot" objects and smaller numbers for the "warm" and "cold" objects. For this purpose we shall make use of the following observations out of everyday experience.

Certain physical properties of matter change with temperature. For example, a rod of steel feels cool to the hand, but when the rod is placed over a Bunsen burner, it feels very hot, that is, its temperature has increased. *Accompanying this increase in temperature is an increase in the length of the rod.* We may take the increase in length as a measure of the temperature of the steel rod. Many temperature-measuring devices, *thermometers*, use this property of expansion. There are a number of other physical properties of matter, however, that also change with temperature. For example, the following changes take place: pressure of confined gases; electrical resistance of conductors; thermoelectric forces; vapor pressure of liquids; magnetic properties and colors of materials. These physical properties that change with temperature we shall call the *thermometric properties* of matter.

Before proceeding to a description of a thermometer scale, it is necessary that we understand the meaning of "temperature equilibrium." An object A which feels cold to the hand and an identical object B which feels hot are placed in contact with each other. After a sufficient length of time, A and B give rise to the same temperature sensation, i.e., are at the same temperature and are said to be in *temperature equilibrium* with each other. If A and B are each in temperature equilibrium with a third object C, then A and B are in temperature equilibrium with each other.

We direct the reader's attention to the way in which the concept of temperature, anthropomorphic in origin (as were all the useful concepts in physics), was placed on an objective basis. We shall develop this train of thought in some detail in the paragraphs that follow.

11.3. Construction of a Thermometer Scale

The construction of a useful thermometer scale necessitates (1) a choice of some thermometric substance whose thermometric properties will indicate changes in temperature; (2) a selection of two fixed temperatures which will determine the *standard temperature interval* (these two standard temperatures are taken as the temperature at which ice melts under normal atmospheric pressure* and the temperature at which water boils under normal atmospheric pressure [normal atmospheric pressure corresponds to a barometer reading of 760 mm of mercury]); (3) a division of this standard interval into subintervals or *degrees;* and (4) a choice of numbers to be associated with each division.

In the centigrade temperature scale the temperature of melting ice is taken as 0° C, while that of boiling water is 100° C. The centigrade degree is taken as exactly $\frac{1}{100}$ of the standard interval. Let us illustrate these ideas with a concrete example—that of the common mercury thermometer. Here the thermometric substances are glass and mercury, and the property used is the differential change in volume of mercury and glass when heated. This volume change is measured by the relative change in volume of the mercury in the thermometer bulb, which results in the mercury's taking different positions in the stem of the thermometer. The mercury thermometer is placed in an ice bath until temperature equilibrium is attained. A scratch is made on the glass to show the mercury level. This mark is labeled 0° C. The thermometer is then placed in the steam bath; and, after temperature equilibrium is attained, another scratch is made and labeled 100° C. The distance between those two marks is divided into 100 equal parts. The distance between consecutive marks corresponds to 1° C *as defined by this particular thermometer.*

11.4. General Definition of Temperature

As we have pointed out, many thermometric properties may be used in the definition of "temperature." Let A_0 be some thermometric property (e.g., length) measured at 0° C; A_{100} the thermometric property at 100° C; and A_t that at some unknown temperature, t. On the centigrade scale $(A_{100} - A_0)/100$ is the size of the unit interval, the degree, while $A_t - A_0$ is the size of the temperature interval from 0° to t° C. By definition, the *temperature number*, or simply the number of degrees

* In 1927, at the International Seventh General Conference of Weights and Measures, it was agreed that 0° centigrade should be defined as the temperature at which *ice and air-saturated water are in equilibrium at 1 atm total pressure.* This temperature is actually 0°.0024 C lower than the equilibrium temperature of ice and pure water at 1 atm total pressure.

temperature, t, is the number of unit intervals, degrees, in the temperature interval $A_t - A_0$. In symbols

$$t^\circ \text{C} = \frac{(A_t - A_0) / (A_{100} - A_0)}{100}$$

or

$$t^\circ \text{C} = \left(\frac{A_t - A_0}{A_{100} - A_0}\right)(100). \qquad (11\text{-}1)$$

Equation (11-1) represents a general definition of temperature.

To illustrate the use of equation (11-1), we shall determine the temperature of a tank of water. For our thermometric substance we shall first use a steel rod and make use of changes in length. The symbol A of equation (11-1) must now be replaced by l, the length of the rod. The length of the rod in melting ice is l_0 cm, in boiling water l_{100} cm, and, when placed in the tank of water whose temperature we want, the rod is found to have a length of l_t cm. By equation (11-1) the temperature number, or the number of degrees temperature, of the tank is

$$t_l^\circ \text{C} = \left(\frac{l_t - l_0}{l_{100} - l_0}\right)(100). \qquad (11\text{-}2)$$

If the change in electrical resistance of the steel rod is used, the temperature would be given by

$$t_R^\circ \text{C} = \left(\frac{R_t - R_0}{R_{100} - R_0}\right)(100), \qquad (11\text{-}3)$$

where R_t is the resistance in ohms of the rod when in the tank, R_0 the resistance at 0° C, and R_{100} the resistance at the boiling temperature.

Another thermometric substance that could be used is a confined gas, like air or hydrogen. Here the increase in pressure of the confined gas kept at a constant volume, or the change in volume of the gas kept at a constant pressure, could effectively be used as a thermometric property. For these two cases we would obtain temperatures t_p and t_v, respectively. The question naturally arises as to whether these various temperatures—t_l, t_R, t_p, and t_v—are numerically the same. *The answer is no.* With the exception of the two fixed points 0° and 100° C, there is no reason to expect them to be the same, since temperature, as defined by equation (11-1), *depends essentially upon the thermometric property* A. Fortunately, between the range 0° and 100° C they do agree within a few tenths of a degree.

11.5. Gas Thermometer

Because different thermometric substances give rise to somewhat different measures of the same temperature, it is, of course, necessary to decide upon some thermometric substance to be used in the construction of a *standard thermometer*. In 1887 the International Committee on Weights and Measures adopted as the standard the constant-volume thermometer, using hydrogen gas (helium is now preferred). A sketch of a constant-volume thermometer is shown in Fig. 11-1. Hydrogen,

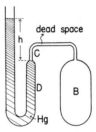

Fig. 11-1.—Constant-volume gas thermometer

carefully dried, is contained in the bulb B, which is connected to a pressure manometer through the capillary C. Sufficient gas is present to make the pressure equal to 1,000 mm of mercury when the bulb is surrounded with ice. The mercury in tube D is adjusted so that the volume of gas always remains constant. For this thermometer, A of equation (11-1) must be replaced by p, the pressure of the gas. Let p_0 be the pressure of the gas at 0° C as measured by the mercury height h, p_{100} the pressure at 100° C, and p_t the pressure when the bulb is immersed in a bath whose temperature is desired. By equation (11-1) the temperature of the bath is

$$t^\circ \text{C} = \left(\frac{p_t - p_0}{p_{100} - p_0}\right)(100). \qquad (11\text{-}4)$$

ALL OTHER THERMOMETERS MUST NOW BE CALIBRATED IN TERMS OF EQUATION (11-4) by comparison of their readings with those of this standard gas thermometer. In the discussions that follow we shall assume that all temperatures on the centigrade scale (whatever the thermometer used to measure these temperatures) are equivalent to those that would have been obtained if equation (11-4) had been used.

The reasons for choosing the constant-volume gas thermometer as an international standard are **171** many. A few of these reasons will now be considered.

Let us define the *average pressure coefficient* of a gas, represented by the Greek letter gamma (γ), by the quantity

$$\frac{p_{100} - p_0}{100 p_0} \equiv \gamma. \tag{11-5}$$

This coefficient represents the average *specific change in pressure*, i.e., *change in pressure per unit pressure per degree centigrade*. Introducing this quantity into equation (11-4), we obtain

$$t^\circ\, C = \left(\frac{p_t - p_0}{p_0}\right) \frac{1}{\gamma}. \tag{11-6}$$

Now it is experimentally found that, as the initial pressure, p_0, is reduced below 1 atm, the average value of the pressure coefficient, γ, *for different gases* approaches the same numerical value—0.00366 per degree centigrade. This result implies that the temperature readings of constant-volume thermometers using *different gases* approach the same value as the initial pressure, p_0, is decreased.

Strictly speaking, it is this extrapolated gas scale that is chosen as the standard gas thermometer, for it depends only upon the general properties of gases and not upon any particular gas.

The use of the resistance thermometer and of thermoelectric effects as thermometers is discussed in some detail in later chapters.

11.6. Fahrenheit Scale

Although the centigrade temperature scale is the one used almost exclusively in scientific work, for engineering purposes and in everyday life another scale is commonly employed—the Fahrenheit scale. In this scale the fixed points are the same as those of the centigrade scale. This standard interval is then divided into 180 parts, and the melting point of ice is labeled 32° F, while that of boiling water is labeled 212° F. The size of the Fahrenheit degree is thus smaller than the centigrade degree. The following useful equation for converting from one scale to another may be verified by the reader:

$$\frac{C}{100} = \frac{F - 32}{180}. \tag{11-7}$$

The two scales are shown in Fig. 11-2 for comparison. Two additional temperature scales that are also employed in engineering and scientific work are shown in Table 11-1. The two scales,

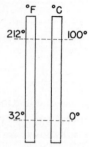

Fig. 11-2.—Comparison of temperature scales

TABLE 11-1

COMMON TYPES OF TEMPERATURE SCALES

Fixed Points	Centigrade	Fahrenheit	Kelvin	Rankine
Steam point.....	100°00	212°00	373°16	671°69
Ice point........	0°00	32°00	273°16	491°69
Absolute zero....	−273°16	−459°69	0°00	0°00

Kelvin and Rankine, are based on considerations that are discussed in later chapters.

EXAMPLE 1: A temperature of 20° C corresponds to what on the Fahrenheit scale?

Solution: By equation (11-7),

$$\frac{20}{100} = \frac{F - 32}{180};$$

$$180 = 5F - 160$$

$$F = 68°.$$

11.7. International Temperature Scale

By international agreement, certain reproducible fixed points, in addition to the ice and steam points, are used for calibrating thermometers. This International Temperature Scale is given in Table 11-2. The constant-volume hydro-

TABLE 11-2

INTERNATIONAL TEMPERATURE SCALE
(STANDARD ATMOSPHERIC PRESSURE)

	° C
Boiling point of oxygen......	−182.970
Melting point of ice.........	0.000
Boiling point of water.......	100.000
Boiling point of sulphur.....	444.600
Melting point of silver......	960.8
Melting point of gold.......	1,063.0

gen thermometer (see § 11.5) agrees very closely with the International Scale. At $-200°$ C the hydrogen thermometer is approximately $0°06$ lower than the International Scale and $0°02$ lower at $400°$ C.

11.8. Linear Expansion of Solids

As we have stated in § 11.2, if a rod of steel is heated uniformly, i.e., so that the temperatures of all parts of the rod are the same, the rod increases in length. We wish to find out whether any simple relationship exists between the initial length of the rod, the final length of the rod, and the change in temperatures. We again emphasize that, whatever be the practical thermometer used to determine t, this temperature is that defined by equation (11-4).

If the length of the rod at $0°$ C is l_0, the final temperature $t°$ C, and the final length l_t, it is found that, with reasonable accuracy, the final length l_t can be represented by

$$l_t = l_0 (1 + at) \qquad (11-8)$$

or

$$a = \left(\frac{l_t - l_0}{l_0}\right)\frac{1}{t}; \qquad (11-9)$$

or, if we use a different notation and call Δl the change in length that accompanies a change in temperature, Δt, then

$$a = \frac{\Delta l}{l_0 \Delta t}. \qquad (11-10)$$

In equation (11-8) a is to be taken as a constant over the range t. From equation (11-9) we see that a represents the fractional change in length per degree and is hence independent of the lengths of the samples but dependent on the size of the degree. The constant a, called the "coefficient of linear expansion," is clearly defined by equation (11-9) as the *fractional or specific change in length per degree change in temperature*. Approximate values of a for some common substances are given in Table 11-3.*

It is important to emphasize the qualifying phrase of the previous paragraph that equation (11-8) represents the change in length of a solid "with reasonable accuracy." If a steel rod were heated over the range of temperature from $-100°$

*Note that a is somewhat dependent upon the temperature range.

to $600°$ C and changes in the length (taking the length at $0°$ C as a standard) were noted, we would find that for this range of temperature equation (11-8) is not at all adequate. It would be necessary, instead, to represent the length l_t by an empirical equation of the type

$$l_t = l_0 (1 + a_0 t + a_1 t^2 + a_2 t^3 + \ldots), \quad (11-11)$$

where the constants a_0, a_1, and a_2 were experimentally determined. For the purposes of this

TABLE 11-3

LINEAR COEFFICIENTS OF EXPANSION

Substances	At or between Centigrade Temperatures of—	$a \times 10^5$ Degrees C^{-1}
Aluminum..............	20°	2.24
	100°	2.35
	300°	2.84
	500°	3.11
Brass (cast)............	0°–100°	1.875
Copper.................	0°–100°	1.68
Fused quartz..........	$-190°$ to $+16°$	0.026
	16°– 500°	0.057
Glass tubing (soft).......	0°– 100°	0.833
Ice....................	$-20°$ to $1°$	5.1
Invar..................	....	0.09
Lead..................	20°	3.12
	100°	2.91
	280°	3.43
Paraffin...............	0°– 16°	10.66
	16°– 38°	13.03
	38°– 49°	47.7
Platinum..............	20°	0.887
Potassium.............	0°– 50°	8.3
Rubber................	....	6.6–7.5
Steel.................	0°–100°	1.10
Tungsten..............	0°–100°	0.45
Wood across grain:		
Beech.............	....	6.0
Pine..............	....	3.4
Wood parallel to grain:		
Ash...............	....	0.9
Beech.............	....	0.2
Zinc..................	0°–100°	3.54

text, however, we shall cut the power series off after the terms $a_0 t$ and simply use equation (11-8). Many problems arise where the initial length is not at $0°$ C. A convenient expression for the increase of length can be obtained from equation (11-8).

For a temperature change from $0°$ to $t_1°$ C, equation (11-8) gives

$$l_1 = l_0 (1 + a_0 t_1); \qquad (11-12)$$

also, for a temperature range from $0°$ to $t_2°$ C, we get

$$l_2 = l_0 (1 + a_0 t_2). \qquad (11-13)$$

Dividing equation (11-13) by equation (11-12), we obtain

$$\frac{l_2}{l_1} = \frac{l_0}{l_0} \frac{(1 + a_0 t_2)}{(1 + a_0 t_1)} ; \qquad (11\text{-}14)$$

also,

$$\frac{1 + a_0 t_2}{1 + a_0 t_1} = 1 + a_0 (t_2 - t_1) - a_0^2 t_1 (t_2 - t_1) + \ldots ;$$

and, since a_0 is of the order of 10^{-5} and a_0^2 is of the order of 10^{-10}, the terms after $a_0(t_2 - t_1)$ may be neglected if $a_0(t_2 - t_1)$ is small. Equation (11-14) then becomes

$$l_2 = l_1 [1 + a_0 (t_2 - t_1)] . \qquad (11\text{-}15)$$

Equation (11-15) is used quite frequently in solving practical problems.

EXAMPLE 2: A steel rod is 1 m long at $50°$ C. What is its length at $75°$ C?

Solution: From equation (11-12)

$$l_t = l_0 (1 + 0.000011 0 t) .$$

Here we do not know l_0 directly. If we write the equation twice, once for each given temperature,

$$l_{75} = l_0 (1 + 0.0000110 \times 75) ,$$

$$100 = l_{50} = l_0 (1 + 0.0000110 \times 50) ,$$

and then divide one equation by the other,

$$\frac{l_{75}}{100} = \frac{l_0 (1 + 0.0000110 \times 75)}{l_0 (1 + 0.0000110 \times 50)} ,$$

the l_0's cancel and the expression can be solved for l_{75}. Hence

$$l_{75} = 100 \left(\frac{1.000825}{1.000550}\right) = 100.027^+ \text{ cm} .$$

From equation (11-15) we have

$$l_{75} = 100 (1 + 0.000011 \times 25)$$

$$= 100.0265 \text{ cm} .$$

11.9. Surface Expansion of Solids

The increase in area of a solid when heated may be calculated in the following fashion. Let us assume that a square sheet of copper has a length of l_0 cm at $0°$ C and that it is uniformly heated and expands into a larger square whose side is l_t. The initial, A_0, area is obviously l_0^2, while the final area, A_t, is l_t^2. If a is the linear expansion coefficient, then

$$A_t = l_t^2 = [l_0 (1 + a t)]^2 = [l_0^2 (1 + 2 a t + a^2 t^2)]$$

or

$$A_t = A_0 (1 + 2 a t) , \qquad (11\text{-}16)$$

where we have neglected the term $a^2 t^2$ in comparison with $2at$. If the initial temperature is not $0°$ C but, say, $t_1°$ C, and the final temperature is $t_2°$ C, then in the same manner that we obtain equation (11-15) we may write

$$A_2 = A_1 [1 + 2 a (t_2 - t_1)] , \qquad (11\text{-}17)$$

where A_2 is the final area and A_1 is the initial area.

It is interesting to note that equation (11-17) is applicable to *the increase in area of a hole in a disk* if the disk is heated.

11.10. Volume Expansion of Solids

It is clear that if a solid cube—copper, for example—were heated, its volume would increase. Let us take a cube whose side is l_0 cm at $0°$ C and heat it uniformly to $t°$ C. If we assume that each side of the cube increased in length by the same amount, so that the side is of length l_t at $t°$ C, then

$$V_t = [l_0 (1 + a t)]^3 = l_0^3 (1 + 3 a t$$
$$\qquad\qquad + 3 a^2 t^2 + a^3 t^3) , \qquad (11\text{-}18)$$

or

$$V_t = V_0 (1 + 3 a t) , \qquad (11\text{-}19)$$

where again $3a^2 t^2$ and $a^3 t^3$ are neglected in comparison with $3at$.

The quantity $3a \equiv \beta$ is called the *coefficient of cubical or volume expansion*. If the initial temperature is not $0°$ C but $t_1°$ C, we may rewrite equation (11-19) to a close approximation as

$$V_2 = V_1 [1 + \beta (t_2 - t_1)] . \qquad (11\text{-}20)$$

This equation is also applicable to increase or decrease in volume of *cavities* in solids. Table 11-4

TABLE 11-4

CUBICAL COEFFICIENTS OF EXPANSION

Solids	At or between Centigrade Temperatures of—	$\beta \times 10^5$ Degrees C^{-1}	Liquids	At or between Centigrade Temperatures of—	$\beta \times 10^5$ Degrees C^{-1}
Fused quartz.	18°	0.09	Ether.......	20°	165.6
Glass (soft)..	20°	2.53	Ethyl alcohol		
Ice.........	−20°– 1°	15.0	99 per cent	20°	112
Invar.......	20°	0.09	Mercury....	20°	18.2
Iron........	0°–100°	3.55	Pentane.....	20°	161
Lead........	0°–100°	8.4	Water.......	20°	20.7
Paraffin....	20°	58.8	Pyrex.......	20°–400°	1.1
Platinum....	....	2.65			

gives the cubical coefficients of a few substances. It will be noted by comparison with Tables 11-2 and 11-3 that β is not always observed to be exactly three times a. Another point that must be brought to the attention of the reader is that the assumption that solids increase by the same amounts in all directions is true for isotropic homogeneous substances but not for anisotropic substances, in which the expansion is different in different directions in the crystalline substance.

EXAMPLE 3: A cube of a metal with a linear coefficient of expansion of 0.00002 per degree centigrade is 10 cm on an edge at 0° C. What will be its volume at 100° C? What percentage of error is made by neglecting the higher terms of equation (11-18)?

Solution: Approximate solution:

$$V_{100} = 10^3 (1 + 3 \times 0.00002 \times 100) = 1,006 \text{ cm}^3 .$$

Exact solution:

$$L_{100} = 10(1 + 0.00002 \times 100) = 10.02 \text{ cm} ,$$

$$V_{100} = (L_{100})^3 = (10.02)^3 = 1,006.012 \text{ cm}^3.$$

Percentage of error:

$$\frac{1,006.012 - 1,006}{1,006.012} \times 100 = 0.001 \text{ per cent .}$$

This shows that the error of the approximation is less than the probable experimental error of the data, for temperature coefficients are not known to this accuracy.

11.11. Applications of the Expansion of Solids

The practical importance and application of the expansion of solids are many. For example, whenever structures such as large bridges or nonwelded rails are subject to temperature changes, expansion or contraction must be allowed for by having some parts of the structure free to slide. Steam pipes have expansion joints; pendulums are compensated; and the bearings of high-speed machines must have proper clearances for expansion.

The fact that various metals have different expansion coefficients has been put to good use in the construction of bimetallic thermal regulators. As illustrated in Fig. 11-3, if a strip of zinc and one

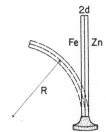

FIG. 11-3.—The bimetallic elements

of iron are riveted together so that at some temperature they are straight, then, if the bimetallic strip is heated, the strip will bend into a curve approximately a circular arc, of radius R, as shown by the dotted lines. It is clear that the solid with the larger expansion coefficient (in this case zinc) is on the convex side of this curve.*

Again from Table 11-3 we observe that platinum and glass have nearly the same expansion coefficient. This makes it possible to seal platinum wires into glass bulbs for electric lights, for example. A less expensive wire material has been developed to replace platinum for this purpose. This wire consists of a solid core of nickel-steel alloy covered with copper. The alloy has an expansion coefficient less than glass, that of copper greater than glass. By controlling the thickness of the copper and nickel-steel, the expansion coefficient of most glasses can be duplicated. Another alloy extensively used for glass-to-metal seals is Kovar—an iron, cobalt, nickel alloy.

11.12. Thermal Stresses

We know from our previous analysis that if a rod is fastened at both ends and not permitted to expand or contract while its temperature changes, tensile or compressive *thermal stresses* will be set up in the rod. We may calculate the magnitude of these stresses by recalling the expression for Young's modulus of equation (7-14)

$$Y = \frac{F/A}{s/l} . \tag{7-14}$$

In this equation s/l is the fractional change in length (strain) due to the stress F/A. If the rod of length l and cross-sectional area A expands an amount s when heated $t°$ C, this change in length is given by

$$s = lat , \tag{11-21}$$

where a is the linear expansion coefficient. The compressive stress necessary to prevent this change is given by equation (7-14). Combining equations (7-14) and (11-21), we obtain

$$F = YatA . \tag{11-22}$$

---

* It can be shown that if $2d$ is the total thickness of the strip and t the temperature rise, then approximately

$$R = \frac{d}{(a_{\text{Zn}} - a_{\text{Fe}}) t} .$$

175

EXAMPLE 4: A steel rod 5 cm² in cross-section is heated to 200° C, and its ends are then fastened to two rigid supports. Find the tension in the rod when the temperature drops to 20° C. Assume that Young's modulus is 22×10^{10} newtons/m^2.

Solution: From equation (11-22),

$$F = (22 \times 10^{10} \text{ newtons}/m^2) \frac{(11 \times 10^{-6})}{°C}$$
$$\times (180°C)(5 \times 10^{-4} \, m^2)$$
$$= 217,800 \text{ newtons .}$$

11.13. Volume Expansion of Liquids

In discussing the volume expansion coefficient of liquids it is necessary to distinguish between the "true" expansion coefficient and the "apparent" or differential expansion coefficient. To illustrate these ideas, consider a glass bulb filled with, say, liquid mercury at a temperature of 0° C. At this temperature let the bulb be filled to the point A (Fig. 11-4). If the temperature is now increased

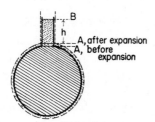

FIG. 11-4.—Volume expansion of liquids

$t°$ C, three effects may take place: the glass bulb may change in volume to a greater degree than the mercury, in which case the mercury level will be below A; the mercury may expand more than the glass, the overflow mercury filling the stem to this point B; or, finally, both the glass and the mercury may expand by the same amount, leaving the mercury level at A. Experimentally* we find that the mercury expands to a greater degree than the glass.

Let β_m represent the fractional change in volume per degree centigrade of the mercury itself, i.e., β_m is the true expansion coefficient; β_g is the volume expansion coefficient of the glass bulb. If V_0 is the volume of the bulb and mercury at 0° C, then, if the temperature is raised $t°$ C, we obtain

$$_gV_t = V_0(1 + \beta_g t) ,$$

as the new volume of the glass bulb ,

and

$$_mV_t = V_0(1 + \beta_m t) ,$$

the new volume of the mercury .

By subtracting $_gV_t$ from $_mV_t$, we obtain

$$(_mV_t - {_gV_t}) = V_0(\beta_m - \beta_g)t ; \quad (11\text{-}23)$$

we may rewrite equation (11-23) as

$$\beta_m - \beta_g = \frac{\Delta V}{V_0}\frac{1}{t} , \quad (11\text{-}24)$$

where we have replaced $_mV_t - {_gV_t}$ by ΔV. Equation (11-24) may be taken as the definition of the "apparent" expansion coefficient of mercury with respect to glass.

In case we have a liquid whose expansion coefficient is less than that of its solid container, as, for example, water in a paraffin bulb, the sign of the right-hand member of equation (11-23) will be reversed. Heating a thermometric device of this kind results in a descent of the recording liquid thread in the stem; conversely, cooling causes the index to rise.

The question naturally arises as to whether it is possible to make an absolute determination of the true expansion coefficient of a liquid without involving a knowledge of the expansion coefficient of the solid that contains the liquid. The answer is contained in an experiment of Dulong and Petit.

This method consists in filling a U-tube with the liquid whose expansion coefficient is wanted, surrounding one arm with a hot bath at temperature t_1 and the other with a cold bath at temperature t_2. A measurement of the heights h_1 and h_2 and the temperatures t_1 and t_2 suffices to give β. The underlying principle of the method is that the pressure at a point within a liquid in hydrostatic equilibrium depends only upon the height of liquid above that point and upon the density of the liquid. Let ρ_1 be the density of the hot liquid at temperature t_1 and let h_1 be its height above the point A (Fig. 11-5). Also, let ρ_2 be the density of the cold liquid at t_2 and h_2 be its height above A. Then, by the principle of hydrostatics,

$$h_1\rho_1 g = h_2\rho_2 g ,$$

where g is the acceleration due to gravity. Rearranging terms, we get

$$\frac{h_1}{h_2} = \frac{\rho_2}{\rho_1} . \quad (11\text{-}25)$$

If v_1 is the volume of a given mass of the liquid at temperature t_1 and v_2 is the volume of the *same mass* of the liquid at t_2, then

$$\rho_1 v_1 = \rho_2 v_2$$

or

$$\frac{\rho_2}{\rho_1} = \frac{v_1}{v_2};$$

also

$$v_1 = v_0(1 + \beta t_1)$$

and

$$v_2 = v_0(1 + \beta t_2).$$

Therefore,

$$\frac{h_1}{h_2} = \frac{(1 + \beta t_1)}{(1 + \beta t_2)},$$

or

$$\beta = \frac{h_2 - h_1}{h_1 t_2 - h_2 t_1}. \qquad (11\text{-}26)$$

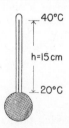

Fig. 11-5.—Apparatus of Dulong and Petit to determine the true expansion coefficients of a liquid.

The reader will observe that the *pressures* of the two arms at point A do not depend upon the cross-section of the arms of the U-tube but only upon the vertical heights h_1 and h_2 and the densities ρ_1 and ρ_2. For the case of mercury, $\beta = 0.000182$ per degree centigrade.

A practical application of equations (11-23) and (11-24) is found in the design of thermometers. We illustrate this application in the following example.

EXAMPLE 5: A bulb-immersion thermometer is to be constructed having a range from $20°$ to $40°$ C. It is desired that the length of the scale be 15 cm. *Approximately* what volume bulb is necessary if the diameter of the capillary hole in the stem is 0.4 mm?

Solution: Let V_1 be the volume of the thin-walled bulb and that of the mercury at the temperature $t_1 =$

$20°$ C. When the temperature is raised to $t_2 = 40°$ C, then the change in volume, as given by equation (11-23), is

$$\Delta V = V_1(\beta_m - \beta_g)(t_2 - t_1).$$

If we assume that this volume of mercury ΔV is forced up the stem to a height h above the bulb, then

$$\Delta V = \pi r^2 h,$$

the volume of the cylindrical hole in the stem. Substituting in this equation, we get

$$V_1 = \frac{\pi r^2 h}{(\beta_m - \beta_g)} \frac{1}{(t_2 - t_1)}$$

$$= \frac{\pi (0.02)^2 (15)}{(0.000182 - 0.000025)(20)} = 6.00 \text{ cm}^3.$$

These same principles may be applied to obtain a correction to barometer readings. Ordinarily, barometric readings are given in terms of the height, h, of a mercury column above some fiducial point, for there the pressure is $h\rho g$, where g is the acceleration due to gravity and ρ is the density of the mercury. For this statement to be fully meaningful, it is necessary to specify the temperature of the mercury column. By convention, barometric pressures are given in terms of the height of the mercury column at $0°$ C. If the height reading is taken at $t°$ C, then a correction must be applied.

Since the pressure is the same for the two temperatures, it follows that

$$h_t \rho_t g = h_0 \rho_0 g. \qquad (11\text{-}27)$$

Also, $\rho_0 = \rho_t(1 + \beta t)$, where β is the true volume expansion coefficient for mercury. Substituting in equation (11-27), we obtain

$$h_0 = h_t \frac{1}{(1 + \beta t)}, \qquad (11\text{-}28)$$

or, approximately,

$$h_0 = h_t(1 - \beta t). \qquad (11\text{-}29)$$

An additional correction must be applied if the scale changes its length appreciably with temperature.

11.14. Anomalous Expansion of Water

If warm water is cooled down to $0°$ C, experiment shows that the volume of a given mass of water decreases with fair regularity until a temperature of $4°$ C is reached. Between $4°$ and $0°$ C the *volume of a given mass of water increases with a decrease in temperature.* This anomalous behavior of water is shown graphically in Fig. 11-6. It follows from the graph that the density (mass/unit volume) of water increases and attains a maxi-

177

mum value at 4° C, after which it decreases. As a matter of fact, 1 cm³ of water at 4° C contains, to within a few parts in a million, 1 gm of water; in other words, the density of water at 4° C is 1 gm/cm³.

An equation which quite closely represents the variation of volume of a given mass of water (identified by its volume V_0 at 0° C) over the range of

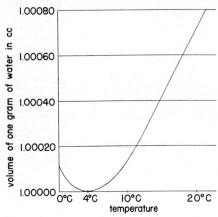

FIG. 11-6.—The anomalous expansion of water

temperature from 0° to 33° C, i.e., an empirical equation of the graph of Fig. 11-6, is

$$V_t = V_0(1 + a_1 t + a_2 t^2 + a_3 t^3),$$

where $a_1 = -6.0 \times 10^{-5}$ deg⁻¹; $a_2 = +8.5 \times 10^{-6}$ deg⁻²; $a_3 = -6.8 \times 10^{-8}$ deg⁻³.

This unusual property of water is used to explain why a pond freezes first at the surface. During prolonged cooling, layers of water near the surface reach 4° C and then sink because of the *increased* density of water. This convection process continues until the entire pond is at 4° C; further cooling of the surface water *decreases* the density of the water, which then remains at the surface. The surface then freezes, and the rate of heat loss from the interior is sharply curtailed.

11.15. Résumé

The following is a brief listing of the definitions, principles, and theories, given in this chapter, which you should know:

The concept and definition of temperature

Measurement of temperature and various temperature scales

Definition of the linear, area, and volume coefficients of expansion

The relationship between these coefficients of expansion

Thermal stresses accompanying expansions

Definitions of true and apparent coefficients of expansion of liquids

Methods of measurement of these coefficients

Queries

1. What factors must be taken into account in the design of a sensitive thermometer?

2. What means would one employ to measure (a) the temperature of a tiny insect? (b) a temperature of 400° C? (c) temperatures around −180° C? (d) temperatures around 3,000° C?

3. What are some of the principal errors of the mercurial thermometer?

4. Would there be any advantage in using a logarithmic scale for temperatures near the absolute zero?

5. On what principle does the common thermostat operate?

6. What would be the units of the expansion coefficients in the mks system?

7. Why is it necessary to distinguish between the coefficient of real and the coefficient of apparent cubical expansion in the case of a liquid?

8. Colored water and colored alcohol are placed in bulbs with long stems and are to be used as thermometers. If temperatures between 1° and 40° C are to be measured, what kind of scales must be placed alongside each stem?

9. Give a description of a simple experiment which shows that a hole in a plate of metal expands just as the metal does.

Practice Problems

1. Change the following thermometer readings from the centigrade to the Fahrenheit scale:

−273° C, −100° C, −40° C, 30° C, 2,000° C.

2. Change the following thermometer readings from the Fahrenheit to the centigrade scale:

−22° F, 0° F, 68° F, 98°.6 F, 932° F.

3. At 25° C a steel rod is 2 cm in diameter. A brass ring has an interior diameter of 1.995 cm at the same temperature. At what temperature, both the rod and the ring being heated, will the ring just slide onto the rod? Ans.: 350° C.

4. A brass rod with a cross-sectional area of 5 mm² has its ends held absolutely fixed. If its temperature is lowered 45° F, what tension will exist in the rod?

5. A steel tape is accurate at 0° C. On a day when the thermometer read 30° C, a surveyor used this tape to measure the width of a building lot without making any correction for temperature. The lot measured 50 ft. What was its true width? Ans.: 50.0165 ft.

6. The volume of a liquid at 0° C is 100 cm³. Find its coefficient of cubical expansion if its volume at 50° C is 102 cm³.

7. In the Dulong-Petit experiment a column of liquid 58.92 cm at 0° C just balances another column 60.01 cm at 100° C. What is the coefficient of absolute expansion of the liquid? Ans.: 0.000185 per degree centigrade.

8. A seconds pendulum (period 2 sec) made of brass keeps correct time at 20° C. How many seconds per day will it lose at 30° C?

9. The mercury of a barometer is 758 mm above the cistern level when the temperature is 23° C. If atmospheric pressure remains constant, calculate the height of the mercury above the cistern when the temperature is 0° C. Ans.: 754.8 mm.

10. Let us assume that the mercury height (758 mm) of problem 9 was measured with a brass scale that was accurate at 0° C. What will the scale read if the temperature of scale and mercury is 0° C?

11. A steel tank is filled with 300 ft³ of water at a temperature of 10° C. Calculate the amount of overflow if the temperature rises to 50° C. Ans.: 2.1 ft³.

12. If, in the bimetallic strip of Fig. 11-3, $d = 0.1$ cm, calculate the radius of curvature of the bent strip if the temperature rise is 20° C. Use steel and brass.

13. At 15° C a copper cylinder of diameter 10 cm exactly fits into a cylindrical hole bored in a steel block. Calculate the width of the gap between copper and steel when both are cooled to −80° C. Ans.: 0.00275 cm.

14. A steel rod 2 in in diameter is heated to 350° F, and its ends are then fastened to two rigid supports. Calculate the tension in the rod when the temperature falls to 50° F. Assume that Young's modulus for the rod is 3×10^7 lbf/in².

15. A brass rod and a steel rod, both of cross-sectional area of 5 mm², are each placed between rigid supports. If the temperature is raised 45° C, compute the force in the rods. Ans.: 3.9×10^7; 5.45×10^7 dynes.

16. The wire of a simple pendulum is made of brass. If the pendulum is 1.40 m long when the temperature is 5° C, compute the change in period of oscillation when the temperature is 25° C.

17. A bulb-immersion mercury thermometer is to be constructed with a range from 0° to 100° C. It is desired that the length of the scale be 10 cm. Approximately what volume bulb is necessary if the diameter of the capillary hole in the stem is 0.2 mm? Ans.: 0.2 cm³.

18. Express the fixed temperatures of the International Scale (Table 11-1) in terms of the Fahrenheit and and Rankine scales.

19. A constant-volume hydrogen thermometer registers a pressure of 1,000 mm of mercury when placed in an ice bath. Compute the temperature of a bath in which the thermometer reads a pressure of 3,010 mm of mercury. Ans.: 549° C.

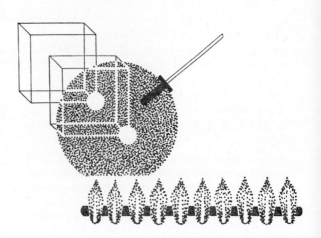

CHAPTER 12

Quantity of Heat: Calorimetry

12.1. Concept of Heat

The reader will recall how we introduced the concept of temperature in § 11.2. We wish to show now that temperature alone is not sufficient to describe all thermal phenomena. We must, in fact, introduce another new concept, that of heat—in particular, quantity of heat. Although the word "heat" is undoubtedly part of the vocabulary of the reader, it is desirable to show by means of an experiment or two the logical necessity for its introduction.

If we have had any experience whatever in heating objects over a fire, we know intuitively (1) that the time required to heat any given amount of material increases with the temperature range through which it is to be heated; (2) that to heat a large mass of material over a given rise of temperature requires a longer time than to heat a smaller mass; (3) that some materials get hot at a much faster rate than others when on equally hot fires.

As another and more readily measurable experiment, let us consider two of the beakers of water described in § 11.2. Let us also assume that the one we called "warm" contains 100 gm of water at 20° C and the one called "hot," 100 gm of water at 90° C. We shall pour the contents of both beakers into a thin-walled thermos bottle. We find that the resulting temperature of the mixture is 55° C. We further experiment by mixing 100 gm of water at 20° C with 200 gm of water at 90° C. Now we observe the temperature to be 66°.7 C. How can we explain the experimental fact

that the final temperature in the first experiment was simply the arithmetic mean of the initial and final temperatures, whereas in the second experiment that is by no means the case?

In order to answer this question, we propose to introduce the concept of quantity of heat and give it certain preliminary properties that will enable us to explain these experiments. We shall assume, for example, that the water in each beaker contains a quantity of heat that depends, among other possible factors, upon the mass of water and its temperature. Thus the 100 gm of water at 90° C contain more heat than 100 gm at 20° C; also, 200 gm at 90° C contain more heat than 100 gm at 90° C. We further assume that when the two are mixed, the *total quantity of heat remains constant*. Also, the amount of heat gained by the colder water is equal to the product of its mass by the rise in temperature; likewise, the heat lost by the warmer water is equal to the product of its mass and the drop in temperature. To place these assumptions in analytical form, let M_1 gm of water at t_1 be mixed with M_2 gm at t_2, where t_2 is greater than t_1, and let t_3 be the resulting temperature. What we are assuming is that

$$M_1 (t_3 - t_1) = M_2 (t_2 - t_3) \qquad (12\text{-}1)$$

is always true. We see immediately that expression (12-1) describes adequately the results of our two experiments, for, if the masses are the same and $t_3 = 55°$ C, $t_1 = 20°$ C, and $t_2 = 90°$ C, equation (12-1) is identically satisfied. Also, if $M_1 = 100$

gm, $M_2 = 200$ gm, $t_3 = 66°.7$ C, $t_1 = 20°$ C, and $t_2 = 90°$ C, the expression is again satisfied.

In order to measure the quantity of heat in the laboratory, a *unit of heat quantity* must be defined.

This unit in the metric system is the calorie (cal), which is defined as the quantity of heat necessary to raise the temperature of 1 gm of pure water from 14°.5 to 15°.5 C.

The calorie is also called the "small calorie," the "gram-calorie," and the "therm." The "large calorie," or "kilogram-calorie," as its name indicates, is equal to 1,000 small calories.

We have been careful to define the calorie in terms of water at an average temperature of 15° C. It has been found experimentally that the amount of heat necessary to raise 1 gm of water 1° C varies somewhat with the average temperature of the water. Thus to raise 1 gm of water 1° C at 0° C requires 1.009 cal; at 25° C, 0.9989 cal; at 100° C, 1.004 cal. This variation is slight and will be disregarded in our computations.

The British Engineering unit of quantity of heat is called the "British thermal unit" (BTU) and is defined as the quantity of heat necessary to raise 1 lb of water from 57°.5 to 58°.5 F. One BTU is equal to approximately 252 cal.

In summarizing this section we may say that we have postulated the existence of a physical entity—quantity of heat—resident in all matter. A body will acquire heat when its temperature rises and will lose heat when its temperature falls. What is measured is a change in the quantity of heat of a body, not the absolute amount of heat which the body has. Second, we defined a standard unit in which heat is measured. Finally, we assumed that when mixtures of different materials are formed, the total amount of heat remains unaltered. In such cases there is automatically transferred a certain quantity of heat from the hotter component of the mixture to the cooler component. The important question as to the nature of heat we shall answer in a later section (§ 12.4).

12.2. Specific Heat; Thermal Capacity

In introducing the concept of heat in the last section we were careful to mix two samples of water and not, say, water and oil. Let us see what happens when 100 gm of water at 20° C and 100 gm of oil at 90° C are thoroughly mixed in the thermos bottle. We find for a particular kind of oil that the resulting temperature is not the arithmetic mean of 55° C but rather 43°.3 C. We explain this experimental fact by extending our heat concept in the following way: Different substances require different amounts of heat to raise 1 gm through 1° C. For example, the 100 gm of water required 100 cal to raise its temperature 1°; the 100 gm of oil, on the other hand, take only 50 cal to raise the temperature 1°. We must ascribe to all substances another thermal property, *specific heat*.

The specific heat of a substance is defined as the number of calories of heat necessary to raise the temperature of 1 gm of the substance 1° C.

We shall use the letter c to represent specific heat.[*] The specific heat is usually not a constant but depends slightly upon temperature. In many of our problems we shall assume that c is constant.

A list of specific heats for some common substances is given in Table 12-1. The reader will observe that, with the exception of one or two cases, water has the largest specific heat, equal to 1 cal/gm ° C at 15° C.

The quantity of heat, Q, in calories necessary to raise m gm of a substance from $t_1°$ C to $t_2°$ C may be expressed mathematically in the following way:

$$Q = (t_2 - t_1) \, cm \text{ cal} . \qquad (12\text{-}2)$$

For the case of water, $c = 1$, and equation (12-2) becomes equation (12-1). Also, the reader may see from equation (12-2) that the units of c are calories per gram per degree centigrade.

The *thermal capacity* of any particular body is defined as the product of its mass in grams and its specific heat in calories per gram per degree centigrade; i.e.,

$$\text{Thermal capacity} \equiv m \, c \text{ cal/deg} . \quad (12\text{-}3)$$

In making measurements of specific heats of materials, a vessel called a *calorimeter* is used. It is usually a cylindrical vessel of highly polished aluminum, fitted with a stirrer and placed within another larger cylinder and insulated from the outer cylinder. If two liquids are poured together in a calorimeter, this inner vessel will, as a rule, absorb a certain amount of heat. We take this into account by assuming the calorimeter to be equivalent, as far as heat exchanges are concerned, to a

---

[*] Small c will be used to represent specific heat of a gram of substance; capital C, for a gram molecular weight of the substance.

mass of water equal to the product of the mass of the cylinder and its specific heat. In other words, this *water equivalent* of the calorimeter is numerically equal to the thermal capacity of the calorimeter.

12.3. Method of Mixtures

As we have seen, when two different objects at different temperatures are placed in contact, heat

cure t_B lower than t_A. Let us further assume that, if the materials are liquids or liquids and solids, no chemical reactions take place and no changes of phase take place; i.e., the solid does not melt or the liquid vaporize, the solid does not go into solution, etc. Under these conditions material A will lose heat and cool, giving it up to B, which will be heated thereby. Ultimately, temperature equilibrium at temperature t_e between t_A and t_B will be reached. Using equation (12-2) to express the fact that the

TABLE 12-1

SPECIFIC HEATS OF SOLIDS, LIQUIDS, AND GASES

Element	At or between Centigrade Temperatures of—	c Specific Heat (Cal/Gm ° C)	Atomic Weight	C Atomic Heat (Cal/Mole ° C)
Al.........	0°–100°	0.2175	26.97	5.83
Au........	0°–100°	.0309	197.2	6.10
B..........	0°–100°	.26	11.0	2.86
C (graphite)	−191° to −79°	.057	12.00	0.68
	−76°–0°	.126		1.51
	0°–100°	.216		2.60
	+896°	.454		5.44
Cu........	0°–100°	.0930	63.57	5.92
	−189°	.0506		3.20
	+600°	.0994		6.30
Fe.........	0°–100°	.110	55.84	6.14
	−256°	.00067		0.04
	+760°	.302		16.8
Hg........	0°	.03346	200.61	6.70
	210°	.0319		6.48
Mg........	0°–100°	.247	24.32	6.00
Ni.........	0°–100°	.1092	58.6	6.41
Pb........	0°–100°	.0310	207.2	6.43
Pt........	0°–100°	.0318	195.2	6.21
Si........	0°–100°	0.177	28.3	5.02

Solids	Temperature (° C)	Specific Heat (Cal/Gm ° C)
Flint glass..................	10–50	0.12
Glass......................	19–100	.20
Ice........................	0	.487
Rubber....................	20–100	.48
Wood......................	20	0.33

Liquids	Temperature (° C)	Specific Heat (Cal/Gm ° C)
Alcohol—ethyl..............	0	0.548
Ammonia...................	20	1.126
Ether......................	0	0.529
Petroleum..................	21–58	0.511
Sea water..................	17.5	0.95
Turpentine.................	0	0.411

Gases	Temperature (° C)	c_p*
Air........................	0–200	0.2375
Argon......................	20–90	0.1233
Bromine....................	83–228	0.0555
CO_2......................	15–100	0.2025
Hydrogen...................	12–198	3.4090
Oxygen.....................	13–207	0.2175
Water vapor...............	0	0.4655
	100	0.421

* c_p = Specific heat under constant pressure.

is transferred from the hotter one to the cooler one until their temperatures are equalized and a condition known as *thermal equilibrium* is attained. This fact supplies us with one of the most convenient methods, although not the most accurate, of determining specific heats—the *method of mixtures*.

Suppose m_A gm of material A at a temperature t_A and a specific heat c_A are mixed with m_B gm of material B having a specific heat c_B and a tempera-

amount of heat lost by A equals that gained by B, we can write the equation as follows:

$$\underbrace{m_A\, c_A\, (t_A - t_e)}_{\text{Heat lost}} = \underbrace{m_B\, c_B\, (t_e - t_B)}_{\text{Heat gained}}. \quad (12\text{-}4)$$

Since temperatures and masses are directly measurable, the ratio, c_B/c_A, of the two specific heats may at once be obtained from this equation.

If the material B is water, $c_B = 1$. If the water is contained in a calorimeter whose mass is m_c, of material of specific heat c_c, the water equivalent of this vessel is $c_c m_c$, and equation (12-4) must be modified as follows:

$$\underbrace{m_A c_A (t_A - t_e)}_{\text{Heat lost}} = \underbrace{(m_B + c_c m_c)(t_e - t_B)}_{\text{Heat gained}}. \quad (12\text{-}5)$$

Solving for c_A, we obtain

$$c_A = \left(\frac{m_B + c_c m_c}{m_A}\right)\left(\frac{t_e - t_B}{t_A - t_e}\right). \quad (12\text{-}6)$$

In making calorimetric measurements, certain obvious precautions must be taken. For example, there must be good heat insulation around the inner calorimeter so as to prevent convection, conduction, and radiation losses to and from the environment. We may partly compensate for such errors by making the initial temperature of water and calorimeter, t_B, such that $(t_r - t_B)$ is slightly greater than $(t_e - t_r)$, where t_r is room temperature. A preliminary experiment is generally necessary to determine the proper value of t_B.

Other interesting applications of the method of mixtures are given in § 14.2.

EXAMPLE 1: A 50-gm block of platinum is taken from a furnace and dropped into 150 gm of water contained in a 250-gm glass beaker. The temperature of the water rises from 10° to 25° C. What was the temperature of the furnace?

Solution: Let t represent the original temperature of the furnace. The platinum lost heat, and the water and beaker gained an equal amount; hence

$$\left[\underbrace{50(t - 25)\,0.0318}_{\text{Heat lost by platinum}}\right]$$

$$= \left[\underbrace{150(25 - 10)\,1 + 250(25 - 10)\,0.2}_{\text{Heat gained by water and beaker}}\right].$$

$$1.59t - 39.75 = (150 + 50)(25 - 10),$$

$$1.59t = 3,039.75,$$

$$t = 1,912° \text{C}.$$

12.4. Heat as a Form of Energy; Mechanical Equivalent of Heat

According to the old "caloric" theory, heat was an intangible, weightless fluid which could mix, or combine (in a chemical sense), with the particles of matter with perfect ease. When mixed with matter, it was sensible, i.e., *perceptible*, heat; when combined, it "disappeared" or became latent. Differences in the *rate* at which different objects become heated (different specific heats) represented different degrees of "affinity" between different materials and caloric. This theory could explain the experimental phenomena that we have discussed thus far without contradiction. However, largely through the observations of Count Rumford (1753–1814) and the classic experiments of James Prescott Joule (1818–89) the caloric theory was abandoned in favor of a dynamical theory of heat.

It was through the brilliant theoretical and philosophical efforts of Clausius (1822–78) in Germany and of Maxwell (1831–79) and Lord Kelvin (1824–1907) in England that this dynamical theory of heat was finally placed on a firm scientific foundation and the concept of thermal energy identified with the mechanical energy of random motion of atoms and molecules on a microscopic scale.

Rumford in 1798 had shown that the amount of heat developed in the boring of guns had no relation to the amount of shavings turned out, as demanded by caloric theory, but only to the amount of mechanical work that had been done. Since with dull tools an indefinitely large amount of heat could be produced, as well as a correspondingly large amount of work used up, he suggested that there was a *thermal equivalent* to work, or a *mechanical equivalent* of heat.

Between 1840 and 1850 Mayer, Helmholtz, and others were expressing general ideas as to the equivalence of heat and work and suggesting that thermal items, if added to mechanical ones in scientific bookkeeping, invariably would put the energy account in balance.

It is fair to state that Joule was the first to prove experimentally the equivalence of energy and heat. His experiments were designed to show that when mechanical energy is expended upon a system, the energy which disappears is exactly equivalent to the quantity of heat produced. In one famous experiment, for example, water was set into motion by means of paddles, and the rise in temperature was compared with the mechanical work done upon it. His results were equivalent to showing that approximately 1 gram-calorie (gm-cal) of heat was equivalent to some 4.2×10^7 ergs of energy. **183**

Stated in other terms, if 4.2×10^7 ergs of energy disappear through friction, the amount of heat produced would raise the temperature of 1 gm of water 1° C.

It remained for Joule to show that if all the work used in churning the water is spent in producing heat, then the same amount of heat should be obtained for a given expenditure of work regardless of the method used to produce the work. This would, indeed, be a critical set of experiments to perform. Joule stirred mercury, contained in an iron vessel, by means of an iron paddle; he produced heat by rubbing together two iron rings under mercury; he measured the heat produced when electrical energy was converted into heat in a wire of known resistance; he produced heat by friction of water in tubes. In all these cases the constant of proportionality between heat produced and work performed agreed to within 5 per cent. This was quite a remarkable result. We may state the results of Joule's experiment in the form of an equation:

$$W = JQ , \qquad (12\text{-}7)$$

where W is the work expended in ergs, Q the resulting heat in gram-calories, and J a universal constant called the *mechanical equivalent of heat.*

The most reliable values of J are obtained from the experiments of Callendar and Barnes (1899), who studied the conversion of electrical energy into heat energy, and the experiments of Jaeger and Steinwehr (1921), who measured the rise in temperature of a large mass of water (50 kg) that was churned by paddles. The accepted value of the mechanical equivalent of heat is

$$\left. \begin{array}{l} J = 4.1855 \pm 0.0004 \text{ joules/cal} \\[4pt] = 778.2 \text{ ft-lbf/BTU} \\[4pt] = 0.4267 \text{ kgf-m/cal.} \end{array} \right\} \qquad (12\text{-}8)$$

In the mks system the specific heat may be expressed in kilocals per kilogram per degree centigrade, or in joules per kilogram per degree centigrade. Thus the specific heat of water is about 4,186 joules/kg° C.

EXAMPLE 2: A 2-kg hammer moving 50 m/sec strikes a 100-gm lead ball lying on an anvil. If half the energy of the hammer goes into heating the lead, by how many degrees will its temperature be raised?

Solution:

$$\text{KE of hammer} = \tfrac{1}{2} \times 2 \text{ kg} \times 50^2 \text{ m}^2/\text{sec}^2$$
$$= 2.5 \times 10^3 \text{ kg-m}^2/\text{sec}^2$$
$$= 2.5 \times 10^3 \text{ joules}$$
$$= \frac{2.5 \times 10^3 \text{ joules}}{4.19 \text{ joules/cal}} = 596.7 \text{ cal} .$$

$$\tfrac{1}{2} \times 597.5 = 100 \, (t - t_0) \, 0.031 ,$$

$$(t - t_0) = 96.4 \text{ C} .$$

12.5. Conservation of Energy; First Law of Thermodynamics

In § 4.8 we discussed the conservation of energy (potential and kinetic) of mechanical systems in which friction was absent. The results of the previous section permit us to extend this principle to include systems in which the total mechanical energy decreases because of the presence of friction or viscous forces. The mechanical energy, W, which disappears because of friction reappears as heat energy, QJ, which is exactly equal to W, i.e., $W = QJ$. In other words, we may state that, in a mechanical system where heat energy appears as a result of motion against friction, the total energy —mechanical energy plus the mechanical equivalent of the heat energy—is constant. Suppose in Fig. 12-1 a block slides down the rough inclined

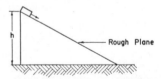

Fig. 12-1.—Illustrating the conservation of mechanical and thermal energy.

plane with constant speed. The kinetic energy remains constant, the potential energy decreases, and therefore the total mechanical energy decreases. According to the conservation principle, an equivalent amount of heat is generated between plane and block. This generalization of the conservation principle to include the equivalence of work and heat and their interconversion is known as the "first law of thermodynamics." This law is our first step in tying together mechanical and thermal phenomena. An analytical formulation of the law is given in the next section.

12.6. Internal Energy

The generalized conservation principle as stated by the first law of thermodynamics permits a logical discussion of the following observations. We know that we may raise the temperature of a beaker of water $t_1°$ by churning the water with a paddle and expending W^1 joules of work in the process. We also know that we may raise the temperature $t_1°$ by adding Q calories of heat to the beaker by means of a Bunsen burner. Or we may stir the water for a while, doing W_1 joules of work, thereby raising the temperature only slightly, and then add enough heat Q_1 to raise the temperature to the required amount, $t_1°$. The first law of thermodynamics tells us that these are all equivalent energy processes and that, quantitatively,

$$W^1 = Q_1 J + W_1 . \qquad (12\text{-}9)$$

It is interesting to observe that if the only information we have is that a given beaker of water formerly at some initial temperature (initial state) is now at a temperature $t_1°$ higher (final state), we have no means of determining how much mechanical work was done on the beaker of water or what amount of heat was added by the burner, except that, whatever their respective values, their algebraic sum $(Q_1 J + W_1)$ must equal a constant by equation (12-9). This analysis suggests that the algebraic sum $(Q_1 J + W_1)$ is the thermodynamic quantity that properly and significantly describes the change that has occurred to the beaker of water.

We shall consider a second more general observation. Suppose some gas or other expansible substance is contained in a cylinder fitted with a piston. We supply Q calories of heat to the substance, which, in addition to other possible effects, expands, thus pushing the piston against atmospheric pressure and thereby doing some external work of amount W joules. This implies that the substance has given up W joules of energy. By the conservation principle, the net gain in energy of the substance $(QJ - W)$ must be absorbed by the substance. This absorption of energy may result from a rise in temperature (the case of the beaker of water) or from a change in phase of the substance from solid to liquid or liquid to vapor without change in temperature or from work in increasing the separation of the molecules against cohesive forces or from any combination of these energy-absorptive processes.

Whatever the process, the energy absorbed by the substance results in a change of *internal energy* of the substance. We shall designate the change in internal energy by $\Delta U = U_1 - U_0$, where U_0 is the internal energy of the initial state and U_1 that of the final state. Although we have no means for measuring the absolute value of the internal energy, U_1 or U_0, we can measure changes in the internal energy of a substance. From our discussion this change in internal energy is

$$\Delta U = QJ - W . \qquad (12\text{-}10)$$

If the energy processes are reversed from those described, the algebraic signs of the quantities QJ and W must also change. Thus, if heat is abstracted from the substance, QJ is negative; if work is done on the substance, W is positive.

We note that the quantity $(Q_1 J + W_1)$ of our first beaker experiment represents the *change* (increase in this case) in *internal energy* of the water. Furthermore, we conclude that the change in internal energy of the water was dependent only on the initial and final states and is *independent* of any particular combination of mechanical or thermal processes used to effect the change in state. Equation (12-10) is a mathematical formulation of the first law of thermodynamics. Further examples of the first law are given in chapters 13 and 16.

EXAMPLE 3: Five kilocalories of heat are supplied to a gas that expands and does 3,000 joules of work. Calculate the change in internal energy of the gas. From equation (12-10),

$$\Delta U = 5 \times 10^3 \times 4.185 - 3,000$$

$$= 17.925 \times 10^3 \text{ joules} .$$

12.7. Continuous-Flow Calorimeter

The specific heat of a liquid may be measured by means of a *continuous-flow calorimeter* sketched in Fig. 12-2. A stream of liquid flows at a constant rate through the tube containing an electrically heated wire. The current through the wires is measured by an ammeter, and the difference in potential across the wire is measured by a voltmeter. When steady-state conditions are attained, the temperature of the liquid as it enters is T_1 and at the outlet after being heated it is T_2. Let **185**

m gm of the liquid whose specific heat is c flow through the calorimeter in t seconds. Then $mc(T_2 - T_1)$ cal of heat are carried away by the liquid in t sec. During this same time interval VIt/J cal of heat are dissipated by the wire,* where V is the potential drop in volts and I the current in amperes; hence

$$m c (T_2 - T_1) + Ht = \frac{V I t}{J}. \quad (12\text{-}11)$$

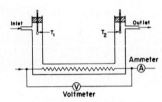

Fig. 12-2.—A continuous-flow calorimeter using electrical heating.

In this equation H represents heat losses from the apparatus. If the rate of flow is changed and the electrical heating adjusted to maintain the same

* See chap. 18.

temperature difference, we have, for steady-state conditions,

$$m_1 c (T_2 - T_1) + Ht_1 = \frac{V_1 I_1 t_1}{J}. \quad (12\text{-}12)$$

From these two equations H may be eliminated and the specific heat, c, determined. This method has been used to measure the specific heat of water at various temperatures. Alternatively, it has been used to measure J with high precision.

12.8. Résumé

The following is a brief listing of the definitions, principles, and theories, given in this chapter, which you should know:

Concept and measurement of quantities of heat

Definition of unit quantities of heat—the calorie and the BTU

Definition and measurement of specific heats of substances

Heat as a form of energy—measurement of the mechanical equivalent of heat

The first law of thermodynamics

The concept of internal energy of substances and how changes in it may be measured

Queries

1. Is the following statement true? "Since Joule's equivalent, J, is a universal constant, it is independent of the units in which the work done and the heat generated are measured." Justify your answer.
2. What is meant by the statement that the mechanical equivalent of heat is 3.0874 ft-lbf/cal?
3. What distinction was drawn between the terms "calorie," "specific heat," "thermal capacity," and "water équivalent"?
4. How would you determine the truth of the statement that the specific heat of ethyl alcohol is 0.548 cal/gm ° C?

Practice Problems

To be used in the problems where needed:

Standard atmospheric pressure
$$= 1,013,250 \text{ dynes/cm}^2$$
$$J = 4.1855 \text{ abs joules/cal}_{15°},$$
$$\text{Cal}_{15°} = 3.0874 \text{ ft-lbf}$$
$$= 0.0011628 \text{ watt-hours}.$$

1. If 100 gm of water at 90° C and 40 gm of water at 10° C are mixed, what is the final temperature? What assumptions are necessary to get a numerical answer? Ans.: 67°.1 C.

2. If 40 gm of water at 10° C are put into 100 gm of water at 90° C contained in a 200-gm glass (specific heat = 0.2 cal/gm ° C) beaker, what will be the final temperature? What assumptions are necessary to get a numerical answer?

3. If 300 gm of a metal at 100° C are put into 260 gm of water at 15° C contained in a 200-gm glass beaker and the final temperature is 20° C, what is the specific heat of the metal? Ans.: 0.0625 cal/gm ° C.

4. A lead bullet with a velocity of 100 m/sec strikes a target. If half the energy of the bullet goes to warming it, how many degrees will its temperature be raised?

5. In the continuous-flow calorimeter 84 watts of power are necessary to maintain a steady-state temperature difference of 5° C with a rate of flow of 10 gm/sec. Assuming negligible heat losses, calculate the specific heat of the liquid. Ans.: $c = 0.40$ cal/gm ° C.

6. The steel projectile of the long-range gun with which the Germans bombarded Paris had a velocity of 3,160 ft/sec at an elevation of 14 miles and a velocity of 2,400 ft/sec when it reached the ground. On the assumption that half the heat generated by friction was kept by the projectile, how many degrees was its temperature raised during this part of the fall? (Do not forget to include the loss of potential energy in your calculation.)

7. A torque of 200 cm-gmf is necessary to turn the paddles of a cake-mixer 300 rev/min. At what rate is heat being generated in the batter? Ans.: 8.82 cal/min.

8. An aluminum calorimeter of mass 100 gm contains 180 gm of a liquid at 10° C. An aluminum ball of mass 50 gm and at a temperature of 100° C is dropped into the calorimeter. If the final temperature of the mixture is 20° C, find the specific heat of the liquid.

9. An electric heating coil raises the temperature of 500 gm of water 15° C in 6 min. What would the rise in temperature be if the water were replaced by 800 gm of a liquid whose specific heat is 0.3 and

the coil allowed in the liquid for 8 min? (Assume that the rate at which the coil gives heat is constant in both cases.) Ans.: 41°7 C.

10. A heating coil is placed in a flask containing 600 gm of water for 10 min and raises the temperature 15° C. If the water is replaced by 300 gm of another liquid and the same rise in temperature results in 2 min, what is the specific heat of the liquid?

11. What is the change in internal energy of a gas that absorbs 200 cal of heat and does 10,000 ergs of work in expansion? Ans.: 8.37×10^9 ergs.

12. A copper calorimeter has a mass of 100 gm and contains 50 gm of water at $t°$ C. A 40-gm specimen of brass, specific heat 0.092 cal/gm °C, at a temperature of 100° C is dropped into the calorimeter. If the room temperature is 20° C, what must t be so that the difference between room temperature and t is the same as the difference between room temperature and the final temperature of the mixture?

13. Two runs with the continuous-flow calorimeter were used to measure the specific heat of a liquid. In the first, a temperature difference of 10° C was maintained for a flow rate of 65 gm/sec, with a power dissipation of 1,407 watts. In the second, a temperature difference of 10° C was maintained for a flow of 45 gm/sec, with a power dissipation of 987 watts. Compute the specific heat of the liquid, taking heat losses into account. Ans.: $c = 0.5$ cal/gm ° C.

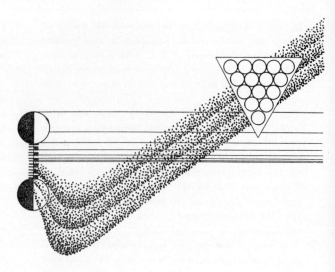

CHAPTER 13

Gas Laws and the Kinetic Theory

13.1. Expansion of Gases

Ordinary changes in pressure have little effect upon the volume of solids and liquids; i.e., solids and liquids are highly incompressible. For this reason, pressure effects were ignored when we discussed expansion coefficients. In the case of gases, however, even a very small variation in pressure changes the volume of a given mass of gas by a

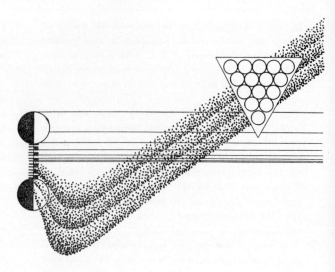

FIG. 13-1.—Volume expansion of gas under constant pressure.

significant amount. The volume of a gas, therefore, *depends upon both the temperature and the pressure.*

We shall discuss the expansion of gases in this section under two quite separate and distinct conditions: (1) that the pressure remains constant while the temperature of the gas is changed and (2) that the volume is constant while the pressure changes with temperature.

Condition 1 may be experimentally realized by confining a given mass of gas in a cylinder fitted with a movable, frictionless piston, as shown in Fig. 13-1 (*a*). The external pressure, P_0, we shall assume to be constant. Let V_0 be the volume of gas at 0° C, and V_t the volume at any higher temperature t. As the gas is heated, the volume increases linearly with temperature, as shown by Fig. 13-1 (*b*). The experimental facts may also be represented by the following equation:

$$V_t = V_0 (1 + \beta t), \qquad (13\text{-}1)$$

where β is a constant. The significance of β may be seen by rewriting equation (13-1):

$$\beta = \left(\frac{V_t - V_0}{V_0} \right) \frac{1}{t}. \qquad (13\text{-}2)$$

From equation (13-2) we see that β is the volume expansion coefficient of a gas under constant pressure and is proportional to the slope of the graph in Fig. 13-1 (*b*). What is, indeed, remarkable about equation (13-2) is that β is practically a constant and is equal approximately to $\frac{1}{273} = 0.00366$ per degree centigrade for all the "permanent gases"— e.g., hydrogen, oxygen, helium, neon, nitrogen, etc.

To proceed in accordance with assumption 2 above, we let the gas be confined to a closed box of invariant volume, as shown in Fig. 13-2 (*a*), and vary the temperature. Then it is found that the pressure of the gas, P, increases with the temperature according to the relation

$$P_t = P_0 (1 + \gamma t), \qquad (13\text{-}3)$$

where P_0 is the initial pressure at $0°$ C and P_t is the pressure at temperature t. The pressures P_0 and P_t may be measured in any convenient system of units—e.g., atmospheres, dynes per square centimeter, pounds-force per square inch, newtons per square meter, etc. If t is in degrees centigrade, γ has the units per degree centigrade.

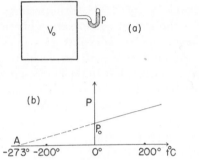

Fig. 13-2.—Variation of pressure with temperature at constant volume.

The graphical representation of the equation is shown in Fig. 13-2 (b) by the solid line. The constant γ is called the "coefficient of pressure change at constant volume." As was pointed out in § 11.5, γ approaches $\frac{1}{273} = 0.00366$ per degree centigrade for all the permanent gases; that is to say, experimentally we find that β is very closely equal to γ. A few expansion coefficients are listed in Table 13-1.

13.2. Absolute Temperature

Since $\beta = \gamma = \frac{1}{273}$ for all the permanent gases, we may introduce a simplification into equations (13-1) and (13-3). Let

$$T = t + 273, \qquad (13\text{-}4)$$

where t is the temperature in degrees centigrade and where T will be called the "absolute temperature" or "degrees Kelvin" (° K). Substituting in equations (13-2) and (13-3) with $\beta = \gamma = \frac{1}{273}$, we obtain

$$V_t = \left(\frac{V_0}{273}\right) T \qquad \text{(pressure constant)}, \quad (13\text{-}5)$$

$$P_t = \left(\frac{P_0}{273}\right) T \qquad \text{(volume constant)}. \quad (13\text{-}6)$$

Equation (13-5) is the analytical expression of Charles's law—*at constant pressure the volume of a given mass of gas varies directly as the absolute temperature.*

Equations (13-5) and (13-6) may be rewritten in a form better adapted to problem-solving. If a given mass of gas at temperature T_1 and volume

TABLE 13-1

THERMAL COEFFICIENTS OF GASES

CONSTANT PRESSURE
(Volume Coefficient, β)

Substance	Pressure (Cm of Hg)	Temperature (° C)	$\beta \times 10^5$ (Degrees C^{-1})
Air...........	76	0–100	367.1
	100	0–100	366.0
H$_2$...........	30,400	0–100	295
	60,800	0–100	242
CO$_2$...........	100	0– 20	376.0
SO$_2$...........	76	....	390.3
H$_2$O (vapor)......	76	0–119	418.7
		0–162	407.1
		0–247	380.0

CONSTANT VOLUME
(Pressure Coefficient, γ)

Substance	Pressure (Cm of Hg)	Temperature (° C)	$\gamma \times 10^5$ (Degrees C^{-1})
Air...........	0.6	0–100	376.66
	25	0–100	365.80
	76	0–100	366.50
	10,000	0–100	410.0
He...........	56.7	....	365.5
H$_2$...........	0.0077	16–132	332.8
	100	0–100	366.3
CO$_2$...........	100	0– 20	372.6
SO$_2$...........	76	....	384.3

V_1 is heated to a temperature T_2 and volume V_2, then, by equation (13-5),

$$\frac{V_1}{T_1} = \frac{V_2}{T_2} \qquad \text{(pressure constant)}, \quad (13\text{-}7)$$

the factor $V_0/273$ canceling out of the ratio. Also, if the volume is kept constant but the pressure is allowed to change, equation (13-6) yields

$$\frac{P_1}{T_1} = \frac{P_2}{T_2} \qquad \text{(volume constant)}. \quad (13\text{-}8)$$

13.3. Boyle's Law

The law of variation of pressure with volume of a given mass of gas when the *temperature* remains **189** constant was discovered by Robert Boyle in 1660. Boyle's law is stated as follows:

The product of pressure and volume of a given mass of gas at constant temperature is a constant.

Stated in symbols, Boyle's law is

$$PV = C = \text{Constant} , \qquad (13\text{-}9)$$

where the numerical value of C will depend upon the temperature and mass of the gas. Equation (13-9) is illustrated graphically in Fig. 13-3. Curve

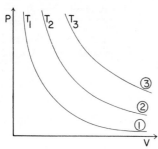

Fig. 13-3.—Graphical representation of $PV = \text{Constant}$

1 represents Boyle's law for a given mass of gas at a temperature T_1 (absolute temperature); curve *2*, for a temperature T_2. These curves are called ISOTHERMAL CURVES and represent, as we have seen, the relation between pressure and volume during the isothermal expansion or compression of a gas.

A convenient statement of Boyle's law for a given mass of gas is

$$P_1 V_1 = P_2 V_2 \text{ (temperature constant)} . \quad (13\text{-}10)$$

Gases whose physical behavior is represented *exactly* by equations (13-5), (13-6), and (13-9) are, by definition, *ideal gases*. We have already noted that such gases as hydrogen, helium, oxygen, nitrogen, etc., may be considered closely to approximate ideal gases. The approximation is increasingly better at lower pressures and higher temperatures. Deviations from the ideal gas are discussed in § 13.19.

Another point that must be emphasized in connection with equation (13-9) is that this equation is applicable only to EQUILIBRIUM STATES OF THE GAS. By "equilibrium state" of a gas we mean this: If a mass of gas is at a constant temperature T throughout, if it fills a definite volume V, and if it is subject to a constant external pressure P, this gas will show no tendency to change this condition. In other words, a mass of gas confined

to a definite volume and at a definite temperature will indicate the same pressure indefinitely. A situation in which equation (13-9) is not applicable at all arises whenever a gas is rapidly expanding.

13.4. Work Done in Isothermal Expansion or Compression of a Gas

It is a common fact of observation that if a gas is compressed in a cylinder, the gas becomes heated. When a gas is compressed, work is done on the gas. If, on the other hand, a gas is allowed to expand, the gas *does work* in pushing the piston of the cylinder out against atmospheric pressure, and hence the gas is cooled. If a mass of air at $0°$ C, confined to a perfectly insulating cylinder so that no heat is lost to the environment, is compressed to one-tenth its original volume, its temperature will rise to $429°$ C. These interesting facts and some of their implications will be discussed in greater detail in § 13.18.

In view of these facts, it is clear that if Boyle's law is to be applicable to compressions or expansions, these expansions or compressions must take place very slowly indeed, to allow the pressure to be the same throughout the gas and to allow sufficient time for the gas to adjust itself to and maintain itself at a constant temperature. This latter involves the transfer of heat to or from the surroundings.

The work done in compressing a gas *isothermally* from a volume V_1 to V_2 may be computed from first principles. Let the gas at pressure P be confined to a cylinder fitted with a piston of cross-sectional area A. If the piston is pushed down a distance dx by an external force F equal at each instant to PA, the force exerted by the gas on the piston, the work done on the gas is

$$\Delta W = F \, dx .$$

Since $F = PA$ and $A dx$ is the volume change dV, we have

$$\Delta W = P \, dV$$

or

$$W = \int_{V_1}^{V_2} P \, dV . \qquad (13\text{-}11)$$

From Boyle's law, $P_1 V_1 = P_2 V_2 = C$ (eq. [13-10]), so that

$$P = \frac{C}{V},$$

and

$$W = \int_{V_1}^{V_2} C\, \frac{dV}{V} = C \log_e \frac{V_2}{V_1} = C \ln \frac{V_2}{V_1}.$$

Therefore,

$$W = P_1 V_1 \ln \frac{V_2}{V_1} = P_2 V_2 \ln \frac{V_2}{V_1}. \quad (13\text{-}12)$$

In this equation $\log_e$ or $\ln$ is the natural logarithm. These results are also shown graphically in Fig. 13-4. The curve AB represents an isotherm at a

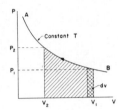

Fig. 13-4.—Work done in isothermal compression of an ideal gas.

temperature T. The shaded area is the work done in isothermally compressing the gas from volume V_1 to volume V_2.

EXAMPLE 1: A gas under atmospheric pressure occupies a volume of 100 liters. Compute the work done in isothermally compressing the gas to half its initial volume.

Solution: The initial conditions are

$$V_1 = 100 \text{ liters} = \tfrac{1}{10}\text{m}^3\ ,$$

$$P_1 = 1.013 \times 10^5 \text{ newtons/m}^2\ .$$

The final condition is $V_2 = \tfrac{1}{2}V_1$. From equation (13-12) we have

$$W = (1.013 \times 10^5 \text{ newtons/m}^2)\,(\tfrac{1}{10}\text{ m}^3)\ln\tfrac{1}{2}$$

$$= (1.013 \times 10^5)\,(\tfrac{1}{10})\,(-0.693)$$

$$= -7,020 \text{ joules}\ .$$

The minus sign indicates that work has been done *on* the gas or that energy is absorbed by the gas.

13.5. The Gas Law

So far we have allowed only two of the three variables P, T, and V to vary on any one occasion. We may inquire as to what relation exists between P, T, and V, if all three are allowed to vary. We may deduce this relationship by using Charles's [eq. (13-7)] and Boyle's [eq. (13-10)]

laws. In Fig. 13-5 (a) a mass of gas is in equilibrium, with initial temperature T_0, pressure P_0, and volume V_0. This is illustrated by point 1 on the PV diagram of Fig. 13-6. Keeping the pressure constant and changing the temperature to T_1, we may apply Charles's law to obtain

$$\frac{V_0}{T_0} = \frac{V'}{T_1}.$$

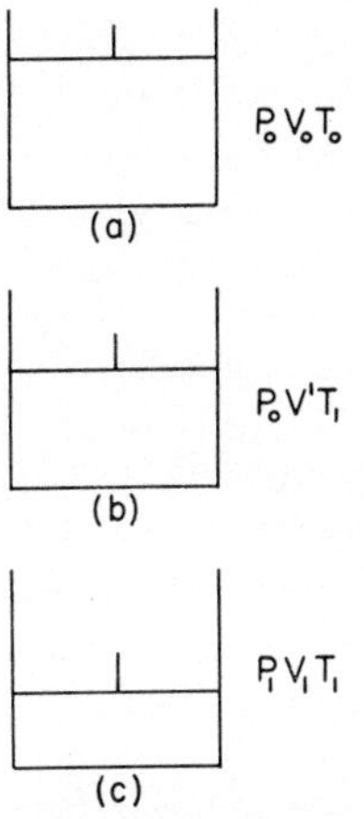

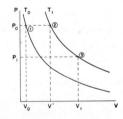

Fig. 13-5.—Derivation of the gas law

Fig. 13-5 (b) and point 2 of Fig. 13-6 represent this state of the gas. By keeping the temperature constant as T_1 and varying the pressure, we obtain, from Boyle's statement,

$$P_0 V' = P_1 V_1\ .$$

This state is represented by point 3 of Fig. 13-6.

Fig. 13-6.—PV diagram, illustrating the gas law

If we eliminate V' from these two equations, we get

$$\frac{P_0 V_0}{T_0} = \frac{P_1 V_1}{T_1},$$

or, leaving off the subscript 1,

$$\frac{PV}{T} = R, \qquad PV = RT, \qquad (13\text{-}13)$$

where $R = P_0 V_0 / T_0$. Equation (13-13) is the gas law, often referred to as the EQUATION OF STATE, **191** for ideal gases. Since the state represented by Fig. 13-5 (b) is any intermediate state, equation (13-13) represents the relationship of P, V, and T for all equilibrium states of an ideal gas. R is called the *gas constant* and depends upon the nature and mass of gases. It is an experimental fact that 1 mole (mass in grams equal numerically to molecular weight) of any ideal gas occupies 22.4 liters at $0°$ C and a pressure of 1 atm (1.013×10^6 dynes/cm²). It follows, then, that R is a universal constant for 1 mole of gas and is equal to

$$R = \frac{P_0 V_0}{T_0} = \frac{1.013 \times 10^6 \times 22.4 \times 10^3}{273}$$
$$= 8.31 \times 10^7 \text{ ergs/deg-mole} \Big\} \quad (13\text{-}14)$$
$$= 8.31 \text{ joules/deg-mole}.$$

The gas law, equation (13-13), may be rewritten in forms convenient for calculations. For the same mass of gas, initially at V_1, P_1, and T_1 and finally at V_2, P_2, and T_2, the gas equation becomes

$$\frac{P_1 V_1}{T_1} = \frac{P_2 V_2}{T_2}. \quad (13\text{-}15)$$

Also, if m_1 gm of a given gas occupy V_1 cm³ at a pressure of P_1 dynes/cm² and at a temperature of $T_1°$ K and, further, if m_2 gm of the *same gas* occupy V_2 cm³ at a pressure of P_2 dynes/cm² and at a temperature of $T_1°$ K, then it can be shown that

$$\frac{P_1 V_1}{T_1 m_1} = \frac{P_2 V_2}{T_2 m_2}. \quad (13\text{-}16)$$

In many problems the pressures and volumes may not be given in dynes per square centimeter and cubic centimeters. The value of R in the case of mixed units is given in the accompanying table.

Units of Pressure	Units of Volume	R per Mole of Gas
Atmospheres................	cm³	82.06
Atmospheres................	liters	0.08206
Mm of mercury.............	cm³	62,370
Lbf/in².....................	in³	18,510
Newtons/m² (mks)..........	m³	8.31

In equation (13-13) V represents the volume of 1 mole of gas. Hence, if M is the molecular weight of the gas, the density, ρ, is equal to M/V, and we may rewrite the gas law as

$$P = \frac{R}{M} \rho T. \quad (13\text{-}17)$$

EXAMPLE 2: Oxygen is contained in a cylinder fitted with a piston. At a temperature of $20°$ C, the volume of oxygen is 100 liters and the pressure 20 atm. What is the pressure when the temperature rises to $30°$ C and the volume becomes 75 liters?

Solution: Initial conditions are: $P_1 = 20$ atm, $V_1 = 100$ liters, $T_1 = 293°$ K. Final conditions are: P_2 is required, $V_1 = 75$ liters; $T_2 = 303°$ K. By equation (13-15)

$$\frac{P_1 V_1}{T_1} = \frac{P_2 V_2}{T_2},$$

$$\frac{20 \times 100}{293} = \frac{P_2 (75)}{303}; \quad P_2 = 27.5 \text{ atm}.$$

13.6. Kinetic Theory of Ideal Gases

The thought has probably occurred to the reader that gases possess some rather interesting properties—properties not shared in general by liquids or solids. For example, gases have no definite shape or volume but fill whatever containers they are placed in. In other words, gases appear able to expand indefinitely. Gases exert on the walls of containers pressures far in excess of what one would expect from the weight of the gas. As we have seen, all the gases have approximately the same pressure and volume expansion coefficients.

Attempts by such men as Daniel Bernoulli (1700–1782), Maxwell (1831–79), Clausius (1822–78), and Boltzmann (1844–1906) to explain these peculiarities have resulted in the development of a highly successful theory of the structure and behavior of gases—*the kinetic theory of gases*.

According to this theory, a gas consists of individual small particles identified as MOLECULES which are in continuous motion, constantly striking one another and the walls of the container. The average velocities of these molecules are high. At a temperature of $0°$ C, under normal conditions, hydrogen molecules have average velocities of the order of 18×10^4 cm/sec. The average distance a molecule travels between impacts is called the *mean free path* and in the case of hydrogen is of the order of 10^{-6} cm at normal pressures. Because of this rather large mean free path, intermolecular forces are too weak to cause molecules to cohere as in a liquid. Thus, as a container is made larger and larger, for a fixed mass of gas, the mean free path becomes greater and greater. It is clear, then, that the apparent volume of this gas, i.e., the volume that it fills *kinetically*, is very much larger than the sum of the molecular volumes.

On the basis of the kinetic theory, the high compressibility of a gas is explained by the small volumes of the molecules relative to the space they fill. The pressure of a gas is explained as due to the rate at which momentum is transferred to the walls by impacts of molecules against one another and against the walls of the container. (The reader will recall that, whenever a change in momentum occurs, a force must be present.) If it is assumed that impacts between the molecules and between the molecules and the walls are perfectly elastic ($\epsilon = 1$ [cf. chap. 3]), one can compute the change in momentum which occurs whenever a molecule strikes the wall and rebounds. Because of the large velocities possessed by the molecules, their enormous numbers, and the resulting high frequency of their collisions with the wall, rather large gas pressures may arise. We shall consider the details of these phenomena in § 13.7.

Another interesting correlation brought out by the kinetic theory is that of temperature and molecular speed. As we have shown in § 13.1, if the temperature of a confined gas is increased, its pressure increases. But, according to the kinetic theory, increase in pressure would, in this case, imply increased molecular speeds. We have here a possibility of an interpretation of temperature in terms of molecular speeds.

13.7. Calculation of Pressure of a Gas

Let us suppose that a cubical vessel of side l cm (Fig. 13-7) contains N molecules, the mass of each

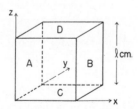

FIG. 13-7.—Coordinate system used for the computation of gas pressure.

molecule being m gm. If u is the velocity of any molecule, then we shall resolve u into components parallel to the x-, y-, and z-axes of the figure: u_x, u_y, and u_z. The reader may verify that

$$u_x^2 + u_y^2 + u_z^2 = u^2 . \quad (13\text{-}18)$$

In order to compute the average pressure exerted by the molecules by elementary methods, we must make certain assumptions as to the behavior of these molecules. Such assumptions we have already indicated in § 13.6, but, for clarity, we shall list them explicitly here.

1. **We assume that all molecules are in rapid motion and move about chaotically in all directions—that is, there is no preferred direction in the box.**
2. **Between collisions, molecules move in straight lines.**
3. **Collisions between molecules and between molecules and the walls are perfectly elastic.**
4. **There are no appreciable intermolecular forces.**
5. **The diameters of the molecules are entirely negligible in comparison with the average distances they travel between collisions.**
6. **The time spent during collisions is much less than the time between collisions.**

Let us consider a molecule of mass m, proceeding from face A toward face B with a space velocity, $_1u$, whose component along the x-axis is $_1u_x$. The component of the momentum of this molecule normal to face B, then, is m_1u_x before it strikes face B. After the molecule has collided with face B and if the collision is perfectly elastic (as assumed), its normal momentum will be $-m_1u_x$.* The *change* in momentum, then, due to m striking face B, is

$$m_1u_x - (-m_1u_x) = 2 m_1u_x . \quad (13\text{-}19)$$

The time for traveling from A to B to A is

$$\frac{2l}{_1u_x} \text{ sec} . \quad (13\text{-}20)$$

The number of collisions per second on face B is, then, for the one molecule,

$$\frac{_1u_x}{2l} \text{ per second} . \quad (13\text{-}21)$$

---

* The reader may wonder to what extent our argument is invalidated in case the molecule, as is most likely, collides with other molecules or with other faces of the inclosure on its way between A and B. It is to be remembered that we are dealing with point molecules; hence all collisions must be central and perfectly elastic. Such collisions between bodies of equal mass result in an exact exchange of velocities. Thus there will be some one molecule that will collide with face B with momentum m_1u_x, which will always correspond to the one that left face A with this momentum. Similarly, any elastic collision with any wall parallel to the x direction leaves the tangential component of momentum unchanged during the impact. Thus neither of these eventualities invalidates our argument.

Therefore, the rate of change of momentum at face B in the x direction is

$$(2 m_1 u_x)\frac{1 u_x}{2 l} \text{ per molecule}. \quad (13\text{-}22)$$

By Newton's Second Law, $F = (mu - mu_0)/t$; and hence the average force, f, exerted by the molecule on face B or A is

$$f = \frac{m_1 u_x^2}{l}; \quad (13\text{-}23)$$

and, for N molecules,

$$F = \frac{m}{l}(_1 u_x^2 + _2 u_x^2 + \ldots + _N u_x^2), \quad (13\text{-}24)$$

$$F = \frac{m}{l}\sum_1^N u_x^2. \quad (13\text{-}25)$$

Let $\overline{u_x^2}$ be the average value of the square of the x component of velocity averaged over all the N molecules, i.e., $N\overline{u_x^2} = \Sigma u_x^2$; then

$$F = \frac{N m \overline{u_x^2}}{l}. \quad (13\text{-}26)$$

By Newton's Third Law the average force exerted by face B on the molecules is also equal to expression (13-26). Since by "pressure" we mean force per unit area, it follows that

$$\frac{F}{l^2} \equiv p_x = \frac{N m \overline{u_x^2}}{l^3},$$

where l^3 is equal to the volume of the cube. Placing $l^3 = V$, we obtain, for the average pressure,

$$p_x = \frac{N m \overline{u_x^2}}{V}. \quad (13\text{-}27)$$

Similar expressions may be obtained for the forces per unit area exerted in the y and z directions:

$$p_y = \frac{N m \overline{u_y^2}}{V} \quad \text{and} \quad p_z = \frac{N m \overline{u_z^2}}{V}.$$

Since there can be no preferred direction for the molecules, it can be shown that

$$\overline{u_x^2} = \overline{u_y^2} = \overline{u_z^2} = \frac{\overline{u^2}}{3}, \quad (13\text{-}28)$$

where $\overline{u^2}$ is the average of the square of the velocities. Also, by Pascal's law, the pressure is the same throughout the volume, $p_x = p_y = p_z = P$.* The

* This is true if the cube is sufficiently small that the weight of the gas may be neglected. Otherwise, the pressure at the top face will be smaller than that at the lower face.

average pressure exerted by the molecules on the container is, therefore,

$$P = \frac{N m \overline{u^2}}{3 V} = \frac{n m \overline{u^2}}{3}, \quad (13\text{-}29)$$

where n is the number of molecules per unit volume. Since the total mass $M = Nm$,

$$P = \frac{1}{3}\frac{M \overline{u^2}}{V}$$

or

$$PV = \tfrac{1}{3} M \overline{u^2}; \quad (13\text{-}30)$$

an alternative way of writing equation (13-30) is

$$P = \tfrac{1}{3}\rho \overline{u^2}, \quad (13\text{-}31)$$

where ρ is the density of the gas.

It should be observed that if the units are those of the cgs system, P must be measured in dynes per square centimeter, ρ in grams per cubic centimeter, and u in centimeters per second.

Using equation (13-31), we may, from observed densities at any pressure and temperature, calculate the root-mean-square velocity $(\sqrt{\overline{u^2}})$ of any gas at any temperature. Table 13-2 gives the re-

TABLE 13-2

VELOCITY OF MOLECULES AND KINETIC ENERGY PER MOLE OF GASES AT 0° C

Gas	$\sqrt{\overline{u^2}}$	$M(O=16,000)$		$\frac{1}{2}M\overline{u^2}$
O_2	4.61×10^4 cm/sec	2×16	$= 32$	3.40×10^{10}ergs
N_2	4.93	2×14	$= 28$	3.39
Air	4.85		28.8	3.28
CO	4.93	$12+16$	$= 28$	3.39
H_2	18.38	2×1.008	$= 2.016$	3.37
He	13.11		4	3.43
Hg	1.84		200	3.39
Ne	5.84		20.1	3.42
CO_2	3.93	$12+(2\times16)$	$= 44$	3.40
H_2O	6.15	$(2\times1)+16$	$= 18$	3.40

sults of these calculations for a number of common gases, calculated for 0° C. Molecular weights also are listed in the third column, and the fourth column gives the total kinetic energy for 1 mole (equal to a mass of 1 gm molecular weight) of gas at 0° C. Example 3 shows in detail how the calculation is made.

We note in Table 13-2 that, although the velocities of different gases differ widely at the same temperatures, the **kinetic energy per mole is the same for all gases at the same temperature.**

EXAMPLE 3: Calculate the root-mean-square velocity for air at 0° C and at 76-cm pressure.

Solution: For air at 0° C and 76-cm pressure,

$$P = 76 \times 13.6 \times 980 \, \text{dynes/cm}^2 ,$$

$$\rho = 0.001296 \, \text{gm/cm}^3 .$$

Therefore, from equation (13-31),

$$\sqrt{\overline{u_{\text{air}}^2}} = \sqrt{\frac{3 \times 1.012 \times 10^6 \, \text{gm-cm/sec}^2\text{-cm}^2}{1.296 \times 10^{-3} \, \text{gm/cm}^3}}$$

$$= 48,500 \, \text{cm/sec} .$$

13.8. Maxwellian Distribution of Molecular Speeds

In the preceding section we were concerned with the root-mean-square (r.m.s.) velocities of the molecules, $\sqrt{\overline{u^2}}$. Actually, the velocities of the molecules range over a wide span. At room temperature and atmospheric pressure there are about 10^9 collisions per second between molecules of a gas. We would expect that some few molecules would attain speeds in excess of the r.m.s. value, others very low speeds, approaching a zero value. Either of these extremes is not very likely, since each would require a large number of preferential collisions. We would expect that the largest number of molecules would encounter collisions of all kinds and therefore have speeds grouped about some average speed.

The problem of the most likely distribution of speeds among a large number of molecules was solved mathematically by Maxwell, based on the reasonable assumption of complete randomness of molecular motions. The result of this computation follows:

$$\frac{\Delta N}{N} = 4\pi v^2 \left(\frac{m}{2\pi kT}\right)^{3/2} \exp\left(\frac{-mv^2}{2k_1}\right) \Delta v . \quad (13\text{-}32)$$

In this equation $\Delta N/N$ represents the fraction of the molecules in the interval Δv; v is the molecular speed; m the mass of the molecule; T the absolute temperature; and k Boltzmann's constant (see § 13.11).

A graphical plot of this Maxwellian distribution of speeds for two different temperatures of a given number of molecules is shown in Fig. 13-8. We observe that the number of molecules in a given speed interval increases, reaching a maximum at the most probable speed, and then decreases asymptotically toward zero value. Calculations show that the arithmetic mean of all the speeds is somewhat larger than the most probable value and that the square root of the sum of the squares (r.m.s.) of the speeds is still larger. As the temperature increases, the most probable speed increases, but the fraction of molecules with that speed decreases. On the other hand, the total number of molecules with speeds greater than a given speed increases rapidly with a rise in temperature. This explains the rapid increase in rates of chemical reactions with rising temperatures.

The experimental evidence for the randomness of molecular motions can be found from the study of *Brownian movements* in liquids and gases. In 1817 the British botanist, Robert Brown, first observed that small particles suspended in a liquid are in a state of continuous random motion. These zigzag motions were ascribed to random molecular collisions. In 1909 Jean Perrin conducted a series of exhaustive experiments in

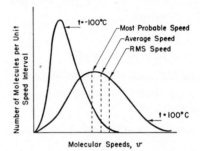

Fig. 13-8.—Maxwellian distribution of molecular speeds at $-100°$ C and $100°$ C.

water and was able to show that the mean kinetic energies of various-sized particles were the same and equal to that of a gas at the same temperature.

13.9. Kinetic-Theory Interpretation of Temperature

The reader will observe that the quantity $\frac{1}{3}M\overline{u^2}$ of equation (13-30) is just $\frac{2}{3}(\frac{1}{2}M\overline{u^2})$, i.e., two-thirds of the total kinetic energy of translation of the molecules. If we make a further assumption that the kinetic energy of the molecules remains constant when the temperature is constant, then equation (13-30) represents Boyle's law:

$$PV = \text{Constant (temperature constant)} . \quad (13\text{-}33)$$

The kinetic theory offers us a deeper insight into the meaning of the temperature concept. As we have indicated previously, equation (13-30) may be written (with M now equal to the gram molecular weight) as

$$PV = \tfrac{2}{3}\left(\tfrac{1}{2}M\overline{u^2}\right) . \quad (13\text{-}30)$$

We introduce the temperature by means of the experimentally determined equation of state for 1 mole of gas,

$$PV = RT . \quad (13\text{-}34)$$

Hence

$$\tfrac{2}{3}\left(\tfrac{1}{2}M\overline{u^2}\right) = RT . \quad (13\text{-}35)$$

This important deduction tells us that the average translational kinetic energy of the molecules is directly proportional to the absolute temperature. **195**

Equation (13-35) gives rise to some rather interesting speculations regarding the nature of temperature. We may, for example, say that, in order to fit the kinetic theory to the gas law, relation (13-35) is a necessary assumption. On the other hand, we may consider equation (13-35) as defining temperature on the kinetic theory. Thus, in either case, we obtain a new and important meaning of temperature.

13.10. The Law of Avogadro

One of the most important conclusions about gases is that, under the same condition of pressure and temperature,

Equal volumes contain equal numbers of molecules.

This is known as the *law of Avogadro*.

This law is derivable from equation (13-29) of the kinetic theory applied to two different gases under the same temperature and pressure and with $V = 1$ cm^3, i.e.,

$$P = \tfrac{1}{3} n_1 m_1 \overline{u_1^2}; \qquad P = \tfrac{1}{3} n_2 m_2 \overline{u_2^2}, \quad (13\text{-}36)$$

and by using the additional assumption given by equation (13-37) below of *equipartition of translational energy* among the degrees of freedom of molecules in gases at the same temperature:[*]

$$\tfrac{1}{2} m_1 \overline{u_1^2} = \tfrac{1}{2} m_2 \overline{u_2^2}, \qquad (13\text{-}37)$$

whence

$$n_1 = n_2 . \qquad (13\text{-}38)$$

A far simpler approach to this conclusion, at the present time, is offered by the fact that modern methods of atomic and molecular analysis (positive-ray mass spectrographs) have provided us with precise knowledge of relative atomic and molecular weights.

Now it is possible to measure with equal precision the absolute densities of gases. The results of such measures are appended in the fourth column of Table 13-3. The *relative* densities of gases, with respect to hydrogen taken as unity, are given in the fifth column of the table. We observe that the relative densities with respect to hydrogen give the SAME SET OF NUMBERS as is given by the

[*] By means of a statistical analysis of the mechanics of myriads of colliding, perfectly elastic molecules it is possible to give a sound deduction of the principle of equipartition of energy.

molecular weights relative to the same element. From this identity of values in the third and fifth columns, one and only one simple conclusion can be drawn, and that is that equal volumes of gases under the same temperature and pressure (calculated in the fourth column for 0° C and 76-cm pressure) contain the same number of molecules.

Thus we have *deduced* the **law of Avogadro** in quite the same direct fashion that we should deduce the fact that two closed boxes contained the *same number of marbles and golf balls, respectively,* if the net weights of the two boxes had exactly the

TABLE 13-3

MOLECULAR WEIGHTS AND DENSITIES

Substance	International Molecular Weight (O = 16)	Molecular Weight (H = 1)	ρ (at 0° C, 76 Cm) $\times 10^3$ Gm/Cm^3	ρ/ρ_{H_1}
H$_2$	2.016	2.000	0.08988	2.0
O$_2$	32.000	31.8	1.4290	31.8
N$_2$	28.016	27.8	1.2505	27.8
CO	28.000	27.8	1.2504	27.8
Air (mixture)	28.8	28.6	1.2929	28.8
He	4.002	3.97	0.1784	3.98
Ne	20.183	20.0	0.9003	20
A	39.944	39.6	1.7837	39.6
X	130.2	129.3	5.851	130
CO$_2$	44.000	43.7	1.9769	43.7

same ratio as the weights of an individual sample marble and golf ball.

One gram molecular weight of a gas under a pressure of 1 atm (1,013,246 dynes/cm^2) and a temperature of 0° C requires a volume of 22,414 cm^3. The number of molecules in a mole is Avogadro's number (N_0).[*] The number of molecules per cubic centimeter at normal pressure and temperature is Loschmidt's number (n). These values are

$$N_0 = 6.0228 \times 10^{23} \text{ molecules/mole}, \quad (13\text{-}39)$$

$$n = 2.687 \times 10^{19} \text{ molecules/cm}^3 . \quad (13\text{-}40)$$

Finally, we deduce **Dalton's law of partial pressures** from the kinetic theory. Dalton's law states that when mixtures of gases or vapors having no chemical interaction are present together in a given space, the pressure exerted by each constituent at a given temperature is the same as it

[*] N_0 is best determined by electrical means (cf. § 19.6).

would exert if it alone filled the whole space and, further, that the total pressure is equal to the sum of the partial pressures due to each gas.

To deduce this law, consider n_1 molecules/cm³ of one gas and n_2 molecules/cm³ of another, each gas confined separately to a volume of 1 cm³. Assuming that the temperatures are the same, the pressure of the first gas on the walls would be

$$p_1 = \tfrac{1}{3} n_1 m_1 \overline{u_1^2} ;$$

that of the second,

$$p_2 = \tfrac{1}{3} n_2 m_2 \overline{u_2^2} .$$

Referring back to equation (13-29), we see that the pressure exerted by a mixture of the two gases in 1 cm³ of volume ($V = 1$ in eq. [13-29]) is simply

$$P = \tfrac{1}{3} n_1 m_1 \overline{u_1^2} + \tfrac{1}{3} n_2 m_2 \overline{u_2^2} ; \quad (13\text{-}41)$$

and, therefore, DALTON'S LAW OF PARTIAL PRESSURES is

$$P = p_1 + p_2 .$$

13.11. Boltzmann Constant; Molecular Energies

If we divide the total translational kinetic energy of the molecules in a mole of gas by the number of molecules in a mole, we have the average translational kinetic energy of any one of them. Let E represent this average energy per molecule. In equation (13-35) we may replace M by its equivalent, $N_0 m$, the product of Avogadro's number and the mass of a molecule. We then have

$$E = \frac{mu^2}{2} = \frac{3RT}{2 N_0}$$

or

$$E = \frac{3 kT}{2} \text{ ergs/molecule}°\text{abs}. \quad (13\text{-}42)$$

The constant $k = R/N_0$ is called the "Boltzmann constant." From values of R and N_0 previously given, we find that

$$\left. \begin{array}{l} k = 1.3805 \times 10^{-16} \text{ ergs/degree} \\ = 1.3805 \times 10^{-23} \text{ joules/degree} . \end{array} \right\} \quad (13\text{-}43)$$

We note that k is a universal constant, since both R and N_0 enjoy this distinction. As we shall see later, this constant appears in many basic physical laws.

The numerical value of E may be computed for any gas that approximates an ideal gas. Thus for an ideal gas (such as oxygen) at 0° C, we have

$$E = \tfrac{3}{2} \times 1.38 \times 10^{-16} \times 273.2$$
$$= 5.656 \times 10^{-14} \text{ ergs/molecule}$$
$$= 5.656 \times 10^{-21} \text{ joules/molecule} .$$

For a mole of gas, the translational kinetic energy at 0° C is

$$N_0 E = 6.0228 \times 10^{23} \times 5.656 \times 10^{-21}$$
$$= 3,407 \text{ joules/mole} .$$

In the gas equation $PV = RT$, V represents the volume occupied by a mole of gas or by N_0 molecules of gas at pressure P and temperature T. Let n be the number of molecules occupying a volume v at the same pressure and temperature. If we multiply the gas equation by the ratio n/N_0, we obtain

$$\frac{PVn}{N_0} = \frac{RTn}{N_0} ;$$

but $R/N_0 = k$, Boltzmann's constant, and Vn/N_0 is the volume v, occupied by the n molecules of gas, so that

$$Pv = nkT . \quad (13\text{-}44)$$

This equation does not contain the molecular weight of the gas and therefore implies that, to the extent that a gas is ideal, equal numbers of any gas molecules confined to a volume v and at a temperature T will exert the same pressure P.

EXAMPLE 4: A gas occupies a volume of 1 cm³ at a pressure of 0.001 atm. Compute the number of molecules present if the gas temperature is − 73° C.

Solution: The pressure $P = 1.013 \times 10^3$ dynes/cm²; $v = 1$ cm³; $T = 200°$ K; and $k = 1.38 \times 10^{-16}$ ergs/degree.

From equation (13-44), we obtain

$$n = \frac{Pv}{kT} = \frac{1.013 \times 10^3 \times 1}{200 \times 1.38 \times 10^{-16}}$$
$$= 3.67 \times 10^{16} \text{ molecules} .$$

13.12. Specific Heats of Gases

According to our definition (§ 12.2), specific heat is the number of calories of heat required to raise the temperature of 1 gm of a substance 1° C. In our discussion of the first law of thermodynamics (§ 12.6) we learned that when heat is added to an expansible substance, part of the heat **197** is used to raise the temperature of the substance, part to do internal work, and part to perform external work if the substance is free to expand. It is clear, therefore, that in measuring specific heats it is necessary to prescribe carefully the experimental conditions under which the measurement is made.

In the case of liquids and solids the specific heats were measured under constant atmospheric pressure. However, since their expansion coefficients are comparatively small, we can neglect the external work done against the atmosphere without incurring serious error. For gases whose expansion coefficients are large, we must not only take into account the external work done by the gas on expansion but also state precisely the conditions under which the heat is added if we are to obtain meaningful results.

Two important cases present themselves: (a) we may measure the specific heat at constant pressure, C_p, by allowing the gas to expand against a constant pressure, or (b) we may measure the specific heat at constant volume, C_v, by confining the gas to a definite volume while heat is added. The capital letters C_p and C_v represent *molar specific heats* equal to the number of calories required to raise 1 mole of gas 1° C. When a gas is heated through 1° C at constant pressure, part of the heat is used to increase the internal energy of the gas and part in doing work against the external pressure. When heated at constant volume, no external work is done, and the heat is used to increase the internal energy of the gas. We expect, therefore, that $C_p > C_v$. If we specify neither constant pressure nor constant volume, specific heat becomes indeterminate, for it is then possible to have an infinite number of specific heats.

The specific heat at constant pressure, C_p, can be accurately measured by means of the continuous-flow calorimeter described in § 12.7. The direct determination of C_v is possible, but, in practice, C_v is deduced from a measurement of C_p and the ratio of the two specific heats, γ,

$$\gamma = \frac{C_p}{C_v}.$$

The ratio γ is accurately measured by a variety of methods, one of the simplest involving a determination of the speed of sound in the gas. As will be

shown in chapter 32, the speed of sound in a gas is given by

$$v = \sqrt{\frac{\gamma P}{\rho}}, \qquad (13\text{-}45)$$

where P is the pressure and ρ the density of the gas.

Table 13-4 lists the experimental facts concerning the molar (molecular) specific heats (C_p and C_v) of a number of gases and vapors, widely different in atomic and molecular characteristics, ranging from the most simple to the complex. We notice that, to some extent at least, the gases can

TABLE 13-4

MOLECULAR HEATS OF GASES AT 20° C

No. of Atoms	Substance	C_p (Cal/Mole °C)	C_v (Cal/Mole °C)	γ
Monatomic	A	4.97	2.98	1.666
	He	4.97	2.98	1.666
Diatomic	N_2	6.95	4.955	1.402
	O_2	7.03	5.03	1.396
	H_2	6.865	4.88	1.408
	CO	6.97	4.98	1.40
	NO	7.10	5.10	1.39
	HCl	7.04	5.00	1.41
	Cl_2	8.29	6.15	1.35
3, 4, or 5 atoms per molecule	H_2O	8.20	6.20	1.32
	NH_3	8.80	6.65	1.31
	CO_2	8.83	6.80	1.299
	SO_2	9.65	7.50	1.29
	CH_4	8.50	6.50	1.31
Polyatomic (5–26 atoms per molecule)	$CHCl_3$	15.2	12.20	1.25
	C_2H_6	12.35	10.30	1.20
	C_2H_5OH	20.9	18.5	1.13
	$C_4H_{10}O$	32.5	31	1.05
	$C_{10}H_{16}$	55	54	1.03

be grouped into categories in which all members have approximately the same molecular heats. In all cases $C_p > C_v$, as we should expect.

EXAMPLE 5: The density of air at 0° C and 1.013 × 10^6 dynes/cm² pressure is 0.001293 gm/cm³. The specific heat of air at constant pressure is about 0.24 cal/gm ° C. Compute the specific heat at constant volume if the velocity of sound in air is 332 m/sec.

Solution: From equation (13-45),

$$\gamma = \frac{v^2 \rho}{P}$$

$$= \frac{332 \times 332 \times 10^4 \times 1.293 \times 10^{-3}}{1.013 \times 10^6} = 1.41,$$

$$c_v = \frac{c_p}{\gamma} = \frac{0.24}{1.41} = 0.17 \text{ cal/gm ° C.}$$

13.13. Difference between the Two Specific Heats of an Ideal Gas

In the previous section we suggested that the difference in specific heats of a gas was a measure of the external work done by the gas on expansion. We shall examine this statement quantitatively, using the first law of thermodynamics as given by equation (12-10). For small changes this equation reads

$$\Delta U = \Delta Q + \frac{\Delta W}{J}, \qquad (13\text{-}46)$$

where all quantities are measured in calories.

If 1 mole of gas is inclosed in a cylinder with rigid walls and heat is supplied, the pressure and temperature of the gas will rise. Since no external work is done, all the heat energy is used to increase the internal energy of the gas. If ΔQ calories of heat are supplied, then the internal energy increase will be

$$\Delta U = \Delta Q ,$$

since $\Delta W = 0$. When the temperature rises by ΔT degrees as a result of absorbing ΔQ calories at constant volume, we have, by definition of C_v,

$$\Delta Q = C_v \Delta T ,$$

and therefore

$$\Delta U = C_v \Delta T . \qquad (13\text{-}47)$$

This important relation is also true for an ideal gas that may change its volume by expansion, since, by definition of an ideal gas, there are no intermolecular forces at play, and therefore no *internal work* is done by the gas on expansion. In other words, the change in internal energy of a gas expanding under constant pressure is also equal to $C_v \Delta T$,

$$\Delta U_P = \Delta U_v . \qquad (13\text{-}48)$$

This result—that the internal energy of an ideal gas is independent of the volume it occupies and is dependent only on its temperature—is very nearly true of all gases (see § 13.19 for exceptions). We now place 1 mole of an ideal gas in a cylinder fitted with a piston so that it may expand against a constant pressure P. If ΔQ calories of heat are added to the gas, its temperature will rise ΔT degrees and, on expanding, will perform $\Delta W/J$ calories of work. The first law requires that

$$\Delta U = \Delta Q - \frac{\Delta W}{J}. \qquad (13\text{-}49)$$

However $\Delta W = P\Delta V$, $\Delta Q = C_p \Delta T$ by definition of C_p, and $\Delta U = C_v \Delta T$ by equations (13-48) and (13-47), so that equation (13-49) becomes

$$C_p \Delta T = C_v \Delta T + \frac{P\Delta V}{J}. \qquad (13\text{-}50)$$

From the ideal gas law we know that, for the initial equilibrium state,

$$PV = RT$$

and that, for the final equilibrium state,

$$P(V + \Delta V) = R(T + \Delta T) .$$

Subtracting, we obtain

$$P\Delta V = R\Delta T ,$$

and equation (13-50) becomes

$$C_p - C_v = \frac{R}{J} = 1.98 \text{ cal/mole degree} . \quad (13\text{-}51)$$

From Table 13-4 we obtain the values given in the accompanying table. Many of the real gases, particularly those having the simpler molecules, agree with the ideal gas deduction.

	$C_p - C_v$ (Cal/Mole Degree)		$C_p - C_v$ (Cal/Mole Degree)
For He	1.99	For HCl	2.04
H_2	1.98	SO_2	2.15
O_2	2.00	C_2H_5OH	2.40
N_2	1.99	$C_{10}H_{16}$	1.00

13.14. Kinetic-Theory Interpretation of Specific Heats

In § 13.11 it was pointed out that the total translational kinetic energy of a mole of gas is given by

$$E = \tfrac{3}{2}RT .$$

The increase in translational energy for a 1° rise in temperature would then be $\tfrac{3}{2}R$. We expect, therefore, that the specific heat at constant volume, C_v, of a gas would be equal to $\tfrac{3}{2}R$ calories, i.e., equal to about 3 calories. Also, since $C_p = C_v + 2$, the ratio of the specific heats, γ, should be $\tfrac{5}{3}$ or 1.66. From an examination of Table 13-4 we find that this is true for the monatomic gases like helium and argon. For diatomic gases, such as hydrogen or oxygen, C_v is greater than $\tfrac{3}{2}R$ and still greater for polyatomic molecules. Let us examine the possible reasons for the discrepancies.

A monatomic gas pictured as a tiny sphere can absorb energy only by increasing its translational kinetic energy. A diatomic molecule pictured as a rigid dumbbell-shaped configuration may absorb not only translational energy but rotational energy as well. When two such molecules collide, their rotational speeds, as well as their translational speeds, must change. Furthermore, it is possible for some of the translational energy to be converted into the rotational energy of another molecule during collision. If such a gas is heated at constant volume, the heat energy is shared between both kinds of motions. However, *only the increase in translational energy can contribute to a rise in temperature.* Therefore, more heat energy is required to raise the temperature of a gas of diatomic molecules 1° C at constant volume than is required for a gas of monatomic molecules, all of whose energy is translational. As an extension of these ideas we note that if the atoms of a complex molecule are able to vibrate with respect to one another, additional energy will be absorbed by the molecule. Therefore, C_v should become progressively greater and the ratio γ smaller as the molecules become more complex. It is clear from Table 13-4 that this quantitative result is essentially correct.

13.15. Degrees of Freedom; Equipartition of Energy

We shall attempt to describe the complexity of the molecule in more quantitative terms. First, let us consider what types of mechanical energy may be possessed by these molecular systems conceived of, in a first approach, as being of simple geometrical shapes. Monatomic molecules we shall continue to consider as tiny spheres. Diatomic molecules we may picture as little dumbbell-shaped arrangements. Associations of three atoms in a molecule might be made up either as a short chain or as a little triangle. More than three atoms might be associated in either linear, plane, or three-dimensional configurations—our pictures of them become somewhat elaborate.

To define the pointlike attributes of a sphere in motion in space, three independent coordinates suffice. Its kinetic energy we will assume to be exclusively translational, since we have no means of describing or of detecting the rotation of a *mona-*

tomic "point molecule." In describing the position and motion of a *diatomic* molecule, on the other hand, we require three coordinates to locate its center of mass and two more to fix its orientation. By the DEGREES OF FREEDOM of motion of a system is meant the number of independent coordinates required to specify or describe the motion.

We have pointed out that a monatomic molecule with three degrees of freedom associated with the three components of velocity has a total translational kinetic energy per mole of $\frac{3}{2}R$. After many random collisions it is reasonable to expect that changes in motion and hence changes in kinetic energy would be the same for the three directions and that therefore the total kinetic energy would be shared equally, on the average, between the three degrees of freedom of motion, i.e., each translational degree of freedom has $R/2$ or about 1 cal of energy associated with it. This is an example of the principle of *equipartition of energy*, or equal division of energy between the three modes of motion of a monatomic gas.

In the case of complex molecules, the equipartition principle predicts that any heat energy supplied to the gas will be equally shared, on the average, by all the possible modes of motion. These modes are described quantitatively in terms of the degrees of freedom of motion of the molecular system and, associated with each degree of freedom, is $R/2$ or about 1 cal of energy per mole of gas.

For the case of 1 mole of a diatomic gas with five degrees of freedom (three translational, two rotational), the equipartition principle would require that each degree of freedom have associated with it 1 cal, so that the molar heat capacity at constant volume would be 5 cal/mole ° C, in agreement with the experiment. Similarly, for polyatomic molecules with six degrees of freedom, the molar heat capacity would be 6 cal/mole ° C.

Heating a gas under constant pressure may result in increasing (*a*) the translational kinetic energy, ΔE_t, as measured by a temperature rise; (*b*) the rotational kinetic energy, ΔE_r; (*c*) the internal energy due to work done against internal forces (such as molecular attractions) while the volume is changing, ΔE_i; and (*d*) the external work of expansion, $P\Delta V$.

On the other hand, heating a gas under constant-volume conditions can result in changing *a* and *b* above. Since the gas does not expand, effects *c* and *d* are excluded.

Therefore, the ratio of the two specific heats, γ, may be expressed symbolically as

$$\gamma = \frac{C_p}{C_v} = \frac{\Delta E_t + \Delta E_r + \Delta E_i + P\Delta V}{\Delta E_t + \Delta E_r} . \quad (13\text{-}52)$$

For all gases, the term ΔE_i is negligible (§§ 13.7, 13.9) and will be dropped from equation (13-52). The term $P\Delta V$ is easily calculated from equation (13-51),

$$P\Delta V = R\Delta T .$$

For $\Delta T = 1°$ C,

$$P\Delta V = 2 \text{ cal/mole} . \quad (13\text{-}53)$$

Thus for a monatomic gas (metallic vapors, helium, argon, etc.) with three degrees of freedom, $\Delta E_t = 3$ cal, and equation (13-52) becomes

$$\gamma = \frac{3+0+2}{3+0} = \tfrac{5}{3} = 1.67 . \quad (13\text{-}54)$$

For a diatomic gas,

$$\gamma = \frac{3+2+2}{3+2} = \tfrac{7}{5} = 1.40 . \quad (13\text{-}55)$$

For a polyatomic gas,

$$\gamma = \frac{3+3+2}{3+3} = \tfrac{8}{6} = 1.33 . \quad (13\text{-}56)$$

These values are in good agreement with experiment (see Table 13-4). At high temperatures the value of γ becomes smaller. This could be explained by assuming that the molecules vibrate. Also, certain diatomic molecules (Cl_2) and nearly all the complex polyatomic molecules (e.g., $C_4H_{10}O$) possess distinctly lower values of γ than those predicted above. Assuming that intramolecular vibrations are present for these molecules, then, since it would be necessary to ascribe two additional degrees of freedom, one each for the kinetic and for the potential energy of vibration, the molar specific heat at constant volume would increase by 2 cal/mole ° C. When these vibrational degrees of freedom are taken into account, the theoretical and experimental values of γ are in better agreement.

13.16. Atomic Heats of Solids

It is of interest to extend the kinetic-theory interpretation of specific heats to solids. In a solid the atoms are not free to move around but are more or less confined to fixed positions and may vibrate about the position only when heat is added. Six degrees of freedom are necessary to describe the motion of an atom in a solid—three for the kinetic vibrational energy and three for the potential energy that the atom possesses by virtue of its position relative to neighboring atoms. According to the equipartition principle, for a dynamical system in thermal equilibrium the energy is equally distributed between various degrees of freedom, and for each of them it is equal to $kT/2$ ($R/2$ for a mole for a $1°$ C rise in temperature). We expect, therefore, that the specific heat or constant volume, C_v, should be $6 \times R/2$ or 6 cal/mole degree. From Table 12-1 we note that many metals between $0°$ C and $100°$ C have atomic heats near 6 cal/mole ° C.

One of the serious drawbacks of the kinetic theory of specific heats is that it predicts that the specific heats will be independent of temperature. Actually, the atomic heats of all solids *decrease* with temperature, approaching a zero value at absolute temperatures near zero, as shown in Fig.

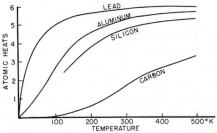

FIG. 13-9.—Variation of atomic heat with absolute temperature.

13-9. Quantum theory, as described in § 13.17, has been successful in accounting for this temperature variation. Apparently the Dulong-Petit law is applicable only at the higher temperatures—the flat portion of the curve of Fig. 13-9. Carbon, whose specific heat is about 2.6 at room temperatures, approaches a value near 6 at 1,000° C.

13.17. Quantum Theory of Specific Heats

An inference that must be made from the foregoing discussion is that some types of molecules in their impacts with one another can receive enough energy to start vibrating. For these molecules, on the basis of classical theory, additional degrees of freedom would be present, and therefore γ would be small. On the other hand, some diatomic molecules, such as hydrogen, apparently do not receive enough energy to start them vibrating, so that for hydrogen, C_v is only 4.88 cal/mole ° C. At very **201** low temperatures $(-233° C)$, where the kinetic energies are small, A. Eucken found in 1912 that $C_v = 2.98$ cal/mole $°C$ for hydrogen, so that it would appear that the *rotational degrees of freedom* are also suppressed and hydrogen behaves like a monatomic gas.

This concept of lost degrees of freedom presents a grave difficulty to the mechanics of classical kinetic theory, for one would expect all possible modes of motion to be present at all temperatures above absolute zero. This conceptual difficulty arising from an application of classical mechanics to molecular and atomic problems, together with other difficulties described elsewhere in this text, led to the formulation of *quantum theory* (see chap. 37). According to this theory of Planck, atoms or molecules can absorb or give up energy only in definite amounts, called *quanta*. As applied to the problem of specific heats by Einstein, a molecule with a natural frequency of vibration, f per second, will not be put into oscillation until it receives a quantum of energy of value hf, where h is a universal constant equal to 6.624×10^{-27} ergs-sec. On the basis of this theory, a diatomic molecule, such as hydrogen, with a high natural frequency of rotation (due to its small moment of inertia) does not receive enough energy by impact at $-233° C$ to set it into rotation. The heavier molecules, such as nitrogen, with low natural frequencies of vibration, require small amounts of energy, hf, for rotation, and therefore we would expect their rotational degrees of freedom to be present at lower temperatures.

An extension of these ideas of Planck and Einstein by Debye and Born has led to a satisfactory theory for the dependence of specific heats of solids on temperature (Fig. 13-9). This theory predicts that, for low temperatures, C_v should vary as the cube of the absolute temperature, that is,

$$C_v \sim \left(\frac{T}{\theta}\right)^3,$$

where θ is a constant different from substance to substance and is referred to as the "characteristic temperature" of the element.

13.18. Adiabatic Expansion or Compression of a Gas

In § 13.4 we discussed the problem of isothermal expansion or compression of an ideal gas. There is another set of experimental conditions for expansion or compression of a gas of great practical and theoretical interest, and that is the case in which *no heat energy is added or subtracted* from the gas. These are known as *adiabatic conditions*. Let us place a mole of gas in a perfectly heat-insulated cylinder fitted with an insulated piston. We shall then compress the gas and perform ΔW joules of work on the gas. Since no heat energy may enter or leave the cylinder, this mechanical work must manifest itself as a rise in temperature of the gas. From the conservation of energy (first law of thermodynamics) we have, from equation (13-46),

$$\Delta U = \Delta Q + \frac{\Delta W}{J}. \qquad (13\text{-}46)$$

However, $\Delta Q = 0$ by our experimental conditions, and $\Delta U = C_v \Delta T$ by equation (13-47). Therefore, equation (13-46) becomes

$$C_v \Delta T = \frac{\Delta W}{J}. \qquad (13\text{-}57)$$

Thus, if work is done on the gas, ΔW is positive, and therefore ΔT is positive—the temperature rises. On the other hand, if the gas expands adiabatically and does work, ΔW is negative, and therefore the temperature of the gas must fall. We observe that heat energy is stored in the gas during compression and, if allowed to expand adiabatically, will perform an equal amount of external work until its temperature is restored. We now inquire as to what relationship exists between P, V, and T during adiabatic processes. It is clear that Boyle's law, $PV = C$, is not applicable. The gas law,

$$PV = RT, \qquad (13\text{-}58)$$

is, however, applicable. The work done on compression, ΔW, is positive and is equal to $-PdV$, since dV is negative. On differentiating equation (13-58), we obtain

$$PdV + VdP = RdT. \qquad (13\text{-}59)$$

If, in equation (13-57) we substitute $-PdV$ for ΔW and the result of equation (13-59) for ΔT replaced by dT, we obtain

$$C_v(PdV + VdP) + \frac{R}{J}PdV = 0. \qquad (13\text{-}60)$$

It was shown in equation (13-51) that $C_p - C_v = R/J$. Substituting in equation (13-60) yields

$$C_v PdV + C_v VdP + C_p PdV - C_v PdV = 0,$$

or, on rearranging terms,

$$\frac{dP}{P} + \frac{C_p}{C_v}\frac{dV}{V} = 0. \qquad (13\text{-}61)$$

With $\gamma = C_p/C_v$, equation (13-61) takes the form of

$$\frac{dP}{P} + \frac{\gamma\, dV}{V} = 0. \qquad (13\text{-}62)$$

Integrating this equation gives

$$\log_e P + \gamma \log_e V = \text{Constant},\, C_1$$

or

$$P\,V^\gamma = \text{Another constant},\, C_2. \quad (13\text{-}63)$$

Equation (13-63) is illustrated graphically in Fig. 13-10. The dotted curves represent isothermal

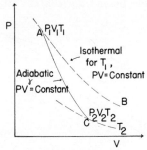

Fig. 13-10.—Comparison of adiabatic and isothermal expansions of a gas.

compression or expansion. Note how much steeper AC is than AB. A gas at C compressed adiabatically is heated and therefore gives rise to a greater pressure for a given volume change than a gas compressed isothermally—hence the steeper slope. Thus, along the dotted curve, $PV = \text{Constant}$ is true; along the solid curve, $PV^\gamma = \text{Constant}$ is the proper relation.

We may rewrite equation (13-63) in other usable forms by substituting for P or V from the equation of state, $PV = RT$. Thus

$$\frac{RT}{V} V^\gamma = \text{Constant}$$

or

$$T V^{\gamma-1} = \text{Constant}\, C_1, \qquad (13\text{-}64)$$

giving the variation of T and V along an adiabat, or

$$P^{(1/\gamma-1)}T = \text{Another constant}\, C_2, \quad (13\text{-}65)$$

giving the variation of P and T along a given adiabat.

EXAMPLE 6: The volume of a flask of air ($\gamma = 1.4$) is suddenly doubled. If the gas was originally at $27°$ C, what is its new temperature?

Solution:

$$V_2 = 2\,V_1,$$

$$T_1 = 273 + 27 = 300°\,\text{K},$$

$$300\,V_1{}^{(1.4-1)} = (t_2 + 273)\,(2\,V_1)\,^{(1.4-1)},$$

$$300 = (t + 273)\,(2)\,^{(1.4-1)}$$
$$= (t + 273)\,(2)\,^{0.4}$$

$$\log_{10} 2 = 0.301030,$$

$$0.4 \times \log_{10} 2 = 0.120412,$$

$$2^{0.4} = 1.3295,$$

$$300 = (t + 273)\,1.3295,$$

$$t + 273 = 227.35,$$

$$t = -45°65\,\text{C}.$$

The work done on an ideal gas in an adiabatic compression is given by

$$W = \int_{V_2}^{V_1} P\,dV. \qquad (13\text{-}66)$$

where, by equation (13-63), $PV^\gamma = \text{Constant} = C$. Therefore, to compress 1 mole of gas adiabatically requires an amount of work given by

$$W = \int_{V_2}^{V_1} \frac{C\,dV}{V^\gamma}$$
$$= \frac{C}{(1-\gamma)}\,[\,V_2{}^{(1-\gamma)} - V_1{}^{(1-\gamma)}\,]. \qquad (13\text{-}67)$$

However, since $P_1 V_1^\gamma = P_2 V_2^\gamma = C$ is a constant along any given adiabat, we have

$$W = \frac{P_2 V_2 - P_1 V_1}{(1-\gamma)}. \qquad (13\text{-}68)$$

On the other hand, we may evaluate the integral (eq. [13-66]) by noting that $\Delta W = PdV$ and, from equation (13-46), $PdV = JC_v dT$,

$$W = \int_{T_2}^{T_1} JC_v dT = JC_v\,(T_2 - T_1). \quad (13\text{-}69)$$

EXAMPLE 7: In Example 6 compute the work done by 1 mole of the gas.

Solution: By equation (13-69),

$$W = (4.185\,\text{joules/cal})\,(0.0375\,\text{cals/mole}°\,\text{C})$$
$$\times (227.35 - 300)\ \text{C}$$
$$= -11.44\,\text{joules/mole}.$$

The negative sign indicates work done by the gas corresponding to the drop in temperature.

EXAMPLE 8: Compute the isothermal and adiabatic bulk modulus of a gas.

Solution: According to § 7.4, the bulk modulus, K, is given by

$$K = -V \frac{dP}{dV}.$$

It is clear that K may be determined, provided that we know the relation between P and V. From our previous analyses there are two important equations that relate P to V, one for isothermal processes and the other for adiabatic ones.

For an isothermal process,

$$PV = \text{Constant},$$

and therefore $dP/dV = -P/V$, so that K_i, the bulk modulus for a gas at constant temperature, becomes

$$K_i = \frac{-V\,dP}{dV} = P. \qquad (13\text{-}70)$$

For an adiabatic process, $PV^\gamma = \text{Constant}$,

$$V^\gamma dP + \gamma V^{\gamma-1} P\,dV = 0$$

or

$$\frac{dP}{dV} = -\gamma \frac{P}{V},$$

and K_a, the bulk modulus for a gas operating under adiabatic conditions, is

$$K_a = \frac{-V\,dP}{dV} = \gamma P. \qquad (13\text{-}71)$$

This last result will be used later in the text to evaluate the velocity of sound in gases.

13.19. Real Gases in Contrast to the Ideal Gas

In the introduction to the kinetic theory of gases which has been presented in the preceding pages, we have, as usual for simplification, assumed conditions that are never realized in nature. Only in the case of hydrogen and helium at ordinary temperatures and pressures are our results approximately correct. The approximation of this simple theory to the truth becomes better and better with all gases and even with vapors at higher and higher temperatures and at lower and lower pressures. As we go to lower temperatures and higher pressures, on the other hand, the relations cease to be even *approximately* characteristic of *any* substance.

We assumed, in the first place, that the volume of the molecules was entirely negligible in comparison with the volume occupied kinetically by the gas. Also, we made a second assumption to the effect that cohesive forces between molecules were entirely absent, or at least that they were entirely ineffective, if present. It was Amagat who first showed experimentally that even Boyle's law was only approximately true, and he indicated how departures from it were to be expected if one took proper account of molecular size and intermolecular forces.

Suppose we plot the *product* of pressure by volume (temperature being constant) against pressure. Boyle's law indicates that the graph, Fig. 13-11, should be a horizontal straight line.

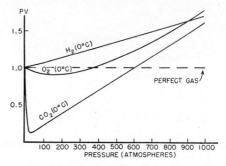

FIG. 13-11.—Plot of PV against P for various gases at constant temperature 0° C.

Actually, in the case of hydrogen this line is approximately straight at low pressures but acquires an upward slope at higher pressures. For air, oxygen, or nitrogen at ordinary temperatures the graph at first descends to a minimum value and then ascends as high pressures are reached. Carbon dioxide is more extreme in this respect than air.

We can easily interpret the *rise* of these curves at high pressures in view of the finite size of molecules. As more and more are crowded together, their total volume becomes a more significant part of the total volume of the container—the free space available for them to fill kinetically is decreased. Twice as many molecules now can be squeezed into what is *actually less than half* the volume only by the exertion of *more than doubled pressure—PV* is no longer constant but *increases with pressure*.

The initial *drop* of the *PV-P* curve in the case of nitrogen and carbon dioxide requires a little more careful analysis. How would the effectiveness of intermolecular forces enter into the situation as one proceeded from low pressures (small density) and large volumes to high pressure (and density) and decreased volume? Intermolecular forces presumably are effective only when molecules are relatively near one another. Furthermore, these forces can produce a significant momentum change only if the molecules are near one another for appreciable intervals of time, since

$$\Delta(mv) = f\Delta t .$$

If these forces are effective, the result will be to draw molecules closer together than they would otherwise be. However, at very low pressures, forces are not effective—molecules are much too far apart. Also, at very high temperatures, even with greater pressures, the forces are not effective. Molecules are moving too swiftly to remain long enough near one another for such forces as exist to produce any appreciable results.

But now let us use comparatively low temperatures and begin to increase pressure, thus compressing the molecules into smaller space. The intermolecular forces then begin to be effective, and the molecules can then, to some extent, be pulled still closer together by their own attractions for one another. They tend thus to condense—density increases in a *greater* ratio than the pressure increase, unassisted by these attractions, would warrant. The product of pressure by volume *declines*, since the volume shrinks faster than the pressure augments. Ultimately, when the molecules begin themselves to occupy a significant part of the free space in the vessel, the opposite effect described in the preceding paragraph occurs, and *PV* increases.

With any gas which, like hydrogen, appears to show no effect of intermolecular forces at ordinary temperatures, i.e., no initial decline of the *PV-P* curve, we might expect that, by going to *lower temperatures*, the forces, entirely ineffective at room temperature, would begin to be rendered effective, because now at lower temperatures the molecules remain longer within the range of these feebler fields and momentum changes are accordingly produced. This is precisely what experiment shows to be the fact. If *PV-P* relations are plotted for hydrogen, keeping the gas very cool during the observations, the same kind of curve is obtained for this gas at these low temperatures as one gets for

air or nitrogen at ordinary temperatures or for carbon dioxide at higher temperatures. Furthermore, if a gas like carbon dioxide, in which the cohesion is large, is strongly heated during such observations, a curve like that of hydrogen at room temperatures (showing no initial decline but only a gradual rise) is obtained.

Thus departures from the simple law of Boyle (and of Charles, too) give definite indication of the existence of molecular forces between molecules of finite size. Qualitatively, indications also of the degree of intensity of these forces in different kinds of molecules are also given.

Thus we see that at sufficiently high pressures and sufficiently low temperatures the cohesive forces assume a dominant role. They are able, in large degree, to overcome the kinetic tendencies of the molecules to rebound to great distances after collisions. A gas then ceases to possess the characteristic kinetic properties of a gas—it condenses into a DIFFERENT STATE OF MATTER and becomes a liquid or a solid. Thus our dynamical picture of gases inevitably leads us to interpretations of liquids and solids. It begins to show a power and a scope originally quite unsuspected of it, and it achieves a status as a physical theory far greater than that of the tentative hypothesis with which it began.

13.20. Résumé

The following is a brief listing of the definitions, principles, and theories, given in this chapter, which you should know:

Definitions of coefficients of expansion of gases at constant pressure and constant volume

Measurement of these quantities

Absolute temperature and a method of measurement

Boyle's law and isothermal expansions

Work done in isothermal expansions

The general gas law

Kinetic-molecular theory of gases, the assumptions and calculations

The interpretation of temperature, of Avogadro's hypothesis by the kinetic theory

The specific heats of gases at constant pressure and at constant volume

The significance of the difference between these two specific heats

Degrees of freedom of the molecules of a gas

Theoretical interpretation of the specific heats of gases

Adiabatic expansions

Contrast of real gases with the ideal gas

Queries

1. What evidence can you give for the molecular theory of matter?
2. What evidence exists that supports the hypothesis that gas molecules are in motion?
3. The barometer gives the pressure of the air at a given place. This pressure, we have said, is due to the *weight of a column of air from the ground to the top of the atmosphere*. A box is opened to the atmosphere and then closed airtight. What is the absolute pressure of the air within the box? Is this air pressure within the box due to the *weight of the air within the box?* Explain away this paradox.
4. How can one explain the phenomenon of gaseous diffusion on the basis of the kinetic theory?
5. On the basis of the kinetic theory explain why a gas should become heated when compressed.
6. Explain why $\Delta U \neq C_p \Delta T$ for a constant-pressure expansion.

Practice Problems

1. What is the volume of 10 gm of O_2 at 27° C and 70-cm pressure? Ans.: 8.35 liters.
2. Without using the numerical value of the gas constant, determine the volume of 3 lb of an ideal gas at 127° C and 70-cm pressure, if 4 lb of the same gas at 290° C and 150-cm pressure have a volume of 100 ft³.
3. Calculate the number of molecules of air in a volume of 1,000 cm³ at 30° C which has been evacuated to 0.0001-mm pressure. Ans.: 3.2×10^{15}.
4. Calculate the root-mean-square velocity of hydrogen at −200° C and at 1,000° C. The velocity of escape from the surface of the earth is approximately 6.95 mi/sec. Can there be much hydrogen in the earth's atmosphere if, as there is reason to believe, temperatures approaching 1,000° C are present in the upper atmosphere?
5. How many grams of hydrogen are necessary to fill a balloon, 10 m in diameter, at 30° C and a pressure of 755 mm? Ans.: 41.8 kg.
6. Five grams of oxygen and 5 gm of nitrogen are placed in a 10-liter flask whose temperature is 0° C. Calculate the total pressure in the flask and the partial pressures of each gas.
7. An automobile tire is heated from 20° to 40° C. Assuming that the tire does not expand appreciably, calculate the new gauge pressure if the original was 30 lbf/in². Ans.: 32.1 lbf/in².
8. An oxygen tank is filled with the gas at a pressure of 2,000 lbf/in². If the tank has a volume of 1 ft³, how many cubic feet of gas at atmospheric pressure does this tank represent? If the temperature of the gas is 20° C, how many grams of oxygen are in the tank?
9. Owing to air above the mercury in a barometer, the barometer reads 746 mm when the actual pressure is 753 mm, and 734 when the actual pressure is 738 mm of mercury. Calculate the reading of the barometer when the true pressure is 760 mm. Ans.: 750.4 mm, assuming a large reservoir of mercury.

10. A U-type mercury manometer shown in (*a*) of the accompanying illustration is connected to a chamber, *c*. When the chamber is highly evacuated (pressure less than 10^{-6} mm of mercury), the mercury levels in the manometer show a difference in height of 1 cm, as shown in (*a*), indicating the presence of some air in the manometer tube. The chamber is now filled with air to a pressure *p*, the manometer now indicating a difference in height of mercury of 3 cm (*b*). Compute the true pressure in centimeters of mercury within the chamber. Draw a calibration curve, plotting as ordinates the true pressure in centimeters of mercury and as abscissas the difference in mercury levels.

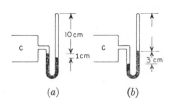

(*a*)   (*b*)

11. The density of air at 0° C and 76-cm pressure is 0.001293 gm/cm³. If 10 gm of air are heated from 0° to 100° C, the pressure remaining constant, how much will the volume change? How many ergs of mechanical work will be done by the gas in expanding? The specific heat of air at constant pressure is about 0.24 cal/gm ° C. From the work done in expanding this gas, determine the specific heat at constant volume and determine γ. Ans.: 2.84×10^3 cm³; 2.87×10^9 ergs; 0.17 cal/gm °C; 1.41.
12. Compute the work necessary to compress a gas isothermally from a pressure of 10 lbf/in² to 50 lbf/in². The initial volume of the gas was 30 ft³. *Hint:* $P_1 V_1 = P_2 V_2$.
13. Determine the mass of air in a room 15 by 30 ft by 10 ft high. The barometric pressure is 745 mm of mercury, and the temperature 20° C. Ans.: 150 kg.

14. Compute the number of molecules of gas in a 1,000-cm³ bottle whose temperature is 15° C and whose pressure is 0.01 mm of mercury.

15. At what temperature will 1 liter of air weigh 1.5 gm with a pressure of 710 mm of mercury? Ans.: 220 K.

16. A cylinder of oxygen of 5-ft³ capacity under a pressure of 6 atm and a temperature of 20° C is connected through a valve of negligible volume to another tank of oxygen of 6-ft³ capacity under a pressure of 8 atm and a temperature of 10° C. Calculate the pressure in the tanks when the valve is opened and the temperatures of the tanks are 24° and 18° C, respectively. *Hint:* Use equation (13-44).

17. Compute the root-mean-square velocity of the gas C_4H_{10} at 100° C. Ans.: Approximately 327 m/sec.

18. A liter of a gas with $\gamma = 1.3$ at 0° C and 76 cm of mercury pressure has its volume suddenly halved. Find the new pressure and temperature.

19. A cubic foot of air ($\gamma = 1.41$) at 0° C and 1 atm pressure has its pressure suddenly doubled. Find the new temperature and the new volume. Ans.: 61° C; 0.612 ft³.

20. If the gas of problem 18 is now cooled back to 0° C, what will be its volume, the pressure remaining constant?

21. If the gas of problem 19 is now cooled back to 0° C, the volume being held constant, what will be the pressure? Ans.: 1.63 atm.

22. The velocity of sound in ammonia gas at 0° C is 415 m/sec. Compute C_v if C_p is 0.523 and the density of ammonia is 0.771 gm/liter at 0° C.

23. The following precise data are available on the gas oxygen: $c_p = 0.2178$ cal/gm ° C; $\gamma = 1.401$ at 273°.16 K. Also 22,414 cm³ of oxygen at 273°.16 K exerts a pressure of $1,013.246 \times 10^3$ dynes/cm². Compute the mechanical equivalent of heat, using these data.

24. Compute the density of nitrogen under a pressure of 20 atm and at a temperature of 27° C.

25. Five moles of CO_2 are heated 10° C at constant volume. Compute the increase in internal energy of the gas. Ans.: 1,421.2 joules.

26. A gram of air at atmospheric pressure and at a temperature of 20° C is compressed adiabatically. Calculate its temperature and its volume when the final pressures are 5 and 20 atm.

27. In problem 24 calculate the number of nitrogen molecules present if the gas is confined to a volume of 3 liters. Ans.: 14.7×10^{23}.

28. Calculate the total kinetic energy of translation of all the molecules of helium in a 15-liter flask at a temperature of 10° C and pressure of 228 cm of mercury.

29. Compute the root-mean-square speed of fog droplets of radius 1 micron (10^{-6} m) at 0° C and atmospheric pressure. Ans.: Approximately 176 cm/sec.

30. One mole of nitrogen gas is confined to a volume of 5 liters at a temperature of 150° C. Compute the isothermal and adiabatic bulk moduli of the gas.

31. Five moles of a monatomic ideal gas are confined to a 10-liter volume at 0° C. Compute the adiabatic bulk modulus of the gas. Ans.: 1.895×10^6 nt/m².

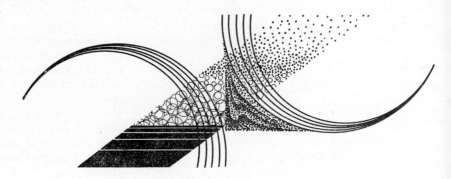

CHAPTER 14

Latent Heats and Changes of Phase

14.1. Latent Heats

One aspect of thermal phenomena was very puzzling to the early experimenters. It had to do with the fact that, under certain circumstances, a very great amount of heat may be applied to a material *without causing its temperature to rise at all*. The energy is absorbed in some manner which in no way affects the average translational molecular agitation. Whenever this happens, it is to be observed also that the material involved is undergoing a CHANGE OF PHASE. By this we most commonly understand that the material, if solid, is melting into a liquid or, if a liquid, is boiling away into a vapor, passing into the gaseous form. It is quite possible for a solid to pass directly into the gaseous condition without the intermediate liquid phase, however.

Heat that is thus absorbed in effecting any change of phase is called LATENT HEAT, and the energy changes which are effected in the material by this absorption are, to a large extent, of the nature of potential, rather than of kinetic, energy changes, i.e., have to do with the third term ΔE_i in the numerator of equation (13-52) (§ 13.15). These phenomena not only give a direct indication of the existence of intermolecular forces—also attested to by the independent facts of cohesion and adhesion universally characteristic of solids and liquids—but also enable us to form some impression even of the order of magnitude of these forces or, more exactly, of the work that must be done against them to overcome their action and drag molecules apart and, to a large extent, away from one another's influence.

Latent heat is measured by the number of calories required to effect a complete change of 1 gm of material from one phase to another.

There are two most common latent heats: one associated with melting, a LATENT HEAT OF FUSION and one with evaporation, a LATENT HEAT OF VAPORIZATION.

The converse processes of condensation from vapor to a liquid and of freezing of a liquid into a solid, of course, give out and return just the same amount of heat as is absorbed when these changes of state occur in the opposite sense.

Table 14-1 gives the values of the latent heats of fusion and of evaporation for some more common substances, together with the melting and boiling points of the same substances. Note the variation of latent heat of vaporization with temperature.

The second and third columns give the melting temperature and, since this is constant over wide variations of pressure, the latent heats of fusion at this temperature. The fourth column gives the boiling point under standard atmospheric pressure. The latent heat of vaporization is given in the fifth column. The latter changes with the temperature, usually decreasing with rising temperature; and therefore the sixth column gives the temperature at which the listed latent heat is determined. This is usually a temperature somewhat higher than the normal boiling point.

The case of carbon dioxide is of special interest. Note that its latent heat of vaporization drops to a very low value at 30°.8 C. This temperature is just a little below the "critical temperature" of this compound, at which latent heats of vaporization might all be expected to vanish, since it is impossible for the liquid state to exist at temperatures higher than the critical temperature. We shall revert to this subject again in Example 2 in the next section, but we shall defer detailed discussion until later in this chapter.

TABLE 14-1

LATENT HEATS OF FUSION AND VAPORIZATION
(ALSO MELTING AND BOILING POINTS)

Elements	Melting Point (°C)	Latent Heats of Fusion (Cal/Gm)	Boiling Point (°C) at Atmospheric Pressure	Latent Heats of Vaporization (Cal/Gm)	Temperature at Which Latent Heat of Vaporization Was Determined
Solids:					
Al......	659.7	94	1,800	....	....
Cu......	1,083	41	2,300	....	....
Fe......	1,535	{ 49 6.56* }	3,000	....	....
Hg......	− 38.87	2.78	356	71	358
Pb......	327.4	5.47	1,620	175	1,170
Zn......	419.7	23	907	475	918
W.......	3,370	....	5,900	....	....
Gases:					
He......	− 272	....	− 268.94	5.6	− 271.3
A.......	− 189.2	6.64	− 185.7	37.6	....
H₂......	− 259.1	14	− 252.8	5.6	− 253
N₂......	− 209.8	6.1	− 195.8	476	− 195.8
O₂......	− 218.4	3.33	− 183	50.9	− 182.9
Compounds:					
H₂O.....	0	79.63†	100	{ 595.4† 538.7 446.9 0.0 }	0 100 220 365
NH₃.....	− 75	108	− 33.5	{ 331.7 301.8 252.6 }	− 40 0 + 40
CS₂.....	− 110	....	46.2	90	0
CO₂.....	− 57	....	− 80	{ 72.23 57.98 44.97 3.72 }	− 25 0 22 30.8
C₆H₆....	5.48	36	80.2	92.9	80.1
C₂H₅OH..	− 117	....	78.5	236	0

* Latent heat of change of magnetic form to nonmagnetic form. This change occurs at about 737° C.

† Because of the difficulty of making exact heat measurements, we use in simple laboratory experiments for the latent heat of fusion of ice, 80 cal/gm and for the latent heat of vaporization of water at 100° C, 539 cal/gm.

14.2. Calorimetric Measurements of Latent Heats

The methods used in determining the numerical facts of latent heats, such as are exhibited in Table 14-1, are calorimetric. By this we mean that some modification of the method of mixtures used for measuring specific heats as described earlier is also used for determining latent heats.

Suppose, for example, M_1 gm of water are cooled from an initial temperature t_1 to a final temperature t_2 by the melting of M_2 gm of material of melting point t_m. We will call the specific heat c_l when the substance is liquid and c_s when solid. Suppose, further, that this material in the solid form is introduced into the water at a temperature $t_3 < t_m$; then the heat that it takes up from the water in being heated to its melting point is

$$M_2 c_s (t_m - t_3) .$$

The heat that it takes up from the water while melting is

$$M_2 L ,$$

where L is the latent heat of fusion in calories per gram. The heat it takes up after being melted, to bring it to the final temperature of the water t_2, is

$$M_2 c_l (t_2 - t_m) .$$

The total heat lost by the water to supply the heat for these three items is

$$(t_1 - t_2)(M_1 c + \text{water equivalent of the calorimeter}) .$$

Hence we have, by applying the principle of conservation of energy, the equation

$$(M_1 c + \text{water equivalent})(t_1 - t_2) = M_2 [c_s (t_m - t_3) + L + c_l (t_2 - t_m)] . \tag{14-1}$$

This equation may then be solved for the single unknown L.

A similar application of the method and equation would, quite obviously, serve to determine any one of the quantities c, c_s, and c_l if all the others, including L, were known.

Ice calorimeters are those in which the melting of a known mass of ice whose constants c_s, c_l, and L are accurately *known* and are used to determine an *unknown specific heat*, c.

Steam calorimeters are those in which a known mass of water is condensed upon an object of unknown specific heat. From the known specific heats of steam and water and the latent heat of vaporization of water, the *unknown specific heat* of the material may be found.

Illustrative examples with numerical data will serve to clarify further these practical procedures.

EXAMPLE 1: How much heat is necessary to change 10 gm of ice at −5° C into steam at 115° C (all at normal pressure)?

Solution:

$[0 - (-5)]0.487 \times 10 = 24.35$ cal is the heat necessary to change 10 gm of ice at −5° C into ice at 0° C.

$10 \times 79.63 = 796.3$ cal is the heat necessary to change 10 gm of ice at 0° C into water at 0° C.

$10 \times (100 - 0)1 = 1,000$ cal is the heat necessary to change 10 gm of water at $0°$ C into water at $100°$ C.

$10 \times 538.7 = 5,387$ cal is the heat necessary to change 10 gm of water at $100°$ C into steam at $100°$ C.

$10(115 - 100)0.45 = 67.5$ cal is the heat necessary to change 10 gm of steam at $100°$ C into steam at $115°$ C.

The sum of these heats, then, is the total heat needed to change 10 gm of ice at $-5°$ C into steam at $115°$ C.

$24.35 + 796.3 + 1,000 + 5,387 + 67.5 = 7,275.15$ cal.

EXAMPLE 2: Ten grams of steam at $120°$ C and 70 gm of ice at $-20°$ C are put into 200 gm of water at $50°$ C. What is the final temperature?

Solution: It is not evident whether the water will warm or cool. Let us assume that the final temperature (t) will be $0 < t < 50$. This implies that the water cools. Then

$$70[0 - (-20)]0.487 + 70 \times 79.63 + 70t$$
$$= 10 \times (120 - 100)0.45 + 10(538.7)$$
$$+ 10(100 - t)1 + 200(50 - t)1$$
$$t = 36°5 \text{ C}.$$

That is,

$$\left\{ \begin{array}{c} 70[0-(-20)]0.487 \\ \text{(Heat to transform ice from} \\ \text{ice at } -20° \text{ to ice at } 0° \text{ C)} \end{array} \right\} + \left\{ \begin{array}{c} 70 \times 79.63 \\ \text{(Heat to transform ice from} \\ \text{ice at } 0° \text{ to water at } 0° \text{ C)} \end{array} \right\}$$

$$+ \left\{ \begin{array}{c} 70t \\ \text{(Heat to transform water at} \\ 0° \text{ to water at } t° \text{ C)} \end{array} \right\} = \left\{ \begin{array}{c} 10 \times (120-100)0.45 \\ \text{(Heat to transform steam at} \\ 120° \text{ to steam at } 100° \text{ C)} \end{array} \right\}$$

$$+ \left\{ \begin{array}{c} 10(538.7) \\ \text{(Heat to transform steam at} \\ 100° \text{ to water at } 100° \text{ C)} \end{array} \right\} + \left\{ \begin{array}{c} 10(100-t)1 \\ \text{(Heat to transform water at} \\ 100° \text{ to water at } t° \text{ C)} \end{array} \right\}$$

$$+ \left\{ \begin{array}{c} 200(50-t)1 \\ \text{(Heat lost by water in cooling} \\ \text{from } 50° \text{ to } t° \text{ C)} \end{array} \right\}.$$

On the basis of the opposite assumption that $50°$ C $< t < 100°$ C, we obtain

$$700[0 - (-20)]0.487 + 70 \times 79.63 + 70t$$
$$+ 200(t - 50)1 = 10(120 - 100)0.45$$
$$+ 10(538.7) + 10(100 - t)1.$$

Therefore,

$$t = 36°5 \text{ C}.$$

In the first trial the water was to lose heat, which would be what the steam also did; also $t < 50$.

In the second trial the water was to gain heat, which would be what the ice also did; hence $t > 50$.

A comparison of these two equations shows that they differ only by the algebraic transposition of $200(50 - t)1$ from one side of the equation to $-200(50 - t)1 =$

$200(t - 50)1$ on the other side. *Hence it was not necessary to know which assumption was correct. It was necessary only to make an assumption and then to write the equation according to the assumption.*

EXAMPLE 3: Into a space filled with saturated steam at $100°$ C is put a 100-gm brass ball at $20°$ C. If 1.4 gm of water condense on the ball, what is the specific heat of the brass ball?

Solution:

$$100(100 - 20) c = 1.4 \times 539,$$

$$c = \frac{1.4 \times 539}{100 \times 80} = 0.0948 \text{ cal/gm} °\text{C}.$$

In problems of this type, where a substance passes through a change of phase, it is well to have in mind a chart such as that given in Fig. 14-1. In

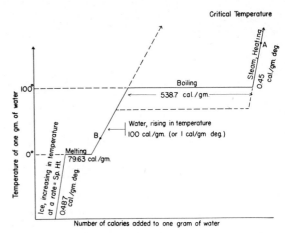

FIG. 14-1.—Temperature changes that accompany the absorption of heat by 1 gm of water from the solid to the gaseous phase.

this, the temperature of 1 gm of water has been plotted against the number of calories added to that water. In any calorimetric equation there will be a separate term for each part of the line connecting the initial and final temperatures of the water. Note that the difference in the specific heats of water and steam results in a change in the heat of vaporization when the boiling point is not $100°$ C. Lowering of the pressure results in a lowering of the boiling point and hence in an increase in the heat of vaporization. It should be noted, however, that the total calories necessary to carry steam at A into water at B is the same, whichever path is followed. This simplifies calorimetric experiments at room pressure.

This chart also implies that if the boiling point were raised sufficiently high, the water and steam lines would intersect and the heat of vaporization would vanish. This actually happens. See the discussion of the "critical point" later in this chapter (§ 14.11). It is to be pointed out, however, that although the lines representing specific heats have been assumed straight, i.e., the specific heats have been assumed constant, this is not quite true. These lines are slightly curved, so that the correct value for the critical-point temperature *cannot* be obtained by extending the steam and water lines until they meet.

14.3. Interpretation on the Basis of the Kinetic Molecular Theory

So far in this chapter we have been dealing entirely with the gross thermal relations that are involved in changes of phase; and we have intimated that these changes in energy imply, in terms of a dynamical theory of heat, interchanges between the potential and kinetic aspects of the average mechanical energy of the atomic and molecular entities that constitute the fine-scale structure of matter in its various phases.

In what follows we shall examine a variety of other experimental phenomena, chiefly relating to the gaseous and liquid phases, which, like latent heats, were exceedingly mystifying to the earlier workers. The adoption of the kinetic molecular hypothesis entirely clarified and correlated these hitherto completely isolated experimental facts.

14.4. Kinetic Mechanism of Changes in Phase

When heat is supplied to a crystalline solid, the temperature of the solid rises; also the average kinetic energy of the molecules held in their crystal lattice increases until their excursions to and fro about their definite and fixed positions, to which they are elastically bound in the crystals, become so energetic that, at the melting temperature, these bonds are ruptured (the fixed configurations disappear), the external geometrical shape dissolves, and the material becomes liquid. The converse of this process—the evolution of heat (which is usually removed by the environment) when a liquid freezes into a solid—accompanies the crystallization of a supersaturated solution.

As heat continues to be poured in, however, the temperature of the liquid now rises. The average kinetic agitation of its freely wandering molecules increases; and, because of the increasing vigor of their jostling one another in their densely packed condition, they are pushed farther away from one another. Thus the liquid expands with rise in temperature, with certain possible exceptions (see § 11.14).

In the continually shifting molecules that are in the region adjacent to the free surface of the liquid, there are those with velocities sufficiently in excess of the average to enable them to break through this region and escape from the liquid. The attraction of the crowd for its members (discussed under "Surface Tension," in chap. 10) constitutes a POTENTIAL (energy) BARRIER—a fence, as it were —which can be leaped over only by those molecules which accidentally, at the moment of their arrival there, may have a superabundant supply of kinetic energy. Thus above the surface of a liquid there are molecules of the material which have now escaped into the third or gaseous state.

As more and more heat is poured into the remaining liquid, the latter's continued expansion further weakens the bonds that hold the individuals together and simultaneously endows larger numbers of them with energies sufficient for escape, so that the change-over from the liquid to the gaseous phase (evaporation) becomes a continually accelerating process.

Often it happens, however, that the "free surface" of the liquid in the vessel is not entirely free. A blanket of another gaseous material (such as air), if present, will smother and impede evaporation by confining escaped molecules near the liquid long enough that they may be jostled back within the range of the attracting surface and then pulled back within the crowded liquid. Under such circumstances evaporation is retarded. As the temperature is raised, however, the swiftly increasing numbers of molecules with velocities high enough to escape ultimately accumulate in sufficient numerical strength and with a sufficient magnitude of momentum transfer per second to force back and drive away the enveloping blanket of alien gas. The vapor pressure has reached a value equal to, or slightly greater than, the atmospheric pressure under which the free surface of the water was confined. Bubbles filled with vapor are formed, and the liquid boils. If there is any reduction of **211** outside pressure now, some water will vaporize almost explosively. An increase in pressure, on the other hand, will stop the boiling instantly.

Finally, after the liquid has all boiled away, converted into a gas, there is, nevertheless, still definite *residual* evidence to be found of the intermolecular attractive forces. Let some of the "vapor" wander into a somewhat cooler region, and these attractive forces will begin to condense the material into liquid again. Measurement of the pressure of the vapor shows it to be much less than that of a perfect gas (which has negligible attractive forces) at the same temperature.

Before the conditions of an "ideal gas" (where the effects of intermolecular forces and the fraction of the total volume filled by the molecules themselves are relatively *negligible*) can be approximated, the newly released "vapor" must be heated much further and be much further expanded, so that its molecules are not only much farther apart on the average but are also endowed with far greater average kinetic energies. Then, at last, they can move so swiftly past one another in their rather infrequent chance encounters that the tendency toward entanglement is quite negligible, and the condition of an ideal gas is approximated.

With this general descriptive background of the kinetic pictures of changes of phase in mind, let us now look more precisely at some of the quantitative relations which not only experimentally confirm the kinetic hypothesis but also enable us to make numerical measures and comparisons with respect to values of intermolecular forces and estimates of molecular cross-section.

It is interesting to contrast this detailed description of molecular processes that kinetic theory affords with the direct operational description of these processes given by the first law of thermodynamics,

$$\Delta U = J\Delta Q - \Delta W . \qquad (12\text{-}10)$$

This generalized conservation principle states that whatever are the detailed processes that absorb heat energy (ΔQ), i.e., rise in temperature, changes of phase, etc., and whatever the mechanism whereby the material does work (ΔW) either in the liquid, solid, or vapor phases, the algebraic sum of ΔQ and ΔW is a measure of the change in internal energy of the material.

14.5. Evaporation

Let us look first at the phenomenon of evaporation. Any liquid freely exposed to the air always EVAPORATES. Because of intermolecular forces, however, any molecule such as the one represented by the dot in Fig. 14-2 within a distance ξ from either side of the surface (where ξ represents the average range of effective molecular attraction) suffers unbalanced forces that tend to draw it back within the body of the liquid. In the figure the crosshatched areas indicate regions within which lie molecules whose attractions for molecules within the surface region are unbalanced. It is seen, moreover, in Fig. 14-2 (*b*) that the unbal-

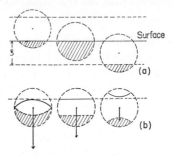

FIG. 14-2.—Intermediate forces present near a molecule or a liquid surface.

anced pull on surface molecules is greater beneath a convex surface than beneath plane surfaces. The later, in turn, provide greater unbalanced forces than do concave surfaces. This is the origin of those "surface forces" responsible for all the mechanical effect which we discussed under the heading of "Surface Tension" in chapter 10.

Also from these same considerations we would expect that the saturation vapor pressure over a concave surface must be smaller than it is over a plane surface and that over a convex surface it must be larger than over a plane surface (see §14.19).

If any given molecule has sufficient kinetic energy ($E_t = \frac{1}{2}mv^2$) to exceed the product of the average force throughout this surface region and the width of the region, i.e., $\overline{f}\xi$, it will escape, that is, with

$$\tfrac{1}{2}\,m\,v^2 > \overline{f}\,\xi .$$

Its escape will be most readily possible, of course, if, upon approaching the region, it receives an impact which sends it off in a direction perpendicular to the surface.

On the other hand, any molecule *in the vapor above the liquid* which wanders into the surface region will, in the vast majority of cases, be entangled in the force field there and pulled back into the liquid.

When a liquid is exposed to the open air, the molecules that evaporate are mixed with molecules of the air and are, to a large extent, carried away by air currents, so that they cannot return to the liquid.

Since only the more swiftly moving molecules escape, the average speed of those remaining behind is lowered. An evaporating liquid is somewhat cooler than its surroundings. It does not become indefinitely cooler, however, since heat begins to flow into it from its surroundings more and more rapidly, the cooler it becomes. The more volatile liquids have, in general, smaller surface forces and evaporate more quickly. Playing a fan upon their surface drives away the closely overlying vapor and greatly reduces the return of the escaped molecules, hence greatly increasing the cooling effect of the evaporation.

In the case of a liquid confined to a closed vessel, from which also any foreign gas other than the "vapor" of the liquid may be removed, we have a distinctly different situation, and we must define a few terms to meet it.

14.6. Vapor Pressure; Properties of Vapors

When a liquid is introduced into a closed, previously evacuated vessel (Fig. 14-3), equilibrium is

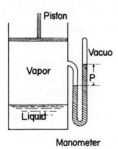

Fig. 14-3.—Illustrating the relation between vapor pressure and temperature.

quickly attained at any temperature between the number of molecules escaping per second and the number returning. Once the density of the vapor (i.e., number of molecules per cubic centimeter) has risen to the point where it furnishes as many molecules going back as the liquid supplies going

out, we call the vapor a SATURATED VAPOR. Its pressure is called the SATURATED VAPOR PRESSURE of the liquid.

Increase in temperature of the liquid, of course, by increasing the average kinetic energy, enables more molecules to escape. Their escape renders the vapor density and pressure greater; the greater density provides more returning molecules; and again equilibrium is reached. Decrease in temperature lowers the rate of escape from the liquid. The rate of return of molecules to it is in excess, and the vapor density and pressure decline until equilibrium again is found. The *saturated vapor pressure* therefore changes rapidly with temperature; its variation can be found experimentally by several methods.

If, in contrast to the preceding paragraph, the closed vessel has not been evacuated before the introduction of the liquid into it, the presence of a foreign group of molecules, such as air mixed in with the molecules of a vapor like water over the surface of water, in no way affects the end results we have been describing, except as to the time taken for the equilibrium to be reached. (The total pressure in the vessel is, of course, the sum of the vapor pressure and that of the foreign gas [Dalton's law of partial pressures].) In an evacuated vessel of about a liter capacity, equilibrium is reached between liquid and vapor in less than a second of time. If the vessel is filled with air at atmospheric pressure, several minutes may be necessary.

The question has probably arisen as to whether saturated vapors obey the same laws as do ideal gases. We shall answer this question by performing a set of experiments using water in the cylinder of Fig. 14-3. Let us fix the volume of the chamber by not allowing the piston to move. With the vapor saturated above the liquid surface, we shall heat the liquid and vapor, observing the change in vapor pressure by means of the manometer. The results are tabulated in Table 14-2 and are shown graphically in Fig. 14-4. If the saturated vapor obeyed the ideal gas laws, we would expect the pressure to vary linearly with temperature, as given by equation (13-3). Figure 14-4 clearly shows that the curve of P versus t is *by no means linear*. The curve is concave upward. This curve above 0° C is called the *steam line*. Below 0° C there is a distinct change in slope of the pressure-temperature curve. Here the curve represents the **213** *vapor pressure above ice;* the curve is called the *hoarfrost line.* It is shown again in Fig. 14-6 on an exaggerated scale.

All the points on the steam line represent equilibrium conditions of pressure and temperature of saturated vapor. Consider, for example, point *A*

TABLE 14-2

SATURATED WATER-VAPOR PRESSURES

$t°$ C	P_{mm} of Hg	ρ (Gm/Cm³)	P_{mb}*
− 60.........	0.008	....	0.011
− 40.........	0.096	....	0.129
− 20.........	0.783	0.894×10^{-6}	1.04
− 15.........	1.252	1.403	1.67
− 10.........	1.964	2.158	2.62
− 5.........	3.025	3.261	4.03
− 2.........	3.887	4.144	5.18
− 1.........	4.220	4.482	5.63
0.........	4.580†	4.847	6.11
1.........	4.924	5.192	6.56
2.........	5.291	5.559	7.05
5.........	6.541	6.797	8.72
10.........	9.21	9.401	12.28
20.........	17.55	17.300	23.40
30.........	31.86	30.371	42.48
40.........	55.40	0.0511×10^{-3}	73.86
50.........	92.6	0.0832	....
60.........	149.6	0.1305	....
70.........	233.9	0.1984	....
80.........	355.4	0.2938	....
90.........	526.0	0.4241	....
92.........	567.2	0.4552	....
94.........	611.1	0.4878	....
96.........	657.8	0.523	....
98.........	707.4	0.560	....
100.........	760.0	0.598	....
110.........	1,074	0.827	....
125.........	1,740	1.299	....
150.........	3,568	2.55	....
200.........	11,650	7.84	....
250.........	29,770	....	....
300.........	64,300	....	....
350.........	123,710	....	....

* One millibar ≡ mb = 10³ dynes/cm².

† In this table the value of the saturated vapor pressure at 0° C over ice and over pure water is given as 4.580 mm of Hg. Actually, these two vapor pressures differ slightly and are identical only at the triple-point temperature, 0°0098 C.

on the curve of Fig. 14-4. That point represents the *saturation vapor pressure* corresponding to a definite temperature; i.e., there is no tendency for increased evaporation of liquid. If we now were to take a point off the curve, such as *B*, new phenomena would occur in our cylinder. The point *B* represents the same vapor pressure as *A* but at a higher temperature. In order to achieve this in the cylinder, we must allow the piston to move upward so that the pressure remains constant as the temperature is increased. On performing this ex-

periment, we find that all the liquid initially in the chamber has now evaporated and that nothing but unsaturated vapor remains. All the points to the right of the steam line thus represent a vapor phase of the material. If, on the other hand, we choose a point, such as *C*, that represents the same temperature as *A* but an increase in pressure over that of *A*, we find that, to achieve this condition in the cylinder, we must compress the vapor. During the process of compression, no appreciable increase in the pressure of the vapor is discernible until all the vapor has been condensed into liquid; then a further compression of the liquid quickly

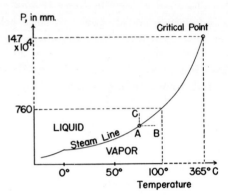

FIG. 14-4.—Vapor pressure of water

brings the pressure to that of *C*. In other words, *Boyle's law does not hold for saturated vapors,* since pressure is independent of volume changes.

We see that points to the left of the steam line represent the substance in the liquid phase only. The steam line, therefore, separates the liquid phase from the vapor phase, and points on it represent conditions of equilibrium of saturated vapor and liquid. This statement holds true up to what is known as the *critical temperature* (365° C for water).

14.7. Triple Point: Case of Water

The melting point of ice (and other solids also) is dependent, to a slight degree, upon the external pressure to which the solid is subjected. In the case of ice, the melting point is lowered by increased pressure. For any particular pressure, ice will melt if above a certain temperature; on the other hand, water will freeze if it is colder than this temperature. The solid curve of Fig. 14-5, called the *ice line,* represents the relation between pressure and temperature at which a mixture of ice and water may be in equilibrium. As in the case of the steam line, to the right of the ice line only liquid may exist; to the left of the ice line, only solid. We wish to emphasize that the slope of the ice line is highly exaggerated, as a pressure increase of approximately *1 atm* lowers the melting point of ice by *only 0°.0072 C.*

As we have stated, the hoarfrost line represents equilibrium conditions between ice and saturated vapor (see Table 14-2 for vapor pressures over ice).

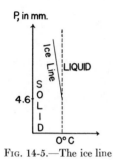

Fig. 14-5.—The ice line

Above the frost line the solid exists; below the line, vapor.

Since the three curves of Figs. 14-4, 14-5, and 14-6 represent relations between pressure and tem-

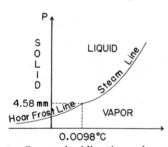

Fig. 14-6.—Curves of sublimation and vaporization

perature, they may be drawn on the same chart. This is done in Fig. 14-7. It was shown experi-

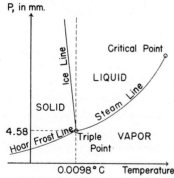

Fig. 14-7.—Triple-point diagram for water (scales exaggerated).

mentally that these three lines must meet in a common point, to which the special name *triple point* is applied.* At the triple point all *three phases* of solid, liquid, and vapor may *coexist in equilibrium with one another;* i.e., at a temperature of 0°.00997 C for water and a partial water-vapor pressure of 4.58 mm, water, ice, and vapor may remain together ad infinitum, without any change in the relative amounts of any one of them.

The triple-point diagram for a solid that *contracts* on solidifying is shown in Fig. 14-8. From a

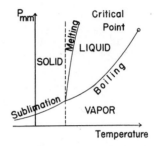

Fig. 14-8.—Triple-point diagram for a solid that contracts on solidifying.

study of the two triple diagrams we see that substances that contract on solidifying cannot exist in the liquid state at temperatures below the triple point. Substances like ice, on the other hand, cannot be *solidified* at temperatures above the triple point.

14.8. Cases of Carbon Dioxide and Nitrogen

Every material with a definite freezing point has a triple-point diagram like that illustrated in Fig. 14-7 for water. The scale is different for every different substance. From such diagrams we can see the reasons for physical properties, often quite puzzling otherwise. For example, carbon dioxide, common in commercial and household use as "dry ice," is always handled in the solid form. Its temperature is −80° C at atmospheric pressure. It never melts under atmospheric pressure, sub-liming directly from the solid to the gaseous state. Less common is liquid air (or nitrogen). Although it is much colder than dry ice (−196° C), we never see it frozen in chunks at atmospheric pressures.

The reasons for this difference between air and

* The triple point is the point of intersection of any vertical (V = Constant) plane with the "triple-point line" shown in the thermodynamic surface models in Pl. 14-1. See § 14.12. **215**

carbon dioxide are easily seen by referring to the triple-point diagrams of these two substances. Atmospheric pressure lies *below* the triple-point pressure (388 cm) for carbon dioxide (Fig. 14-9). Warming the solid converts it directly to vapor or gas. No liquid phase is possible unless pressures higher than atmospheric, viz., above 5.1 atm (388 cm of Hg), are attained.

Fig. 14-9.—Triple-point diagram for CO_2 (scales exaggerated).

The case for nitrogen is different. Atmospheric pressure lies higher than the triple-point pressure for this liquid, just as it does for water. If the material is at room temperature, far to the right in the diagram (Fig. 14-10), it cannot be liquid. If cooled, however, and kept thermally insulated, as in thermos flasks, it is readily kept in liquid form. For it to be solidified, it is necessary only to *cool the liquid*

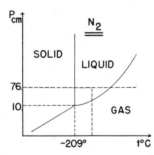

Fig. 14-10.—Triple-point diagram for N_2 (scales exaggerated).

further. This is readily done by forced evaporation. If, by means of a fast-working pump, liquid air is put under a somewhat reduced pressure, the resulting accelerated evaporation cools the liquid further, and solid nitrogen is produced within it, appearing as a white precipitate, which settles out. If this precipitate is filtered out and kept under a partial vacuum of less than 10-cm pressure, it does not disappear by melting first into the liquid form and evaporating: it sublimes directly into gas, as does carbon dioxide.

Conversely, liquid carbon dioxide may be obtained from the solid by dropping some of the latter into a heavy walled tube, kept chilled below

216

−80° C, and then sealing the tube and allowing it to warm up to room temperature. So high a pressure will be accumulated by conversion of the solid to gas that soon the triple-point pressure will be exceeded and the solid will be seen melting directly into the liquid form. Needless to say, THIS EXPERIMENT IS HIGHLY DANGEROUS and must not be attempted except by those fully experienced in high-pressure work—and, even then, with extreme caution.

14.9. Intermolecular Forces Responsible for Changes of Phase

Let us next look at some other types of experiment, which will be even more revealing as to the nature and magnitude of those intermolecular forces that are responsible for the possibility of the condensation of matter out of the gaseous phase into the liquid and solid condition. It is the existence of these intermolecular forces, as well as the finite sizes of the molecules, that causes *real* gases to behave somewhat differently from ideal gases; and so it might be well if the reader would, at this point, return and reread our remarks contrasting the behavior of real and ideal gases (§ 13.19).

14.10. Behavior of a Real Gas when Greatly Compressed

Having these facts in mind, now let us imagine that we take a given mass of some gas like carbon dioxide and inclose it in a cylinder fitted with a piston, so that the gas may be compressed (Fig. 14-11) (Andrews' experiment [1863]). We will start

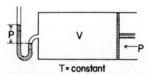

Fig. 14-11.—Isothermal compression of a real gas

with a relatively large volume and small pressure. Since the temperature will be altered by the very fact of compression, we will suppose that some means are provided for removing the heat as produced and for keeping the temperature uniform, at about room temperature, during compression; i.e., we are to imagine an ISOTHERMAL COMPRESSION. At the beginning, if the pressure is sufficiently low, the molecules are so far apart that the effects of their mutual attractive forces are negligible.

Furthermore, the total volume of the molecules is negligible, compared with that of the cylinder. Boyle's law is closely followed:

$$PV = \text{Constant}$$

or

$$P \propto \frac{1}{V};$$

and a graph of P against V begins as an equilateral hyperbola, such as section AB of the curve T_1 in Fig. 14-12.

However, as the compression continues, molecules become more closely crowded together, remain within one another's influence for a longer

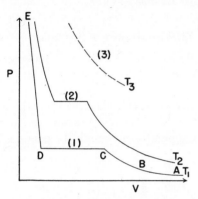

FIG. 14-12.—Isotherms of a real gas

time, and the intermolecular forces begin to make themselves felt. Furthermore, the total volume of the molecules is no longer entirely to be neglected, in comparison with the now greatly reduced volume of the vessel. For these two reasons the P-V curve begins to depart more and more from the simple inverse proportionality. We reach the portion BC of the diagram, which we will call the "vapor" portion, meaning by VAPOR a gas that is far from "ideal." Soon, on reaching point C (Fig. 14-12), molecules begin to be ensnared within the meshes of their own attractive forces and held with sufficient strength that the kinetic impacts can no longer separate them to great distances. The saturated vapor has begun to CONDENSE into a liquid. Drops of fog may appear throughout the cylinder, or condensed vapor may begin to trickle down its walls. Now, as the volume is further reduced, the *pressure remains constant*. More and more vapor is being condensed.

The latent heat of vaporization is now given up to the inclosing cylinder; but our assumed heat-conducting mechanism takes it away as fast as it

is given up, and the temperature remains the same. Our experimental curve is now the horizontal portion CD in Fig. 14-12.

Finally, the gas is all condensed into liquid after the free volume, which it was kinetically filling, is all removed. The point D is reached, and the volume of the cylinder is now simply the volume of the condensed liquid. Actually, the liquid is not solidly packed with molecules, since some interstices remain between them within the liquid; but the molecules now strongly resist further reduction in total volume. The liquid is highly incompressible, and any further reduction in volume sends the pressure soaring up along the segment DE.

Next let us repeat the entire process, but at a higher temperature, T_2. Departure from Boyle's law and condensation appear again, but at smaller volumes and at higher pressures than before, since the greater speed of molecular agitation at any given volume results in higher external pressure on the walls and internally gives any pair of molecules a shorter time to remain within the range of each other's forces. Thus the effects of the forces are now less than formerly. Ultimately, however, the vapor will again begin to condense out into liquid. When all the material is in the liquid state, its total volume, because of the higher temperature, will be a little greater than before, so that the steep portion of the curve is reached more quickly, i.e., at a larger volume than before.

14.11. The Critical Temperature

It is now conceivable that if we should repeat the experiment yet another time, it might be possible to go to a temperature sufficiently high that, whatever the reduction of volume and the resulting molecular density, the thermal agitation might be sufficiently rapid to prevent any condensation into the liquid state whatsoever (cf. dotted curve *3*, Fig. 14-12). This is found to be borne out by experiment. That temperature above which condensation* is impossible by pressure only is called the CRITICAL TEMPERATURE.

The existence of such a critical temperature may be beautifully illustrated in another way: Suppose we have a partly condensed vapor sealed

* Condensation may be recognized only by the presence of effects due to a surface or "interface" between the liquid and vapor states. Aside from this, there is no simple way of deciding whether a substance under high pressure above its critical temperature is vapor or liquid.

up in a transparent tube sufficiently strong that it may be heated without danger of bursting.* The volume of vapor and liquid are given by V_v and V_l, respectively. At low temperatures we observe the condensed liquid V_l in the lower half of the tube. This condition would be represented by the point P on the isothermal, T_1, for that temperature (Fig. 14-13). If we now heat the tube, we move successively to points P', P'', etc., on higher isothermals, the horizontal portions of

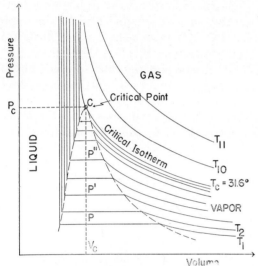

Fig. 14-13.—Isotherms of carbon dioxide

which are becoming shorter and shorter, as we have seen. Ultimately, we observe that the line of demarcation between liquid and vapor in the tube becomes more and more tenuous; finally it disappears altogether. Whatever there is in the tube becomes entirely invisible as the critical temperature is exceeded. In the diagram we have passed upward to isothermals, T_c, T_{10}, etc., that have no flat portion whatsoever. The horizontal portion of the curve represents the region where liquid and vapor are in equilibrium. The "critical point," C, of Fig. 14-13 is located at that temperature, pressure, and critical volume, T_c, P_c, V_c, at which all the properties of a liquid and its vapor become identical. Some theoretical bases for these phenomena are given in § 14.13.

From our discussion it is clear that at the critical point the surface tension becomes zero and the heat of vaporization is zero, as previously noted.

In Fig. 14-14 is shown the variation in the densities of a liquid and its saturated vapor with tem-

* Demonstration of these effects is highly dangerous without proper technique on the part of the experimenter. The novice should never attempt them.

perature changes up to the critical temperature (C.T.). As the critical temperature is approached, the densities of liquid and vapor approach each other.

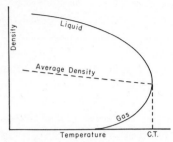

Fig. 14-14.—Variation with temperature of the density of a liquid and its saturated vapor.

14.12. Thermodynamic Surfaces

If a three-dimensional graph is made in which the different isothermal lines of the P-V plane are arranged one behind the other, using the direction perpendicular to the plane of the page as a temperature axis (lower temperatures in front of, and higher temperatures behind, the page), a continu-

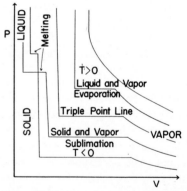

Fig. 14-15.—P-V section of a thermodynamic surface

ous surface is developed, all points on which represent possible values of P, V, and T for a given initial mass of material. Such surfaces are called THERMODYNAMIC SURFACES. Photographs of several of these are shown in Plate 14-1, a and b. Frequent references should be made to them.

If one carries out the experiments described in the preceding pages (§ 14.10) at temperatures *below the triple point*, the isothermal relations in the case of water form a family of curves, which, when carried to sufficiently high pressures, *change from solid back to the liquid state*, as the ice condensed from the vapor melts under pressure alone when the pressure is very great (Fig. 14-15). These changes, represented by the projecting abutment

PLATE 14-1
THERMODYNAMIC SURFACES

If *at any fixed temperature*, T, one plots, for a nearly saturated vapor, the pressure against volume for a range of pressures that carries the vapor through condensation and on into the liquid phase, a curve similar to *dcba* of Plate 14-1, *a*, is obtained. This curve, then, represents a series of equilibrium states of the vapor as the pressure changes at constant temperature. (See also Fig. 14-12.)

At some higher temperature, T', the curve $d'c'b'a'$ represents another series of equilibrium states. Note that for this temperature the straight portion of the curve, $c'b'$, is shorter than cb. At temperatures above the critical temperature, liquefaction is impossible—no straight portion whatsoever exists for any curve drawn for temperatures above the critical temperature.

These facts are made vivid by plotting pressures, volumes, and temperatures along three rectangular axes and constructing a *surface* on which lie all the points that represent *equilibrium states* of the given substance. Such a surface for carbon dioxide is shown in Plate 14-1, *a;* the surface is cut off by a P-V plane at a temperature far *above* the critical temperature.

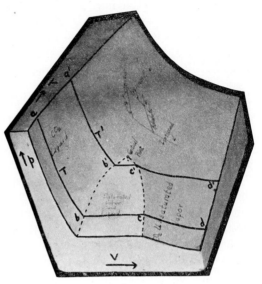

a) CARBON DIOXIDE

A much more extensive surface with several other very interesting characteristics is obtained if one plots experimental data for carbon dioxide taken at temperatures below its freezing point (Pl. 14-1, *b*). Since liquid and vapor are in equilibrium on the dotted surface *m* and solid and vapor on a lower adjoining surface, all three phases must be in equilibrium on the line common to both surfaces. This line is the "triple-point line." Its projection on the P-T plane is the triple point of Fig. 14-9.

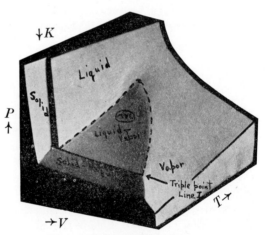

b) CARBON DIOXIDE

The thermal properties of water are exceptional—different from those of most all other substances in many respects. The P-V-T relations at and below the freezing point, when represented on a surface (Pl. 14-1, *c*), show some of these exceptional aspects of water quite clearly.

The "solid" wall or abutment, K', of the water surface projects in front of the wall that represents the liquid phase. Contrast this abutment, K', of Plate 14-1, *c*, with the receding "corner," K, of Plate 14-1, *b*. This difference in the models represents the well-known fact that water *expands* on freezing.

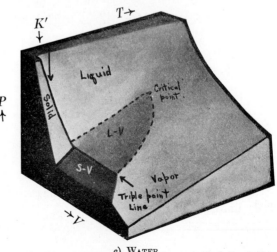

c) WATER

in Plate 14-1, *c*, are very difficult to visualize from the plane drawings alone. Since, even in these regions, the curves may be imagined as ruled on a continuous surface, the extended thermodynamic surface becomes very helpful. Such surfaces below the triple point contain additional features best visualized from photographs of such models if the models themselves are not available.

Furthermore, if one cuts such surfaces by planes of constant *V*, the *PT* sections so obtained are found to be precisely the *steam* and *hoarfrost lines* lying in parallel constant-volume (isometric) planes that were discussed earlier.

To see additional significance with respect to the *ice line*, we must consider a third family of curves lying in planes of constant pressure (isopiestic). These are sections cut from the surface by horizontal (*P* = Constant) planes and giving *T-V* curves. (Pl. 14-1, *a*, *b*, for carbon dioxide, and 14-1, *c*, for water, illustrate these sections.)

14.13. The More General Equation of State: van der Waals' Equation

The reader will recall that our former equation of state for 1 mole of gas,

$$PV = RT ,\qquad (13\text{-}13)$$

represents a series of rectangular hyperbolas. For temperatures sufficiently above the critical temperature, this simple equation represents the facts fairly well. Below the critical temperature, however, we have seen that the isothermals are quite different in form.

The first successful attempt to extend the simple gas equation (13-13) to include the effect of molecular interactions and to take account of the finite dimensions of molecules was made by J. D. van der Waals (1837–1923), who proposed the following equation:

$$\left(P+\frac{a}{V^2}\right)(V - b) = RT,\qquad (14\text{-}2)$$

to represent the behavior of 1 mole of a real gas. The term a/V^2, where a is a constant for a given mass of gas and V is the volume of this mass of gas, represents a correction for the attraction exerted between neighboring molecules; the term $(-b)$, where b is equal to four times the volume of the molecules in 1 mole of gas, is the correction for finite size of the molecules.

We may rewrite equation (14-2) as follows:

$$P = \frac{RT}{V - b} - \frac{a}{V^2}.\qquad (14\text{-}3)$$

The set of isothermal curves representing this equation is shown in Fig. 14-16. Each curve corresponds to a constant temperature, the upper curves to larger values of the temperature. Note that the isothermal marked T_c contains no maximum or minimum but merely a point of inflexion.

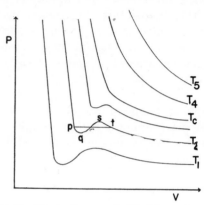

FIG. 14-16.—Illustrating van der Waals' equation of state

Let us compare the theoretical curves of Fig. 14-16 with the experimental ones of Fig. 14-13. It is natural to identify the isotherm T_c of Fig. 14-16 with the critical isotherm T_c of Fig. 14-13. If we consider the curves below T_c, we see that they are distinctly suggestive of the experimental facts, except that these curves are ∼-shaped rather than straight horizontal lines. As a matter of fact, the portions *st* and *pq* of Fig. 14-16 may be ascribed to a supercooled vapor and a superheated liquid, respectively. The portion *qs* represents an unstable condition. In view of the simplicity of van der Waals' corrections, the agreement between experiment and theory is remarkably good. Since solidification is not taken into account in van der Waals' theory, this equation must be considered only as a first approximation.

One can obtain from van der Waals' equation an expression for the critical constants T_c, V_c, and P_c in terms of the constants a, b, and R:

$$T_c = \frac{8a}{27Rb}\ {}^\circ\text{K},\qquad (14\text{-}4)$$

$$V_c = 3b\ \text{cm}^3 ,\qquad (14\text{-}5)$$

$$P_c = \frac{a}{27b^2}\ \text{atm} .\qquad (14\text{-}6)$$

Equations (14-4), (14-5), and (14-6) have been among the most useful in the field of molecular physics. A list of critical constants for some of the more common substances is given in Table 14-3. In this table the critical temperature is given in degrees centigrade, the pressure in atmospheres, and the critical density in grams per cubic centimeter.

TABLE 14-3

CRITICAL CONSTANTS

Substance	$t_c°$ C	P_c (in Atmospheres)	Critical Density (Gm/Cm³)
N₂	−147	34	0.3110
CO₂	31.1	73	.460
H₂	−239.9	12.8	.0310
H₂O	365.0	195	.40
SO₂	157	78	.052
He	−269.9	2.26	.0693
Air	−140.7	37.2	.35
O₂	−118.8	49.7	.430
CCl₄	283.15	45	.5576
Pentane	197.2	33	0.232

14.14. Free Expansion: The Joule-Kelvin* Experiment of the Porous Plug

We are now in a position to make more precise our distinction between a real gas like carbon dioxide and an ideal gas. In § 13.4 we pointed out that a gas expanding against a pressure does work and hence cools. The question arises: If the gas were to expand *without doing work against a piston, i.e., expand freely into a vacuum, would the gas cool?* In other words, does the internal energy of a gas change on free expansion? A definite answer to this question would give some evidence as to the magnitude of intermolecular forces; for, if attractive forces exist between molecules, then, during expansion, work would be done to separate the molecules, this work appearing as increased potential energy of the molecules at the expense of the kinetic energy of translation. Hence the gas should cool. If, on the other hand, no intermolecular forces exist, as we assumed for an ideal gas, there should be no change in internal energy and hence no cooling. In 1845 Joule performed his celebrated experiment (shown in Fig. 14-17) to find out the facts.

* Often referred to as the "Joule-Thomson (Kelvin) experiment."

Two flasks, *A* and *B*, were immersed in a tank of water, care being taken to see that the tank was isulated against external thermal effects. Flask *A* contained gas under high pressure, which was allowed to expand into flask *B*, which was initially evacuated. In such an expansion *no external work* is done by the expanding gas. Joule found no detectable change in the temperature of the bath and concluded from this experiment that no change in the internal energy of the gas took place.

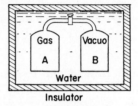

FIG. 14-17.—Free expansion of a gas

These results were not convincing, because of the fact that small changes in temperature could not be detected. Lord Kelvin proposed a modification of Joule's experiment, substituting a porous plug for the stopcock and allowing gas of a known pressure and temperature to enter at *A* and leave the plug at atmospheric pressure (Fig. 14-18).

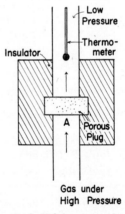

FIG. 14-18.—Joule-Kelvin porous-plug experiment

After correcting for changes in temperature due to the net work done on the gas in going from the high-pressure side to the low-pressure side, Kelvin and Joule found that in all gases except hydrogen there was a fall in temperature whenever the gases passed through the plug, showing that work was done against intermolecular forces. In the case of hydrogen a slight *increase* in temperature was found. But if the hydrogen was first cooled below −80° C, it behaved like the other gases. On the **221** other hand, if the other gases were initially above a certain temperature known as the *temperature inversion*, they, too, showed heating rather than cooling. The inversion temperature of hydrogen is $-80°$ C; that of helium, $-238°$ C.

As a result of these experiments we may say that *an ideal gas is one that obeys the simple gas law,* $PV = RT$, *and does not cool on free expansion.* We may go a step further and state that *the internal energy of an ideal gas is independent of volume and pressure changes*, provided that the temperature does not change. This latter statement is sometimes called "Joule's law" (see § 13.13).

14.15. Liquefaction of the Permanent Gases

As was shown by the Joule-Kelvin experiments, all gases will cool on expansion if they are initially below the inversion temperature.

We shall briefly describe one commercial apparatus used for the liquefaction of air—the Linde process (1895). A rough sketch of the apparatus is shown in Fig. 14-19. Air is allowed to expand

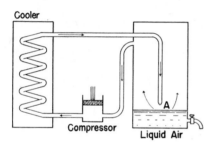

Fig. 14-19.—Linde apparatus for liquid air

through a small orifice at A and is cooled as a consequence of the Joule-Kelvin effect. The cooled air is then allowed to circulate about the inner tube, so that the approaching air becomes progressively cooler before expansion. By the continuous operation of this cycle, the expanding air is finally cooled to the point where it liquefies and collects in the trap below. Using the same technique with pre-cooled hydrogen, Dewar in 1898 liquefied hydrogen.

14.16. The Production of Very Low Temperatures

In 1908 Kamerlingh Onnes succeeded in liquefying helium by passing pure compressed helium gas through liquid hydrogen boiling at reduced pres-

sure and a temperature of $15°$ K, or $-258°.16$ C. The helium was then passed on through a nozzle similar to that of the Linde apparatus of Fig. 14-19 and allowed to expand. Liquid helium was produced at a temperature of $4°.2$ K, or $-269°$ C, a temperature very close to absolute zero.

By subjecting helium, confined to a tube, to a pressure of 130 atm and immersing the tube in liquid helium, Keeson succeeded, in 1926, in solidifying helium. Helium has been solidified at pressures of only 86 atm and at temperatures of $3°.2$ K. By rapidly evaporating liquid helium, temperatures as low as $0°.7$ K have been attained.

Using a method proposed by Debye (1926) and by Giauque (1927), whereby certain magnetic materials initially at the temperature of boiling helium were demagnetized adiabatically, physicists (in 1949) have been able to reach temperatures as low as $0°.0029$ K. It has recently been suggested that if the nuclei of substances were adiabatically demagnetized rather than molecules

TABLE 14-4

VERY LOW TEMPERATURES

	° C	° K
Ice point	0	273.16
Oxygen, boiling point	−182.97	90.19
Nitrogen, boiling point	−196	77
Hydrogen, boiling point	−252.7	20.5
Helium, boiling point	−269.0	4.2
Demagnetization	−273.14	0.003

or atoms, it should be theoretically possible to reach a temperature of 10^{-6} ° K, very near absolute zero indeed. But, as has been pointed out by Lindemann and others, such nearness to absolute zero is apparent rather than real, since the use of a geometric or logarithmic temperature scale, rather than an arithmetic scale, would place absolute zero an infinite distance away! Only by the use of such a scale can one represent the significant changes in physical phenomena which occur, for example, between temperatures of 10^{-6} and 10^{-3} degrees absolute. Table 14-4 lists some of the low temperatures attained in the laboratory.

14.17. Properties of Dilute Solutions

In the discussions of the previous sections attention was directed to the thermal and mechanical properties of pure liquids. In this section we shall briefly study some of the important properties of dilute solutions. By a "solution" we mean a homogeneous mixture of two or more chemical substances which has the same chemical composition and physical properties in every part. In a solution the *solvent* is that component which is present in the larger proportion, and the *solute* the other component. In the paragraphs that follow we shall discuss only a few of the properties associated with dilute solutions.

a) Lowering of vapor pressure.—It has been known for a long time that the vapor pressure of a solvent is lowered with the addition of a nonvolatile solute, such as a salt. This phenomenon can be understood from a kinetic point of view—the solute molecules reduce the effective concentration of solvent molecules and, by their motions, reduce the chances that a fast-moving molecule of liquid will leave the surface. On this basis we would expect the lowering of the vapor pressure to be proportional to the number of dissolved molecules. This conjecture was experimentally verified by Raoult in 1887.

b) Lowering of the freezing point.—The freezing point of water or any other liquid is lowered by the addition of a solute. Here again the magnitude of the lowering depends on the total number of dissolved particles. It is found that when a dilute solution, for example, begins to freeze, the pure solvent crystallizes out. The solution therefore must become more concentrated. The more concentrated the solution, the greater the lowering of the freezing point. A stage will ultimately be reached at which the solution becomes saturated. Further cooling will cause precipitation of both solvent and solute, so that the concentration remains constant. The temperature at which this occurs is called the *eutectic point*.

c) Elevation of the boiling point.—The boiling point of a pure liquid is raised if a salt is dissolved in the liquid. Since the vapor pressure is lowered by the addition of a solute, a solution must be heated above the boiling point of the pure solvent before the vapor pressure is equal to that of the atmosphere. Raoult found here also that in dilute solutions the elevation of the boiling point is proportional to the concentration of the solute.

d) Osmosis.—A phenomenon closely related to the lowering of the freezing point and vapor pressure and elevation of the boiling point is that of *osmosis*. In 1748 Abbé Nollet found that if a membrane separated a dilute solution from a more concentrated one, the solvent would pass through the membrane from the dilute solution to the more concentrated one. For example, if a long tube closed at one end with a membrane made of pig's bladder is partly filled with a sugar solution and then placed in a vessel of pure water, it is found that water diffuses through the membrane into the solution but that the sugar does not pass out into the water. The water will rise in the tube until a rather high hydro-static pressure is attained (about 3 atm for a solution containing 2 gm of sugar dissolved in 50 cm³ of water at 25° C). Membranes with this property are called *semipermeable* membranes.

The hydrostatic pressure associated with osmosis is called *osmotic pressure*, and its value is found to depend on the nature of the solute and the solvent, the concentration of the solution, and the absolute temperature. It can be demonstrated that osmotic pressure must be independent of the nature of the membrane employed. The nature of the membrane and its area determine how rapidly osmosis will occur.

The quantitative features of osmosis were discovered by the botanist Pfeffer in 1877. He found that for non-electrolytic solutions the osmotic pressure was directly proportional to the concentration of the solution and, further, that the pressure for a given concentration was approximately proportional to the absolute temperature. Van't Hoff pointed out in 1887 the striking similarity between these osmotic properties of solutions and the properties of gases. He propounded a kinetic theory of very dilute solutions that correlated, in part, the experimental results of Pfeffer.

It is important to point out that the osmotic pressures of aqueous solutions of most electrolytic substances are much greater than would be expected from the osmotic pressures of solutions of sugars for corresponding concentrations. This difference is due to the dissociation of the electrolytic substances and hence the corresponding increase in the total number of particles present in the solution.

Osmosis plays an important role in many organic functions of both animal and plant life. The surfaces of roots of plants act as semipermeable membranes, permitting soil water to pass into the more concentrated sap solution.

14.18. Measurements of Vapor Content of Atmosphere

Perhaps the most important group of phenomena that we encounter often in daily life, which are to be classified under studies of saturated vapors, are those that have to do with mixtures of water vapor and the air. Aside from relatively small amounts of the so-called "noble gases," carbon dioxide, and the main constituents of nitrogen and oxygen, a most vital constituent of the atmosphere is its constantly changing content of water vapor. Surrounding, as it does, a globe whose surface of 200,000,000 square miles is 71 per cent water, much of which, in equatorial regions, is continually warm and evaporating, the atmosphere contains an amount of water vapor that is prodigious.

The amount of water vapor in the atmosphere is of great importance to the comfort and well-being of human life as well. Only in recent times has it been fully realized that humidity, as well as temperature, must be controlled for the maximum of comfort of indoor living when the outdoor environment is inclement. For all these reasons the study of the simple facts and measurements that have to do with atmospheric humidity are of prime importance. This field of study is called HYGROMETRY; the measuring instruments, HYGROMETERS.

The water vapor in the atmosphere may be treated—to a first approximation, at least—as a perfect gas, provided that no condensation is taking place. If p is the water-vapor pressure, M_w its molecular weight (18), ρ_w the density of the water vapor, and T the absolute temperature, then equation (13-17) can be written as

$$\rho = \frac{p}{RT} M_w, \qquad (14\text{-}7)$$

where R is the universal gas constant. Also, if ρ' is the density of *dry air* at a pressure P' and temperature T',

$$\rho' = \frac{P'}{RT} M_a, \qquad (14\text{-}8)$$

where M_a is the average molecular weight of dry air (28.97).

If ρ is the density of *moist air*, it is equal to the sum of the densities of dry air and the water vapor contained therein. Also, if P is the total pressure of the moist air, then $P - p$ is the pressure of the dry air and p that of the water vapor. Under these conditions we may express the gas equation as

$$\begin{aligned} \overset{\text{moist air}}{\rho} &= \frac{(P-p)}{RT} \overset{\text{dry air}}{M_a} + \frac{p}{RT} \overset{\text{vapor}}{M_w} \\ &= \frac{PM_a}{RT}\left\{1 - 0.379\,\frac{p}{P}\right\}, \end{aligned} \qquad (14\text{-}9)$$

where $0.379 = (M_a - M_w)/M_a$. Equation (14-9), together with equation (14-8), shows that the density of *moist air* is less than the density of *dry air* of the same temperature and pressure.

The water-vapor content of the atmosphere is designated frequently by a number of different, but interdependent, parameters—absolute humidity, relative humidity, specific humidity, mixing ratio, and dew-point temperature. We shall briefly define and discuss these parameters.

a) *Absolute humidity.*—The absolute humidity or density of the water vapor is defined by equation (14-7). As is clear from the nature of this equation, ρ_w varies with the vapor temperature. This quantity is not in widespread use.

b) *Relative humidity* (h).—The relative humidity, h, is defined as the ratio of the actual vapor pressure, p, of the moist air to the saturation vapor pressure, p_m, corresponding to the temperature of the mixture:

$$h = \frac{p}{p_m},$$

or, on a percentile basis,

$$h = \frac{p}{p_m}\,100 \text{ per cent}. \qquad (14\text{-}10)$$

This quantity is widely used in practice, though it, too, varies with air temperature.

c) *Specific humidity* (q).—The specific humidity, q, is defined as the ratio of the density of water vapor to the density of the *moist air*. It follows from equations (14-7) and (14-9) that

$$q \equiv \frac{\rho_w}{\rho} = 0.621\,\frac{p}{P - 0.379p}, \qquad (14\text{-}11)$$

where $0.621 = M_w/M_a$. Because q is a rather small quantity, it is generally expressed in terms of grams of water vapor per kilogram of moist air. On a cool, dry winter day, q may have values of from 1 to 3 gm/kg, whereas on a warm, moist day q may vary from 12 to 25 gm/kg. Specific humidity is not sensitive to temperature variation, provided that no water vapor is added to, or removed from, the air.

d) *Mixing ratio* (w).—The humidity mixing ratio, w, is defined as the ratio of the density of water vapor to that of *dry air*. From equations (14-7) and (14-8), with $P - p$ substituted for P', the pressure of dry air, we have

$$w = \frac{0.621p}{P - p}. \qquad (14\text{-}12)$$

Since P is of the order of one hundred times p, it is convenient to replace $P - p$ by P, the barometric pressure, so that w becomes

$$w = 0.621\,\frac{p}{P}. \qquad (14\text{-}13)$$

If this approximation is made use of in equation (14-11), then $q = w$, and the two terms can be used interchangeably. Equation (14-13) is widely used. In the practice of air-conditioning, w is usually expressed in pounds of water vapor per pound of dry air. Parameters b, c, and d are ratios, and therefore any convenient system of units may be used for the pressures.

e) Dew-point temperature (t).—The dew-point temperature, t, is defined as the temperature to which air must be cooled, at constant pressure, in order to become saturated. It follows from this definition that from the dew-point temperature and a table of saturated water vapor versus temperature, the actual water-vapor pressure in the air may be computed. If, therefore, the initial air temperature is also known, relative humidities may immediately be computed. Provided that no water vapor is added to, or condensed from, a sample of air, the dew-point temperature is not affected by temperature changes.

EXAMPLE 4: It is observed that the temperature of a sample of air is 20° C, the dew point 10° C, and the barometric pressure 29.50 mm of Hg or 990 mb. Compute the following quantities: Actual vapor pressure, p; h; q; ρ_w; and w.

Solution: From the dew-point temperature and Table 14-2, the actual vapor pressure is

$$p = 12.28 \text{ mb} .$$

Also the saturation pressure for an air temperature of 20° C is

$$p_m = 23.4 \text{ mb} .$$

Therefore

$$h = \frac{p}{p_m} 100 = \frac{12.28}{23.4} 100 = 53 \text{ per cent} .$$

Further,

$$q = \frac{0.621p}{P - 0.379p} = \frac{12.28 \times 0.621}{990 - 0.379 (12.28)}$$

$$= 7.7 \times 10^{-3} \text{ gm vapor/gm moist air}$$

$$= 7.7 \text{ gm vapor/kg moist air}$$

$$w = 0.621 \frac{P}{p} = 7.7 \text{ gm vapor/kg dry air} ;$$

$$\rho_w = \frac{p}{RT} M_w = \frac{12.28 \times 18 \times 10^3}{8.31 \times 10^7 \times 293}$$

$$= 9.07 \times 10^{-6} \text{ gm/cc}$$

or

$$9.07 \text{ gm/m}^3 .$$

The experimental problem of determining accurately the water-vapor content of the atmosphere is, oddly enough, difficult. No *practical meth-* od comparable in accuracy and convenience to the methods of measuring temperature and pressure is at present available. The absolute humidity may, of course, be determined accurately by drawing a known amount of moist air through an apparatus which will chemically absorb all the water vapor it contains. Weighing the chemicals before and after the air is drawn through determines the mass of water vapor. Such a method, though capable of good results, is obviously not suitable as a routine method.

The WET- AND DRY-BULB HYGROMETER (or SLING PSYCHROMETER) is widely in use today. The instrument consists of the two thermometers attached to a sling in such a way that the thermometers may be whirled around at a rate of about 120 rev/min. The bulb of one thermometer is surrounded by a moistened cloth from

TABLE 14-5

VAPOR PRESSURES IN MM OF HG FROM WET-DRY BULB TEMPERATURE READINGS*

$t_d°$ C	$t_d - t_w$										
	0	1	2	3	4	5	6	7	8	9	10
0........	4.6	3.7	2.9	2.1	1.3	...	...	...	...	...	...
5........	6.5	5.5	4.5	3.5	2.6	1.8	1.0	...	...	...	...
10........	9.2	8.0	6.8	5.7	4.6	3.5	2.5	1.5	0.5	...	...
15........	12.8	11.3	9.9	8.6	7.4	6.1	5.0	3.8	2.7	1.6	0.5
20........	17.5	15.7	14.1	12.6	11.1	9.7	8.3	6.9	5.6	4.3	3.1

* $t_d \equiv$ dry-bulb temperature; $t_w \equiv$ wet-bulb temperature. Barometric pressure $= 760$ mg of Hg.

which water is free to evaporate. The other thermometer is freely exposed to the air. Because of evaporation, the temperature indicated by the wet-bulb thermometer will be somewhat lower than the other. When the air is saturated, the two thermometers will indicate the same temperature, while, if the air is very dry, the wet-bulb thermometer will indicate a temperature many degrees lower than the other. Tables are available giving the actual vapor pressure from a knowledge of the wet-dry bulb readings and the barometric pressure. The relation connecting the vapor pressure with the wet-dry bulb reading is, however, empirical. A portion of such a table is given in Table 14-5. The first vertical column is the dry-bulb reading; the first horizontal line the difference between the wet-dry bulb readings; the remaining columns give the actual vapor pressure in millimeters of mercury.

Thus, if the dry-bulb reading is 15° C and the wet-bulb reading is 12° C, the actual vapor pressure from Table 14-5, p, is 8.6 mm of mercury; and from the second column the saturation pressure at 15° C is $p_m =$

12.8 mm of mercury. By equation (14-10) the relative humidity is 67 per cent.

The HAIR HYGROMETER consists of several strands of stretched human hair which are arranged in such a way that changes in the length of the hair due to humidity changes can be recorded on a calibrated dial. The hair hygrometer has the advantages of being simple in construction and of giving direct readings. However, the instrument shows considerable lag and needs frequent recalibration. Both the hair hygrometer and the sling psychrometer become quite unreliable when the air temperatures reach approximately −20° C.

14.19. Atmospheric Condensation

When the temperature of the water vapor in the air is lowered, a temperature (dew point) will be reached at which saturation is present; i.e., we say that the air is saturated with water vapor. The question may arise as to what will happen if the temperature is lowered further. If the air is in contact with surfaces such as rocks, vegetation, buildings, etc., the water vapor will condense as *dew* on the surfaces if the dew point is above 0° C and as *frost* if the dew point is lower than 0° C.

However, if we are considering water vapor not in contact with surfaces, will the water-vapor molecules join to form water molecules at the instant the dew point is exceeded? It has been found experimentally that condensation into droplets will not occur in pure water vapor unless the vapor pressure is many times that of the saturated vapor pressure, i.e., unless the air is supersaturated with water vapor. A discussion of the fundamental reasons for this would be beyond the scope of this text. (Suffice it to say that the vapor pressure over a curved surface is greater than that over a plane surface; also, the greater the curvature, the greater will be the difference in vapor pressure. Thus a water droplet over a liquid plane surface would evaporate unless the vapor pressure over the plane surface was the same as that over the spherical surface of the droplet; i.e., the vapor must be supersaturated.)

However, if suspended particles of matter are present, condensation appears when the saturated vapor pressure is attained and in some isolated cases when the air is somewhat less than saturated. In other words, a nucleus must be present for water vapor to condense upon.

It is interesting to point out that not all particles in the atmosphere can act as nuclei for condensation. Experiments show, for example, that ordinary dust particles, such as are on the floor of the coal shed, will not serve as nuclei unless the air is considerably supersaturated. On the other hand, water-soluble salts, such as nitric, phosphoric, and sulphuric compounds, to name a few, are well suited as nuclei. Near the seashore, salt particles in the air are effective nuclei. In industrial areas products of combustion generally contain many water-soluble compounds. Nuclei which can serve as condensation centers in the atmosphere are known as *hygroscopic nuclei*.

In the atmosphere, it has been estimated, there are from 2,000 to 50,000 hygroscopic nuclei per cubic centimeter of air near the surface of the earth. The radii of these nuclei range in size from 10^{-4} to 10^{-6} cm. There is good evidence that condensation under normal conditions takes place on nuclei whose radii are of the order of 10^{-5} cm. The precise physicochemical processes involved in the initial condensation process are not yet clearly understood.

When the dew point is reached, the water vapor condenses on the hygroscopic nuclei, forming droplets whose radii range in size from approximately 4×10^{-4} to 3×10^{-3} cm. Because of their minute size, their terminal velocity (Stokes's law, § 9.23) is quite small, and so a minor vertical air current is sufficient to keep them afloat. This aggregate of tiny water droplets is called either a *fog* or a *cloud*. It is a fog if the suspended water droplets are on or very near the surface of the ground, and a cloud if some distance above the ground. Thus the distinction between a fog and a cloud is really one of position.

14.20. Saturation Vapor Pressure over Ice and Supercooled Water

In our discussion of the vapor-liquid phase of water in § 14.6 we stated that each point on the steam line represented the saturation vapor pressure over water for a given temperature. In a similar fashion each point on the hoarfrost line or sublimation curve represents the saturation vapor pressure over ice for a given temperature. In Fig. 14-20 the solid line represents the equilibrium conditions of water vapor over ice. We note that the sublimation curve ends at the triple point, since ice cannot be heated above this temperature.

Water, however, can be cooled below the triple

point and, although thermodynamically unstable, frequently exists in the atmosphere at temperatures as low as $-30°$ C. The dotted curve below the triple point in Fig. 14-20 is an extension of the steam line. Points on this dotted curve represent saturation vapor pressure over supercooled water. It is to be noted that these pressures are a little larger than those over ice for corresponding tem-

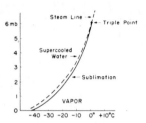

Fig. 14-20.—Saturation vapor pressure over ice and supercooled water.

peratures, i.e., the vapor is *supersaturated* with respect to ice. An ice crystal, therefore, placed near supercooled water droplets at the same temperature, will grow at the expense of the water droplets.

This phenomenon has several important applications in meteorology. For example, a supercooled fog that moves over a snow-covered surface will tend to dissolve because of a differential in vapor pressure. Also ice crystals present in a supercooled cloud in the atmosphere will grow rapidly and become sufficiently large to overcome the buoyant forces of the ice and fall, reaching the ground as snow or rain, depending on the air temperatures near the ground. The Swedish meteorologist Bergeron was the first to suggest that most rains in the temperate zones first have their origin as ice crystals that grow at the expense of any supercooled water droplets present.

Recently Langmuir and Schaeffer discovered that a pellet of solid carbon dioxide dropped into a cloud of supercooled water droplets generates tiny ice crystals of the order of tens of millions along the path of the pellet. Then, if these ice crystals are spread throughout a cloud of supercooled water droplets and permitted to grow, a possibility exists for dissipating the cloud. By dropping about 10 lb of small pellets per mile from

Fig. 14-21.—A photograph taken from an airplane at 25,000 ft, showing the transformation of supercooled water droplets into ice crystals after seeding the cloud system with dry-ice pellets. The seeded area is about 20 mi long.

an aircraft, Langmuir has been able to clear large areas of supercooled stratus-type clouds. Figure 14-21 is a photograph showing the transformation of supercooled water droplets into ice crystals. after seeding the cloud system with dry-ice pellets.

14.21. Résumé

The following is a brief listing of the definitions, principles, and theories, given in this chapter, which you should know:

The concept of latent heat of fusion and evaporation
The method of measurement of latent heats

The kinetic-theory interpretation of latent heats and changes of phase
Saturation vapor pressures and boiling
The freezing point of liquids and its dependence on pressure
The triple point and the three phases—solid, liquid, and vapor—in equilibrium
The critical temperature of a gas
Thermodynamic *P-V-T* surfaces and their significance
Interpretation of these phenomena by van der Waals
Free expansion of gases and internal forces between molecules
Humidity of the atmosphere

Queries

1. How could one separate oxygen from the nitrogen of the air? *Hint:* Note the boiling points in Table 14-4.
2. To what extent does van der Waals' equation afford a satisfactory explanation of the experimental results of Amagat (Fig. 13-11)?
3. What is meant by the popular skating phrase "the ice is slick"?
4. In connection with Fig. 14-7, is it possible to account, in part at least, for the movement of glaciers?
5. In the Joule experiment on free expansion (Fig. 14-17), tank A is found to cool slightly and B to become somewhat heated when the valve is opened. Can you give reasons why this should be so? Would

this phenomenon occur if the gas were an ideal gas, as defined in § 14.14?
6. On the basis of the discussion in §§ 14.10 and 14.11, would you say that it is possible to take a substance from the unsaturated vapor state to the gaseous state and then to the liquid state without encountering a free surface? Justify your answer by means of Fig. 14-13.
7. Why would the geometric temperature scale mentioned in § 14.16 place the absolute zero beyond reach?
8. What is the essential difference between the specific heat of a substance and its latent heats in terms of the kinetic-molecular theory?

Practice Problems

1. Of the 538.7 cal/gm necessary to change water at 100° C to steam at 100° C when the pressure is 76 cm, how many are used to push back the surroundings; how many go into the steam? The specific volume of steam at 100° C is 1,671 cm³/gm. Ans.: 40.3 cal; 498.3 cal.
2. Of the 79.63 cal/gm involved in freezing water, how many actually come from the water? Density of water at 0° C is 0.999841 gm/cm³, and ice, 0.917 gm/cm³.
3. How far must a block of ice fall to just melt if all the energy goes to melting ice? (Air at 0° C.) Ans.: 34.0×10^5 cm.
4. If 10 gm of steam at 150° C are run into 100 gm of water at 20° C contained in a 200-gm glass beaker (specific heat = 0.2 cal/gm ° C), what will be the final temperature?
5. If 50 gm of ice at −30° are put into 100 gm of

water at 80° C contained in a 200-gm glass beaker, what will be the final temperature? Ans.: 34°.14 C.
6. If 10 gm of steam at 150° and 50 gm of ice at −30° C are put into 100 gm of water at 50° C, contained in a 200-gm glass beaker, what will be the final temperature?
7. If 1,000 gm of steam at 150° are combined with 10 gm of water at 50° C, what will be the result? Ans.: Steam at 136°.5 C.
8. If 100 gm of steam at 150° are combined with 10 gm of water at 50° C, what will be the result?
9. The temperature of the air is 27°, and the dew point 21° C. What is the relative humidity? Ans.: 69.9 per cent.
10. Calculate the pressure due to dry air on a day when the relative humidity is 90 per cent, the temperature 20° C, and the barometric reading 757.3 mm of mercury.

11. The relative humidity in a room $20 \times 10 \times 6$ m is 80 per cent, and the temperature 25° C. How many grams of water are present? How many grams of water must be removed to lower the relative humidity to 50 per cent? Ans.: 22.8×10^3; 8.5×10^3.

12. Ten liters of air at 50 per cent relative humidity and temperature of 10° C expand isothermally to 40 liters. How many grams of water must be added to the air to maintain a constant relative humidity?

13. If in problem 12 the air is now compressed isothermally back to 10 liters, calculate the number of grams of water that will condense. Ans.: 0.094.

14. Compute the number of calories of heat necessary to change 30 gm of ice at −10° into steam at 200° C at atmospheric pressure.

15. Using Plate 14-1, a, draw a V-T curve at constant p. Explain the physical significance of this curve.

16. Repeat for water, using Plate 14-1, c.

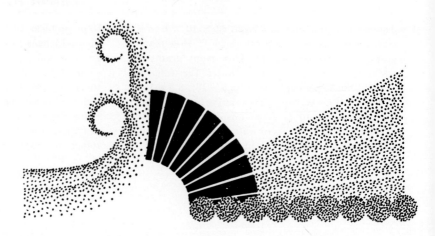

CHAPTER 15

The Transmission of Heat

We have, in what has preceded, frequently used such expressions as "when the temperature is raised," "as heat is poured into," "dissipated into heat," etc., implying both deliberate and inevitable transmission or transfer of thermal energy (a quantity of heat measured in calories) from one location to another—a transfer involving a finite time for its accomplishment, at least by implication. A brief account of the various means by which heat is transmitted is therefore in order. *These means are three.* They may be separately involved or associated; i.e., heat may be transferred from one location to another by CONVECTION, by CONDUCTION, or by RADIATION, singly or in various degrees of association.

15.1. Convection

Convection is involved only when fluids—either liquids or gases—are themselves heated in definitely circumscribed regions or when they are in contact with heated solid objects. *Convection is a mechanical process involving the bodily transfer of heated substances.* Heat transfer caused by the natural motion of the fluid due to density differences is known as *natural convection*. On the other hand, if the heated fluid is made to move by means of a fan or pump, the transfer is called *forced convection*. We shall treat natural convection in this section.

Whenever crystals of any colored substance that are somewhat soluble in a liquid are present on the bottom of a transparent container over a fire or any other source of heat, the *rising convection currents* may be seen. These are due, of course, to the fact that the liquid expands and becomes less dense when heated, and hence rises through the adjacent denser, cooler portions. The circulation so established tends to keep the liquid more uniformly heated throughout than would otherwise be the case. In heavy, dense, or highly viscous liquids, however, there may be, in spite of convection, great temperature differences and gradients.

Convection currents in liquid have a direct practical application in hot-water furnaces. A sche-

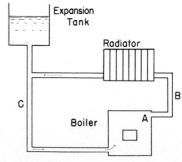

Fig. 15-1.—A natural convection hot-water heating system

matic drawing of a hot-water system is shown in Fig. 15-1. As the boiler is fired, the water at A becomes heated, rises in pipe B to the radiator, where it cools, becomes more dense, and then returns to the boiler through pipe system C. The expansion tank allows for volume changes in the circulating system and so prevents pipes from bursting because of water expansion.

Gases, being more tenuous than liquids, have

much more vigorous convection currents within them. The convection currents rising from a shaded parlor lamp (if the shade is open at the top) are readily seen if a puff of tobacco smoke is sent in the direction of the lamp and beneath the shade. Indeed, great pains must be taken to have the walls of a vessel containing a gas at the same temperature everywhere if local differences in density and the convection currents produced thereby are to be avoided. For the transmission of beams of light in rectilinear paths across any considerable distance in air, freedom from local variation in density and from convection currents is necessary.

Ocean currents and atmospheric winds have their origin in the heating of some areas in excess of others. The resulting circulation, in addition, is affected by its streaming into higher or lower latitudes, where the linear velocity of the earth's surface is altered by its changing distances from the axis of rotation.

Local land and sea breezes are accounted for by convection currents in the atmosphere. As shown in Fig. 15-2 (a), during the day the land becomes

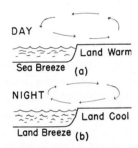

FIG. 15-2.—Illustrating a land and sea breeze

hotter than the adjoining sea, owing in part to the higher specific heat of water, as compared with sand and rock. The air above the earth expands, rises, and is replaced by the cooler sea air, the general pattern of circulation being that of the diagram. During the night the land loses heat more rapidly by radiation than does the sea. The air over the land thus becomes cooler than that over the sea; and local convection currents, as shown in Fig. 15-2 (b), are set up.

Calculating in detail how much heat is carried away by convection from a heated surface is, in general, very difficult, if not impossible. Some special cases involving small plane surfaces and wires have been solved, but the details are beyond our present scope.

15.2. Conduction

By far the most common means of heat transference, at least between small terrestrial objects and in the most common examples in everyday life, is by conduction. *Conduction is a molecular process of heat transfer through adjacent layers of any material medium by the diffusion of the high kinetic energy of the random-moving molecules from the heated regions into successively cooler ones.* Since liquids and gases are relatively poor conductors of heat in this sense, compared to many solids, especially the metals, and since, furthermore, conduction is complicated by convection in fluids in general, we shall discuss such details as are fundamental to solids. Quantitatively, different solids differ enormously in their ability to conduct heat energy. Rods of silver, copper, iron, and glass, placed each with one end in a furnace, will differ greatly in the length of time taken for the other end to show a definite rise in temperature. The copper and silver rods quickly become too hot to hold; the iron will probably never become too hot to handle with a cloth holder; the glass rod may be incandescent at one end while the other end will show almost no rise in temperature. These facts are given quantitative recognition in ascribing to all substances a *coefficient of thermal conductivity.*

Consider the thin slab of material shown in Fig. 15-3, which has a thickness of d cm; the area of

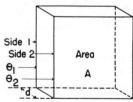

FIG. 15-3.—Conduction of heat through a thin slab of material.

one face, A cm^2; and the temperature of the rear face $\theta_1°$ C and of the front face $\theta_2°$ C, with $\theta_2 > \theta_1$. If the temperatures θ_1 and θ_2 are maintained constant, then the experiment shows that the total number of calories of heat transferred from side *2* to side *1* depends only upon the following factors: (a) the nature of the material, which we shall represent by k; (b) the surface area, A; (c) the thickness, d; (d) the elapsed time, t; and (e) the temperature difference, $\theta_2 - \theta_1$. Analytically, the total heat, Q, transported is given by

$$Q = \frac{k A (\theta_2 - \theta_1) t}{d} \text{ cal};$$

(15-1)

or

$$k = \frac{Q}{t\,[\,(\theta_2 - \theta_1)\,/d\,]\,A}\ \text{cal/cm-sec}\,^\circ\text{C}\,,\quad (15\text{-}2)$$

where k is the coefficient of thermal conductivity, Q the number of calories transferred, t the time in seconds, $(\theta_2 - \theta_1)/d$ the TEMPERATURE GRADIENT, and A the surface area. The units of k from equation (15-2) are calories per centimeter per second per degree centigrade (see Table 15-1 for values).

TABLE 15-1

COEFFICIENTS OF THERMAL CONDUCTIVITY

Substance	k (Cal/Cm Sec ° C)	Substance	k (Cal/Cm Sec ° C)
Ag..........	1.006	Pb..........	0.083
Cu..........	0.918	Glass........	.0024
Au..........	0.700	H₂O.........	.00143
Al..........	0.480	Rubber......	.00045
Graphite.....	0.315	Asbestos.....	.0004
Hg..........	0.145	Pine........	.0002
Steel........	0.11 and 0.15	Air.........	0.000055

Because there are no perfect *nonconductors* of heat, a precise measurement of thermal conductivity is one of the most difficult problems in laboratory physics. Thermal conductivities of metals in the form of rods have been measured by the method of Forbes. This, in essence, consists of (*a*) heating one end electrically, providing a well-known heat input, and (*b*) cooling the other end with circulating water, thus enabling the output to be measured. This will be somewhat less than the input, and therefore surface losses between the two ends may be found. Thermocouples are used to measure the temperature at different points along the bar, whence the thermal gradient may be established.

In many practical applications the heat lost through a cylindrical pipe is needed. We give the following formula for the number of calories per second transferred radially through a pipe of unit length, Fig. 15-4:

$$q = \frac{2\pi k\,(\theta_1 - \theta_2)}{\log_e (r\,/\,r_1)}\ \text{cal/sec-cm}\,,\quad (15\text{-}3)$$

Fig. 15-4.—Cross-section of a cylindrical pipe of radii r_1 and r_2.

where k is the thermal conductivity of the pipe, r_1 the inner radius, r_2 the outer radius, θ_1 the temperature within the pipe, and θ_2 that outside the pipe. The proof of this is asked for in problem 13.

15.3. Radiation

A third way in which heat may be transmitted away from an object is one which involves *no material means* of transfer, either convective or conductive. Heat is transferred as an ELECTROMAGNETIC RADIATION. This "radiant" heat is familiar to everyone. The life-giving and life-maintaining sunlight that bathes our planet comes to us thus. We can *feel* its *longer infrared heat waves* as well as *see its light.*

Blackened bulbs filled with water, alcohol, or ether and sealed after removal of the air form excellent crude thermometers for detecting the radiant heat from flatirons, hot plates, toasters, etc. These radiations, like ordinary light, are transmitted in straight lines, are reflected from polished surfaces, may be refracted by prisms made from rock salt, which is transparent to them, or diffracted by gratings in just the same manner as visible light. They can likewise be polarized and follow the inverse-square law of attenuation with distance (see chap. 34).

There are great variations in the transparency of different materials to these heat radiations (a phenomenon called "diathermacy"). A deep-black solution of iodine in carbon disulphide, which transmits practically no light, is almost perfectly transparent to heat. A weak solution of cuprous chloride in water—quite transparent optically—absorbs all heat radiations.

15.4. Prevost's Theory of Exchanges

If a thermocouple or any other sensitive detector of heat is placed near a block of ice, it will register an effect opposite to that which it gets from a hot body. This does not imply, however, that "cold" is radiated, except in the sense that some radiant heat is cut off. Prevost, in 1792, first explained this effect. When a number of objects at different temperatures are placed in an inclosure into which no heat can be conducted, all objects will ultimately come to the same temperature. This will happen even in the absence of any circumambient medium through which convection or conduction can occur. Even though all the objects are at the same temperature, they are continuously emitting radiant energy. This loss of energy lowers the temperature of the radiating body, but radiations are received by each body from all the others, thus gaining energy; and, since the temperature of all remains the same, the amount received by any one from all the others must equal what it itself emits.

These ideas are contained in the following statement, known as the *Prevost law of exchange:*

All objects are continuously radiating heat energy; in a state of thermal equilibrium the amount of energy radiated per second from an object is equal to the energy absorbed by it in the form of radiations from surrounding objects.

A direct consequence of great importance follows: Suppose two objects, one with a highly polished surface and the other blackened, are placed within a constant-temperature inclosure. They, too, will reach the same temperature. But now it is clear that the polished surface reflects and does not absorb most of the radiations that fall on it. Therefore, unless it emits its own radiation at a rate correspondingly smaller, in accordance with its large reflecting power, it would lose more than it received and would fall to a lower temperature than the blackened body.

The validity of this argument may be demonstrated experimentally. The emitting power and absorptive power of various types of surfaces vary in the same ratio. Black surfaces absorb more readily than white ones, rough surfaces more readily than polished ones. Consequently, we should expect the blackened and roughened portions of the surface of any heated object to emit more radiation than the white and polished portions. This, too, is readily demonstrated.

15.5. Quantitative Laws of Radiation

It is found in the laboratory that the rate at which an object emits thermal radiation depends, first, upon its temperature and, second, upon the nature of the surface. By definition, the TOTAL EMISSIVE POWER, e, is equal to the total radiant energy emitted per second per unit area; i.e.,

$$e \equiv \text{Emissive power} = \text{Total radiant energy per second per square centimeter .} \quad (15\text{-}4)$$

When radiant energy falls on an object, some of the radiation is absorbed, some transmitted

through the object, and some reflected. By definition, the fraction of the total incident radiation that is absorbed by an object is its TOTAL ABSORPTIVITY, a. The factor a depends upon the temperature and nature of the absorbing surface and has values ranging from 0 to 1; i.e.,

$$a \equiv \text{Absorptivity} = \text{Fraction of total incident radiation absorbed .} \quad (15\text{-}5)$$

As we have stated, the absorptivity differs widely among substances. Some substances, such as lampblack, have absorptivities very near unity; i.e., practically all the radiation that falls on them is absorbed. A body that *does* absorb all the incident radiation we shall call a *black body;* thus, for a black body,

$$a_B = 1 . \quad (15\text{-}6)$$

It was first suggested by Stefan in 1879 and subsequently predicted theoretically by Boltzmann that the total emissivity of a black body is proportional to the fourth power of the absolute temperature. This law, known as the "Stefan-Boltzmann law," may be written as

$$e_B = \sigma T^4 , \quad (15\text{-}7)$$

where σ is a universal constant, the "Stefan-Boltzmann constant," and is equal to

$$\left. \begin{aligned} \sigma &= 5.735 \times 10^{-5} \text{ erg/sec-cm}^2\text{-deg}^4 \\ &= 5.735 \times 10^{-8} \text{ watts/m}^2\text{-deg}^4 . \end{aligned} \right\} \quad (15\text{-}8)$$

The relation between the emissive power and the absorptivity of any object not necessarily a black body may be obtained from the following experiment. Two objects, one of which we assume to be a black body, are introduced into a box whose interior walls are at a temperature T. The temperature of the two objects is also T. The amount of radiation arriving in unit time on a unit area of each object is the same. Let us call this amount of energy per second per square centimeter, E. For the black body the energy absorbed per second per square centimeter is, by definition of a black body, E. The energy emitted per second per square centimeter is e_B; and, since the black body is in equilibrium,

$$E = e_B . \quad (15\text{-}9)$$

For the other body, not a black body, the amount of energy absorbed per second per square centimeter is equal to aE, while the radiation emitted **233** per second per square centimeter is e. Again, since the object is also in equilibrium with the walls and with the black body,

$$e = aE . \qquad (15\text{-}10)$$

Since, however, $E = e_B$, it follows that

$$e = e_B a . \qquad (15\text{-}11)$$

This equation represents the contents of Kirchhoff's law, which states that *the total emissive power of any object at any temperature is equal to a fraction of the emissive power of a black body at that temperature*. The fraction is the absorptivity of the object at the given temperature.

From equation (15-7) it follows that equation (15-11) may be written as

$$e = a\,\sigma T^4 \ \text{erg/sec-cm}^2 . \qquad (15\text{-}12)$$

Kirchhoff's law, in the form of equation (15-11), states:

At any given temperature a body which is a good absorber of radiation is also a correspondingly good radiator of radiation.

Many interesting phenomena have their quantitative explanations in this statement. We leave it to the reader to explain (1) why a thermometer whose bulb is coated with lampblack reads higher in the sun than one not coated (*Hint:* What about convection currents?); (2) why steam radiators are purposely made rough and dull.

Other theoretical aspects of radiation are treated in § 37.3.

15.6. Newton's Law of Cooling

From our discussions thus far it is clear that, in general, a hot object will lose heat to its surroundings by conduction, radiation, and convection, natural or forced. The degree to which these processes contribute to the *rate of cooling* of the object will depend on experimental conditions. Several empirical laws have been proposed. One of the most useful, particularly for calorimetric measurements, is that proposed by Newton. His law of cooling, an approximate statement of conditions when the difference in temperature between a heated object and its surroundings is only a few degrees, states that the rate of loss of heat

is proportional to the difference in temperature. For the case of the calorimeter we can write

$$\text{Rate of loss of heat} = -\,b\,(T - T_1) . \,(15\text{-}13)$$

In this equation T is the instantaneous temperature of the calorimeter; T_1 is the constant temperature of the room; and b, called the "radiation constant," is a constant of the calorimeter and depends primarily on the size, material, and condition of the outer surface of the calorimeter. The temperatures are in degrees centigrade. If mc is the heat capacity of the calorimeter and contents, then

$$\text{Rate of loss of heat} = \frac{dT}{dt}\,m\,c ,$$

and equation (15-13) becomes

$$\frac{dT}{dt} = \frac{-\,b}{m\,c}\,(T - T_1) \qquad (15\text{-}14)$$

or

$$\frac{dT}{T - T_1} = \frac{-\,b}{m\,c}\,dt .$$

Integrating, we have

$$\log_e (T - T_1) = \frac{-\,bt}{m\,c} + \text{constant} .$$

If $T = T_0$, for $t = 0$, then the constant is equal to $A \log_e(T_0 - T_1)$, and hence

$$\log_e (T - T_1) - \log_e (T_0 - T_1) = \frac{-\,b}{m\,c}\,t, \,(15\text{-}15)$$

or the temperature $T\,°\,C$ at any time t of a calorimeter having a temperature $T_0\,°\,C$ at time zero, and cooling into a room at $T_1\,°\,C$, is

$$T = T_1 + (T_0 - T_1)\,e^{-(b/mc)\,t} . \qquad (15\text{-}16)$$

Figure 15-5 is a plot of equation (15-16).

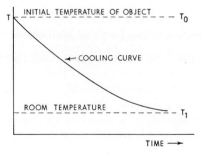

Fig. 15-5.—Cooling curve

EXAMPLE 1: Calculate the time in minutes required for a calorimeter initially at a temperature of 30° C to cool to within 1° of room temperature of 20° C. The heat capacity of the calorimeter and contents is 400 cal/degree, and the radiation constant has been measured to be 20 cal/min-degree.

Solution: This is a straightforward application of equation (15-15):

$$\log_e(1) - \log_e(10) = \frac{-20}{400} t,$$

$$0 - 2.3 = \frac{t}{20} \quad \text{or} \quad t = 46 \text{ min}.$$

15.7. Résumé

The following is a brief listing of the definitions, principles, and theories, given in this chapter, which you should know:

Definitions of the processes of heat transfer by convection, conduction, and radiation

The Stefan-Boltzmann law for the radiation emitted by a black and by a nonblack body

Radiation equilibrium and Prevost's law of exchange

Queries

1. An airplane flies over terrain that includes a small lake, forest, sandy beach, and newly plowed fields and encounters considerable "bumpiness of the air." Explain, in terms of the principles of this chapter, what is happening.

2. On the basis of Kirchhoff's law (eq. [15-11]), how would you expect a piece of green glass and a piece of red glass to appear when just removed from a furnace in which they had been heated sufficiently to radiate visible light?

3. What units must be given to b of equation (15-14)?

4. What evidence can you give that thermal radiations have the same general properties that visible light possesses?

5. The thermal conductivity of air is far less than that of cotton or wool. Why, then, use cotton or woolen blankets to keep warm?

6. On what principle does the greenhouse operate?

7. Why is it that ground fogs appear more frequently during nights when there are no overhanging clouds than when the sky is overcast?

8. A slab of wood and one of copper remain in contact until a temperature equilibrium of 0° C is attained. If one touches the wood and then the copper with one's finger, the copper "feels" colder than the wood. Explain.

9. In view of query 8, what criticism can be made of the definition of temperature equilibrium in § 11.2?

10. If in equation (15-15) we were to plot $\log_e(T - T_1)$ against time, what kind of curve would result? What is the value of the slope of this curve?

Practice Problems

1. A plate of glass having an area of 500 cm² is maintained with a temperature difference of 10° C between its surfaces. If the thickness is 0.5 cm, how much heat passes through it in 1 sec? What is the temperature gradient? Ans.: 24 cal/sec; 20° C per centimeter.

2. A steam pipe made of steel carries steam at 110° C. The inner radius of the pipe is 2 cm; the outer, 3 cm. Calculate the number of calories of heat lost radially per second and per unit length if the room temperature is 30° C.

3. A vessel is divided into parts by an aluminum partition 1 cm thick, 20 cm high, and 30 cm wide. One compartment contains ice; the other has steam at 100° C passing through it. How much ice is melted per minute? Ans.: 21,700 gm/min.

4. An evaporating-pan made of steel has 2 m² of heating surface. The iron is ½ cm thick, and its outer heating surface is kept at 200° C. Calculate the amount of water that evaporates per hour while the water is boiling at 100° C.

5. The temperature of a tungsten filament is 2,775° K. Assuming the filament to be a black body, compute the number of watts per square centimeter of surface radiated by the filament. Ans.: 340 watts.

6. A small hole in a furnace may be taken to approximate the black body in so far as emitted radiation is concerned. If the temperature is 1,700° C, compute the number of calories per second per square centimeter radiated through the hole.

7. Two very large parallel planes are 1 cm apart in an atmosphere of air. One plane is maintained at a temperature of 0° C; the other, at 2° C. (a) Calculate the number of calories per second per square centimeter transferred through the air by conduction from the hot to the cold plane. (b) Assuming the two planes to be black bodies, calculate the net amount of calories per second per square centimeter transferred by radiation from the hot to the cold plane. Ans.: 1.1×10^{-4}; 2.25×10^{-4}.

8. Repeat the calculation of problem 7, assuming the cold plane to be 0° C but the hot plane 200° C.

9. Calculate the number of watts of electrical power necessary to take care of radiation losses of an incandescent filament if its temperature is 2,000° C and that of the bulb 100° C. The filament is 20 cm in length and 0.002 cm in diameter. Assume that the filament and bulb radiate as black bodies. Ans.: 19.2.

10. Calculate the total number of calories of heat conducted radially outward from an aluminum pipe 10 m long if the liquid within is at a temperature of 200° C and the room temperature is 30° C, the inside diameter is 8 cm and the outside 10 cm. Ans.: 23×10^5 cal/sec.

11. Derive Newton's law of cooling from the Stefan-Boltzmann law, assuming that the temperature T of the hot body is close to the constant temperature T_0 of the surroundings.

12. Calculate the time required for a calorimeter at a temperature of 36° C to cool to within 0°.1 C of the room temperature of 16° C. The heat capacity of the calorimeter and water is 340 cal/degree, and the radiation constant is 17 cal/min-degree.

13. Derive equation (15-3). (Note that the quantity of heat, q, transmitted through a thin ring of the pipe of radius r and thickness dr, temperature difference $d\theta$, is $Q = - k2\pi r d\theta/dr$ per unit length.) Explain the presence of the negative sign in this differential equation.

CHAPTER 16

Thermodynamics and Heat Engines

16.1. The Science of Thermodynamics: First Law

The science of thermodynamics deals with the quantitative relations between heat and other forms of energy, particularly mechanical energy. We showed in § 12.4 that all machines convert some part of their energy into heat energy through ever present friction. It is easy to arrange conditions to convert any desired fraction of the available mechanical energy of a machine into heat energy by controlling the amount of friction. (As a matter of fact, it is possible to convert all forms of available energy—electrical, magnetic, chemical, electromagnetic, etc.—into heat energy.)

The inverse process of converting heat energy back into, say, mechanical energy is accomplished in practice by means of a *heat engine*. The heat engine may be a steam engine, wherein the working substance, steam under pressure, does work, or a gasoline engine, hot-air engine, or diesel engine, to name the more practical devices.

However, we found that whatever device was used, be it a heat engine that converted heat energy Q (calories) into mechanical energy W (joules) or a mechanical machine that converted mechanical energy W into heat energy Q, the two forms of energy were interconvertible and that a definite, quantitative relation existed between them, i.e.,

$$W = QJ , \qquad (16\text{-}1)$$

where J is the mechanical equivalent of heat in joules per calorie. By means of this relation we were in a position to generalize the law of conservation of energy to include heat energy. Thus, *in a closed system, the total amount of energy is constant.* The phrase "in a closed system" implies that no energy is allowed to enter or leave the system. This general conservation principle is also known as the *first law of thermodynamics.* In § 12.6 we formulated the first law in mathematical form as

$$\Delta U = QJ + W , \qquad (16\text{-}2)$$

which states that the change in internal energy of a system is equal to the heat energy absorbed by the system measured in mechanical units plus the work done on the system. The validity of the first law is based on a vast amount of experimental evidence. There are no known exceptions to it.

16.2. Second Law of Thermodynamics

The early designers of heat engines quickly found that such devices were notoriously inefficient. Only a fraction of the heat energy given up by the working substance could be converted into useful work. This difficulty was basically not due to poor engineering design but to a fundamental limitation given by thermodynamic theory. No one has ever constructed a heat engine that converted all the available heat into mechanical energy, and there is reason to believe that no one ever will. This limitation is not one imposed by the first law.

From our definition of efficiency (§ 4.11) we have

$$\text{Efficiency} \equiv E = \frac{\text{Useful work output}}{\text{Total work expended}}.$$

If an amount of heat Q_1 is put into a heat engine which, during the course of its operations, fails to convert all of this into useful mechanical work and returns an amount Q_2 to its environment (but not to the source of heat), the useful work, according to the first law, will be $J(Q_1 - Q_2)$. The total heat energy expended by the source of heat is JQ_1, and hence the thermal efficiency of the heat engine is

$$E = \frac{Q_1 - Q_2}{Q_1}. \tag{16-3}$$

For an amount of heat energy Q_1 given to the engine, the only way to make the thermal efficiency 100 per cent is to make Q_2 zero. In practice this is impossible. In principle it is possible under the special conditions of absolute zero, as shown in § 16.8. An example will illustrate the point.

When we burn 1 kg of coal, approximately 8,000 cal/gm or some 8×10^6 cal are released. Let us suppose that we wish to convert this heat energy into mechanical energy by means of a steam engine. Let us further assume that the heat is used to maintain steam in the boiler at 250° C and that we discharge the steam at 100° C. Then, neglecting all heat losses due to friction in bearings and all heat losses by conduction, convection, etc., we find that the maximum amount of this energy that we can convert into mechanical energy is only 29 per cent. This does not mean that we have destroyed any heat energy; rather, it means that 71 per cent of the total energy remains in the thermal form of exhaust steam, Q_2, of equation (16-3).

This example points up the basic characteristics of all heat engines that operate cyclically (§ 16.3), namely, that heat is absorbed by the engine at some high temperature and part of the heat is necessarily ejected at a lower temperature. The studies of the French engineer, Sadi Carnot, in 1824 first clearly brought this inherent limitation of heat engines to the attention of engineers. The statement that embodies these ideas constitutes the *second law of thermodynamics*. The second law is, therefore, an inference from experience. A useful statement of the law follows:

It is impossible in principle to construct an engine, operating in a cycle, that will produce mechanical work by extracting heat from a single reservoir and not return any heat to a reservoir at a lower temperature.

The second law of thermodynamics has far-reaching implications in all branches of science—in physics, chemistry, biology, cosmology—for it turns out that in any of these disciplines where heat-transfer problems are involved, it is possible, at least in principle, to reduce the problem to that of a heat engine that converts heat energy to mechanical energy. For such a process the statement of the second law is directly applicable.

There have been many formulations of the second law by such masters as Clausius, Kelvin, Boltzmann, and Planck, each of the statements, of course, logically equivalent, though emphasizing another facet of this broad generalization of nature. For example, the second law enables us to determine the directions of thermal processes. A simple experiment will illustrate the point. Consider two blocks of copper, one at a temperature of 100° C, the other at 20° C. When these blocks are brought together, we know from experience that heat flows from the hot to the cold block. Why does it not flow from cold to hot? Surely, the first law of thermodynamics would not be violated, since whatever heat energy the hot block might acquire, the cold block would lose. It is clear that the first law must be supplemented by a statement that determines the direction in which heat changes may occur. This statement is the second law of thermodynamics. Its formulation by Clausius is useful for this purpose:

It is impossible for any self-acting machine, unaided by any external agent, to convey heat from one body to another at a higher temperature.

This simple statement implies that to convey heat from a cold to a hot object it is necessary to do external work, as in the case of a refrigerator. In the sections that follow we shall develop these ideas more fully, leading to a precise mathematical formulation of the second law.

16.3. Cycles; Thermodynamics Diagram

Let us consider a mole of gas confined in a cylinder with a movable piston. As we have stated in § 13.5, the gas law, $PV = RT$, is the equation of state of an ideal gas. This means that the gas whose volume is V and temperature T throughout and which is subjected to a pressure P is in equilibrium. The variables P, V, and T are said to define the thermodynamic state of the system, here a mole of gas in the cylinder. If we change from one state, P_1, V_1, T_1, to another state, P_2, V_2, T_2, we have effected a thermodynamic transformation. This may be represented diagrammatically in many ways, one of the most useful for our purpose being a P versus V plot, as shown in Fig. 16-1. The point A represents the initial state,

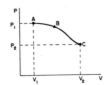

Fig. 16-1.—Thermodynamic *PV* diagram

P_1, V_1, T_1. We may allow the gas to expand and do useful work, the pressure, volume, and temperature in general changing; each point on the curve ABC represents the values of P, V, and T at each instant during the expansion. When we stop the expansion, the final state of the gas will then be represented by the point C, whose value is $P_2V_2T_2$. We have seen that the work done by a gas in expanding is W, given by

$$W = \int_{V_1}^{V_2} P\,dV , \qquad (16\text{-}4)$$

which is the area under the curve ABC of Fig. 16-1. According to the first law, if U_A is the internal energy of the gas at the initial state and U_C its final state, then

$$U_C - U_A = QJ - W . \qquad (16\text{-}5)$$

Here Q represents any heat that was absorbed or given up by the gas.

The gas is now brought back to its initial state, P_1, V_1, T_1, along the curve CDA by a series of

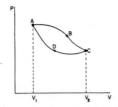

Fig. 16-2.—A thermodynamic cycle

compressions. We have completed *a cycle of operations* on the gas in returning it to its initial state (Fig. 16-2). The work done *on the gas* during com-

pression is given by the area under the curve CDA. The net work done in the cycle is, therefore, the area inclosed by the curve $ABCDA$. We note that this area, in general, is not zero, for the gas received heat during the cycle and hence did work.

Applying the first law (eq. [16-5]) to the completed cycle, we observe that, since $U_C = U_A$, we have

$$W = QJ ;$$

that is, the net work done by the gas, W, is exactly equal to the amount of heat, QJ, absorbed by the gas during the cycle. If we had traversed the cycle in the opposite direction, $ADCBA$, the work done on the gas would be the area under the curve ADC and that done by the gas would be the area under the curve CBA. The work represented by the area inclosed by the curves now represents work done *on the gas* rather than by the gas.

There are an infinite number of possible cycles through which the gas, the working substance, may be subjected. In § 16.5 we shall consider a few simple cycles, though ones of great practical interest.

16.4. Reversible and Irreversible Processes

Let us consider a mole of gas confined in a cylinder with a movable piston. The mole of gas in the cylinder is maintained in thermal equilibrium with respect to its environment by the external agencies: the cylinder confines the gas; the piston and cylinder walls exert the necessary external pressure equal to the gas pressure; and the surrounding atmosphere, via the cylinder walls, maintains the gas at the proper temperature. If these various external agencies change slowly, i.e., the temperature of the cylinder is lowered or raised very slowly and the external pressure is increased or decreased by infinitesimal amounts, *then the gas will pass through a continuous series of equilibrium states.* Such a slow change is called a *reversible process*, because the gas will be passed through the same series of equilibrium states in reverse order if the external agencies are changed slowly in the reverse direction.

If infinitesimal changes in pressure, for example, are not applied to the piston, so that the gas may expand infinitely slowly, kinetic energy in the form of eddies will develop within the gas and will be converted into heat. In the reverse process, not only does reconversion of this dissipated heat into **239** mechanical energy not take place, but more mechanical energy (in the form of eddies) is converted into heat. Such processes are called *irreversible*.

Since, in actual practice, we cannot get rid of at least some eddy formation and also of friction in the piston, we may state that all actual cycles in practice are irreversible. Nevertheless, the ideal concept of a reversible cycle proves to be of great theoretical value in the study of thermodynamics. One very important reversible cycle is the Carnot cycle, discussed in some detail in the next section.

16.5. The Carnot Cycle

The thermodynamic studies of Carnot were based on an idealized type of heat engine. Imagine a cylinder fitted with a frictionless piston and containing some working substance which expands with increasing temperature and decreasing pressure (Fig. 16-3). For simplicity *only*, we shall as-

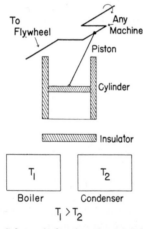

Fig. 16-3.—Schematic drawing of a simple heat engine used to illustrate the Carnot cycle.

sume that our working substance is an ideal gas. We propose to take a mole of gas through a complete cycle made up of two adiabatics (see § 13.18) and two isotherms, that is, the closed curve *ABCDA* of Fig. 16-2 will have four portions to it, representing an isothermal expansion, an adiabatic expansion, then an isothermal compression, and, finally, an adiabatic compression, bringing the gas back to its original state.

Let T_1 and T_2 be the temperatures in degrees absolute of the two sources of heat which can be maintained at these temperatures (Fig. 16-3). Suppose the piston is near the bottom of the cylinder and the gas is in thermal contact with

the hot source, the boiler. The working substance is therefore at a temperature T_1. In a *PV* diagram (Fig. 16-4) made to correspond to the constants of this engine, the pressure and volume of the working substance at this instant are indicated by the point *A*. The substance now absorbs heat of an amount Q_1 at temperature T_1 as it expands isothermally to *B*. Next we remove the cylinder from the hot source and place it on the insulating

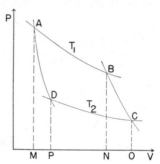

Fig. 16-4.—The Carnot cycle

pad. Now let the substance expand adiabatically until its temperature has dropped to T_2 (*C* in Fig. 16-4). Next we place the cylinder in contact with the condenser at temperature T_2 and compress isothermally to point *D*, removing an amount of heat Q_2. Finally, we again insulate the cylinder and compress adiabatically until it has been heated to the initial point *A*. We have now completed a Carnot cycle.

As we have seen in § 16.3, the area *ABNMA* is the work W_1 done by the gas while absorbing the quantity of heat JQ_1. This work is equal to

$$W_1 = \int_{V_A}^{V_B} P\,dV \,,$$

and for an ideal gas it is given by (§ 13.4)

$$W_1 = RT_1 \log_e \frac{V_B}{V_A}. \tag{16-6}$$

The area *BCONB* represents the work W_1' done by the gas during the adiabatic expansion. No heat was absorbed. By equation (13-69),

$$W_1' = JC_v(T_2 - T_1). \tag{16-7}$$

The area *CDPOC* represents work W_2 done on the gas while a quantity of heat JQ_2 is ejected to the condenser. For an ideal gas,

$$W_2 = RT_2 \log_e \frac{V_D}{V_C} = -RT_2 \log_e \frac{V_C}{V_D}. \tag{16-8}$$

Finally, area $DAMP$ represents work W_2' done on the gas during the adiabatic compression. No heat is given up. The work is given by

$$W_2' = -JC_v(T_2 - T_1). \qquad (16\text{-}9)$$

The total work W done around the cycle is

$$\left. \begin{array}{l} W = W_1 + W_2 + W_1' + W_2' \\[1mm] \quad = RT_1 \log_e \dfrac{V_B}{V_A} - RT_2 \log_e \dfrac{V_C}{V_D}. \end{array} \right\} \quad (16\text{-}10)$$

By the first law of thermodynamics, this net work done by the gas is

$$W = J(Q_1 - Q_2).$$

Also the heat energy given up by the boiler is $Q_1 J$, which is

$$Q_1 J = RT_1 \log_e \frac{V_B}{V_A}.$$

Since efficiency is, by definition,

$$E = \frac{\text{Useful work output}}{\text{Total heat energy input}},$$

we have, for an ideal gas,

$$E = \frac{Q_1 - Q_2}{Q_1} =$$

$$\frac{RT_1 \log_e(V_B/V_A) - RT_2 \log_e(V_C/V_D)}{RT_1 \log_e(V_B/V_A)}. \qquad (16\text{-}11)$$

The student can show, using $P_1 V_1 = P_2 V_2$ for isothermal processes and $P_1 V_1^\gamma = P_2 V_2^\gamma$ for the adiabatic process, that

$$\frac{V_B}{V_A} = \frac{V_C}{V_D},$$

so that, for the ideal gas,

$$\frac{Q_1 - Q_2}{Q_1} = \frac{T_1 - T_2}{T}, \qquad (16\text{-}12)$$

and

$$E = \frac{T_1 - T_2}{T_1}. \qquad (16\text{-}13)$$

The temperatures T_1 and T_2 are those measured on the gas thermometer scale described in § 11.5. We shall discuss the significance of equations (16-12) and (16-13) in §§ 16.7 and 16.8.

16.6. The Second Law of Thermodynamics and the Efficiency of Engines

As we have stated, the second law of thermodynamics may be expressed in many ways. One useful expression is that of § 16.2, that no self-contained system can cause heat to flow from a cooler to a warmer region indefinitely.

On the basis of this law, one can state that the efficiency of a Carnot engine is independent of the working substance in the engine and is the maximum efficiency that any thermal engine may have in working between the same temperature limits. Let us suppose, for the moment, that we have two Carnot engines such that the work done per cycle $(Q_1 - Q_2)$, when working between the same two temperatures, is the same for the two engines. But let it further be assumed that, owing to different working substances, one engine is more efficient than the other. For example,

$$\frac{Q_1' - Q_2'}{Q_1'} > \frac{Q_1'' - Q_2''}{Q_1''}; \qquad (16\text{-}14)$$

and, since the work per stroke is adjusted to be the same,

$$Q_1' - Q_2' = Q_1'' - Q_2''. \qquad (16\text{-}15)$$

Let the more efficient one drive the less efficient one backward, which (in the ideal case of no friction or other losses) it could just do if equal amounts of work were done by the two engines per cycle. The more efficient engine takes its heat (Q_1') from the heat source, T_1, and gives heat to the heat reservoir, T_2, while the less efficient engine running backward takes its heat Q_2'' from T_2 and gives heat Q_1'' to T_1.

From equations (16-14) and (16-15) it follows directly that $Q_1'' > Q_1'$ and also, then, that $Q_2'' > Q_2'$. Thus the hot source gains heat $Q_1'' - Q_1'$, and the cool reservoir loses heat $Q_2'' - Q_2'$; or, in other words, heat is transferred from T_2 to T_1. This is contrary to the second law of thermodynamics, and hence our assumption that one of these Carnot engines may be more efficient than the other is contrary to experience. As a result of this type of argument, we may state that the efficiency of a *Carnot* engine depends only upon the temperatures and not at all upon the working substance.

16.7. Kelvin Thermodynamic Temperature Scale

Nearly all the phenomena we have so far discussed on heat have involved the nature of the substance being examined. The efficiency of heat engines, we have just seen, is an exception to this rule and one in which the properties of the material in the apparatus are not involved. This leads us to an important definition of temperature, which is likewise independent of any thermometer substance.

In chapter 11 on temperature, attention was **241**

called to the dependence of the definitions of temperature therein employed upon the material of the thermometer. Lord Kelvin recognized the possibility of defining a temperature scale suggested by equation (16-12). He proposed that

$$\frac{Q_1 - Q_2}{Q_1} = \frac{\theta_1 - \theta_2}{\theta_1} \quad \text{or} \quad \frac{Q_1}{Q_1} - \frac{Q_2}{Q_1} = \frac{\theta_1}{\theta_1} - \frac{\theta_2}{\theta_1}, \quad (16\text{-}16)$$

which may be transformed into

$$\frac{Q_2}{Q_1} = \frac{\theta_2}{\theta_1}, \quad (16\text{-}17)$$

where θ is the temperature on this new scale. This equation states that any two temperatures are in the same ratio as the quantities of heat which are absorbed and ejected, respectively, in a Carnot cycle operating between these two temperatures.

Imagine a Carnot engine working between 0° and 100° C (Fig. 16-5). There is then a certain

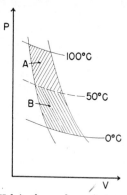

Fig. 16-5.—Kelvin thermodynamic temperature scale

area of useful work. Now we can define 50° C as a temperature such that a Carnot engine working between 0° and 50° C will do the same work, measured by area B (Fig. 16-5), as one working between 50° and 100° C, measured by area A (Fig. 16-5). As we have seen, this engine is independent of the working substance and hence gives a *thermodynamic temperature scale independent of any material.*

The equivalence of Kelvin's thermodynamic definition of temperature, θ (eq. [16-16]), with equation (16-12) derived for an ideal gas indicates that the gas thermometer, in so far as the gas is ideal, gives temperatures on the thermodynamic scale (§§ 11.5 and 13.2). With real gases obeying van der Waals' equation, for example, discrepancies arise between the Kelvin thermodynamic temperature and the measured real gas temperature.

16.8. The Efficiency of Heat Engines in General

By an argument similar to the one given previously for two Carnot engines, it can be shown that *no other type of heat engine* can ever be more efficient than a Carnot engine—in fact, no actual engine *will ever be so efficient*, since we have made no allowance for friction and heat losses. Hence, for actual engines, we have an upper limit of efficiency, given by

$$E < \frac{T_1 - T_2}{T_1} = 1 - \frac{T_2}{T_1}. \quad (16\text{-}18)$$

This conclusion came as a great surprise to the early investigators, who had imagined that in heat engines, operating by the expansive properties of gases or vapors, the physical properties of the working substance would undoubtedly affect the efficiency of the engine. This is seen not to be so—the efficiency of any thermal motor at its best is proportional only to the temperature range over which it operates and inversely proportional to the absolute temperature of the source of its heat supply. In the limiting case this ratio $(T_1 - T_2)/T_1$ could be unity (100 per cent efficiency) for $T_2 = 0°$ K. For the best efficiency, a large temperature range and a low-temperature condenser are indicated. It is obvious that this combination is one most difficult to achieve.

EXAMPLE 1: A simple steam engine takes steam from the boiler at 200° C (approximately 211 lbf/in² gauge pressure) and exhausts directly into the air at 100° C (i.e., 0 gauge pressure or 14 lbf/in² actual pressure). What is the upper limit of its efficiency?

Solution:

$$\text{Efficiency} = \frac{(200 + 273) - (100 + 273)}{200 + 273}$$

$$\times 100 = 21.1 \text{ per cent}.$$

Modern multiple-expansion engines using condensers to lower the exhaust temperature may have an ideal efficiency of 30–35 per cent and may, in very favorable conditions, have an actual efficiency of 0.6 of the ideal, or 18–20 per cent. Efficiencies of 13–15 per cent are far more common.

16.9. The Concept of Entropy

The free expansion of an *ideal gas* (Joule experiment, § 14.14) is a good example of an *irreversible process.* There is no change in temperature of the gas and hence no change in the internal energy of the gas; i.e., the gas possesses the same amount of energy after the expansion as before. Yet, if we wish to bring this gas back to its *initial* state, *we must do work on the gas.* We may say that, depending upon the extent of the free expansion, the gas has undergone a certain amount of *thermodynamic degradation.* By this word, in a technical sense, we mean nothing more than that the gas has lost some of its capacity for doing useful work. Again, if a hot object is brought into contact with a cold object, so that they reach temperature equilibrium, no net loss of thermal energy has occurred. However, before being placed in contact, the hot and cold objects could do useful work by means of a suitable heat engine; after contact, on the other hand, the two objects can do no work, since no temperature difference now exists between them. Here, again, a certain amount of thermodynamic degradation has been involved. As a final example, the diffusion of two gases is an irreversible process involving thermodynamic degradation. As a matter of fact, *in all irreversible processes* thermodynamic degradation is involved.

It is possible to find some quantity a measure of which would be a gauge of the thermodynamic degradation of an irreversible process. This physical quantity is called *entropy.* (The reader should recall how the necessity for introducing the concepts of work and energy arose from the need for a measure of the space effect of a force or of momentum as a measure of the time effect of a force.) The entropy of a substance may be used in descriptions of thermal phenomena in the same manner that we use the temperature or pressure or volume or internal energy and, further, is a measurable property of a substance, as are temperature, etc. Entropy, however, is defined by means of reversible processes. Its application to irreversible processes will be shown in the discussion that follows.

We have seen that, for a Carnot cycle,

$$\frac{Q_1}{T_1} = \frac{Q_2}{T_2}$$

or

$$\frac{Q_1}{T_1} - \frac{Q_2}{T_2} = 0. \qquad (16\text{-}19)$$

We may interpret this equation as follows: The sum of the quantities Q/T is zero for a Carnot cycle between any two adiabatics. Here Q is the

heat absorbed (negative if ejected) at temperature T.

If we now consider any general reversible cycle, such as is shown in Fig. 16-6, we may replace the general reversible cycle by a series of infinitesimal Carnot cycles. Two such Carnot cycles are shown in the figure. If ΔQ is the heat given to a substance at temperature T (ΔQ is negative if it represents

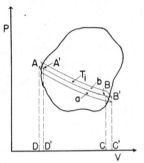

Fig. 16-6.—Representation of a general reversible cycle by elementary Carnot cycles.

heat given up by the working substance), then for each Carnot cycle we may write the following equation:

$$\frac{\Delta Q_1}{T_1} + \frac{\Delta Q_2}{T_2} = 0.$$

And for all the Carnot cycles that approximate the reversible cycle we obtain

$$\sum_{i=1}^{i=n} \frac{\Delta Q_i}{T_i} = 0; \qquad (16\text{-}20)$$

that is, if the algebraic sum of the quotients $\Delta Q/T$ is taken for a substance that undergoes a reversible cycle, this sum is always zero. This important result suggests the introduction of a new thermodynamic quantity for a *reversible process,* the ratio of heat absorbed to the absolute temperature. Thus, if dQ is the heat absorbed at the absolute temperature T, then the entropy, S, is given by

$$S_B - S_A = \int_A^B \frac{dQ}{T} \qquad (16\text{-}21)$$

or

$$dS = \frac{dQ}{T}. \qquad (16\text{-}22)$$

Equation (16-21) states that the difference in entropy of a substance in going from state A to state B along any reversible path is equal to the sum of the quantities dQ/T evaluated along the path. Equation (16-22) states that for a reversible **243** process an infinitesimal change in entropy is equal to the ratio of the infinitesimal heat absorbed to the absolute temperature. Furthermore, we observe that, like the internal energy of a substance, only *changes in entropy* are significant. Entropy has the same dimensions as heat capacity and may be expressed as calories per degree.

It remains to be shown that the entropy concept defined for *reversible processes* is useful as a measure of thermodynamic degradation of an *irreversible process*. For this purpose we return to our experiment of free expansion of an ideal gas. The gas has an initial volume V, pressure P, temperature T, and entropy S_1. Allow the gas to expand freely to a volume $V + \Delta V$. The temperature is still T; but, since the gas is now in a different state, its entropy is different. Call it S_2. As we have emphasized, such an expansion is an irreversible process and involves a certain amount of thermodynamic degradation.

We shall now bring the gas back to its initial state by the *reversible process of very slowly compressing the gas* and maintaining the gas at constant temperature T by removing from it a quantity of heat ΔQ. We observe that this reversible process involves no degradation and, further, that it brings the gas back to its initial state with temperature T, pressure P, volume V, and entropy S_1. Thus, in going from the final state to the initial state by the reversible process of compression, the entropy changed from S_2 to S_1; and, since we *removed* an amount of heat ΔQ at the temperature T from the working substance during the reversible process, by definition, the entropy of the working substance *decreased* by an amount $\Delta Q/T$. It follows, therefore, that the entropy of the gas *after free expansion is greater than that before expansion;* and, furthermore, this increase in entropy ΔS during the irreversible process is known by measuring the entropy change involved in a reversible process that brings the substance back to its initial state. The fundamental reason why this statement is true is that every substance possesses, when in thermal equilibrium, a definite amount of entropy.

Our measure of thermodynamic degradation or the degree of irreversibility of a process is a measure of the increase in entropy of the substance. If T_0 is the lowest temperature attained by our system that undergoes an irreversible process whose entropy change is ΔS, the *amount of unavailable energy* due to the irreversible process is $T_0\Delta S$. This does not mean, of course, that energy is lost, for the first law of thermodynamics is applicable to these processes.

Since all natural processes are irreversible and involve increases in entropy, we may generalize and, following Clausius, state that the *second law of thermodynamics is equivalent to the statement that the entropy of the universe is increasing* and that the *first law is equivalent to the statement that the total energy of the universe is constant.*

A final remark on entropy: As we indicated previously, the state of a substance can be determined from a knowledge of the variables U, S, P, V, and T, only two of which are independent. In the case of a gas, we chose P and V, the temperature being determined by the gas law. In the Carnot cycle of Fig. 16-4, P and V were the variables. Let us now choose the absolute temperature T and the entropy S as the variables. During an adiabatic expansion such as BC, no heat is lost by the working substance, so that the entropy must be a constant during the change. During the isothermal expansion AB, heat was absorbed by the substance at *constant temperature*, so that the entropy increased. Thus in Fig. 16-7 the isothermals

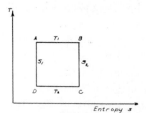

Fig. 16-7.—The Carnot cycle

are the straight lines parallel to the axis of entropy, and the adiabatics are straight lines parallel to the temperature axis. These lines of constant entropy are also called *isentropics*.

EXAMPLE 2: Compute the change in entropy when 100 gm of ice at 0° C are melted and converted into water at 0° C, assuming the latent heat of ice to be 79.63 cal/gm.

Solution: Since the temperature remains constant at $273°.16$ K, we may write equation (16-21) as

$$S_{\text{water}} - S_{\text{ice}} = \frac{1}{T} \int_0^Q dQ = \frac{Q}{T}$$

or

$$S_{\text{water}} - S_{\text{ice}} = \frac{100 \times 79.63}{273.16} = 29.3 \text{ cal/°K}$$

$$= 122.4 \text{ joules/°K}.$$

This represents an increase in entropy of the system.

EXAMPLE 3: One hundred grams of copper, whose specific heat $c = 0.093$ cal/gm degree, are heated from $0°$ to $100°$ C. Compute the change in entropy of the copper.

Solution: In equation (16-21), $dQ = mcdT$, so that

$$S_2 - S_1 = m\,c \int_{T_1}^{T_2} \frac{dT}{T} = m\,c \log_e \frac{T_2}{T_1}$$

or

$$S_2 - S_1 = (100 \text{ gm}) (0.093 \text{ cal/gm degree})$$
$$\times \log_e \tfrac{373}{273} = 2.9 \text{ cal/degree},$$

representing an increase in entropy. Note that temperatures T_1 and T_2 must be in degrees absolute.

16.10. Applications of the Laws of Thermodynamics

One interesting application of the principles discussed in this chapter is the calculation of the depression of the boiling point produced by a given reduction in pressure. We shall obtain our desired result by taking 1 gm of saturated steam around a Carnot cycle. We start with 1 gm of saturated steam at a temperature T, pressure P, and volume V. The state of this sample of "working substance" is represented in the PV diagram of Fig. 16-8 by point a. The first step in the cycle is to

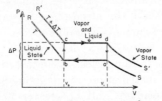

FIG. 16-8.—Carnot cycle for saturated water vapor

compress the vapor to a volume V_2 (point b) and to condense all but an infinitesimal amount. During this compression we shall take L units of heat from the condensing vapor and thus keep the compression isothermal. The amount of work done on the vapor is

$$(V_1 - V_2)P . \qquad (16-23)$$

We continue the compression adiabatically by an infinitesimal amount, thus condensing all the rest of the vapor into water and, since the compression is adiabatic, raising the temperature to $T + \Delta T$ (point c in the figure). The next step is to vaporize all the water except an infinitesimal amount, expanding the volume from V_2 to V_1 and supplying a sufficient amount of heat to maintain the temperature at $T + \Delta T$; i.e., we maintain an isothermal expansion. The working substance does an amount of work equal to

$$(V_1 - V_2)(P + \Delta P) . \qquad (16-24)$$

Finally, we vaporize the remaining liquid by allowing the vapor to expand adiabatically to point a, our original starting point.

The net work done *by the vapor* during the cycle, obtained by subtracting equation (16-23) from equation (16-24), is equal to

$$(V_1 - V_2)\Delta P . \qquad (16-25)$$

Since L units of heat were absorbed, owing to evaporation, the *total heat absorbed by the vapor* at temperature $T + \Delta T$ is

$$L + (V_1 - V_2)\Delta P . \qquad (16-26)$$

In this equation L must be measured in ergs per gram. The heat rejected by the vapor during the cycle is L units at temperature T. Hence, by equation (16-17), placing Q_2 equal to $L + (V_1 - V_2)\Delta P$, T_2 equal to $T + \Delta T$, Q_1 equal to L, and T_1 equal to T, we obtain

$$\frac{L + (V_1 - V_2)\Delta P}{T + \Delta T} = \frac{L}{T} \qquad (16-27)$$

or

$$\Delta T = \frac{\Delta P (V_2 - V_1) T}{L} . \qquad (16-28)$$

Equation (16-28) is known as "Clausius-Clapeyron's equation." Applying this equation to the depression of the boiling point of pure water by a diminution of pressure of 1 mm of mercury, we obtain

$$\Delta P = -1 \text{ mm of mercury}$$
$$= -1.34 \times 10^3 \text{ dynes/cm}^2$$

$V_2 = 1,674$ cc, volume of 1 gm of vapor at $100°$ C

$V_1 = 1.04$ cc, volume of 1 gm of water at $100°$ C,

$T = 373°$ K,

$L = 537 \times 4.2 \times 10^7$ ergs/gm .

Equation (16-28) yields

$$\Delta T = -0.036° \text{ C},$$

in good agreement with the experimental value of $0°.037$ C. Our calculations show that if water is boiling at $100°$ C under a pressure of 760 mm, the boiling point will be $99°.964$ C if the pressure drops to 759 mm.

16.11. Thermodynamics of the Atmosphere

The atmosphere may be considered to be a huge thermodynamic engine, with the heat energy that drives this engine derived from the sun's radiation, which is absorbed in the atmosphere and at the ground. As a very rough approximation, we may consider the region about the equator as a primary **245** heat source and that about the North Pole (for the Northern Hemisphere) as a cold source. The difference between the heat absorbed at the heat source and that given up at the cold source is converted into work. It appears as the kinetic energy of a wind system which on a nonrotating earth blows to the north aloft and to the south near the ground. Unless brakes of some kind are applied, the winds must constantly increase in speed. The brakes supplied are due to friction between the winds and the ground or sea. The energy of the winds is thus converted back into heat, which is finally re-radiated back into space as thermal radiation. We have here another example of the degradation-of-energy principle.

Actually, atmospheric processes are far more complicated than we have outlined. Heat sources are scattered over a wider geographical area; the distribution of land and the oceans is an important factor that affects the distribution of heat sources. Finally, the rotation of the earth profoundly modifies the wind fields described. Although the heat-engine concept is still valid, it does not provide us with the details we wish to know about our atmosphere. It will be of interest to see what happens to the energy received from the sun.

16.12. Heat Transfer in the Atmosphere

An area exposed at right angles to the sun's rays just outside the earth's atmosphere receives about 1.94 cal/cm²-min of radiant energy. This radiation from the sun is approximately that of a black body with an effective temperature of 6,000° K. Some of this radiation is reflected back into space from the upper surface of clouds, and some is lost through scattering by dust and air molecules. Roughly 39 per cent is lost through these processes. It is the back-scattered radiation that determines the earth's whiteness or albedo as viewed from outer space. The earth's albedo as a planet is 0.39. For comparison the albedo of Mars is about 0.15 and that of Venus 0.6.

The remaining 61 per cent of the radiation passes through the atmosphere without appreciable absorption and is finally absorbed by the surface of the earth without much loss due to reflection. (A snow surface is an exception.) Since the mean temperature of the earth does not change much, the earth must re-radiate the absorbed radiation. For this purpose the earth may be considered as a black body at a temperature of about 300° K. The radiant energy, therefore, is largely in the long-wave-length or infrared region (see § 34.6). This radiation is readily absorbed principally by the water vapor in the lower parts of the atmosphere and a small part by the ozone present in the upper layers. The atmosphere, in turn, radiates infrared upward to outer space and downward to the ground. If radiative processes alone controlled the temperature of the atmosphere, it would have to emit as much as it absorbed; but, since it emits in two directions but absorbs radiation only from the ground, it follows that the mean temperature of the atmosphere must be lower than that of the ground. Also, since the ground receives not only solar radiation but radiation in the infrared from the atmosphere, the surface temperature of the earth must be higher than might be expected from solar radiation alone. The atmosphere, therefore, provides in essence a "greenhouse effect," in which the temperature inside a glass greenhouse is higher than outside because of the differential absorption of glass for the visible and infrared radiations.

It can be shown that the temperature decrease with height near the ground for an atmosphere controlled by radiative processes alone would be so great as to make the atmosphere unstable. As a result, violent currents would be set up, tending to equalize the temperature discrepancies. It is believed that radiation processes control the vertical temperature distribution above about 8 km, i.e., in the stratosphere and at greater heights. In these regions the distribution of such gases as ozone, water vapor, and carbon dioxide that strongly absorb infrared radiations have a marked influence on the temperature. In the lower atmosphere, the troposphere, radiation plays a lesser role, in that the day-to-day temperature variations are caused by the motion of the air, particularly vertical motions, wherein parcels of air are cooled and heated by adiabatic processes. Let us see what the vertical temperature distribution actually is.

16.13. Vertical Distribution of Temperature in the Atmosphere

Direct measurements of atmospheric temperatures have been made with balloons carrying instruments to heights of 42 km (26 mi). By means of rockets, temperatures up to 160 km have been recorded. Estimates of temperatures to 500 km have been attempted, based on indirect evidence. The best available data (1955) on atmospheric temperatures are presented in Fig. 16-9. The measurements of temperatures to 30 km are now made on a daily basis by the U.S. Weather Bureau. Data above this height are still meager

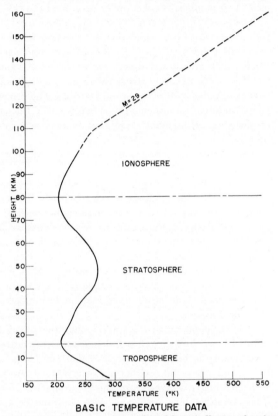

FIG. 16-9.—Temperature obtained from balloon and rocket flights at White Sands, New Mexico. The dotted curve above 100 km represents the temperature deduced from pressure measurements based on the assumption that the molecular weight of air is 29 gm.

and are dependent on infrequent rocket flights. A study of this curve reveals several interesting and rather important points. From the ground to about 10–15 km, the temperature decreases at approximately 6° C per kilometer. This region is called the *troposphere*, and it is here that most of the weather that affects us occurs. From a minimum temperature of −70° C at 15 km the temperature gradually rises, reaching a maximum temperature of about +10° C at 50 km. This high value is due to the absorption of ultraviolet light by ozone. The maximum concentration of ozone,

however, occurs at 25 km, but because of the large absorption coefficient, a large fraction of the radiation is absorbed at higher altitudes, hence the maximum heating. From this maximum the temperature then drops to a minimum of −80° C at 80 km and then sharply rises to values in excess of several hundred degrees. From studies of aurorae and from analysis of the light emitted by the atmosphere during the night, it has been estimated that the temperature at 600 km is about 500° C and at 800 km it rises to about 1,000° C.

The temperatures between 30 and 125 km were deduced from pressure measurements made from a rocket in flight. The pressure measurements taken from rockets are shown in Fig. 16-10. The slope

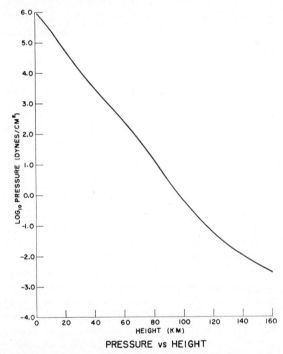

FIG. 16-10.—Pressure measurement made from rocket flight above White Sands, New Mexico. The logarithm of the atmospheric pressure is plotted against height in kilometers. If the atmosphere is isothermal, the curve should be a straight line.

of this curve is proportional to the absolute temperature. At 82 km a pressure of 0.007 mm of Hg was found; at 60 km a pressure of 0.17 mm.

16.14. Formation of a Thunderstorm Cloud

With this preliminary discussion of the temperature distribution, we are prepared to explain the formation of a typical thunderstorm cloud, *cumulus nimbus*. Let us **247** assume that at 7:00 A.M. on a day in July a sounding balloon is sent aloft in Chicago and that the temperature-height curve of the air above Chicago has been drawn from the collected data. The unbroken line of Fig. 16-11 (*a*) represents this temperature distribution. As is characteristic of such early-morning soundings in summer, a marked temperature inversion is present, *AB* of the curve; i.e., the air temperature near the ground *increases* with height. The dot-dash curve, *BKJ*, represents the moist adiabatic curve, i.e., the curve giving the rate at which moist air will cool if it rises and expands adiabatically.

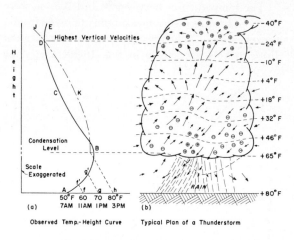

FIG. 16-11.—Formation of a thunderstorm: (*a*) typical temperature-height curve; (*b*) vertical plan of a mature thunderstorm.

As we see from the curve at 7:00 A.M., the air temperature near the ground is 50° F. At 11:00 A.M. the air has been heated to 60° F, owing to the sun's rays heating the ground. This shallow mass of air near the ground becomes somewhat unstable and rises, cooling, according to the dotted curve *ff'*, until it is as cool as the surrounding air at this height and so ceases to rise. Finally, at 3:00 P.M. the air somewhat below *B* has reached 80° F and, rising along *hB*, cools at a rate of the dry adiabatic of 5°.5 F per 1,000 ft, so that the dew point is reached at *B*. The vapor condenses on the hydroscopic nuclei present and forms the base of the cumulus cloud. This height indicated by the horizontal dotted line through *B* is called the *condensation level*.

As the water vapor condenses, it gives up its heat of condensation (some 539 cal/gm) to the rising air, and this liberated heat immediately raises the temperature of the rising air above that of the surrounding air. From point *B* the air is thus accelerated upward by the buoyant force of the more dense air of the environment. The rapidly rising moist air cools along the curve *BKJ*, so that between *B* and *D* the rising air is warmer than the surrounding air, as given by the heavy line *BCD*.

At *D* the temperature of the rising air becomes equal to that of the surrounding air, making their densities equal and reducing the accelerating force to zero. The air, however, has attained its greatest upward velocity at *D* and so, because of its momentum, continues to rise beyond *D*, cooling at the rate indicated by *KDJ*. Now the density of the rising air is much *greater* than the surrounding air and so forces acting downward come into being and bring the air to rest at some point such as *J*. This point marks the top, or mantle, of the cloud.

The top of this cumulus cloud is now sufficiently high that ice crystals form, which then act as nuclei for more vapor to condense. The crystals fall into the warm levels, melt into liquid drops, on which more vapor condenses, the resulting heavy raindrops finally falling to the ground in spite of the upward current. The mantle of the cumulus nimbus cloud is an indication of the ice-crystal stage. If no mantle is present, there is, as a rule, no rain.

The vertical plan of a mature summer thunderstorm cloud is shown in Fig. 16-11 (*b*). The direction of the winds near the ground and the distribution of vertical currents within the cloud are shown by arrows. The beginning of the heavy rain and the initial appearance of a downdraft are nearly simultaneous. The downdrafts appear to originate near the freezing level and increase somewhat in magnitude as the bottom of the cloud is reached. The updrafts, as we have noted, continue to increase as the cloud matures and may reach vertical velocities as high as 60–80 ft/sec near the level marked *D* in the diagram.

The temperature distribution within the cloud is shown by the dotted curves. A departure of the dotted curve from the horizontal indicates that the temperature within the cloud is either higher or lower than the corresponding temperature of the environment. Thus lower temperatures are found with downdrafts and higher temperatures with updrafts.

The probable distribution of negative and positive charges within the cloud structure is also shown. There is evidence for the existence of three regions of charge concentration—a large positive charge near the top of the cloud, a main center of negative charge more or less distributed near the bottom of the cloud, and a small pocket of positive charge associated with the strong downdraft and heavy rain. There is still some uncertainty as to whether this small positive center imbedded in the general negative charge is typical of thunderstorms.

16.15. Résumé

The following is a brief listing of the definitions, principles, and theories, given in this chapter, which you should know:

The first law of thermodynamics
The various statements of the second law of thermodynamics
Reversible and irreversible processes

The theory and efficiency of a Carnot cycle
The Kelvin thermodynamic temperature scale
The concept of entropy
Heat relationships in the atmosphere

Queries

1. What examples can you give of irreversible processes in nature?
2. Why was it necessary to emphasize that reversible processes must take place very slowly?
3. Is it possible to operate a steam engine continuously in the tropics, using the temperature difference between the surface of the ocean and a point some distance below?
4. How could the Carnot cycle be used for refrigeration?
5. Is it possible to cool a closed room by placing an electric refrigerator in the room, opening the door of the refrigerator, and allowing a fan to blow the cool air from the coils into the room?
6. Why should the moist adiabatic lapse rate be less than the dry adiabatic lapse rate?
7. How would you explain the formation of hail in a thunderstorm?
8. Why do aviators avoid thunderclouds? Is it on account of lightning risks?
9. Are the units of equation (16-28) correct? Explain.

Practice Problems

1. Calculate the maximum theoretical efficiency of a steam engine which is supplied steam at a temperature of 450° C and exhausted at a temperature of 210° C. Ans.: 33.2 per cent.
2. A heat engine is 39 per cent efficient. If 10 lb of coal are burned per hour, compute the horsepower developed. (One pound of coal gives rise to approximately 3.6×10^6 cal.)
3. Compute the theoretical efficiency of a heat engine working between 0° and 150° C. Ans.: 35.4 per cent.
4. In a heat transfer, 1,000 cal of heat are taken from a heat source whose constant temperature is 110° C and given to a cold source of constant temperature 0° C. Calculate the change in entropy of the system.
5. Calculate the over-all efficiency of a plant that develops 1.2 kwh for each kilogram of coal burned. (Use the data of problem 2.) Ans.: 13 per cent.
6. A gram of water at 0° C occupies very closely a volume of 1 cm³, while 1 gm of ice at 0° C occupies a volume of 1.087 cm³. Using equation (16-28), compute the change in the melting point of ice if subjected to an increase in pressure of 1 atm (1.013×10^6 dynes/cm²). Ans.: -0.007 C.
7. Carnot's engine (§ 16.5) is an ideal heat engine which loses no energy by conduction of heat and no energy by friction. It carries a constant amount of the working substance through a Carnot cycle. This means that the working substance in the cylinder takes in energy as heat by an isothermal expansion at a high temperature, gives out part of that energy as heat while it suffers an isothermal compression at a lower temperature, and changes its temperature by adiabatic processes, during which work is done upon the material, or by it, but no exchange of heat takes place. Each of the four steps of the cycle is reversible and could be performed in the opposite direction, so as to restore the reservoir and the working substance to their original states, by use of just as much mechanical work as was developed during that step in the original direction. Accordingly, the cycle is reversible *as a whole.* Suppose that such an engine takes 12 units of energy as heat from the hot reservoir into the working material during the isothermal expansion at the high temperature. The efficiency of the engine is $\frac{1}{3}$. How much work is done, and what is done with the rest of the 12 units of energy, during one reversible cycle?
8. An engine like that of the foregoing problem may be set to work backward so that the working material is carried around the Carnot cycle in the sense opposite to that given. It will then take in a certain amount of energy from a cold reservoir, for the working material will expand isothermally while in contact with the cold reservoir and will restore a greater amount of heat to a hotter reservoir, for the material will be compressed isothermally while in contact with the hot reservoir. The term "efficiency," as applied to this engine, has the same meaning as in problem 7. If a second engine has an efficiency of $\frac{1}{4}$ and 4 units of energy are supplied as mechanical work to drive the engine around the cycle backward, what will happen to the two reservoirs during the cycle? How much energy will each reservoir gain or lose?

9. If the 4 units of work necessary for the backward operation of the second engine are supplied during one cycle of operation by the first engine and if both engines work between the same two reservoirs, what is the net result upon the working substance and upon the two reservoirs of the combined operation of the two engines? What is the net gain or loss in mechanical work?

10. Does the operation of the first engine violate the law of conservation of energy? Does the operation of the second engine (through the cycle backward) violate it? Does the combined operation violate that law? Is the net result of the two operations, as described, possible? Is it probable? What practical advantage could be secured if the two ideal engines, working between the same two reservoirs—one forward and one backward—could have different efficiencies? What may be said in general about the efficiency of an engine working between reservoirs at given temperatures?

11. The Carnot engine described in problem 7 develops 4 units of mechanical work when, in the course of one cycle, it removes 12 units of energy as heat from a hot reservoir and restores 8 units to a colder reservoir. A second similar engine is driven backward, and during one of its cycles it restores 12 units of heat to the same hot reservoir. Its efficiency is $\frac{1}{4}$. How much mechanical work does it require? How much heat does it remove from the colder reservoir? What is the net result of the combined operation of the two? What is the credit balance of mechanical work? Is this complete result contrary to the law of conservation of energy? Is it possible? What is the specific characteristic of these two engines which experience forbids?

12. The Carnot engine described in problem 11 develops 4 units of mechanical work when going through one cycle, its efficiency being $\frac{1}{3}$. A second engine is worked backward between the same two reservoirs, and during one cycle it removes 8 units of energy from the colder reservoir. Its efficiency is $\frac{1}{4}$. How much mechanical work is necessary to operate it? How much heat energy does it restore to the hotter reservoir? What is the net result of the combined operation through one cycle of each of these two engines between the same two reservoirs? Is this result in contradiction to the law of conservation of energy? Is it possible? Is it probable? These two engines have what specific characteristic which experience forbids?

13. Compute the change in entropy when 400 gm of mercury at $-38°.87$ C are melted and converted into liquid at $-38°.87$ C. Ans.: 19.86 joules/° K.

14. Five hundred grams of aluminum are heated from 70° to 300° C. Compute the change in entropy of the aluminum.

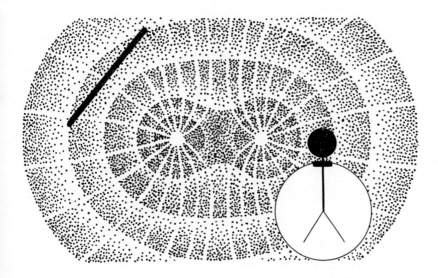

CHAPTER 17

Electrostatics

17.1. Early History of Electricity and Magnetism

The story of the development of the science of electricity and magnetism is practically the story of the development of modern civilization for the last century and a half. It has been a remarkable story of the union of experiment, theory, and application. Whereas man's knowledge and application of mechanics had their beginnings in prehistoric time, the history of electricity and magnetism is distinctly modern. Except for the observations ascribed to Thales of Miletus (sixth and seventh centuries B.C.), the use of natural magnetic iron ore and its semimagical properties by the Phrygians, a matter that was legendary to Plato in the fourth century B.C., and the possibility that Chinese navigators and explorers may have used suspended rocks containing magnetic ore as compass needles by the second century A.D., systematic studies of magnetism were first recorded by one Peter Peregrine in the thirteenth century.

During the reign of Elizabeth I of England (1533–1603) her court physician, William Gilbert, of Colchester, published a systematic work and critical treatise on magnetism, *De magnete*. This work, remarkable for its scientific spirit in an era prior to the time of Galileo, is indeed quite competent.

It is probable that some knowledge of the compass needle filtered into western Europe before the fifteenth century and was a factor in the rapid development of world-wide exploration, which started in the latter part of that era. In this case, of course, the tendency of a compass needle to lie north and south could be put to immediate use without any need of an explanatory theory. However, the behavior of a compass needle could be explained on the assumption that the earth was itself a huge magnet. It was this assumption that was made by Gilbert. As to why the earth is a magnet, however, we still are in doubt.

The observation that amber rubbed with fur acquires the ability to attract light objects to itself, likewise ascribed to Thales of Miletus, has been known since the days of the early Greeks. The force of this attraction is so small, however, that no practical application for it was suggested; consequently, it commanded no interest. Furthermore, the early natural philosophers were not given to increasing their knowledge of any phenomenon by experimentation. Bacon's *Novum organum*, which appeared in 1620, was the first publication to recognize a change in this earlier attitude. Indeed, it was undoubtedly one of the causes of that change. A school of keen observers and experimenters, headed by the same William Gilbert previously mentioned, had come into existence. About 1570, along with the discoveries in magnetism, it was found that many substances besides magnets and amber could be caused to attract objects. It was nearly seventy years later, however, before Von Guericke of Magdeburg (famous for his spectacular experiments on air pressure) discovered that there was sometimes a force of *repulsion*, as well as one of attraction, between electrified bodies.

Stephen Gray, about 1729, discovered electrical conductivity and thus explained the failure of Gilbert to entrap electricity upon certain materials like the metals. In 1747 Benjamin Franklin introduced the words "positive" and "negative" into the science as an improvement in nomenclature for the two kinds of electrification (vitreous and resinous) that had been shown by Dufay to be coexistent. It was also Franklin who suggested that lightning was but another example, on a more grandiose scale, of the phenomenon of electrification. Franklin was able to prove his theory experimentally by means of his famous kite and was more fortunate than his Russian contemporary, Richmann, who was killed performing the same experiment.

17.2. Fundamental Phenomena

It is a fact of common experience that a stick of amber or sealing wax, when rubbed with any dry woolen cloth or with cat's fur, will attract bits of paper, sulphur dust, and other small bodies. We say that the amber or sealing wax has become *electrified* or is in a state of *electrification* or possesses a *"charge"* of electricity. Glass rubbed with silk exhibits a similar effect. If sufficient care is exercised, all bodies can be placed in a state of electrification or receive a charge of electricity.

The effect of rubbing two bodies together can best be studied by means of a pith ball (carved out of raw potato or bamboo heart and dried) suspended by a string. If the pith ball is stroked with an electrified amber rod (thus receiving a portion of the charge of electricity from the amber) and then released, upon bringing the amber near it we find that the pith ball is strongly repelled (Fig. 17-1). If, however, the fur with which the amber was rubbed is brought near, the pith ball is attracted to the fur. By allowing the pith ball to touch the fur and hence receive part of its electrification, we find that the amber rod attracts the pith ball but that the fur repels it.

The experiment we have described suggests the presence of two distinct kinds of electrification—

that on the rubbed amber and that on the fur. We shall say arbitrarily that the amber rod acquires a negative $(-)$ charge and the fur a positive $(+)$ charge. In the case of a glass rod and silk, the glass rod acquires positive charges and the silk negative charges.

We may summarize this experiment by the following statements: (1) When electrification is brought about by rubbing, equal *negative* and *positive* charges of electricity are made apparent. (2) Like charges repel each other; unlike attract.

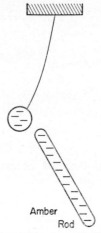

Amber
Rod

FIG. 17-1.—Repulsion between like charges

How is this electrification related to matter? Let us look for the answer on the basis of present-day knowledge of the structure of matter.

17.3. Matter and Electricity

The physicist subdivides the gross structure of matter into molecules and atoms. He has experience relating to a hundred elements or different kinds of atoms, combinations of which yield the great variety of compounds used in our everyday life. Hydrogen, oxygen, copper, silver, lead, and iron are examples of well-known elements.

According to the present universally accepted view, atoms are electrically neutral systems consisting of a central massive nucleus with a net positive charge and a number of light negative charges, called "electrons," which surround the nucleus in a manner that is superficially analogous to planets surrounding the sun. The analogy is *not* too good—atomic mechanics is different from celestial mechanics.

These negative charges, *electrons*, constitute the smallest subdivision of negative electricity that has ever been observed. All electrons have the same mass when at rest and identical charges, regardless of their origin. The mass of the electron is exceedingly small—about 9.107×10^{-31} kg.

The central nucleus contains very nearly all the mass of the atom ($\sim 10^{-27}$ kg). Recent research has shown the nucleus to be a very complex structure, consisting of positively charged *protons* (hydrogen nuclei) and *neutrons which are uncharged* and have about the same mass as the proton. The mass of the proton or neutron is roughly eighteen hundred times that of the electron. Thus hydrogen consists of a proton and one external electron. This pair is held together by their mutual force of attraction. As the atoms become more complex, the nucleus becomes more massive, containing many protons and neutrons, while the number of electrons in the surrounding cloud increases in direct ratio to the nuclear charge, so that the entire atom remains neutral. Chromium, with an atomic weight of 52, can be considered as having a nucleus of 24 protons and 28 neutrons and as having 24 electrons surrounding the nucleus (for further discussion see § 38.37).

With this brief description of the constitution of matter before us, we may view the process of electrification by rubbing as due to the transference of electrons from one material to another *at the surface of contact*. The rubbing serves to bring the two surfaces into intimate contact, thus facilitating the transference of electrons. Certain experiments, which are not convenient to describe here for lack of space, indicate that certain atoms lose electrons quite readily and that others gain electrons with corresponding ease. We may say, then, that a positively charged atom implies a loss of electrons from its external swarm; a negative charge, an excess or acquisition of electrons therein. In neither case is there any electricity *produced*. Electricity of one kind is simply separated from its close association with the other kind. This is known as the *conservation of electricity*. This leads us to assume that electricity is something which remains unchanged, irrespective of the phenomena in which it participates.*

17.4. Conductors and Insulators

If a long brass rod is held by a glass handle and rubbed with a dry cloth, the brass rod acquires

* A possible exception to this statement may be seen in the union of positron and electron (cf. chap. 38).

a charge of electricity. We observe that the charge of electricity instantly spreads over the entire rod; furthermore, that if the rod is touched at any point along its length, either by the hand or by another metal rod held in the hand, the charge (or the state of electrification) on the rod either disappears entirely or becomes negligibly small. If the brass rod is replaced by any other metal, such as copper, silver, etc., the same phenomena will occur.

On the other hand, if the brass rod is replaced by an amber rod and the same experiment performed, we would find, first, that the electrification remains localized in patches along the amber; it does not spead readily over the surface. Second, if the amber is touched by the hand at some point, such as the end of the rod, the electrification or charge of the rod is not appreciably altered.

We have here a means of classifying substances. Those that behave like the brass rod we shall call *conductors*, for they actually do conduct the electricity. Substances like amber, which may retain their charges in patches, we shall call *insulators* or *dielectrics*. It must be emphasized, however, that the difference between conductors and insulators is one of degree only.

The metals are the best conductors; solutions of salts, acids, and bases are next; while the poorest conductors or insulators are the organic oils, waxes, glass, amber, and sulphur (see Table 20-1).

According to the atomic picture presented in the last section, good conductors are substances in which the outermost electrons of the swarm about the atomic nucleus are very loosely bound to this massive core and hence are free to move about, exchanging positions with electrons in near-by swarms in such a way that, *on the average*, each atom remains electrically neutral. The ease with which the conduction electrons may move about differs with various substances, thus accounting for the varying degrees of conductivity. In the case of insulators, there are no free electrons. Each electron is tightly bound to its atomic core. In this chapter we shall be concerned with the two idealized extremes of conductivity—the "perfect conductor" and the "perfect insulator."

17.5. Electroscopes

For the detection of the presence of electric charge and for quantitative measurement of quantity of charge, an *electroscope*, as shown in Fig. 17-2, may be used. The electroscope consists essentially of two strips of gold leaf, b and b, hanging from the end of an insulated metallic rod. The upper end of the rod has a metal disk, d, attached to it. Two large metal disks, a and a, are placed on each side of the gold leaf, the disks being connected by wire to the earth. The reason for this

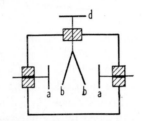

FIG. 17-2.—Gold-leaf electroscope

will be clear when we discuss induction phenomena (§ 17.20). If an amber rod previously rubbed with cat's fur touches the disk d, the leaves fly apart, each having acquired a small portion of the charge and repelling the other thereby. The degree to which they are separated depends upon the charge on the leaves.

In Fig. 17-3 one of the gold strips is removed, and the other is attached to the central rod, as

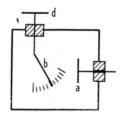

FIG. 17-3.—Simple electrostatic voltmeter

shown. A scale is provided with the instrument to facilitate quantitative measurement.

17.6. The Coulomb Law of Force between Electric Charges

The law of the force of attraction of unlike charges or repulsion of like charges was first brought to the attention of the scientific world by Coulomb[*] in 1785. The apparatus he used was similar in principle to the torsion balance of Cavendish (see § 2.18) and is illustrated schematically in Fig. 17-4. The force of repulsion be-

[*] It appears that Priestley and Cavendish were aware of this inverse-square law some years before Coulomb published his results.

tween two spheres carrying like charges, q_1 and q_2, will twist the suspension, S, so that the distance between the centers of the spheres, r, increases. The distance between centers can be varied by turning the knob A through the proper angle. From a knowledge of the elastic constants of the suspension and the angle of twist, the electrical force of repulsion F between q_1 and q_2 can be measured for any distance of separation, r. Coulomb showed that the *force was inversely propor-*

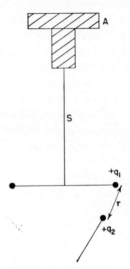

FIG. 17-4.—Schematic drawing of a torsion balance for measuring the force of repulsion between like charges.

tional to the square of the distance r (provided that r is large compared with the radius of the spheres), i.e., $F \propto (1/r^2)$. The force of repulsion between charges of varying size but at a constant separation was found to be proportional to the product of the charges, i.e., $F \propto (q_1q_2)$. Thus the law of interaction of point charges becomes

$$F = \frac{1}{k} \frac{q_1q_2}{r^2}, \qquad (17\text{-}1)$$

where k depends upon the units chosen for F, q, and r and also on the medium between the charges. The force F is a vector quantity, directed along the line joining the point charges. We may interpret F as the force acting on q_1 due to the presence of q_2 and directed away from q_2 if q_1 and q_2 are of like sign and toward q_2 if the charges are of unlike sign. Also, by Newton's Third Law, the force on q_2 due to the presence of q_1 is that given by equation (17-1). This point is illustrated in Fig. 17-5. Equation (17-1), the Coulomb law of force for electric charges, is one of the fundamental equations of electricity and magnetism. Many of the concepts that we shall develop in this chapter are either definitions or logical deductions from equation (17-1).

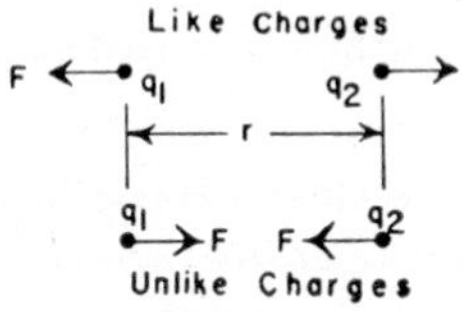

Fig. 17-5.—When q_1 and q_2 are of like signs, the force F is positive and is directed as shown at the top of the figure; when q_1 and q_2 are of unlike signs, the force is negative and directed as shown at the bottom of the figure.

17.7. Units of Charge and the mks System

Equation (17-1) affords us the opportunity of defining an appropriate set of electrical units and of determining a value for the constant k. In principle, at least, we have at our disposal a wide choice of units, depending on the selection we make for F, r, q, or k. If we make $q_1 = q_2$, $F = 1$ dyne, $r = 1$ cm, and $k = 1$ (vacuum), then this defines the unit of charge in the *electrostatic system of units* (esu). In words:

The electrostatic unit of charge, or statcoulomb, is that quantity of electricity which will repel with a force of 1 dyne an equal and similar charge when at a distance of 1 cm from it in vacuum.

Although we placed $k = 1$ for vacuum, its units from equation (17-1) are

$$k = \frac{1 \text{ statcoulomb}^2}{\text{dyne-cm}^2}. \qquad (17\text{-}2)$$

Coulomb's law for this case becomes

$$F = \frac{q_1 q_2}{r^2}. \qquad (17\text{-}3)$$

Because of its simplicity, the electrostatic system of units is still used in the literature to describe electrostatic phenomena. However, for studies of the magnetic effects of charges in motion, i.e., electric currents, this system of units is not useful, and it has been customary to introduce a second set of units, the electromagnetic units (emu). There is still a third set of units, the practical units, widely used in circuit analysis. Many attempts have been made to invent a single uni-

form system of electrical units that would be universally accepted and applied to all electrical and magnetic measurements. The system that is gaining wide acceptance by scientific workers throughout the world and appears destined to become the standard system is the *rationalized mks*—meter, kilogram, second system. This is the system that we shall emphasize in this text.

We have already used the mks system in our studies of mechanical phenomena (§§ 2.5, 4.2), where, in addition to the meter, kilogram, and second, we introduced the newton as a unit of force and the joule as a unit of work. To extend the mks units to electrical phenomena, we must introduce a new unit of charge, the *coulomb*. This unit of charge is not defined in terms of equation (17-1), as was the case for the esu system but rather is defined in terms of unit current, the ampere. *The coulomb* is the quantity of electricity that is carried across a section of a conductor in 1 sec by a constant current of 1 amp* (see § 18.1).

The question of the so-called "rationalization" of units has to do with whether the factor 2π or 4π is used in those equations concerned with phenomena in which there is circular or spherical symmetry. It is not possible to get rid of π in one situation without its appearing in another. About 1890 Oliver Heaviside, an eminent British electrical engineer wrote: "The unnatural suppression of 4π in the formulas of central force, where it has a right to be, drives it into the blood, there to multiply itself, and afterward break out all over the body of electromagnetic theory."

The force between two point charges depends on the separation r and not on the direction, that is, there is spherical symmetry in this instance.

Thus Coulomb's law for two point charges, q_1 and q_2 coulombs, separated by a distance r meters in a vacuum is written in the rationalized system of units as

$$F = \frac{q_1 q_2}{4 \pi \epsilon_0 r^2} (\text{newtons}), \qquad (17\text{-}4)$$

where ϵ_0 is the permittivity of free space. As will appear in the later discussions, the value of ϵ_0 is

$$\epsilon_0 = 8.85 \times 10^{-12} \text{ coulomb}^2/\text{newton}(\text{m}^2). \quad (17\text{-}5)$$

This number assumes a very important role in the theory of electricity and will appear often, so that it is recommended that the student remember it.

* A coulomb of charge is equal to about 3×10^9 electrostatic units of charge.

Using the value of ϵ_0 given in equation (17-5), it follows that

$$4\pi\epsilon_0 = 1.11 \times 10^{-10} \qquad (17\text{-}6)$$

and that Coulomb's law in the rationalized mks system may be written for point charges in a vacuum very approximately as

$$F = \frac{q_1 q_2}{1.11 \times 10^{-10} \times r^2}$$

$$= 9.0 \times 10^9 \frac{q_1 q_2}{r^2} \text{ newtons} . \qquad (17\text{-}7)$$

If the charges are immersed in a uniform medium, such as oil or paraffin, etc., then the force between the charges is smaller than in a vacuum. In order to take into account the effect of the medium, a numerical κ is introduced into Coulomb's equation. This quantity κ is variously known as the "specific inductive capacity," "dielectric constant," or "relative permittivity" of the medium. For any homogeneous medium of dielectric constant κ, Coulomb's law may be written as follows:

$$F = 9.0 \times 10^9 \frac{q_1 q_2}{\kappa r^2}, \qquad (17\text{-}8)$$

in which F is in newtons, q_1 and q_2 are in coulombs, r is in meters, and κ has no units. Values of the dielectric constant are given in Table 17-1 (§ 17.26).

The unit of charge associated with atoms, namely, the charge on the proton, $+e$, or the negative charge on the electron, $-e$, has a numerical value of

$$e = 1.602 \times 10^{-19} \text{ coulomb} .$$

In terms of the electrostatic unit of charge, the value of e is 4.803×10^{-10} statcoulomb or esu. The ratio of these numbers is very approximately 3×10^9 or ten times the velocity of light in meters per second. The significance of this will appear later.

EXAMPLE 1: Compute the force of repulsion between two protons whose distance of closest approach is 10^{-14} m. Assume that the dielectric constant for the medium is unity.

Solution: From equation (17-8) we have

$$F = \frac{9 \times 10^9 \times (1.602)^2 \times 10^{-38}}{(10^{-14})^2} = 2.31 \text{ newtons} .$$

17.8. Electrostatic Field

From our previous discussion we have seen that the region that surrounds a charge was character-

ized by the fact that forces act on another charge brought into the region. This observation permits us to introduce the useful concept of an *electrostatic field* as a region in space wherein a charge—a test charge—experiences an electrostatic force. We shall take as the test charge a positive charge sufficiently small that it does not disturb the existing field.

The intensity of the electrostatic field at a point is defined as the force per unit charge on any positive charge placed at the point.

We shall use the symbol $\mathcal{E}$ to represent the magnitude and direction of the electrostatic field intensity or *electric intensity*. Thus if a small test charge of q coulombs experiences a force of f newtons, then the magnitude of the field $\mathcal{E}$ in the mks units is:

$$\text{Electric intensity } \mathcal{E} \equiv \frac{f}{q} \text{ newtons/coulomb} \quad (17\text{-}9)$$

(in the esu system the units are dynes per statcoulomb).

The electric field $\mathcal{E}$ is a vector quantity having the direction of the force that acts on a *positive charge* placed at the point in question. The existence of an electrostatic field implies the existence of charges. Let us see what kind of field is obtained from an isolated positive charge q_1. At a field point P (Fig. 17-6), distant r_1 from the point charge, the

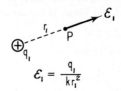

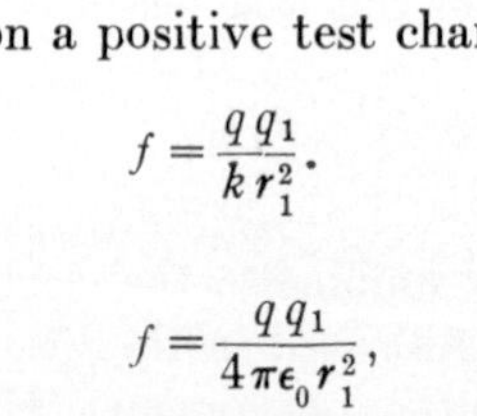

$$\mathcal{E}_1 = \frac{q_1}{\kappa r_1^2}$$

FIG. 17-6.—Electric field due to a single positive charge

force acting on a positive test charge q would be

$$f = \frac{q q_1}{\kappa r_1^2} .$$

In mks units:

$$f = \frac{q q_1}{4\pi\epsilon_0 r_1^2},$$

so that

$$\frac{f}{q} = \mathcal{E}_1 = \frac{q_1}{4\pi\epsilon_0 r_1^2} . \qquad (17\text{-}10)$$

That is, the field intensity varies inversely as the square of the distance r_1. The direction of the field is along the radius vector r_1, pointing away from q_1. If q_1 were negative, $\mathcal{E}_1$ would point toward q_1.

If another charge, q_2, is placed r_2 m from P (Fig. 17-7), then the field due to q_2 at P will be

$$\mathcal{E}_2 = \frac{q_2}{4\pi\epsilon_0 r_2^2}. \qquad (17\text{-}11)$$

The resultant electric intensity or field $\mathcal{E}$ at P we may calculate by adding the separate fields $\mathcal{E}_1$ and $\mathcal{E}_2$ in accordance with the parallelogram law. In general, the resultant intensity due to a dis-

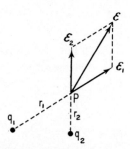

FIG. 17-7.—Electric field due to two point charges

tribution of charges positive and negative is obtained by adding vectorially the fields due to each charge. For numerical computation of field intensity due to charges in a vacuum (or in air), it is convenient to represent the field intensity in the rationalized system of units as

$$\mathcal{E} = 9 \times 10^9 \frac{q}{r^2} \text{ newtons/coulomb}. \quad (17\text{-}12)$$

EXAMPLE 2: A positive charge of 20×10^{-6} coulomb (20 μcoul) is placed 2 m away from a negative charge of 20 microcoulombs, as shown in the diagram. Compute the electric field intensity at P, 2 m from each charge.

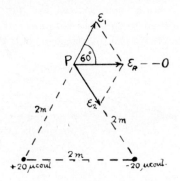

Solution: The electric field at P due to $+20$ microcoulombs is

$$\mathcal{E}_1 = \frac{9 \times 10^9 \times 20 \times 10^{-6}}{4}$$
$$= 4.5 \times 10^4 \text{ newtons/coulomb}$$

and is directed as shown in the diagram. Also the field due to -20 microcoulombs is

$$\mathcal{E}_2 = \frac{9 \times 10^9 \times 20 \times 10^{-6}}{4}$$
$$= 4.5 \times 10^4 \text{ newtons/coulomb},$$

directed as shown. If we resolve each vector representing the field intensity into components along PO and at right angles to PO, it is clear that the components at right angles to PO cancel and those along PO add. The direction of the resulting field $\mathcal{E}_R$ is therefore along PO from P to O, as shown, and the magnitude is given by

$$\mathcal{E}_R = 4.5 \times 10^4 \cos 60° + 4.5 \times 10^4 \cos 60$$
$$= 4.5 \times 10^4 \text{ newtons/coulomb}.$$

EXAMPLE 3: Given the charges in microcoulombs shown in the figure. Compute the electrostatic field at X.

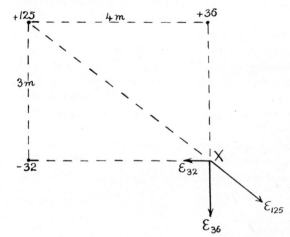

Solution:

$$\mathcal{E}_{36} = \frac{9 \times 10^9 \times 36 \times 10^{-6}}{9}$$
$$= 36 \times 10^3 \text{ newtons/coulomb}$$

$$\mathcal{E}_{125} = \frac{9 \times 10^9 \times 125 \times 10^{-6}}{25}$$
$$= 45 \times 10^3 \text{ newtons/coulomb},$$

$$\mathcal{E}_{32} = \frac{9 \times 10^9 \times 32 \times 10^{-6}}{16}$$
$$= 18 \times 10^3 \text{ newtons/coulomb}.$$

Resolve $\mathcal{E}_{125}$ into horizontal and vertical components; thus the horizontal component directed to the right of X is 36×10^3, and the vertical component directed downward is 27×10^3. The total horizontal component is $36 \times 10^3 - 18 \times 10^3$ to the right, and the total **257** vertical component is $27 \times 10^3 + 36 \times 10^3 = 63 \times 10^3$ downward. Thus

$$\mathcal{E}_R = \sqrt{(18 \times 10^3)^2 + (63 \times 10^3)^2}$$

$$= 65.5 \times 10^3 \text{ newtons/coulomb}$$

$$\tan \theta = \tfrac{7}{2}.$$

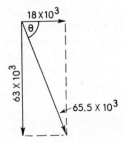

17.9. Lines of Force and Field Intensity

It is clear from our previous analysis that the electrostatic field is a *vector quantity* because associated with each point of the field is a vector whose magnitude represents the force per unit charge on a test charge placed at the point and whose direction is the direction that the positive test charge will move if released with zero velocity. In order to aid in visualizing such a field, we introduce the concept of a "line of force." By definition, a line of force is a curve so drawn that the tangent to the curve at any point will give the direction of the field. In Fig. 17-8 the vector field

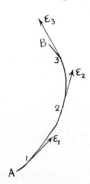

FIG. 17-8.—Electrostatic line of force

intensity at point *1* is given by $\mathcal{E}_1$, at *2* by $\mathcal{E}_2$, and at *3* by $\mathcal{E}_3$. The curve AB is drawn tangent to the vectors $\mathcal{E}_1$, $\mathcal{E}_2$, and $\mathcal{E}_3$ and is the line of force.

As suggested by Fig. 17-6, the electric field from an isolated positive charge decreases radially outward; the lines of force for this case would be

represented by a bundle of straight lines diverging

from the charge. For a negative charge the pattern would be the same, except that the arrows representing the direction of the field would point toward the negative charge, i.e., the lines converge toward the charge. Figure 17-9 represents the lines

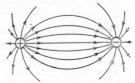

FIG. 17-9.—Lines of force between two equal but unlike charges.

of force for two equal but unlike charges; Fig. 17-10 that of two equal positive charges. The

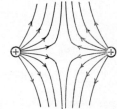

FIG. 17-10.—Lines of force between two equal and like charges.

electric field must have a definite direction and magnitude at each point in space. Therefore, lines of force are continuous and cannot cross one another.

Lines of force start from positive charges and end on negative charges. They cannot start or stop in space, nor can electrostatic lines of force form closed loops.

Lines of force may be given a quantitative aspect by spacing them so that the number per unit area normal to the field is numerically equal to the field. Where the lines are concentrated, the

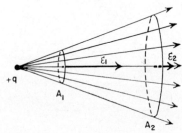

FIG. 17-11.—Lines of force representing field intensity

field is intense. We may illustrate this point by considering the distribution of lines of force from a point charge as shown in Fig. 17-11. Areas A_1 and A_2 are drawn normal to the lines of force, i.e., normal to the field, with A_2 greater than A_1. Since, by construction, the same number of lines of force enter A_1 as enter A_2, the number of lines per unit area at A_1 is greater than at A_2, so that intensity $\mathscr{E}_1$ is greater than $\mathscr{E}_2$, as we should expect. Note that the total number of lines $N = \mathscr{E}_1 A_1 = \mathscr{E}_2 A_2$ and represents the "flux" of lines of force through A_1 or A_2. A collection of lines of force such as that bounded by A_1 and A_2 is also called a "tube of force."

Let us next determine the number, N, of lines of force radiating from q units of positive charge. If we draw a sphere of radius r centered at the charge q, the total number of lines of force or flux crossing the spherical surface will be equal to the product of the surface area by the field intensity; thus

$$N = (4\pi r^2)\left(\frac{q}{4\pi\epsilon_0 r^2}\right) = \frac{q}{\epsilon_0}, \quad (17\text{-}13)$$

that is, the total flux through the spherical surface is directly proportional to the charge q and is independent of the size of the spherical surface. It follows that $1/8.85 \times 10^{-12}$ or 1.13×10^{11} lines of force radiate outward from 1 coulomb of positive charge.

17.10. Gauss's Law for Electric Lines of Force

Although the results of equation (17-13) were deduced for the special case of an isolated charge, it is possible to generalize the results into the following theorem, first stated by Gauss (1777–1855). If charges

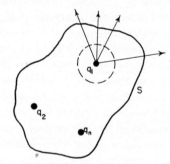

Fig. 17-12.—Gauss's theorem for charges within a closed surface S.

$q_1, q_2, \ldots, q_n$ are completely surrounded by some surface S, as shown in Fig. 17-12, the net flux, N, through the surface is given by

$$N = \frac{1}{\epsilon_0}(q_1 + q_2 + \ldots + q_n)\,\text{mks units}. \quad (17\text{-}14)$$

A mathematical proof of Gauss's law may be found in more advanced texts. The reasonableness of the result can be seen by observing that if we draw a sphere around some charge q_1 within the surface S, as shown in Fig. 17-12, the flux q_1/ϵ_0 cuts the sphere and must also cut the surface S. Lines of force that leave S are counted positive, and those entering are counted negative. If the charge resides outside the closed surface S, as in Fig. 17-13, then by Gauss's law the net flux of lines of

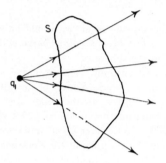

Fig. 17-13.—Gauss's theorem for charges outside a closed surface S.

force through S must be zero. We see from Fig. 17-13 that a line of force must cut the surface twice, and therefore the algebraic sum of the negative, inward flux and the positive, outward flux is zero.

It is important to realize that Gauss's law as given by equation (17-13) or (17-14) is a direct mathematical consequence of the validity of Coulomb's inverse-square law for point charges.

17.11. Applications of Gauss's Law

The intensity of the electric field due to a uniformly charged spherical conductor may be readily calculated by means of Gauss's law. Let the radius of the spherical conductor be a and the charge on it q. As shown in Fig. 17-14, we shall draw a spherical surface, S, of radius r

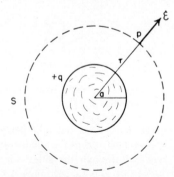

Fig. 17-14.—Electric field intensity outside a uniformly charged spherical conductor in a vacuum.

concentric with the charged sphere and through the point P at which the field intensity is desired. If $\mathcal{E}$ is the field intensity at P, then the total flux through S is

$$N = 4\pi r^2 \mathcal{E},$$

and, by Gauss's theorem, $N = q/\epsilon_0$, so that

$$4\pi r^2 \mathcal{E} = \frac{q}{\epsilon_0}$$

or

$$\mathcal{E} = \frac{q}{4\pi\epsilon_0 r^2} \text{ newtons/coulomb}. \quad (17\text{-}15)$$

This is precisely the result we would have obtained if the charge q had been a point charge concentrated at the center of the sphere.

As a second example we shall derive a relation between the surface charge density of a conductor and the field intensity normal to the conductor. The student will recall from our discussion in § 17.4 that electrons are relatively free to move about in a conductor and if an electric field were present, the electrons would move until they distributed themselves in such a way as to attain some equilibrium condition. In general, the resulting charge distribution on the surface of the conductor is complicated and would depend on the shape of the conductor and on the presence of other conductors in the field. But, whatever the distribution, the resulting field within the conductor must be zero, and the field at the surface must be normal to the conductor. These deductions, then, are simply consequences of the definition of a conductor and the statement of static equilibrium between charges on the conductor.

To compute the field, $\mathcal{E}$, just outside the conductor, we construct a small pillbox surface, B, partly inside the conductor and partly outside, as shown in Fig. 17-15. Let the surface area normal to $\mathcal{E}$ be ΔA and the

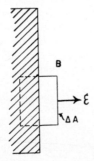

Fig. 17-15.—Use of Gauss's theorem to deduce the field intensity near a conductor.

surface charge density, σ coulombs per square meter. Since there is no field within the conductor and no lines of force leak out through the sides, the total flux N through the surface B is that through ΔA and is given by

$$N = \mathcal{E}\Delta A.$$

The total charge q within the box is $\sigma\Delta A$, so that, by Gauss's law,

$$\mathcal{E}\Delta A = \frac{\sigma}{\epsilon_0}\Delta A$$

or

$$\mathcal{E} = \frac{\sigma}{\epsilon_0} \text{ newtons/coulomb}. \quad (17\text{-}16)$$

EXAMPLE 4: The field intensity near a conducting surface is known to be 5×10^4 newtons/coulomb. Compute the surface charge density.

Solution: From equation (17-16),

$$\sigma = \mathcal{E}\epsilon_0 = 5 \times 10^4 \times 8.85 \times 10^{-12}$$

$$= 0.4425 \text{ microcoulomb/m}^2.$$

17.12. Electrostatic Induction

Let us suppose that an uncharged conductor (copper) is mounted on an insulated stand (glass) and that a charged glass rod is brought near, but not touching, the conductor (Fig. 17-16 [a]). Since

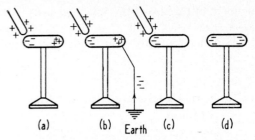

Fig. 17-16.—Charging a conductor by induction

the conductor has free electrons present, these electrons will be attracted toward the end of the conductor nearest the charging rod, leaving the other end with a deficiency of electrons and hence positively charged. The drift of electrons toward the charged glass rod quickly stops, owing to the equilibrium of three forces: an attraction by the charge on the glass rod, a strong repulsion by the excess electrons at one end of the conductor, and a strong attraction by the positive charge left in excess at the other end of the conductor.

If the conductor is now connected to the earth by a wire, as shown in Fig. 17-16 (b) (the symbol $\equiv$ represents the earth as a conductor), electrons flow from the earth to the copper conductor, neutralizing the positive charge. By breaking the ground connection (Fig. 17-16 [c]) these extra electrons are bound on the conductor, leaving it, as a whole, with excess electrons. On removing the glass rod, the extra electrons distribute themselves about the conductor, leaving it negatively charged.

By this sequence of operations we have charged the conductor negatively, using a positive charge. Such a process of producing electrification is known as "charging by induction."

If the uncharged conductor were suspended by a string, we would observe a marked attraction between the conductor and the inducing charge. This effect is strictly in accordance with Coulomb's law, for it can be shown that, although the induced plus charge repels the glass rod, the force due to the induced negative charge is greater because of the inverse-square power. In § 17.20 we shall explain the process of induction from the point of view of electrostatic potential.

The exact laws of electrostatic induction were discovered by Michael Faraday (1791–1867), a brilliant English experimenter, whose classic "ice-pail experiments" we shall describe. Let us suspend a positively charged conductor, B, in a metal pail that almost completely surrounds the charge (Fig. 17-17). The pail is connected to a gold-leaf

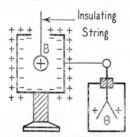

Fig. 17-17.—Faraday ice-pail experiments

electroscope. We observe that the leaves diverge to some angle θ and remain at this angle regardless of the position of the ball within the pail. The leaves are found to be positively charged, as shown in the figure. On withdrawing the ball, the leaves collapse.

If the ball is again introduced, the leaves attain the same separation, θ. The ball is now allowed to touch the side of the pail. *We observe no change in the deflection.* Also, the ball is found to be completely discharged.

We conclude that the magnitude of the induced negative charge on the inside of the pail was exactly equal to the inducing charge.

As a further proof of the equality of the charge, we allow the positively charged ball to enter the pail as in the first experiment and observe the deflection, θ. We now ground the electroscope, keep-

ing the ball inside. The leaves immediately collapse, since electrons from the earth neutralize the positive charge on the leaves. On breaking the ground connection and removing the charged ball, the leaves again diverge *to the same angle,* θ, though the charge on the leaves is now negative.

Thus a charge completely surrounded by a conducting shield induces on the inside of the conductor a charge exactly equal in magnitude to the inducing charge but of opposite sign. On the outside of the metallic shield a charge identical to the inducing charge is present. The charge on the inside surface of the conductor remains only as long as it is bound by the inducing charge; if the inducing charge is removed, no charge resides on the inner surface.

17.13. Electrostatic Machines

In the process of charging by induction we noted that a charge may be placed on a conductor without diminishing the inducing charge. We may store the induced charge in some convenient device (capacitor) and, by repeating the induction process, again obtain an induced charge. This may be carried on indefinitely, obtaining a very large charge from a small initial inducing charge.

In the ELECTROPHORUS (Fig. 17-18) a slab of resin, B, mounted on a metallic plate, D, is charged

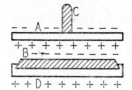

Fig. 17-18.—The electrophorus

with negative electricity by rubbing it with a dry woolen cloth. A metal disk, A, held by an insulating handle, C, is brought very near the resin. A positive charge is induced on the lower portion of the disk, A, while a negative charge resides on the upper side. The operator touches the disk, allowing the electrons to escape to earth and leaving the disk with a net positive charge. The disk is removed (work is done by removing the disk), and the charge is stored in a capacitor. The process may be repeated at will, thus storing up a positive charge with no loss of any charge from the resin slab.

In the older forms of electrostatic machines, **261**

Wimshurst's and Toepler-Holtz's, the foregoing routine is accomplished mechanically by turning a handle and allowing two charged disks to rotate in opposite directions. The resulting induced charges are removed by conducting brushes to capacitors for storage.

The modern form of electrostatic machine is the Van de Graaff generator, shown diagrammatically in Fig. 17-19. Here, again, the basic principle is

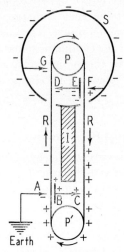

Fig. 17-19.—Van de Graaff generator

essentially that of a continuous induction process. The operation of the machine is as follows: An insulating silk belt, R, passes continuously over two pulleys, P and P', pulley P being inside a hollow conducting sphere, S, mounted on an insulating stand, I. The letters A, C, D, F, and G represent sharp points, while E and B are flat metal disks. Let us assume that the conducting system BC (disk plus the sharp point) has initially a small positive charge. The positive charge induces a negative charge on the conductor, A, which is connected to the earth. Because of the sharp point, the negative charge leaves A and is literally "sprayed" on the belt.* This net negative charge is carried to the upper collecting rod, G, connected to the sphere, giving the sphere a negative charge. The point D also collects a negative charge, so that the disk E, to which it is connected, induces a positive charge on F. The point F discharges positive electricity to the belt, which is carried down and collected by C, reinforcing the positive charge on B. This completes the cycle, which results in a steady transference of electrons from the ground

262   * See § 17.18.

to the outside of the sphere. In a more recent design the entire mechanism of Fig. 17-19 is inclosed in a high-pressure chamber, which has the effect of increasing the discharge potential, thus preventing loss of charge. From this compact electrostatic machine, voltages in excess of 2,000,000 volts are readily attained. Such high voltages are being used to accelerate charged particles for use in research in nuclear physics.

17.14. Electrostatic Potential

In our studies of electrostatics thus far, we have introduced the idea of an electrical charge with its associated electric field and showed how this field could be described in terms of the vector quantity, electric field intensity, $\mathscr{E}$. Another important concept in electricity widely used is that of *potential* and *difference in potential* in an electric field. This concept is concerned with the problem of determining the amount of work done in moving unit charges about in an electric field. As we shall see, we may also describe an electric field in terms of potential, with the added advantage that potential is a scalar quantity.

We shall start by computing the work done in bringing a positive test charge q from a very great distance (infinity) to a distance r from an isolated positive charge q_1. We know from our previous analysis that a force $\mathscr{E}q$ will tend to push the test charge away from q_1. Therefore, some external agency must provide a force equal and opposite to $\mathscr{E}q$ and do work in pushing the test charge toward q_1. The work done in pushing the test charge a distance Δr toward q_1 is, therefore,

$$\Delta W = -\mathscr{E}q\Delta r \,,$$

where $\mathscr{E}$ is the electric field intensity at Δr, a distance r from q_1. From equation (17-10) this field is given by

$$\mathscr{E} = \frac{q_1}{4\pi\epsilon_0 r^2} \text{ newtons/coulomb} \,,$$

so that

$$\Delta W = \frac{-q_1 q \Delta r}{4\pi\epsilon_0 r^2} \text{ joules} \,, \qquad (17\text{-}17)$$

and therefore the total work is

$$\left. \begin{aligned} W &= -\int_{\infty}^{r} \frac{q_1 q}{4\pi\epsilon_0 r^2} \, dr \text{ joules} \\[2mm] &= \frac{q_1 q}{4\pi\epsilon_0 r} \text{ joules} \,. \end{aligned} \right\} \qquad (17\text{-}18)$$

The work W was supplied by the external agency and represents the potential energy of the test charge, q coulombs, at a distance r meters in the electrostatic field of q_1. The useful quantity for our purposes is *potential energy per unit charge*, or *potential*, and is given by

$$\frac{W\ (\text{joules})}{q\ (\text{coulombs})} \equiv V = \frac{q_1}{4\pi\epsilon_0 r} \text{ volts } (17\text{-}19)$$

or

$$V = 9\times 10^9 \frac{q_1}{r} \text{ volts}.$$

By definition,

The potential at a point in any electrostatic field is the work per unit charge necessary to bring any positive charge from infinity to the point in question. Potential is a scalar quantity.

In the mks system *the unit of potential, joule per coulomb, is called the "volt."*

We are now prepared to compute the *difference in potential* between two points in an electrostatic field. In particular, we shall compute the difference

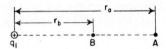

FIG. 17-20.—Difference in potential between two points in an electrostatic field.

in potential between the two points A and B of Fig. 17-20. The potential at A is

$$V_A = \frac{q_1}{4\pi\epsilon_0 r_A} \text{ volts},$$

which represents the work per unit charge in bringing the test charge from infinity to r_A, and, at B,

$$V_B = \frac{q_1}{4\pi\epsilon_0 r_B} \text{ volts},$$

which is the work per unit charge in bringing the test charge from infinity to r_B, so that the difference in potential $V_B - V_A$ is given by

$$V_B - V_A = \frac{q_1}{4\pi\epsilon_0}\left\{\frac{1}{r_B} - \frac{1}{r_A}\right\} \text{ volts},$$

which is the work per unit charge in bringing the test charge from A to B.

In general, a unit difference of potential, called the *volt*, exists between two points if it requires 1 joule of work per coulomb charge to carry any positive test charge from the point of low potential to that of higher potential.

Let us next consider the application of potential to the special case of a *uniform field*. By "uniform field" we mean one in which the force on the unit charge is constant in magnitude and direction for every point in the field; i.e., $\mathcal{E}$ treated as a vector is constant. We may obtain a uniform field by charging two large parallel plates with equal but opposite kinds of charge, as shown in Fig. 17-21.

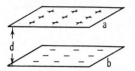

FIG. 17-21.—Potential difference in a uniform field

The field $\mathcal{E}$ is uniform except near the edges, which we avoid.

A positive charge, q, lifted parallel to $\mathcal{E}$ from the lower plate, b, to the upper plate, a, a distance d has work done *on* it, by some outside agent, of an amount

$$W = -q\mathcal{E}d, \qquad (17\text{-}20)$$

since the *force on* q *at each point of its path is* CONSTANT. By definition, the difference of potential between the two points a and b, which we designate as $V_a - V_b$, is the work per unit charge:

$$\frac{W}{q} \equiv V_a - V_b = -\mathcal{E}d. \qquad (17\text{-}21)$$

If the charge moves from a to b, and equal amount of work is done *on* the charge *by the field*.

We may rewrite equation (17-21) thus:

$$\mathcal{E} = -\left(\frac{V_a - V_b}{d}\right) \text{ volts/m}. \quad (17\text{-}22)$$

The quantity $-([V_a - V_b]/d)$ is called the "potential gradient of the field" and is a measure of the electric field intensity.

Instead of moving the positive q charge along ab, let us move it along the straight line ca, making an angle θ with the vertical (Fig. 17-22). The

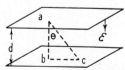

FIG. 17-22.—Work done in an electrostatic field is independent of the path.

force along ca is $q\mathcal{E} \cos \theta$, and the work done is $q\mathcal{E}\ (ca) \cos \theta$. The quantity $ca \cos \theta$ is equal to d, and hence the work done is $q\mathcal{E}d$. The path ca is **263** equivalent to the path cb at right angles to the field and the path ba parallel to the field. No work is done in moving a charge along cb. Since any arbitrary path between a and c may be broken up into paths parallel and perpendicular to $\mathcal{E}$, *the work done in moving a charge from* c *to* a *is independent of the path connecting* c *and* a. This result may be generalized to state that the work done in moving a charge from one point to another in any electrostatic field is independent of the path. Thus the electrostatic field, like the gravitational field, is a conservative force field.

17.15. Electrostatic Potential of Several Charges

Potential, like work, whence it is derived, is a scalar quantity. It is determined by its magnitude alone. To obtain the potential at a point due to several point charges, we must add *algebraically* the potentials due to each charge. Thus, if charges $q_1, q_2, \ldots, q_n$ are present at distances $r_1, r_2, \ldots, r_n$ from a given point P in the field, then, for each charge, equation (17-19) is applicable, and the total potential at P is

$$V_P = \sum_{i=1}^{n} \frac{q_1}{4\pi\epsilon_0 r_i} = 9 \times 10^9 \sum \frac{q_1}{r_i} \text{ volts}. \quad (17\text{-}23)$$

If the result is positive, we interpret this as meaning that work must be done by some external agency to bring the test charge from infinity to the point in question. If V_P is negative, then the field does the work on the charge.

EXAMPLE 5: Given the charges shown in Example 3, compute the potential at the point x.

Solution: By equation (17-23) we have

$$V_x = \frac{9 \times 10^9 \times 125 \times 10^{-6}}{5}$$

$$+ \frac{9 \times 10^9 \times 36 \times 10^{-6}}{3} - \frac{9 \times 10^9 \times 32 \times 10^{-6}}{4}$$

$$= 261 \times 10^3 \text{ volts}.$$

EXAMPLE 6: For the same distribution of charge, compute the work required to carry $+3$ coulombs of charge from x to the midpoint of the diagonal by any path which does not actually intersect any of the given charges.

Solution: $V_x = 261 \times 10^3$ volts from Example 5. The potential, V, at the midpoint is

$$V = \frac{9 \times 10^9 \times 125 \times 10^{-6}}{2.5}$$

$$+ \frac{9 \times 10^9 \times 36 \times 10^{-6}}{2.5} - \frac{9 \times 10^9 \times 32 \times 10^{-6}}{2.5}$$

$$= 464.4 \times 10^3 \text{ volts};$$

$$W = q\,(V - V_x) = 3\,(464.4 - 261) \times 10^3$$

$$= 610.2 \times 10^3 \text{ joules}.$$

17.16. Equipotential Surfaces

In the case of an isolated point charge q, the potential at a point r m from it is $q/4\pi\epsilon_0 r$. If we were to draw a sphere of radius r about the charge, then every point on the surface of the sphere would have the same potential, $q/4\pi\epsilon_0 r$. We call such a surface over which the potential is constant an "equipotential surface." It follows that no work is required to move charges about an equipotential surface. This implies that the *electric field intensity* $\mathcal{E}$ *must be at right angles to the surface* (Fig. 17-23);

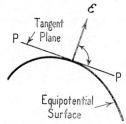

FIG. 17-23.—Electric field intensity is always normal to an equipotential surface.

otherwise, there would be a component parallel to the surface, and work would be done in moving a charge. This, of course, would contradict our definition of an equipotential surface.

What the geometrical shape of an equipotential surface will be depends upon the distribution of the charges that give rise to the field. As we have shown, a single isolated point charge gives rise to spherical equipotential surfaces. In the case of a uniform field between parallel conductors, the equipotential surfaces are planes parallel to the conductors.

The surface of a homogeneous conductor, uncharged or with charges at rest upon it, is *necessarily* an equipotential surface. If this were not so, any charges on it would move from places of high potential to those of lower, and we would not have a *static* distribution of charge. It follows that the electric field due to the surface charges on a conductor is *normal* to the surface of the conductor (Fig. 17-23). In other words, the lines of force leave or end normal to the conductor.

A quantitative picture of the electric field intensity $\mathcal{E}$ surrounding a charged conductor may be obtained by drawing a series of equipotential surfaces each differing from the other by a constant potential, say 5 volts, as shown in Fig. 17-24.

FIG. 17-24.—The field intensity is large where the equipotential surfaces are closely spaced.

In this figure traces of equipotential surfaces are drawn for potentials of 100 volts, 95 volts, etc. The field is most intense in the region where the surfaces are most crowded. We observed from equation (17-22) that, for the case of a uniform field, $\mathcal{E} = -\Delta V/\Delta s$, that is, the field intensity was equal to the negative potential gradient.

It can be shown that this result is of general validity, provided that equation (17-22) is replaced by

$$\mathcal{E} = -\frac{dV}{ds} \text{ volts/m} \qquad (17\text{-}24)$$

and interpreted to mean that the field intensity $\mathcal{E}$ at any point is equal to the *maximum negative potential gradient at that point and, as a vector, points in the direction of that gradient.* In Fig. 17-24 the field $\mathcal{E}$ at P is normal to the 100-volt equipotential surface and points in the direction of the maximum space change of potential. We see that a positive test charge placed at P would move in the direction of the arrow, from high to low potential, in accordance with our definitions.

Another important property of equipotential surfaces is that we may replace any equipotential surface by a conductor of the same geometrical shape without altering the field or potential outside the conductor. The potential of the conductor would be that of the equipotential surface it replaced.

By comparing equations (17-9) and (17-24), the student will observe that

1 newton/coulomb = 1 volt/meter .

The preferred unit for field intensity is the volt per meter.

17.17. Potential of a Charged Spherical Conductor

Consider a conducting sphere in a vacuum far removed from all other objects and with a surface charge of $+q$ coulombs. Our problem is to compute the potential at some distance R from the center of the sphere (Fig. 17-25). In § 17.11 we proved by

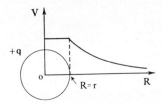

FIG. 17-25.—Potential of a charged spherical conductor

means of Gauss's theorem that the field intensity at some point outside the sphere was the same as that of a point charge $+q$ at the center of the sphere. In the previous section we showed that the equipotential surfaces of a point charge were spherical surfaces whose potentials were given by $q/4\pi\epsilon_0 R$, where R is the distance from the point charge. The potential at a distance R from the center of a sphere charged with q coulombs is therefore

$$V = \frac{q}{4\pi\epsilon_0 R} \text{ volts}. \qquad (17\text{-}25)$$

The potential of the sphere itself whose radius is r and charge q is given by

$$\left. \begin{array}{l} V = \dfrac{q}{4\pi\epsilon_0 r} \text{ volts} \\[2mm] \quad = 9 \times 10^9 \dfrac{q}{r} \text{ volts}. \end{array} \right\} \qquad (17\text{-}26)$$

The potential within the sphere is constant and given by equation (17-26). The electric field intensity is zero within the sphere and falls off inversely at the square of the distance from the center of the sphere for points outside the sphere. **265**

EXAMPLE 7: It has been experimentally determined that in dry air the maximum potential to which a sphere 1 cm in radius may be raised without causing corona discharge is about 30,000 volts. (a) Compute the surface charge density on the sphere. (b) What is the maximum field intensity at the surface of the sphere? (c) What is the maximum potential to which the sphere of the Van de Graaff machine may be raised if the diameter of the sphere is 2 m?

Solution:

a) Since $q = 4\pi r^2\sigma$, it follows from equation (17-26) that

$$\sigma = \frac{\epsilon_0 V}{r} = \frac{8.85 \times 10^{-12} \times 30,000}{10^{-2}}$$

$$= 2.65 \times 10^{-5} \text{ coulomb/m}^2 .$$

b) From equation (17-16),

$$\mathcal{E} = \frac{\sigma}{\epsilon_0} = \frac{\epsilon_0 V}{\epsilon_0 r} = \frac{V}{r} = \frac{30,000}{10^{-2}}$$

$$= 3 \times 10^6 \text{ volts/m} .$$

c) The maximum potential of the Van de Graaff sphere is V', where

$$V' = \frac{q}{4\pi\epsilon_0 R}; \qquad q = 4\pi R^2\sigma ,$$

so that

$$V' = 4\pi R^2\epsilon_0 \frac{V}{r} \frac{1}{4\pi\epsilon_0 R} = \frac{VR}{r} = \frac{3 \times 10^4}{10^{-2}}$$

$$= 3 \times 10^6 \text{ volts} .$$

17.18. Distribution of Charges on Conductors

That the surface density of charge of a conductor depends upon the shape of the conductor may be shown experimentally in the following way. By touching a conductor with a "proof plane" (a tiny conductor with an insulating handle), part of the charge of the conductor is removed. If the proof plane is brought into contact with an electroscope, the divergence of the leaves can be taken as a measure of the charge on the plane and hence the charge density of the conductor. In this way we learn that the density of charge is greatest in the regions where the curvature is greatest. This is illustrated in Fig. 17-26 (a), (b), and (c), by the distribution of plus (+) signs. In the case of an isolated sphere, the charge is distributed uniformly over the surface. The reason why the charge density varies as the curvature is suggested by the following analysis. The density of charge σ on a spherical surface is the quotient of the total charge divided by the total area, i.e.,

$$\sigma = \frac{Q}{4\pi r^2};$$

also, since $V = Q/4\pi\epsilon_0 r$, therefore

$$\sigma = \frac{\epsilon_0 V}{r}. \qquad (17\text{-}27)$$

For a given potential V of a spherical surface, equation (17-27) tells us that the density of charge is proportional to the curvature of the surface. If the conductor is not spherical but pear-shaped, as

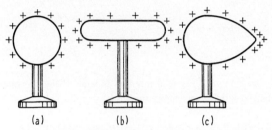

FIG. 17-26.—Charge distribution on conductors

in Fig. 17-26 (c), it has, nevertheless, the same potential at all points, being a conductor. Since its curvature is variable, we would expect the charge density to increase with an increase in the curvature.

The fact that the surface density is greatest where the curvature is greatest is of some practical importance, for it tells us that fine wires and sharp points are regions of intense electrification and so tend to discharge. The reason for this tendency to discharge is revealed by a detailed study of gaseous phenomena, which shows that the air always contains some free ions (electrically charged particles), owing to radioactive materials in the earth and atmosphere or due to cosmic rays. An intense field anywhere accelerates the ions toward or away from the point, depending upon the sign of the charge, i.e., upon the direction of the field. These ions in a short space gain large velocities—large enough to ionize more molecules by collision, and the charge is carried away by the ions. If a sufficient number of molecules is ionized, we observe a faint glow, called a "corona discharge" (see § 38.1).

17.19. Electrostatic Screening

If we take a hollow conductor with a tiny hole in it and explore the inner surface of the conductor, we find that, regardless of the amount of charge on the outer surface, *no charge resides on the inner surface* (Fig. 17-27). This implies that there is no electric field within the hollow conductor.

Another variation of this experiment is interesting. Two proof planes, originally uncharged, are placed within the hollow conductor in contact with each other (Fig. 17-27). If there is an electric field present, then, by electrostatic induction, one proof plane would be negatively charged, the other positively. If the planes are separated within the

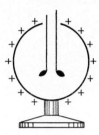

Fig. 17-27.—The use of a proof plane to demonstrate the absence of an electric field within a hollow conductor.

charged conductor, these charges would be trapped on each proof plane. We find, however, no measurable charge. There can therefore be no field within the hollow conductor. Hence the potential everywhere within the cavity is the same and equal to that at the surface.

This unique property of a closed conductor is the basis of *electrostatic shielding*. Delicate apparatus may be shielded against electrostatic disturbances by simply surrounding the apparatus with a conductor. Wire screening serves very well.

Within a solid conductor under static conditions there cannot be an electrostatic field or a potential difference, although the potential of the conductor need not be zero. If there were any field within the conductor, there would be a force acting on the free electrons, and they would be in motion. Conditions would not be static.

This property of zero electrostatic field within a conductor provides a test for the validity of the inverse-square law far more sensitive than the direct measurement of forces between charges. It can be shown by a mathematical analysis that if the exponent in Coulomb's law is not exactly 2, then there should be a field within the charged conductor. Experiments using modern techniques indicate that the exponent does not differ from 2 by more than 1 in 10^9, i.e., one part in 1,000,000,000.

17.20. Electrostatic Induction from the Point of View of Potential

It is instructive to describe the process of electrostatic induction, taking the *changes of potential* of the conductor into consideration. We shall use the diagrams of Fig. 17-28.

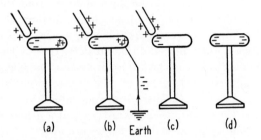

Fig. 17-28.—Charging a conductor by induction

The positively charged glass rod gives rise to an electric field in the region that surrounds it. The potential of the region is positive; i.e., work must be done on a positive charge to bring it from infinity to any point in the region about the rod. We now introduce the uncharged copper bar of zero potential, as in Fig. 17-28 (*a*). Since no field can exist within the conductor, the electrons on the bar redistribute themselves in such a way that the field within the bar is reduced to zero. The bar also acquires the positive electrostatic potential of the field. What its numerical value may be is of no consequence here. All we know is that the potential is some constant positive value.

The next step is to connect the bar to the earth via a wire. The bar immediately comes to the potential of the earth, since the bar, wire, and earth now constitute a single conductor. For practical purposes the potential of the earth is taken as zero; so the potential of the bar must decrease to zero. This may happen only if additional negative charges—electrons—are placed on the bar, for the electrons give rise to a negative potential, which, added algebraically to the positive potential (due to the inducing charge), brings the resultant potential to zero (Fig. 17-28 [*b*]).

In the next step the ground connection is broken, the potential remaining unchanged (Fig. 17-28 [*c*]). When the inducing positive charge is withdrawn, the positive potential of the field disappears, and hence the bar is left with a negative potential—excess of electrons (Fig. 17-28 [*d*]). It is clear from this discussion that it *makes no difference what part of the copper bar is earthed.*

267

17.21. Measurement of Potential Difference

In view of what we have learned about potential, we now may modify the statement of § 17.5 that a gold-leaf electroscope measures *charge* directly. If in Fig. 17-2 the metal disks are made very large so that *all* the lines of force from the leaves terminate on the disks and if, further, the disks d and aa are connected with a wire and a charge is then placed on d, no deflection of the leaves is observed. The leaves, disks d and aa, being a single conductor, are at the same potential. When, however, the disks aa are insulated from the leaves, the leaves, when charged, are at a *different potential* from aa, and the leaves are observed to diverge. We conclude, then, that the divergence of the leaves is a measure of the *difference of potential* between the leaves and disks aa. If the disks aa are connected to the ground, which, by convention, is at zero potential, then the divergence of the leaves may be taken as a measure of the potential of the region surrounding d or of any conductor in contact with d.

Instruments of the type of Fig. 17-3, with the gold leaf replaced by an aluminum rod, can be made to measure potentials of thousands of volts and are called ELECTROSTATIC VOLTMETERS.

For the measurement of potential differences as small as of the order of 0.001 volt, the QUADRANT ELECTROMETER is used. The instrument consists essentially of a light, paddle-shaped needle (N, Fig. 17-29), suspended by a fine conducting fiber.

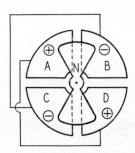

FIG. 17-29.—Electrical connections of the quadrant electrometer.

The needle hangs within a metal cylindrical box divided into four quadrants, A, B, C, and D, supported on insulating rods. The diagonally opposite quadrants are electrically connected. With the needle kept at some constant potential, usually 100 volts, potential differences between two bodies

(conductors) are measured by connecting each conductor to one pair of quadrants. The angular deflection of the needle is a measure of the potential difference.

EXAMPLE 8: (*a*) An insulated ball having a diameter of 6 cm has a positive charge of 0.027 microcoulomb (μcoulomb). What is the potential of the sphere?

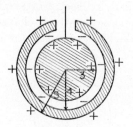

Solution: By equation (17-26),

$$V = 9 \times 10^9 \frac{q}{r} = \frac{9 \times 10^9 \times 0.027 \times 10^{-6}}{3 \times 10^{-2}}$$
$$= 8,100 \text{ volts .}$$

b) If the ball of part *a* is placed inside an uncharged hollow sphere, so that their centers coincide, compute the potential of the ball and sphere if the outside diameter of the sphere is 10 cm and the inside diameter is 8 cm (see the accompanying figure).

Solution: By electrostatic induction (§ 17.12) a -0.027-μcoulomb charge is on the inside surface of the hollow sphere and a $+0.027$-μcoulomb charge on the outside. Each charged surface makes a contribution to the potential of the ball and hollow sphere, thus

$$V_{\text{ball}} = \frac{9 \times 10^9 \times 0.027 \times 10^{-6}}{3 \times 10^{-2}}$$
$$- \frac{9 \times 10^9 \times 0.027 \times 10^{-}}{4 \times 10^{-2}}$$
$$+ \frac{9 \times 10^9 \times 0.027 \times 10^{-6}}{5 \times 10^{-2}}$$

$$= 6,885 \text{ volts}$$

Note that the potential of the ball has decreased from the value calculated in *a*, owing to the presence of the second conductor:

$$V_{\text{H. sphere}} = \frac{9 \times 10^9 \times 0.027 \times 10^{-6}}{5 \times 10^{-2}}$$
$$- \frac{9 \times 10^9 \times 0.027 \times 10^{-6}}{5 \times 10^{-2}}$$
$$+ \frac{9 \times 10^9 \times 0.027 \times 10^{-6}}{5 \times 10^{-2}}$$

$$= 4,860 \text{ volts ,}$$

since (§ 17.17) the potential at the outer surface of the sphere may be calculated, assuming the charges to be point charges at the center; therefore,

$$V_{ball} - V_{H.\,sphere} = 6{,}885 - 4{,}860$$
$$= 2{,}025 \text{ volts} .$$

c) Let the outside shell be grounded so that its potential is zero. Then

$$V_{ball} = \frac{9 \times 10^9 \times 0.027 \times 10^{-6}}{3 \times 10^{-2}}$$
$$- \frac{9 \times 10^9 \times 0.027 \times 10^{-6}}{4 \times 10^{-2}} = 2{,}025 \text{ volts},$$

$$V_{H.\,sphere} = 0 .$$

Therefore,

$$V_{ball} - V_{H.\,sphere} = 2{,}025 \text{ volts} .$$

We shall use this last result in § 17.24.

17.22. Capacitance: Capacitors or Condensers

An isolated conducting sphere of radius r has a potential V, whose value, we have seen (§ 17.17), is directly proportional to the charge Q:

$$V = \frac{1}{C} Q$$

or

$$Q = CV . \tag{17-28}$$

Whatever the size or shape of the conductor, provided only that it be away from the influence of other conductors, the relationship of charge and resulting potential is still given by equation (17-28). The value of the constant, C, however, depends upon the size and shape of the isolated conductor and upon the medium in which it is situated. The constant C is called the *capacity* or *capacitance* of the conductor. In the case of an isolated sphere *in vacuo*,

$$V = \frac{Q}{4\pi\epsilon_0 r} .$$

Therefore,

$$C = 4\pi\epsilon_0 r , \tag{17-29}$$

or the capacitance of an isolated sphere *in vacuo* is equal to its radius times the constant $4\pi\epsilon_0$.

If several conductors are present, we cannot associate any definite *capacitance* with each conductor, because the potential on any one conductor depends also upon the potentials of the other conductors (see Ex. 8, § 17.21).

By having two geometrically similar conductors present, so that they have equal charges but of opposite sign, we are able to assign a capacitance to the combination. *Such a combination of similar conductors is called a capacitor or condenser.* Capacitors are used in practice for the storage of charge. In the case of a capacitor, its capacitance is defined as the ratio of the charge on one of the conductors to the difference in potential between the conductors:

$$C = \frac{Q}{V} . \tag{17-30}$$

We wish to emphasize that capacitance *does not mean* "the amount of electricity the capacitor can hold" but rather the amount that it requires to bring its two conductors to unit difference in potential. The capacitance of a capacitor depends upon the geometry of the conductors and the material of the dielectric between them. There is, however, a practical limit to the amount of charge that a capacitor may hold, a limit dictated by the breakdown potential of the dielectric (see Ex. 7, § 17.17).

In addition to their uses as simple storage devices for electrical charges, capacitors play a basic role in electronics as tuning elements in radios, as filtering units in rectifying circuits, and as coupling devices to radio tubes, to name a few of the more important applications. We shall treat some of these applications in later chapters.

17.23. Units of Capacitance

In the mks or practical system of units, charge is measured in coulombs and potential in volts, so that equation (17-30) defines the unit of capacitance, which is called the *farad*.

$$1 \text{ farad} = \frac{1 \text{ coulomb}}{1 \text{ volt}} . \tag{17-31}$$

The farad is a large unit for practical work, and therefore the microfarad (μf), 10^{-6} farad, and the micromicrofarad ($\mu\mu$f), 10^{-12} farad, are commonly used. The capacitance of a sphere of 1-cm radius is given by equation (17-29) as

$$C = 4\pi \times 8.85 \times 10^{-12} \times 10^{-2} \text{ farad} = 1.1 \ \mu\mu\text{f} .$$

We may also note that, from equation (17-29), the units of ϵ_0 are farads per meter. Thus it follows from equation (17-5) that the units of farads per meter are the same as coulombs^2 per newton-meter^2.

17.24. Capacitance of Simple Capacitors

As the first example of a simple capacitor, we shall discuss the case of two concentric spheres of radii r_1 and r_2, the inner sphere having a positive charge of q units, while the inner surface of the outside sphere has a charge of $-q$ units (Fig. 17-30). The outside sphere is connected to the

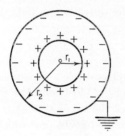

Fig. 17-30.—Spherical capacitor

ground. The resultant potential at the center o is that due to $+q$ equal to $q/4\pi\epsilon_0 r_1$ and that due to $-q$ equal to $-q/4\pi\epsilon_0 r_2$ (see worked Ex. 8 [c]). The resultant potential is then

$$V_1 = \frac{q}{4\pi\epsilon_0 r_1} - \frac{q}{4\pi\epsilon_0 r_2},$$

which is thus the potential over the entire inner sphere. Since the potential V_2 of the outer sphere is zero, the difference in potential between the spheres is simply

$$V_1 - V_2 = \frac{q}{4\pi\epsilon_0}\left(\frac{1}{r_1} - \frac{1}{r_2}\right). \quad (17\text{-}32)$$

The capacitance C becomes

$$\frac{q}{V_1 - V_2} = C = \frac{4\pi\epsilon_0 r_1 r_2}{r_2 - r_1} \text{ farads}. \quad (17\text{-}33)$$

The student will note that the magnitude of the capacitance of the spherical capacitor is greatly influenced by the size of the air gap between the conductors, $r_2 - r_1$ (see Ex. 9 for a numerical calculation).

As a second example of a capacitor we take the case of a parallel-plate capacitor in a vacuum, whose surface area is A and where the distance between plates is d (Fig. 17-21). We shall assume that the separation is small compared to the linear dimensions of the capacitor, so that the field between plates is essentially uniform. If q is the charge on either plate, the surface density of charge σ is q/A coulombs/m². From § 17.11, equa-

tion (17-16), it follows that the electric field intensity just outside a plane conductor is

$$\mathcal{E} = \frac{\sigma}{\epsilon_0} \text{ volts/m}.$$

Since the electric field is uniform, the potential difference V between the plates is

$$V = \mathcal{E}d$$

$$= \frac{\sigma d}{\epsilon_0} = \frac{q\,d}{A\epsilon_0}.$$

Also

$$C = \frac{q}{V},$$

so that

$$C = \frac{A\epsilon_0}{d} \text{ farads}. \quad (17\text{-}34)$$

That is, the capacitance of a parallel-plate capacitor is proportional to the area of the plates and inversely proportional to the distance between the plates. It should be noted that equations involving situations with spherical symmetry contain the factor 4π—equation (17-33)—while those not having such symmetry do not contain the factor 4π—equation (17-34).

EXAMPLE 9: Compute the capacitance of a spherical capacitor in air if the outer and inner radii of the sphere are, respectively, (a) 10 and 9 cm; (b) 10 and 9.9 cm; (c) 10 and 9.99 cm; (d) 100 and 9 cm; (e) $r_1 = 9$ cm and $r_2 \rightarrow \infty$.

Solution: From equation (17-33) with $4\pi\epsilon_0 = \frac{1}{9} \times 10^{-9}$, we have

$$a) \quad C = \frac{90 \times 10^{-4}}{9 \times 10^9 \times 10^{-2}} = 10^{-10} \text{ f} = 100 \ \mu\mu\text{f}.$$

$$b) \quad C = \frac{99 \times 10^{-4}}{9 \times 10^9 \times 10^{-3}} = 11 \times 10^{-10} \text{ f}$$
$$= 1{,}100 \ \mu\mu$$

$$c) \quad C = \frac{99.9 \times 10^{-4}}{9 \times 10^9 \times 10^{-4}} = 111 \times 10^{-10} \text{ f}$$
$$= 11{,}100 \ \mu\mu\text{f}.$$

$$d) \quad C = \frac{900 \times 10^{-4}}{9 \times 10^9 \times 91 \times 10^{-2}} = 0.11 \times 10^{-10} \text{ f}$$
$$= 11 \ \mu\mu\text{f}.$$

$$e) \quad C = 4\pi\epsilon_0 r = \frac{9 \times 10^{-2}}{9 \times 10^9} \text{ f} = 10^{-11} \text{ f} = 10 \ \mu\mu\text{f}.$$

17.25. The Dielectric

In our discussion of electrostatics so far, we have neglected to say exactly what effect an electrostatic field has on an insulator or dielectric All our formulas were deduced for charges and conductors in a vacuum. Our next problem is to determine what changes occur if charges are immersed in a dielectric and what effect a dielectric placed between the plates of a capacitor has on its capacitance.

According to § 17.4, an insulator or dielectric is characterized by the fact that there are no free electrons present; i.e., the electrons are tightly bound to their respective atomic nuclei. If, therefore, an insulator is placed in an electric field, free charges no longer redistribute themselves in such a way that the field within the body is zero, as is the case for conductors. *There will be a field within the insulator* and, in general, a component of the field parallel to the surface of the dielectric. However, dielectrics placed in electric fields do exhibit surface charges, so that some sort of distribution of charge does take place.

A mechanism to explain this is suggested by the following experiment. If a dielectric in an electric field is cut into two pieces by a plane normal to the field, the pieces separated and then withdrawn from the field, we find that the net charge on each is zero. This is not the case, however, for conductors, for here each half has electrification of the opposite kind (§ 17.12).

We may explain the action of an insulator in an electric field from the point of view of atomic structure. An atom is made up of electrons and a positive nucleus. When atoms are placed in the electrostatic field, a force acts on the charges in

at the other side. In this way the dielectric would exhibit *surface charges*. An atom so disturbed by an electric field is called an *induced dipole*. When the field is removed, the atom reverts to its normal state. *All atoms* are polarizable or acquire an induced dipole in an electric field.

A dielectric thus adjusts its surface charge so that the electric field that passes through it is diminished to some extent, the field going to zero only in the case of conductors. The reduction of the field is experimentally found to be approximately constant for any given dielectric. We designate this physical property of an insulator by the letter κ. The dielectric constant of a medium, then, is a measure of the extent to which the dielectric may be polarized.

In order to obtain a quantitative measure of the effect of a dielectric in an electric field, we shall examine the problem of a dielectric placed between the plates of a parallel-plate capacitor, as illustrated in Fig. 17-32. Let us assume that the

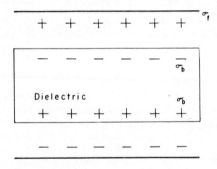

Fig. 17-32.—Dielectric between the plates of a condenser, showing free and bound charges.

surface charge density on the conductor is σ_f coulombs/m². Owing to the presence of the electric field, surface charges appear on the dielectric as shown. We shall designate the surface charge density of this bound charge by σ_b. It has been found experimentally that, for most substances, the magnitude of the bound charge is proportional to the applied field. Thus

$$\sigma_b = \chi \mathcal{E} \text{ coulombs/m}^2 . \qquad (17\text{-}35)$$

The constant χ is called the *electric susceptibility*, and its units are coulombs per volt-meter. Values for χ are given in Table 17-1.

With both free and bound charges present at the conductor-dielectric interface, the total field-producing charge density is then $\sigma_f - \sigma_b$, and the

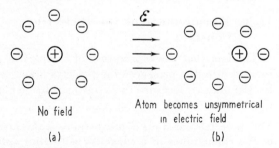

No field

Atom becomes unsymmetrical in electric field

(a)   (b)

Fig. 17-31.—Polarization of a dielectric due to an electric field.

such a way that the negative charges and positive charges are pulled slightly away from each other, as shown in Fig. 17-31. The dielectric is said to be *polarized*. For atoms near the surface, the electrons, for example, would be pushed toward the surface at one side of the dielectric and pulled in

271

resulting electric field between the capacitor plates is given by equation (17-16):

$$\mathcal{E} = \frac{\sigma_f - \sigma_b}{\epsilon_0} \text{ volts/m}. \qquad (17\text{-}36)$$

If we replace σ_b in this equation with its value given by equation (17-35), we obtain

$$\mathcal{E} = \frac{\sigma_f - \chi\mathcal{E}}{\epsilon_0}$$

or

$$\mathcal{E} = \frac{\sigma_f}{\epsilon_0(1 + \chi/\epsilon_0)}. \qquad (17\text{-}37)$$

TABLE 17-1
DIELECTRIC CONSTANTS

Material	Dielectric Constant κ	$\epsilon = \kappa\epsilon_0$ Permittivity, ϵ Farads/M	$\chi = \epsilon_0(\kappa-1)$ Electric Susceptibility, χ Farads/M
Vacuum	1	$\epsilon_0 = 8.85 \times 10^{-12}$	0
Air (NTP)	1.0006	8.85×10^{-12}	53.1×10^{-16}
Paraffin	2.0–2.3	17.7–20.4×10^{-12}	8.85–11.6×10^{-12}
Dry paper	2.0–2.4	17.7–21.2×10^{-12}	8.85–12.4×10^{-12}
Rubber	2.1–2.3	18.6–20.4×10^{-12}	9.75–11.6×10^{-12}
Wood	2.5–7.7	22.1–68.1×10^{-12}	13.3–59.3×10^{-12}
Shellac	3.1	27.4×10^{-12}	18.6×10^{-12}
Sulphur	4.0	35×10^{-12}	26.2×10^{-12}
Paraffin oil	4.7	41.6×10^{-12}	32.8×10^{-12}
Mica	5.7–7.0	50.4–62×10^{-12}	41.6–53.2×10^{-12}
Flint glass	7–10	62–88.5×10^{-12}	53.2–79.7×10^{-12}
Ethyl alcohol	27.0	2.39×10^{-10}	2.3×10^{-10}
Distilled water	81.07	7.18×10^{-10}	7.1×10^{-10}
Polystyrene	2.6	2.26×10^{-11}	1.4×10^{-11}

Since χ and ϵ_0 have the same dimensions, the quantity $(1 + \chi/\epsilon_0)$ is dimensionless, and we shall designate it with the letter κ:

$$\kappa = \left(1 + \frac{\chi}{\epsilon_0}\right). \qquad (17\text{-}38)$$

The quantity κ is called the *relative dielectric constant* or specific inductive capacity of the material. From equation (17-37) we see that the relative dielectric constant κ is the ratio of the field with no dielectric present to that when a dielectric is present (see § 17.26 for an equivalent definition). Since the resulting electric field is given by equation (17-37),

$$\mathcal{E} = \frac{\sigma_f}{\kappa\epsilon_0},$$

and the difference in potential by

$$V = \mathcal{E}d$$

$$= \frac{\sigma_f d}{\kappa\epsilon_0} = \frac{qd}{\kappa\epsilon_0 A},$$

it follows that the capacitance C of the capacitor is

$$C = \frac{q}{V} = \frac{A\kappa\epsilon_0}{d} \text{ farads}. \qquad (17\text{-}39)$$

In a vacuum,

$$C = \frac{A\epsilon_0}{d} \text{ farads}. \qquad (17\text{-}40)$$

The capacitance of the capacitor has been increased by the factor κ. If equation (17-38) is multiplied through by ϵ_0, we obtain

$$\kappa\epsilon_0 = \epsilon_0 + \chi.$$

The quantity $\kappa\epsilon_0$ is called the *permittivity*, ϵ, of the dielectric or the absolute dielectric constant and ϵ_0 the permittivity of free space.

By a similar analysis we may show that if two small conductors with charges q_1 and q_2 are imbedded in a dielectric, the electric field intensity at q_2 due to q_1 will be decreased by a factor κ because of the presence of bound surface charges on the dielectric in contact with the conductor. The field at q_1 is similarly reduced. It follows, therefore, that the force between the charges will be diminished by the factor κ, and Coulomb's law of force between charges in a dielectric becomes

$$F = \frac{q_1 q_2}{4\pi\kappa\epsilon_0 r^2} \text{ newtons}. \qquad (17\text{-}41)$$

Also the electric field and potential of a point charge in a dielectric are reduced by κ,

$$\mathcal{E} = \frac{q}{4\pi\kappa\epsilon_0 r^2} \text{ volts/m}; \quad V = \frac{q}{4\pi\kappa\epsilon_0 r} \text{ volts}. \quad (17\text{-}42)$$

If a dielectric is placed between the concentric spheres of a spherical capacitor, the capacitance is increased by κ,

$$C = \frac{4\pi\kappa\epsilon_0 r_1 r_2}{(r_2 - r_1)} \text{ farads}. \qquad (17\text{-}43)$$

17.26. Measurement of the Dielectric Constant

The fact that the capacitance of a capacitor is increased by the presence of a dielectric was used by Faraday to measure this constant. By definition:

The dielectric constant of a substance is the ratio of the capacitance of a capacitor with that substance present as a dielectric to the capacitance of the capacitor with a vacuum as a dielectric.

Thus, if C is the capacitance of a capacitor (parallel plate, for example) with a vacuum as a dielectric, and C_κ the capacitance with another dielectric between the plates, then

$$\kappa = \frac{C_\kappa}{C}. \qquad (17\text{-}44)$$

From Table 17-1 it is clear that the dielectric constant for air and that for a vacuum are nearly equal. The particular advantages of dielectrics in capacitors are twofold—first, the presence of the dielectric permits the capacitor to store a greater charge per unit difference of potential than a capacitor in a vacuum, and, second, the potential to which a capacitor may be charged can be substantially increased by the use of dielectrics. The critical electrical intensity $\mathcal{E}_C$ at which breakdown occurs is called the *dielectric strength* of the dielectric or insulator. The dielectric constant, κ, and the dielectric strength, $\mathcal{E}_C$, should not be confused. They are not basically related.

In Example 7 of § 17.17 we pointed out that the dielectric strength of dry air is about 30,000 volts/cm. Transformer oil has a dielectric strength of about 100,000 volts/cm; mica, values up to 1,000,000 volts/cm; and glass, about 300,000 volts/cm.

17.27. Capacitors in Series and Parallel

Capacitors, arranged as shown in Fig. 17-33, are connected in series. A charge $+q$ placed on the left-hand plate of C_1 will induce a charge $-q$ on

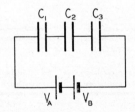

FIG. 17-33.—Capacitors connected in series

the right-hand plate. This charge comes from the left of C_2, leaving the left of C_2 with a $+q$ charge. This will, in turn, induce $-q$ on the right of C_2, leaving the left of C_3 with $+q$. The right of C_3 then has a $-q$ charge on it. Evidently, each capacitor has the same charge q. The potential difference between the terminals is $V_A - V_B$, which we call V. This constant potential difference may be maintained, for example, by a battery. The total work done per unit charge on a test charge taken across the three capacitors is, by definition, V, and it must be equal to the work done per unit charge in carrying the charge across each capacitor separately. Since $V = q/C$,

$$V = V_1 + V_2 + V_3 .$$

Therefore,

$$\frac{q}{C} = \frac{q}{C_1} + \frac{q}{C_2} + \frac{q}{C_3}$$

or

$$\frac{1}{C} = \frac{1}{C_1} + \frac{1}{C_2} + \frac{1}{C_3} \qquad (17\text{-}45)$$

or, in general,

$$\frac{1}{C} = \sum_{i=1}^{n} \frac{1}{C_i} . \qquad (17\text{-}46)$$

In words, the reciprocal of the capacitance of a number of capacitors connected in series is equal to the sum of the reciprocals of the separate capacitances. It is well to remember that equation (17-45) shows that the capacitance of the group is less than the smallest individual capacitance. However, the combination may withstand a greater potential difference, without breaking down, than any individual capacitor, since the total potential is divided among the individual capacitors.

Suppose the three capacitors are arranged in parallel, as in Fig. 17-34. The total quantity of

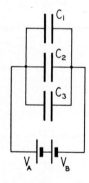

FIG. 17-34.—Capacitors connected in parallel

charge, q, will distribute itself between the capacitors, so that

$$q = q_1 + q_2 + q_3 .$$

Each capacitor, however, has the same potential difference:

$$V_A - V_B = V ,$$

$$V = V_1 = V_2 = V_3 .$$

Since $q = CV$, therefore,

$$CV = C_1V + C_2V + C_3V ,$$

or

$$C = C_1 + C_2 + C_3 , \qquad (17\text{-}47)$$

273

or, in general,

$$C = \sum_{i=1}^{n} C_i. \qquad (17\text{-}48)$$

The total capacitance of a set of capacitors connected in parallel is equal to the sum of the individual capacitances.

17.28. Energy of a Charged Capacitor

If we have a capacitor with no charge on it and hence no difference of potential between the plates, no work would be needed to carry a charge from one plate to the other. However, after a finite but small charge has been placed on the capacitor, a finite but small amount of work,

$$dw = V\, dq, \qquad (17\text{-}49)$$

must be done to carry a charge dq from one plate to the other, whose difference of potential is now V. The V, proportional to q, increases linearly as the charge increases, and hence the work done changes for a given dq, the more the total q. If we wish to calculate the total work done in placing q units of charge on a conductor whose final potential is V_1, we may compute the work as follows:

Plot the potential of the conductor against the total charge on it and thus obtain a straight line, $q = CV$, in which the slope is $1/C$ (Fig. 17-35).

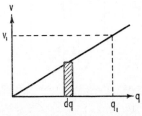

Fig. 17-35.—Change in potential of a capacitor due to a charge dq.

The work done in increasing the charge on the conductor by an amount dq against a potential V is given analytically by equation (17-49) and graphically by the small rectangle of Fig. 17-35. The total work done in charging the conductor to a potential V_1 by the transference of q_1 units of charge is given by the area under the curve,

$$W = \tfrac{1}{2} q_1 V_1. \qquad (17\text{-}50)$$

Also, since $q = CV$,

$$W = \tfrac{1}{2} C V_1^2, \qquad (17\text{-}51)$$

and

$$W = \frac{1}{2}\frac{q_1^2}{C}. \qquad (17\text{-}52)$$

More simply, using the calculus, we obtain

$$dW = V\, dq$$

$$= \frac{q}{C}\, dq,$$

$$W = \frac{1}{C}\int_0^{q_1} q\, dq = \frac{q_1^2}{2C}.$$

In these equations q is measured in coulombs, V in volts, C in farads, and W in joules.

EXAMPLE 10: Given two capacitors, as indicated in the accompanying figure (a), where

$$C_1 = 100 \ \mu\text{f}, \qquad C_2 = 50 \ \mu\text{f},$$

$$Q_1 = 1{,}000 \ \mu\text{coulombs}, \quad Q_2 = 2{,}500 \ \mu\text{coulombs}$$

$$V_1 = 10 \text{ volts}, \qquad V_2 = 50 \text{ volts}.$$

Suppose a wire is attached between two of the plates. There will be no movement of electricity, since the charges on the lower plates are "bound" by the charges on the upper plates, and these upper charges have no opportunity to move.

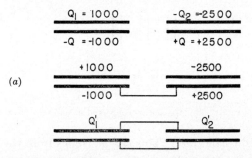

(a)

Now suppose a second wire is attached between the upper plates. Almost instantly there will be a rearrangement of the charges. What will be the new charge on each capacitor? On the upper pair of plates, +1,000 and −2,500 μcoulombs will combine to give a total of −1,500 μcoulombs. The total capacitance is $100 + 50 = 150 \ \mu$f. The new potential difference V', which must be the same for each capacitor, will then be

$$V' = \tfrac{1500}{150} = 10 \text{ volts}.$$

Then

$$Q_1' = 100 \ \mu\text{f} \times 10 \text{ volts} = 1{,}000 \ \mu\text{coulombs},$$

$$Q_2' = 50 \ \mu\text{f} \times 10 \text{ volts} = 500 \ \mu\text{coulombs}$$

as shown in figure (*b*).

(*b*)

What has happened to the energy stored on the capacitors? Before the arrangement,

$$W = \tfrac{1}{2} \times 1{,}000 \times 10^{-6} \times 10 + \tfrac{1}{2} \times 2{,}500 \times 10^{-6}$$
$$\times 50 = 675 \times 10^{-4} \text{ joule}.$$

After the arrangement,

$$W' = \tfrac{1}{2} \times 1{,}000 \times 10^{-6} \times 10 + \tfrac{1}{2} \times 500 \times 10^{-6}$$
$$\times 10 = 75 \times 10^{-4} \text{ joule}.$$

Hence there has been a loss of energy of $(675 - 75) \times 10^{-4} = 600 \times 10^{-4}$ joule. This energy has gone to heat the connecting wires and to supply energy for the light and sound and heat of a spark, and perhaps has been partly radiated as radio waves.

17.29. Résumé

The following is a brief listing of the definitions, principles, and theories, given in this chapter, which you should know:

How electric charges are inferred from forces

Coulomb's law of force for point charges

Definition of unit charges, the electostatic unit, the coulomb

The significance of ϵ_0, the permittivity of free space

Definition of the dielectric constant κ

Definition of electric field

Lines of force and their relationship to electric field strengths

Gauss's theorem

Definition of potential difference; of the volt

Calculation of potential differences for simple charge distributions

Definition of capacitance

Calculation of capacitance for some simple geometrical figures

Measurement of dielectric constant

Capacitances placed in series and in parallel

Energy of a charged capacitor

Queries

1. Calculate the force of repulsion between 1 coulomb (6.24×10^{18} electrons) placed 1 m away from an equal charge. Assuming each electron to have a mass of 9×10^{-28} gm, compute the gravitational force between the electrons. What is so striking about the results?

2. In the Cottrell process for smoke or small-particle precipitation, a long, negatively charged wire is suspended vertically within the column of rising particles. Explain in a general way the mechanism of precipitation.

3. In gravitational phenomena the attraction of masses follows an inverse-square law, mathematically the same as that for electrostatic charges. Can you give a possible reason why we may talk of a nearly uniform gravitational field near the surface of the earth, ye in electrostatics all the fields, with few exceptions, vary rapidly over very short distances? *Hint:* In $f = \gamma m_1 m_2 / R^2$, what is the effect of a change of a few centimeters in R?

4. Is it possible for the potential at a point to be zero but the electric field at the point not to be zero?

5. Is it possible for the potential at a point to be some value other than zero but the electric field intensity at the point to be zero?

6. Can both the electric field and the potential at a point be zero for some distribution of charge?

7. Discuss the charging of an insulated conductor (Fig. 17-16) from the point of view of potential.

Practice Problems

Useful constants for the solution of the following problem:

$$\epsilon_0 = 8.85 \times 10^{-12} \text{ coulomb/volt m};$$

$$4\pi\epsilon_0 = 1.10 \times 10^{-10}; \quad \frac{1}{4\pi\epsilon_0} = 9 \times 10^9;$$

$$\frac{1}{\epsilon_0} = 1.13 \times 10^{11}.$$

1. Two small spheres, hung from insulating strings from the same point, carry charges of 4 μcoulombs each. The separation of the charges is 1 m. What force does each exert on the other? Ans.: 0.144 newtons.

2. The two balls hung from strings from the same point have a mass of 1 gm each and are 2 cm apart. Calculate the charges on each, assuming them to be the same. The strings are each 50 cm long.

3. Two α-particles of charges $+2e$ approach each other to a distance of 10^{-8} cm. Calculate the force of repulsion in newtons. What is the electric field intensity in volts/meter at 10^{-8} cm from an α-particle? Ans.: 9.2×10^{-8} newton; 2.9×10^{11} newtons/coulomb or volts/m.

4. Two charges of $+12$ μcoulombs and -4 μcoulombs are 10 cm apart. Where must a third charge be placed so that the force on it is zero? Is this a position of stable equilibrium for the test charge?

5. Two equal charges of $+25$ μcoulombs are placed 1 m apart. Compute (a) the force exerted by each charge on a $+2$-μcoulomb charge placed midway between them; (b) the resulting force on the test charge. Ans.: (a) 1.8 newtons (direction?); (b) 0.

6. What would happen to the test charge of problem 5 if it were pushed slightly to the left or right of the mid-point? What would happen if it were pushed up or down?

7. Compute the electrostatic field at x due to these charges in microcoulombs. Ans: $\mathcal{E}_x = -54 \times 10^3$ volts/m; $\mathcal{E}_y = -63 \times 10^3$ volts/m.

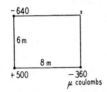

8. A regular hexagon is 10 m on an edge. Compute the electrostatic field at A and at B due to the given charges in microcoulombs.

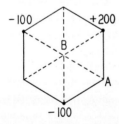

9. In problem 7 compute the electrostatic potential at x. Ans.: -810 kv.

10. In problem 8 compute the electrostatic potential at A and at B.

11. In problem 8 how much work would be needed to carry $+3 \times 10^{-7}$ coulomb from B to A? Ans.: 0.0135 joule.

12. Charges of $+0.002$ coulomb each are placed at the vertices of an equilateral triangle whose sides are 1 m long. Compute the force (magnitude and direction) acting on any one of the charges due to the presence of the others.

13. How much work is necessary to bring a 1-coulomb negative charge from a great distance to the center of the triangle of problem 12? Ans.: -93.6×10^6 joules.

14. Compute the force acting on a 1-coulomb negative charge placed at the center of the triangle of problem 12.

15. A charge of 1 μcoulomb is distributed uniformly along a wire bent into the form of a circle of radius 10 cm. What is the potential and electric field intensity at the center? Ans.: $\mathcal{E} = 0$; $V = 90$ kv.

16. Two point charges of 1,000 and 500 μcoulombs are spaced 15 m apart in a vacuum. Compute the work required to reduce the separation to 10 m.

17. The following series of diagrams represents a partially drawn "animated cartoon" of a continuous series of events similar to the diagrams of Fig. 17-17. You are to complete the drawings by putting the proper plus ($+$) and minus ($-$) signs to indicate the presence of electrostatic charges. Also complete the leaves and mark the charges on the gold-leaf electroscope. The symbol "$\nearrow \infty$" implies a distance so great as to be considered infinite. (In diagram [11] the ball touches the cup.)

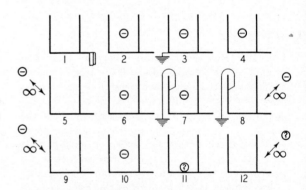

18. The potential gradient at a point in an electrostatic field is 350 kv/m. What is the force on a $+8$-μcoulomb charge placed at this point?

19. A plane conductor has a surface charge density of -0.3 μcoulomb/m². Calculate the force and acceleration on an electron that comes within 10^{-5} cm of the surface. Use $e = 16 \times 10^{-20}$ coulomb; $m = 9 \times 10^{-31}$ kg. Ans.: 5.4×10^{-15} newton; 6×10^{15} m/sec².

20. In the electron gun of a cathode-ray tube an electron falls through a constant potential gradient of 10^4 volts/m and remains in the field for 10^{-9} sec. (a) Calculate the force in newtons on the electron. (b) With what velocity does the electron leave the gun?

21. Given two charged capacitors, as indicated. Connect them together ($+$) to ($+$) and ($-$) to ($-$).

Compute the new charge on each and the apparent loss of energy. Ans.: 100 μcoulombs; 400 μcoulombs; 0.00125 joule lost.

$$
\begin{array}{cc}
C_1 & C_2 \\
+200 & +300 \\
\hline
-200 & -300
\end{array}
$$

MICROCOULOMBS

$C_1 = 5\ \mu f$

$C_2 = 20\ \mu f$

22. In problem 21 connect the capacitors (+) to (−) and (−) to (+). Compute the new charge on each and the apparent loss of energy.

23. Two spheres of radii 12 cm each have charges of $+12 \times 10^{-8}$ and -6×10^{-8} coulomb. If the centers of the spheres are placed 2 m apart, compute the potential on each sphere and the difference in potential between spheres. It is assumed that the charges are uniformly distributed over each sphere. Ans.: 8,730 volts; −3,960 volts; +12,690 volts.

24. The dielectric constant of a phenolic material is 6.3. Calculate the permittivity and electrical susceptibility of the material.

25. Two parallel metal plates are permanently fixed 1 cm apart. The space between is filled by a glass plate of dielectric constant 5. The area of each metal plate is 2,000 cm², and of the glass plate it is 3,000 cm². Compute the potential difference between the plates if the capacitor is charged with 2-μcoulomb charges. If the glass plate is removed, the metal plates remaining in place, what will be the new potential difference? Compute the electrical energy in each case. Where did the extra energy come from? Ans.: 2.26 kv; 11.3 kv; $w_1 = 0.00226$ joule; $w_2 = 0.0113$ joule.

26. Two spheres of radii 10 and 4 cm are connected by a long wire. If 1 μcoulomb of charge is given to the system, how will the charges distribute themselves? Which sphere will have the greatest charge density?

27. How much work is necessary to charge an isolated sphere of radius 10 m with 2 μcoulombs of charge? Ans.: 18×10^{-4} joule.

28. A sphere of diameter 4 cm has a uniformly distributed charge of 2×10^{-7} coulomb. Calculate the field intensity at the surface of the sphere. What is the total flux of lines of force from the sphere?

29. Compute the surface charge density on a parallel-plate capacitor whose plate separation is 0.1 mm and whose potential difference is 1,500 volts. Ans.: 13.3×10^{-5} coulomb/m².

30. Calculate the capacitance of the earth in farads. Mean radius of the earth is 6,371 km.

31. All the energy of a capacitor of capacitance 0.0021 farad and charged to a potential of 200 volts is used to heat a metal. How many calories of heat are evolved? Ans.: 10 cal.

32. The radii of the concentric spheres of a capacitor are 12 and 14 cm. Compute the capacitance of the capacitor if a dielectric of value $\kappa = 6$ fills the space between the spheres.

33. If the dielectric strength of a sample of mica ($\kappa = 7$) is 500,000 volts/cm, calculate the maximum free surface charge density that a parallel-plate capacitor may hold if the plate separation is 0.1 mm. Ans.: 3.1×10^{-3} coulomb/m².

34. A parallel-plate capacitor of area 500 cm² has a dielectric of mica ($\kappa = 6$) whose thickness is 0.001 cm. Calculate the capacitance of the capacitor. If the capacitor is connected across a battery of 12-volt potential, calculate the charge on each plate in coulombs.

35. If in Example 8 the charge on the ball is $+0.08$ μcoulomb, compute (a) the potential of the ball isolated from other conductors; (b) the potential of ball and sphere when the ball is inside the sphere; (c) the potential of ball and sphere if the sphere is grounded; and (d) after disconnecting the ground, the potential of the sphere and ball if the ball is allowed to touch the insulated sphere and then brought back to its original position. Ans.: (a) 24 kv; (b) 20.4 kv, 14.4 kv; (c) 6 kv, 0; (d) 0, 0.

36. A parallel-plate mica capacitor is to be designed for a potential of 9,000 volts. How thick must the mica dielectric be?

37. Four capacitors are connected as shown in the diagram. What is the equivalent capacitance of the system from A to B? Ans.: 4.85 μf.

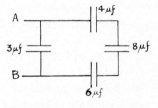

38. If a 100-volt battery is connected between A and B of problem 37, what will the charges be on each of the capacitors?

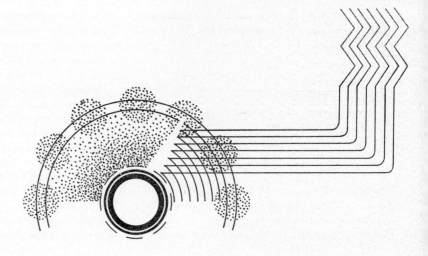

CHAPTER 18

Ohm's Law: Resistance

18.1. Meaning of an Electric Current

One of the fundamental observations in electrostatics was the absence of an electric field *within a conductor* placed in an electrostatic field and the constancy of potential throughout the conductor. For example, if the conductor A, charged positively (deficiency of electrons), is connected to B, charged negatively (excess of electrons), by a copper wire C (Fig. 18-1), there immediately takes

A   C   B
$\oplus$————————$\ominus$
              ← e

Fig. 18-1.—Flow of electrons from low to high potential

place a transference of electrons from C to A and from B to C to bring the field within C to zero. When this occurs, the transference ceases, and the initial difference in potential $(V_A - V_B)$ no longer exists.

Let us assume, however, that A and B are connected to some suitable mechanism (such as an electrostatic machine or preferably a battery) which maintains a constant difference of potential between A and B. *In this case an electric field, $\mathcal{E}$, does exist within the conductor.* A force $q\mathcal{E}$ acts on the mobile electrons, each of charge q, causing a steady movement of the electrons through C from points of low potential to higher potential. Static conditions no longer exist, for charges are in motion.

Charges in motion constitute an electric current. If

q units of charge pass a cross-section of a conductor in time t, the average current, I, is

$$I = \frac{q}{t}; \qquad (18\text{-}1)$$

i.e., the magnitude of the current, I, is equal to the time rate of flow of charge. If equal quantities of charge q pass a given section of the conductor in successive equal intervals of time t, the current I is said to be constant in magnitude or "steady." Equation (18-1), then, defines the magnitude of the current. By taking q equal to 1 esu and t equal to 1 sec, we could define an electrostatic unit of current. Such a unit is so small as to be of no practical value. If q equals 1 coulomb of charge (3×10^9 esu) and t equals 1 sec, then equation (18-1) defines an *ampere* (amp) of current strength. In a later section (§ 21.12) we shall define current strength in terms of its magnetic properties.

In the case of currents in metallic conductors, we know that the electrons are the charges that move, going from points of lower to points of higher potential. Hence it would be logical to represent the direction of the current in a conductor by an arrow pointing in the direction of electron flow. Unfortunately, the convention of current direction was adopted before the electron theory of matter was developed. The direction of the current in a conductor is conventionally adopted as the direction of flow of positive charges, i.e., high to low potential. Because this convention has become so widespread in its use, it is convenient to adopt it in this book also. The convention is illustrated in Fig. 18-2.

An electric current gives rise to three fundamentally important effects. The first is the presence of a magnetic field about the current. The second is the heating that takes place in the con-

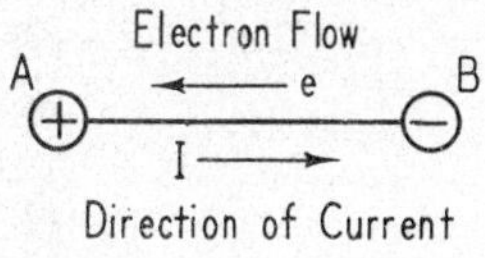

FIG. 18-2.—Distinction between electron flow and conventional current direction.

ductor through which the current passes. Finally, an electric current, when passed through certain solutions—"electrolytes"—gives rise to a chemical activity technically known as *electrolysis*. All three effects may take place simultaneously. However, in this chapter we shall emphasize the heating effects in conductors through which a current passes.

18.2. Difference in Potential: The Volt

In the previous section we pointed out that if a difference in potential were continuously maintained between the two points A and B of the conductor C (Fig. 18-1), a current would pass from the point of higher potential to that of lower. This is purely by the convention originally set up by Ampère. From the definition of potential difference as work per unit charge (§ 17.14), this means that energy must be continuously dissipated in the conductor C at a rate that depends upon the difference of potential and the strength of the current. We shall study this important point in some detail.

In the mks system of units the unit of work is taken as the joule and the charge as the coulomb. It follows that the unit of potential difference in the mks system—the volt—is the difference of potential between two points if a joule per coulomb is done by any charge in moving from the point of high potential to that of lower potential. Stated analytically,

$$V = \frac{W}{Q} \text{ volts} \quad \text{or} \quad W = QV \text{ joules} . \quad (18\text{-}2)$$

From the definition of power as the time rate of doing work, we have

$$P = \frac{W}{t} = \frac{QV}{t} \text{ watts} . \quad (18\text{-}3)$$

Since $Q/t = I$,

$$P = VI \text{ watts} \quad (18\text{-}4)$$

and

$$W = VIt \text{ joules} . \quad (18\text{-}5)$$

The preferred definition of the volt is based on equation (18-4).

By definition, the volt is the difference of electrical potential between two points (or two equipotential surfaces) in a conductor carrying a constant current of 1 ampere when the power dissipated between the two points is 1 watt.

(In the electromagnetic system of units or emu system the unit of potential difference is the abvolt, equal to 0.1 erg per coulomb. 1 volt = 10^8 abvolts.) The potential difference between the poles of common batteries is from 1 to 2 volts and that of the common house-lighting circuits 115 or 230 volts.

We could measure the potential difference (PD) between two points on a conductor by means of the electrometers discussed in chapter 17. However, in practice, potential differences associated with electric currents are measured by the voltmeter, an instrument whose operation and construction we shall discuss in detail in chapter 22.

18.3. Electromotive Force and Potential Difference

In our study of electrostatics we noted that the production of electrification or the isolation of free charge could be obtained only by separating the positive and negative electricity which is present in all matter and in all uncharged bodies in equal amounts. It was necessary to do work in order to separate positive from negative electricity because of the attraction between these unlike charges. When unlike charges are permitted to reunite, the energy which was stored in the charges owing to their separation again becomes available to do work. This energy appears as a heating of the conductors through which the charges pass. It is inherent in part in the magnetic field which accompanies a current and partly in other forms, which we will study later.

Any device which separates positive and nega- **279**

tive charges must supply energy to the electrical circuit. The efficacy of such a device may be described by designating the amount of energy supplied for each unit quantity of electricity which has been separated from its equivalent of opposite sign.

Definition of electromotive force: In a circuit in which current is present, the total rate at which energy is drawn from the source of the current and dissipated in the circuit per unit current is defined as the electromotive force (emf) in the circuit.

This is a quite general definition of electromotive force. Primary electric cells or batteries supply energy to an electric circuit in this manner, the energy coming from chemical reactions in the cells (§ 19.8). Other examples of devices with an electromotive force are dynamos and thermal junctions, details of which will be discussed in later chapters. A *volt of electromotive force* is supplied by a device if 1 joule of energy per coulomb is supplied to the circuit.

It is important to note the following precise relation between electromotive force, as just defined, and a potential difference between two points on a current-carrying metallic conductor.

Definition of potential difference: The potential difference (PD) between two points on a conductor is the work per unit charge done by a charge in moving from a point of higher potential to that of lower.

Here electricity is doing work. A *volt of potential difference* exists between the two points if 1 joule of work per coulomb is done by the charge.

The reader will note that, since both electromotive force and potential difference are concerned with energy transformations per unit charge, they are measured in the same unit, the *volt*.

18.4. Ohm's Law

We wish next to determine whether any definite relationship exists between the potential difference V of two points on a *metallic conductor* (A and B of Fig. 18-3) and the resulting current I that passes between these two points. We may measure the potential difference by means of an electrometer suitably calibrated and the current by a galvanometer (§ 21.4). *It is found that the current is strictly proportional to the potential difference.* If the

potential difference is increased, the current also increases. We may represent the result of the experiment by the equation

$$\frac{V}{I} = R, \text{ a constant} \quad \text{or} \quad V = IR. \quad (18\text{-}6)$$

This is another fundamental equation in electricity. The constancy of the ratio V/I represented by equation (18-6) was discovered in 1827 by the

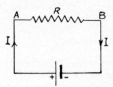

Fig. 18-3.—Elementary circuit containing a resistance and a battery.

German physicist, George Simon Ohm (1789–1854). It is known as *Ohm's law*.

If we plot V against I, the straight line of Fig. 18-4 results. The ratio V/I is, by definition, the resistance of a conductor between two given points and, for the case of a metallic conductor, a constant measured by the slope of the straight line of Fig. 18-4. *The most delicate tests have failed*

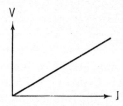

Fig. 18-4.—Ohm's law. The slope of the curve is the resistance in ohms.

to reveal any departure from the strict linearity of equation (18-6). It holds for the feeblest currents that have been measured up to current densities in excess of 15,000 amp/mm². For this reason we are justified in calling the resistance of a metallic conductor a "measure of a definite physical property of the conductor." Like most physical constants, the resistance is constant only if certain other conditions are constant. In this case we assume that the physical properties, such as temperature, tension, and pressure of the conductor, do not change. In the diagrams we shall designate the resistance of a wire by the symbol —⋀⋀⋀⋀⋀—, whereas a straight line will indicate a wire of zero resistance.

Furthermore, equation (18-6) allows us to define the unit of resistance: *If* V *is in volts and* I *in amperes, then* R *is measured in ohms.*

Definition of ohm: The ohm is the electric resistance between two points of a conductor when a constant difference of potential of 1 volt, applied between these two points, produces in this conductor a current of 1 ampere, provided that between the two points there is no electromotive force.

Some idea of the size of this unit may be gained from the fact that 1 mi of copper wire 0.76 cm in diameter has a resistance of approximately 0.6 ohm.

Ohm's law, i.e., the constancy of V/I, is not applicable to all conducting media.* In what follows, our statements are true only for constant, unvarying currents in metallic conductors and also, we shall see, in electrolytes, under appropriate experimental conditions.

18.5. Energy and Power Relations: Joule's Law

When a charge of electricity moves from a point of higher potential to one of lower potential, work is done by it of an amount given by equation (18-2):

$$W = QV ; \qquad (18\text{-}2)$$

and from $Q = It$, this may be written as

$$W = VIt . \qquad (18\text{-}5)$$

In accordance with the principle of conservation of energy, the work expended by the charge is not lost but gives rise to an equivalent amount of energy of some kind.

It was shown in 1845 by Joule (1818–89) that the work spent in "driving the current" through a simple resistance (such as *ab* of Fig. 18-5) was

Fig. 18-5.—Electrical energy is transformed into heat energy in the resistance *ab*.

converted into heat. In fact, Joule showed experimentally that the heat generated in a conductor per unit time was proportional to the square of

---

* Exceptions are electric arcs, vacuum tubes, and tubes in which conduction is due to gaseous ionization.

the current. This statement is known as "*Joule's law* of electrical heating":

$$W = RI^2t \text{ joules} , \qquad (18\text{-}7)$$

where the constant R is the resistance measured in ohms, I is in amperes, and t in seconds. In terms of heat units, since 4.19×10^7 ergs (4.19 joules) equals 1 cal, we have

$$H = 0.239RI^2t \text{ cal} . \qquad (18\text{-}8)$$

Note that this joulean heating is independent of the direction of the current through the resistance.

We may deduce equation (18-7), using Ohm's law. From equation (18-5),

$$W = VIt \text{ joules} ,$$
$$V = RI \text{ (Ohm's law)} .$$

Therefore,

$$W = RI^2t \text{ joules} ,$$

or

$$W = \frac{1}{4.19} RI^2t \text{ cal} .$$

Let us assume that the conductor connecting the two points *a* and *b* (Fig. 18-5) contains a source of electromotive force, E, i.e., a device capable of putting electrical energy into the circuit (§ 18.3). We shall also assume that the electromotive force opposes the flow of electricity. This would be the case if there were a charging storage battery connected into the line (Fig. 18-6).† The

Fig. 18-6.—Electrical energy is absorbed by the cell and part transformed into heat in the resistance.

difference of potential between *a* and *b*, as measured by an electrometer or voltmeter, implies that energy is expended by a charge in moving from *a* to *b*. Where does the energy go? Part of the energy is converted into heat in the resistance between *a* and *b*, and part of the energy is absorbed by the cell. Thus

$$W = RI^2t + EIt \text{ joules} , \qquad (18\text{-}9)$$

where RI^2t is the heat produced and EIt is the work done in charging the battery. If we divide

---

† An electric motor connected at E would also furnish a counter electromotive force in the circuit (see chap. 27, § 27.7).

equation (18-9) by It, the charge, we obtain V for the difference in potential between a and b,

$$V = RI + E \text{ volts} . \qquad (18\text{-}10)$$

Also, since power is W/t,

$$P = RI^2 + EI \text{ watts} , \qquad (18\text{-}11)$$

where again RI^2 is the rate at which heat is produced and EI is the rate at which other work is being done. Equations (18-9)–(18-11) are simply statements of the conservation of energy as applied to electrical circuits. We shall return to this point again in chapter 20.

EXAMPLE 1: The current in Fig. 18-6 is 10 amp, and the resistance shown is 0.1 ohm. What is the value of the electromotive force of the battery if 32 watts of power are absorbed in the circuit between points a and b?

Solution: From equation (18-11),

$$E = \frac{P - RI^2}{I} = \frac{32 - (0.1)(100)}{10} = 2.2 \text{ volts} .$$

18.6. Nature of Resistance

Experiment shows that for wires of different sizes the resistance R increases with the length and diminishes as the cross-section becomes larger. Different materials of the same physical dimensions vary in their resistance. These facts may be described by the equation

$$R = \rho \frac{l}{A} , \qquad (18\text{-}12)$$

where l is the length (in meters), A is the cross-sectional area (in square meters), and ρ is a constant (called the RESISTIVITY) depending upon the material of the conductor. If we substitute the units of R, l, and A in this formula and solve for the units of ρ, we find that ρ is measured in ohm-meters.

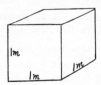

FIG. 18-7.—Resistivity is equivalent to the resistance of a meter cube.

If one had a solid cube 1 m on an edge (Fig. 18-7), then between two opposite faces $l = 1$ m and $A = 1$ m², so that $R = \rho$ numerically. From this the resistivity of a substance is sometimes de-fined as "the resistance between opposite faces of a cube of the substance, 1 m on an edge." For this reason resistivity is sometimes called the "resistance per meter cube." It should *not* be called the "resistance per cubic meter," since a conductor $\frac{1}{2}$ m² in cross-section and 2 m long would also contain 1 m³ of the material of the conductor, but its resistance would not be its resistivity.

The reciprocal of the resistance $1/R$ is called the *conductance* of the conductor and is generally designated by the letter G. The unit of conductance is called the *mho*. The reciprocal of the resistivity $1/\rho$ is called the *conductivity*, σ. The

TABLE 18-1

THE RESISTIVITY OF VARIOUS SUBSTANCES

Material	Temperature (° C)	Resistivity in Ohm-Meters
Silver............	18	1.629×10^{-8}
Copper............	20	1.7241×10^{-8}
Aluminum.........	20	2.828×10^{-8}
Nickel............	20	7.8×10^{-8}
Iron..............	20	10×10^{-8}
Platinum..........	20	10×10^{-8}
Advance ⎱ Constantan ⎰	0	$47-49 \times 10^{-8}$
Chromel ⎱ Nichrome ⎰	0	$70-110 \times 10^{-8}$
Mercury...........		94.07×10^{-8}
Germanium........		8×10^{-4}
Tellurium.........		0.2×10^{-2}
Graphite..........		8
Water (distilled).....	18	5×10^{3}
Wood.............		$3 \times 10^{8} - 4 \times 10^{11}$
Glass.............	20	9×10^{11}
Paraffin...........		1×10^{14}
Amber............		5×10^{14}
Sulphur...........		1×10^{15}

unit of conductivity is the *mho per meter*. The conductivity of copper is 5.8×10^7 mho/m; that of sulphur 10^{-19} mho/m. In many tables the resistivity is given in mixed units, ohm-cm, and σ in mho/cm.

While great progress is being made, year by year, in our knowledge of the structure of matter, still we cannot, as yet, completely correlate the resistivity of a substance with its other properties. We have a fairly clear picture, nevertheless, of the mechanism by which electricity is conducted through a metal (see §§ 18.9 and 37.14). Resistivity as a purely empirical quantity, however, is to be determined only by experiment. A short list of the resistivities of various common metals is given in Table 18-1. For a more complete table the reader is referred to any handbook of physics.

In Table 18-1 it is interesting to note that the resistivity of the better insulators is of the order of 10^{23} times as great as is the resistivity of the better conductors. *The fact that substances exhibit such enormous differences in their electrical resistivity is an important factor in making possible very exact electrical measurements.*

It is of interest to write Ohm's law in terms of the conductivity σ of a conductor and the electric field intensity $\mathcal{E}$ present. From equation (18-12) it follows that

$$R = \frac{l}{\sigma A}. \qquad (18\text{-}13)$$

If the current in the conductor of cross-sectional area A is I amp, then the average current density J in the conductor is

$$J = \frac{I}{A} \quad \text{or} \quad I = JA. \qquad (18\text{-}14)$$

If these quantities are substituted in equation (18-6), we have

$$V = IR = (JA)\left(\frac{l}{\sigma A}\right)$$

or

$$\frac{V}{l} = \frac{J}{\sigma}.$$

But V/l is the potential gradient, $\mathcal{E}$, along the conductor, so that

$$J = \sigma \mathcal{E}. \qquad (18\text{-}15)$$

Equation (18-15) is also Ohm's law and states that the current density in a conductor is proportional to the electric field. In this equation J is in amp/m², σ in mho/m, and $\mathcal{E}$ in volts/m.

EXAMPLE 2: An electric field intensity of 0.0344 volt/m exists between two points on a copper conductor. Calculate the current density in the conductor. If the cross-sectional area is 1 mm², what is the current in the conductor?

Solution: From equation (18-15),

$$J = \sigma \mathcal{E} = \left(\frac{1}{1.72 \times 10^{-8}} \text{ mhos/m}\right)$$

$$\times (0.0344 \text{ volts/m}) = 2 \times 10^6 \text{ amp/m}^2.$$

Since 1 mm² $= 10^{-6}$ m²,

$$I = JA = (2 \times 10^6 \text{ amp/m}^2)(10^{-6} \text{ m}^2) = 2 \text{ amp}.$$

18.7. Resistance and Temperature

The resistivity of a substance and hence the resistance of a wire of that substance depend upon the temperature. The relation between resistivity and temperature for lead is given in Fig. 18-8.

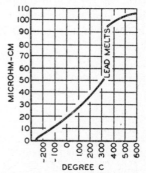

FIG. 18-8.—The resistivity of lead as function of temperature.

Lead melts at 327°.5 C. Between $-260°$ C and the melting point the graph shows a smooth curve. Any such type of curve may be represented in a purely empirical fashion by a power series of the form

$$R_t = R_1 [1 + a_1(t - t_1) + \beta_1(t - t_1)^2$$
$$+ \gamma_1(t - t_1)^3 + \ldots]. \qquad (18\text{-}16)$$

Here R_t is the resistance at $t°$ C and R_1 is the resistance at the temperature of reference, t_1. *The values of the coefficients depend upon the reference temperature.* For most metals in the temperature range ordinarily used, β_1, γ_1, and higher coefficients are so small as to be negligible for many purposes. This corresponds to assuming that within the range of temperatures under discussion, usually relatively small, the curve of Fig. 18-8 is a straight line. Then we may write

$$R_t = R_1 [1 + a_1(t - t_1)]. \qquad (18\text{-}17)$$

If we also choose 0° C as the reference temperature (this is not always done), we may write

$$R_t = R_0 (1 + a_0 t). \qquad (18\text{-}18)$$

Since the value of a depends upon the reference temperature, we have given it a subscript.

The values of these coefficients can be determined only by measuring the resistance of a wire at various temperatures. Our knowledge of the structure of matter is not yet sufficiently complete that these coefficients can be predicted or calculated from other properties of the substances.

283

If we substitute the known units for R_t, R_0, and t in equation (18-18) and solve for the units of a, we find that a has the units of deg^{-1}. This signifies that the numerical value of a depends upon the temperature scale used.

Approximate values of the temperature coefficient of resistance of a number of substances are given in Table 18-2. In this table it is to be noted

TABLE 18-2

Temperature Coefficient of Resistance

Material	a_0 in $(^\circ C)^{-1}$	Material	a_0 in $(^\circ C)^{-1}$
Aluminum.....	0.0039	Carbon.......	−0.0005
Copper........	.0039	Manganin....	.000000
Iron..........	.0062	Constantan⎫	.000008
Lead..........	.0043	Advance ⎬	
Platinum......	.0037	Nichrome⎫	.0004
Silver........	.0038	Chromel ⎬ ...	
Tungsten......	0.0045	Thermistor...	−0.044

that the temperature coefficients of resistance for various pure metals are all of about the same order of magnitude. Carbon has a negative but small coefficient. *Certain nonmetals, like porcelain, have large negative coefficients;* but these substances have an entirely different mechanism of conduction from that involved in metals.

It is of especial interest to note that the temperature coefficient of alloys is frequently very different from that of the constituent metals. Small traces of impurities in most metals will greatly change the values of these coefficients. Hence the values given in different tables may differ widely and may be quite different from the values one finds on testing commercial samples of the substances.

The temperature coefficient of the alloy manganin (84 per cent copper, 12 per cent manganese, 4 per cent nickel) is practically negligible. For this reason manganin is used extensively in the manufacture of standard resistors and of other electrical instruments where constancy of resistance is essential.

If the relation between the resistance and temperature of a wire is known, then the measurement of temperature may be made by measuring a resistance. For this purpose pure platinum wire is best. Values of $a = 0.00392$ $(^\circ C)^{-1}$ and $\beta = -0.000000588$ $(^\circ C)^{-2}$ represent well the performance of platinum for temperatures between -40° and $1,000^\circ$ C. The platinum resistance thermome-

ter is sufficiently good to be made the basis of the International Temperature Scale, between 0° and 66° C, using the three-constant formula

$$R_t = R_0 (1 + at + \beta t^2). \qquad (18\text{-}19)$$

For rough work a two-constant formula is adequate:

$$R_t = R_0 (1 + at) \qquad (18\text{-}20)$$

or

$$t = \frac{R_t - R_0}{aR_0}. \qquad (18\text{-}21)$$

18.8. Thermistors

There are materials known as "semiconductors" that are assuming increasing importance as resistive elements because of their unusual temperature-resistance characteristics. These resistors with large negative temperature coefficients are called *thermistors*.

Thermistors are made of various mixtures of oxides of manganese, nickel, cobalt, iron, zinc, titanium, and magnesium and come in the shapes of beads, cylinders, and thin disks. These oxides are mixed in the proper proportion to provide the required resistivity and temperature coefficient and then are sintered under controlled temperatures. A thermistor made of the oxides of manganese and nickel has a temperature coefficient of -0.044 per $^\circ$ C at 25° C and a resistivity that ranges from 10^7 ohm-cm at -100° to 200 ohm-cm at 100° C. Over this same range the resistivity of platinum changes by a factor of about 2.

A plot of the variation in resistivity with temperature for a manganese-nickel mixture is shown in Fig. 18-9.

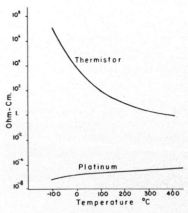

Fig. 18-9.—The variation of resistivity with temperature of a thermistor compared with platinum.

The resistivities are plotted logarithmically. It is found experimentally that the resistivity of thermistors obeys a law of the form

$$\rho = \rho_0 e^{A(1/T - 1/T_0)}, \qquad (18\text{-}22)$$

where ρ is the resistivity at the absolute temperature T, ρ_0 at T_0, and A a constant which depends on the material.

Because of their unusually wide range in temperature resistance, good electrical stability, and small compact sizes, thermistors are being widely used in measurement equipment. Temperatures can be easily measured with a standard Wheatstone bridge (see § 22.6) to an accuracy of ± 0.01 C over a temperature range of $-80°$ to $+200°$ C. Thermistors have been used as compensators for resistors with positive temperature coefficients, as power-measuring elements at radio frequencies, and as a time-delay element in electrical circuits.

18.9. Theory of Metallic Conduction

According to present ideas concerning the structure of matter discussed in §§ 17.3 and 38.22, a metallic conductor consists of atoms or, more exactly, ions forming a crystal lattice and a cloud of electrons free to move about within the interstices of the lattice, in such a way, however, that each small volume of the conductor is electrically neutral. At a given temperature, T, the ions are thought of as vibrating about their positions of equilibrium. The electrons in their random motion about the conductor are supposed to collide frequently with the ions and thus acquire a kinetic energy characteristic of the temperature.

When a potential difference V is set up between two points in a conductor, the following phenomena occur: Under the influence of the field the electrons acquire a motion in the direction of the field. This motion is superimposed on the random motion already possessed by the electrons. The electron moves a short distance freely, then collides with an ion of the lattice, giving up the energy it gained. It again moves freely, making another collision, and so the cycle is repeated. The electron, however, moves progressively in the direction of the field. The energy given up increases the mean kinetic energy of the ion, thus increasing the amplitude of motion of the ion and making collisions with electrons more frequent. This explains the joulean heat of equation (18-7).

The drift velocity of the electrons along the conductor is not large; in fact, it is much smaller than the random motion of the electron. We may roughly estimate this drift velocity as follows: Let us assume that a conductor of cross-sectional area A has n free electrons per cubic meter. If the drift

velocity in the direction of the field is v, then the current I is

$$I = Anev. \qquad (18-23)$$

If we take the reasonable values $I = 1$ amp; $e = 1.6 \times 10^{-19}$ coulomb, the charge on the electron; $A = 10^{-6}$ m²; and assume that n for copper is equal to the number of atoms per cubic meter and hence equal to about 8.3×10^{28}, we obtain, for v,

$$v = \frac{1}{10^{-6} \times 1.6 \times 10^{-19} \times 8.3 \times 10^{28}}$$
$$= 7.5 \times 10^{-5} \text{ m/sec}.$$

18.10. Superconductivity

All metals show a resistivity that diminishes as temperature decreases. The rate of diminution, when observed at higher temperatures, is one which is roughly proportional to absolute temperature. The temperature coefficients of resistance are of the same order of magnitude, and some of them, indeed, are not far, numerically, from the temperature coefficients of pressure or volume expansion of gases. This would imply, if extrapolation were allowable, zero resistance at $0°$ K, i.e., at $-273°$ C.

However, in 1911 Kamerlingh Onnes discovered that some metals, notably mercury, drop in resistance quite suddenly from values that are about 0.001 their resistance at $0°$ C to practically nothing at all, i.e., resistances that are of the order of 10^{-20} of their resistance at $0°$ C. The results obtained by Onnes in 1911 for mercury are shown in Fig. 18-10. So nearly perfect does the conductivity be-

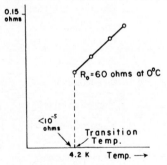

Fig. 18-10.—The resistance of mercury as a function of absolute temperature showing the discontinuity in resistance at the transition temperature of 4.2 K.

come below the "transition temperature" that currents inductively started in them decrease less than 1 per cent per hour and may not fall to half-value before the better part of a day has elapsed. By such studies of the depletion of persistent currents, the resistivity below the transition temperatures has been established to be less than 10^{-20} ohm-cm.

TABLE 18-3

ELEMENTS EXHIBITING SUPERCONDUCTIVITY

Element	Transition Temp. (° K)	Element	Transition Temp. (° K)
Aluminum 13....	1.2	Lanthanum 57..	4.7
Titanium 22.....	0.5	Hafnium 72.....	0.3
Vanadium 23....	4.3	Tantalum 73....	4.4
Zinc 30.........	0.9	Rhenium 75.....	0.9
Gallium 31......	1.1	Mercury 80.....	4.2
Zirconium 40....	0.7	Thallium 81....	1.2
Niobium 41......	9.2	Lead 82........	1.2
Cadmium 48.....	0.6	Thorium 90.....	1.4
Indium 49.......	3.4	Uranium 92....	0.8
Tin 50..........	3.7		

Table 18-3 shows the elements found to possess a superconducting state. Certain compounds, such as niobium nitride and columbium nitrite, also possess a superconductivity state. As a matter of fact, the highest transition temperature is that of columbium nitrite at 14° K. Other metals acquire, as these very low temperatures are reached, a *zero temperature coefficient of resistance* but *not a zero resistance*. Such metals as gold, silver, copper, and bismuth remain normal below 1° K.

An interesting property of the superconducting state is that at each temperature below the transition temperature there exists a critical value of a magnetic field that will destroy superconductivity. At present there is no good accepted explanation of superconductivity.

18.11. Summary of the mks Units

It is convenient to summarize here the mks absolute system of electrical units which we shall

TABLE 18-4

RELATION OF THE mks TO OTHER SYSTEMS OF UNITS

Symbol	Quantity	Definition	Unit	Equivalents
l..........	Length	....	Meter	10^2 cm
m..........	Mass	....	Kg	10^3 gm
f..........	Force	$f = ma$	Newton	10^5 dynes
W.........	Work	$dW = f \cos \theta ds$	Joule	10^7 ergs
P..........	Power	$P = dW/dt$	Watt	10^7 ergs/sec
$\mathcal{L}$..........	Torque	$\mathcal{L} = fl \sin \theta$	Newton-m	10^7 dyne-cm
i..........	Electric current	See text, § 18.11	Amp	0.1 emu, 3×10^9 esu
q..........	Electric charge	$q = it$	Coulomb	0.1 emu, 3×10^9 esu
PD, V	Potential difference	$V = W/Q = P/i$	Volt	10^8 emu, 1/300 esu
R..........	Resistance	$R = PD/i$	Ohm	10^9 emu, $1/9 \times 10^{11}$ esu
$\mathcal{E}$..........	Electric field strength	$\mathcal{E} = f/q = -dV/dr$	Newtons/coulomb or volts/m	10^6 emu, $1/3 \times 10^4$ esu
C..........	Capacitance	$C = q/V$	Farad	9×10^{11} esu
ϵ_0..........	Permittivity of free space	$f = q_1 q_2 / 4\pi \epsilon_0 r^2$	Farads/m	$4\pi \times 9 \times 10^9$ esu
κ..........	Relative dielectric constant	$\kappa = C_\kappa / C_0$	....	$\kappa = 1$ *in vacuo*
L..........	Inductance	$Ldi/dt = -V$	Henry	10^9 emu
F.........	Magneto-motive force	$F = Ni$	Amp-turn	0.4π gilbert
H.........	Magnetic field intensity	$dH = i \sin \theta ds / 4\pi r^2$ $H = Ni/l$ (long solenoid)	Amp-turn/m or amp/m or newtons/weber	$4\pi/10^3$ oersted
M........	Magnetic moment	$MH \sin \theta = \mathcal{L}$	Weber-m or newton-m²/amp	$10^{10}/4\pi$ emu
p.........	Pole strength	$p = M/l$	Newton-m/amp or weber	$10^8/4\pi$ emu
ϕ.........	Magnetic flux	$V = -N(d\phi/dt)$	Weber	10^8 maxwell
$\mathcal{R}$.........	Reluctance	$\mathcal{R} = F/\phi$	Rowland	$4\pi/10^9$ emu
B..........	Magnetic induction or flux density	$df = Bi \sin \theta ds$ $BA = \phi$	Webers/m² or newtons/amp-m	10^4 gauss
μ..........	Permeability	$\mu = B/H$	Webers/amp-m or newtons/amp²	$10^7/4\pi$ emu
μ_r..........	Relative permeability	$\mu_r = \mu/\mu_0$	....	$\mu_r = 1$ *in vacuo*
I..........	Intensity of magnetization	$I = M/V = p/A$ $(B = \mu_0 H + I)$	Webers/m²	$10^4/4\pi$ emu

use extensively in our study of electrical circuits. These are the definitions adopted by the International Electrotechnical Commission and were officially adopted in this country on January 1, 1948.

Ampere.—The ampere is an electric current of such magnitude that when maintained in two straight parallel conductors of infinite length and of negligible cross-section, at a distance of 1 meter from each other in a vacuum, it would produce between the conductors a force of 2×10^{-7} newton per meter of their length (see § 21.12).

Volt.—The volt is the unit of electromotive force or of potential difference. It is the difference of electrical potential between two points (or two equipotential surfaces) in a conductor carrying a constant current of 1 ampere when the power dissipated between these points is 1 watt (see § 18.2).

Ohm.—The ohm is the electric resistance between two points of a conductor when a constant difference of potential of 1 volt maintained between these points produces a current of 1 ampere in the conductor, the conductor not being the source of any electromotive force (see § 18.4).

Coulomb.—The coulomb is the quantity of electricity transported in 1 second across any cross-section of a circuit by a current of 1 ampere (see § 17.7).

Farad.—The farad is the capacitance of an electric capacitor in which a charge of 1 coulomb produces a difference of potential of 1 volt between the plates of the capacitor (see § 17.23).

Henry.—The henry is the inductance of a circuit in which an electromotive force of 1 volt is produced when an electric current in the circuit changes uniformly at the rate of 1 ampere per second, or the inductance between two circuits when a uniform change of 1 ampere per second in one of them produces an electromotive force of 1 volt in the other (see chap. 23).

18.12. Résumé

The following is a brief listing of the definitions, principles, and theories, given in this chapter, which you should know:

Definition of electric current
Definition of electromotive force
Distinction between an electromotive force and a potential difference
Ohm's law
Definition of the ohm
Energy and power relationships in a current-carrying circuit
Resistivity and conductivity of a conductor
Variation of resistance with temperature
Superconductivity and transition temperatures

Queries

1. What does a negative temperature coefficient of resistance imply?
2. How many of the following electrical quantities must you know in order to derive the others: (*a*) energy in joules, (*b*) potential difference in volts, (*c*) power in watts, (*d*) current in amperes, (*e*) resistance in ohms, and (*f*) charge in coulombs? What interrelations exist between these quantities?
3. What distinction exists between the resistance of a conductor and the resistivity of a conductor?
4. Is it possible to deduce Ohm's law from Joule's law? What other information is necessary?
5. What change would be necessary in equation (18-9) if the battery *E* of Fig. 18-6 were reversed? What would the modified equation mean physically?

Practice Problems

1. A wire has a cross-section of 2 mm² and is 500 cm long. Its resistance is 0.03 ohm. What is the resistivity of the material of the wire? Ans.: 1.2×10^{-8} ohm-m.
2. A carbon lamp has a resistance of 108.9 ohms at 20° C and of 82.5 ohms at 500° C. Compute its temperature coefficient of resistance.
3. Each of the cables connecting a powerhouse and factory has a resistance of $\frac{1}{3}$ ohm/mi. The factory needs 60 hp (44.76 kw). If the potential difference

287

between the wires at the entrance to the factory is 100 volts, compute the power which the powerhouse must supply. Ans.: 179 kw.

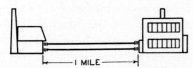

4. In problem 3, if the potential difference between the wires at the entrance to the factory is 10,000 volts, compute the power which the powerhouse must supply.

5. How much energy is necessary to run a 50-watt lamp for half an hour? Ans.: 9×10^4 joules.

6. The resistance of a platinum resistance thermometer at $0°$ C in 20 ohms. When placed in an oil bath, the resistance is found to be 30 ohms. Compute the temperature of the bath, using the two-constant formula.

7. A standard resistance of 1 ohm is to be made, using constantan wire of 1 mm in diameter. What length of wire is required? Ans.: 163.6 cm.

8. A heating coil is to be constructed of nichrome wire of diameter 0.02 in. If the coil must dissipate 500 watts when connected to the 110-volt line, what length of wire in feet is required? Use a resistivity of 90×10^{-6} ohm-cm.

9. If energy costs 5 cents per kilowatt-hour, how much does it cost to operate an electric motor that takes 6 amp at 110 volts, assuming the motor runs for 8 hr? Ans.: 26.4 cents.

10. What is the difference in potential of the adjacent turns of a solenoid that takes 60 amp when 120 volts are applied? The solenoid has ninety turns of wire. Ans.: 1.33 volts.

11. A current of 10 amp passes through a resistance of 20 ohms for 5 min. What is the potential difference across the resistance? How much work is done in joules? What is the power dissipation? Ans.: 200 volts; 6×10^5 joules; 2,000 watts.

12. What is the efficiency of a 6-hp motor that draws 25 amp at 220 volts?

13. Predict the outcome of the connections of the two lamps as shown herewith. The normal operating

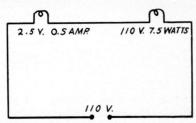

ratings are shown. Ans.: The 2.5-volt lamp would not light.

14. A standard string of eight Christmas tree lights attached to a 110-volt line (all go out if one lamp burns out) could be rated at about 24 watts. Therefore, each bulb could be rated as a .... watt .... volt bulb. In normal operation each bulb has a resistance of .... ohms and passes .... amp. (*Note:* The change in resistance with temperature is neglected. Assume the resistance of each lamp to be the resistance under normal operation.)

15. Show that a standard 550-watt (110-volt) radiant heater element in series with a 32-candle-power (cp) (approximately 32-watt), 6-volt microscope lamp, will offer the necessary extra resistance to allow operation directly from the 110-volt household lines.

16. A parallel-plate capacitor whose area is 900 cm² and whose separation is 0.1 cm has a dielectric of value 12. If the capacitor is charged at the rate of 10 volts per microsecond (μsec), compute the charging current in amperes.

17. An excellent capacitor has a discharge rate of 1 volt per 24 hours. If the capacitor has a capacitance of 864 $\mu\mu$f, compute the discharge current in amperes. What power does the capacitor deliver when its potential is 1,000 volts? Ans.: 10^{-14} amp; 10^{-11} watt.

18. The drop in potential along an aluminum power transmission line 300 km long is 10,000 volts. What is the current density in the line?

19. Compute the conductance of a copper wire 10^3 m long and 10^{-4} m² in cross-sectional area. Ans.: 5.8 mhos.

20. The resistance of a coil of nichrome wire at $20°$ C is 80 ohms. What is its resistance at $130°$ C?

21. The electric field between two points on a silver conductor is 3.26 volts/m. What is the current density in the conductor? Ans.: 2×10^8 amp/m².

22. A tungsten resistor whose resistance at $20°$ C is 100 ohms is connected to a constant potential of 10 volts. The temperature of the resistor rises to $1,000°$ C. Compute the current in the resistor at temperatures of $20°$, $100°$, and $1,000°$ C. Assume that equation (18-17) is applicable.

23. The temperature of a water bath of 1,000-liter capacity is raised $60°$ C. Compute the cost of heating the water if electrical energy is 4 cents/kwh. Ans.: $2.79.

24. Calculate the drift velocity of the electrons in a conductor if the current density is 10^7 amp/m² and the moving charge per unit volume is 8×10^9 coulombs/m³.

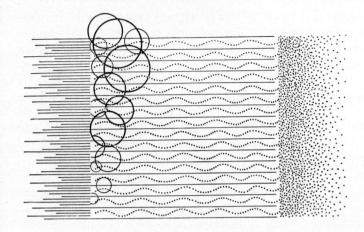

CHAPTER 19

Conduction of Electricity through Liquids: Electric Cells

A. Electrolysis

In the previous chapter we confined our attention to the phenomena associated with the flow of electricity through metallic conductors. The characteristics of metallic conduction were found to be (1) the presence of resistance, which generally increases with a rise in temperature; (2) the constancy with temperature of the ratio of V/I, which is called "Ohm's law"; (3) a conduction process, which consisted in the flow of electrons, i.e., *no matter* (e.g., no copper) was transported, nor was there any chemical change discoverable in the conductor anywhere.

19.1. Conduction through Liquids

We come next to the discussion of the phenomena that take place when electricity flows through liquids. Let us look at the following experiment: Two platinum electrodes connected in series with an ammeter and a source of electromotive force, such as a battery or power mains, are dipped first into a beaker containing liquid mercury (Fig. 19-1 [a]). The ammeter reads a sizable current. Although mercury is liquid, nevertheless we find that electricity is conducted through it just as it is conducted through a copper wire, so that we need to consider metallic liquids no further.

If the electrodes are dipped into freshly distilled

water, we note that no appreciable amount of electricity is conducted through the water. Liquids of this type—distilled water, oils, etc.—have so great a resistivity that they are essentially nonconductors, and we shall exclude them from further discussion (see Table 18-1).

In case (c) of Fig. 19-1, however, we find that a pinch of sodium chloride (NaCl) or a small

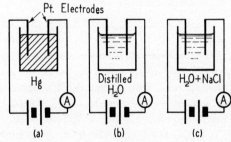

Fig. 19-1.—Conduction of electricity through liquids; appreciable currents pass between the electrodes of cells (a) and (c).

amount of copper sulphate ($CuSO_4$) added to the distilled water renders the solution a fairly good conductor. In general, solutions of acids, bases, and salts in water or similar solvents are good conductors. However, to make a solution a conductor of electricity, we must choose both a suitable solvent and a suitable solute. Thus solutions of alcohols and sugar in water are poorer condnctors than water, whereas hydrogen chloride (HCl) in water is a good conductor. A solution of hydrogen chloride in benzene is not a good conductor.

An aqueous solution of hydrogen chloride or of sodium chloride belongs to an important class of solutions known as *electrolytes*.

19.2. Electrolysis

When a current is passed through an electrolyte, two important phenomena take place. At the electrodes chemical reactions occur. The chemical reaction that occurs at the anode (the electrode at which the current enters) is, in general, different from that which takes place at the cathode (the electrode at which the current leaves). For example, in the electrolysis of an aqueous solution of hydrogen chloride, chlorine gas is evolved at the anode, while hydrogen gas is evolved at the cathode.

The second important observation is that *matter is actually transported through the solution*. The conduction of electricity in electrolytes, technically called ELECTROLYSIS, is due to an entirely different mechanism from that responsible for the conduction of electricity through metals. We shall outline this new mechanism in the following paragraphs.

19.3. The Arrhenius Theory of Electrolytic Dissociation

The Swedish chemist Svanté Arrhenius (1859–1927) was the first to suggest that when a hydrogen chloride molecule dissolves in water, the molecule immediately splits into two parts, or, as we say, "dissociates," into a positive hydrogen ION H^+ and a negative chlorine ION Cl^-. By placing platinum electrodes having a difference in potential between them into the hydrogen chloride solution, we introduce an electric field within the electrolyte. Each ion, therefore, is acted upon by a force equal to $\mathscr{E}q$, where $\mathscr{E}$ is the effective electric field and q is the charge on the ion. This force causes the ion to migrate or drift, the positive ions going to the cathode, the negative ions to the anode.

According to Arrhenius, the molecule dissociates into ions the instant it dissolves. *The dissociation is not due to the presence of the electrodes or of any applied field.*

As we have already pointed out, modern theories of atomic structure consider the atom as an electrical system with a positive nucleus and a number of electrons circulating about the nucleus. If an atom should lose an electron from its outer region, the remaining structure, called a *positive ion*, would have one unit excess of positive charge. If two electrons are lost, the positive ion would have an excess of two positive charges on the nucleus, and the ion would be *doubly ionized*.

The electrons lost in this process attach themselves to other atoms, which thereby acquire excess negative charges and become *negative ions*. Thus, in the dissociation of a hydrogen chloride molecule, the hydrogen atom loses one electron and becomes a H^+ ion, while the chlorine gains one electron and becomes a Cl^- ion.

It is at once obvious, then, that the *chemical properties of an ion* are very different from the properties of the corresponding neutral atom or molecule. In the case of a solution of sodium chloride, the sodium ion (Na^+) is stable in an aqueous solution of sodium chloride. A neutral sodium atom (Na) would react violently.

In the case of aqueous solutions in general, it is believed that the water molecule also undergoes a very slight dissociation into H^+ and OH^- ions. It is not believed that H^+ ions exist in solution but that they are attached to a neutral water molecule to form the complex ion $(H_3O)^+$. This distinction is not important to our analysis.

19.4. Typical Electrolytic Reactions: Effect of the Electrodes

We shall apply the ideas outlined here to the description of a few typical cases of electrolysis.

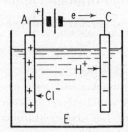

FIG. 19-2.—Electrolysis of hydrogen chloride

1. *Electrolysis of hydrogen chloride.*—Two platinum electrodes, *A* and *C*, are immersed in a solution of hydrogen chloride, *E* (Fig. 19-2). The electrodes are connected to a battery or some other source of electromotive force. In Fig. 19-2, *A* rep-

resents the anode, while C is the cathode. Although the current enters A and leaves C by convention, actually the electrons flow oppositely, as indicated.

Owing to the electric field between A and C, the H^+ ion will be attracted toward the cathode C, while the Cl^- ions are repelled by the cathode and attracted to the anode, A. The H^+ ion, on touching the cathode, receives an electron from the cathode and escapes as neutral gaseous hydrogen. We may represent the reaction at the cathode as

$$2\,e + 2H^+ \rightarrow H_2 \ .$$

The letter e represents a negative electron. At the anode the reaction is

$$2Cl^- \rightarrow Cl_2 + 2\,e \ ,$$

with the chlorine gas escaping.

2. Electrolysis of silver nitrate.—In the electrolysis of silver nitrate ($AgNO_3$), using *silver* electrodes, the following phenomena occur. Because of the electric field, the ions begin to migrate, the Ag^+ toward the cathode and the NO_3^- toward the anode. At the cathode the silver ion takes on an electron and deposits on the electrode as a silver atom:

$$Ag^+ + e \rightarrow Ag \ .$$

At the anode, instead of the NO_3^- ion discharging, a silver ion from the electrode goes into solution:

$$Ag - e \rightarrow Ag^+ \ .$$

In this way the number of NO_3^- and Ag^+ ions in the solution remains constant. This reaction is essentially the one used in silver electroplating.

If platinum electrodes are used instead of silver, the Ag^+ and NO_3^- ions migrate as before. At the electrodes the following reactions are found to take place:

Cathode: $4Ag^+ + 4e \rightarrow 4Ag$,

Anode: $4OH^- \rightarrow 2H_2O + O_2 + 4e$.

Platinum is chemically inert and does not go into solution. At the anode the hydroxide ion is discharged, giving rise to oxygen gas. The discharge of the OH^- ion is a *secondary reaction*.

3. Electrolysis of acidulated water.—As a final example of electrolysis, we may cite the decomposition of water as a *secondary reaction* again in the electrolysis of dilute sulphuric acid (H_2SO_4). With platinum electrodes immersed in the acid,

the H^+ of the dissociated sulphuric acid migrates toward the cathode, discharging as hydrogen gas,

$$4H^+ + 4\,e \rightarrow 2H_2 \ .$$

On reaching the anode, the sulphate ion ($SO_4^=$) is not discharged, but rather the hydroxide ion (OH^-) of the water:

$$4OH^- \rightarrow O_2 + 2H_2O + 4\,e \ .$$

From this analysis we may conclude that in the electrolysis of solutions the ions that migrate through the solution and thus carry the current from one electrode to the other *are not necessarily* the ones that are discharged at the various electrodes. If the ions that migrate are also discharged, the reaction is called a *primary reaction*. If other ions are discharged, the reaction is called *secondary*.

What the discharge products will be depends upon many factors. In general, the prediction of the products requires extensive chemical knowledge. Some typical reactions are listed in Table 19-1.

TABLE 19-1

TYPICAL ELECTROLYTIC REACTIONS

Electrolyte	Electrodes	Cathode Reaction	Anode Reaction
$AgNO_3$..	$\begin{cases} Ag \\ Pt \end{cases}$	$Ag^+ \ + e \rightarrow Ag$ $4Ag^+ \ +4e \rightarrow 4Ag$	$Ag \ - e \rightarrow Ag^+$ $4OH^- - 4e \rightarrow O_2 + 2H_2O$
H_2SO_4..	Pt	$4H^+ \ +4e \rightarrow 2H_2$	$4OH^- - 4e \rightarrow O_2 + 2H_2O$
NaI....	Ag	$2H^+ \ +2e \rightarrow H_2$	$2I^- \ -2e \rightarrow I_2$
NaCl...	Pt	$2H^+ \ +2e \rightarrow H_2$	$2Cl^- \ -2e \rightarrow Cl_2$
$\left.\begin{matrix} FeCl_2 \\ FeCl_3 \end{matrix}\right\}$..	Pt	$Fe^{+++} + \ e \rightarrow Fe^{++}$	$Fe^{++} - \ e \rightarrow Fe^{+++}$

19.5. Faraday's Laws of Electrolysis

Michael Faraday was the first to investigate systematically the phenomena attending the passage of electricity through liquids. By a series of brilliant experiments he was able in 1833 to correlate these complex phenomena into two simple laws.

1. The mass of any material released or deposited out from any electrolyte by the passage of electricity is proportional to the QUANTITY *of electricity that passed through the solution.*

2. The masses of substances liberated at the electrodes, when the same quantity of electricity passes through solutions of different electrolytes, are proportional to their CHEMICAL EQUIVALENTS.

The first law may be stated mathematically, as follows:

$$M = Zq = Zit , \qquad (19\text{-}1)$$

where M is the mass in grams, q the charge in coulombs, i the current in amperes, t the time in seconds, and Z, the *electrochemical equivalent*, is a constant of proportionality characteristic of the element deposited and the compound from which it was deposited. The units of Z are grams per coulomb.

In the statement of the second law it is necessary to define the term *chemical equivalent* (CE), which is not to be confused with "*electro*chemical equivalent" in the preceding paragraph. The chemical equivalent of any element or radical is defined as the ratio of the atomic or group weight (W) in grams to the valence v. That is, $CE = W/v$. It is the mass of an element that will replace or combine with 1.00813 gm of hydrogen in any chemical reaction. Experimentally it is found that liberating 1.00813 gm of hydrogen by electrolysis requires the passage of 96,519 coulombs of electricity. This quantity of electricity is called the *faraday* (F). This same quantity of electricity will liberate half the atomic weight of a bivalent element or one-third the atomic weight of a trivalent element.

To illustrate these ideas, let us send 96,519 coulombs of charge through four cells connected in series and determine the masses deposited at the electrodes (Fig. 19-3). The first cell consists of silver electrodes in a solution of silver nitrate; the second, copper electrodes in copper sulphate; the third, platinum electrodes in hydrochloric acid; while the last cell consists of platinum electrodes in dilute sulphuric acid. The products of electrol-

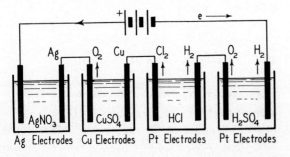

Fig. 19-3.—The masses deposited on the electrode per 96,519 coulombs are equal to the chemical equivalents of the elements in solution.

ysis are indicated next to the electrodes in the diagrams.

The results of this experiment are summarized in Table 19-2. The electrochemical equivalent Z is found by equation (19-1), using 96,519 as q and the mass deposited as M. The chemical equivalent

TABLE 19-2

Some Values Used in Electrolysis

Element	Mass Deposited (Gm)	Atomic Weight	Va-lency	Chemical Equiva-lent	Electrochemical Equivalent (Gm/Coulomb)
Ag.....	107.909	107.909	1	107.909	0.0011180
Cu.....	31.8	63.57	2	31.8	.0003294
H₂.....	1.0083	1.0083	1	1.0083	.00001046
Cl₂.....	35.46	35.46	1	35.46	.0003675
O₂......	8.00	16.00	2	8.00	0.00008291

is known from the systemized data of chemical combination.

Faraday's laws of electrolysis may be *summarized* in the following *single law;*

In electrolysis, 1 faraday of electricity (96,519 coulombs) liberates one chemical equivalent (atomic weight/valence) of an element from solution, and other quantities liberate proportional amounts.

Faraday's law, as stated here, has been subjected to severe experimental tests and is found to be one of the most exact laws in physical science. It holds at low and high temperatures, in dilute and concentrated solutions, at various pressures, and with different solvents. Faraday's law applies to those chemical changes that involve electron transfer.

In this connection it is important in very accurate work to distinguish between the two scales of atomic weights—the chemical scale, in which the atomic weight of naturally occurring oxygen (with all its isotopes [see § 38.4]) is taken by definition to be 16, and the physical scale, in which the O^{16} isotope alone is given a weight of 16. The accuracy with which the physicist has been able to measure the relative masses of different isotopes is much greater than has been achieved by chemical techniques. On the basis of the physical scale, where the mass of O^{16} is exactly 16, the average of the chemical to physical scales of atomic masses is 1.000272. The atomic weights in Table 19-2 are corrected to the physical scale. The best value for the faraday has been determined by the National Bureau of Standards by the electrolytic oxidation of sodium oxalate; $F = 96,519.4 \pm 0.7$ coulombs/mole, physical scale.

19.6. Faraday's Laws and the Atomicity of Electricity

If Faraday's laws of electrolysis are considered from the point of view of the modern theory of atomic structure, they lead to a remarkable prediction about the nature of electricity. It was pointed out by the great German physicist, Hermann von Helmholtz (1821–94), in an address on the researches of Faraday, that "if we accept the hypothesis that the elementary substances are composed of atoms, we cannot avoid concluding that electricity also is divided into definite elementary portions which behave like atoms of electricity."

From Faraday's laws we know that 96,519 coulombs of electricity will deposit 107.909 gm of silver. The atomic theory tells us that 1 gm atomic weight (107.909 gm) contains 6.0254×10^{23} atoms of silver (Avogadro's number). Therefore, during the electrolysis of silver nitrate each silver ion Ag^+ transports, on the average, a charge e, given by

$$e = \frac{96,519 \text{ coulombs}}{6.0254 \times 10^{23} \text{ atoms}}$$
$$= 1.602 \times 10^{-19} \text{ coulomb/atom}$$
$$= 4.802 \times 10^{-10} \text{ esu/atom}.$$

The NO_3^- molecular group also carries 1.602×10^{-19} coulomb of electricity. Similarly, Cu^{++} of copper sulphate carries $2 \times (1.602 \times 10^{-19})$ coulomb, since 96,519 coulombs liberate half the gram atomic weight of copper.

The fact that ions in electrolytes are found invariably associated with $1 \times$, $2 \times$, $3 \times$, etc., 1.602×10^{-19} coulomb of charge and never with any other multiple or any fraction of this amount was strikingly paralleled by experiments in an entirely different field of study—that of conductivity of electricity in gases (chap. 38). This elementary charge e is identical with that of the electron whose existence we postulated in the earlier chapters.

Since e, the charge on the electron, and F, the faraday, can both be experimentally determined

with precision, we use these values to compute Avogadro's number (N):

$$N = \frac{F}{e}. \text{ *} \qquad (19\text{-}2)$$

From the outline given previously, we have a clue as to the reason why Faraday's law is so exact. *It represents essentially a counting process:* for each univalent positive ion that touches the cathode and is discharged, an electron is transferred from the cathode to the ion.

19.7. Ohm's Law: Resistance of an Electrolyte

When a potential difference V is applied across the electrodes of an electrolytic cell, a current I is found to pass. The question arises as to whether Ohm's law is applicable to electrolytic conduction. It is found that the current I is not proportional to V, but rather to $V - E'$, where E' is a constant for a particular cell. This result suggests that a cell acts like a battery, giving rise to an electromotive force that opposes the flow of electricity. We shall discuss such "polarization" electromotive forces in greater detail in § 19.11.

When polarization is taken into account, we find that

$$\frac{V - E'}{I} = \text{Constant} = R, \qquad (19\text{-}3)$$

where the constant, R, is the resistance of the cell.

In order to learn how the resistance varies with the geometry of the electrolytic conductor, let us consider the resistance of an electrolyte confined to a cylinder. We find that, for a given concentration of silver nitrate and at a given temperature, the resistance, as defined by equation (19-3), is directly proportional to the length and inversely proportional to the cross-sectional area of the electrolyte, as in metallic resistance. Unlike the metallic case, however, we discover that the resistance *decreases* with *increase in temperature;* that is to say, an electrolyte has a *negative temperature coefficient* of resistance. This, it will be recalled, is true also of such solid and rather poor conductors as carbon and porcelain (§ 18.8), which also had negative temperature coefficients.

B. Electric Cells or Batteries

19.8. Conversion of Chemical Energy into Electrical Energy

We have already described methods of maintaining differences of potential by means of electrostatic machines. Here mechanical energy was converted into electrical energy. There are, at present, more convenient sources of electrical energy, known as "galvanic" or "voltaic" cells, whose operation depends upon the continuous conversion of chemical energy into electrical energy.

The discovery of the voltaic cell was made accidentally in 1792 by an Italian professor, Luigi Galvani, of the University of Bologna, who noticed that a frog's leg twitched when touched with two dissimilar metals. The same phenomenon occurred if the frog's leg was attached to an electrostatic machine. Galvani attributed the effect to "animal electricity." However, another Italian professor, Alessandro Volta, of the University of Pavia, took issue with this conclusion and showed, in 1800, that the essential phenomena of Galvani did not depend upon the presence of animal tissue but was due rather to the presence of the two dissimilar metals. By placing moistened paper sheets between strips of zinc and copper alternately, Volta constructed the first practical voltaic cell, called at that time a voltaic "pile." Soon after the announcement of Volta's discovery was made, investigators in other countries rapidly devised many types of cells, a few of which are still used today. Most of them, however, are now obsolete.

A simple voltaic cell may be constructed by immersing a zinc and a copper rod in dilute sulphuric acid. We find that immediately on immersion a potential difference is established between the zinc and copper rods. This observation is one of fundamental importance.

The existence of this potential difference may be demonstrated by means of a capacitor and a gold-leaf electroscope. As shown in Fig. 19-4, we connect the upper plate of the capacitor to the earth, while the lower one is connected to the electroscope. With the plates very close together (large capacitance), the cell is connected to the capacitor, as shown. The lower plate receives a positive charge and the upper plate a negative charge. Because of the small difference in potential between the zinc and copper (about 1 volt), the

electroscope shows no appreciable effect. However, on disconnecting the cell and separating the plates by a large distance, the potential difference is greatly increased (why?) and may be sufficient to deflect the gold-leaf electroscope. This experiment shows that the electric charges associated with batteries and those obtained in electrostatics

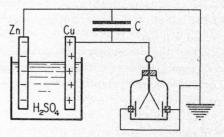

Fig. 19-4.—The potential difference between the electrodes can be demonstrated by the electroscope and capacitor C.

from electrified glass or wax are fundamentally identical.

When the two electrodes are joined by a resistance, a current, I, passes from copper to zinc outside the cell and from zinc to copper within the cell. The presence of the current is detected by the galvanometer, G, of Fig. 19-5. Coincident with the

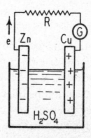

Fig. 19-5.—The simple $Zn(H_2SO_4)Cu$ cell

flow of electricity around the circuit, we find chemical activity present within the cell.

If zinc dissolves in sulphuric acid, a definite amount of heat, known as the "heat of reaction," is given to the environment. In the voltaic cell this energy, in general, appears as electrical energy rather than as heat-of-reaction energy.

We may ascribe 1 volt of *electromotive force* to the voltaic cell for every joule of work per coulomb transported that is done by the cell in lifting a charge from the potential of the zinc to that of the copper.

19.9. Standard Electrode Potentials; Electromotive Series

What is found to be characteristic of all voltaic cells is the presence of an electric field in the *boundary layer* between an electrode and an electrolyte. This electric field makes itself known by the presence of a difference in potential between the electrode and the electrolyte.

In the case of the zinc-copper cell described previously, the copper is at a lower potential than the acid, which, in turn, is at a higher potential than the zinc. We may conveniently represent the relative values of these potentials by means of the diagram of Fig. 19-6. We note a sharp discon-

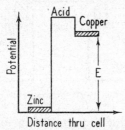

Fig. 19-6.—Variation of potential through a Zn(H$_2$SO$_4$)Cu cell. The potentials are not absolute values but relative to the zinc electrode.

tinuity in the potential on going from zinc to the acid. This is due to zinc atoms dissolving in the acid as Zn^{++} ions, leaving the zinc with an excess of electrons. Another discontinuity in potential occurs at the acid-copper junction, the copper being at a lower potential than the acid.

We see from the diagram that the function of the cell is to lift charges of electricity from zinc to the copper potential. The total difference in potential between zinc and copper on open circuit is numerically the electromotive force of the cell. *In general, we may say that the electromotive force of a voltaic cell is equal to the algebraic sum of all the abrupt changes in potential through the cell on open circuit.*

However, if the circuit is closed and a current passes, then certain changes in potential take place that modify our diagram profoundly. These are shown in Fig. 19-7. The difference in potential between zinc and copper is no longer the electromotive force of the cell. The importance of knowing the value of the potential difference between an electrode and its electrolyte is thus evident.

It is found experimentally that the magnitude of potential difference between an electrode and an electrolyte depends not only upon the material of the electrode and the composition of the electrolyte but also upon the concentration of ions in the electrolyte. A direct measurement of this potential difference is difficult, if not impossible, since any measuring device necessarily introduces a second electrode and therefore measures the resultant potential difference between two electrodes immersed in an electrolyte. For this reason the potential difference between an electrode and a solution of its

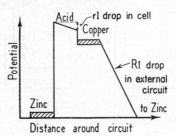

Fig. 19-7.—Variation of potential around a complete cell, as shown in Fig. 19-5, including the drop in potential in the external resistance R.

own ions is measured with respect to a reference electrode whose potential is arbitrarily taken to be zero. The value given by such a measurement is called the *standard electrode potential*.

The reference electrode generally taken in practice is the hydrogen electrode, which consists of a platinum wire coated with platinum black, i.e., colloidal platinum. The electrode is immersed in a solution containing 1 gm-equivalent of hydrogen ion per liter (a normal hydrogen-ion solution). Hydrogen gas is then allowed to bubble past the blackened electrode (Fig. 19-8).

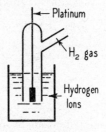

Fig. 19-8.—Hydrogen electrode

The procedure followed in determining the standard electrode potential of a metal is as follows: A voltaic cell is constructed with its two electrodes in separate containers. One electrode consists of the metal immersed in a normal solution of its ions, while the other electrode is the hy-

Analytical Experimental Physics

drogen electrode. A conducting path is formed between the two solutions by means of a U-tube filled with a concentrated solution of salt (KCl). The reason for its use is that the difference of potential at the liquid junctions is thus rendered negligibly small (see Fig. 19-9). The resultant po-

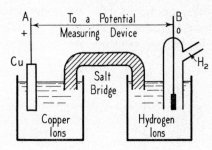

Fig. 19-9.—Method for measuring standard electrode potentials. *AB* is connected to potentiometer.

tential difference of the copper-hydrogen cell is then measured by means of a suitable potential measuring device, generally a potentiometer (§ 22.8).

In the case of copper (Fig. 19-9), the potential difference is $+0.34$ volt, the copper being at the higher potential. This number, then, is the *single or standard electrode potential* of copper. Zinc has a potential of -0.76 volt, meaning that zinc is at a lower potential than hydrogen. In Table 19-3 the

TABLE 19-3

STANDARD ELECTRODE POTENTIALS
ELECTROMOTIVE SERIES

Elements	Voltage	Elements	Voltage
Li.............	-2.96	Sn (Sn^{++})	-0.13
Na.............	-2.71	H.............	0.00
Mg.............	-2.40	Cu (Cu^{++})	$+0.34$
Al.............	-1.70	Hg.............	$+0.79$
Zn.............	-0.76	Ag.............	$+0.86$
Cd.............	-0.40	Br.............	$+1.23$

standard electrode potentials of some of the common elements are given. This arrangement of the elements is called the *electromotive series*. The standard electrode potentials measure the relative tendency of the element to form ions. Thus zinc has a greater tendency to form positive ions than hydrogen or copper, while magnesium forms ions even more readily. We shall use the information of Table 19-3 in our discussion of cells.

19.10. Daniell's Cell

One of the oldest and simplest of voltaic cells is the one invented by J. Daniell. Because of its theoretical simplicity, ease of construction, and useful relative constancy of electromotive force, we shall describe it in some detail. The cell consists of a zinc rod dipping into a solution containing zinc ions (Zn^{++}) and a copper rod dipping into cupric ions (Cu^{++}), as shown in Fig. 19-10. Let us assume, for the moment, that neither the external circuit nor the salt bridge is present.

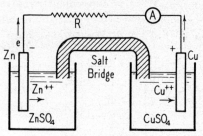

Fig. 19-10.—Daniell cell

When the zinc rod dips into a solution of zinc sulphate, zinc ions go into solution, leaving the negative electrons on the zinc. This process continues until it is stopped by the accumulation of charges on the boundary. The equilibrium reaction at the zinc electrode is

$$Zn \rightleftarrows Zn^{++} + 2e.$$

At the copper rod a similar reaction takes place, cupric ions going into solution until equilibrium is set up:

$$Cu \rightleftarrows Cu^{++} + 2e.$$

From the position of the two elements in the electromotive series, it can be seen that zinc has a greater tendency to go into solution as positive ions than copper has; hence the zinc electrode has a greater density of negative charge upon it than that on the copper electrode, and a difference of potential is established between zinc and copper. On connecting the electrodes by a wire R, there is an electron flow from zinc to copper, where they are taken up by the cupric ions to form copper on the electrode. Soon, however, the electron flow stops, since there is an accumulation of SO$_4^-$ in the copper cell and Zn^{++} in the zinc cell. By placing a salt bridge between the cells, as shown in Fig. 19-10, the SO$_4^-$ ions can pass to the zinc to form zinc sulphate. In Fig. 19-10 the zinc electrode is called negative and the copper positive

296

because electrons (e) flow from zinc to copper, and therefore the current (i) passes from copper to zinc.

Faraday's law applies; and so, for each gram-equivalent of copper ions that is deposited from among the many ions that strike the copper electrode, 1 gm-equivalent of zinc is ionized, and 96,519 coulombs, or 6.0254×10^{23} electrons, travel along the wire from zinc to copper.

In practice, a simplified technique eliminating the salt bridge is used. The zinc electrode with its solution is placed in a *porous cup*. The cup is then suspended within the copper sulphate solution. The porous cup replaces the salt bridge and allows $SO_4^=$ ions to diffuse from the copper sulphate to the zinc.

If the copper and zinc sulphates are normal solutions, then the electromotive force of the cell on open circuit is the difference between the standard electrode potentials of zinc and copper. From Table 19-3, $E = 0.34 - (-0.76) = 1.10$ volts. The potential difference of the liquid junctions may be neglected if a salt (KCl) bridge is used.

19.11. Polarization of Cells

In the operation of the copper-zinc–sulphuric acid cell of Fig. 19-4 (§ 19.8) we noted that zinc went into solution as Zn^{++}, while hydrogen gas was liberated at the copper electrode because of its position in the electromotive series. In practice we find that the current supplied by this cell is rapidly reduced in magnitude because of the presence of the hydrogen gas on the surface of the copper electrode. The effect of the presence of the hydrogen gas is, first, to increase enormously the internal resistance of the cell and, second, to introduce an electromotive force that opposes the electromotive force of the cell proper. This back electromotive force arises from the fact that a monomolecular layer of hydrogen surrounds the copper electrode and acts like a hydrogen electrode. The electromotive force of the zinc-hydrogen cell is less than that of a copper-zinc cell. Another effect that may cause the original electromotive force of the cell to decrease is the presence of internal "concentration cells," owing to the change in concentration of the ions about the electrodes. These effects, taken together, give rise to the phenomenon of *polarization*.

Gases other than hydrogen may cause polariza-

tion. For example, if a difference in potential of 1 volt is applied to two platinum electrodes immersed in hydrochloric acid, the strength of the current passing in an external circuit approaches zero very quickly. This is due to the accumulation of hydrogen at the cathode and chlorine at the anode, the two gases setting up an opposing electromotive force of roughly 1.35 volts. If the applied electromotive force is increased to, say, 1.5 volts, the counter electromotive force is no longer sufficient to reduce the current to zero. At a voltage above 1.35 volts, a continuous current passes. The minimum voltage at which a continuous current will pass through an electrolyte is known as the *decomposition potential*. The relationship between the applied electromotive force, E, and the counter electromotive force, E', the total resistance of the circuit R, and the current I is given by

$$E - E' = RI . \qquad (19-4)$$

Since no gas is formed in the Daniell cell (§ 19.10), this cell does not polarize. In many of the voltaic cells used at present in the laboratory, however—e.g., dry cells—hydrogen gas is generally evolved. These cells have present in the electrolyte some oxidizing agent (called a "depolarizer") whose function is to eliminate the hydrogen. A cell with one electrode of mercury, one of zinc, and an electrolyte of very dilute hydrochloric acid (HCl) will polarize so completely as to give almost no current. A minute fragment of mercurous chloride (HgCl) dropped into the cell will cause the electromotive force to jump up to its original value. Table 19-4 lists some of the more important cells, together with their electromotive forces.

19.12. Standard Cell

The electromotive force of most cells, depending, as it does, upon a chemical reaction, is altered by a change in the temperature of the cell. Researches primarily by Weston have shown that if zinc, which is common to most primary cells, is replaced by cadmium, the temperature effect is much reduced. These cells are constructed of an H-shaped glass tube having platinum lead-in wires sealed into the bottom of each leg, as shown in Fig. 19-11. The anode is the mercury seen at the bottom of one leg. This is covered with a layer of mercurous sulphate (Hg_2SO_4) paste, which acts as **297** a depolarizer. The cathode is a liquid amalgam of 1 part of cadmium in 7 parts of mercury.

These are made in two types. In the "normal cell," or saturated cell, the electrolyte is a saturated solution of cadmium sulphate ($CdSO_4$) with added crystals of the solute, so that it is saturated saturated at room temperatures. This type has an electromotive force of about 1.0186 volts, which remains practically constant between 4° and 40° C. However, the electromotive force varies slightly between different cells, so that each needs to be calibrated.

TABLE 19-4

COMPOSITION, RESISTANCE, AND ELECTROMOTIVE FORCE OF VOLTAIC CELLS

Name	Resistance in Ohms	Negative Pole	Solution	Positive Pole	Solution	Electromotive Force in Volts
Double-fluid cells:						
Bunsen..........	0.1 −0.2	Zn	$1(H_2SO_4)+12(H_2O)$	C	Fuming HNO_3	1.9
Bichromate......	.08 − .4	Zn	$1(H_2SO_4)+12(H_2O)$	C	$12(K_2Cr_2O_7)+100(H_2O)$	2.03
Daniell..........	.85	Zn	$1(H_2SO_4)+4(H_2O)$	Cu	Saturated solution of $CuSO_4$	1.08
Grove...........	.1 − .2	Zn	$1(H_2SO_4)+12(H_2O)$	Pt	Fuming HNO_3	1.93
Single-fluid cells:						
Dry cell........	.05 − .1	Zn	NH_4Cl		C with MnO	1.53
Edison-Lalande...	.03	Zn	KOH and H_2O		Cu and CuO	0.70
Storage cells:						
Lead (Planté)....	0.004−0.02	Pb	H_2SO_4, density 1.1		PbO_2	2.2
Edison...........	....	Fe	20 per cent KOH		Nickel oxide	1.1

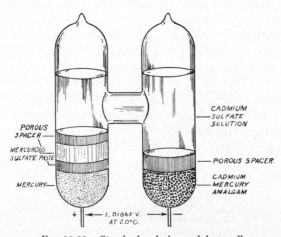

FIG. 19-11.—Standard cadmium sulphate cell

at all temperatures. This cell has a definite electromotive force of 1.01830 volts at 20° C, reproducible to a few parts in 100,000. Its temperature variation is given by

$$E_t = E_{20°C} - 4.06 \times 10^{-5}(T-20) - 9.5 \times 10^{-7}(T-20)^2 . \tag{19-5}$$

In the "unsaturated" cell the solution of cadmium sulphate is saturated at 4° C, but no additional crystals of the solute are added, so that it is not

19.13. Secondary Cells: Storage Batteries

Most of the primary cells are used for intermittent discharge purposes only. When the electrolyte is depleted, the cell is discarded. There is a class of cells known as "secondary cells" or "storage batteries," in which, on discharge, the electrolyte is re-formed by sending a reverse current through them. Indeed, the Daniell cell can have its action reversed by sending a current through it in the opposite direction, when zinc will be deposited and copper will be removed from those electrodes, respectively. The concomitant mixing of the two solutions limits the action, however.

The Planté lead storage battery and the Edison alkaline cell are the two most commonly used. We shall describe the operation of the lead cell in some detail.

By immersing two lead plates in a dilute solution of suphuric acid, a trace of lead sulphate is formed on the surface of each plate. When a current of electricity is passed through the solution, the lead sulphate on the cathode is reduced to metallic lead, and on the anode the lead sulphate is oxidized to lead peroxide. Such a cell will give rise to an electromotive force of about 2.1 volts.

However, the amounts of lead and lead peroxide produced in this way are so small that very little total quantity of electricity (ampere-hours) can be furnished by this action. Systematically repeated charging and discharging are needed when the cell is newly made, to roughen and so to increase the plates' surfaces, that the cells may have a somewhat larger capacity.

The capacity of the cell may be enormously increased by making the electrodes in the form of "grids" and by filling the grids with a paste of red lead and lead oxide. On sending a current through the solution, electrolysis takes place, forming spongy lead at the cathode and lead peroxide at the anode. By connecting the two electrodes with a wire, electrons flow from the lead plate to the peroxide plate, forming lead sulphate at each electrode.

The total chemical change taking place at charge and discharge may be written as follows:

$$PbO_2 + Pb + 2H_2SO_4 \underset{charge}{\overset{discharge}{\rightleftarrows}} 2PbSO_4 + 2H_2O . \quad (19\text{-}6)$$

The fundamental ionic reactions that give rise to the flow of electrons are as follows:

At the negative pole: $Pb \rightarrow Pb^{++} + 2e$,

At the positive pole: $Pb^{++++} + 2e \rightarrow Pb^{++}$.

According to equation (19-6), as discharge of the lead cell progresses, the heavy sulphuric acid molecule is replaced by the lighter water molecule.

The density of the solution thus changes and may be taken as an indication of the degree of discharge. A density of about 1.29 gm/cm^3 corresponds to a fully charged battery, 1.15 gm/cm^3 to one in need of charging.

In the Edison alkaline cell the electrodes are of iron and nickel oxide (Ni_2O_3) when charged, the solution being potassium hydroxide. The chemical reactions taking place on discharge and charge are

$$Fe + Ni_2O_3 + 3H_2O \underset{charge}{\overset{discharge}{\rightleftarrows}} Fe(OH)_2 \quad (19\text{-}7)$$
$$+ 2Ni(OH)_2 .$$

Although the electromotive force of this cell is only 1.1 volts, its usefulness lies in its ruggedness and light weight. An Edison cell may be completely discharged or overcharged without permanent damage. This is, unfortunately, not true with the Planté cell, in which complete discharge means the formation of hard lead sulphate crystals, which have high electrical resistance. Cells so "sulphated" cannot be restored to usefulness.

19.14. Résumé

The following is a brief listing of the definitions, principles, and theories, given in this chapter, which you should know:

The definition of electrolysis
The dissociation theory of electrolytes
A proof of the atomicity of electricity
The theory of the voltaic cell—electrode potentials
The theory of some practical cells, both primary and secondary

Queries

1. Is it possible to construct a voltaic cell of two copper electrodes in a copper sulphate solution? Explain.
2. What important factors determine the electromotive force of a voltaic cell?
3. Does a voltaic cell store electrical charge? What does it do?
4. What is the fundamental difference between a primary and a secondary reaction in electrolysis?

Practice Problems

1. A circuit consists of a heating coil of 100-ohm resistance and an electroplating bath with copper electrodes immersed in copper sulphate; both coil and bath are in series. If 0.99 gm of copper is deposited in 10 min, calculate (a) the average strength of the current; (b) the amount of heat dissipated in the coil in calories per second. Ans.: (a) 5 amp; (b) 599 cal/sec.
2. A current of 20 amp passes through a silver nitrate and a copper sulphate cell in series. What are the amounts of silver and copper deposited in ½ hr?
3. In the silver plating of a disk, 0.1 amp is allowed to

pass for 2 hr. What is the thickness of the silver plate deposited if the total area of the thin disk is 150 cm²? Ans.: 5.11×10^{-4} cm.

4. A current of 5 amp passes through acidulated water for 20 min. How many cubic centimeters of hydrogen are liberated, assuming standard conditions: 0° C, 76 cm mercury pressure? (Density of H_2 = 0.000089 gm/cm³, at 0° C.)

5. A cell of silver nitrate with silver electrodes is connected in series with a copper sulphate cell with copper electrodes. If 1,000 coulombs of charge are sent through the cells, how many atoms of silver and of copper are deposited on the electrodes? Ans.: silver, 6.28×10^{21}; copper, 3.14×10^{21}.

6. A current of 5 amp, as read by an ammeter, is sent through a silver nitrate cell for 600 sec. The increase in weight of the cathode is found to be 3.341 gm. Does the ammeter read correctly? What correction, if any, must be applied?

7. A potential difference of 3 volts is across a copper sulphate electroplating bath. How much does it cost to deposit 500 gm of copper if the price of electrical energy is 1.5 cents/kwh? Ans.: 1.9 cents.

8. From the table of standard electrode potentials calculate the potential of a cell made up of lead electrodes in equilibrium with Pb^{++} and silver electrode in equilibrium with Ag^+. Which is the positive electrode?

9. What volume of oxygen at standard conditions is liberated if 0.05 amp passes through acidulated water for 20 min? Ans.: 3.48 cm³.

10. A Daniell-type cell is made of silver and cadmium electrodes in normal solution of their salts. What electromotive force will this cell have? Which electrode is negative?

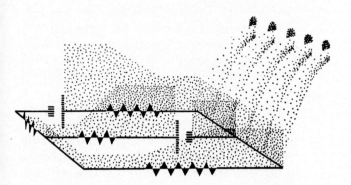

CHAPTER 20

Circuit Theory: *Applications of Ohm's Law*

In chapter 18 we had a simple statement of Ohm's law to the effect that the ratio of potential difference to the resulting current is constant for a given conductor at any specific temperature. This is one of the most important electrical laws, especially in connection with practical calculations concerning electric circuits carrying currents in general and involving, in addition, resistances, differences of potential, and electromotive forces. The law itself is quite simple, but the circuits to which it may be applied are not necessarily so. Therefore, we must devote some time to discussing the application of Ohm's law to circuits of varying complexity.

20.1. Ohm's Law Applied to an Undivided Part of a Circuit Not Containing an Electromotive Force

Suppose we have a current, I, through a circuit of resistance, R. In order to determine the potential difference across this resistance, a voltmeter (see § 22.3) may be connected to its terminals. A VOLTMETER is usually a galvanometer with a high resistance. It gives readings proportional to the difference of potential between its terminals. While the actual resistance of most voltmeters is not infinite, it should be so great that, even if it were greater, the readings of the instrument would not be appreciably different.*

* Few, if any, actual infinities occur in practical experience; but the physicist regards anything to be, to all intents and purposes, infinite if making it larger does not make a measurable change in the experiment. To the student of atomic structure an electron taken 1 cm away from an atom could well be considered as removed to infinity. In electrostatic experiments a

For use with direct currents, that terminal of the voltmeter which is to be placed at the higher potential is marked plus (+). In Fig. 20-1, for a *simple resistance*, we see that the current external to the applied electromotive force is directed from the high-potential side to the low-potential side; hence the voltmeter should be connected as indicated.

FIG. 20-1.—The resistance R is equal to the ratio of the voltmeter reading to the ammeter reading.

To measure this current, an ammeter may be inserted in the circuit. An AMMETER is likewise a galvanometer with an auxiliary (shunt) circuit such that the combination is one of *nearly zero resistance*. This gives readings proportional to the current through the instrument (see § 22.2). The electricity enters at the positive (+) terminal and leaves at the negative (−) terminal of the ammeter. The flow of electrons through the instrument is, of course, the reverse of the preceding conventions.

In this simple situation Ohm's law gives directly

$$\frac{\text{Voltmeter reading of potential difference}}{\text{Ammeter reading of current}}$$

$$= \text{Resistance}$$

charged ball carried into the next room might be considered as carried to infinity.

In a similar way, the physicist often calls a quantity zero when it is so small that to make it smaller would not change the experiment in a measurable way.

or

$$\frac{V}{I} = R. \qquad (20\text{-}1)$$

This method is sometimes called the "voltmeter-ammeter method" for measuring resistance. It is not a reliable method unless one is *sure* that conditions referred to in footnote * of § 20.1 are satisfied.

20.2. Ohm's Law Applied to a Complete Undivided Circuit

Let us consider the circuit of Fig. 20-2. The source of energy, an electromotive force, is here indicated to be a primary cell represented by

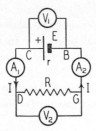

Fig. 20-2.—Application of Ohm's law to the complete undivided circuit.

$(+\,|\,\blacksquare\,-)$. Every cell must itself offer some resistance to the passage of current through it. Let this INTERNAL RESISTANCE be represented by r. The positive terminal of the electromotive force of the cell is that toward which negative electrons flow in the external circuit. Hence, outside the cell, positive electricity conventionally flows away from this electrode, as indicated (see § 18.1).* Will the current in the two ammeters be the same? Realizing that by "current" we mean the actual flow of negatively charged electrons in the direction opposite to the "positive" direction of the current, it becomes evident that, unless the number of electrons passing one point is the same as the number passing another point, there must be an excess or a deficiency of them accumulating between these points. Since currents may exist for days or even years, such accumulations are evidently impossible. Hence the two ammeters must read the

---

* While the resistance of the connecting wires cannot well be made absolutely zero, it can be made so small as to be completely negligible compared to R and r, at least when these are not too small; i.e., the resistance of connecting wires must be so small that to make it smaller alters nothing else by a measurable amount.

same. A circuit of this sort, in which the current is everywhere the same, we call an *undivided* circuit.

Our next question is: Will the two voltmeters read the same? Since we have assumed that the resistance between D and C is negligible, i.e., A_1 is of negligible resistance, D and C must be at the same potential. The same is true for B and G, A_2 being of negligible resistance. Hence the two voltmeters are, to all intents and purposes, connected to the same points electrically and thus will read the same.

When Ohm's law is applied to such a circuit as this, it takes the form

$$\text{Current} = \frac{\text{Total electromotive force}}{\text{Total resistance}}$$

or

$$I = \frac{E}{R + r}. \qquad (20\text{-}2)$$

Let us note, however, that between D and G we have just the undivided part of a circuit not containing an electromotive force which we have previously discussed. Then, since the same current is present in one part of an undivided circuit as in another, the same I is given by

$$I = \frac{V}{R},$$

as well as by

$$I = \frac{E}{R + r}. \qquad (20\text{-}3)$$

We can eliminate R from these two equations. From the first we get $IR = V$, and from the second $IR + Ir = E$. Replacing IR by V, we have $V + Ir = E$, and solving for I gives

$$I = \frac{E - V}{r}. \qquad (20\text{-}4)$$

This shows *that* $E = V$ *only when* $I = 0$, since the internal resistance of a cell r is *never zero*. In other words, when a current exists, a voltmeter connected to the terminals of a cell does *not* give the electromotive force of the cell. If, however, $I = 0$, then $E = V$, and the voltmeter does give the electromotive force of the cell. Also, to the extent that I approaches zero, the voltmeter reading approaches the electromotive force.

The total electromotive force of equation (20-2) needs interpretation. If a circuit contains more than one electromotive force, the total electromotive force that goes into equation (20-2) is the algebraic sum of the separate electromotive forces.

Whether an electromotive force is to be taken as positive or negative is determined by the following convention. If the current in the circuit enters a battery at the negative pole and leaves by the positive pole, that electromotive force is to be considered positive $(+)$. On the other hand, if the current enters the positive pole and leaves by the negative pole, its electromotive force is to be considered negative $(-)$. Thus a positive electromotive force in a circuit implies that the electromotive force supplies energy from its chemical store to the circuit, while a negative electromotive force abstracts energy from the circuit and thereby adds to its chemical store. In Fig. 20-3 (a), if the cur-

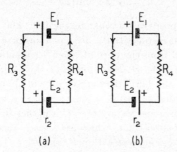

FIG. 20-3.—In (a) E_1 provides energy to the circuit, E_2 absorbs energy; in (b) both batteries supply energy to the circuit.

rent, I, circulates as indicated, then the electromotive force E_1 is larger than the opposing one, E_2, and, in time t sec, E_1 gives E_1It joules of energy to the circuit, while E_2 absorbs E_2It joules of energy. If R is the *total resistance of the circuit*, then RI^2t joules of energy are converted into heat. By the law of conservation of energy, in Fig. 20-3 (a),

$$E_1It - E_2It = RI^2t ,$$

or

$$E_1 - E_2 = RI ,$$

$$I = \frac{E_1 - E_2}{R} .$$

In Fig. 20-3 (b) both electromotive forces are positive, or the batteries are in series:

$$I = \frac{E_1 + E_2}{R} .$$

EXAMPLE 1: Let us calculate the current I in the circuits of Fig. 20-3 (a) and (b) if

$E_1 = 10$ volts ; $r_1 = 1$ ohm ; $R_3 = 5$ ohms ,

$E_2 = 5$ volts ; $r_2 = 2$ ohms ; $R_4 = 2$ ohms .

Solution: In case (a), $I = (10 - 5)/10 = \frac{1}{2}$ amp. In case (b), $I = (10 + 5)/10 = 1.5$ amp.

20.3. Ohm's Law Applied to an Undivided Part of a Circuit Containing an Electromotive Force

Let us next consider that part of the circuit of Fig. 20-4 lying between C and D, on the assumption that $E' > E$. Then the current, I, will have the direction shown. We must combine the poten-

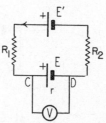

FIG. 20-4.—The voltmeter reads $(Ir + E)$

tial changes due to an electromotive force with the potential changes due to a current through a resistance, r. If we confine our attention to the resistance only, there would be a potential difference of Ir, with the high-potential side on the left, i.e., on the side the current enters. Again, if we ignore the resistance and consider only the electromotive force, E, there would be a resulting potential difference of E, with the high-potential side again on the left, as given in the diagram. These two potential differences may be represented as

$$\begin{array}{c} + I\,r - \\ + E - \\ \hline + (I\,r + E) - \end{array}$$

and the voltmeter will read $(Ir + E)$ and have its positive terminal on the left.

If we now reverse E, a somewhat different situation is presented (Fig. 20-5). As before, we have

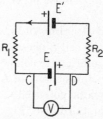

FIG. 20-5.—The voltmeter may read $(Ir - E)$ with C the positive terminal, or may read zero, or may read $(E - Ir)$ with C the negative terminal.

the high-potential side of Ir on the left, since the direction of the current has not changed and since there is no direction to resistance; but now the right side of E is its high-potential side.

Three cases present themselves (Figs. 20-6, 20-7, and 20-8). Here the voltmeter may give a direct reading, may read zero, or may even have its polarity reversed from that of the electromotive force of the cell across which it is connected. We may interpret these three cases in terms of the energy transformations that take place as a charge moves from C to D.

$E > Ir$:

$$\begin{array}{c} - E + \\ + I r - \\ \hline - (E - I r) + \end{array}$$

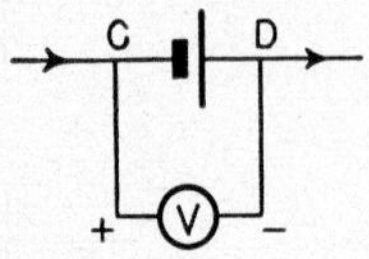

FIG. 20-6

$E = Ir$:

$$\begin{array}{c} - E + \\ + I r - \\ \hline 0 \end{array}$$

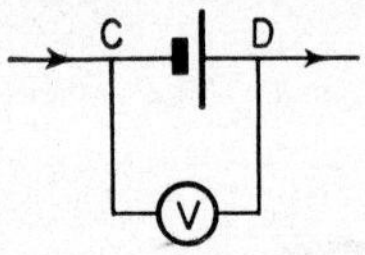

FIG. 20-7

$E < Ir$:

$$\begin{array}{c} - E + \\ + I r - \\ \hline + (I r - E) - \end{array}$$

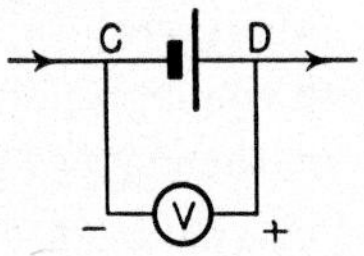

FIG. 20-8.—Voltage readings across cells with different currents in them.

In the first case (Fig. 20-6), a charge of q coulombs, in moving from C to D, loses qIr joules of energy in the internal resistance of the cell.* We interpret this net loss of energy in the resistance as meaning that D is at a lower potential than C. The cell E, however, does an amount of work, Eq joules, on the charge, which not only compensates for the loss of Ir but supplies, in addition, $(Eq - Irq)$ joules of energy to the rest of the circuit. Hence the point D must be at a *higher potential* than C.

In the second case (Fig. 20-7), when the voltmeter reads zero, the cell E does an amount of work, Eq joules, just sufficient to compensate for the loss qIr in the internal resistance of E. Hence C and D must be at the same potential.

Finally, in the third case (Fig. 20-8), the loss of qIr joules in the internal resistance of E *exceeds* the energy contribution Eq of cell E, so that $(Irq - Eq)$ joules must be supplied by E' (see Fig. 20-5) if the current I is to pass through E. Thus the potential at D is *less* than that at C.

* This energy is not really lost but is merely transformed into heat energy.

EXAMPLE 2: Let us calculate the voltmeter readings of Fig. 20-4, given the following data: $E' = 6$ volts; $E = 4$ volts; $r = 1$ ohm (internal resistance of E), and $R_1 = R_2 = 0.5$ ohm.

Solution: By equation (20-2),

$$I = \frac{6 - 4}{2} = 1 \text{ amp}.$$

The Ir drop through the cell E is $+ (1 \text{ volt}) -$, with the higher potential on the left. The electromotive force of the cell is $+ (4) -$, with the high potential also on the left. The voltmeter will read as follows:

$$\begin{array}{c} + \ \ (1 \text{ volt}) - \\ + \ \ (4 \text{ volts}) - \\ \hline + \ \ (5 \text{ volts}) - \end{array}$$

with the positive terminal to the left. The interpretation of this result is that, in 1 sec, 5 joules of energy are abstracted from the current when 1 coulomb of charge goes from C to D. Four joules of this energy are absorbed by the cell E in charging it or in producing other chemical changes, while 1 joule is lost in heat in the internal resistance; the battery E' supplies to the circuit 6 joules of energy, 5 of which are already accounted for; 1 joule is lost as heat within R_1 and R_2.

EXAMPLE 3: What will the voltmeter read if E is reversed (as in Fig. 20-5)?

Solution: By equation (20-2),

$$I = \frac{6 + 4}{2} = 5 \text{ amp},$$

where the Ir drop is $\qquad + (5 \text{ volts}) -$
the electromotive force of the cell is $\quad - (4 \text{ volts}) +$
and the voltmeter reading is $\qquad + (1 \text{ volt}) -$

i.e., the point C is at a higher potential than the point D. The interpretation of this result is that 25 watts of power are lost as heat in the internal resistance of the cell E but that only 20 watts of power are supplied to the circuit by E. The battery E' must supply the extra 5 watts of power if the current of 5 amp is to be maintained through E. As a matter of fact, E' supplies 30 watts; 5 watts are used in battery E; and 25 watts are dissipated in resistances R_1 and R_2.

20.4. Ohm's Law Applied to Divided Parts of Circuits Not Containing Electromotive Forces

Suppose several resistances are connected in PARALLEL, as shown in Fig. 20-9. Here there are several possible paths for the electrons; and the currents in the different paths are not, in general, the same. Hence the circuit is called DIVIDED. Furthermore, the current passing through the undivided part of any circuit must be the sum of the various currents passing through the divided parts; otherwise charges would be accumulating (or disappearing) somewhere. Therefore,

$$I = I_1 + I_2 + I_3 . \qquad (20\text{-}5)$$

Since each resistance joins in a common terminal at B, the drop in potential must be the same across each; i.e.,

$$V = V_1 = V_2 = V_3 . \qquad (20\text{-}6)$$

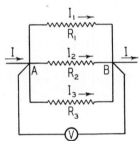

Fig. 20-9.—Resistances connected in parallel

Combining these two equations with $I = V/R$, where R is the total resistance of the three resistances in parallel, gives

$$\frac{V}{R} = \frac{V_1}{R_1} + \frac{V_2}{R_2} + \frac{V_3}{R_3}, \qquad (20\text{-}7)$$

$$\frac{1}{R} = \frac{1}{R_1} + \frac{1}{R_2} + \frac{1}{R_3}. \qquad (20\text{-}8)$$

The reciprocal of the total resistance of several resistances in parallel is the sum of the reciprocals of the individual resistances.

Also, equation (20-8) shows that

The total resistance of several resistances in parallel is less than that of the smallest individual resistance.

In the case of resistances in series (Fig. 20-10),

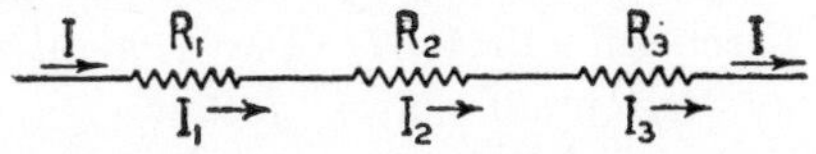

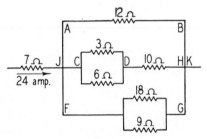

Fig. 20-10.—Resistances connected in series

the current in each must be the same, since there are no side paths through which current may be deflected, and so

$$I = I_1 = I_2 = I_3 .$$

However, the total drop in potential is the drop in the first resistance plus the drop in the second resistance plus the drop in the third resistance; or the total work done per unit of electricity is the work done in passing through the first resistance plus the work done in passing through the second resistance, etc.; i.e.,

$$V = V_1 + V_2 + V_3 . \qquad (20\text{-}9)$$

Combining these two equations with $V = IR$ gives

$$IR = I_1 R_1 + I_2 R_2 + I_3 R_3 , \qquad (20\text{-}10)$$

$$R = R_1 + R_2 + R_3 . \qquad (20\text{-}11)$$

The total resistance of several resistances in series is the sum of the individual resistances and is greater than that of the largest individual resistance.

For more complicated combinations of resistances free from electromotive forces, definite formulas can hardly be derived; but an example may show a method which, by repeated application of the rules we have given, will determine the current in any part.

EXAMPLE 4: How much of this current passes through the 6-ohm coil? (See Fig. 20-11.)

Fig. 20-11

Solution:

$$\frac{1}{R_{CD}} = \tfrac{1}{3} + \tfrac{1}{6} ,$$

$$R_{CD} = 2 \text{ ohms} .$$

Therefore,

$$R_{CH} = 2 + 10 = 12 \text{ ohms} .$$

Also,

$$\frac{1}{R_{FG}} = \tfrac{1}{9} + \tfrac{1}{18}.$$

Therefore,

$$R_{FG} = 6 \text{ ohms} .$$

Moreover,

$$\frac{1}{R_{JK}} = \tfrac{1}{12} + \tfrac{1}{12} + \tfrac{1}{6}.$$

Therefore,

$$R_{JK} = 3 \text{ ohms} .$$

Thus

$$V_{JK} = 24 \times 3 = 72 \text{ volts} ,$$

305

and

$$I_{CH} = \tfrac{72}{12} = 6 \text{ amp};$$

also,

$$V_{CD} = 6 \times 2 = 12 \text{ volts},$$

$$I_6 = \tfrac{12}{6} = 2 \text{ amp}.$$

This might have been shortened a little by recognizing that the 12-ohm coil, the 9-ohm coil, the 18-ohm coil, and R_{CH} are in parallel. Then

$$\frac{1}{R_{JK}} = \tfrac{1}{12} + \tfrac{1}{12} + \tfrac{1}{9} + \tfrac{1}{18},$$

$$R_{JK} = 3 \text{ ohms}.$$

20.5. Ohm's Law Applied to Certain Divided Circuits Containing Similar Electromotive Forces

When divided circuits contain scattered electromotive forces, the calculation of the currents may be rather difficult. However, there are certain cases in which a number of similar cells* are so connected that the computation of the current is not difficult. Consider Fig. 20-12, in which n cells,

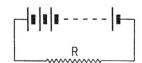

Fig. 20-12.—Cells connected in series

each with an electromotive force e and an internal resistance r, are connected to an external resistance R. This is not a divided circuit and is covered by equation (20-2):

$$I = \frac{ne}{R + nr}. \qquad (20\text{-}12)$$

The total electromotive force is the sum of the individual electromotive forces, since they are in series; and also, according to equation (20-11), the total internal resistance is the sum of the individual resistances.

Next consider Fig. 20-13. Suppose m similar cells are connected in parallel. In this case, any electron goes through one cell or through another but not through more than one in going once around the circuit. Hence the energy given to any electron is the energy given by *but one cell;* or, in other words, the total electromotive force of the

* Only to the degree that the cells are *identical in electromotive force and internal resistance* are the results of this section exact.

battery is the same as the electromotive force of one cell. However, since the electron streams have several possible paths, the total internal resistance is less than the internal resistance of one cell. From equation (20-8) the sum of several equal resistances in parallel is given by

$$\frac{1}{R_{\text{total}}} = \frac{1}{r_1} + \frac{1}{r_2} + \frac{1}{r_3} + \ldots = \frac{m}{r}, \quad (20\text{-}13)$$

or

$$R_{\text{total}} = \frac{r}{m},$$

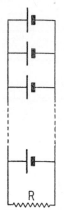

Fig. 20-13.—Cells connected in parallel

since the r's, m in number, are all equal. Then equation (20-2) takes the form†

$$I = \frac{e}{R + r/m}. \qquad (20\text{-}14)$$

20.6. Network Problems: Kirchhoff's Laws

It is quite apparent from the preceding discussion that the application of the methods we have discussed to divided circuits, including different electromotive forces, may be very long and confusing unless some definite and systematic procedure is applied to the solution. A general method of attack was formulated by the German physicist G. H. Kirchhoff (1824–87). Two generalizations, known as "Kirchhoff's laws," follow:

1. From our understanding of the nature of the electric current we know that in any steady state

† Certain practical suggestions should be made here. When cells are connected in series, one bad cell with a high internal resistance may spoil the effect of all the other good cells.

When cells are connected in parallel, it is essential that they be of exactly the same electromotive force. Otherwise, when the external circuit is open and the cells are apparently not supplying current, there will be local currents forward through the cell of higher e and backward through the cell of smaller e.

of current flow there can be no accumulation or disappearance of charge at any point. Thus, at point P of Fig. 20-14, several conductors carrying current are joined. The sum of the currents that enter P must be equal to the sum of the currents

that leave P, i.e., there is conservation of electric charge at P, or

$$I_1 + I_2 = I_3 + I_4 .$$

If we call currents that enter P "positive" and those that leave "negative," then, in general, Kirchhoff's first law is stated as follows:

LAW I:

$$\Sigma I = 0 . \qquad (20\text{-}15)$$

2. To obtain the second generalization, let us recall, first, that a potential difference is the energy per unit charge being given up by the electricity as it moves through the resistance and that an electromotive force is the energy per unit charge being supplied to the electric circuit. Then the conservation of energy requires that, around any closed path in an electric circuit, the algebraic sum of the differences in potential must be equal to the algebraic sum of the electromotive forces. This is Kirchhoff's second law:

LAW II:

$$\Sigma RI = \Sigma E . \qquad (20\text{-}16)$$

In the application of these generalizations to a circuit network, we formulate the following procedure:

1. *Assume and mark down* a direction for the current in each conductor of the circuit. It is quite *unimportant* whether the assumed direction is the *correct* one or not.

2. In going around a closed path (the arrow ⌐⌐ represents the direction around the path), the IR is taken as positive $(+)$ if one traces through the resistance in the *direction assumed for the current in 1;* it is negative $(-)$ if the current flows in the opposite direction.

3. Any electromotive force encountered will be considered positive if one first encounters the nega-

tive terminal and then the positive terminal of the electromotive force. This is the same convention as that used formerly (§ 20.2).

4. One must get as many *independent equations* as there are unknowns.

These rules can best be illustrated by an example. Given the network of Fig. 20-15, what are the values of the currents in the various resistances?

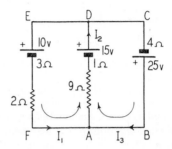

Fig. 20-15.—Illustrating the application of Kirchhoff's law of circuit analysis.

We have arbitrarily assumed a direction for currents I_1, I_2, and I_3. Applying Law I to point A, we obtain

$$I_1 + I_3 - I_2 = 0 . \qquad (a)$$

In applying Law II we shall take three closed paths: *ADEFA*, *ADCBA*, and *AFEDCBA*. The direction of travel around the three paths is indicated by the order of the letters or by the curved arrows.

Around path *ADEFA* we obtain

$$9 I_2 + I_2 (1) + 3 I_1 + 2 I_1 = 15 - 10 = 5 ; (b)$$

around path *ADCBA*,

$$9 I_2 + I_2 (1) + 4 I_3 = 15 + 25 = 40 ; \qquad (c)$$

and, finally, around *AFEDCBA*,

$$- 2 I_1 - 3 I_1 + 4 I_3 = 10 + 25 = 35 . \qquad (d)$$

We have here four equations but only three unknowns. These four equations are not all independent. By subtracting equation (b) from equation (c) we obtain equation (d). The useful equations, then, are (a), (b), and (c). On simplifying these equations, we have

$$I_1 = I_2 - I_3 , \qquad (a)$$

$$10 I_2 + 5 I_1 = 5 , \qquad (b)$$

$$10 I_2 + 4 I_3 = 40 . \qquad (c)$$

Since I_1 appears but once in the last two equations, let us substitute its value from equation (a):

$$10 I_2 + 5 (I_2 - I_3) = 5 , \qquad (a), (b)$$

or

$$15 I_2 - 5 I_3 = 5 . \qquad (e)$$

307

Multiplying equation (*e*) by 2 and equation (*c*) by 3, we obtain

$$30 I_2 - 10 I_3 = 10 ,$$

$$30 I_2 + 12 I_3 = 120 ,$$

or

$$22 I_3 = 110 ,$$

$$I_3 = 5 \text{ amp} .$$

From equation (*c*) we get

$$I_2 = 2 \text{ amp} ,$$

while equation (*a*) gives

$$I_1 = -3 \text{ amp} .$$

Currents I_2 and I_3 are in the directions assumed, while I_1 has a direction opposite to that assumed.

There is another manner of applying Kirchhoff's laws, in which the currents are assumed to be in closed loops. Let the current in the loop *EFADE* be I_1, in the counterclockwise direction shown by the arrow in Fig. 20-15. In the loop *DCBAD* there is the current I_3 in a clockwise direction, as shown by the arrow. In the arm *AD* the current is the sum of I_1 and I_3, since the currents are both in the direction from *A* to *D*.

Applying Kirchhoff's second law to the loop *EFADE*, starting from the point *A*, we have

$$9 (I_1 + I_3) + 1 (I_1 + I_3) + 3 I_1 + 2 I_1 = 15 - 10$$

or

$$15 I_1 + 10 I_3 = 5 . \tag{f}$$

Similarly, for the loop *ADCBA*, starting from *A*, we have

$$9 (I_1 + I_3) + 1 (I_1 + I_3) + 4 I_3 = 15 + 25$$

or

$$10 I_1 + 14 I_3 = 40 . \tag{g}$$

Multiplying equation (*f*) by $\frac{7}{5}$, we have

$$21 I_1 + 14 I_3 = 7 . \tag{h}$$

By subtraction of equations (*g*) and (*h*), it follows that

$$-11 I_1 = 33 \quad \text{or} \quad I_1 = -3 \text{ amp} .$$

Substituting for I_1 in equation (*f*) or equation (*g*) gives $I_3 = 5$ amp. The current $(I_1 + I_3)$ in the arm *AD* is then 2 amp. The two methods given here are entirely equivalent, and which one is chosen is a matter of personal preference.

20.7. Résumé

The following is a brief listing of the definitions, principles, and theories, given in this chapter, which you should know:

Ohm's law as applied to a part or to the whole of a circuit

An understanding of the difference between the electromotive force of a cell and the potential difference across the cell

The electromotive force and internal resistance of a number of cells connected in series or parallel

Kirchhoff's laws—their meaning and application

Queries

1. What advantages and disadvantages are there in connecting cells in parallel?
2. Given two lengths of wire of the same diameter, one of copper and one of nichrome, which wire would you use if you wished to make one of them as hot as possible, using a battery of a given electromotive force and negligible resistance?
3. A wire of nichrome glows when a current passes through it. A portion of this wire is cooled, which results in the remaining wire glowing brightly and finally melting. What has happened?
4. How many different values of resistance can one obtain from three resistors whose values are R_1, R_2, and R_3 ohms?

Practice Problems

1. See Fig. 20-2. For a given cell, when *R* equaled 3 ohms, the current was 3 amp. When *R* was reduced to 1 ohm, the current increased to 5 amp. Compute the electromotive force and internal resistance of the cell. Compute V_2 and *I* when $R \rightarrow 0$ and when $R \rightarrow \infty$. Tabulate your answers.

E	r	V_2	I	R
		....	3	3
		....	5	1
		....	....	0
		....	....	∞

2. See Fig. 20-2. For a given cell, when $I = 2$ amp, $V = 4$ volts. By changing R, I was reduced to 1 amp, and then $V_2 = 7$ volts. Compute the electromotive force and internal resistance of the cell and the external resistance readings. Then compute V_2 and I when $R \to 0$ and when $R \to \infty$. Tabulate your answers.

E	r	V_2	I	R
		4	2	....
		7	1	....
		....	....	0
		....	....	∞

3. Compute the current in the 6-ohm coil. Ans.: 16 amp.

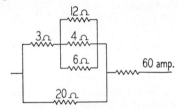

4. Compute the current in the 3-ohm coil.

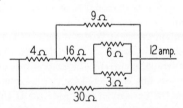

5. Compute the reading of each voltmeter and mark its high-potential side with a plus ($+$) sign. Draw a graph above each cell, showing the rise and fall of potential difference due to E and Ir. Ans.: $+ 11 -$; $- 0.5 +$; 0; $- 3 +$; $- 1.5 +$.

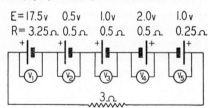

6. Compute the potential drop across each cell and indicate which side of each cell is at the higher potential.

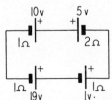

7. Twenty-four cells, each with an internal resistance of 9 ohms and an electromotive force of 1.5 volts, are connected in series through a resistance of $R = 2$ ohms. Compute (a) the current through the external resistance; (b) the power dissipated in the batteries and in the external resistance; (c) the power dissipated in cells and external resistance when the external resistance is equal to the internal resistance of the cells. Ans.: (a) 0.165 amp; (b) 5.89 watts, 0.05 watts; (c) 1.50 watts, 1.50 watts.

8. The above cells are connected in parallel through an external resistance of $R = 2$ ohms. Compute (a) the current through the external resistance; (b) the power dissipated in the cells and in the external resistance.

9. Compute the current in each part of this circuit. Ans.: 6 amp to right through 5-ohm resistor; 2 amp upward through 8-ohm resistor; 4 amp upward through 7-ohm resistor.

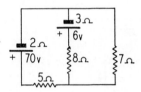

10. A 100-watt, 110-volt lamp is connected in parallel with a 40-watt, 110-volt lamp. Find their combined resistance.

11. A lamp bank used for charging batteries consists of a group of 60-watt, 120-volt bulbs connected in parallel, as shown in the accompanying figure. What current passes through the 20-volt storage battery if a lamp bank of three bulbs is connected to a 120-volt line? In order to increase the current, should the number of lamps be increased or decreased? (Assume negligible resistance for the battery.) Ans.: 1.25 amp; increased.

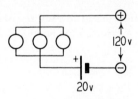

12. A storage battery is charged, using a lamp bank similar to that in problem 11. If the line voltage is 120 volts (assumed constant) and a current of 5 amp passes through the battery for 10 hr, calculate the amount of energy transformed into chemical energy, the amount wasted as heat, and the total

energy abstracted from the mains. The electromotive force of the battery is 20 volts.

13. Given two resistances, R_1 and R_2, in parallel, as shown in the figure herewith, compute the equivalent resistance between A and B. The current I splits into the components I_1 and I_2. Find I_1 and I_2 in terms of I and the resistances R_1 and R_2. Ans.: $R_1R_2/R_1 + R_2$; $R_2I/R_1 + R_2$; $R_1I/R_1 + R_2$

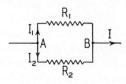

14. It is desired to replace the complicated circuit of figure (a) by the simple circuit of figure (b). What must be the electromotive force of battery E with $r = 0$?

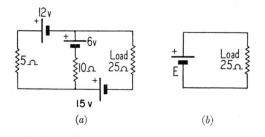

(a) (b)

15. Two circuits are joined by a conductor AB (see accompanying figure). Compute the current in each part of the circuit. *Hint:* Apply Kirchhoff's first law to A or B. Ans.: 2.22 amp; 0.423 amp.

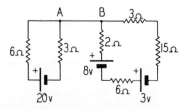

16. Compute the current in each resistance of the circuit in the accompanying figure.

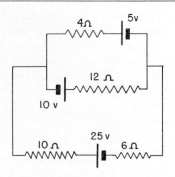

17. In Fig. 20-4, $E' = 9$ volts, $E = 6$ volts, $R_1 = 1$ ohm, $R_2 = 2$ ohms. The internal resistance of E' is negligible, that of E is $R = 2$ ohms. Calculate the voltmeter reading and interpret the results in terms of energy changes per coulomb. Ans.: 7.2 volts with point C at the higher potential.

18. In problem 17 the polarity of cell E is reserved, as in Fig. 20-5. What does the voltmeter read? What is the interpretation of this result?

19. One-hundred-ohm resistors form the sides of a hexagon. What is the resistance between any two symmetrical points? Ans.: 150 ohms.

20. Five rods of resistance 50 ohms each make up a square and one diagonal. Compute the resistance between the two points not joined by a rod.

21. In problem 20 what is the resistance between the two points connected by the rod? Ans.: 25 ohms.

22. In problem 20 a current of 8 amp enters and leaves the corners not connected by the diagonal rod. (a) What is the current in each rod? (b) If the current enters the corner joined by a rod and leaves by the opposite corner, calculate the current in each of the rods.

23. In problem 20 a current of 8 amp enters the corner where the three rods join but leaves at the corner where only two rods join. What is the current in each rod? Ans.: 5 amp in the rod joining entrance and exit points; 2 amp along the diagonal. What are the other currents?

24. In problem 23, what is the equivalent resistance between the two points at which the current enters and leaves?

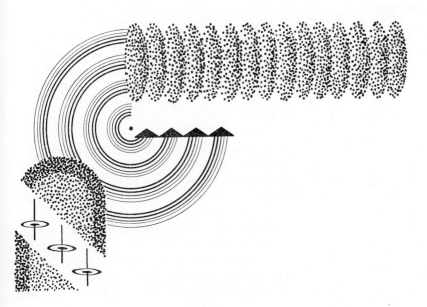

Electromagnetism

21.1. Some Phenomena in Magnetism

The beginnings of the science of magnetism go far back into history, when it was found that a freely suspended piece of natural iron oxide, called *magnetite*, would point approximately along a north-south line. This formed a primitive magnetic compass. Later it was found that bars of iron could acquire this property of magnetism, and these magnetized iron bars became known as *magnets*. Magnets could also attract iron filings, and the regions where the iron filings were concentrated were called the *magnetic poles*. For each magnet there were usually two magnetic poles, which were called a *north-seeking* or N pole and a *south-seeking* or S pole.

It can easily be shown that like magnetic poles repel each other while unlike poles attract each other, and if a compass needle is placed near a magnet, forces will act on the poles of the compass needle. The region about the magnet is said to be a *magnetic field*, and the compass needle will align itself along the direction of the field. As in the case of an electrostatic field, a magnetic field may be represented by lines of force. The magnetic field associated with iron bars or so-called "permanent magnets" will be left to a later chapter (chap. 25). At present we are interested in the phenomena associated with electric currents.

About 1820, Oersted discovered that a magnetic

compass needle was deflected when placed near a current-carrying wire. This was of fundamental importance, for it immediately suggested a connecting link between the sciences of electricity and magnetism.

By exploring the field around a very long conductor with a tiny compass needle, it is found that the lines of force are circles with their centers on the wire. A convenient rule for determining the

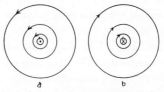

Fig. 21-1.—Magnetic field surrounding a long conductor. (*a*) Current passing out of paper, (*b*) current into the paper.

direction of the field about a wire is the "right-hand rule."

Right-hand rule: If you grasp the wire with the right hand, holding the extended thumb in the direction of the current, the fingers encircle the wire in the direction of the magnetic field.*

In Fig. 21-1 (*a*) the tiny circle with a dot in it represents a current passing normally out of the

---

* The reader will note that this is the same rule that we used in mechanics to denote a positive rotation, i.e., one which, when viewed in the direction of the vector (thumb), was clockwise in direction.

paper, while the cross (×) of Fig. 21-1 (*b*) represents a current into the paper. The larger circles with arrows are the lines of force about the current.

In the case of a circular loop carrying a current *I*, the lines of force are as shown in Fig. 21-2.

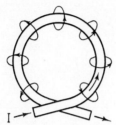

Fig. 21-2.—Magnetic field surrounding a current-carrying loop.

The magnetic field due to a solenoid is shown in Fig. 21-3. A solenoid is formed by winding an

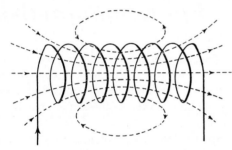

Fig. 21-3.—Magnetic field about a solenoid

insulated wire on a cylinder. On exploring the field, either by a compass needle or with iron filings, it is found that the field is quite uniform at the center but that near the ends the lines of force diverge (Fig. 21-3). Each line of force, however, is a closed loop. The similarity between the field produced by a solenoid and that of a permanent magnet is striking (see § 21.13). As a matter of fact, a few weeks after Oersted's discovery, the brilliant French physicist, André Ampère (1775–1836), showed experimentally that a *permanent magnet may be replaced by a suitable solenoid as far as magnetic effects exterior to the magnet are concerned* (see § 24.3). This is another suggestion as to what is probably responsible for the magnetic field of a permanent magnet.

A further proof that magnetic fields are produced by *motion of charges* is furnished by the classic experiment of H. A. Rowland (1848–1901), an American physicist. He proved in 1876 that isolated electric charges, when rotated mechanical-

ly, deflected a compass needle in exactly the same way that a charge moving in a conductor as an electric current would deflect the needle. His experiment consisted in rotating a charged vulcanite disk at very high angular velocities.

On the basis of a vast amount of experimental data, we are justified in making the following important generalization: Regardless of the origin of the charge—whether it be conduction electrons in metals, charges on insulators, or ions in solution—charges in motion, relative to the experimenter, invariably give rise to magnetic fields.

In Fig. 21-3 it is seen that the field near the center of the current-carrying solenoid is uniform. Now it is possible to measure the magnetic effect at the center in terms of the total number of lines of force, or magnetic flux, crossing the central region. The magnetic flux ϕ is measured in the mks system in terms of a unit called a *weber*.[*] The magnetic field strength or flux density is measured by the flux crossing the unit area and is measured in webers/m² in the mks system.

If the area of the central portion of the solenoid is $A(\text{m}^2)$ and the flux density B (webers/m²) is uniform across this area, then the total flux ϕ (webers) is given by

$$\phi = B A .$$

We shall see in the next section how these quantities can be measured, at least in theory.

21.2. Magnetic Flux Density *B* and Forces on Moving Charges

Among Ampère's experiments in 1830, he showed that there was a force of attraction between two parallel wires carrying currents in the same direction and a repulsion between two parallel wires carrying currents in the opposite direction. This is shown in Fig. 21-4, in which the symbol ⊙ represents a current out of the paper and ⊕ a current into the paper. Similarly, if one of the current-carrying wires is replaced by a

[*] The unit of the weber for magnetic flux is named in honor of Wilhelm Eduard Weber (1804–91), a German physicist who did pioneer work in electromagnetism. Together with Karl Friedrich Gauss (1777–1855), he first devised an *electromagnetic system of units*. In this system the unit of flux density, *B*, is the gauss or the maxwell per square centimeter. The relationship of the electromagnetic (emu) to the mks system of units gives, for the magnetic flux, 1 weber = 10^8 maxwells and, for the flux density, 1 weber/m² = 10^4 maxwells/cm² = 10^4 gauss.

beam of electrons in a vacuum tube (to minimize scattering by gas molecules), then the electron beam is deflected or has a force exerted on it. We may thus state that a *magnetic flux density exists at a point whenever a force acts on a moving charged particle* but that no force is exerted by the magnetic field on a stationary charged particle at the point. An electrostatic field can exert a force on a

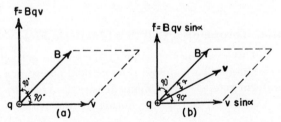

Fig. 21-4.—Forces of attraction and repulsion between parallel wires carrying currents.

stationary or moving charged particle, whereas a magnetic field exerts a force on a moving charged particle, not on one at rest.

If a charge of q coulombs is moving with a velocity of v m/sec at right angles to a flux density of B webers/m², then the moving charge experiences a force of f newtons at right angles to both B and v, which is experimentally given by

$$f = B v q . \qquad (21\text{-}1)$$

Equation (21-1) is a defining equation for the flux density B. The directions of B, v, and f are shown in Fig. 21-5 (a). If the velocity of the

Fig. 21-5.—Diagrams illustrating the directions of the vectors B, v, and f of a charged particle in a magnetic field.

positively charged particle makes an angle a with the flux density B, then the component $v \sin a$ is at right angles to B, and the force f is at right angles to the plane containing B and v (Fig. 12-5 [b]). The force an the moving charged particle is

$$f = qB v \sin a . \qquad (21\text{-}2)$$

Let us apply equations (21-1) and (21-2) to a current-carrying wire. Suppose a wire of uniform

cross-section A and length l which is carrying a current I is placed at right angles to a magnetic field of flux density B, as shown in Fig. 21-6. If

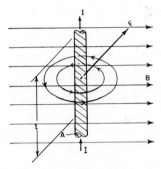

Fig. 21-6.—A current-carrying conductor at right angles to a magnetic field.

there are n charges per unit volume in the wire of area A and each charge q has a velocity v, then, from equation (18-23),

$$I = A n q v .$$

The number of charges N in a length l is

$$N = nl A ,$$

and on each of these moving charges is a force f, given by equation (21-1). Thus the total force F on the length l is

$$\left. \begin{aligned} F = N f &= (nl A)(B v q) \\ &= (A n v q)(Bl) \\ &= IBl , \end{aligned} \right\} \qquad (21\text{-}3)$$

where F is in newtons; I is in amperes; l is in meters; and B is in webers per square meter. From equation (21-3) it follows that the unit of B may be expressed as webers per square meter or as newtons per ampere-meter, i.e.,

$$1 \text{ weber} / \text{m}^2 = 1 \text{ newton} / \text{amp-m.}$$

The direction of the force F is perpendicular to both B and l and is directed into the plane of the paper. From the rule given in § 21.1, it is seen that the magnetic field due to the current in the wire has circular symmetry about the wire, thus adding to the field B in front and opposing it behind. Thus the resultant field is increased in front of the current-carrying wire and weakened behind it. Now the force on the wire is exerted in the **313** direction from the stronger to the weaker resultant field and is therefore at right angles to the plane of I and B. This is shown in Fig. 21-7.

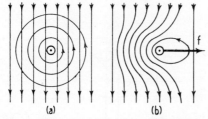

Fig. 21-7.—(a) Superposition of magnetic fields due to B and current-carrying conductor; (b) resulting field and force on the conductor.

If the current-carrying wire of length l makes an angle a with the direction of the magnetic field B, then the force F on the wire is at right angles to B and I into the paper (Fig. 21-8) and is given by

$$F = BIl \sin a . \qquad (21\text{-}4)$$

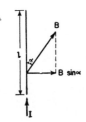

Fig. 21-8.—Illustrating the relationship between B and I for the case in which the conductor makes an angle a with the field.

The component of the flux density $B \cos a$ is along the wire, and there is no force exerted on the wire due to this component. The student should show that the direction of the forces in Fig. 21-4 between two parallel current-carrying wires is correct and in the direction from the stronger to the weaker resultant magnetic field. The relationship between the directions of F, I, and B is also shown* in Fig. 21-9 (a) and (b).

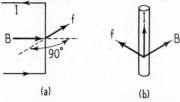

Fig. 21-9.—Illustrating the force on a current in a magnetic field.

* The direction of F is that in which a right-handed screw would progress when turned in the direction of I into B.

In order to determine the direction of the force on the current, we shall describe an apparatus due to Ampère, whose classical experiments with current-carrying conductors laid the foundations of modern electromagnetism. The apparatus consists essentially of a section of a conductor, ab, bent into the arc of a circle (Fig. 21-10), the two ends

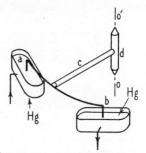

Fig. 21-10.—Ampère's experiment

dipping into mercury troughs. The wire, ab, is rigidly attached to a rod, c, which in turn is rigidly attached to the rod d. The rod d may freely rotate about the axis OO'; i.e., the conductor, ab, may move freely in the direction ab but cannot move at right angles to ab. Ampère found that, whatever the position of a magnet and hence direction of the magnetic field relative to ab, *there was no detectable movement of the wire in the direction* ab even when very large currents were sent through ab. Hence the force on the conductor carrying the current was always at right angles to the conductor, regardless of the direction of the magnetic field.

21.3. Torque on a Coil in a Magnetic Field

Let us examine the value of the torque acting on a rectangular coil with sides a and b placed in a uniform magnetic field of flux density B. As shown in Fig. 21-11, NN' is the normal to the plane of

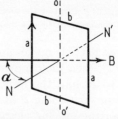

Fig. 21-11.—Rectangular coil in a magnetic field. Vector B makes an angle a with the normal NN' of the coil.

the coil, while a is the angle between NN' and B and is in a horizontal plane. If current I passes around the coil, the force on the sides of the coil is given by equation (21-4). The forces on the horizontal sides, b, are equal and opposite and in the same plane, while the forces on the vertical sides, a, are also equal and opposite *but not in the same plane*. These latter forces constitute a torque tending to rotate the coil about the axis OO'. The value of the force is BIa, the lever arm is $b \sin a$, and the torque is

$$L = BIab \sin a \text{ newton-meters}$$

or

$$L = BIA \sin a . \qquad (21\text{-}5)$$

Here A is the area of the coil. We see from equation (21-5) that the coil tends to turn so that it is at right angles to the field. The torque has a maximum value when $a = 90°$; i.e., the coil is parallel to the field and is zero when $a = 0$, with the coil perpendicular to the field. Since the expression for L contains only the area, it might well be that L is independent of the shape of the coil. A general analysis verifies this prediction.

21.4. The D'Arsonval Galvanometer

A large number of the direct-current measuring instruments in use today have a permanent magnet and a moving coil. In these the weak and variable field of the earth is replaced by the magnetic field of a permanent magnet so strong that the effect of the earth's field is quite negligible. The permanent magnet, being very heavy, is held fixed, and the current coil moves. Such a galvanometer is known as a D'Arsonval type.

A light rectangular coil is hung by a fine conducting wire, usually phosphor-bronze, between

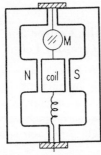

Fig 21-12.—Elevation drawing of the D'Arsonval galvanometer.

the poles of the magnet. The coil ends in a loose spiral wire, which conducts the current out at the bottom (Fig. 21-12). When no current passes, the

plane of the coil is parallel to the field. Now, if a current is passed through the coil, there will be a torque acting on the coil, tending to rotate the coil normal to B. From equation (21-5),

$$L = BIAN \cos \theta , \qquad (21\text{-}6)$$

where N is the number of turns in the coil and θ is the angle between the plane of the coil and the field (θ is the complement of a of eq. [21-5]).

As the coil turns, the supporting wire is twisted; and the mechanical torque necessary to twist the wire through an angle θ, measured in radians, is

$$L_m = L_0 \theta ,$$

where L_0 is the moment of torsion of the wire. When the coil is at rest, the mechanical torque, L_m, is equal to the magnetic torque:

$$L_0 \theta = BIAN \cos \theta$$

or

$$I = \frac{L_0}{BNA} \frac{\theta}{\cos \theta}. \qquad (21\text{-}7)$$

We may simplify this equation by modifying the design of the instrument so that the motion of the coil is always normal to the field. The $\cos \theta$ is then always unity, and equation (21-7) becomes

$$I = \frac{L_0 \theta}{BNA}, \qquad (21\text{-}8)$$

where I is in amperes; L_0 is in newton-meters per radian; B is in webers per square meter; A is in square meters; and θ is in radians. It is not possible to achieve this change perfectly, so that equation (21-8) is not strictly valid over very large angular deflections.

In actual practice it would be difficult to determine the constants B, N, and A with precision. It is better to group the constants together and write

$$I = K\theta , \qquad (21\text{-}9)$$

where

$$K = \frac{L_0}{BNA}, \qquad (21\text{-}10)$$

and determine K by passing a known current through the instrument and observing the deflection.

In the ordinary type of galvanometer a beam of reflected light from the mirror M is intercepted by a curved scale, d. If the mirror turns through an **315** angle θ, the beam of light turns through 2θ (Fig. 21-13). If D is the distance from mirror to scale and d is the deflection, then

$$2\theta = \frac{d}{D}$$

and

$$I = \frac{Kd}{2D} = kd . \qquad (21-11)$$

With $D = 1,000$ mm and d measured in millimeters, k is called the "current sensitivity" of the galvanometer. This sensitivity, k, varies from 10^{-7} to 10^{-10} amp/mm for ordinary instruments.

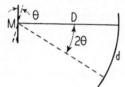

Fig. 21-13.—Relationship between mirror rotation θ and the scale reading d.

A more rugged type of instrument used for measuring heavy currents is called an AMMETER. The modifications necessary are discussed in chapter 22.

EXAMPLE 1: A long straight conductor carrying a current of 150 amp is placed between the pole pieces of the electromagnet of the cyclotron (§ 21.16), where the flux density is 2 webers/m². If the field is confined to the region between the pole pieces of 1-m diameter, calculate the force necessary to keep the wire from moving.

Solution: We shall assume that the conductor is perpendicular to B. By equation (21-4),

$f = (2 \text{ webers/m}^2) (150 \text{ amp}) (1 \text{ m}) (\sin 90°)$

$= 300 \text{ newtons} .$

21.5. The Magnetic Field B Produced by Currents

In § 21.1, Figs. 21-1, 21-2, and 21-3 show the direction of the magnetic field produced by a current. It is our purpose here to calculate the magnitude of the flux density B at a distance r from a current-carrying wire. From the results of experiments, it was shown by analysis by Biot and Ampère that a short length, dl, of wire carrying a current I (called a current element $I dl$) produces in vacuum a flux density dB at a point distant r, making an angle θ, as shown in Fig. 21-14:

$$dB = K \frac{I dl \sin \theta}{r^2}, \qquad (21-12)$$

where the quantity K depends on the units chosen.* In the rationalized mks system the value of K is $\mu_0/4\pi$ and has a value of 10^{-7} weber per ampere-meter. The quantity $\mu_0 = 4\pi \times 10^{-7}$ weber/ampere-meter is called the *permeability* of a vacuum. Thus Ampère's law (eq. [21-12]) for the

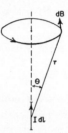

Fig. 21-14.—Magnetic field due to a current element $I dl$

flux density due to a current element is given in the rationalized mks system of units as

$$dB = \frac{\mu_0 I dl \sin \theta}{4\pi r^2} = \frac{I dl \sin \theta}{10^7 r^2}, \quad (21-13)$$

where μ_0 is in webers per ampere-meter; B is in webers per square meters; I is in amperes; dl is in meters; and r is in meters.

For any real circuit, equation (21-13) must be integrated to give the total flux density, and this is usually difficult in any but the simplest cases.

From Fig. 21-14 it is seen that the flux density is the same around the circle shown and that lines of flux density are closed. Unlike lines representing an electric field, which begin and end on electric charges, lines of magnetic flux density or *induction*, as they are sometimes called, always form closed loops, for there is no such thing as isolated magnetic charges.

21.6. Field at the Center and along the Axis of a Circular Loop

Let us apply Ampère's law (eq. [21-13]) to the calculation of the magnetic field at the center O due to a current in a circular loop of radius r (Fig. 21-15). The distance of each current element $I dl$ is r, and the angle θ in each case is 90°, so that $\sin \theta = 1$ and

$$dB = \frac{\mu_0 I dl}{4\pi r^2} = \frac{I dl}{10^7 r^2}.$$

* In the electromagnetic system, K is set equal to unity, the current I in amperes/10, r and dl in centimeters, and the flux density dB in gauss.

The net field at O due to all the current elements at the same distance r is

$$B = \Sigma dB = \frac{I}{10^7 r^2} \Sigma dl.$$

For this case, $\Sigma dl = 2\pi r$, the circumference of the circle; hence

$$B = \frac{2\pi I}{10^7 r} \text{ webers/m}^2 \qquad (21\text{-}14)$$

and is perpendicular to the plane of the page and into the page.

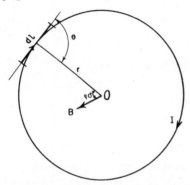

FIG. 21-15.—Magnetic field at the center of a circular loop

EXAMPLE 2: Find the flux density at the center of a loop of wire of radius 20 cm in which there is a current of 2 amp.

Solution: From equation (21-14), we obtain

$$B = \frac{2\pi I}{10^7 r} = \frac{2\pi \times 2}{10^7 \times 0.2} = \frac{2\pi}{10^6} \text{ weber/m}^2.$$

We shall now find the flux density at a point along the axis of a circular loop of wire (Fig. 21-16). Let the radius of the loop be r and have a

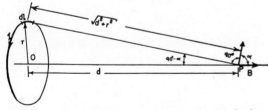

FIG. 21-16.—Magnetic field along the axis of a circular loop

current I in it. It is clear from symmetry that the resultant field B must be along the axis OP, and we shall therefore sum up only the components along OP. The contribution of the current element Idl to the flux density along OP is

$$dB_{OP} = dB \cos \alpha = \frac{\mu_0 I \, dl \cos \alpha}{4\pi (r^2 + d^2)},$$

where $\cos \alpha = r/\sqrt{d^2 + r^2}$. Hence

$$dB_{OP} = \frac{\mu_0 I \, dl \, r}{4\pi (r^2 + d^2)^{3/2}} = \frac{I \, dl \, r}{10^7 (r^2 + d^2)^{3/2}},$$

and the resultant flux density along OP is

$$B = \Sigma dB_{OP} = \frac{I r}{10^7 (r^2 + d^2)^{3/2}} \Sigma dl$$

$$= \frac{I r \, 2\pi r}{10^7 (r^2 + d^2)^{3/2}}.$$

The area A of the loop is πr^2, so that

$$B = \frac{2 A I}{10^7 (r^2 + d^2)^{3/2}} = \frac{2 \mu_0 A I}{4\pi (r^2 + d^2)^{3/2}}. \quad (21\text{-}15)$$

If $d \gg r$, then

$$B = \frac{2 \mu_0 A I}{4\pi d^3} = \frac{2 A I}{10^7 d^3}, \qquad (21\text{-}16)$$

where B is in webers per square meter; μ_0 is in webers per ampere-meter; A is in square meters; I is in amperes; and d is in meters.

If the coil consists of N closely wound loops, then the field is N times that given for a single loop, or the field at P is

$$B = \frac{2 N A I}{10^7 d^3}. \qquad (21\text{-}17)$$

EXAMPLE 3: Suppose the radius of the loop in Fig. 21-16 is 1 mm. Compute the flux density B at point P, 2 m from O, if the loop carries a current of $\frac{1}{2}$ amp.

Solution: From equation (21-16) the flux density is

$$B = \frac{2 A I}{10^7 d^3} = \frac{2\pi (10^{-3})^2 \times 0.5}{10^7 2^3}$$

$$= \frac{\pi}{8} \times 10^{-13} \text{ weber/m}^2.$$

21.7. The Flux Density B Due to a Current in a Straight Conductor

Suppose a wire of length l is carrying a current I. It is desired to find the flux density at a point

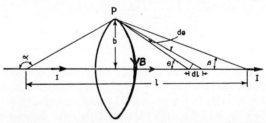

FIG. 21-17.—Diagram for calculating the magnetic field due to a straight conductor.

P distant b from the wire and subtending angles α and β from the ends (Fig. 21-17). Consider a **317** current element Idl a distance r from P subtending an angle θ at P. The contribution of Idl to the flux density at P is given by the following equation:

$$dB = \frac{\mu_0 I\,dl}{4\pi}\frac{\sin\theta}{r^2} = \frac{I\,dl\,\sin\theta}{10^7 r^2}.$$

From the figure it is seen that

$$-\,dl\,\sin\theta = r\,d\theta$$

(the minus sign appears because as θ increases, l decreases); also $b = r\sin\theta$; hence

$$dB = -\frac{I r\,d\theta}{10^7 r^2} = -\frac{I}{10^7 b}\sin\theta\,d\theta.$$

The net flux density B due to the wire subtending the angles α and β is given by

$$B = \int dB = -\frac{I}{10^7 b}\int_\alpha^\beta \sin\theta\,d\theta$$

$$= \frac{I}{10^7 b}(\cos\beta - \cos\alpha). \qquad (21\text{-}18)$$

Suppose the wire is very long compared to the distance b. Then in the limit for infinitely long wire, $\alpha = \pi$ and $\beta = 0$, and the flux density at a distance b from the wire carrying a current I is

$$B = \frac{2I}{10^7 b} = \frac{\mu_0 I}{2\pi b}. \qquad (21\text{-}19)$$

An experimental confirmation of the fact that the flux density varies inversely as the distance b from a very long current-carrying wire was made by Biot and by Maxwell (see § 21.13).

EXAMPLE 4: A very long, straight, thin strip of copper is 10 cm wide and carries a current of 10,000 amp. What is the flux density, B, at a point 10 cm away from an edge of the strip and in the plane of the strip?

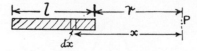

Solution: We shall first solve this problem analytically by making use of the accompanying figure and equation (21-19). In the diagram showing a cross-section of the long, straight strip, l is the width, r the distance to the field point, and x the variable distance from the infinitesimal current-carrying filament dx. The current dI in dx is

$$dI = \frac{I\,dx}{l}.$$

By equation (21-19) the flux density at P due to the straight, infinitely long current filament dI is

$$dB = \frac{2I\,dx}{10^7 l x},$$

and

$$B = \frac{2I}{10^7 l}\int_r^{r+l}\frac{dx}{x} = \frac{2I}{10^7 l}\log_e\left(\frac{r+l}{r}\right) \text{ webers/m}^2;$$

with $r = 10$ cm, $l = 0.1$ m, and $I = 10^4$ amp, the value of the flux density B at the point P is

$$B = \frac{2\times 10^4}{10^7\times 10^{-1}}\log_e\frac{20}{10} = 2\times 0.693\times 10^{-2}$$

$$= 0.01386 \text{ weber/m}^2$$

21.8. The Line Integral of B

As we have repeatedly emphasized, the lines of magnetic flux density form closed loops. These closed loops encircle currents, and we shall show for the special case given above that the value of B multiplied by an element of length tangential to B and summed over the closed loop is equal to

$$\frac{4\pi I}{10^7} = \mu_0 I.$$

In Fig. 21-17, it is seen that the value of B is everywhere the same around the circle of radius b, which is concentric with the infinitely long current-carrying wire. Also the flux density is everywhere tangential to this circle, and for the closed loop the summation of the elements of lengths tangential to B is $2\pi b$. Thus the value of B multiplied by the elements of length tangential to B or the line integral of B around the closed loop of radius b is

$$B2\pi b = \frac{4\pi I}{10^7} = \mu_0 I \text{ webers/m}. \quad (21\text{-}20)$$

This line integral around a closed loop, such as is shown in Fig. 21-18, is written as

$$\int_{\text{closed loop}} B\cos\alpha\,ds,$$

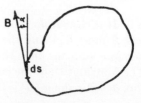

FIG. 21-18.—Calculation of the line integral of B around a closed curve.

where a is the angle between B and ds. In our special case $a = 0$ and $\cos a$ and B have a constant value around the circle of radius b, so that

$$\int_{\text{closed loop}} B\,ds \cos a = B\int ds = B\,2\pi b.$$

Substituting from equation (21-19), $B = 2I/10^7 b$, we obtain

$$\int_{\text{closed loop}} B\,ds \cos a = \frac{4\pi I}{10^7} = \mu_0 I. \quad (21\text{-}21)$$

This result is true for a closed loop of any shape, encircling a current, and is proved for the general case in more advanced texts. From this result the flux density can be found relatively simply for a few important current-carrying circuits.

21.9. The Magnetic Flux Density B within a Very Long Solenoid

A solenoid* consists of uniform turns of wire closely wound on a cylinder. If the length of the solenoid is some ten times its diameter or greater, then the magnetic field within the solenoid, produced by a current in the windings, is very uniform near the center. In order to determine this field, consider a closed loop a, b, c, d (Fig. 21-19).

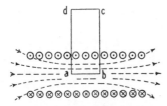

Fig. 21-19.—Magnetic field within a very long solenoid

The lines of flux density form closed loops and emerge from within the solenoid to join up with those entering at the other end, as partially shown in Fig. 21-3. Suppose the closed loop a, b, c, d is so drawn that along the sides bc and ad the flux density is perpendicular to them and that the angle a (eq. [21-21]) is 90°. Then along these sides $\Sigma B ds \cos a$ is zero. Also let cd be so far away that the flux density is zero in that region. Then the only contribution to the line integral (eq. [21-21]) around a, b, c, d is along the side ab.

Suppose the solenoid has n turns per meter length and that a current of I amp in the windings

* "Solenoid" comes from the Greek word *solen*, meaning "tube."

produces a flux density B in the region about ab. Along ab there are $n \times ab$ current windings, or the closed loop a, b, c, d encircles a total current of $n \times ab \times I$. Applying equation (21-21) to the closed loop a, b, c, d, we have

$$\int_{abcd} B \cos a\,ds = \int_{ab} B\,ds = B\,a\,b = \mu_0 n\,a\,b\,I$$

and

$$B = \mu_0 n I = \frac{4\pi n I}{10^7}, \quad (21\text{-}22)$$

where B is in webers per square meter; I is in amperes; and n is turns per meter.

This equation is valid for a region near the center of a very long solenoid. If the solenoid is not long, then the flux density at some point P (Fig. 21-20) can be found from Ampère's law to be

$$\left.\begin{aligned} B &= \frac{\mu_0 n I}{2}(\cos \beta_1 - \cos \beta_2) \\ &= \frac{2\pi n I}{10^7}(\cos \beta_1 - \cos \beta_2). \end{aligned}\right\} \quad (21\text{-}23)$$

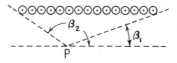

Fig. 21-20.—Magnetic field within a solenoid of finite length.

For an infinitely long solenoid, the case discussed above, β_1 approaches 0 and β_2 approaches π, so that equation (21-23) becomes that given in equation (21-22), or

$$B = \frac{\mu_0 n I}{2}[1 - (-1)] = \mu_0 n I = \frac{4\pi n I}{10^7}. \quad (21\text{-}22)$$

At one end of a very long solenoid, β_1 approaches $\pi/2$ and β_2 approaches π, so that

$$B_{nd} = \frac{\mu_0 n I}{2} = \frac{2\pi n I}{10^7}. \quad (21\text{-}24)$$

21.10. The Magnetic Flux Density B within a Toroidal Solenoid

Suppose several turns of wire are wound in the manner shown in Fig. 21-21 to form a toroidal solenoid. If there is a current in the windings, then, if the windings are closed together, the flux **319** density is confined to within the cross-section of the windings, and there is no flux density outside. The closed loop of B is formed by circles whose center is at the center of the toroid. Let us consider the closed loop marked l in Fig. 21-21, which has a total length of $2\pi r$. If the toroid has n windings per meter, then the closed loop l encircles $2\pi rn$ windings and $2\pi rnI$ current windings if I is the current in the windings. For the line integral

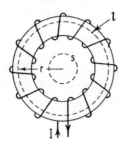

FIG. 21-21.—Magnetic field within a toroidal solenoid

around the closed loop l we have, from equation (21-21), since B is everywhere along the path l or $a = 0$,

$$\int_{\text{Path } l} B \cos a \, dl = B \, 2\pi r = \mu_0 2\pi rnI ,$$

$$B = \mu_0 nI = \frac{4\pi}{10^7} \, nI , \qquad (21\text{-}25)$$

where μ_0 is in webers per ampere-meter; B is in webers per square meter; and nI is in ampere-turns per meter.

If a path such as S in Fig. 21-21 is taken, then the line integral around this path is zero, since no currents or turns are encircled. Hence the toroidal solenoid has no external field.

EXAMPLE 5: A toroidal solenoid has a current of 10 amp in the windings, which have 8 turns/cm. Find the value of B within the solenoid.

Solution: From equation (21-25), with $n = 800$ turns/m, we have

$$B = \frac{4\pi \times 800 \times 10}{10^7} = 3.2\pi \times 10^{-3} \text{ weber/m}^2 .$$

21.11. Magnetic Flux φ in Webers

The magnetic flux ϕ over any area is the summation of the normal component of B over each element of area times the area, or

$$\phi = \int B_n \, dA = \int B \cos a \, dA , \quad (21\text{-}26)$$

where a is the angle between the normal to the element of area dA and the direction of B. If the flux density is uniform, then the flux through an area A perpendicular to B is

$$\phi = AB , \qquad (21\text{-}27)$$

where ϕ is in webers; B is in webers per square meter; and A is in square meters.

The flux density in a closely wound toroid is sensibly uniform over a cross-section of the core. Hence, from equation (21-25), for the flux density within the core, the total flux across the core is

$$\phi = B \, A = \mu_0 NIA = \frac{4\pi NIA}{10^7} \text{ webers}, \quad (21\text{-}28)$$

where A is the cross-sectional area of the core and N is the total number of turns of wire.

EXAMPLE 6: Suppose the core of the toroid in example 5 has an area of 10^{-2} m^2. Find the flux ϕ within the core.
Solution:

$$\phi = B \, A = 3.2\pi \times 10^{-5} \text{ weber} .$$

21.12. The Force between Parallel Currents in Wires; the Ampere

Let us return to the problem given in § 21.1, in which we showed that two wires having currents in them attracted or repelled each other. We shall now find the force F per unit length between two infinitely long parallel wires having currents I_1 and I_2 in them, as shown in Fig. 21-22. The flux

FIG. 21-22.—Force between two parallel currents

density B_1 at a distance r from it is given by equation (21-19) for an infinitely long wire as

$$B_1 = \frac{\mu_0 I_1}{2\pi r} . \qquad (21\text{-}29)$$

This flux density is everywhere perpendicular to the wire having current I_2 in it, so that the force per unit length of wire 2 is, from equation (21-3),

$$F = I_2 B_1 = \frac{\mu_0}{2\pi} \frac{I_1 I_2}{r} = \frac{2 I_1 I_2}{10^7 r}, \quad (21\text{-}30)$$

where F is in newtons per meter; I_1 and I_2 are in amperes; r is in meters; and μ_0 is in webers per ampere-meter.

From the symmetry of the problem, it is seen that the force per unit length on wire *1* is the same as that on wire *2*. For the currents in the same direction, the forces are those of attraction, whereas if the currents are in opposite directions, the forces are repulsive, as previously discussed.

Equation (21-30) is made the basis for the definition of the ampere in the mks system of units. The apparatus incorporating this idea for the purpose of measurement is called a *current balance* and is described in the National Bureau of Standards bulletins. In the mks system:

The ampere is that constant current which, if maintained in each of two parallel conductors of infinite length of negligible circular sections and placed 1 meter apart in a vacuum will produce a force on these conductors of exactly 2×10^{-7} newton per meter of length.

It follows from equation (21-30) and the definition of the ampere that μ_0 in the rationalized mks system is exactly equal to $4\pi \times 10^{-7}$ weber per ampere-meter.

EXAMPLE 7: It is instructive to compare the magnitude of the electrostatic forces of stationary charges with that of electromagnetic forces of charges in motion. As a specific example we shall compute the electrostatic force between two stationary point charges, each of 1 coulomb and separated by 1 m, and also compute the electromagnetic force between the same charges moving with a speed of 1 m/sec along parallel paths 1 m apart.

From equation (17-8) the electrostatic force is given by

$$f_{elec} = 9 \times 10^9 \frac{q_1 q_2}{r^2} = 9 \times 10^9 \text{ newtons},$$

an enormous force!

To compute the electromagnetic forces, we shall modify equation (21-13), giving the magnetic field due to Ids,

$$dB = \frac{\mu_0}{4\pi r^2} I \sin a\, ds \qquad (a)$$

to deduce an expression for the field due to a charge in motion with a speed v_1. From equation (18-13) we have $I_1 = qv_1 n_1$, where n_1 is the number of charges per unit length in conductor 1, in which the speed of the charges is v_1:

$$I\, ds = q\, v_1 n\, ds . \qquad (b)$$

But $qnds$ is the charge dq in the element ds, so that $Ids = v_1 dq$ (note the dimensions of each side of this equation). Substituting in equation (*a*) above, we get

$$dB = \frac{\mu_0}{4\pi r^2} v_1 \sin a\, dq \qquad (c)$$

for the field due to the charge dq. Thus, for a charge q_1, we have

$$B = \frac{\mu_0}{4\pi r^2} v_1 q_1 \sin a , \qquad (d)$$

where a is the angle between v_1 and r and B is perpendicular to the plane of v and r. If a second charge q_2 is r meters from q_1 and moving with a speed v_2, then, by equation (21-2), the force on the charge q_2 is

$$f_{mag} = B q_2 v_2 \sin \beta , \qquad (e)$$

where β is the angle between v_2 and B. Substituting for B from equation (*d*), we obtain the electromagnetic force between moving charges,

$$f_{mag} = \frac{\mu_0}{4\pi r^2} q_1 v_1 \sin a \times q_2 v_2 \sin \beta \text{ newtons} . \qquad (f)$$

For our problem we have $\sin a = \sin \beta = 1$, $\mu_0 = 4\pi \times 10^{-7}$ weber/amp-m, $q_1 = q_2 = 1$ coulomb, $r = 1$ m, $v_1 = v_2 = 1$ m/sec, so that equation (*f*) becomes

$$f_{mag} = 10^{-7} \text{ newton} . \qquad (g)$$

The ratio of the electrostatic to the electromagnetic forces is

$$\frac{f_{elec}}{f_{mag}} = \frac{9 \times 10^9}{10^{-7}} = 9 \times 10^{16} . \qquad (h)$$

This is numerically equal to the square of the velocity of light c^2 ($c = 3 \times 10^8$ m/sec), i.e., the electrostatic forces are of the order of c^2 times as large as the electromagnetic forces of comparable charges moving with unit velocity. Equation (*d*) is of basic importance, for it reveals the true origin of magnetic fields—charges in motion*—and also establishes B, the magnetic flux density, as the primary magnetic vector of field strength.

21.13. Permanent Magnets

It is not the purpose here to discuss the theory of permanent magnets, for this will be done in chapter 25. Rather we shall consider magnets as sources of magnetic flux. We are all familiar with the fact that a uniformly magnetized bar of iron will attract iron filings in small regions about the ends. These regions from which the lines of flux

---

* This statement will require some modification, inasmuch as the neutron, an electrically neutral particle within the nucleus of an atom, has a magnetic moment. At present there is **no simple explanation** for this phenomenon.

density arise are called *magnetic poles*. If a uniformly magnetized bar is placed at some angle a with a uniform magnetic field of flux density B, then a torque is required to hold the magnet in position. There is a force F exerted on each end of the magnet, and these are equal and opposite, as shown in Fig. 21-23. If the distance between the

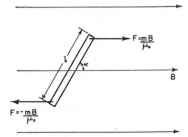

Fig. 21-23.—Torque on a bar magnet in a uniform magnetic field.

points of application of the forces on the magnet is l, then the torque L, tending to make the magnet become parallel with the field, is

$$L = Fl \sin a . \qquad (21\text{-}31)$$

Now in § 21.3 we showed that a torque L on a loop of wire of area A, having a current I in it and placed at an angle a with the field of flux density B, is

$$L = BI A \sin a .$$

The quantity $\mu_0 IA$ is defined as the *magnetic moment* M of the loop, so the torque on the coil may be written as

$$L = \frac{MB \sin a}{\mu_0}. \qquad (21\text{-}32)$$

If the magnet and the loop of wire having the current I experience the same torque at an angle a in a field B, then

$$\frac{MB \sin a}{\mu_0} = Fl \sin a$$

or

$$M = \frac{\mu_0 Fl}{B}.$$

The quantity $\mu_0 F/B$, or the force exerted on the poles in a field of unit flux density, is called the *pole strength* m. Hence the magnetic moment of a bar magnet can be written as

$$M = ml . \qquad (21\text{-}33)$$

Though magnetic poles are fictitious, nevertheless they are a useful concept. In chapter 24 it will be

seen how the magnetism of a bar magnet can be explained in terms of the motion of electrons in atoms. In chapter 25 the concept of a magnetic pole is developed from another point of view.

A solenoid having a current in it has an external magnetic field very similar to that of a bar magnet. Thus the right-hand end of the solenoid (Fig. 21-24 [a]) would behave very much like the N pole

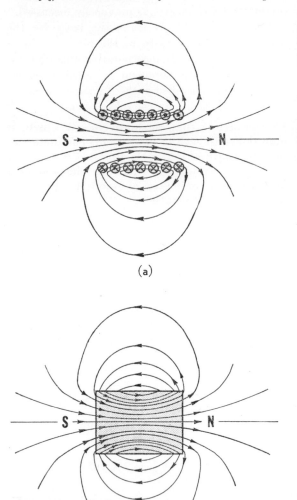

(a)

(b)

Fig. 21-24.—Diagrams showing similarity of the external fields of a solenoid and of a bar magnet.

of a magnet. If the solenoid having a current in it is freely suspended, then it will align itself approximately along a north-south line.

An experimental confirmation of the fact that the field from a long wire having a current I in it varies inversely as the distance was supplied by Clerk Maxwell. If a series of bar magnets is arranged radially on a light circular disk which is suspended so as to rotate freely (Fig. 21-25), not the slightest tendency toward rotation can be discovered when even the greatest currents are sent through the wire, AB. This means that the torques on the north poles tending to cause rotation were exactly balanced by the torques on the

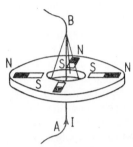

Fig. 21-25.—Bar magnets arranged radially about a current-carrying conductor show no tendency to rotate.

south poles. If r_N and r_S are the distances of the poles from the wire, and B_N and B_S the fields at r_N and r_S due to the current-carrying wire AB, then

$$\frac{B_N}{\mu_0} m r_N = \frac{B_S}{\mu_0} m r_S$$

or

$$\frac{B_N}{B_S} = \frac{r_S}{r_N}, \qquad (21\text{-}34)$$

in agreement with equation (21-19).

21.14. Motion of Charged Particles in Electric and Magnetic Fields

We may gain valuable information as to the properties of electrons and other electrically charged particles by studying their motion through electric and magnetic fields combined in suitable ways.

For simplicity, let us first assume that a uniform electrostatic field $\mathcal{E}$ is parallel to the X-axis and that a particle of mass m and positive charge q has a velocity V_A at A along the X-axis (Fig. 21-26). By definition of the electric field, the force f acting on q is

$$\mathcal{E}q = f \, .$$

According to Newton's Second Law, the constant acceleration a along X is

$$f = ma$$

or

$$a = \frac{\mathcal{E}q}{m} \, .$$

Also for this case the work done by the field on the charge in accelerating it from A to B, a distance d, is measured by the difference in kinetic energies at B and A,

$$\mathcal{E}qd = \tfrac{1}{2} m V_B^2 - \tfrac{1}{2} m V_A^2 \, . \qquad (21\text{-}35)$$

Since the difference in potential between B and A for a uniform field is $V = \mathcal{E}d$, equation (21-35) may be written as

$$Vq = \tfrac{1}{2} m V_B^2 - \tfrac{1}{2} m V_A^2 \, , \qquad (21\text{-}36)$$

indicating that a charged particle falling through a potential difference V acquires an increase in kinetic energy. Although this result was deduced for a uniform electrostatic field, it is valid for any field distribution. If an electron of charge 1.602×10^{-19} coulomb and $m = 9 \times 10^{-31}$ kg (a value to be deduced later) and initially at rest falls through a potential difference of only 1 volt, it acquires a speed of 5.9×10^5 m/sec.

Consider, next, the motion of a charged particle projected between the plates of a parallel-plate capacitor, as shown in Fig. 21-26 (b). The constant force $\mathcal{E}q$ acts

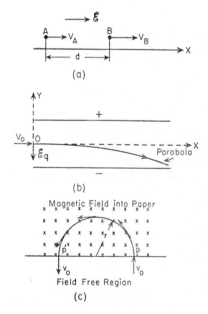

(a)

(b)

(c)

Fig. 21-26.—Trajectories of charged particles in electric and magnetic fields: (a) electrostatic field parallel to v; (b) electrostatic field at right angles to v; (c) magnetic field at right angles to v.

continuously at right angles to the initial speed v_0, and the particle will describe a parabola between the plates, as shown. The motion of the particle is similar to that of the projectile discussed in § 1.23. The equation of the parabola is

$$v = \left(\frac{1}{2} \frac{\mathcal{E}q}{v_0^2 \, m} \right) x^2 \, . \qquad (21\text{-}37)$$

The particle will continue in a straight line in the field-free region outside the plates.

Consider, now, a negatively charged particle of mass m, moving with constant velocity v_0 and entering a uniform magnetic field of flux density B, as shown in Fig. 21-26 (c). The force acting on the particle is constant in magnitude and always at right angles to the velocity, v_0, so that the charged particle moves in a circular path of radius r. From § 2.14 it follows that the centripetal force on the particle is

$$f = \frac{m v^2}{r};$$

and, since the force is Bqv, it follows that

$$B q v = \frac{m v^2}{r}$$

or

$$r = \frac{m v}{q B}; \qquad (21\text{-}38)$$

also, since $\omega = v/r$ and $T = 2\pi/\omega$, we have

$$\left. \begin{aligned} \omega &= \frac{qB}{m}, \\[2mm] T &= \frac{2\pi m}{qB}, \end{aligned} \right\} \qquad (21\text{-}39)$$

where T is the time required for the charge to traverse a complete circle. We observe from these equations that, for m, q, and B constant, (1) the radius of the path is directly proportional to the speed and (2) the period and angular velocity are independent of speed or radius, which means that faster-traveling particles traverse larger circles in the same time that slower particles traverse smaller circles.

For discussions of the energies of high-speed particles it is convenient to introduce a new unit of energy called the *electron-volt* (ev). This unit, by definition, is the amount of energy acquired by an electron in falling through a difference of potential of 1 volt, i.e.,

$$\left. \begin{aligned} 1 \text{ electron-volt} &= \text{Electronic charge} \\ &\quad \times 1 \text{ volt} \\ &= 1.602 \times 10^{-19} \\ &\quad \text{coulomb} \times 1 \text{ volt} \\ &= 1.602 \times 10^{-19} \text{ joule} . \end{aligned} \right\} \quad (21\text{-}40)$$

This unit of energy is widely used in measurements of all types of energy, especially in nuclear work.

21.15. Relativistic Variation of Mass with Velocity

According to the theory of the electron proposed by Lorentz and also according to the special theory of relativity of Einstein, the mass of an electron (or any other moving object) should vary with its velocity, in accordance with the following equation:

$$m = \frac{m_0}{\sqrt{1 - \beta^2}}, \qquad \beta = \frac{v}{c}, \qquad (21\text{-}41)$$

where m_0 is the rest mass of the electron, v its velocity, and c the velocity of light. Since $c = 3 \times 10^8$ m/sec, the mass m differs only slightly from m_0, except for values of v approaching c.

Experiments by Kaufmann in 1906 showed definitely that e/m varies with the velocity of the electron and approaches zero as v approaches c. If an electron falls through a difference of potential of about 3,000 volts, it attains a speed of about $\frac{1}{10}c$. For this speed, m is about 0.5 per cent larger than m_0. For speeds in excess of $\frac{1}{10}c$, the relativistic change in mass should be taken into account in the equations of motion. Newton's Second Law of Motion is valid for high-speed particles if expressed as $d(mv)/dt = F$, i.e., the time rate of change of momentum is equal to the applied force.

21.16. The Cyclotron

The principles that we have considered in § 21.14 (eqs. [21-38] and [21-39]) were employed in a most ingenious way by Lawrence in 1932 to develop the apparatus known as the *cyclotron*. By means of this device it is possible to produce particles possessing very high energies. This apparatus has produced proton beams (hydrogen nuclei) of about 400 million ev (Mev) and deuterons (ions of the heavy hydrogen isotope) of about 200 Mev at average currents of the order of 1.0 μamp.

The principle of operation can best be seen from Fig. 21-27. The cyclotron consists essentially of two

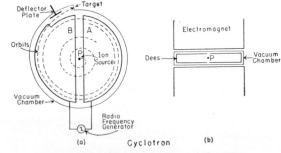

Fig. 21-27.—Schematic drawing of a cyclotron

hollow electrodes shaped in the form of the letter D, as shown in the figure. These two dees, slightly separated, are placed in a vacuum chamber, and the entire structure is placed between the pole faces of a large electromagnet capable of producing a field of about 1.2–1.8 webers/m², or 12,000–18,000 gauss. The plane of the dees is perpendicular to this uniform magnetic field.

The dees are coupled to a powerful source of radio-frequency voltage, so that between the dees (A and B

of Fig. 21-27) an alternating electric field is produced. A frequency between 2 and 20 megacycles/sec is generally used. A source of positive ions is placed at P. Let us assume that at the moment that a positive ion is present at P, dee A is at the negative voltage peak and B at the positive voltage peak. Under the influence of this potential difference, the ion will be accelerated toward A and enter the dee. Within it, the electric field is zero, but under the influence of the magnetic induction field the ion will move in the arc of a circle of radius r, given by equation (21-38),

$$r = \frac{m\,v}{q\,B}.$$

The magnetic field is so adjusted that the angular velocity, v/r, of the ion in the field is equal to the angular velocity, ω, of the radio frequency employed, i.e.,

$$\frac{v}{r} = \frac{B\,q}{m} = \omega = 2\,\pi\,f.$$

Thus, by the time that the ion has completed a semicircle, the field between the dees has reversed, and the ion receives another increment of energy, entering dee B with increased velocity. The radius of the semicircle in B is larger than in A. However, as shown by equation (21-39), the *time required* to traverse this longer path is unaltered if m is constant. Thus the ion appears in the gap at exactly the correct moment to receive an acceleration by the electric field. In this way the ion travels in circles of increasing radius until it passes out through a window to a suitable target.

If the cyclotron imparts 50,000 ev of energy each time the ion passes the gap, in order to attain energies of the order of 5 Mev, the ion must pass the accelerating gap 100 times, i.e., make 50 complete circles. The frequency of the oscillator must be the reciprocal of the time required for an ion to make one complete revolution, as given by equation (21-39), for "ionic resonance"

to take place. For a magnetic field of 1.2 webers/m² or 12,000 gauss and deuterons—heavy hydrogen nuclei (see § 38.31)—the frequency of the oscillator is about 10,000,000 cycles/sec.

The cyclotron must be modified if it is to be used to accelerate electrons, since, even for moderate accelerating fields, the velocity of the electrons approaches that of the speed of light and hence the electron mass increases in accordance with equation (21-41). Under these conditions the simple resonance condition of equation (21-39) is no longer applicable. For the acceleration of electrons, the betatron, described in a later section, has proved remarkably successful.

21.17. Résumé

The following is a brief listing of the definitions, principles, and theories, given in this chapter, which you should know:

The expression for the magnitude and also the direction of the force acting on a moving charge or a current when placed in a magnetic field

The torque on a current-carrying coil placed in a magnetic field

The D'Arsonval galvanometer theory and its current sensitivity

Ampère's law for the flux density due to a current element

Calculation of the flux density due to a current in (a) a circular loop, both at the center and along the axis; (b) a long, straight wire; (c) a solenoid; and (d) a toroid

The force between two parallel current-carrying wires

The definition of the ampere

The equations of motion of charges moving in electric and magnetic fields

The basic theory of the cyclotron

Queries

1. A straight wire carrying a current passes vertically through a hole in the center of a disk which floats freely on the surface of a liquid. A magnet is placed radially on the disk. What happens to the magnet and disk?

2. A long helical spring is arranged so that one end is fixed and the other end just dips into a dish of mercury. If a current is passed through the spring to the mercury, what will happen?

3. On what factors does the sensitivity of the D'Arsonval galvanometer depend?

Practice Problems

1. If two long parallel bars are 10 cm apart and each carries a current of 1,000 amp, what will be the force between the bars per meter of their length? Ans.: 2 newtons/m.

2. A very long, straight solenoid has a diameter of 10 cm. There are 20 turns of wire per centimeter of length of the solenoid. Compute the magnetic flux density at the center of this solenoid when a current of $\frac{1}{2}$ amp exists in the wire.

3. Compute the magnetic flux density at one end of the

solenoid of problem 2. Ans.: $2\pi \times 10^{-4}$ weber/m².

4. If the solenoid of problem 2 is but 20 cm long, compute the magnetic flux density at the center. (Same number of turns per centimeter as in problem 2.)

5. Compute the magnetic flux density at 5 cm from a long, straight wire in which there is a current of 100 amp. Ans.: 4×10^{-4} weber/m².

6. Compute the intensity of the magnetic field at the center of a circular coil of 50 turns having a mean radius of 10 cm. The current in the coil is 0.5 amp. This coil is not a solenoid.

7. A toroidal solenoid carries a current of 1,000 amp. If the solenoid is wound with 4 turns per centimeter, compute the flux density within the solenoid. If the cross-sectional area is 100 cm², what is the value of the flux within the toroid? Ans.: 0.16π weber/m²; 0.0016π weber.

8. A wire 50 cm long is held horizontally at right angles to the earth's magnetic field. If the horizontal component is 2.5×10^{-5} weber/m² and the angle of dip is 72°, calculate the force acting on the wire when a current is 60 amp.

9. Two long parallel wires, 8 in apart, carry currents in different directions, of value 40 amp. Compute the flux density at a point midway between them. Ans.: 1.58×10^{-4} weber/m².

10. Compute the torque on a square coil of 50 turns, 10 cm on a side, placed in a uniform field of strength 0.05 weber/m². The normal to the coil makes an angle of 30° with the field. The current is 100 amp.

11. An electron, $q = 1.602 \times 10^{-19}$ coulomb, moves in a circular path of radius 10^{-8} cm at a speed of 10^7 cm/sec. What is the average current around the path? Ans.: 0.26×10^{-4} amp.

12. An electron with a velocity of 2×10^7 cm/sec is projected into a uniform magnetic field of 0.3 weber/m². If the electron moves at right angles to the field, what is the radius of the circle it describes?

13. Some cosmic-ray particles have energies of the order of 3×10^{10} ev. What kinetic energy in joules does this represent? Ans.: 4.8×10^{-9} joule.

14. An electron is projected into an electric field, as shown in the accompanying diagram. If the electron is deflected $\frac{1}{2}$ cm in traveling 15 cm, calculate its initial velocity. Assume that the field strength is 5 volts/cm.

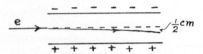

15. How must a magnetic field be arranged, and what must its strength be, in order to prevent the $\frac{1}{2}$-cm deflection of problem 14? Ans.: 3.53×10^{-5} weber/m².

16. A proton has a speed of 10^5 m/sec perpendicular to a magnetic field of 0.02 weber/m². What is the radius of curvature of the path?

17. At a given instant a proton has a speed of 10^6 m/sec and is moving at 45° to a magnetic field of 1.2 webers/m². Calculate the instantaneous force on the proton. What is the acceleration of the proton at this same instant? Ans.: 1.27×10^{-13} newton; 7.6×10^{13} m/sec²

18. Calculate the flux density within a long solenoid of 10 turns/cm with a current of 100 amp present.

19. What is the flux density on the axis of a small circular loop 1 m from the center, if the area of the loop is 0.1 cm² and the current in the loop is 1 amp? What is the magnetic moment of the loop? Ans.: 2×10^{-12} weber/m²; $4\pi \times 10^{-12}$ weber-m.

20. A circular loop of radius 1 m and 100 turns has a current of 5 amp present. Calculate the flux density at a distance of 1 m, 50 cm, and 10 cm from the center of the loop, as shown in Fig. 21-16.

21. A second identical and coaxial loop is placed 1 m away from the loop of problem 20. Calculate the flux density at the mid-point between the coils, assuming that the fields are additive. Also calculate the flux densities at points 10 cm on either side of the center point. What is the percentage variation in field?

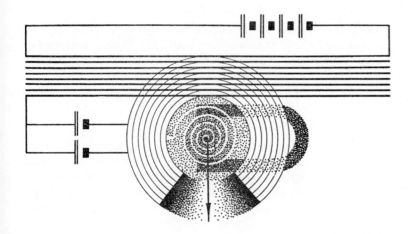

CHAPTER 22

Direct-Current Instruments
and Measurements

In the last few chapters we have been discussing a number of electrical quantities without giving practical methods for their measurement and computation. This was necessary because most electrical measurements depend upon the determination of currents in more or less complicated circuits. Now that we have learned the principles of such circuits, we are ready to make use of them.

It should be noticed that most electrical measurements consist of the determination of one quantity in terms of other quantities, which are given by an *apparatus suitably calibrated;* that is to say, our measurements are *not absolute* in the sense that the defining quantities are directly used. It is especially significant to observe the degree to which the successful operation of the various instruments depends on the validity of Ohm's law.

We will assume that our instruments are accurate. Actual electrical equipment ought to be checked occasionally against the standards kept at the National Bureau of Standards in Washington, D.C.

22.1. Galvanometer Constant

In equation (21-9) the current sensitivity constant of a D'Arsonval galvanometer was given by $I = K\theta$, where I is the current in amperes and θ is the deflection in radians. If the deflection is meas-

ured in any other unit than radians, as, for example, in millimeters of scale deflection, d, we can write

$$I = kd \, ,$$

where d is the deflection in these new units. The new factor of proportionality, k, differs from the former K. In order to determine k, a known current, I, is passed through the galvanometer, and the resulting deflection is observed. A convenient way to obtain a known current is shown by Fig. 22-1, where G is the galvanometer whose k is de-

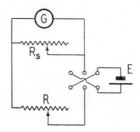

FIG. 22-1.—Measurement of the galvanometer constant k

sired, R is a high resistance (5,000–100,000 ohms), and R_s (called the "resistance of the galvanometer SHUNT") is of much lower resistance (1–5 ohms). The application of the methods given in chapter 20 results in

$$I_G = \frac{E}{R + [1/(1/R_G + 1/R_s)]} \times \left(\frac{R_s}{R_s + R_G}\right), \quad (22\text{-}1)$$

where E is the electromotive force of the cell of negligible resistance and R_G is the galvanometer resistance. The resistance R is so much larger than the resistance of the *galvanometer and shunt* that it is a satisfactory approximation to write

$$I_G = \frac{E}{R} \times \frac{R_s}{R_s + R_G}. \qquad (22\text{-}2)$$

Then $k = I_G/d$ gives

$$k = \frac{E}{R} \times \frac{R_s}{R_s + R_G} \times \frac{1}{d}. \qquad (22\text{-}3)$$

If d is the deflection in millimeters on a scale 1 m away, then equation (22-3) gives k in amperes per millimeter.

22.2. The Direct-Current Ammeter

Galvanometers are always made with many turns of fine wire in the moving coil. For the measurement of a large current, a galvanometer must be modified by increasing its current-carrying capacity. This is done *not* by changing the coil of the instrument but by putting a low-resistance *shunt* across the galvanometer. Let us illustrate this by an example. Suppose we wish to measure the current in an ordinary 100-watt lamp bulb. This current will be of the order of magnitude of 1 amp. Suppose we have a galvanometer with a resistance, R_G, of 100 ohms and of such sensitivity that a current of 20×10^{-5} amp causes full-scale deflection. Now we want the full-scale deflection of the galvanometer to correspond to 3 amp in the bulb. Therefore, we will connect a low resistance in series in the lamp circuit and use this as a shunt across the galvanometer, as in Fig. 22-3. Since the

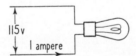

FIG. 22-2.—Measurement of current in the lamp bulb

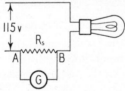

FIG. 22-3.—The use of a shunt to convert a galvanometer into an ammeter.

total current in the divided circuit is to be 3 amp when that in the galvanometer is to be 20×10^{-5}

amp, the difference, 2.9998 amp, must be the current through the shunt, R_s. Then, equating the potential drop between A and B through the two different paths, we have

$$2.9998 R_s = 100 \times 20 \times 10^{-5},$$

$$R_s = 0.00667 \text{ ohm}.$$

Here we have constructed an ammeter for measuring currents of several amperes out of a galvanometer designed to measure only a few hundred thousandths. Note that this result agrees with our previous assumption that the resistance of an ammeter should be negligibly small compared to the other resistances of the circuit. This is likewise true, since the resistance of the bulb is

$$\tfrac{115}{1} = 115 \gg 0.006667.$$

When the normal current of 1 amp passes through the lamp, the galvanometer, so shunted, reads one-third of its full-scale deflection.

In practice, ammeters have coils that are pivoted on jeweled bearings and are fitted with a pointer that moves across a previously calibrated scale (Fig. 22-4). The pointer is brought back to

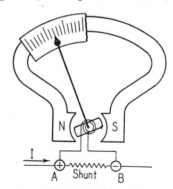

FIG. 22-4.—Schematic diagram of an ammeter

its zero reading by a small spiral spring, which, in addition, serves to lead the current to the coils and also supplies the equilibrant torque. The coil, whose resistance is generally high, is shunted by a low resistance, which results in a very tiny current passing through the coil. For a given coil mechanism, the range of the instrument is determined by the value of the shunt. Instruments are available with interchangeable shunts, whose ranges are from 10^{-6} amp per scale division to 1 amp per scale division.

EXAMPLE 1: A direct-current milliammeter has a full-scale reading of 20 millamp (0.020 amp). The total resistance between its terminals is 5.0 ohms. What value of shunt must be used with this millammeter so that it will read 100 amp full scale?

Solution: Let R_m be the resistance of the millammeter between terminals and including terminal leads, R_s the resistance of the shunt, I_m the full-scale current through the meter, I_s the shunt current, and I the total current to be measured. Following the foregoing analyses, the student can show that

$$R_s = \frac{I_m R_m}{I_s}$$

and

$$R_s = \frac{R_m}{(I/I_m) - 1}.$$

For our problem,

$$R_s = \frac{5}{(100/0.02) - 1} = 0.001 \text{ ohm}.$$

The ratio I/I_m is known as the *multiplying factor* of the shunt. In our problem the multiplying factor is 5,000. Manganin, with a very low temperature coefficient, is generally used for standard shunts.

22.3. The Direct-Current Voltmeter

In this same lamp circuit we can measure the potential drop across the lamp, making use of this same galvanometer, provided that we modify it in a different way (Fig. 22-5). If we put a suf-

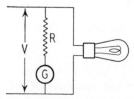

Fig. 22-5.—The use of a high resistance, R, to convert a galvanometer into a voltmeter.

ficiently large resistance, R, in series with the galvanometer, we may connect it across the *bulb*, deflecting only a negligible current away from the bulb. Let us use a series resistance such that a potential difference of 150 volts gives full-scale deflection on the galvanometer. In other words, make R of such size that a potential drop of 150 volts across the resistance and galvanometer will produce a current of 20×10^{-5} amp in each. Then, from Ohm's law,

$$20 \times 10^{-5} = \frac{150}{R + 100}, \qquad (22\text{-}4)$$

$$R = 749,900 \text{ ohms}.$$

Note that this resistance is very great, compared to the resistance of the lamp.

The voltmeter, as used in practice, is identical in design to the ammeter, the essential difference being that the movable coil is connected in series with a very high resistance, R (Fig. 22-6). The in-

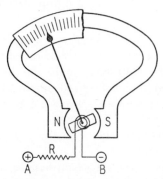

Fig. 22-6.—Schematic diagram of a voltmeter

strument may be connected directly across the voltage mains, so that only a small current will pass through the coil. For a given coil mechanism, the range of the instrument is determined by the resistance R.

EXAMPLE 2: A moving-coil, direct-current millivoltmeter has a full-scale reading of 100 millivolts. It is desired to increase the range of the meter to a full-scale reading of 1,000 volts. If the resistance between terminals is 10 ohms, what value of resistance must be placed in series with the millivoltmeter?

Solution: Let V be the voltage drop to be measured, V_m the maximum voltage drop across the meter, R_s the required series resistance, and R_m the resistance of the meter. The student can show that

$$R_s = R_m \left(\frac{V}{V_m} - 1 \right).$$

For our problem,

$$R_s = 10 \left(\frac{1,000}{0.1} - 1 \right) = 99,990 \text{ ohms}.$$

Commercial-grade voltmeters have internal resistances of the order of 100 ohms per volt deflection. Precision voltmeters may have resistances as high as 1,000 ohms per volt deflection.

22.4. Use of Ammeters and Voltmeters

Suppose an inexperienced experimenter made the mistake of connecting the voltmeter in Fig. 22-6 into the line in series with a 330-watt lamp, as an ammeter should be connected. Will the instrument be damaged, will the lamp burn, will the instrument give a reading? If it gives a read-   *329* ing, will this be nearly 110 volts (on a 110-volt direct-current line) or nearly zero? The current in the lamp and voltmeter would evidently be

$$\frac{110}{750,000 + 36\frac{2}{3}} = 0.000146 \text{ amp};$$

but, since it would require a current of 0.0002 amp to cause the voltmeter to give full-scale deflection, the voltmeter would not be damaged. The current of only 0.000146 amp in a lamp which normally carries 3 amp would not be sufficient to heat the filament. The reading of the voltmeter would be less than 110 volts by only $33\frac{1}{3}$ parts in 750,000.

Now consider what would happen if the same inexperienced experimenter were to connect the ammeter across the line as a voltmeter should be connected. The current in the ammeter would rise toward $110 \div 0.00667 = 16,500$ amp! Before this current was reached, the instrument would be destroyed.

When testing wires, a voltmeter should always be used, not an ammeter. *One further precaution is necessary:* Some voltmeters are made with two different voltage ranges, i.e., with two different series resistances. It is nearly as destructive to connect a voltmeter with a 3-volt range across 110 volts as it is to connect an ammeter across such a line.

22.5. Measurement of Resistance with Ammeter and Voltmeter

An approximate way to measure a resistance is shown in Fig. 22-7. From Ohm's law,

$$R = \frac{V}{I}. \qquad (22\text{-}5)$$

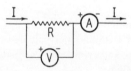

Fig. 22-7.—Ammeter-voltmeter method for measuring resistance. Ammeter reading must be corrected for current through the voltmeter.

This method is precisely what we used in § 20.1 by assuming that the resistance of the voltmeter was so great that it drew negligible current. Nevertheless, the ammeter shows the current in the resistance coil plus the current in the voltmeter, however small the current in the voltmeter

330

may be. Therefore, the correct form of the equation would be

$$R = \frac{V}{I - (V/R_V)}, \qquad (22\text{-}6)$$

where R_V is the resistance of the voltmeter and I is the current in the ammeter. Hence the current through the voltmeter, V/R_V, is deducted from the total current, and the denominator is the current in R of Fig. 22-7.

This difficulty could be avoided by connecting the instruments, as in Fig. 22-8; but now the volt-

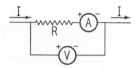

Fig. 22-8.—Voltmeter reading must be corrected for potential drop across the ammeter.

meter gives the potential drop across the resistance coil plus that across the ammeter, and the true equation would be

$$R = \frac{V}{I} - R_A, \qquad (22\text{-}7)$$

where R_A is the resistance of the ammeter.

EXAMPLE 3: A resistance of 1,000 ohms is to be measured by the ammeter-voltmeter method. The following data are available: the constant voltage that drives the current in either Fig. 22-7 or Fig. 22-8 is 100 volts; the voltmeter has a full-scale range of 200 volts and an internal resistance of 40,000 ohms, i.e., 200 ohms per volt; the milliammeter has an internal resistance of 5 ohms.

If the method of Fig. 22-7 is used, then the current through the voltmeter is $100/40,000 = 0.0025$ amp. The current through R is $100/1,000 = 0.1$ amp. The current measured by the milliammeter is, then, $0.1 + 0.0025 = 0.1025$ amp, and the apparent resistance is $100/0.1025 = 976$ ohms or 2.4 per cent less than the true value.

If the method of Fig. 22-8 is used, the total resistance across the voltmeter is 1,005 ohms. The current read by the milliammeter is $100/1,005 = 0.1$ amp to an accuracy of 0.5 per cent, and the resistance is $100/0.1 = 1,000$ ohms.

22.6. The Wheatstone Bridge

Far more accurate than an ammeter-voltmeter method for measuring a resistance is a special form of divided circuit known as a WHEATSTONE

BRIDGE* (Fig. 22-9). Four resistances, R_1, R_2, R_3, R_4, one of them the one to be measured, form the divided circuit. A battery, with a resistance R in series to protect the rest of the circuit from excessive currents, is connected across one side of the divided circuit; the galvanometer, across the other. At least one of the known resistances must be adjustable. This variable resistance is adjusted until there is no current in the galvanometer.

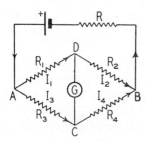

FIG. 22-9.—Wheatstone bridge for the measurement of resistance.

Then the points C and D must be at the same potential. If this condition is true, the potential drop from A to C must be the same as from A to D, or

$$I_1 R_1 = I_3 R_3 \; ;$$

and also

$$I_2 R_2 = I_4 R_4 \; .$$

Dividing one equation by the other gives

$$\frac{I_1 R_1}{I_2 R_2} = \frac{I_3 R_3}{I_4 R_4}. \qquad (22\text{-}8)$$

However, if there is no current in the galvanometer, $I_1 = I_2$ and $I_3 = I_4$; and the equation reduces to

$$\frac{R_1}{R_2} = \frac{R_3}{R_4}. \qquad (22\text{-}9)$$

Suppose R_4 is the unknown resistance; then

$$R_4 = R_3 \frac{R_2}{R_1}. \qquad (22\text{-}10)$$

Note that it would be sufficient to know R_3 and the ratio of R_2 to R_1.

In one common form of the instrument the resistance R_3 is arranged so as to be variable from 0 to 9,999 ohms in steps of 1 ohm by means of a

---

* Sir Charles Wheatstone (1802–75), whose name has always been associated with this device, while doubtless having independently invented it himself, acknowledges that he was anticipated in it by ten years by a Mr. S. H. Christie.

"dial box" in which appropriate coils are connected by turning marked dials. The ratio R_2/R_1 is adjusted either by a single dial or by two separate dial boxes.

In another form of instrument, known as the SLIDE-WIRE BRIDGE, if R_4 is unknown and R_3 is the variable dial box, the "ratio arms," R_1 and R_2, consist of a uniform wire equipped with a sliding contact which can be adjusted to a balance in which no current passes through the galvanometer. This contact divides the wire into two lengths, L_1 and L_2, proportional (if the wire is of uniform cross-section) to R_1 and R_2, thus satisfying equation (22-9):

$$\frac{R_1}{R_2} = \frac{L_1}{L_2} = \frac{R_3}{R_4}$$

or

$$R_4 = R_3 \frac{L_2}{L_1}.$$

Highest precision is obtained when all the resistances, R_1 to R_4, are of the same order of magnitude. Binding posts are present in the slide-wire bridge for inserting the unknown resistance, R_4, the battery, and the galvanometer. Notice that the method consists in comparing an unknown with a known resistance.

EXAMPLE 4: A pair of identical telephone wires of resistance 10 ohms/mi connects two stations, 30 mi apart, as shown in the diagram. One of the wires is

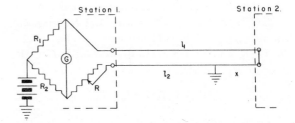

accidentally grounded x miles from station 2. The problem is to locate the fault. A modified Wheatstone bridge is set up as shown. The wires at station 2 are joined, and the resistance R for a balance determined. For a balance:

$$\frac{R_1}{R_2} = \frac{300 + 10x}{300 - 10x + R}.$$

If $R_1 = R_2 = 1,500$ ohms and $R = 24$ ohms, then $x = 24/20 = 1.2$ mi from station 2. This is known as the *Varley loop test*.

22.7. Ideal Measurement of Potential Difference

We have seen that a voltmeter measures potential difference by passing a small but finite amount of current. A perfect device for measuring potential difference ought to pass no current at all; then it would in no way change the circuit to which it was attached. The electrostatic voltmeter or quadrant electrometer discussed in chapter 17 is such an instrument, but it is seldom used in elementary work because of experimental difficulties. However, it is possible to measure potentials by *special forms of circuits* arranged to draw no current from the potential source under consideration. Such a device is the POTENTIOMETER.

22.8. The Potentiometer

The potentiometer is a device for comparing a potential difference with a known electromotive force. The arrangement is as shown in Fig. 22-10.

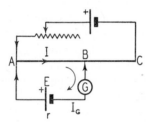

FIG. 22-10.—Basic electrical circuit of the potentiometer

A steady current, I, is passed through a uniform wire ABC. A cell and galvanometer are connected in parallel with part of this wire. Let R_x represent the resistance of the uniform wire between A and B. Then, applying Kirchhoff's second law to the closed circuit $ABGEA$, we have

$$IR_x + R_g I_g + r I_g = E_x . \qquad (22\text{-}11)$$

If we now adjust R_x until $I_g = 0$, the equation reduces to

$$IR_x = E_x . \qquad (22\text{-}12)$$

Suppose we now replace E_x with another cell, E_s, and find a new R_s, which reduces the galvanometer current to zero:

$$IR_s = E_s . \qquad (22\text{-}13)$$

Then, dividing one equation by the other gives

$$\frac{R_x}{R_s} = \frac{E_x}{E_s} . \qquad (22\text{-}14)$$

If E_s is known, the other is determined. For the known E_s, a Weston standard cadmium cell is ordinarily used (§ 19.12). Notice that in this instrument there is no current in the cell when the final balance is obtained.

It is necessary that the direction of the main current and the direction of the electromotive force of the cell, E_x or E_s, be related, as indicated in the diagram, in order that a balance may be obtained. For instance, if the cell were reversed, the equation around the closed circuit would read

$$IR_x + R_g I_g + r I_g = -E_x , \qquad (22\text{-}15)$$

and there would be no possible positive value of R_x which would cause I_g to equal zero.

Because of its precision, the potentiometer has been adapted to many purposes. If the unknown electromotive force of our description is replaced by any potential difference, the potentiometer becomes essentially a voltmeter. Resistances may be built which are very exact. By passing an unknown current through a known resistance and measuring the drop in potential by means of a potentiometer, one has, from $I = V/R$, an "ammeter" far more accurate than the ordinary type of ammeter (Fig. 22-11).

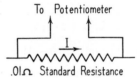

FIG. 22-11.—Use of the potentiometer for the measurement of current.

In the commercial form of the instrument, the resistance, AC, consists of a set of precision-wound resistors controlled by a dial and a highly uniform wire of manganin wound on a drum. By suitable calibration the instrument may be made to read volts directly.

22.9. Résumé

The following is a brief listing of the definitions, principles, and theories, given in this chapter, which you should know:

The theory of the measurement of the sensitivity of a galvanometer

The basic electrical requirements of an ammeter and a voltmeter and the limitations of the measurement of resistance with them

The theory of a Wheatstone bridge

The theory of a potentiometer

Queries

1. How can a potentiometer be used to measure currents? Resistances?
2. Why is it desirable for a voltmeter to have a large resistance per scale division?
3. If a Wheatstone bridge is balanced, will interchanging the battery and galvanometer affect this balance? Explain.

4. Explain why the potentiometer can be used to measure the electromotive force of cells.

Practice Problems

1. See Fig. 22-7. The voltmeter reads 25 volts, and the ammeter reads 5 amp. The resistance of the ammeter is 0.1 ohm; of the voltmeter, 5,000 ohms. What percentage of error is made in the value of R if we use equation (22-5) instead of equation (22-6)? Ans.: 0.1 per cent.
2. See Fig. 22-8. With the same data as in problem 1, what percentage of error is made if we use equation (22-5) instead of equation (22-7)?
3. A galvanometer has a resistance of 198 ohms and is of such sensitivity that a current of 0.002 amp causes a deflection of one scale division. Show how to make this into an ammeter such that five scale divisions indicate 1 amp. (Diagram should be included.) Ans.: Use a shunt of about 2 ohms.
4. Show how to make the galvanometer of problem 3 into a voltmeter such that one scale division means 20 volts. (Diagram should be included.)
5. See Fig. 22-10 of the potentiometer. Suppose the galvanometer is at zero when the sliding contact is at B. Show on the diagram the direction of the current in the galvanometer if the sliding contact be moved to a new position B', nearer to A than to B. Prove your answer.

6. For a D'Arsonval galvanometer, if K is the current in amperes necessary to cause a deflection of 1 rad and k is the current in amperes needed to cause a deflection of 1 mm on a scale 1 m from the mirror, what relation exists between K and k? *Hint:* When a mirror turns, a reflected ray of light turns through twice as great an angle as the mirror itself.
7. A direct-current milliammeter has a full-scale reading of 5 milliamp. The total resistance between its terminals is 2.0 ohms. What value of shunt must be used with this milliammeter to extend the range to 50 amp full scale? Ans.: About 2×10^{-4} ohm.
8. A voltmeter has a full-scale reading of 10 volts. It is desired to increase the range to a full-scale reading of 1,000 volts. If the resistance between the terminals is 2,000 ohms, what value of resistance must be placed in series with the voltmeter?
9. In the Wheatstone bridge of Fig. 22-9, $R = 0$, R_1 and R_2 have values of 20 ohms, $R_3 = 40$ ohms, $R_4 = 44$ ohms, and the galvanometer has a resistance of 60 ohms. The electromotive force of the battery is 10 volts. What is the value of the current through the galvanometer? What value of R_4 will balance the bridge? Ans.: 2.6×10^{-3} amp; 40 ohm.

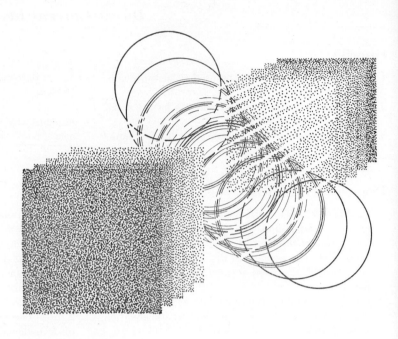

Electromagnetic Induction

23.1. The Discovery of Electromagnetic Induction

In 1820 Oersted discovered that a magnetic field is always associated with an electric current (§ 21.1). This discovery led the physicists of that day to ask the question: "If a steady electric current produces a steady magnetic field, should not a steady magnetic field produce a steady electric current?" All efforts to find such an electric current were unsuccessful, although both Ampère and Arago observed the phenomenon but failed to recognize it. However, Joseph Henry in 1830 discovered current circulating in a coil around the "keeper" or iron bridge across the poles of one of his electromagnets when current was *started* or *discontinued* in the coils of the magnet itself. Michael Faraday, likewise, had been one of those seeking this phenomenon without success. He probably was the first to recognize that the question should be rephrased somewhat as follows: "If moving electric fields (those of the moving charges that constitute the current) produce *magnetic* fields, should we not expect that moving magnetic fields will produce *electric* fields and thus be able to cause charges to circulate, i.e., cause electric currents?" In 1831 Faraday found current produced in a coil of wire wound on an iron core when, and only when, another current was started or stopped in

another coil on the same core. His prompt publication of this result and his rapid clarification of the phenomenon by other shrewd experiments gave him prior claim over Henry,* who, indeed, did not publish his observations until stimulated to do so by seeing Faraday's publication.

The importance of Faraday's discovery in connection with the development of the modern industrial aspects of civilization and the effects of these upon the social life in which we find ourselves can hardly be overestimated. The electric generators which supply all our electrical energy and the transformers that enable this energy to be transmitted far and wide, to turn the wheels of industry and transportation, are direct applications of Faraday's discovery of electromagnetic induction.

23.2. Faraday's Experiments on Induced Currents

If the north pole of a permanent magnet is moved toward a coil of wire containing a galvanometer, G, a current is found to pass along the wire in the direction indicated in Fig. 23-1. If there is no relative motion of the magnet and coil,

---

* For an interesting account of Henry's work see the *Scientific American*, July, 1954, p. 72.

no current is present. The faster the magnet moves toward the coil, the greater the deflection of the galvanometer, but the deflection lasts for a shorter time. When the magnet is withdrawn, a current in the opposite direction passes around the circuit. *The presence of the current in the wire implies the existence of an electromotive force along the*

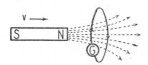

FIG. 23-1.—Electromotive force induced in the galvanometer circuit due to motion of the bar magnet.

wire. This induced electromotive force is present only while there is relative motion of the magnet and coil, i.e., while the magnetic flux within the coil or linking the coil is varying with time.

If the magnet is replaced by a current-carrying solenoid, as in Fig. 23-2, then a movement of the

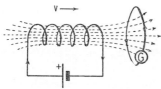

FIG. 23-2.—Electromotive force induced by the relative motion of the current-carrying solenoid and the galvanometer circuit.

solenoid with respect to the coil also produces an electromotive force in the coil. Here, again, the magnetic flux through the coil varies with time.

A third experiment performed by Faraday is illustrated in Fig. 23-3. Two circuits, P and S, are

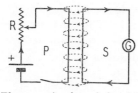

FIG. 23-3.—Electromotive force induced in the secondary circuit while the current in the primary changes.

placed side by side. Circuit P, called the "primary," consists of a battery, a variable resistance, and a key, while S, the secondary, consists of a loop of wire containing a galvanometer. Faraday found that an electromotive force was induced in S at the instant the key was closed, i.e., during the time that the current, and hence the magnetic field, started to increase in circuit P. When the

current reached a steady value, no effect was observed in circuit S. On opening the key, the galvanometer again indicated an induced electromotive force in S, which quickly decreased to zero. Faraday found that, by varying the magnitude of the current in P by means of the variable resistance R, he could induce an electromotive force in circuit S. Again in this experiment we note that an electromotive force was induced only during the time that the magnetic flux linking S was varying.

23.3. Faraday's Law of Electromagnetic Induction

From experiments such as we have described, Faraday concluded that whenever there is a change in *magnetic flux* linking a coil, there is an electromotive force induced in that coil. Also, the magnitude of the electromotive force is proportional to the number of turns of wire N in the coil and proportional to the *time rate of change* of that flux.

The magnetic flux, ϕ, through a given area we had previously defined as the product of an area, A, by the component of the flux density vector B normal to the area. In symbols

$$\phi = B A \cos \alpha \text{ webers}, \qquad (23\text{-}1)$$

where ϕ is the magnetic flux in webers, A is the area in square meters, and α is the angle between B and the normal to the area. If B is perpendicular to the area, then

$$\phi = B A . \qquad (23\text{-}2)$$

We may now formulate Faraday's law of electromagnetic induction in analytical form, for a coil having N turns, as

$$E = - k N \frac{d\phi}{dt}. \qquad (23\text{-}3)$$

The reason for the negative sign is discussed in the next section. The proportionality constant, k, depends on the units chosen for ϕ and E. In our mks system, $k = 1$, E is measured in volts, and ϕ in webers. Thus we have

$$E = - N \frac{d\phi}{dt} \text{ volts}, \qquad (23\text{-}4)$$

where E is in volts; ϕ is in webers; t is in seconds; and N is a numeric. *This is another fundamental equation of electricity.*

If $\bar{E}$ is the average electromotive force induced in a coil of N turns when the magnetic flux through the coil changes from ϕ_1 to ϕ_2 webers in time t sec, then we may write, for this average electromotive force,

$$\bar{E} = -\frac{N(\phi_2 - \phi_1)}{t} \text{ volts}. \qquad (23\text{-}5)$$

In terms of flux density, B, equation (23-4) becomes for the instantaneous electromotive force

$$E = -NA\frac{dB}{dt}, \qquad (23\text{-}6)$$

where E is in volts; B is in webers per square meter; A is in square meters; and t is in seconds, if the size and orientation of A are fixed and only B changes with time. From the foregoing it is seen that

$$1 \text{ volt} = 1 \text{ weber per second}$$
or
$$1 \text{ volt-sec} = 1 \text{ weber}.$$

23.4. Lenz's Law

The direction of the induced electromotive force is always such that the resulting magnetic field of the *induced current* will *oppose* the change of flux which gave rise to the electromotive force. This experimental law was discovered in 1834 by the Russian physicist, Emil Lenz (1804–65), who, as a matter of fact, had independently discovered the phenomena of electromagnetic induction without knowing anything of Henry's work and but little of that of Faraday. The negative sign of equations (23-4) and (23-5) represents the mathematical expression of Lenz's law. This law is a result of the conservation of energy, for if the flux from the induced current were in the same direction as that of the inducing flux, then each would aid the other, and an infinite current would be set up. This would imply an infinite amount of energy produced by a finite amount expended. This is impossible. (Why?)

23.5. Faraday's Law and the Cutting of Lines of Flux

In § 23.2 we stated that (*a*) an electromotive force was induced when the flux linking a circuit changed with time and there was no motion between circuits (Fig. 23-3); and (*b*) an electromotive force was also induced if there was relative motion between a conductor in which the electromotive

force was induced and the current-carrying circuit that produced the field (Fig. 23-2), i.e., when a conductor "cuts lines of flux." The transformer which we shall describe in § 28.12 is a practical example of *a*, and the electric generator (see § 27.2) an example of *b*. In this section we shall examine in some detail the case of a conductor that moves across a magnetic field and "cuts" lines of flux.

Let us consider a conductor of length, l, moving with constant velocity, v, across a magnetic field of uniform flux density, B, directed outward from the paper, as shown in Fig. 23-4. We know from

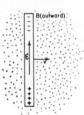

Fig. 23-4.—Induced electromotive force due to flux "cutting."

our previous analysis in § 21.2 that each electron and positive charge of the conductor will experience a force given by equation (21-1):

$$f = q\,vB.$$

As a result, the free negative electrons will move to the top end of the conductor and the positive charges will remain fixed, causing a separation of the charges, as shown in the figure. This separation will continue until the electric field intensity, $\mathcal{E}$, due to the charge separation is just sufficient to counteract the effects of the magnetic fields or the resultant force on the charges within the conductor is zero. The constant electric field vector, $\mathcal{E}$, along the conductor is directed upward, and its magnitude is

$$\mathcal{E} = \frac{f}{q} = vB \text{ volts/m}, \qquad (23\text{-}7)$$

and the induced electromotive force E between the ends of the wire of length l is

$$\text{Induced emf } E = \mathcal{E}l = vBl \text{ volts}. \quad (23\text{-}8)$$

This electromotive force is known as a motional electromotive force. In this case v, B, and l are perpendicular to one another. If v or B changes direction, then the electric field vector and the induced electromotive force change sign. Also if v

is not perpendicular to the field but makes an angle α with B, then equation (23-8) becomes

$$E = vBl \sin \alpha \text{ volts} . \qquad (23\text{-}9)$$

Equation (23-9) represents Faraday's law of induction for the case of moving conductors in magnetic fields. If v is in meters per second, l in meters, and B in webers per square meters, then E is in volts. Equation (23-9) is the basic equation of the electric generator.

We may extract energy from the moving conductor by the arrangement of Fig. 23-5. In this

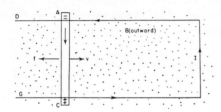

FIG. 23-5.—Induced current due to flux "cutting"; principle of the electric generator.

figure a U-shaped piece of bare copper wire, DG, is placed with its plane normal to a uniform magnetic field of flux density B. Let AC be another bare copper wire which can be moved along DG without breaking the electrical contact. The induced electromotive force starts a current, I, around the circuit, as shown. The moving wire is equivalent to a battery of electromotive force given by vBl, with the negative polarity at the top and the positive polarity at the bottom of the wire. In accordance with equation (21-3), a force acts on the conductor, directed to the left and of magnitude

$$f = BIl \text{ newtons} .$$

The mechanical power expended in pushing the wire to the right with a speed v is

$$fv = BIlv . \qquad (23\text{-}10)$$

However, by equation (23-8), the induced electromotive force E is Blv. Therefore, mechanical power is transformed into electric power, P, given by

$$P = EI \text{ watts} . \qquad (23\text{-}11)$$

It is interesting to observe that, in pushing the wire AC a distance dx in time dt, the area swept out is

$$dA = l \, dx$$

and the flux $d\phi$ through this area is

$$d\phi = Bl \, dx ,$$

so that the change in flux per unit time is

$$\frac{d\phi}{dt} = Bl \frac{dx}{dt} = Bl \, v . \qquad (23\text{-}12)$$

Since Blv is the induced electromotive force E in the conductor AC, we may state that

$$E = -\frac{d\phi}{dt} \text{ volts} ; \qquad (23\text{-}13)$$

that is, when a conductor moves across a magnetic field, an electromotive force is induced whose value is equal to the rate at which the magnetic flux is "cut." From Fig. 23-5 $d\phi/dt$ is negative, and therefore E is positive or produces counterclockwise current.

EXAMPLE 1: Suppose a coil of 1,000 turns of wire is wound around a book, as indicated in the accompanying figure, and this book is lying on the table. The vertical component of the earth's magnetic field is about 0.6×10^{-4} weber/m². The area of the coil is $0.15 \times 0.20 = 0.03$ m². If this book is turned over once (from position A to position D) about a horizontal axis in $\frac{1}{10}$ sec, what average electromotive force will be induced in the coil?

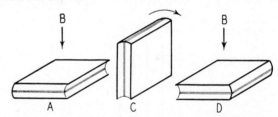

Solution: In position A, $\phi_A = 0.03 \times 0.6 \times 10^{-4} = 1.8 \times 10^{-6}$ weber. In position D, $\phi_D = 0.03 \times 0.6 \times 10^{-4} = 1.8 \times 10^{-6}$ weber. Notice, however, that the direction of the flux through the coil has changed. At an intermediate position, as at position C, the plane of the coil is parallel to the flux. Then

$$\phi_D - \phi_A = 1.8 \times 10^{-6} - (-1.8 \times 10^{-6})$$
$$= 3.6 \times 10^{-6} \text{ weber} ,$$

$$\bar{E} = \frac{(1,000)(3.6 \times 10^{-6})}{1/10} = 0.036 \text{ volt} .$$

23.6. Induced Current and Charge

Using Ohm's law, we may calculate the current and charge that pass around a circuit of N turns in which an electromotive force has been induced. Considering magnitudes of the quantities only, we **337** obtain for the average current, $\bar{I}$, in a circuit of resistance R measured in ohms $(\bar{E} = \bar{I}R)$:

$$I = \frac{N(\phi_2 - \phi_1)}{Rt} \text{ amp} \qquad (23\text{-}14)$$

or instantaneously

$$I = \frac{N}{R}\frac{d\phi}{dt} \text{ amp}. \qquad (23\text{-}15)$$

The induced charge Q is equal to $\int I dt$ or to $\bar{I}t$; hence

$$Q = \frac{N(\phi_2 - \phi_1)}{R} \text{ coulombs}, \qquad (23\text{-}16)$$

where Q is in coulombs; ϕ_2 and ϕ_1 are in webers; R is in ohms; and N is a numeric. *We note that the total charge that passes across any section of the conductor is independent of the time during which the flux changed from ϕ_1 to ϕ_2 and dependent only upon the total amount of the change in flux.*

Equations (23-14)–(23-16) assume that a conductor is present in the region where the flux is changing. If no conductor is present, we have excellent reason to believe that an electric field $\mathcal{E}$ is present in the region where the flux is changing. The electric lines of force are closed curves encircling the changing flux. This is crudely illustrated in Fig. 23-6, where the flux density B is

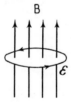

B

$\mathcal{E}$

Fig. 23-6.—Electric lines of force encircle a changing magnetic flux.

considered to be decreasing with time and $\mathcal{E}$ is the induced electric field. When a conductor is so placed as to coincide with $\mathcal{E}$, forces $\mathcal{E}e$ act on the electrons, producing the induced currents (e is the charge on the electron).

Maxwell made the assumption that associated with a changing electric field there is present a magnetic field whose magnitude is proportional to $d\mathcal{E}/dt$. Ampère's law for the magnetic field due to a conduction current (eq. [21-12]) requires modification to take into account the contributions due to a changing electric field. In our earlier discussions we assumed that the electric field was constant, $d\mathcal{E}/dt = 0$. However, if there are no conduction currents, $i = 0$, but the electric field

is changing (for example, in the region between the plates of a capacitor that is being charged or discharged), then Maxwell's assumption states that a magnetic field must be present. The lines of magnetic induction are closed curves encircling the changing electric flux. By symmetry, the vectors B and $\mathcal{E}$ of Fig. 23-6 would be interchanged, with a change in sign. The entire development of the theory of light and radio propagation is based on the validity of this assumption.

EXAMPLE 2: If the ends of the coil of Example 1 are connected and the total resistance of the coil is 20 ohms, what average current was produced while the book was being turned? What charge passed through the coil?

Solution: By equation (23-14),

$$\bar{I} = \frac{0.036}{20} = 0.0018 \text{ amp},$$

$$Q = \bar{I}t = (0.0018)(\tfrac{1}{10}) = 0.00018 \text{ coulomb}.$$

EXAMPLE 3: A horizontal wire, 1 m long, with its length east and west, is permitted to fall freely. In so doing, it sweeps over an area whose normal is parallel to the horizontal component of the earth's magnetic field, which is 2×10^{-5} weber/m². (a) What is the average electromotive force induced during the first 3 sec? (b) What is the induced electromotive force 3 sec after the wire started to fall? (c) What is the average electromotive force during the third second?

Solution:

$$V_0 = 0 \; ; \; V_2 = V_0 + 9.80 \times 2 = 19.6 \text{ m/sec at end of second sec},$$

$$V_3 = V_0 + 9.80 \times 3 = 29.4 \text{ m/sec at end of third sec}.$$

$$\bar{V}_{\text{over 3 sec}} = \frac{0 + 29.4}{2} = 14.7 \text{ m/sec},$$

$$\bar{V}_{\text{during third sec}} = \frac{19.6 + 29.4}{2} = 24.5 \text{ m/sec},$$

$$\bar{E}_{\text{over 3 sec}} = 2 \times 10^{-5} \times 14.7$$
$$= 0.000294 \text{ volt}, \qquad (a)$$

$$E_3 = 2 \times 10^{-5} \times 29.4 = 0.000588 \text{ volt}, \qquad (b)$$

$$\bar{E}_{\text{over third sec}} = 2 \times 10^{-5} \times 24.5$$
$$= 0.000490 \text{ volt}. \qquad (c)$$

EXAMPLE 4: In 1831 Faraday described an experiment to show that an electromotive force can be induced in a conductor that "cuts" lines of flux but through which the total flux remains constant. Faraday arranged a copper disk to rotate in a magnetic field of a large electromagnet, with the rotational axis of the disk coincident with the direction of the field. An electromotive force was induced between the periphery and the axis of the rotating disk, even though the flux linkages through the disk were constant. Our problem is to calculate the induced electromotive force if a 20-cm-radius disk rotates with angular speed of 100 rad/sec in a uniform field of flux density 1 weber/m².

Solution: We shall compute the flux cut per second by a radius of the disk. In time δt a radius of the disk sweeps out a sector of $\delta\theta$ radians, whose area is $r^2\delta\theta/2$. The flux through this area is $Br^2\delta\theta/2$, and hence the rate at which the flux is cut is $Br^2\delta\theta/2\delta t$. But $\delta\theta/\delta t$ is, by definition, the angular speed ω, so that

$$E = \frac{Br^2\omega}{2} \text{ volts},$$

with $B = 1$ weber/m², $r = \frac{1}{5}$ m, $\omega = 100$ rad/sec, $E = 1 \times \frac{1}{25} \times \frac{1}{2} \times 100 = 2$ volts. If the rotation is clockwise as viewed along the direction of B, then the direction of E is from the center to the rim.

23.7. Magnetic Damping: Foucault Currents

Whenever a conductor is moved across a magnetic field, there is induced in that conductor an electromotive force which tends to cause currents whose magnetic fields retard the motion. If the magnetic field is *not uniform over the whole conductor*, there will be closed paths within that conductor along which there will be an electromotive force not zero, and hence currents will be established along these paths. If the magnetic field is perfectly uniform over the whole conductor, there will be no *closed* path along which the total electromotive force is not zero. Suppose a pendulum with a bob of any conducting material (preferably not iron) swings between two poles of an electromagnet (Fig. 23-7). Upon energizing the magnet, the

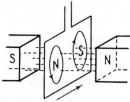

Fig. 23-7.—Foucault or eddy currents induced in a plate moving across a magnetic field.

pendulum will be damped as if it had been immersed in some viscous fluid. In Fig. 23-7 the plate, moving into the paper as it swings like a

pendulum bob, cuts the magnetic flux so as to cause currents to flow, as indicated. These currents produce magnetic fields and poles on the front of the moving plate, as marked, and also produce opposite poles on the back of the plate. The repulsion between the north pole on the front of the plate approaching the fixed north pole, the attraction between the south pole, on the front of the plate, moving away from the fixed north pole (and two similar forces between the poles on the back of the plate and the fixed pole) each acts to retard the motion. An interesting check on this explanation is to use a bob in which slits have been cut, as indicated in Fig. 23-8. In case A the slits

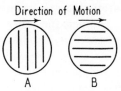

Direction of Motion

A   B

Fig. 23-8.—In A the presence of the vertical slits does not greatly reduce the eddy currents; in B the horizontal slits do reduce the eddy currents, since they are across the direction of the current.

do not greatly reduce the currents, and the damping is not much reduced by cutting the slits. In case B, however, where the slits have been cut across the direction of the current, the amount of damping is very much reduced. These currents induced in the body of a conductor (as distinguished from currents induced in wires) are sometimes called "Foucault" (or "eddy") currents.

The cores of the armatures of motors and generators, the cores of the field magnets of alternating-current motors, and the cores of transformers are ordinarily made from thin sheets of iron placed together rather than from a single block of iron, in order to reduce the Foucault currents, which are induced in the iron. By having the planes of the laminations normal to the direction of the induced electromotive forces, the resistance of the oxide on the surface of the laminations greatly reduces the resulting currents and helps to keep down undesired heat losses.

In the Duncan-type watt-hour meter a disk of aluminum is made to rotate between the poles of a permanent magnet and an iron armature. The Foucault currents induced in this disk retard its motion. The amount of electromagnetic damping may be varied and the meter thus adjusted to run at the correct speed.

23.8. Mechanical Forces between Conductors where the Current in One Induces Current in the Other

We have seen that when a conductor is moved across a magnetic field, current is induced in it and also that the magnetic field of the induced current is such as to oppose the inducing field and thus exert a force opposite to the direction of motion of the moving conductor.

If the fields are strong, these forces are by no means small. Suppose, for example, we have two closed circuits in two independent conductors in close physical proximity, and we cause a large current to circulate in one of these circuits. As the magnetic field around this rising current builds up, the flux passes through the other circuit and so builds up a current in it. Around this secondary circuit, then, there arises a second magnetic field whose direction is opposed to that of the primary field. These fields repel, and the seat of the forces lies in the conductors in which the current flow (electron motion) is confined; hence the conductors repel each other. If either or both are free to move, they do so. If they are not free to move, additional tensions are produced in their supports.

23.9. Mutual Induction

In general, whenever two coils are in the same neighborhood, a part of the flux produced by a current in one coil will cut through the other coil (Fig. 23-9). Now, if the current in the first coil

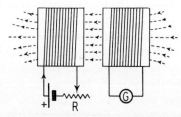

Fig. 23-9.—Mutual induction between two coils

changes by varying R, the flux through the second will change, and an electromotive force will be induced in the second coil. The magnitude of this induced electromotive force will depend on the rate of change of current in the first coil, on the number of turns of wire in each coil, on the geometry of the coils, and perhaps on other factors. Except for certain arrangements of the coil, the exact calculation of this induced electromotive force is too difficult for us to undertake. However, we can say, in general, that

$$E_s \propto \frac{dI_p}{dt}, \qquad (23\text{-}17)$$

where E_s is the induced electromotive force in the second or *secondary* coil and dI_p/dt is the rate of change of current in the first or *primary* coil, P. We may write this as an equation by introducing a constant of proportionality,

$$E_s = -M\frac{dI_p}{dt}, \qquad (23\text{-}18)$$

where M is the COEFFICIENT OF MUTUAL INDUCTION of the two circuits.

Two coils have a coefficient of mutual induction of 1 henry if current *changing* at the rate of 1 amp/sec in the primary coil induces an electromotive force of 1 volt in the secondary coil.

The value of E_s is also given by equation (23-4), and equating the two equations for E_s gives

$$E_s = -M\frac{dI_p}{dt} = -N_s\frac{d\phi}{dt}, \qquad (23\text{-}19)$$

which may be written

$$MdI_p = N_s d\phi_s \qquad (23\text{-}20)$$

or

$$M = N_s\frac{d\phi_s}{dI_p}\text{ henrys},$$

with I_p expressed in amperes. The quantity $N_s d\phi_s$ is called the total flux "linkages" between the primary and secondary coils produced by the current dI_p in the primary. This total number is independent of the rate of cutting; hence mutual inductance is often defined as "the number of linkages occurring when unit current flows in *either* coil" (since the role of primary and secondary may be reversed).

Mutual inductance between two coils is constant only when their positions are fixed and when no iron is present to alter (by its own inductive effect) the flux between the two coils.

In radio work, mutual inductance is a subject of much importance; and one speaks of "loosely coupled" and "closely coupled" coils, depending upon whether little or much of the flux of one coil links through the other.

In the special case where all the flux of the primary passes through all the turns of the secondary, we would say that there is no "magnetic leakage"; equation (23-20) may be put in a more simple form, which avoids the use of differentials. In this case $\phi_p = \phi_s$, and equation (23-20) reduces to

$$N_s \phi_p = M I_p . \qquad (23\text{-}21)$$

23.10. Self-induction

A coil of many turns of wire could be considered as many neighboring coils of a single turn each and connected in series. The magnetic flux produced by one turn would pass through that turn and, in part at least, pass through other turns. Hence a change of current in the coil would result in a change in flux through the coil, with a resulting induced electromotive force. Since this electromotive force opposes the "change" which causes it, i.e., would retard the building-up of a current in the coil by an applied potential difference, it is commonly spoken of as a "back electromotive force." In a manner similar to the development of equation (23-18) we may obtain

$$E_B = -L \frac{di}{dt}, \qquad (23\text{-}22)$$

where L is the *coefficient of self-induction*.

A coil has a coefficient of self-induction of 1 henry if current in the coil *changing* at the rate of 1 amp/sec induces a back electromotive force of 1 volt in the coil.

The value of E_B is also given by equation (23-4); hence, on equating, we obtain

$$N d\phi = L di$$

or

$$N\phi = L I , \qquad (23\text{-}23)$$

where I represents the total *change* of current in amperes, ϕ the total *change* in flux in webers, and L the self-inductance of the coil in henrys. The quantity $N\phi$, as we have seen, is called the *flux linkage*.

A coil has a self-inductance of 1 henry if a current of 1 ampere establishes one flux linkage (1 weber-turn) through it.

In the case of a long solenoid of length l and total turns N,

$$B = \mu_0 \frac{N I}{l} \text{ webers/m}^2 .$$

If the area of the coil is A m², the flux ϕ through the coil is

$$\phi = AB = \frac{\mu_0 A N I}{l} \text{ webers} .$$

Substituting in equation (23-23), we have, for the self-inductance,

$$L = \mu_0 \frac{A N^2}{l}, \qquad (23\text{-}24)$$

where L is in henrys; μ_0 is in webers per ampere-meter; A is in square meter; l is in meters; and N is a numeric.

In many cases it is desirable to construct coils of wire with very little self-inductance. For example, the resistance coils of Wheatstone bridges and potentiometers must have negligible inductance. This is accomplished in practice by taking the insulated wire of the given resistance, doubling it on itself, and then winding the double wire on a spool or cylinder. In this way the magnetic flux due to the current in one wire neutralizes the flux due to the current in the second wire. Such resistances are known as "noninductive."

In equation (23-24) the self-inductance is measured in henrys, A in meters squared, and l in meters. It follows that the units of μ_0 may be determined in terms of these quantities, thus; μ_0 can be measured in units of henrys per meter. This is the preferred unit for μ_0 rather than square webers per square meter-newton or webers per ampere-meter.

23.11. Growth of Current in an Inductive Circuit

In our discussion of Ohm's law (chap. 18) it was stated that the current must be constant. Let us see what happens when the key of the circuit shown in Fig. 23-10 is closed. The circuit consists

Fig. 23-10.—Circuit to determine growth of current in a coil.

of a coil of self-inductance L henrys, resistance R ohms, and a battery of electromotive force E. When the key is closed, the current begins to build up, and hence a back electromotive force is induced in accordance with Faraday's law,

$$E_B = -L \frac{dI}{dt}.$$

341

The direction is such as to oppose the flow of electricity through L. The resultant instantaneous electromotive force at any moment is, then, $E - L(dI/dt)$. Applying Ohm's law, electromotive force = RI,

$$
\left.
\begin{aligned}
E - L\frac{dI}{dt} &= RI \\
E &= L\frac{dI}{dt} + RI \; ;
\end{aligned}
\right\} \qquad (23\text{-}25)
$$

or

that is, the battery must overcome the RI drop in the circuit and also the back electromotive force set up while the current is increasing from zero to its maximum value. In order to find the value of the current at any instant of time, we rearrange terms in equation (23-25) and divide both sides by RdI, so that

$$
\frac{E - RI}{RdI} = \frac{L}{R\,dt},
$$

and change signs and invert, so that

$$
\frac{-R\,dI}{E - RI} = -\frac{R}{L}\,dt.
$$

Integration gives

$$
\log_e (E - RI) = -\frac{R}{L}\,t + \text{constant}.
$$

If $I = 0$ when $t = 0$, the constant equals $\log_e E$. Taking antilogs,

$$
\frac{E - RI}{E} = e^{-(R/L)t},
$$

$$
1 - \frac{RI}{E} = e^{-(R/L)t},
$$

and

$$
I = \frac{E}{R}\left[1 - e^{-(R/L)t}\right]. \quad (23\text{-}26)
$$

In these equations e is the base of natural logarithms, $e = 2.7183. \ldots$ If we plot I against t for

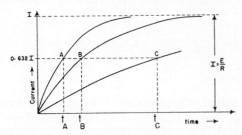

Fig. 23-11.—The growth of current in an inductive circuit; curve C represents a greater self-inductance than curves B and A.

various values of L, we obtain the curves of Fig. 23-11. Curve A represents a very small inductance,

while curve C represents a very high self-inductance. The horizontal dotted line is the maximum current given by $I = E/R$, i.e., the current after sufficient time has elapsed for $e^{-(R/L)t}$ to have become inappreciable ($e^{-\infty} = 0$).

The *time constant* of such a circuit is, by definition, L/R (inductance divided by resistance). This is the time required for the current to rise to $(1 - 1/e)$th $= 0.632$ of its maximum E/R value. These values are marked t_A, t_B, and t_C in Fig. 23-11.

23.12. Energy in the Magnetic Field

From equation (23-25) we see that, while a current is being established in the inductive circuit, the battery supplies E joules of energy for every coulomb of change that traverses the circuit. Part of the energy goes into heat and part into establishing a current against the back electromotive force. To calculate the amount of work done against the back electromotive force, $L(dI/dt)$, we note that in time dt a charge $dq = Idt$ passes around the circuit. The work done is, then,

$$
dw = dq\,\frac{L\,dI}{dt} = LI\,dI.
$$

In order to obtain the total work done, we must integrate the expression $dw = LIdI$ from zero current to I amp, or

$$
W = \int dw = \int_0^I LI\,dI = \tfrac{1}{2}LI^2.
$$

If L is in henrys and I in amperes, W is in joules. This energy is not lost but resides in the magnetic field that surrounds the coil.

If we have two coils, A and B, of self-inductance L_1 and L_2 and carrying currents I_1 and I_2, mutually coupled with mutual inductance M, then a precisely similar argument shows that the energy in the field due to the flux from A cutting coil B is $\tfrac{1}{2}MI_1I_2$, and due to the flux from B cutting coil A is $\tfrac{1}{2}MI_2I_1$.

Hence the magnetic energy due to the coupling is MI_1I_2, and, in general, the *total magnetic energy* W for two such coupled coils is

$$
W = \tfrac{1}{2}L_1I_1^2 + \tfrac{1}{2}L_2I_2^2 + MI_1I_2 \text{ joules}. \quad (23\text{-}27)
$$

23.13. The Induction Coil

The induction coil is an apparatus for producing a high electromotive force in a secondary S by allowing a current to increase and decrease periodically in the primary circuit, *P*. The apparatus consists essentially of a laminated iron core, *R*, around which are wound a few turns of heavy insulated copper wire, connected to a battery, as shown in Fig. 23-12. The secondary con-

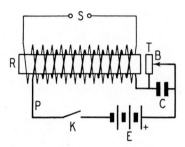

FIG. 23-12.—The electrical circuit of the induction coil

sists of a very large number of turns of insulated wire closely coupled to the primary. When the key is closed, a current passes to the contact, *B*, along the heavy wire encircling the primary back to the battery. The core is then magnetized and pulls the armature, *T*, breaking the contact at *B*. With the contact broken, the flux decreases to zero, and the armature springs back, again again making contact. In this way the flux through the secondary is periodically changed. By Faraday's law of electromagnetic induction, the electromotive force in-

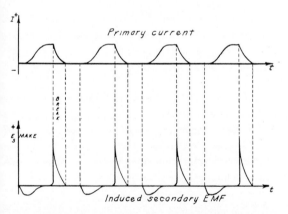

FIG. 23-13.—The primary current and induced secondary electromotive force of the induction coil.

duced in the secondary depends upon the number of conductors cut by the flux and also upon the rate at which the flux changes. A capacitor, *C*, placed across the contact, *B*, avoids sparking at this point and also makes it possible to reduce the flux in the core from its maximum value to zero in a short interval of time. In this way a very large electromotive force, of the order of a few thousand volts, is induced in the secondary. The changes in current that take place in the primary and the resulting electromotive force in the secondary are shown in Fig. 23-13.

23.14. Electromotive Force Induced in a Rotating Coil

By far the most interesting case of induced electromotive force is that of a coil rotating with constant angular velocity, ω, in a uniform magnetic field, B, the axis of rotation being in the plane of the coil.

In Fig. 23-14 (*a*) a rectangular coil of area *A* cm² is rotated about the axis *OO'* between the

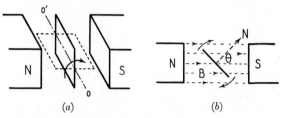

(*a*)   (*b*)

FIG. 23-14.—The elementary alternating-current generator

poles of a magnet. The magnetic flux density, *B*, is directed from north to south. Let θ be the angle between *B* and a normal, *N*, to the plane. This angle is shown in Fig. 23-14 (*b*), which represents a view of Fig. 23-14 (*a*) along the axis *OO'*.

When $\theta = 0°$, the flux through the coil is

$$\phi = B A \text{ webers} ,$$

with $\theta = 90°$, $\phi = 0°$.

For any general angle θ,

$$\phi = B A \cos\theta .$$

If the coil is rotating with uniform angular velocity of ω rad/sec, then the angle θ is given by

$$\theta = \omega t .$$

The flux through the coil is changing in time in accordance with the following equation:

$$\phi = B A \cos \omega t . \qquad (23\text{-}28)$$

If we plot the flux through the coil as ordinates and the time, *t*, as abscissas, we obtain the heavy

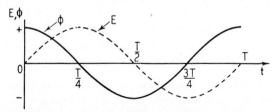

FIG. 23-15.—Variation of electromotive force and flux in the rotating loop.

curve of Fig. 23-15. We start to measure time when the coil is in the position shown in Fig. **343**

23-14 (*a*). The time for one revolution is T sec. We note the following characteristics of this curve. At $t = 0$, $t = T/2$, and $t = T$ sec, the flux is a maximum, and the curve is horizontal, indicating that the time rate of change of flux (or the slope of the curve) at these points is zero; and hence by Faraday's law the induced electromotive force is zero. At $t = \frac{1}{4}T$ and $t = \frac{3}{4}T$ sec, however, the flux is zero, but the time rate at which the flux is changing through the coil is a maximum (slope a maximum). The induced electromotive force is a maximum. In this way we plot the dotted curve of Fig. 23-15, which represents the variation of the induced electromotive force in the coil. The student may show by Lenz's law that the electromotive force is reversed when the coil turns from $\theta = 180°$ to $\theta = 360°$, i.e., from $t = \frac{1}{2}T$ to $t = T$ sec.

We may obtain a mathematical expression for E by introducing ϕ of equation (23-28) into the fundamental equation of electromagnetic induction, equation (23-4). For a coil of N turns,

$$E = -NBA\,\frac{d}{dt}(\cos \omega t)\ \text{volts}$$

or

$$E = NBA\,\omega \sin \omega t\ \text{volts} \qquad (23\text{-}29)$$

or

$$E = E_M \sin \omega t\ , \qquad (23\text{-}30)$$

where

$$E_M = BA\omega N\ , \qquad (23\text{-}31)$$

where E is in volts; B is in webers per square meters; A is in square meters; ω is in radians per second; N is a numeric; and E_M of equation (23-31) is the maximum electromotive force induced in the coil. Such changing electromotive forces, called *alternating electromotive forces*, give rise to changing currents, *alternating currents*. The problems that arise when alternating currents are set up in circuits that contain resistance, inductance, and capacitance will be treated in chapter 28.

EXAMPLE 5: A flat coil of 10 turns and area 100 cm² is rotated with an angular speed of $\omega = 12$ rad/sec in a magnetic field of flux density 0.5 weber/m². Calculate the average electromotive force induced in the coil for a quarter of a period starting from $t = 0$, the maximum electromotive force in the coil, and the ratio of the average and maximum values (see Fig. 23-15).

Solution: At $t = 0$, the flux is $\phi_1 = BA$. At $t = T/4 = \pi/2\omega = \pi/24$ sec, the flux $\phi_2 = 0$. By equation (23-5),

$$\phi_1 = AB = (5 \times 10^{-1}\ \text{weber/m}^2)(10^{-2}\ \text{m}^2)$$
$$= 5 \times 10^{-3}\ \text{weber}\ ,$$

$$E_{av} = \frac{N(\phi_1 - \phi_2)}{t} = \frac{10 \times 5 \times 10^{-3}}{\pi/24}$$
$$= \frac{12}{\pi} \times 10^{-1}\ \text{volt}\ ;$$

$$E_{max} = NBA\omega = (10)(5 \times 10^{-3})(12)$$
$$= 6 \times 10^{-1}\ \text{volt}\ ;$$

and

$$\frac{E_{av}}{E_{max}} = \frac{2}{\pi}\ .$$

23.15. The Betatron and the Synchrotron

In 1941, D. W. Kerst constructed a device capable of imparting very high energies to the electron (20 Mev). This device, called a "betatron" or "electron induction accelerator," utilizes an electric field which is induced by a changing magnetic flux (§ 23.6, Fig. 23-6). In some respects the betatron is similar to a transformer (§ 28.12), in that (*a*) both use a varying magnetic flux, which in some cases is a sinusoidal alternating flux and in other cases might be a pulsed flux; (*b*) the secondary winding of a transformer consists of many turns of a conductor, which encircle the magnetic field, while in a betatron this function is performed by many electrons traveling in a circular orbit; (*c*) the conductor of the secondary winding of a transformer has an extremely large number of free electrons, which move very slowly, while the betatron has relatively few electrons, which move with a velocity near that of light.

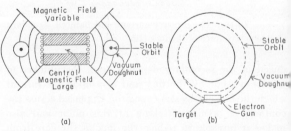

FIG. 23-16.—Principle of operation of the betatron, showing the magnetic flux lines (*a*) and stable orbit (*b*).

A diagrammatic sketch, showing the principle of operation, is given in Fig. 23-16. Between the poles of a magnet is placed a vacuum chamber in the form of a doughnut. Within the chamber are the electron gun and the targets. During operation, the magnetic flux density at the orbit is increased in proportion to the magnetic flux within the orbit. It was shown by Kerst and Serber that radial magnetic focusing is attained if the magnetic field decreases with the radius in a certain fashion at the orbit. Also if the field falls off less rapidly than $1/r$, the orbits are stable.

The conditions for proper operation may be stated analytically as follows: If the magnetic flux through the circuit of the stable orbit is ϕ, then the induced electromotive force around this circuit is $d\phi/dt$, and the field strength $\mathcal{E}$ is equal to the quotient of the electromotive force divided by the orbit path length. Thus, if the orbit is a circle of radius r,

$$\mathcal{E} = \frac{1}{2\pi r}\frac{d\phi}{dt}. \qquad (23\text{-}32)$$

According to Newton's Second Law, the time rate of change of momentum is equal to the applied force. For the case of an electron of charge e, we obtain

$$\mathcal{E}e = \frac{d}{dt}(mv) = \frac{e}{2\pi r}\frac{d\phi}{dt}$$

or

$$\Delta(mv) = \frac{e}{2\pi r}\Delta\phi, \qquad (23\text{-}33)$$

indicating that the change in momentum of the electron is proportional to the change of flux through the orbit. It is to be noted that equation (23-33) takes into account the relativistic increase of mass of the electron. Also from equation (21-38) we have

$$mv = erB_0, \qquad (21\text{-}38)$$

where B_0 is the flux density at the orbit, so that

$$\Delta(mv) = er\Delta B_0,$$

and therefore

$$\Delta B_0 = \frac{\Delta\phi}{2\pi r^2}. \qquad (23\text{-}34)$$

It follows that the flux density at the orbit must be weaker than in the interior. This change in flux density is indicated in Fig. 23-16 (a) by the spacing of the lines of force.

A model of a betatron has been constructed that accelerates electrons to energies of the order of 200 Mev. There is good reason to believe that energies of the order of 500 Mev can be attained. These high-speed electrons, on striking the target, will produce X-rays of extremely short wave length.

It has been possible to accelerate electrons to energies in excess of the limits of the betatron by combining the principle of the betatron with that of the cyclotron (see § 21.16 for a discussion of the cyclotron). This device, invented by McMillan and Veksler and called the *synchrotron*, combines in one apparatus the essential principles of construction of the betatron and the cyclotron. The electrons are held in an orbit of constant radius by the guiding magnetic field and obtain their initial acceleration through the betatron principle. Further acceleration to the maximum energies is obtained by a cyclotron-type alternating electric field. It is essential that the changing magnetic field and the frequency of the accelerating electric field occur synchronously with the acceleration of each injected electron group. The synchrotron may also be used to accelerate heavy particles such as the proton, provided that the particles are injected into the synchrotron with large initial energies (about 4 Mev). The largest instrument now under design is expected to accelerate protons to energies in excess of 100 Bev, i.e., a hundred billion electron volts!

The various devices that we have described for accelerating charged particles are used to provide high-speed particles for nuclear disintegration experiments (see § 38.25).

23.16. Résumé

The following is a brief listing of the definitions, principles, and theories, given in this chapter, which you should know:

Faraday's law for induced electromotive forces

Lenz's law for the direction of the induced current

The value of the induced charge which is moved in a flux change

Induced currents and magnetic damping

The meaning of the mutual inductance of two or more coils and the self-inductance of a single coil

The growth of a current in an inductive circuit

The energy of a magnetic field

The variation of the electromotive force in a coil rotating in a uniform field

The basic theory of the betatron and the synchrotron   **345**

Queries

1. What factors determine the magnitude of induced electromotive forces?
2. What determines the direction of this induced electromotive force?
3. Why are resistance coils used as standards wound noninductively?
4. What factors determine the coefficient of self-inductance of a coil?
5. What factors determine the coefficient of mutual inductance?
6. What would happen if Lenz's law were not true?
7. Two coils are near each other. The current starts to vary in one of them, the primary. A sensitive galvanometer in the secondary coil shows no appreci-

able current. How could you explain this, assuming the secondary circuit to be continuous and its resistance not too large?
8. The roles of the two coils of Fig. 23-9 are interchanged; i.e., the right-hand coil is made the primary, the left-hand coil the secondary. Will the coefficient of mutual inductance be the same?
9. Give the validity in terms of equations in the text for the following statements:

$$\text{webers} \equiv \text{volt-sec}$$
$$\text{webers} \equiv \text{coulomb-ohms}$$
$$\text{webers} \equiv \text{henry-amp}$$
$$\text{joules} \equiv \text{henry-amp}^2$$

10. Show that the quantity Rt/L has no units.

Practice Problems

1. A flat, circular coil of 1,000 turns and area of 100 cm² is placed in a uniform magnetic field of flux density 0.02 weber/m² and is rotated 100 times/sec about an axis in the plane of the coil and normal to the magnetic field. What is the induced electromotive force at the instant the plane of the coil makes an angle whose cosine is $\frac{3}{5}$ (i.e., $\cos^{-1}\frac{3}{5}$) with the direction of the field? What is the maximum electromotive force induced in this coil? What is the average induced electromotive force over a quarter of a rotation, starting when the induced electromotive force is zero? Ans.: 24π volts; 40π volts; 80 volts.

2. Two coils are placed as indicated in the accompanying figure. The resistance of coil No. 1 is 5 ohms; of

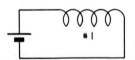 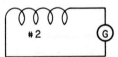

coil No. 2, 10 ohms; of the galvanometer, 90 ohms. When the current in No. 1 changes from 2 to 7 amp, the galvanometer shows that 0.002 coulomb has passed through it. Compute the coefficient of mutual induction between the two coils. If the same change in current could be produced more quickly in No. 1, would more or less electricity pass through the galvanometer?

3. A coil of wire having a resistance of 10 ohms and a coefficient of self-induction of $\frac{1}{100}$ henry is connected to a constant source of potential difference of 100 volts. Compute the current in the coil 0.0001 sec, 0.001 sec, and 0.01 sec after the circuit is closed. Ans.: 0.95 amp; 6.32 amp; 10.00 amp.

4. A wire 150 cm long has a vertical velocity of 30 m/sec. Calculate the electromotive force induced in the wire if the total intensity of the earth's field is 0.7 gauss and the angle of dip is 69°. The wire is horizontal and oriented east-west. (1 gauss = 10^{-4} weber/m².)

5. Calculate the self-inductance of a coil of 100 turns if a current of 2 amp gives rise to a magnetic flux of 5×10^{-5} weber through the coil. Ans.: 2.5×10^{-3} henry.

6. A coil having a self-inductance of 10 henrys has a current of 0.1 amp passing through it. What is the energy in joules in the magnetic field about the coil?

7. A long solenoid is wound, 20 turns to the centimeter, on an iron core whose cross-sectional area is 5 cm² and whose permeability is $\mu = 1,000$. A current of 2 amp is established in the solenoid. If the iron core is withdrawn, calculate the charge in coulombs that flows in a secondary coil of 50 turns wound tightly about the solenoid. The total resistance of the secondary is 8 ohms. Ans.: 0.0157 coulomb.

8. A coil of 200 turns and a diameter of 24 cm is placed with its plane normal to a field of 1 weber/m². What is the electromotive force induced if the field is uniformly decreased to zero in 5 sec?

9. A coil of 150 turns and 20 cm in diameter is horizontal. If the coil is turned through 180° about a north-south axis in $\frac{1}{10}$ sec, compute the number of coulombs of charge that circulate in the coil if its resistance is 110 ohms. The total flux density of the earth's field is 7×10^{-5} weber/m², and the angle of dip is 69°. Ans. 5.61×10^{-6} coulomb.

10. The coil of problem 9 is placed with its plane per-

pendicular to the horizontal component of the earth's field and turned about a horizontal axis through π rad in $\frac{1}{10}$ sec. Compute the average electromotive force induced and the charge in microcoulombs that circulate in the coil.

11. Calculate the electromotive force induced in a square coil 10^{-1} m on a side that rotates at $2,000\pi$ rad/min in a uniform field of flux density 10^{-1} weber/m^2. The axis of rotation is through the midpoints of the sides and is perpendicular to the flux density. Ans.: $0.1 \sin (105t)$ volt.

12. A coil of 10 cm^2 in area and 50 turns is removed from the center of a solenoid whose field is 10^{-2} weber/m^2. If the resistance of the coil is 8 ohms, calculate the amount of charge that flows in the coil.

13. What is the self-inductance of a solenoid in air 2 m long, 10 cm in diameter, and wound with 800 turns of wire? Ans.: $32\pi^2 \times 10^{-7}$ henry.

14. A coil has a self-inductance of 4 millihenrys and a resistance of 2 ohms. What time is required for the current to rise to 90 per cent of the value given by Ohm's law when a constant electromotive force is applied?

15. A conductor moves normally across a magnetic field, cutting lines of flux at a rate of 40 webers/sec. What electromotive force is induced in the conductor? Ans.: 40 volts.

16. A loop is at rest in the xy plane. The z-axis is normal to the plane and is positive upward. Using Lenz's law, determine the direction of the induced electromotive force for the following four cases: (*a*) the flux density points along the positive z-axis and the flux is decreasing; (*b*) the same source as *a*, but flux is increasing; (*c*) the flux density points along the negative z-axis, and the flux is decreasing; (*d*) the same as *c*, but the flux is increasing. Draw diagrams illustrating the results.

17. A disk 5 cm in radius rotates with an angular speed of 200 rad/sec in a uniform field and generates an electromotive force of 1 volt between the rim and the axis. Calculate the flux density normal to the disk. Ans.: 4.0 webers/m^2.

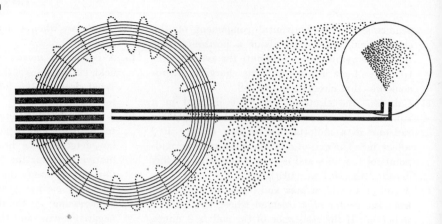

The Magnetic Properties of Iron

All motors and dynamos, transformers, induction coils—in fact, all types of electrical machinery for the production, distribution, and use of electric current, both alternating and direct—contain "magnetic circuits," essential in their operation, in which the magnetic fields of current-carrying coils are greatly enhanced by the addition of iron cores. Therefore, we digress for a space to discuss these magnetic circuits and some of the magnetic properties of materials in general.

24.1. Magnetic Substances

In order to classify a substance as magnetic or nonmagnetic, we must determine whether or not forces or torques act on the substance when brought into a magnetic field. If such an experiment were performed on all available substances, we would find, first, that all *matter is magnetic*, as we have defined the term, and, second, that there are two fundamentally *different types of magnetic behavior*. We find, for example, that certain substances are repelled by magnets and, when made into long rods and placed between the poles of a strong magnet with a nonuniform field, align themselves at right angles to the field. Such substances are called *diamagnetic* and exhibit the phenomenon of *diamagnetism*. Bismuth, phosphorus, antimony, flint glass, and mercury are examples of diamagnetic substances.

The second type of magnetic behavior is known as *paramagnetism*. Paramagnetic materials are attracted by magnets, and rods of such materials align themselves parallel to the magnetic field. Aluminum, erbium, and ferric chloride are examples of paramagnetic substances. In the next section we shall discuss the origin of para- and diamagnetism.

A third class of elements—iron, nickel, cobalt, and certain alloys, Permalloy, Supermalloy, and Heusler alloys—exhibit paramagnetism to a phenomenal degree. They are strongly attracted toward magnets. Because of these very conspicuous magnetic properties, these substances are classed together as *ferromagnetic substances* and treated separately.

It is indeed fortunate that ferromagnetic materials exist and in plentiful amounts, for the successful operation of practically all electrical machinery depends directly upon the unusual magnetic properties of iron and allied materials.

The words "magnetic substance," as used in practical life, refer to ferromagnetic materials. Since the magnetic effects of dia- and paramagnetic materials are negligibly small in general, we call them "nonmagnetic." Thus copper, glass, and wood are nonmagnetic, implying negligible magnetic effects. Nevertheless, the study of paramagnetism and diamagnetism has great scientific importance in aiding our understanding of the electrical structure of matter.

24.2. Magnetism and the Electron Theory

According to the view now accepted, the magnetic properties of matter are to be explained in terms of the motions of electrons within atoms. As long ago as 1823, Ampère had suggested that magnetism was due to electric currents circulating within matter. However, the identification of these "Amperian currents" with the motion of electrons is a rather recent achievement, due principally to J. J. Thomson, Lord Rutherford, and Niels Bohr.

An electron may give rise to a magnetic field in two ways: (1) an electron revolving in an orbit about the nucleus of an atom is equivalent to a tiny current loop, which gives rise to a magnetic field; (2) the electron gives rise to a magnetic effect which may be interpreted as due to the spin of the electron about an axis through its center, somewhat analogous to the spin of the earth on its axis. In an atom containing many electrons, it is possible that the orbits of the electrons and their spins may be so oriented that the atom as a whole possesses a resultant magnetic field. The atom then possesses a *permanent magnetic moment*. That some atoms, such as Fe, Co, and Ni, should have large magnetic moments, while others do not, has been explained by quantum mechanics. If such atoms are placed in a magnetic field, two things occur. First, the external field tends to orient the magnetic moments parallel to the field, i.e., turn the planes of the orbits at right angles to the field, thus adding the field of the electronic currents to the external field. This phenomenon we have called *paramagnetism*. This tendency to align parallel to the external field is opposed by the thermal agitation of the atoms. We therefore expect paramagnetism to be temperature-dependent.

The second effect of an external magnetic field on an atom is associated with induced electromotive forces in the atomic circuits. An electron moving in an atomic orbit is equivalent to an elementary current circuit. If an external field increases from zero to its full value of B webers/m², an electromotive force will be induced in this circuit in such a way as to oppose the field that caused it. This is in accordance with Lenz's law. Depending on the direction of the electron in its orbit relative to the direction of B, the electron will either speed up or slow down and in either case will contribute to a magnetic flux opposed to B. This effect is called *diamagnetism*. Diamagnetic effects are usually quite small and are generally masked by the larger paramagnetic effects if they are present. Many of the orbits that give rise to diamagnetism are deep within an atom and are

therefore not greatly affected by the thermal motions of the atoms.

We shall make use of this concept of atomic electron currents in discussing the bulk magnetic properties of matter. The theory of ferromagnetics is treated in § 24.9.

24.3. Intensity of Magnetization and Equivalent Surface Currents of a Magnet

In the previous section we stated that the magnetic properties of matter can be explained in terms of the equivalent current loops of the atoms. We shall now show how, for a bar magnet, the magnetic field, and flux density, outside and inside the magnet, are basically due to the combined effects of the oriented electronic currents. In Fig. 24-1 is shown a cut of a section of width δl from

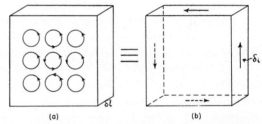

Fig. 24-1.—The magnetic field due to the Amperian currents (a) is equivalent to that of a current around the periphery (b).

the bar magnet in Fig. 24-2. The equivalent current loops are shown with all the currents circulat-

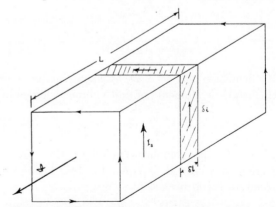

Fig. 24-2.—The intensity of magnetization $\mathscr{I}$ of a bar magnet is equivalent to that of a surface current sheet I_s of the magnet.

ing in the same direction (Fig. 24-1 [a]). In the interior of the magnet each loop has a neighboring loop, in which the current circulates in the op- **349** posite direction for part of the loop, thereby canceling each other's magnetic fields. Thus only those current elements on the surface of the magnet will contribute to the resultant field. The net effect of the atomic current loops is that due to an equivalent current around the periphery of the section δl, as shown in Fig. 24-1 (b). The flux density inside and outside the magnet can therefore be considered as due to an equivalent surface current I_s circulating around the magnet of length L m. We are thus replacing the magnet by a surface current sheet which may be thought of as an infinitely closely wound solenoid in its magnetic effects.

The basic physical quantity that describes the state of magnetization of a material is the magnetic vector, *magnetic moment per unit volume* or *intensity of magnetization* $\mathcal{J}$. We may compute its value in terms of the equivalent surface current I_s by referring to the diagram of Fig. 24-2. For this computation we assume that $\mathcal{J}$ has the same value throughout the specimen. In Fig. 24-2, a surface current δi equal to $(I_s/L)\delta l$ circulates around the hatched portion of cross-sectional area A m². We stated in § 21.13 that the magnetic moment M of a current loop of area A and carrying a current I is given by $M = A\mu_0 I$ in weber-meters. The magnetic moment is normal to the area A. For the elementary current loop of Fig. 24-2, the magnetic moment is

$$\delta M = A\mu_0 \delta i , \qquad (24\text{-}1)$$

or, since

$$\delta i = \frac{I_s}{L} \delta l ,$$

$$\delta M = A\mu_0 \frac{I_s}{L} \delta l . \qquad (24\text{-}2)$$

By definition, the *intensity of magnetization* $\mathcal{J}$, or the magnetic moment per unit volume, is given by

$$\mathcal{J} = \frac{\delta M}{\delta V} \text{ webers/m}^2 , \qquad (24\text{-}3)$$

where δV, the volume occupied by the current section of width δl, is equal to $A\delta l$. Therefore, equation (24-3) becomes

$$\mathcal{J} = \frac{\delta M}{\delta V} = A\mu_0 \frac{I_s}{L} \frac{\delta l}{A\delta l} = \mu_0 \frac{I_s}{L} \text{ webers/m}^2 . \ (24\text{-}4)$$

This result should be compared with equation (21-22) for the flux density within a long solenoid placed in a vacuum,

$$B = \mu_0 n I \text{ webers/m}^2 . \qquad (21\text{-}22)$$

Here n is the number of turns per meter, so that nI, representing ampere-turns per meter, is equivalent to the total surface current around the solenoid per meter. In other words, we expect the field of a solenoid and that of a bar magnet to be similar. As a matter of fact, from equations (21-22) and (24-4) the flux density, B_i, within the bar magnet and away from the ends is given by

$$B_i = \mathcal{J} \text{ webers/m}^2 . \qquad (24\text{-}5)$$

The flux density within the magnet is equal in magnitude and direction to the intensity of magnetization within the magnet.

24.4. Magnetization of Iron

The magnetic properties of iron or any other ferromagnetic material may be conveniently studied by means of the apparatus sketched in Fig. 24-3. The apparatus, known as the Rowland

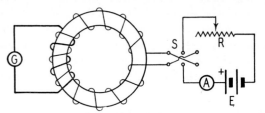

FIG. 24-3.—The Rowland ring

ring, consists of a toroidal solenoid (see § 21.10) connected to a battery, E, through a variable resistance. The secondary circuit consists of a few turns of insulated wire connected to a BALLISTIC GALVANOMETER, an instrument that measures the quantity of electricity (§ 29.4).

If the solenoid is in a vacuum (or has an air core), then the magnetic flux density, B_0, within the toroid is due to the current-carrying conductor and is given by equation (21-25),

$$B_0 = \mu_0 N_p \frac{I}{L} \text{ webers/m}^2 ,$$

where N_p is the total number of turns, I the current in amperes, and L the mean circumference of the solenoid. The flux through the coil is

$$\phi_0 = B_0 A \text{ webers} ,$$

where A is the cross-sectional area of the solenoid measured in square meters.

By changing the primary current, an electromotive force will be induced in the secondary winding. If R is the total resistance of the secondary and N_s the number of turns in the secondary, then the charge Q_0 in coulombs that passes through the galvanometer is, by equation (23-16),

$$Q_0 = \frac{N_s B_0 A}{R} \text{ coulombs}. \qquad (24-6)$$

Now, if the solenoid has a core of some magnetic material, the flux through the core is given by

$$\phi = BA, \qquad (24-7)$$

where B is the total flux density within the magnetic material, due to conduction currents in the coil and induced effects of the material. The charge Q induced in the secondary for this case is

$$Q = \frac{N_s B A}{R} \text{ coulombs}. \qquad (24-8)$$

If we divide equation (24-8) by equation (24-6), we obtain

$$\frac{Q}{Q_0} = \frac{B}{B_0} = \mu_r, \qquad (24-9)$$

where the ratio μ_r is the *relative permeability* of the magnetic substance and is dimensionless.

If the material within the core is diamagnetic, then $B < B_0$; that is, the flux density is less than that in an air core for the same current in the toroid. For diamagnetic substances the relative permeability μ_r is less than unity.

For a paramagnetic material such as aluminum, $B > B_0$, and the relative permeability μ_r is slightly greater than unity. On the other hand, if the core is iron or any other ferromagnetic substance, $B \gg B_0$ and μ_r is much larger than unity, ranging up to many hundreds of thousands. We shall discuss this special case in the next section.

It is clear from this experiment with the Rowland ring that the total flux density B is made up of two parts, that contributed by the conduction current I of the toroid, $\mu_0 NI/L$, and that due to the equivalent surface current sheet, I_s, of the electronic currents of the atoms of the magnetic material, $\mu_0 I_s/L$, as shown in § 24.3. That is,

$$B = \mu_0 \frac{NI}{L} + \mu_0 \frac{I_s}{L}. \qquad (24-10)$$

It is customary to introduce the magnetic intensity vector H, which arises only from the conduction or magnetizing currents around the endless toroid:

$$H = \frac{NI}{L} \text{ amp-turns/m}.$$

Also $\mu_0 I_s/L$ represents the intensity of magnetization vector $\mathcal{I}$ in the magnetic material, as shown by equation (24-4),

$$\mathcal{I} = \mu_0 \frac{I_s}{L},$$

so that equation (24-10) may now be rewritten in terms of H and $\mathcal{I}$,

$$B = \mu_0 H + \mathcal{I}. \qquad (24-11)$$

In this equation, $\mu_0 H$ and $\mathcal{I}$ are vectors whose vector sum is the resultant flux density within the magnetic material. Although this relation between $\mu_0 H$ and $\mathcal{I}$ was deduced for the special case of a toroid, actually it represents a general relation between these magnetic vectors and is true for all magnetic materials. In free space, $\mathcal{I} = 0$ and $B_0 = \mu_0 H$, where the subscript denotes free space.

Experimentally it is found that, for many magnetic materials, $\mathcal{I}$, the intensity of magnetization, is proportional to the magnetizing force H, that is,

$$\mathcal{I} = \chi_m H, \qquad (24-12)$$

where χ_m is called the *magnetic susceptibility* of the material. If we substitute in equation (24-11), we obtain

$$B = (\mu_0 + \chi_m) H \text{ webers/m}^2,$$

and the quantity

$$\mu = \mu_0 + \chi_m \qquad (24-13)$$

is called the *absolute permeability* of the magnetic material. It follows that

$$B = \mu H \text{ webers/m}^2. \qquad (24-14)$$

The quantities μ_0, μ, and χ_m are measured in the same units, henrys per meter. For diamagnetic ma-

TABLE 24-1

MAGNETIC SUSCEPTIBILITIES

	Temp. (° C)	χ_m (Henrys/M)
Antimony........	18	-6×10^{-11}
Bismuth..........	18	-21×10^{-11}
Copper...........	18	-1.18×10^{-11}
Argon (NTP).....	0	-1.2×10^{-14}
Aluminum........	18	2.7×10^{-11}
Ferric chloride....	20	380×10^{-11}
Oxygen liquid.....	-220	504×10^{-11}
Oxygen (NTP)....	0	22.6×10^{-13}

terials, χ_m is negative; for paramagnetic material, χ_m is positive. Typical numerical values of χ_m are given in Table 24-1.

The relative permeability μ_r, which we introduced into equation (24-9), is given by

$$\mu_r = \frac{\mu}{\mu_0} \text{ (dimensionless)} \qquad (24\text{-}15)$$

and

$$B = \mu_r \mu_0 H. \qquad (24\text{-}16)$$

The relative permeability is the quantity widely used in engineering practice (see Table 24-2).

TABLE 24-2

RELATIVE PERMEABILITIES μ_r OF MAGNETIC
SUBSTANCES GIVEN IN TERMS OF $(\mu_r - 1)$

Diamagnetic Substances		Paramagnetic Substances	
Antimony......	-4.8×10^{-5}	Aluminum......	$+0.21 \times 10^{-4}$
Bismuth........	-17×10^{-5}	Liquid oxygen...	$+40 \times 10^{-4}$
Copper.........	-0.94×10^{-5}	Platinum.......	$+3 \times 10^{-4}$
Mercury........	-3.2×10^{-5}	Air (NTP)......	$+3.6 \times 10^{-7}$
Water..........	-0.8×10^{-5}	Oxygen (NTP)..	$+17.9 \times 10^{-7}$
Argon (NTP)....	-0.95×10^{-8}	Dy$_2$O$_3$..........	$+0.0225$

FERROMAGNETIC MATERIALS

Material	μ_{ri}	μ_{max}	$^r B$ Saturation (Webers/M²)	Resistivity (Ohm-Meters)	Curie Temp. (° C)
Cobalt..........	70	240	1.8	35×10^{-8}	1,120
Nickel..........	110	1,000	0.6	8×10^{-8}	360
Transformer iron.	800	6,000	2.0	60×10^{-8}	690
Silicon iron......	250	3,750	2.1	24×10^{-8}	760
Permalloy.......	9,000	100,000	1.07	16×10^{-8}	580
Supermalloy.....	100,000	1,000,000	0.8	60×10^{-8}	400
Perminvar......	360	1,800	1.6	20×10^{-8}	710
Cu-Zn ferrite....	1,500	....	....	$\sim 10^4$	90

EXAMPLE 1: The core of a Rowland ring is made of aluminum of cross-sectional area 10^{-4} m² and mean circumference 0.5 m. Compute the intensity of magnetization and total flux density within the core for a magnetizing field of 2,000 amp-turns. What is the equivalent surface current?

Solution:

$$\vartheta = \chi_m H = 2.7 \times 10^{-11} \times 2 \times 10^3$$
$$= 5.4 \times 10^{-8} \text{ weber/m}^2,$$

$$B = \mu_0 H + \vartheta = 4\pi \times 10^{-7} \times 2 \times 10^3 + 5.4 \times 10^{-8}$$
$$= 2.5 \times 10^{-3} \text{ weber/m}^2,$$

$$(NI_s) = \frac{\vartheta L}{\mu_0} = \frac{5.4 \times 10^{-8}}{4\pi \times 10^{-7} \times 2} = 0.022 \text{ amp-turns},$$

which is only 10^{-5} that of the value of (NI) of the coil.

24.5. Magnetization of Ferromagnetic Materials

In the case of nonferromagnetic materials, the flux density B within the Rowland core and the magnetizing force H are linearly related, that is, μ is a constant, independent of B and H. However, for ferromagnetic materials the flux density B and the permeability μ depend on the magnetizing force H and also on the previous magnetic history of the specimen. The actual dependence of B on H for any substance made into the form of a Rowland ring can be determined experimentally by means of the Rowland ring of Fig. 24-3. By increasing the current in the primary of the circuit, the value of H in ampere-turns per meter can be increased, and thus the corresponding value of B can be measured by means of the induced charge that flows through the galvanometer (eq. [24-8]). If the values of B in webers per square meter are plotted as ordinates and H in ampere-turns per meter as abscissas, the magnetization curves shown in Fig. 24-4 are obtained. These curves,

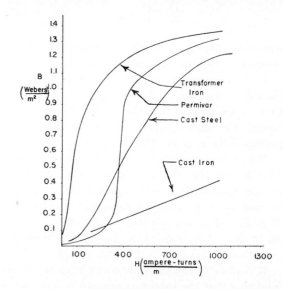

FIG. 24-4.—*B-H* curves for different magnetic materials

known as "*B-H*" curves, are of importance to the designer of electrical equipment, for they give valuable information about the magnetic behavior of materials. Note the different shapes of curves for different magnetic materials. It is also evident from some of these curves that, after a sufficiently high value of H is reached, there is a tendency for the core to saturate, that is, there is little or no further increase in B.

We observe from these *B-H* curves that the permeability of a sample is not a constant but depends on the value of the magnetizing field H.

The permeability μ can be found for each point on the curve by taking the ratio of B to H, that is,

$$\mu = \frac{B}{H}.$$

For example, in the case of cast steel (Fig. 24-4), for a magnetizing field of $H = 500$ amp-turns/m, the flux density B is 0.75 weber/m², so that

$$\mu = \frac{0.75}{500} = 1.5 \times 10^{-3} \text{ henrv/m}.$$

The relative permeability,

$$\mu_r = \frac{\mu}{\mu_0} = \frac{1.5 \times 10^{-3}}{4\pi \times 10^{-7}},$$

is approximately 1,200.

The variation of relative permeabilities with the magnetizing force H for transformer iron and cast steel is shown in Fig. 24-5. The maximum value of

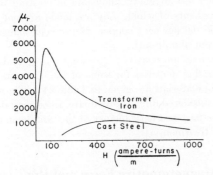

FIG. 24-5.—Curves showing the variation of relative permeabilities of transformer iron and cast steel.

μ_r for transformer iron is about 5,750 for a magnetizing force of 70 amp-turns/m and a flux density of 0.5 weber/m². When the flux density within the iron reaches 1.4 webers/m², the relative permeability drops to about 1,200. The initial relative permeability μ_{ri} is the value of the relative permeability at the origin where B approaches zero. From Fig. 24-5 the initial relative permeability of transformer iron is about 800.

EXAMPLE 2: The flux density within an iron sample of relative permeability 1,000 in the form of a Rowland ring is 1.0 weber/m². If the cross-sectional area is 10^{-3} m² and the mean circumference is 1 m, calculate the total flux, the number of ampere-turns on the core, the intensity of magnetization within the core, and the equivalent surface current of the core:

Solution:

$$\mu = \mu_r \mu_0 = 4\pi \times 10^{-7} \times 1,000$$

$$= 4\pi \times 10^{-4} \text{ henrv/m},$$

$$\phi = BA = (1)(10^{-3}) = 10^{-3} \text{ weber}, \qquad (a)$$

$$H = \frac{B}{\mu} = \frac{1}{4\pi \times 10^{-4}} = 796 \text{ amp-turns/m}, \qquad (b)$$

$$B_0 = \mu_0 H = 4\pi \times 10^{-7} \times \frac{10^4}{4\pi} = 10^{-3} \text{ weber/m}^2,$$

$$\mathfrak{I} = B - \mu_0 H = 1 - 0.001 = 0.999 \text{ weber/m}^2. \quad (c)$$

Equivalent current sheet,

$$(NI_s) = \frac{\mathfrak{I}L}{\mu_0} = \frac{0.999 \times 10^7 \times 1}{4\pi} \qquad (d)$$

$$= 795,000 \text{ amp-turns},$$

which is about 1,000 times that of the (NI) of the coil.

24.6. Hysteresis Loop

If we now start decreasing the current in the Rowland ring, shown in Fig. 24-3, by finite steps and continue determining the change in flux by observing the quantity of electricity induced in the secondary, we find that the magnetic condition of the iron, as determined by its B-H values, does not follow the curve OA but follows, instead, the curve from A to C (Fig. 24-6). When the current is

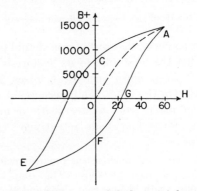

FIG. 24-6.—B-H curve and the hysteresis loop

again zero, i.e., $H = 0$, there is "residual magnetism" or retentivity, OC, left in the iron. To remove this residual magnetism, we must reverse the current. When the field H has reached the negative value indicated at D, the magnetization is again zero. The field OD, needed to demagnetize the sample, is known as the *coercive force*. If we continue to increase this reversed field, the magnetic 353 condition will follow the curve *DE*. Then decreasing the current to zero and applying it in the original direction will give the curve *EFGA*. This entire curve, *ACEFGA*, is known as the *hysteresis loop*. The area of a hysteresis loop is a measure of the amount of energy which is transformed into heat when the magnetic condition of a piece of iron is carried through this cycle. An approximate value for this energy is given by an empirical equation due to the American engineer, Charles F. Steinmetz:

$$W = aB_m^{1.6} \text{ joules/m-cycle}, \quad (24\text{-}17)$$

where B_m is the maximum flux density and a is a constant dependent upon the kind of iron or steel. Its value ranges from 130 for transformer iron to several thousand for hard tool steels.

The hysteresis curves for two different materials are given in Fig. 24-7. For the cores of alternating-

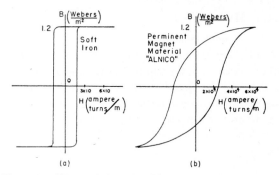

Fig. 24-7.—Hysteresis loops for different magnetic materials

current magnets, that sort of iron should be chosen which has the least area to its hysteresis loop, in order to keep down such heat losses (see Fig. 24-7 [a]). On the other hand, permanent magnets must be made from a steel with a maximum coercive force and high retentivity (see Fig. 24-7 [b]).

There has been much recent progress made in the metallurgy of magnetic alloys. The alloy known as "Permalloy" with 78.5 per cent nickel and 21.5 per cent iron has very large relative permeabilities. A magnetizing force of only 2 amp-turns/m (the earth's magnetic field is equivalent to about 56 amp-turns/m) gives rise to a flux density of 0.05 weber/m², corresponding to $\mu_r = 20,000$, while 4 amp-turns/m corresponds to a flux density of 0.5 weber/m², yielding $\mu_r = 100,000$. Permalloy is extensively used as a core material for small inductance coils. An even more amazing alloy is "Supermalloy" with 79 per cent nickel, 5 per cent molybdenum, and 16 per cent iron. A magnetizing force of

only 0.2 amp-turn/m gives rise to a flux density of 0.3 weber/m², corresponding to a relative permeability $\mu_r \sim 10^6$!

The alloy "Perminvar," 45 per cent nickel, 25 per cent cobalt, and 30 per cent iron, has a low permeability for low fields, but it has the special quality that the *B-H* curve is practically a straight line and the relative permeability a constant up to flux densities of 0.08 weber/m². The hysteresis losses are negligibly small. This material is extensively used in telephone equipment, in which magnetic response should be directly proportional to the excitation current for faithful reproduction of sound.

The alloy "Alnico I," 20 per cent nickel, 12 per cent aluminum, 5 per cent cobalt, and 63 per cent iron, has especially high retentivity, which is an essential property for strong permanent magnets. Its coercive force is about 36,000 amp-turns/m and its residual induction (retentivity) 0.72 weber/m². Alnico magnets are useful for their ability to withstand rather strong demagnetizing fields and continued mechanical vibration. Alnico magnets are used in the design of sensitive magnetometers, phonograph magnetic pickups, and loud-speakers.

Heusler ferromagnetic alloys are interesting in that the ingredients of which they are made are not ferromagnetic (60 per cent copper, 24 per cent manganese, 16 per cent aluminum).

The maximum relative permeability, μ_{max}, and corresponding flux densities for some of the common ferromagnetic materials are given in Table 24-2. The values can be considered only approximate, since they depend upon the past magnetic history of the sample and upon thermal treatment.

24.7. Magnetomotive Force and the Magnetic Circuit

Many problems arise, especially in electrical engineering, where it is desirable to know the flux in a closed magnetic circuit due to a given magnetizing field. The problem is analogous to that in current electricity, where we calculate the current, I, that passes around a closed circuit with resistance R and a battery of E volts. Ohm's law, $I = E/R$, is applicable. In the case of a toroidal solenoid we are able to calculate ϕ for a given H. Using the solenoid of Fig. 24-3, we obtain, for the flux ϕ,

$$\phi = \frac{\mu N I A}{l} \text{ webers}, \quad (24\text{-}18)$$

where I is in amperes, N the total number of turns, A the cross-sectional area in square meters, μ the absolute permeability of the iron, and l the mean circumference in meters. We may rewrite equation (24-18) as

$$\phi = \frac{NI}{l/\mu A}. \qquad (24\text{-}19)$$

The denominator, $l/\mu A$, may be considered as a kind of "resistance" offered to the establishment of the flux, with $1/\mu$ corresponding to ρ resistivity in the corresponding equation for $R = \rho l/A$ (see § 18.6). This denominator is called the *reluctance*:

$$\text{Reluctance } \mathcal{R} = \frac{l}{\mu A}. \qquad (24\text{-}20)$$

The units of reluctance are ampere-turns per weber.

The numerator NI may be considered the "cause" of the flux, just as the numerator of Ohm's law $I = E/R$ may be considered the cause of the electric current. The quantity NI is thus called the *magnetomotive force* (mmf) and is measured in ampere-turns. Equation (24-19) may be written in the following way:

$$\text{Flux} = \frac{\text{Magnetomotive force}}{\text{Reluctance}},$$

$$\phi \text{ (webers)} = \frac{\text{mmf (amp-turns)}}{\mathcal{R} \text{ (amp-turns/weber)}}. \qquad (24\text{-}21)$$

The analogy of equation (24-21) with Ohm's law is only formal. In the first place, the reluctance of a magnetic circuit is not a constant but varies, depending upon the amount of flux through the circuit. The reason for this behavior is clear, for, according to the curves of Fig. 24-5, the permeability is not constant. As a matter of fact, the value of μ that goes into equation (24-20) must be obtained from the curves of Fig. 24-5. Second, in the electrical case, charges actually move along the circuit, whereas there is no movement of magnetic flux around the magnetic circuit. A third important difference is that the flux ϕ is not the same throughout the magnetic circuit, for a certain amount of "leakage" takes place. In the case of an electrical circuit, the same charge passes each cross-section of the circuit. Magnetic leakage occurs because there are no magnetic insulators.

In engineering practice, equation (24-19) is applied to magnetic circuits that contain different materials and are only approximately closed, i.e., possess small air gaps. In this case the flux is not entirely confined to the magnetic circuit, and con-

siderable "leakage" occurs. We may rewrite equation (24-19) for the case of two materials as

$$\phi = \frac{NI}{(l_1/\mu_1 A_1) + (l_2/\mu_2 A_2)}, \qquad (24\text{-}22)$$

where the subscripts refer to iron and air. This formula, though only approximately true, is valuable in designing electrical machinery. An empirical correction factor is applied to the formula to take care of magnetic leakage. In the useful application of equation (24-22) the circuit need not be a circular ring. It is, however, necessary that the length of the magnetic field be definite. We could not use the equation, for instance, to determine the flux produced by a straight solenoid of finite length, since the length of the field outside the solenoid is indefinite.

EXAMPLE 3: The length of the iron core is 40 cm, and the air gap is 5 mm. The cross-sectional area of the iron is 2 cm². There are 700 turns of wire with a current of 6 amp. If the relative permeability of the iron is 200, what is the flux density?

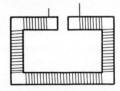

Solution: Magnetomotive force = $700 \times 6 = 4,200$ amp-turns. The absolute permeability is $\mu = \mu_r\mu_0 = 200 \times 4\pi \times 10^{-7}$ henry/m.

$$\text{Reluctance} = \frac{0.4}{200 \times 4\pi \times 10^{-7} \times 2 \times 10^{-4}}$$

$$+ \frac{5}{10^3 \times 4\pi \times 10^{-7} \times 2 \times 10^{-4}}$$

$$= \frac{35}{4\pi} \times 10^7 \text{ amp-turns/weber};$$

$$\phi = \frac{\text{mmf}}{\mathcal{R}} = \frac{4,200 \times 4\pi \times 10^{-7}}{35}$$

$$= 480\pi \times 10^{-7} \text{ weber}$$

$$B = \frac{\phi}{A} = \frac{480\pi \times 10^{-7}}{2 \times 10^{-4}}$$

$$= 2,400\pi \times 10^{-4} \text{ weber/m}^2$$

24.8. Stern-Gerlach Experiment

That a single atom may be magnetic and possess a definite and measurable magnetic moment was brilliantly demonstrated by Stern and Gerlach in **355**

1921. The experiment consists in allowing a beam of, say, silver atoms to pass between the poles of an electromagnet. If the field is homogeneous, then an atom of silver, even though it possessed a magnetic moment, would pass undeflected through the field. The reason for this is clear, for if the atom is thought of as a tiny magnet, the forces on the north and south poles would be the same, and the only effect would be slight rotation (see § 21.13). If, however, the field is nonhomogeneous, the magnet will be deflected, for now the force on the north pole is different from that on the south pole.

In order to obtain an appreciable effect, however, it is necessary to have the field change markedly within the distance of the elementary magnet. Stern produced the necessary inhomogeneity by means of specially designed pole pieces —one pole piece was flat, the other shaped like a knife edge. In passing through such a field, a silver atom is deflected, as shown by traces of these atoms on a photographic plate. In the case of silver, two distinct lines were found.

The classical theory of magnetism would predict of such an experiment merely a broadening of the beam. These remarkable experiments are taken as proof of the magnetic character of silver atoms and as a direct demonstration that the silver atom orients itself in the field in only two directions— along the direction of the field or opposite to the field.

24.9. Domain Theory of Ferromagnetism

Ferromagnetism is not a property of the individual atom alone, as are para- and diamagnetism but is, rather, a property of the crystals that constitute a ferromagnetic material. Evidence for this is provided by the fact that iron vapor as well as iron ions in a solution exhibit only paramagnetism, whereas certain mixed crystals of nonferrous materials, such as the Heusler alloys, are strongly ferromagnetic.

According to the modern theory of ferromagnetism, it is possible for the electron spins of the conduction electrons of an aggregate of atoms of a microcrystal to align themselves spontaneously with all the spin magnetic moments parallel to one another. The sizes of these aggregates or "domains" vary but are of the order of 10,000 atoms in diameter or about 10^{12} atoms in a domain. At the absolute zero, $T = 0°$ K, it is assumed that the magnetic spin moments of all the atoms are parallel and hence that the intensity of magnetization of the microcrystal is at a maximum. At room temperature, $T = 300°$ K, a certain small fraction of the atoms of a domain are randomly aligned because of thermal agitation. As the temperature is increased, more and more of the atoms get out of alignment until at a temperature known as the "Curie temperature" (see Table 24-2) the thermal agitation is so great that all the spins are randomly distributed within a domain and the microcrystal becomes paramagnetic. The Curie temperature for iron is about 1,000° K; that for nickel, 633° K; and that for cobalt, 1,393° K. At room temperature a large number of the spins are aligned, so that the domain is still strongly magnetic.

In a bulk specimen of a ferromagnetic material, the magnetic moments of the many microcrystal domains are themselves randomly oriented, so that the specimen appears unmagnetized. If such a specimen is placed in the form of a Rowland ring (Fig. 24-3) and subjected to a magnetizing field H, the domain theory predicts that, for a small magnetizing force, some of the domains, nearly parallel to H, will be rotated parallel to the field, causing an increase in the flux density B. In addition, there is a translation of domain boundaries, with domains whose magnetization is parallel to H increasing in size at the expense of those whose magnetization is not parallel to H. This accounts for portion oa of the B-H curve of Fig. 24-8. As

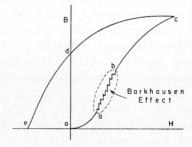

Fig. 24-8.—Domain theory explanation of the hysteresis loop.

the magnetizing field is continuously increased, the magnetization within the specimen increases discontinuously, in discrete jumps, as some of the favorably oriented domain walls move over the domain walls which are unfavorably oriented.

The discontinuous nature of the magnetization is shown in exaggerated form in Fig. 24-8, curve ab. This effect is known as the "Barkhausen effect" and can be easily detected experimentally. By a further increase in H, the domains are rotated into parallelism with H, and the sample becomes magnetically saturated. This explains the shape of the curve bc. When the field is removed, the domains reorient only slightly, leaving a residual magnetization, od. To bring the sample back to a condition approximating the initial magnetic state requires a demagnetizing force, oe.

One of the triumphs of modern quantum theory was to demonstrate quantitatively that spontaneous magnetization in a microcrystal was possible. Qualitatively two general conditions must be fulfilled: (1) The atoms of the crystal must not form closed shells, and the electron in these shells should be in orbits of high quantum number (see § 37.9). (2) The radius of these electron shells must be small compared to the lattice size.

These two conditions are fulfilled rather ideally by elements of the iron group—iron, cobalt, and nickel—and to a lesser degree by the rare earths. In the case of the rare earths, the Curie temperatures are very low (Gd, $16°$ C; Dy, $-168°$ C), so that, at room temperature, thermal agitation is sufficient to upset the tendency toward spontaneous magnetization. The rare earths are ferromagnetic near absolute zero but are paramagnetic at room temperatures. It has been possible through metallurgical research to increase in effect the lattice size of certain nonmagnetic materials and produce ferromagnetism.*

24.10. Ferrites and Eddy-Current Losses

In our discussion of energy losses in magnetic materials in § 24.6, we took into account only hysteresis losses due to magnetic cycling of the material. The value of this energy loss was given by the Steinmetz equation (24-17). Another energy loss, particularly serious at high frequencies, is the *eddy-current* loss, first discussed in § 23.7. As we have shown, changing mag-

* For an excellent qualitative account of the theory of magnetism see R. M. Bozorth, "Magnetic Materials," *Scientific American*, January, 1955, p. 68.

netic fluxes set up induced eddy voltages and hence currents that are dissipated as heat in the material. For a given size of material, these eddy losses in watts per cubic meter are proportional to the square of the maximum flux density, the square of the frequency of flux change, and the conductivity of the material. For very high frequencies approaching 100 kc/sec, eddy losses become prohibitive. A contributing factor to these losses is the high conductivity of magnetic materials whose values range about 10^8 mhos/m.

A major advance has come from the study of certain new materials known as "ferrites," which are nonmetallic (oxides) and have conductivities 10^4–10^{-5} mhos/m. The relative permeabilities of the ferrites range from 10 to over 1,000. Eddy-current effects in some ferrites are practically nonexistent up to frequencies of 1,000 kc/sec.

Ferrites have the general formula $MOFe_2O_3$, in which M stands for a bivalent metal ion. The important ferrites for magnetic effects are those containing the bivalent ions Mg, Zn, Ni, Cu, Fe, Co, and Mn. Of these, a mixture of the copper and zinc ferrites has important practical applications as transformer cores and cores for inductance coils. The relative permeability ranges from 200 to 1,000. One practical advantage of ferrite is that mixtures of powdered ferrites can first be pressed into almost any shape or form and then sintered in an oxygen atmosphere at temperatures of about $1,200°$ C. The final product is very hard and not malleable. It must be ground to final dimensions.

24.11. Résumé

The following is a brief listing of the definitions, principles, and theories, given in this chapter, which you should know:

The basic properties of the three magnetic types of matter—diamagnetic, paramagnetic, and ferromagnetic

The electron theory of magnetism

The measurement of the relative permeability of magnetic material

The relationship of the magnetic intensity vector H, the flux density B, and the intensity of magnetization $\mathcal{J}$

The significance of the magnetization or B-H curves for iron, the hysteresis loop, and coercive force

The magnetic circuit and the meaning of magnetomotive force

Queries

1. In what way is the analogy between the magnetic circuit and the electric circuit faulty?
2. What shape of hysteresis curve would be desirable for an electromagnet? Transformer? Permanent magnet?

Practice Problems

Note: In these problems the lengths given are to be considered as the effective lengths, and all leakage is to be neglected.

1. An iron rod having a cross-sectional area of 10 cm² is bent as indicated in the accompanying figure. One thousand turns of wire are wound on as primary, and 10 turns as secondary. The relative permeability of the iron is 300. A steady current of 20 amp is passed through the primary.

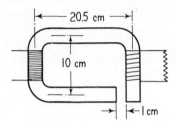

a) Compute the magnetomotive force of the primary. Ans.: 20,000 amp-turns.

b) Compute the reluctance of the magnetic circuit.

c) Compute the flux density in the air gap.

d) Compute the coefficient of self-induction of the primary.

e) Compute the coefficient of mutual induction between the two coils. Ans.: 0.00105 henry.

f) The total resistance of the closed secondary circuit is 5 ohms. If the secondary coil is slipped off its iron core, what quantity of electricity will be induced in the secondary? Ans.: 0.00419 coulomb.

2. A toroidal solenoid made of transformer iron is wound with 350 turns of wire. The mean radius of the ring is 20 cm, and the cross-sectional area of a section is 3 cm². Calculate the current required to set up a flux of 1.5×10^{-3} weber. Use $\mu = 4\pi \times 10^{-4}$ henry/m.

3. In problem 2 the solenoid is broken by a gap 3 mm in width. Calculate the current required to produce the same flux. Ans.: 48.3 amperes.

4. A ring of transformer iron has a mean circumference of 2 m and a cross-sectional area of 1.6×10^{-3} m². Determine from Fig. 24-4 the flux density corresponding to a magnetizing field of 500 amp-turns/m. What is the flux in the core? What is the equivalent surface current of the core?

5. In problem 4, compute the hysteresis loss per cubic meter of material per cycle if the maximum flux density is that calculated in problem 4. Use $a = 130$ for transformer iron. Ans.: About 180 joules.

6. An Alnico II permanent bar magnet is uniformly magnetized to a flux density of 0.8 weber/m². If the magnet is 6 cm long and has a cross-sectional area of 1 cm², calculate the intensity of magnetization, the total magnetic moment of the bar magnet, and the equivalent surface current in ampere-turns.

7. Plot a *B-H* curve and a μ_r-*H* curve from the following data on Armco iron:

B (Webers /M²)	H (Amp-Turns/M)	B (Webers /M²)	H (Amp-Turns/M)
0.05	50	1.35	300
0.85	100	1.40	450
1.15	150	1.45	800
1.25	200	1.48	1,200

8. From the data of problem 7, compute the intensity of magnetization and magnetic susceptibility for each of the given points on the *B-H* curve.

9. In problem 6, what is the pole strength of the bar magnet? If the magnet is placed at right angles to a magnetic field of 0.2×10^{-4} weber/m², compute the torque of the magnet. Ans.: 8×10^{-5} weber; $2.4 \times 10^{-4}/\pi$ newton-m.

10. An electromagnet is to be designed with the following specifications: field in the air gap, 1 weber/m²; $\mu_r = 1,000$; effective length of magnet, 8 m (iron portion); length of air gap, 0.3 m; effective diameter of magnet, 1 m. Compute the number of ampere-turns necessary.

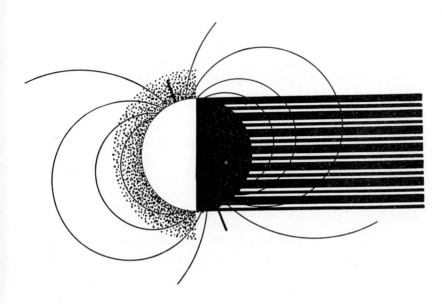

CHAPTER 25

Magnetostatics

25.1. Permanent Magnets; Coulomb's Law

In the previous chapter, we saw how a piece of
iron or other ferromagnetic material retains much
of its magnetic properties after the magnetizing
field is reduced to zero. Such a bar of iron is then a
"permanent" magnet, although its state of magnetization can be changed by subjecting it to other
strong magnetic fields or to high temperatures.
The magnetic properties of the magnetized bar
were interpreted in terms of the surface currents
(§ 24.3) and ultimately in terms of atomic currents. However, many of the gross properties of
bar magnets and the magnetic fields that surround
them can be conveniently correlated in terms of
the concepts of magnetic poles and magnetic moment of the bar magnet.

Let us suppose we have a bar magnet like that
shown in Fig. 25-1, which is uniformly magnetized.

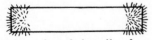

Fig. 25-1.—Poles at the end of a uniformly magnetized bar

If iron filings are sprinkled over the magnet, they
become magnetized by the external magnetic field
and become concentrated at or near the ends. The
effective centers of these regions are called *poles*.

If the bar magnet is suspended by a string or
mounted on a pivot, it is found that the bar lines
up roughly north and south, *the same end of the bar*
magnet always pointing north. For this reason we
shall call the pole that tends to seek the geographic
north, the *north pole* (N), and the other, which
tends to seek the geographic south, the *south pole*
(S). In this way we may mark the ends of all bar
magnets as north poles or south poles.

By mounting bar magnets on pivots or suspensions so that they are free to move, we find that
the like poles of two magnets repel, whereas unlike
poles attract each other. By using long, slender, magnetized needles, the north and south poles of a
magnet or magnets may be sufficiently separated
from each other that the magnitude of the forces
between individual, approximately isolated poles
may be measured. This was first done by the
French physicist Charles A. Coulomb (1736–
1806) in 1785. By suspending a long needle from
a brass wire attached to a graduated scale, as
shown in Fig. 25-2, he was able to determine what
mechanical torque had to be applied to hold the
needle in its zero position when another long magnetic needle was brought near. Coulomb found,
first, that the force of repulsion between two north
poles was inversely proportional to the square of
the distance between N and N'. By experimenting
with various magnets, he was able to demonstrate
that different magnets, when at the same distance,
r, from the suspended magnet, produced different
forces. He was able to ascribe to each magnet a

Analytical Experimental Physics

definite *pole strength*, *m*, relative to the standard magnet. He then expressed the results of his investigation quantitatively as

$$f = \frac{1}{k} \frac{m_1 m_2}{r^2},\qquad(25\text{-}1)$$

where m_1 and m_2 are the strengths of the poles of the two magnets, r the distance between the poles, and k a constant of proportionality which depends upon the units chosen and also upon the medium between the poles.

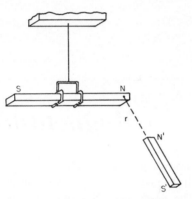

Fig. 25-2.—Principle of Coulomb's experiment for measuring the forces between magnetic poles.

In the rationalized mks system of units the constant k of equation (25-1) is replaced by $4\pi\mu_0$, where μ_0 is the permeability of a vacuum (or air) and is equal to $4\pi \times 10^{-7}$ (12.57×10^{-7}) henry per meter. The Coulomb law may be written as

$$f = \frac{1}{4\pi\mu_0} \frac{m_1 m_2}{r^2},\qquad(25\text{-}2)$$

where f is in newtons; m is in webers; r is in meters; and μ_0 is in henrys per meter. From equation (25-2) it follows that the unit magnetic pole, the weber, may be defined as one which, when placed 1 m from an equal and similar pole in a vacuum, repels it with a force of $10^7/16\pi^2 = 63,325$ newtons. This is a very large force, much too large for practical bar magnets. We expect to find only fractional pole strengths in webers in actual magnets. A good cobalt-steel bar magnet may have a pole strength of the order of 10^{-5}–10^{-4} weber. For purposes of calculation, Coulomb's law may be written very approximately as

$$f = 6.33 \times 10^4 \frac{m_1 m_2}{r^2}\ \text{newtons}.\quad(25\text{-}3)$$

This equation is really only an approximation to actual conditions, for it assumes point poles which do not exist. In general, the poles of a uniformly magnetized bar are not at the ends of the bar, but the distance between them is approximately five-sixths the length of the magnet. It follows from the earlier discussion that the strength of the N pole is equal to that of the S pole. This can easily be proved experimentally. Can you suggest such an experiment?

EXAMPLE 1: As an illustration of these ideas, let us compute the forces acting on two bar magnets 15 cm between poles and with their N poles 10 cm apart and S poles 40 cm apart. We shall take the pole strength of each magnet to be 4×10^{-4} weber.

Solution: The force of repulsion between N poles by equation (25-3) is

$$f_1 = 6.33 \times 10^4 \frac{m_1 m_2}{r^2}$$

$$= \frac{6.33 \times 10^4 \times (4 \times 10^{-4})(4 \times 10^{-4})}{10^{-2}}$$

$$= 1.013\ \text{newtons}.$$

The force of repulsion between S poles is

$$f_2 = \frac{6.33 \times 10^4 \times (4 \times 10^{-4})(4 \times 10^{-4})}{(0.4)^2}$$

$$= 0.063\ \text{newton}.$$

The force of attraction between pairs of N and S poles is

$$f_3 = \frac{6.33 \times 10^4 \times 2 \times (4 \times 10^{-4})(4 \times 10^{-4})}{(0.25)^2}$$

$$= 0.324\ \text{newton}.$$

The net force is one of repulsion, or

$$f_1 + f_2 - f_3 = 1.013 + 0.063 - 0.324$$

$$= 0.752\ \text{newton}.$$

If the magnets were connected by a cord, the tension in the cord would be 0.752 newton.

25.2. Magnetostatic Field

From our previous discussion we have seen that the region surrounding a bar magnet is characterized by the fact that forces act on other bar magnets and magnetic materials brought into the

360

region. This observation may be used to introduce the concept of a magnetostatic field as a region wherein magnetic forces act. This is analogous to the concept of an electrostatic field, as described in § 17.8.

By definition, the direction of the magnetic field is that of the force acting upon an isolated north pole. The magnitude of the field, called the *magnetic field intensity, H,* is defined as the force per unit pole acting on a test pole placed at the point in question. What we mean by "isolated pole" is one end of a long, needle-like magnet; the other pole is assumed so far away as to have negligible effect. Thus the magnetic vector H, the magnetic field intensity, is given by

$$H = \frac{f}{m} \text{ newtons/weber} \quad \text{or} \quad \text{amp-turns/m}.$$

The preferred units for H are ampere-turns per meter, as used in chapter 24, § 24.4.

From equation (25-2) and the foregoing definition, it follows that the magnetic field intensity, H, a distance r meters from an isolated pole m, is

$$H = \frac{f}{m} = \frac{m}{4\pi\mu_0 r^2} \text{ amp-turns/m}^2. \quad (25\text{-}4)$$

We showed in § 24.4 that the flux density vector B of free space and H are related by

$$B = \mu_0 H,$$

so that the force on a pole of strength m webers in a field of flux density B is

$$f = \frac{B}{\mu_0} m. \quad (25\text{-}5)$$

Also, the flux density a distance r meters in a vacuum or air from an isolated pole of strength m webers is given by

$$B = \frac{m}{4\pi r^2} \text{ webers/m}^2. \quad (25\text{-}6)$$

Let us determine the number of lines of flux ϕ radiating from a north pole of strength m webers. If we draw a spherical surface of radius r meters centered at the isolated N pole of m webers or at the pole of a long, thin magnet but not including the S pole, the total number of lines of flux ϕ in webers crossing the spherical surface will be equal to the product of the surface area and the flux density:

$$\phi = 4\pi r^2 \frac{m}{4\pi r^2} = m \text{ webers}; \quad (25\text{-}7)$$

that is, the total flux through the surface is equal to the pole strength m, and each is measured in webers, or one line of flux is associated with each unit pole.

The student will observe that in describing the magnetostatic field in free space surrounding a bar magnet we may employ the magnetic vector, H—the magnetic field intensity—or the flux density vector, B. The flux density vector B is preferred.

As a second illustration, we shall use equation (25-6) to compute the flux density B, at a point P, a distance d meters in a vacuum or air from the center of a bar

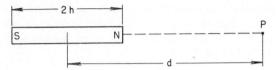

FIG. 25-3.—Magnetic field along the axis of a bar magnet

magnet of length $2h$ and of pole strength m (Fig. 25-3). From equation (25-6) the flux density at P is

$$
\begin{aligned}
B &= \frac{m}{4\pi(d-h)^2} - \frac{m}{4\pi(d+h)^2} \\
&= \frac{m}{4\pi}\left[\frac{(d^2 + 2dh + h^2) - (d^2 - 2dh + h^2)}{(d^2 - h^2)^2}\right] \\
&= \frac{4hdm}{4\pi(d^2 - h^2)^2}.
\end{aligned}
$$

If we assume that $d \gg 2h$, the field is that of a magnetic dipole, and B is given by

$$B = \frac{hm}{\pi d^3}. \quad (25\text{-}8)$$

Thus, at points far from the magnet, the flux density varies *inversely as the cube of the distance*. This is in contrast to the inverse-square law for isolated poles.

As another example, let us find the flux density B at the point P shown in Fig. 25-4. At point P there will

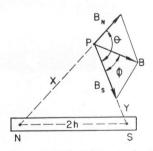

FIG. 25-4.—Magnetic field off the axis of a bar magnet

be a field due to the north pole B_N, and also a field due to the south pole B_S. These being vector quantities, the resultant field at P is obtained by the parallelogram law.

Since

$$B_N = \frac{m}{4\pi X^2}; \qquad B_S = \frac{m}{4\pi Y^2},$$

therefore,

$$B = \sqrt{B_N^2 + B_S^2 + 2B_N B_S \cos \theta} \, .$$

Also

$$\cos \phi = \frac{B^2 + B_S^2 - B_N^2}{2BB_S} \, . \qquad (25\text{-}9)$$

If X, Y, and $2h$ were known, θ could be found, and it would be possible to evaluate expression (25-9). As we obtained the field at P, so we may calculate the field at any other point about the magnet. We also see from this analysis that the magnetic field that surrounds a bar magnet is far different from that of an isolated pole. This is illustrated in Fig. 25-5.

EXAMPLE 2: To determine the magnetic flux density at point 5 cm and 3 cm from the poles of the magnet shown, with $m = 2.25 \times 10^{-4}$ weber.

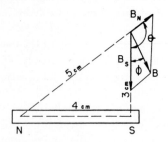

A

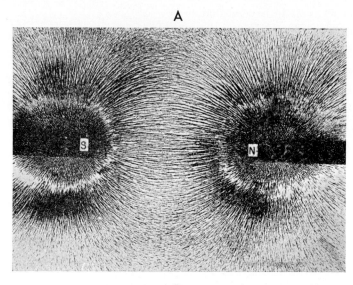

B

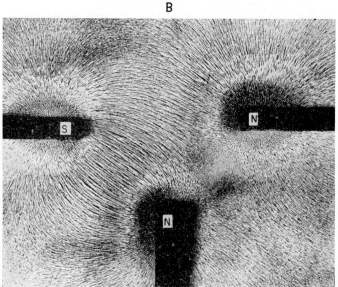

FIG. 25-5.—Magnetic fields surrounding bar magnets. In A, the magnetic field is due to two unlike poles. Note how intense the field is near the poles, as evidenced by the high concentration of iron filings. A close scrutiny of B will reveal a tiny region about halfway between the north poles where the field is zero.

Solution:

$$B_N = \frac{2.25 \times 10^{-4}}{4\pi \times 25 \times 10^{-4}} = \frac{0.09}{4\pi} \text{ weber/m}^2;$$

$$B_S = \frac{2.25 \times 10^{-4}}{4\pi \times 9 \times 10^{-4}} = \frac{0.25}{4\pi} \text{ weber/m}^2;$$

$$\cos\theta = -\tfrac{3}{5}.$$

Therefore,

$$B = \frac{1}{4\pi}$$

$$\times \sqrt{(0.09)^2 + (0.25)^2 + 2(0.09)(0.25)(-0.6)}$$

$$= \frac{0.2088}{4\pi} \text{ weber/m}^2;$$

$$\cos\phi = \frac{(0.21)^2 + (0.25)^2 - (0.09)^2}{2(0.21)(0.25)}$$

$$= 0.938, \quad \text{and} \quad \phi = 20^\circ.2.$$

25.3. Torque on a Magnet in a Uniform Magnetic Field; Magnetic Moment

If a bar magnet of length $2h$ is placed in a uniform magnetic field of flux density B, there will be a force acting on each end, as shown in Fig. 25-6.

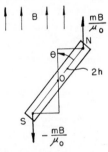

Fig. 25-6.—Torques on a magnet in a uniform magnetic field of flux density B.

This pair of equal and opposite forces $(mB/\mu_0, -mB/\mu_0)$ constitutes a couple tending to rotate the bar magnet until it is parallel to B. The value of this torque is

$$\left.\begin{aligned} L &= \frac{mB}{\mu_0} \times \text{lever arm} \\ &= \frac{mB}{\mu_0} 2h \sin\theta \\ &= (2mh)\frac{B}{\mu_0}\sin\theta. \end{aligned}\right\} \quad (25\text{-}10)$$

The product $(2mh)$ is an important characteristic of the bar magnet and is called the *magnetic*

moment of the magnet. It is a vector quantity whose direction is from the south to the north pole of the magnet. Its importance resides in the fact that neither the pole strength nor the distance between the poles is an accurately determined quantity, though the product is. If we designate moment by M, then, from equation (25-10),

$$L = \frac{MB}{\mu_0}\sin\theta \text{ meter-newtons}. \quad (25\text{-}11)$$

The units of M are usually taken as weber-meters. Typical values of magnetic moments of magnets range from 10^{-6} to 10^{-8} weber-meter.

In chapter 24, we stated that the intensity of magnetization $\mathcal{I}$ equal to the magnetic moment per unit volume is also an important magnetic quantity of a magnet. In the special case of a uniformly magnetized bar magnet of length l and cross-sectional area A, we have

$$\mathcal{I} = \frac{M}{V} = \frac{ml}{Al} = \frac{m}{A} \text{ webers/m}^2,$$

or the intensity of magnetization is a measure of the pole strength per unit area.

In terms of magnetic moment M, the field at a great distance from a small magnet, a magnetic dipole (eq. [25-8]) may be written as

$$B = \frac{M}{2\pi d^3}. \quad (25\text{-}12)$$

In § 21.6 we calculated the magnetic field at a distance d from a small loop of wire having a current I in it and found

$$B = \frac{AI\mu_0}{2\pi d^3}.$$

The field due to the magnetic dipole and that of the small loop become the same if we identify $AI\mu_0$ as the equivalent magnetic moment of the loop, i.e.,

$$M = AI\mu_0.$$

In § 21.13 we used this expression to introduce the idea of a magnetic pole.

25.4. The Earth as a Magnet

As we have already remarked, it has been known almost from the time of the discovery of magnetism that the earth itself behaves like a large magnet. At first, only those magnetic effects existing at the surface of the earth were observed. **363**

Later, magnetic effects were found a short distance within the earth's interior. Furthermore, phenomena related to the aurora borealis and certain aspects of cosmic-ray studies show that the magnetic field of the earth extends to great distances from the earth's surface—many times as far as any appreciable atmosphere can be traced. The knowledge of the character of the earth's magnetic field at or near the earth's surface has been gradually accumulated by routine observations during the course of navigation, by scientific exploration to all parts of the globe, including especially both polar regions, and by such organizations as the Department of Terrestrial Magnetism, Carnegie Institute.

The present state of our knowledge of the magnetic field of the earth indicates that the earth is equivalent to a large magnet, the poles being located as shown in Fig. 25-7, some distance away

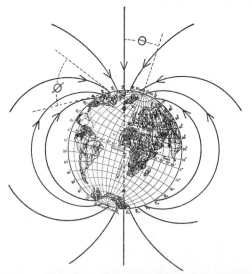

FIG. 25-7.—The earth as a magnet

from the axis of rotation. Indeed, the equivalent "poles" of the magnet that is our earth are located *deep in its interior*. The line joining the poles—the magnetic axis—makes an angle θ, which is approximately equal to 20°, with its axis of rotation. This axis *intersects the earth's surface* in Boothia Peninsula in the north and in South Victoria Land. Thus the *geomagnetic* latitude and longitude are very different from the *geographical*. Also, the earth's magnetic axis does not pass through the center of the earth. Therefore, the magnitude of the magnetic field at the surface is greater in one hemisphere than in the other. It has long been

known that the magnetic field in the vicinity of the Indian Ocean is greater than at the opposite side of the earth. Recent measurements of cosmic rays have confirmed this same fact as being true of the earth's field at great distances above the surface.

A factor which disturbs the earth's magnetic field is the presence of large bodies of iron ore in which inductive effects are produced.

At any point on the surface of the earth the magnetic field, B, will be directed at an angle, ϕ, with the horizontal plane, as shown in Figs. 25-8

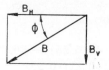

FIG. 25-8.—The components of the earth's magnetic field, showing the angle of dip.

and 25-7. This angle is called the "dip" because a needle free to rotate in a vertical plane will dip that much below the horizontal.

Only the horizontal component, B_H, will affect magnetic needles free to rotate in the horizontal plane. Such a needle will set itself parallel to the horizontal component of the earth's field. The needle will not, in general, point to the geographic north. The deviation from the true north, which at places may be very large, is called the *magnetic declination* at the station. This must be known for all parts on the earth where navigation takes place.

From Fig. 25-8 it follows that

$$B_H = B \cos \phi \, ,$$

$$B_V = B \sin \phi \, .$$

For Washington, D.C., $B_H = 0.184 \times 10^{-4}$ weber/m², $B_V = 0.541 \times 10^{-4}$ weber/m², and $\phi =$ angle of dip $= 71°.2$.

25.5. The Gauss Magnetometer

A convenient way to measure the strength of an approximately uniform field, such as the earth's field, for a tiny region of space, and also to determine the magnetic moment, M, of a bar magnet was suggested in 1833 by Karl Friedrich Gauss (1777–1855), an eminent German mathematician and physicist. Let a bar magnet whose magnetic moment is M be placed normal to the earth's field,

B_E, as shown in Fig. 25-9. A *tiny* magnetic needle whose magnetic moment is M_1 is suspended by a fiber at a point P. The suspended magnet and housing are called a *magnetometer*. In Fig. 25-10 is

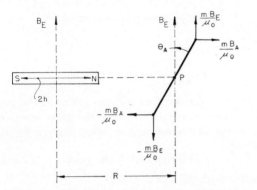

FIG. 25-9.—Gauss's method for measuring the earth's magnetic field and the magnetic moment of a magnet.

shown a typical magnetometer in which the magnet, made of a special alloy known as "Alnico," is damped in oil. The suspension for this magnetometer is an almost torsionless fiber such as silk, or a jeweled steel needle-point bearing, supporting the needle with a minimum of friction.

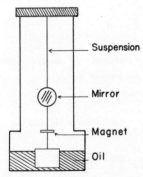

FIG. 25-10.—Principle of a magnetometer

The point P lies on the line joining the poles of the bar magnet, at a distance R from the midpoint between these poles. The experimental arrangement is known as Gauss's "A-position." The magnetometer needle is in the presence of two magnetic fields at right angles to each other, B_E, the earth's field, and B_A, the field of the bar magnet. The needle will set itself so as to be in equilibrium. The torques due to both fields become equal and oppositely directed. If θ is the angle between B_E and the axis of the needle, the torque due to the earth's field is, by equation (25-11),

$$L_E = \frac{B_E M_1 \sin \theta}{\mu_0};$$

that due to the bar magnet is

$$L_A = \frac{B_A M_1 \cos \theta}{\mu_0}.$$

Since the two torques are equal and any torque in the suspension is made zero,

$$B_E \tan \theta = B_A . \qquad (25\text{-}13)$$

By equation (25-8),

$$B_A = \frac{h\,m}{\pi R^3} = \frac{M}{2\pi R^3},$$

assuming $R \gg 2h$. Substituting in equation (25-13), we get

$$\frac{M}{B_E} = 2\pi R^3 \tan \theta . \qquad (25\text{-}14)$$

To obtain a second relation between M and B_E, we shall mount the *bar magnet* so that it is free to rotate about a vertical axis. The torque, L, due to the earth's field will give the magnet an angular acceleration a, where $L = Ia$. Here I is the moment of inertia of the magnet about the axis of suspension. This acceleration is opposite to the direction of the angular displacement, so that if at some instant the magnet makes an angle ϕ with the magnetic field and has an angular acceleration a in this position, then

$$I a = -L ,$$

$$I a = -\frac{M B_E \sin \phi}{\mu_0} \quad (\text{by eq. } [25\text{-}11]) ,$$

so that

$$a = -\frac{M B_E \sin \phi}{\mu_0 I} .$$

If ϕ is sufficiently small, $\sin \phi \to \phi$, measured in radians, and

$$a = -\frac{M B_E}{\mu_0 I} \phi .$$

We have here a case of angular simple harmonic motion. The period of vibration is

$$T = 2\pi \sqrt{-\frac{\phi}{a}} = 2\pi \sqrt{\frac{I \mu_0}{M B_E}} \qquad (25\text{-}15)$$

or

$$M B_E = \frac{4\pi^2 I \mu_0}{T^2} . \qquad (25\text{-}16)$$

By combining equations (25-14) and (25-16), we get

$$\left. \begin{array}{c} M^2 = \dfrac{8\pi^3 \mu_0 I R^3 \tan \theta}{T^2} , \\[2mm] B_E^2 = \dfrac{2\pi I \mu_0}{T^2 R^3 \tan \theta} . \end{array} \right\} \qquad (25\text{-}17)$$

The reader will observe that we obtain B_E and M in terms of easily measured mechanical quantities.

EXAMPLE 3: What will be the period of oscillation of a compass needle with a moment of inertia of 2.5×10^{-7} kg-m² and a magnetic moment of $2\pi \times 10^{-8}$ weber-m in a magnetic field of 2×10^{-5} weber/m²?

Solution:

$$T = 2\pi\sqrt{\frac{I\mu_0}{MB}}$$

$$= 2\pi\sqrt{\frac{2.5 \times 10^{-7} \times 4\pi \times 10^{-7}}{2\pi \times 10^{-8} \times 2 \times 10^{-5}}} = \pi \text{ sec}.$$

25.6. Comparison of Magnetic Units

Although we have emphasized the rationalized mks units in our analysis of electrical and magnetic effects, it is useful to examine another system of units, the *electromagnetic units* (emu), which is used to some extent to describe magnetic phenomena. We shall summarize the more important formulas for future reference.

In the emu system the constant k of equation (25-1) is placed equal to unity, so that Coulomb's law becomes

$$f = \frac{m_1 m_2}{r^2}. \qquad (25\text{-}18)$$

From this equation we define a unit magnetic pole as one which, when placed 1 cm from a like and equal pole, repels it with a force of 1 dyne. There is no special name for the emu unit pole. The numerical relation between the mks unit pole, the weber, and the emu pole can be deduced by noting that 1 weber placed 1 m from an equal pole repels it with a force of $10^7/16\pi^2$ newtons or $10^{12}/16\pi^2$ dynes. The number, m, of emu poles that can exert such a force at this distance of 1 m or 100 cm is

$$\frac{m^2}{10^4} = \frac{10^{12}}{16\pi^2} \text{ dynes}; \quad \text{or} \quad m = \frac{10^8}{4\pi} \text{ unit emu poles};$$

i.e., $\quad$ 1 weber (mks)

$$= \frac{10^8}{4\pi} \text{ unit poles (emu)},$$

or $\quad$ 1 unit pole (emu)

$$= \frac{4\pi}{10^8} \text{ webers (mks)}.$$

$$(25\text{-}19)$$

The magnetic field intensity H in the emu system is the *oersted*, defined as the dyne per unit emu pole. The numerical relation between the ampere-turn per meter

of the mks system and the oersted of the emu system is given by using equation (25-4), $H = f/m$, or

$$1 \text{ amp-turn/m} = \frac{10^5 \text{ dynes}}{10^8/4\pi \text{ unit poles}}$$

$$= \frac{4\pi}{10^3} \text{ dynes/unit pole or oersted},$$

$$1 \text{ oersted (emu)}$$

$$= \frac{10^3}{4\pi} \text{ amp-turns/m (mks)}.$$

$$(25\text{-}20)$$

The flux density B in the emu system is the gauss. The numerical relation between the gauss (emu) and the weber per square meter (mks) is given by $B = \mu_0 H$, or

$$1 \text{ weber/m}^2 = 10^4 \text{ gauss}; \qquad (25\text{-}21)$$

also for magnetic flux $\phi = BA$,

1 weber of magnetic flux (mks)
$$= 10^8 \text{ gauss} \times \text{cm}^2 \text{ or maxwells (emu)}. \qquad (25\text{-}22)$$

The units of magnetic moment M in emu are (unit pole-cm):

$$1 \text{ weber-m (mks)} = \frac{10^8}{4\pi} \text{ unit poles}$$

$$\times 10^2 \text{ cm} = \frac{10^{10}}{4\pi} \text{ unit pole-cm (emu)},$$

$$1 \text{ unit pole-cm (emu)}$$

$$= \frac{4\pi}{10^{10}} \text{ weber-m (mks)}.$$

$$(25\text{-}23)$$

The units of intensity of magnetization $\mathcal{I}$ in emu are unit poles per square meter:

$$1 \text{ weber/m}^2 \text{ (mks)}$$

$$= \frac{10^8}{4\pi} \text{ unit poles}/10^4 \text{ cm}^2$$

$$= \frac{10^4}{4\pi} \text{ unit poles/cm}^2 \text{ (emu)},$$

$$1 \text{ unit pole/cm}^2 \text{ (emu)}$$

$$= \frac{4\pi}{10^4} \text{ webers/m}^2 \text{ (mks)}.$$

$$(25\text{-}24)$$

25.7. Résumé

The following is a brief listing of the definitions, principles, and theories, given in this chapter, which you should know:

The Coulomb law for magnetic poles
The magnetic flux from a magnetic pole
The magnetic moment of a magnet
The torque on a magnet in a magnetic field
The magnetic field of the earth
Method of measurement of the earth's magnetic field by the Gauss magnetometer

Queries

1. If a soft piece of iron is brought near a permanent magnet, the iron itself becomes a magnet while in the field. This is known as *magnetic induction* or *induction of magnetization*. Explain the phenomenon.
2. Given two identical steel bars, one a permanent magnet, the other unmagnetized, without using any apparatus, is it possible to determine which bar is magnetized?
3. Explain why the following experiment suggests that the poles of a magnet are equal in strength. A bar magnet placed on a cork and floating on water is observed to remain at rest.
4. In equation (25-2), using the units for f, r, and μ_0, show that the unit of pole strength, m, is the weber.
5. In equation (25-10), using the units for m, h, B, and μ_0, show that the quantity mhB/μ_0 has the units of newton-meters.

Practice Problems

1. Two bar magnets, one 20 cm long and of pole strength $8\pi \times 10^{-5}$ weber and another 10 cm long with a pole strength of $2\pi \times 10^{-5}$ weber, are placed in a straight line with their north poles 5 cm apart. Assuming the poles to be at the end of the magnets, compute the force between the magnets. Ans.: 0.348 newton.
2. The pole strength of a magnet is 36×10^{-6} weber. Compute the magnetic flux density at x due to this magnet. *Note:* The length of the magnet given is the distance between the poles.

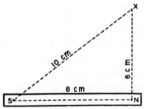

3. The pole strength of a magnet is 7.24×10^{-6} weber. Compute the magnetic field intensity at x due to this magnet. Ans.: 146 newtons/weber.

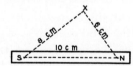

4. The pole strength of a magnet is 42.25×10^{-6} weber. Compute the magnetic flux density at x due to this magnet.

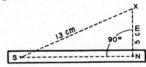

5. Each pole has a strength of 6×10^{-6} weber. Compute the total force between these two magnets. Ans.: 9.93×10^{-4} newton.

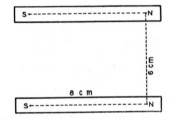

6. At Chicago the horizontal and vertical components of the earth's field are 1.92×10^{-5} and 5.86×10^{-5} weber/m², respectively. Find the total intensity and dip of the earth's field at this point.
7. A cylindrical compass needle is 10 cm long, 4 mm in diameter, and weighs 24 gm. What will be its period of oscillation if set up where the earth's magnetic field has a horizontal component of 2×10^{-5} weber/m²? *Suggestion:* Assume that the distance between the poles is five-sixths the length of the magnet. The pole strength of the needle is 2×10^{-6} weber. Ans.: 17.1 sec.
8. What torque would be necessary to hold the magnet of problem 7 normal to the field?
9. Each of two cobalt-steel magnets has a pole strength of about 11.3×10^{-5} weber. If each magnet weighs 200 gmf, at what distance above it will either magnet support the other? The magnets are 6 cm long. Neglect attractive forces and assume that the poles are separated by five-sixths the length of the magnet. See figure of problem 5. Ans.: 2.88 cm.

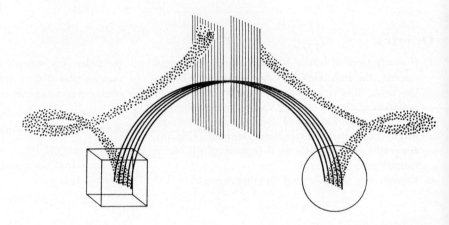

CHAPTER **26**

Thermoelectricity

26.1. Seebeck Effect

In 1822 Thomas Johann Seebeck (1770–1831), of Berlin, discovered that an electric current can be set up in a circuit of two different metals without the presence of any electrolyte, provided that the two metals are joined to form a closed circuit and the junctions are kept at different temperatures. Such a device, known as a THERMOCOUPLE, is illustrated in Fig. 26-1. The metals chosen are

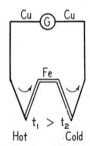

FIG. 26-1.—Copper-iron thermocouple

iron and copper. If the cold junction, t_2, is kept at 0° C and the other junction is heated, a current, as registered by the galvanometer, G, passes around the closed circuit, from copper to iron at the hot junction. The presence of the current implies the existence of a resultant electromotive force, the Seebeck electromotive force. We shall discuss the possible origin of this electromotive force in § 26.8. Thermoelectric electromotive forces are very small, the copper-iron couple giving rise to an electromotive force of the order of 16×10^{-5} volt for 1° C difference in temperature of the junctions. (However, if the resistance of the

thermocouple is made very small by having the copper and iron in the form of short rods of large cross-section, very large currents of the order of amperes may exist in the circuit.)

It is found that the insertion of a third metal in the circuit does not affect the electromotive force, provided that the ends of the third metal are at the same temperature. This makes it possible to introduce within the thermoelectric circuit such measuring devices as galvanometers, voltmeters, and potentiometers without affecting the Seebeck electromotive force.

For a rough determination of the Seebeck electromotive force, E, at various temperatures, we may make use of Ohm's law,

$$E = IR \ ,$$

where I is the current, as measured by a galvanometer and R is the total resistance of the circuit. It is obvious that the resistance of the wires will change with temperature; hence measurements of this kind would ordinarily be unreliable. However, in practice we may place such a large resistance in series with the thermocouple that the change in resistance of the wires will be negligible. A precision method, using the potentiometer, is discussed in § 26.3.

26.2. The Variation of the Seebeck Electromotive Force with Temperature

If we keep one junction of the copper-iron thermocouple at 0° C and gradually heat the other junction, we observe that the current and hence the electromotive force increase in magnitude, reaching a maximum value at about 275° C. On increasing the temperature above 275° C, the electromotive force decreases in value, reaching zero at 550° C. For temperatures in excess of 550° C, the electromotive force reverses direction, the current passing from iron to copper at the hot junction. A plot of the Seebeck electromotive force, as ordinates, against temperature differences of the two junctions is shown in Fig. 26-2. In the case of

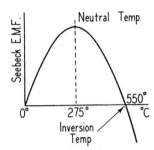

FIG. 26-2.—Variation of Seebeck electromotive force with temperature.

many pairs of metals, this curve is well represented by a parabola. The temperature at which the curve reaches a maximum value is called the *neutral point*, while the temperature at which the electromotive force changes direction is called the *inversion point*. In the case of the copper-iron couple, these two points are 275° and 550° C, respectively. Other pairs of metals have their characteristic neutral and inversion points. For the constantan-alumel thermocouple, the neutral point is beyond the melting point of these metals.

26.3. Uses of Thermocouples

Since the electromotive forces of thermocouples are so small, they are not useful as sources of electrical energy. Their practical use is confined to the measurement of temperature and temperature differences and the measurement of alternating currents of high frequency in radio devices. Because thermojunctions can be constructed of very fine wire, the heat capacity of the junction can be made very small indeed; and hence such devices would be useful in measuring temperatures of objects whose heat content is correspondingly small or where rapid small changes in temperature require a thermometric device of small heat capacity and consequently small lag.

For a given pair of metals the dependence of the Seebeck electromotive force upon temperature may be experimentally determined and a curve of E against temperature plotted; or some empirical equation between E and t (eq. [26-1]) may be used, its constants being determined by experiment. Thus, if one junction is kept at a constant temperature, usually 0° C, the temperature of the other may be determined by a measurement of the Seebeck electromotive force and a reference to a calibration curve, such as that in Fig. 26-2. The alternative method is to determine an empirical equation between E and t. The curve of Fig. 26-2 may be approximately represented by

$$E = Bt + \frac{Ct^2}{2}. \qquad (26\text{-}1)$$

The cold junction is at 0° C, and the hot junction at t° C. The two constants of this equation may be computed by observing the electromotive forces for two temperatures of the hot junction. If, then, any other electromotive force is observed, the corresponding temperature can be computed.

A precision method for measuring E for different temperatures by means of the potentiometer is shown in Fig. 26-3. The thermojunction is made

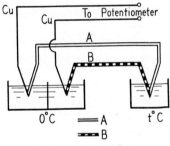

FIG. 26-3.—The use of thermocouples to measure temperatures.

of metals A and B. Since it is not practical to have the leads to the potentiometer made of the junction metals, copper leads are used. To eliminate any possible effect due to the copper, the junctions of copper and A and B are kept at 0° C, while the A-B junction is at temperature t.

Many pairs of metals may be used for thermocouples. A couple with one wire of platinum and the other of an alloy of 90 per cent platinum and 10 per cent rhodium is satisfactory for a wide range of temperatures, being permanent up to 1,500° C. Another thermocouple uses the alloys known by the trade-names of Alumel and Chro- **369** mel. While this couple is not permanent to quite so high a temperature as the platinum-rhodium couple, it is satisfactory up to 1,000° C and has the advantage of being far cheaper and giving a much greater electromotive force.

26.4. Thermoelectric Power

Instead of plotting equation (26-1) for every available pair of metals, it proves more convenient to plot a quantity called the *thermoelectric power* against the temperature of the hot junction. The thermoelectric power of a thermocouple is the change in E per degree of change in temperature of the hot junction; i.e., it is the slope of the graph of Fig. 26-2, or it is equal to dE/dt, the rate of change of the Seebeck electromotive force with temperature. By equation (26-1) it follows that

$$\frac{dE}{dt} = B + Ct. \qquad (26\text{-}2)$$

Thus there is a linear dependence of thermoelectric power on the temperature, and such a curve is a straight line. It should be pointed out that thermoelectric power is *not* "power" in the technical sense in which we have been using the word heretofore, i.e., as the time rate of doing work. The units of thermoelectric power are volts per degree.

Table 26-1 gives the values of B and C for a number of metals with respect to lead. Lead was

TABLE 26-1

THERMOELECTRIC POWER WITH RESPECT TO LEAD

Substance	B (Micro-volt/° C)	C (Micro-volt/° C
Alumel..........................	−17.48	+0.00144
Antimony.......................	+35.58	+ .1450
Bismuth........................	−43.69	− .4647
Chromel........................	+24.40	.0
Constantan (60 per cent Cu, 40 per cent Ni).....................	−38.105	− .0888
Copper.........................	+ 1.34	+ .0094
Iron...........................	+17.15	− .048
Nickel.........................	−23.3	− .008
Platinum.......................	− 0.60	− .0109
90 per cent platinum; 10 per cent rhodium......................	+ 6.413	+ .0064
Zinc...........................	+ 3.0	−0.01

chosen as the reference metal because the Thomson effect (§ 26.6) is negligible for this metal. The

signs of B are assigned so that, when B is positive,

the current passes in the metal considered from the hot junction to the cold junction. With values of B and C taken from Table 26-1, the thermoelectric power, dE/dt, of equation (26-2), is measured in microvolts per degree.

We may illustrate the use of Table 26-1 by determining the equation that represents the temperature dependence of E for the iron-copper junction:

	B	C
Fe-Pb	17.15	− 0.048
Cu-Pb	1.34	+ 0.0094
Fe-Cu	15.81	− 0.0574

so that, for the iron-copper couple,

$$E = [15.81t − \tfrac{1}{2}(0.0574)\,t^2]\ \text{microvolts} . \quad (26\text{-}3)$$

Also, the thermoelectric power is

$$\frac{dE}{dt} = (15.81 − 0.0574t)\ \text{microvolts/° C} . \quad (26\text{-}4)$$

At the neutral temperature, E is a maximum and $dE/dt = 0$, so that

$$15.81 = 0.0574t\,,$$

$$t = 275°\ \text{C (neutral temperature)} .$$

Also, the temperature at which E of equation (26-3) is zero gives the inversion temperature,

$$15.81t − \tfrac{1}{2}(0.0574)\,t^2 = 0\,,$$

$$t = 550°\ \text{C (inversion temperature)} .$$

Fig. 26-4 shows graphs of the thermoelectric powers of the substances given in Table 26-1. Here equation (26-2) has been plotted, using the tabular values for B and C. In a thermocouple composed of any two of these metals, in that metal which is the higher (greater positive value of dE/dt) on the graph the current will be from the hot junction toward the cold junction. *An area* on this graph bounded by the two lines given for the two metals and the ordinates through the temperatures of the junctions *gives the electromotive force produced by those metals at those temperatures.*

As an example of this, notice the lines for iron and copper—say between the temperatures 0° and 275° C, where the lines cross. Here is a triangle with its base (vertical on this graph) about 17 − 1 = 16 microvolts/° C and of altitude (horizontal) 275° C. Then the area of this triangle is $\tfrac{1}{2} \times 16 \times 275 = 2{,}200$ microvolts = 0.0022 volt. Since the iron line is higher than the copper line, the current will be directed from the hot junction toward the cold junction *in the iron.*

Now examine the same lines between 275° and 400° C. Again we have a triangle, this time of base $5 - (-2) = 7$ microvolts/° C and of altitude $400 - 275 = 125$. Since the electromotive force of this thermocouple between 275 and 400 is the area of this triangle, $E = \frac{1}{2} \times 7 \times 125 = 437.5$ microvolts = 0.0004375 volt. Since the copper line

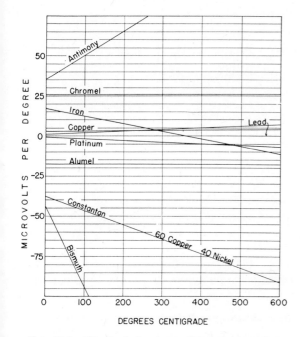

Fig. 26-4.—Graph of the thermoelectric powers of some common substances.

is the higher, the current is directed from the cold junction to the hot junction *in the iron.* Lastly, for an iron-copper thermocouple between 0° and 400° C, the electromotive force will be the algebraic sum of the two electromotive forces we have computed, or $E = 0.0022 + (-0.0004375) = 0.00176$ volt. And, since the first was larger than the second, the current will be directed as given for the first. In this way the thermal electromotive force for any pair of metals and any two temperatures given on the graph may be determined with the accuracy with which the graph may be read.

Since the numerical values of the thermal electromotive forces depend upon the exact crystalline state of the metals and since they are so greatly changed by minute traces of impurities, the values given here must be considered typical values for these substances, and it must not be expected that

these values will be correct for every sample of these substances which may be measured. An examination of the data given in such a source as the *International Critical Tables* (Vol. VI) will show a wide divergence in the values for any one substance.

The electromotive force for various temperatures of certain thermocouples is given in Table 26-2. The cold junction is at 0° C.

It is customary to calibrate a new thermocouple at a number of different temperatures and then plot an error curve, showing the difference between the observed electromotive force and that given by a table like Table 26-2.

TABLE 26-2

ELECTROMOTIVE FORCE FOR CERTAIN THERMOCOUPLES
(In Millivolts)

Temperature (° C)	Platinum-Rhodium (90% Pt; 10% Rh)	Chromel-Alumel	Copper-Constantan	Iron-Constantan
0........	0	0	0	0
100......	0.64	4.08	4.28	5.28
200......	1.43	8.17	9.29	10.77
300......	2.31	12.32	14.86	16.30
400......	3.24	16.50	....	21.83
500......	4.22	20.76	....	27.41
600......	5.22	25.03	....	33.13
700......	6.26	29.22	....	39.19
800......	7.33	33.31	....	45.49
900......	8.43	37.32	....	51.83
1,000......	9.57	41.16	....	58.17
1,100......	10.74	44.89	....	64.50
1,200......	11.93	48.49	....	70.84
1,300......	13.13	51.96	....	....
1,400......	14.33	....	....	....
1,500......	15.55	....	....	....
1,600......	16.75	....	....	....
1,700......	17.95	....	....	....

A number of convenient temperatures for calibration are as follows:

°C
0...........Freezing point of water
32.38........Transition of sodium sulphate
100.0........Boiling point of water
217.96........Boiling point of naphthalene
231.85........Freezing point of tin
444.60........Boiling point of sulphur
960.5........Freezing point of silver

The boiling-point temperatures must be corrected for variation from standard atmospheric pressure.*

* See *Smithsonian Physical Tables* for more details with regard to calibration temperatures.

26.5. Peltier Effect

In connection with thermoelectricity, the following important question has to be raised: What is the source of the energy of the Seebeck electromotive force? A partial answer to this question was given in 1834 by the French physicist, Jean C. Peltier (1785–1845), who observed that heat energy was absorbed at one junction and liberated at the other when a current passed around the circuit of two dissimilar metals. *This was a fundamental discovery.*

We may show the existence of this heating and cooling by the following simple experiment. The thermocouple formed of copper and iron is sealed in an H-tube, as shown in Fig. 26-5. A drop of

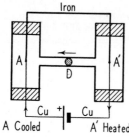

Fig. 26-5.—Illustrating the Peltier effect

oil, D, in the fine capillary tube separates junction A from A'. When a current is passed from copper to iron, junction A is found to cool while A' becomes heated, thus causing the air pressure about A' to exceed the pressure about A. The drop moves toward A. If the current is reversed, the drop moves toward A', indicating that A is heated and A' is cooled. Thus the Peltier effect, as this phenomenon is called, is *reversible;* i.e., the same amount of heat is evolved when a current passes in one direction as is absorbed when the current is reversed. It is found that the heat developed is proportional to the *first power* of the current. There is an essential difference between this Peltier heating and ordinary heat. It will be recalled, of course, that heat produced when a current is present in a *resistance* is proportional to the *square* of the current (Joule's law). Furthermore, joulean heat is irreversible; i.e., heat is always evolved, irrespective of the direction of the current.

The fact that energy is absorbed at A when a current passes from copper to iron suggests that at the junction there is an electromotive force, E_1, directed from copper to iron, and also, since

energy is released at A', it is suggested that at A' an electromotive force, E_2, is directed from copper to iron. When the current passes along E_1, i.e., in the direction of the electric field, work is done on the current, heat is then absorbed, and the junction cools. When the current passes across E_2, the current does work and develops heat, warming this junction. This similarity with batteries is illustrated in Fig. 26-6 (*b*). The value of E_1 and E_2 de-

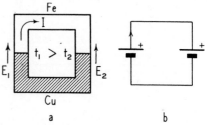

Fig. 26-6.—Similarity of Peltier electromotive forces with batteries.

pends upon the two metals that form the junction and also upon the absolute temperature of the junction. If, therefore, one junction is maintained hotter than the other, and in case of iron-copper junctions both are less than 550° C, E_1 is greater than E_2, and a current passes, as in Fig. 26-6 (*a*). These Peltier electromotive forces are small, of the order of 50 microvolts.

26.6. Thomson Effect

In 1854 William Thomson (later Lord Kelvin) showed that, if the Peltier electromotive force at the junction were the only source of energy in the thermoelectric circuit, then the resultant electromotive force, which we have called the "Seebeck electromotive force," would depend on the first power of the temperature. Figure 26-2 indicates that this is by no means the case.

Thomson predicted that this actual behavior implied another source of electromotive force in the thermoelectric circuit and said that this electromotive force must be due to the differences in temperature (temperature gradient) along the individual metals themselves. We may demonstrate the existence of this electromotive force as follows: By heating a copper rod at its center to a temperature T_2 and keeping the ends at a low temperature, T_1, a symmetrical temperature gradient along the rod may be maintained, as shown in Fig. 26-7 (*a*). The height of the ordinates at each point is a measure of the temperature of the rod at that point.

Now, if a current, I, is passed along the rod, the temperature distribution along the rod is changed somewhat, as shown in Fig. 26-7 (*b*).

We may interpret this as meaning that an electromotive force, e_T, is distributed along the rod. At one end of the rod the current is in the direction of e_T, and hence energy is absorbed, while at the other end energy is supplied.

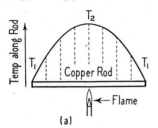

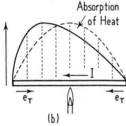

FIG. 26.7.—(*a*) Symmetrical distribution of temperature along the copper rod with no current present; (*b*) asymmetry in the temperature distribution when a current is present in the rod due to Thomson electromotive forces along the rod.

Experiment and theory show that

$$e_T = \int_1^{T_2} \sigma \, dT \,,$$

where σ is the electromotive force per unit temperature difference between two points at temperatures T_1° and T_2° K. The factor σ is called the *Thomson coefficient* and has values of a few microvolts per degree. The Thomson coefficient of lead is practically zero.

As we have seen, Thomson electromotive forces are not located at a single point but are distributed along the conductor wherever a temperature gradient is present. The magnitude of e_T depends upon the metal and the temperature gradient along the metal.

26.7. Complete Thermoelectric Circuit

From the preceding analysis it is evident that the observed Seebeck electromotive force, E, is really the sum of four electromotive forces—two at the junctions E_1 and E_2 (Peltier electromotive forces), one along the iron, e_T, and another in copper, e'_T (Thomson electromotive forces). The Seebeck electromotive force, then, is

$$E = E_1 - E_2 + e_T + e'_T \,. \qquad (26\text{-}5)$$

These various electromotive forces are shown in Fig. 26-8. The current for t_1 less than 550° C is

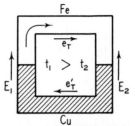

FIG. 26-8.—The complete thermoelectric circuit

also indicated. The reader should remember that the electromotive forces of equation (26-5) are functions of temperature.

26.8. Electronic Explanation of Thermoelectricity

The density of the conduction electrons in a metal A is different, in general, from the density of the electrons in another metal, B. If the metals are brought into contact, there will be a drift of electrons. In the case of copper and iron, electrons are found to drift from iron to copper at each junction.

The difference of potential thus produced will quickly stop any further drift. When the temperature of one junction is increased, the drift begins again, since the increased temperature does not affect the density and velocity of the electrons in the two metals identically. The energy which maintains this drift comes from heat supplied at the warmer junction. This accounts for the presence of the Peltier electromotive forces.

The electron explanation of the Thomson electromotive force is as follows: On heating one end of a copper rod, for example, the electrons diffuse away from the region where the thermal agitation is greatest, leaving the hotter region positive with respect to the cooler region. The Thomson electromotive force would then be directed from the cool to the hot region. In this way the simple electron theory predicts that all metals should have positive Thomson electromotive forces, i.e., electromotive forces directed as in the case of copper.

In some metals, notably iron, bismuth, and **373**

nickel, the Thomson electromotive force is directed in the opposite sense, indicating that here electrons tend to drift from the cool to the hot region. A partial explanation of this anomaly has been recently given by applying the theory of quantum mechanics to the solid state.

26.9. Résumé

The following is a brief listing of the definitions, principles, and theories, given in this chapter, which you should know:

The conditions required for producing a Seebeck or thermoelectric electromotive force

The variation of the Seebeck electromotive force with temperature

The meaning of thermoelectric power

Neutral and inversion temperatures for pairs of metals in a thermocouple

Peltier electromotive force; the evolution and absorption of heat at thermoelectric junctions

Thomson electromotive force due to temperature gradient in metal

The Seebeck electromotive force as the sum of the Peltier and Thomson electromotive forces

Practice Problems

1. For a copper-constantan thermocouple between 100° and 300° C, what will be the Seebeck electromotive force and the direction of the current in copper? Use Fig. 26-4. Ans.: 11.9 millivolts.

2. Calculate the electromotive force of a copper-constantan couple between 0° and 300° C by using equation (26-1).

3. Calculate the electromotive force of a chromel-alumel couple between 0° and 400° C by using equation (26-1). Ans.: 16.4 millivolts.

4. What is the thermoelectric power of an alumel-constantan couple if one junction is at 0° and the other at 30° C?

5. Compute the neutral and inversion temperatures of a zinc-iron thermocouple. Ans.: 372° C, 744° C.

6. One junction of a copper-alumel thermocouple is kept at 0° C. If the electromotive force of the couple, as measured by a potentiometer, is 10 millivolts, what is the temperature of the hot junction?

7. A thermopile consists of a series of junctions connected as shown in the accompanying figure. The junctions b are kept at a constant temperature, while

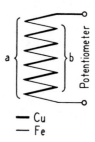

junctions a are exposed to heat radiation. If b is maintained at 0° C while a attains a temperature of 1° C, compute the number of junctions in series that will be required to generate an electromotive force of 10 millivolts. Ans.: 1,268.

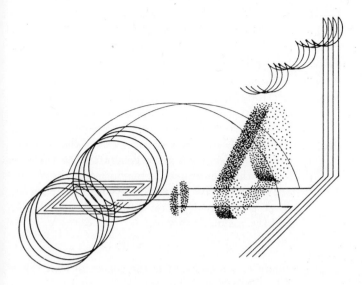

CHAPTER 27

Direct-Current Motors and Generators

We now resume the train of thought we were following in chapter 23 with reference to electromagnetic induction, carrying it on to a more detailed discussion as to the workings of electrical machinery employing "direct" current, in this chapter, and to "alternating" current, in the next one (chap. 28).

27.1. Commutation

When rotating coils are to be connected to stationary external circuits, it is necessary that some

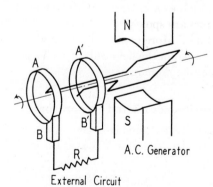

FIG. 27-1.—Simple alternating-current generator

special form of electrical connection between the two be devised if we are to make use of the induced electromotive force (§ 23.14).

These connections can be made in two essentially different ways. In Fig. 27-1 the **ends** of the rotat-

ing coil are attached to two SLIP RINGS, A and A', mounted on the axle of the coil. A sliding contact between fixed BRUSHES, B and B', and these rings connects the coil to the external circuit. In this case, one brush is always connected to the same terminal of the coil, and the current in the external circuit is identically the same as the current in the rotating coil.

In Fig. 27-2, on the other hand, the ends of the coil are connected to the two halves (SEGMENTS) of

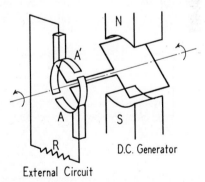

FIG. 27-2.—Direct-current generator with commutator rings.

a divided ring. The sections of this ring are insulated from each other. Such a divided ring is called a COMMUTATOR. One brush makes contact with each segment of the commutator. The brushes are so oriented with respect to the magnetic field that at the instant the current passes through zero the **375** brushes change segments. The result is that the current in the external circuit, instead of alternating, as it does in the rotating coil, remains always in the same direction. Such currents are known as "pulsating" or *unidirectional* currents (Fig. 27-3).

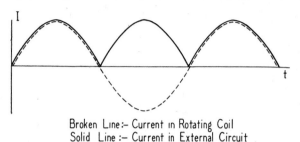

Broken Line :– Current in Rotating Coil
Solid Line :– Current in External Circuit

FIG. 27-3.—Currents produced in the rotating coil of Fig. 27-2 and the unidirectional current (*heavy line*) in the external circuit.

27.2. Direct-Current Generator

In order to reduce the amount of fluctuation in the external current, the moving part of direct-current generators is composed of a large number of coils oriented at different angles about the axis. These coils are all connected in series. The commutator has as many segments as there are coils, and each junction between adjacent coils is connected to a segment of the commutator. The whole group of coils is known as the ARMATURE. The two brushes divide the armature windings into two parallel sets of coils. The electromotive forces in the different coils of one set are represented in Fig. 27-4. Since the coils of one set are in series

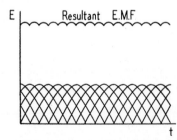

FIG. 27-4.—Effect of many coils connected to appropriate segments of the commutator.

with one another, the total electromotive force of the generator would be the sum of the individual electromotive forces. Since there can be but a finite number of coils, there will be a little RIPPLE in the total electromotive force. However, this will be far less than the change which results from the use of a single coil (Fig. 27-3). Furthermore, the

self-induction of the circuit helps to reduce the amount of this ripple, so that it can barely be detected in commercial direct current.

From equation (23-31) we learned that the electromotive force induced in a rotating coil placed in a magnetic field was proportional to the flux, $\phi = BA$, the number of conductors cutting the flux, N, and the angular speed of the rotating coil, ω. Hence, for a given generator, the variable factors that determine the magnitude of the resultant electromotive force, E, are the angular speed, ω, or its equivalent, revolutions per minute, n, and the magnetic flux, ϕ,

$$E = K\phi n \text{ volts },\qquad (27\text{-}1)$$

where K is a constant of the generator that takes into account such factors as the number of conductors in the armature and the number of poles of the generator. This simplified form of the electromotive-force equation is basic to our study of the performance of direct-current generators.

27.3. The Field Magnets

The magnetic field in which the coil rotates can be produced either by permanent magnets or by electromagnets. Only in small machines known as MAGNETOS are permanent magnets used. It is possible to produce electromagnets which are much stronger than any permanent magnet; these are used for all large generators and most small ones. The electric current necessary to excite the FIELD magnets is occasionally supplied by some outside source, such as a battery or other generator. Such generators are spoken of as "SEPARATELY EXCITED generators." More commonly, current from a generator is used in the field coils of such a generator.* This may be done in several ways.

27.4. Series-connected Generators

In the SERIES-connected generators all the current from the generator is passed through the field windings (see Fig. 27-5). Since there is ordinarily a rather large current, only a few turns of large wire are used for the field coils. Generators are ordinarily turned at a constant rate of speed, $n =$ Constant (eq. [27-1]).

* Obviously, only in the case of a direct-current generator can the current from it be used to excite the field coils. All alternating-current generators must have fields separately excited from an auxiliary direct-current generator.

Suppose such a generator were to be used to supply lighting current. In the middle of a bright day few lamps would be turned on, and the resistance of the LOAD* would be high. Hence the current in the line and in the field coils would be small. This would result in a small magnetic flux ϕ, and the induced electromotive force would be small (see eq. [27-1]). The line difference of potential being small, the few lamps turned on would be

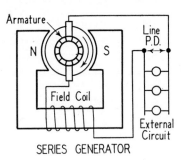

Armature
Line
P.D.

N   S

Field Coil

External
Circuit

SERIES GENERATOR

FIG. 27-5.—Electrical circuit of the series-wound generator

very dim—if, indeed, they lighted at all. As more lamps were turned on toward evening, the current would increase, the flux increase, the induced electromotive force increase, the line potential difference increase, and the lamps burn brighter. If enough lamps were turned on, some might be overheated and burn out. This change of line potential difference with load current is indicated in Fig. 27-6. With no load current, the residual

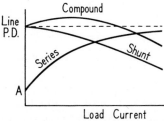

Compound

Line
P.D.

Series   Shunt

A

Load Current

FIG. 27-6.—Variation in line voltage with load current for different types of generators.

magnetism in the cores of the field magnets is ordinarily sufficient to produce a small potential difference, as indicated at A.

27.5. Shunt-connected Generators

Another way to excite the field magnets is to connect them in parallel with the load, i.e., use SHUNT connections (see Fig. 27-7). With this

* By "load" is meant the energy usefully employed in that part of the circuit external to the generator.

method the current in the field windings is wasted, as far as the *useful load* current is concerned; hence the field current is made as small as possible. So that the field magnets may be sufficiently excited by a small current, they are wound with many turns of wire. The characteristics of this type of generator in terms of a load composed of incandescent lamps are a little more involved than was the case for the series generator. The resistance of the armature must be considered. The potential difference at the brushes is the induced electromotive force less the IR drop in the armature:

$$PD = E - (I_{\text{armature}} \times R_{\text{armature}}) . \quad (27\text{-}2)$$

Since the armature supplies all the current used,

$$I_{\text{armature}} = I_{\text{load}} + I_{\text{field}} . \quad (27\text{-}3)$$

The current in the field windings and hence the magnitude of the magnetic flux ϕ depend upon the

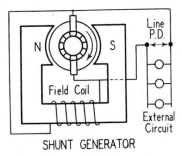

Line
P.D.

N   S

Field Coil

External
Circuit

SHUNT GENERATOR

FIG. 27-7.—Electrical circuit of the shunt-wound generator

potential difference across the field magnets. If only a few lamps were turned on, i.e., load resistance high and load current small, then the potential difference would be large, and the induced electromotive force would be large. As more lamps were turned on, the load current would increase, the potential difference decrease, the magnetic flux become weaker and the induced electromotive force less. Hence all the lamps would be less bright (see Fig. 27-6).

27.6. Compound-connected Generators

Since neither of these methods of excitation is satisfactory and since they seem to be opposite in their characteristics, a better method than either is a combination of the two (see Fig. 27-8). By having a few series turns and also many shunt turns, the line potential difference cannot be held absolutely constant, but it can be held much more **377** nearly constant than by either method alone (see Fig. 27-6).

Aside from these fundamental principles, a more detailed theory of generators is far too complicated to be discussed here. The descriptions we have given represent the general types of generators. Starting from these types, the electrical en-

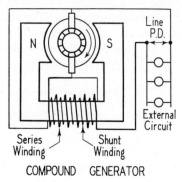

COMPOUND GENERATOR

FIG. 27-8.—Electrical circuit of the compound generator

gineer has developed many modifications for particular purposes and in efforts to increase efficiency.

27.7. Direct-Current Motors

Electric generators are turned by some outside source of energy, such as a steam, gas, or oil engine or a water wheel. It is important to remember that *work must be done* to push the wires of the armature, in which there are currents, across the magnetic field. This work is the source of the energy, which appears as a current driven by the electromotive force induced in the armature coils. If, on the other hand, we supply current through the coils from some outside source, there would be the same force on the armature wires, and the machine would turn itself. This is but a description of an electric MOTOR. Whereas

$$E = -N \frac{d\phi}{dt} \qquad (27\text{-}4)$$

is the fundamental equation for a generator,

$$F = IBl \qquad (27\text{-}5)$$

is the fundamental equation for a motor. However, in a rotating motor, conductors are cutting across lines of magnetic flux, and hence in the motor there is also *an induced or back electromotive force*. Once the motor is in motion, equation (27-1),

$$E_{\text{back}} = K\phi n , \qquad (27\text{-}1)$$

applies to its operation. Since the current in the armature is the cause of the motion of the armature and since an induced electromotive force is always in such a direction as to oppose the change which causes it (Lenz's law, § 23.4), this back electromotive force opposes the current. In the case of a shunt-wound motor (Fig. 27-9), these relations are

$$V_{\text{line}} = E_{\text{back}} + (I_{\text{armature}} \times R_{\text{armature}}) . \quad (27\text{-}6)$$

Actually, this equation is true for any motor regardless of the speed or load conditions, provided that the proper value for the resistance is

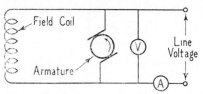

FIG. 27-9.—Circuit diagram of a shunt-wound motor

used. For a shunt-wound motor the resistance is simply the armature resistance; for a series-wound motor the resistance is the sum of the armature and field-coil resistance, which we can consider as part of the armature circuit. With this qualification in mind, we may repeat the basic motor equation as

$$V_{\text{line}} = E_{\text{back}} + (I_{\text{armature}} \times R) \text{ volts} . \quad (27\text{-}7)$$

Note that equation (27-7) represents a situation electrically similar to that of Fig. 20-2. The back electromotive force of equation (27-7) is analogous to E of equation (20-4), the armature resistance similar to the internal resistance of the cell.

As we have noted, to produce rotation of the armature, a torque must be supplied. The origin of this torque can be clearly traced to the force acting on a current-carrying conductor in a magnetic field (eq. [27-5]). We showed in § 21.3 that the torque on a coil in a magnetic field is given by

$$L = BIA \sin \alpha ,$$

where α is the angle between B and the normal to the coil. For the case of motors the current I is the current in the armature. The quantity BA is the magnetic flux ϕ that permeates the armature. For a given motor, therefore, we may express the torque exerted on the armature by the interaction of current-carrying conductors and magnetic field as

$$L = K_t \phi I_a \text{ meter-newtons} , \qquad (27\text{-}8)$$

where ϕ is the flux in webers, I_a the armature current in amperes, and K_t a factor that accounts for the number of conductors, magnetic paths, and other factors which are constant for a particular motor.

The three basic equations necessary for the study of the performance of direct-current motors are equations (27-1), (27-7), and (27-8).

EXAMPLE 1: Suppose the shunt-wound motor of Fig. 27-9 draws 52 amp when connected to a 100-volt line and running at normal speed. For such a motor the resistance of the field coils might well be 50 ohms, and the resistance of the armature 0.05 ohm. Then the current in the field magnets is $100/50 = 2$ amp. Since the armature and field are in parallel, $I_{line} = I_{field} + I_{armature}$; and the armature current is 50 amp. At first, one might have expected the armature current to be $100/0.05 = 2,000$ amp, but this is neglecting the back electromotive force. By equation (27-6),

$$V_{line} = E_{back} + (I_{armature} \times R_{armature}),$$

which in this case gives

$$100 = E_{back} + (50 \times 0.05)$$

and

$$E_{back} = 97.5 \text{ volts}.$$

27.8. Power Relations in a Motor; Efficiency

The power delivered to a motor armature is the product of the line voltage, V, by the armature current,

$$P_a = VI_a \text{ watts}. \qquad (27-9)$$

By substituting from equation (27-7), we obtain

$$P_a = E_{back}I_a + I_a^2 R \text{ watts}. \qquad (27-10)$$

The quantity $I_a^2 R$ represents the power lost as heat in the total armature circuit. The remainder $E_{back}I_a$ represents the electrical power output converted into equivalent mechanical power.

If the total load current drawn by the motor is I amp, with a line voltage of V volts, then the total electrical power absorbed by the motor is

$$P = VI \text{ watts},$$

or

$$P = E_{back}I_a \text{ (mechanical power)}$$
$$+ \text{sum of all electrical losses}. \qquad (27-11)$$

Part of this power, $E_{back}I_a$, is converted into mechanical power; the rest represents electrical losses in the conducting wires of the armature and field

coils. The maximum electrical efficiency of the motor, then, is

$$\text{Efficiency} = \frac{E_{back}I_a}{P} \times 100 \text{ per cent}. \qquad (27-12)$$

In this calculation we have provided for only the heat loss in the copper conducting wires. Actually, there are certain other losses inside every motor, so that not all the power $E_{back}I_a$ is available at the shaft to do useful work. In § 24.6 we discussed hysteresis loss. The magnetism of the iron of the armature is constantly changing, and hence the armature core will be heated because of the resulting hysteresis. The manufacturer uses iron of a sort which has as small a hysteresis loss as possible, but this loss cannot be entirely avoided. Second, the iron of the armature core is itself conducting and in motion in a magnetic field; hence there will be induced, in the iron, eddy or Foucault (§ 23.7) currents. These can be reduced by forming the armature core from thin sheets of iron, but there will always be some loss. And, last, there is always friction—air friction due to motion through the air and sliding friction in the bearings. It is very difficult to separate all these different losses from one another. Their total can be obtained when the output is measured by a Prony brake. The total over-all efficiency, therefore, is less than that given by equation (27-12).

EXAMPLE 2: To continue the preceding example: The power loss in the field windings, due to heating, is

$$P_{field} = VI_{field} \text{ or } R_{field} \times I_{field}^2$$

$$= 100 \times 2 = 50(2)^2 = 200 \text{ watts}.$$

The power loss in the armature is

$$P_{armature} = R_{armature} \times I_{armature}^2 = 0.05 \times (50)^2$$

$$= 125 \text{ watts}.$$

Hence the total power loss is $200 + 125 = 325$ watts. The total power input to the motor is

$$P_{input} = VI$$

or

$$5,200 \text{ watts} = 100 \text{ volts} \times 52 \text{ amp}.$$

Then the "electrical" output is $5,200 - 325 = 4,875$ watts. We could have computed this in other ways. The total power input to the armature is

$$(97.5 + 2.5) 50 = (4,875 + 125) \text{ watts},$$

but we have found that the 125 watts went to producing **379** heat. Hence the 4,875 watts, a quantity which is evidently $E_{back} \times I_{armature}$, are left to do the work.

The computed electrical output (4,875 watts in our example) minus the measured output (suppose it is 4,500 watts) is the sum total $(4,875 - 4,500 = 375$ watts) of the hysteresis, eddy current, and frictional losses.

The electrical efficiency, then, is

$$\frac{4,875}{5,200} \times 100 = 93.75 \text{ per cent},$$

while the total efficiency is

$$\frac{4,500}{5,200} \times 100 = 86.53 \text{ per cent}.$$

27.9. Motor Characteristics

Electric motors may be connected with their field windings in series with the armature, in parallel (shunt) with the armature, or with a combination of the two, in ways similar to generator connections. Each of these has its own running characteristics.

Motors, in general, are started with a certain amount of overload, i.e., with a greater current than their normal running current. With a series motor an excess current provides increased magnetic field and also increased current in the armature, so that the starting torque is proportional to the product of the armature current and the flux through the armature (see eq. [27-8]) is thus roughly proportional to the square of the current. When motors must start a heavy load, this "square" effect makes series motors desirable. On the other hand, shunt motors have no increased field strength when starting, and hence their starting torque is proportional only to the first power of the current.

Direct-current motors accelerate until the increasing back electromotive force reduces the current to such an extent that the power input $(I_{total} \times V_{line})$ is just sufficient to supply the losses of the motor and to do the external work to which the motor is attached, as shown by equation (27-11). Let us consider a series motor which is not connected to any external load, so that the power input goes only to supplying the losses, including friction, and, at first, to increasing the speed of the motor. In this case the back electromotive force must be quite large—nearly as large as the line potential difference—in order that the

current may be small. As the current in a series motor is reduced, the magnetic field weakens; and the motor must run faster and faster to build up the necessary back electromotive force (eq. [27-1]). For this type of motor the speed will increase unless the losses are quite large—until the mechanical bonds of the armature can no longer supply the necessary centripetal force and the motor destroys itself. For this reason, a series motor should never be started without a load and should never be connected to its load by a connection which can fail, as, for instance, a belt, which might break or slip off. A streetcar which has a great mass to be accelerated and hence needs a motor with a large starting torque and which always, when running, has a motorman in attendance, would be a good place to use a series motor, the motor being geared to the wheels (see Ex. 4).

With a shunt motor, where the field windings are connected directly to the line, the magnetic field is essentially constant. The increasing back electromotive force and decreasing armature current do not weaken the field; and hence in such a motor the speed reaches a steady state before becoming excessive. The starting torque, however, is not so large as that of a series motor.

A compound motor has the virtues of both of these and the defects of neither, and hence is most commonly used.

EXAMPLE 3: A 15-hp, 220-volt shunt-wound motor has a no-load speed of 1,800 rpm and a current input at no load of 6.0 amp. The full-load current input is 60 amp, the field-coil resistance is 160 ohms, and the armature resistance is 0.15 ohm. Calculate the full-load speed.

Solution: By equation (27-1) the back electromotive force E is given by

$$E = K\phi n.$$

For the shunt motor ϕ is independent of the load condition, so that if E_1 and E_2 are the back electromotive forces at no load and full load, then

$$E_1 = K\phi n_1; \qquad E_2 = K\phi n_2,$$

and

$$n_2 = n_1 \frac{E_2}{E_1}.$$

For no load,

$$I_f = \tfrac{220}{160} = 1.38 \text{ amp},$$

$$I_a = 6.0 - 1.38 = 4.62 \text{ amp};$$

$$E_1 = 220 - (4.62 \times 0.15) = 219.31 \text{ volts}.$$

For full load,

$$I_f = 1.38 \text{ amp},$$

$$I_a = 60 - 1.38 = 58.62 \text{ amp};$$

$$E_2 = 220 - (58.62 \times 0.15) = 211.2 \text{ volts},$$

and

$$n_2 = \frac{211.2}{219.31} \times 1{,}800 = 1{,}734 \text{ rpm}.$$

EXAMPLE 4: A 30-hp, 500-volt series-wound motor draws 20 amp of current at a speed of 1,200 rpm. The resistance of the field coil is 0.5 ohm and that of the armature 0.5 ohm. Calculate the speed of this motor when drawing a current of 40 amp under the assumption (a) of a constant flux and (b) that the flux is 50 per cent greater when the current is 40 amp.

Solution: The total resistance in the armature plus field coil is 1.0 ohm. The back electromotive force for $I = 20$ amp is

$$E_1 = 500 - (1 \times 20) = 480 \text{ volts}.$$

For $I = 40$ amp,

$$E_2 = 500 - (1 \times 40) = 460 \text{ volts}.$$

For constant flux, since $E = K\phi n$,

$$n_2 = n_1 \frac{E_2}{E_1} = 1{,}200 \times \tfrac{460}{480} = 1{,}150 \text{ rpm}.$$

For a 50 per cent increase in flux, $\phi_2 = 1.5 \phi_1$,

$$n_2 = n_1 \frac{E_2}{1.5 E_1} = \frac{460}{1.5 \times 480} \times 1{,}200 = 767 \text{ rpm}.$$

27.10. Speed Regulation

The speed of a direct-current motor may be regulated by changing the potential difference of its terminals. This, in general, involves the use of rheostats and the loss of energy by its conversion into heat. In the case of shunt motors, the speed can be *increased* by reducing the current in the field windings by adding resistance to the shunt field. Here the heat loss is small, since the field current is small. The one danger of running a large shunt motor unloaded is that a connection in the field windings might become disconnected. This would so weaken the field that only at an excessive speed could the back electromotive force be sufficient to keep down the current so that the input power would be equal to the losses only.

27.11. Reversing the Direction of Rotation

The direction of rotation of a motor can be reversed by changing the direction of the current in either the field or the armature, but *not* in both. Reversing the direction of the magnetic field and also reversing the direction of the current in the armature would leave the direction of the torque unchanged. For this reason a direct-current motor will "run" on alternating current. However, the self-induction of the coils makes a large-sized motor, designed for direct current, quite inefficient when used on alternating current. The "universal motors" used on many household appliances and which run on either direct or alternating current are simply small direct-current (generally series) motors. As we shall see later, many motors designed for alternating current will not run at all and may be destroyed if connected to terminals at constant potential difference.

27.12. Résumé

The following is a brief listing of the definitions, principles, and theories, given in this chapter, which you should know:

The role of the commutator in the direct-current generator

The advantages and disadvantages of series-, shunt-, and compound-wound generators

The characteristics of a direct-current motor; speed regulation and reversal of direction of rotation

Queries

1. Is there any fundamental difference in design between a simple generator and a motor?
2. Does an electric generator create electricity? What does it do?
3. Why does the series-wound motor have such a large starting torque?
4. If the load on a shunt-wound motor is increased, how does the motor respond in taking up the increased load?
5. Can you draw a circuit with a battery and resistances that is electrically equivalent to the series-wound motor? Shunt? Compound?

Practice Problems

1. A shunt-wound, direct-current motor has a field resistance of 22 ohms and an armature resistance of 0.05 ohm. When directly connected to 110-volt mains and running under normal load (900 rpm), the motor draws 65 amp. Compute the back electromotive force and the "electrical" efficiency of the motor. Ans.: 107 volts; 89.8 per cent.

2. When the motor in problem 1 is connected to a Prony brake (see § 4.12), a torque of $(2/\pi) \times 10^9$ cm-dynes slows it down to its normal speed of 900 rpm. Compute the total efficiency of this motor. How many watts go to hysteresis, eddy current, and friction combined?

3. If this motor were loaded until it stalled, what total current would it take from the mains? Ans.: 2,205 amp.

4. A series-wound, direct-current motor has a field resistance of 0.012 ohm and an armature resistance of 0.05 ohm. When directly connected to 110-volt mains and running under normal load (900 rpm), the motor draws 65 amp. Compute its "electrical" efficiency and its current when stalled.

   Note: It is wrong to draw any conclusions as to the relative efficiencies of these two types of motors from problems 1–4.

5. The armature of a shunt-wound motor, when at rest, passes 50-milliamp current when a dry cell of 1.5 electromotive force is connected directly to it. The shunt field coil, likewise with the same electromotive force, passes 1.8-milliamp current. What is the back electromotive force of this motor when running on a 115-volt line under a load that requires a total current consumption of 1 amp? Of 3.5 amp? Ans.: For 1 amp, 89.14 volts.

6. A motor with a resistance of 0.2 ohm is connected to a battery of 250 volts and internal resistance of 0.05 ohm. If the back electromotive force is 225 volts, find the current that the motor draws, the mechanical power delivered, and the potential difference across the motor.

7. A shunt-wound generator has an armature resistance of 0.1 ohm and a field-coil resistance of 120 ohms. If the terminal voltage delivered by the generator is 120 volts when drawing 40 amp, calculate the mechanical power required to drive the generator. Ans.: 5.08 kw.

8. If the armature of problem 7 is turning at 1,500 rpm, what speed is required to maintain the terminal voltage of 120 volts with a load current of 60 amp?

9. A 10-hp, 220-volt shunt-wound motor draws an armature current of 40 amp when turning at a speed of 600 rpm at its rated loading. (*a*) Calculate the speed if the field flux is halved by changing the field excitation and the armature current remains the same. (*b*) What are the torque and horsepower at the new speed? Ans.: (*a*) 1,200 rpm; (*b*) 43.7 lb-ft, 10-hp.

10. A shunt-wound motor of 10 hp and 220 volts has a no-load speed of 2,000 rpm and a current input at no load of 10 amp. The full-load current input is 70 amp, the field resistance is 200 ohms, and the armature resistance 0.22 ohm. Calculate the full-load speed.

11. A 50-hp, 500-volt series motor draws 40 amp of current when operating at a speed of 1,100 rpm. The resistance of the armature is 0.24 ohm and that of the field coil 0.56 ohm. Calculate the speed of this motor when drawing 90 amp under the assumption of (*a*) constant flux and (*b*) that the flux is 65 per cent greater when the current is 90 amp. Ans.: (*a*) 1,006 rpm; (*b*) 610 rpm.

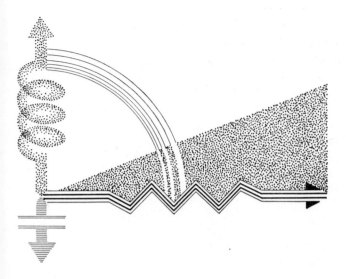

CHAPTER 28

Alternating Currents

28.1. Introduction

In nearly all the domestic uses and a vast majority of the industrial needs which today find electricity almost universally available in America, it is alternating currents that are employed. This is, in large part, due to the ease with which alternating voltages can be stepped up to high voltages for efficient transmission and distribution and then back to low voltages for ordinary use.

We saw in chapter 23, equation (23-30), that when a coil of wire is rotated at a uniform rate in a magnetic field, there was induced in that coil a changing electromotive force of the form

$$e = E_M \sin \omega t \ , * \qquad (23\text{-}30)$$

where e is the instantaneous value of the electromotive force, ω is the angular velocity of rotation of the coil, and E_M is the maximum value of the electromotive force. We may represent equation (23-30) graphically, as in Fig. 28-1. The abscissas may represent either the time or the angle θ through which the coil has turned in time t, since $\theta = \omega t$.

In the figure, o–c represents *one cycle*. The period, T, is the time required for the completion of one cycle, while the frequency, f, is the number of cycles per second. A "60-cycle electromotive

*It is convenient, for the work of this chapter, to have small letters represent instantaneous values of the electrical quantities.

force" is one in which the impressed electromotive force goes through 60 complete cyclic variations in 1 sec. The frequency in this case is said to be 60 cycles per second.

If an alternating electromotive force is impressed on a circuit, such as is shown in Fig. 27-1 (§ 27.1), it is reasonable to assume that a current will be set up which alternates with the same frequency as the applied electromotive force. Throughout our discussion of alternating currents,

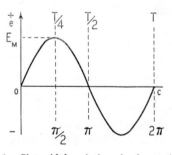

Fig. 28-1.—Sinusoidal variation of voltage with time

we shall assume that the current in the circuit is given by the expression

$$i = I_M \sin \omega t = I_M \sin \frac{2\pi t}{T} = I_M \sin 2\pi f t, \quad (28\text{-}1)$$

where again i is the instantaneous value of the current, I_M the maximum value of the current, and ω the angular velocity equal to $2\pi f$.

28.2. Effects of Alternating Currents

The effects of alternating currents and alternating voltages in circuits are quite different from direct currents and direct voltages. We shall list some of the more important differences:

1. The current, in general, is not given by E/R; i.e., the simple form of Ohm's law must be modified to take other factors, such as inductance and capacitance, into consideration.

2. It is possible for the applied electromotive force and the resulting current not to reach their maximum values at the same instant.

3. Continuous direct currents cannot be set up in circuits containing capacitors, but we shall see that this is not the case for alternating currents.

4. Because of 2, the simple algebraic sum of voltmeter readings or voltage drops across the components of a circuit carrying alternating current may not be equal to the electromotive force impressed on the circuit. To illustrate, in Example 2 of this chapter, 100 volts are impressed across a circuit containing inductance, capacitance, and resistance. The algebraic sum of the voltmeter readings in 120 volts. However, the algebraic sum of the instantaneous potential differences across each component must equal the impressed electromotive force.

In the following sections we shall try to understand the reasons for these effects of alternating currents.

28.3. Phase Relations

As we pointed out in the previous paragraph, the maximum value of the current in a circuit may not be attained at the same instant that the applied electromotive force is a maximum. In other words, the current in the circuit may either lead or lag behind the applied electromotive force; or, to use a technical expression, *a phase difference* may exist between them. This difference in phase may exist between various electromotive forces applied to a circuit, as well as between the current and the applied electromotive force.

We shall illustrate the concept of phase difference for the case of voltages by means of two generators connected in series and giving rise to electromotive forces of the same frequency. We shall assume that generator No. 1 gives rise to the electromotive force

$$e_1 = E_M \sin \omega t$$

and that the electromotive force of generator No. 2, in phase with No. 1, is

$$e_2 = E_M' \sin \omega t .$$

The resultant *instantaneous* electromotive force, then, is

$$e_R = e_1 + e_2 = (E_M + E_M') \sin \omega t . \quad (28\text{-}2)$$

This situation is illustrated in Fig. 28-2 (*a*), where the two dashed curves represent the indi-

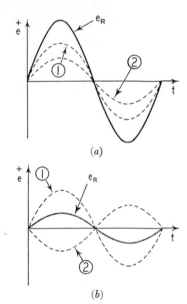

(*a*)

(*b*)

Fig. 28.2.—(*a*) Resultant voltage of two generators of the same frequency and phase but different amplitudes; (*b*) resultant voltage of two generators of the same frequency but out of phase by π rad.

vidual electromotive forces of the generators, while the heavy curve is the resultant electromotive force. Note that in this case the resultant *maximum* electromotive force is the sum of the individual maxima.

If the two generators are *out of phase* by π rad or $T/2$ sec, the resultant instantaneous electromotive force is again obtained by adding the individual instantaneous electromotive forces:

$$e_R = e_1 + e_2 = E_M \sin \omega t + E_M' \sin (\omega t - \pi)$$

or

$$e_R = (E_M - E_M') \sin \omega t . \quad (28\text{-}3)$$

Note, however, that the resulting maximum value is the difference between the individual maxima (Fig. 28-2 [*b*]).

Now, if the two generators are out of phase by ϕ rad, the resultant instantaneous value is

$$e_R = e_1 + e_2 = E_M \sin \omega t + E'_M \sin (\omega t - \phi) , \quad (28\text{-}4)$$

but the maximum electromotive forces cannot be added algebraically. We may always obtain an answer by plotting the two curves and adding ordinates, as shown in Fig. 28-3. The heavy curve is

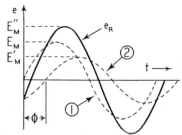

Fig. 28-3.—The resultant voltage of two generators of the same frequency but out of phase by ϕ rad.

the required resultant, with E''_M as the maximum value.

By means of trigonometric formulas it is possible to reduce equation (28-4) to the form

$$e_R = E''_M \sin (\omega t - \delta) , \quad (28\text{-}5)$$

which is the equation of the solid curve of Fig. 28-3. The angle δ is the difference in phase between the resultant electromotive force and the electromotive force of generator No. 1. In general, such trigonometric methods are long and tedious.

A third method, in common use today for obtaining the resultant maximum electromotive force of two or more generators with the same fre-

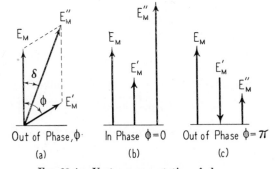

Out of Phase, ϕ In Phase $\phi=0$ Out of Phase $\phi=\pi$
(a)   (b)   (c)

Fig. 28-4.—Vector representation of phase

quency but out of phase with each other, is to represent the maximum electromotive forces of the individual generators as if they were vectors and then add them according to the parallelogram law. This is illustrated in Fig. 28-4 (a).

By the parallelogram law,

$$E''_M = \sqrt{(E_M)^2 + (E_M)^2 + 2E_M E_M \cos \phi} , \quad (28\text{-}6)$$

and

$$\cos \delta = \frac{(E''_M)^2 + (E_M)^2 - (E'_M)^2}{2E_M E_M} . \quad (28\text{-}7)$$

By means of these equations we may obtain E''_M and angle δ of equation (28-5).

This vector representation of electromotive force proves to be very useful in the discussion of circuit problems. In Fig. 28-4, (b) and (c) are vector representations of the generator electromotive forces in phase and out of phase by π rad.

Example 1: If the two generators give rise to the same electromotive force and are out of phase by 60°, compute the resultant electromotive force.

Solution: Using equations (28-6) and (28-7), we have

$$E''_M = E_M \sqrt{2 + 2 \cos 60} = E_M \sqrt{3} ,$$

$$\cos \delta = \frac{3}{2\sqrt{3}} = \frac{\sqrt{3}}{2} \text{ or } \delta = 30° \text{ or } \frac{\pi}{6} \text{ rad} ,$$

$$e_R = E_M \sqrt{3} \sin \left(\omega t - \frac{\pi}{6}\right) .$$

28.4. Effective Currents

An alternating current varies periodically in strength from zero to some maximum value (Fig. 28-1). What, then, is meant by an alternating current of strength of 1 amp? In our earlier discussion of electricity in motion we found that currents could be measured by one of three effects they produced: heat, magnetic fields, or chemical action. Only the heating effect in a resistance which is proportional to the square of the current is independent of the direction in which the current is circulating. Hence the heating effect of alternating current is taken as the basis for the definition of an ampere.

An alternating current is said to have an effective value of 1 ampere when it will develop the same amount of heat in a given resistance as would be produced by a direct current of 1 ampere in the same resistance at the same time. In other words, the same amount of energy is dissipated per second by both currents.

Since we are assuming that the alternating currents are sinusoidal, it is interesting to find out **385** what relationship exists between the effective value of the current, as defined here, and the maximum value of the current. For this analysis we shall use one theorem from trigonometry which we shall not prove. This theorem is as follows: *The value of the* $\overline{\text{sine}^2}$ *or* $\overline{\text{cosine}^2}$ *over one complete period is numerically equal to* $\frac{1}{2}$, *with the bar* (———) *meaning "average."*

From our discussion of direct currents we learned that the power dissipated when a current passed through a resistance of R ohms was $P = Ri^2$ watts if i is in amperes, direct current.

If we designate the *effective alternating current* by I, then, from our definition, we want the same value of the power, P, to be given by

$$P = RI^2 , \qquad (28\text{-}8)$$

where RI^2 is the average power dissipated in the resistance by the alternating current.

At any instant we write the alternating-current formula for the instantaneous power as

$$p = R\,(I_M \sin \omega t)^2 = RI_M^2 \sin^2 \omega t ; \quad (28\text{-}9)$$

the average power for any continuous half-cycle or any integral number of half-cycles is

$$P = RI_M^2 \; \overline{\sin^2 \omega t} .$$

But the average value of $\sin^2 \omega t$ is $\frac{1}{2}$. Hence

$$P = \tfrac{1}{2}RI_M^2 ;$$

and, since from equation (28-8) the average power is also RI^2, by definition we have

$$RI^2 = \frac{RI_M^2}{2}$$

or

$$I = \frac{I_M}{\sqrt{2}} = 0.707 I_M \qquad (28\text{-}10)$$

or

$$P = RI^2 . \qquad (28\text{-}11)$$

In an analogous way we define the *effective value of voltage*, V,

$$V = \frac{V_M}{\sqrt{2}} = 0.707 V_M . \qquad (28\text{-}12)$$

Thus, in speaking of current or voltage, we may use the instantaneous values, the maximum values, or the effective values, as the case may be. In our discussion we shall designate instantaneous values by small letters, effective values by capital letters, and maximum values by capital letters

with the subscript M. Ammeters and voltmeters are calibrated to read effective values.

The effective values of the current and voltages are also called the "root-mean-square" values. It is important to emphasize that equations (28-10) and (28-11) are valid only if these quantities vary sinusoidally.

Equation (28-11), $P = RI^2$, gives the rate at which heat is produced in a circuit. For alternating currents this equation must be considered as defining "alternating-current resistance," since this is the only equation we have for R in which all other quantities are defined. If the resistance of a coil, particularly of a coil with an iron core, is measured by means of a wattmeter and ammeter, the resistance given by equation (28-11) will be greater than the resistance of that coil measured by a direct-current Wheatstone bridge. This is caused by the heat loss due to hysteresis in the core and by any eddy currents which may be induced in near-by conductors. It is this greater resistance, as given by equation (28-11), *not* the Wheatstone bridge resistance, which should be used in equation (28-39). For an air-core coil far from other conductors, the two resistances are essentially the same.

28.5. Phase Relationships between Current and Electromotive Force

As we pointed out in § 28.2, the current and impressed electromotive force will not, in general, be in phase. What we do assume is that both the current and the electromotive force vary harmonically with time, i.e., can be represented by sine or cosine functions. If, then, we represent the applied electromotive force by

$$e = E_M \sin \omega t , \qquad (28\text{-}13)$$

the current circulating must be represented by

$$\imath = I_M \sin(\omega t + \phi) , \qquad (28\text{-}14)$$

where ϕ is the phase angle. If the current leads the electromotive force, then ϕ is positive; if it lags, then ϕ is negative; if it is in phase, then $\phi = 0$.

On the other hand, if we choose to represent the current by

$$\imath = I_M \sin \omega t , \qquad (28\text{-}15)$$

then the impressed electromotive force must be written as

$$e = E_M \sin(\omega t + \phi) . \qquad (28\text{-}16)$$

The *two sets* of equations are physically equivalent. For example, if in a given circuit the current lags behind the electromotive force by 60°, we may represent this either by setting $\phi = -60°$ in equation (28-14) or by setting $\phi = +60°$ in equation (28-16). Whether we use the set of equations (28-13) and (28-14) or the set (28-15) and (28-16) will depend upon which is convenient for the particular problem at hand.

28.6. Circuits with Resistance Only

When an alternating-current voltage is impressed on a circuit composed wholly of noninductive resistance, the current is obtained by an application of Ohm's law, $E = RI$. When the voltage is zero, so is the current; when the voltage is at a maximum, the current is also at a maximum. In other words, the current and the applied electromotive force are in phase. This state of affairs is

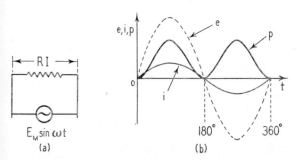

FIG. 28-5.—Voltage, current, and power in phase in a purely resistive circuit.

illustrated graphically in Fig. 28-5 (a) and (b). The following relations hold:

$$i = \frac{E_M}{R} \sin \omega t \ (\text{instantaneous current}) ,$$

$$e = E_M \sin \omega t \ (\text{instantaneous emf}) ,$$

$$I = \frac{E}{R} \ (\text{effective current}) ,$$

$$P = EI \ (\text{average power}) ,$$

$$p = ei = E_M I_M \sin^2 \omega t \ (\text{instantaneous power}) .$$

$$(28\text{-}17)$$

28.7. Circuit with Inductance Only

The second ideal circuit we shall discuss is one containing an inductance alone. We assume that the resistance of the inductive coil is zero. If the current in the circuit of Fig. 28-6 (a) is varying sinusoidally, i.e.,

$$i = I_M \sin \omega t , \qquad (28\text{-}18)$$

we know from our studies of induced currents in chapter 23 that a self-induced electromotive force will be set up in the coil whose value depends upon the magnitude of the self-inductance and also upon

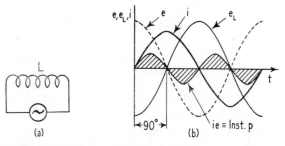

FIG. 28-6.—Current lagging behind voltage by 90° phase angle.

the rate at which the current is varying. Stated analytically,

$$e_L = -L \frac{di}{dt} . \qquad (28\text{-}19)$$

We know, moreover, how the current varies with time (eq [28-18]). By differentiating, we obtain

$$e_L = -L\omega I_M \cos \omega t . \qquad (28\text{-}20)$$

Since we are assuming that the resistance of the coil is negligible, the applied electromotive force, e, has simply to overcome the self-induced electromotive force, e_L, in order to cause the current, $i = I_M \sin \omega t$, to pass; i.e.,

$$e = -e_L ,$$

or

$$e = L\omega I_M \cos \omega t ,$$

or

$$e = L\omega I_M \sin (\omega t + 90°) , \qquad (28\text{-}21)$$

which is the value of the applied electromotive force necessary to drive the current given by equation (28-18) through the pure inductive circuit. Note that the current and applied electromotive force are out of phase by 90°, with the *current lagging behind the applied electromotive force or the applied electromotive force leading the current*. In other words, when the voltage across the inductance is a maximum, the current through it is zero and starting to increase. The current leads the self-induced electromotive force by 90°. Figure 28-6 (b) summarizes the information contained in equa- **387** tions (28-18), (28-20), and (28-21). Observe that the applied electromotive force and self-induced electromotive force are out of phase by 180°.

A very simple way of showing the phase relation of Fig. 28-6 (b) is to represent the various quantities as vectors, as discussed in § 28.3. In Fig. 28-7

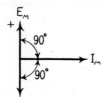

FIG. 28-7.—Vector representation of phase relations

let the horizontal line represent the current; then the applied electromotive force can be represented by a line at right angles to this, the arrow pointing up, while the induced electromotive force points down. If the circuit contains appreciable resistance, the lag is less than 90° (see eq. [28-36]).

From equation (28-21) the maximum value of the applied electromotive force will occur when $\cos \omega t = 1$:

$$E_M = I_M \omega L . \qquad (28\text{-}22)$$

The effective values are

$$E = L\omega I ,$$

since $E = E_M/\sqrt{2}$ and $I = I_M/\sqrt{2}$, or

$$\frac{E}{I} = L\omega . \qquad (28\text{-}23)$$

The quantity $L\omega$ is known as the *inductive reactance* and is defined as the ratio of the component of the applied electromotive force which overcomes the self-induced electromotive force to the current, I, set up in the inductance. The inductive reactance varies directly as the frequency and is measured in ohms.

We see from the preceding analysis that the presence of an inductance in a circuit produces two distinct effects: (1) it introduces a phase lag between the impressed electromotive force and the current of 90°, with the current lagging; (2) it exerts a choking effect upon the current, reducing by its presence the value of the current that may pass for a given applied electromotive force (eq. [28-23]).

A highly inductive coil having only a very small resistance is generally called a "choking coil," or simply a "choke," and is used to cut down the value of the current. It has an advantage over a

rheostat, in that much less power is wasted in heating with the coil than with the resistance of a rheostat. Since the choking effect increases with frequency, it is very efficient as a choke for high frequencies, such as are encountered in radio work.

The instantaneous power dissipated in the circuit is

$$p = ei = E_M I_M \sin \omega t \cos \omega t . \qquad (28\text{-}24)$$

In order to obtain the average power over a complete cycle, it is necessary to know the average value of $\sin \omega t \cos \omega t$ for one cycle. From trigonometry we know that this average is zero. In other words, *the average power dissipated in a pure inductive circuit is zero.* In Fig. 28-6 (b) the shaded area is the energy stored in the magnetic field. The reader will note that for one cycle the average power is actually zero.

EXAMPLE 2: In Fig. 28-6 L is 3 henrys, $E_{\text{eff}} = 10$ volts, and $f = 1,000$ cycles/sec. What is the effective current through the circuit?

Solution: Inductive reactance $= L\omega = 2\pi f L = 6,000\pi$ ohms,

$$I = \frac{E}{6,000\pi} = \frac{10}{6,000\pi} = 0.000531 \text{ amp} ,$$

and the current will be 90° behind the impressed electromotive force.

28.8. Circuit with Capacitance Only

If an alternating electromotive force is applied to a circuit containing only a capacitance, an alternating current, as detected by an ammeter, is found to pass continuously. This phenomenon does not take place in direct-current circuits. We shall see what relation must exist between the applied electromotive force and the resulting current.

If the electromotive force applied to the capacitor of the circuit represented by Fig. 28-8

FIG. 28-8.—Simple capacitor circuit

is $e = E_M \sin \omega t$, then at each instant the capacitor will have a charge sufficient to raise the plates to a potential difference equal to the applied electromotive force:

$$q = Ce \qquad (28\text{-}25)$$

When e changes at its greatest rate, we see from equation (28-25) that q also varies at its greatest rate. Since current is equal to the time rate of change of charge, the preceding statement implies that i is at a maximum value. The electromotive force changes most rapidly just as it passes zero. Hence we conclude that the applied electromotive force and consequent current are out of phase by 90°; thus for a capacitor, *the current leads the applied electromotive force by 90°*.

Since $i = dq/dt$ by definition, equation (28-25) can be written as

$$i = \frac{dq}{dt} = C \frac{de}{dt}. \qquad (28\text{-}26)$$

We know how the electromotive force varies with time, so, by differentiating, we obtain

$$i = C \frac{d}{dt}(E_M \sin \omega t) = C\omega E_M \cos \omega t$$

or

$$i = C\omega E_M \sin(\omega t + 90°); \qquad (28\text{-}27)$$

and, for $\cos \omega t = 1$,

$$I_M = C\omega E_M, \qquad (28\text{-}28)$$

$$I = C\omega E,$$

or

$$E = \frac{I}{C\omega}. \qquad (28\text{-}29)$$

The equations for the applied electromotive force and resulting currents are plotted in Fig. 28-9 (a).

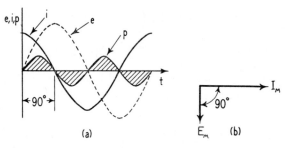

(a)   (b)

Fig. 28-9.—Current leading the voltage by 90° phase angle

In Fig. 28-9 (b) we show the phase difference by a vector representation.

Since the current and applied electromotive force are out of phase by 90°, no power is dissipated in the circuit. The energy accumulated by the capacitor, represented by the shaded area of Fig. 28-9 (a), is given back to the source periodically.

The quantity $1/C\omega$ is called the *capacitance re-*

actance and is measured in ohms. We note that, as the frequency increases, this term becomes smaller, in contrast to the inductive reactance. Capacitance in a circuit has two distinct effects: (1) it introduces a phase shift in which the current leads the applied electromotive force by 90° (eq. [28-27]); and (2) it also limits the value of the current, the effectiveness for a given frequency increasing as the value of the capacitance decreases (eq. [28-29]).

EXAMPLE 3: In Fig. 28-8, C is 2 μf, $E_{\text{eff}} = 10$ volts, and $f = 1{,}000$ cycles/sec. What is the effective current through the circuit?

Solution:

$$\text{Capacitance reactance} = \frac{1}{C\omega}$$

$$= \frac{1}{2 \times 10^{-6} \times 2\pi \times 1{,}000} = 79.5 \text{ ohms},$$

$$I = \frac{E}{79.5} = \frac{10}{79.5} = 0.126 \text{ amp},$$

and the current leads the impressed electromotive force **y** 90°.

28.9. Resistance, Capacitance, and Inductance

In all actual circuits used in practice there is always present a small amount of capacitance. This capacitance, of course, may be so small as to be negligible for frequencies of 60 cycles/sec; but for higher frequencies, such as are encountered in radio work, this may not be negligible. Likewise, inductance is present to varying degrees in cir-

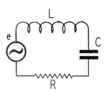

Fig. 28-10.—Series circuit with inductor, capacitor, and resistor.

cuits, as is also resistance. We shall want to investigate the relation existing between current and applied electromotive force in a circuit containing a resistance, a capacitance, and an inductance; R will be measured in ohms, C in farads, L in henrys. The diagram of Fig. 28-10 shows these elements explicitly.

Let us assume that the current circulating in this series circuit is given by

$$i = I_M \sin \omega t . \quad (28\text{-}30)$$

As this same current passes through each element of the circuit, we know from our previous studies that the voltage drop across the resistor is in phase with this current, that the voltage drop across the capacitor lags behind the current by 90°, and, finally, that the voltage drop across the inductor leads the current by 90°. The values of these instantaneous voltage drops are given by the following equations:

$$\left.\begin{array}{l} v_R = I_M R \sin \omega t \text{ (across the resistor)}, \\[1em] v_C = \dfrac{I_M}{C\omega} \sin(\omega t - 90°) \\[0.5em] \qquad \text{(across the capacitor)}, \\[1em] v_L = L\omega I_M \sin(\omega t + 90°) \\[0.5em] \qquad \text{(across the inductor)}. \end{array}\right\} \quad (28\text{-}31)$$

The applied electromotive force, e, necessary to drive the given current around the circuit, must at each instant be equal to the sum of these voltages:

$$e = v_R + v_C + v_L$$
$$= I_M \left[R \sin \omega t + \frac{1}{C\omega} \sin(\omega t - 90°) \right.$$
$$\left. + L\omega \sin(\omega t + 90°) \right].$$

The problem that confronts us in this: Can we simplify this expression, i.e., put it in the form

$$e = E_M \sin(\omega t + \phi) ? \quad (28\text{-}32)$$

If the reader will refer back to § 28.3 on phase shifts, he will find that there it was shown that, although trigonometric methods for simplifying such expressions are available, the methods are involved.

The alternative method is to consider the maximum voltage drops of equations (28-31) as vectors and then add them vectorially. The magnitude of the resultant will be the E_M of equation (28-32), while the angle that the resultant makes with the direction chosen for i will be the phase angle ϕ of equation (28-32).

In Fig. 28-11 (a) are shown the voltage drops with their proper phase relations to i; in (b) these

390 voltages are added vectorially to obtain the re-

sultant E_M. From the diagram, called the *electromotive-force triangle*,

$$E_M = \sqrt{(I_M R)^2 + \left(L\omega I_M - \frac{I_M}{C\omega}\right)^2}, \quad (28\text{-}33)$$

and

$$\phi = \tan^{-1}\left(\frac{L\omega - 1/C\omega}{R}\right). \quad (28\text{-}34)$$

These are the values that must be used in equation (28-32). If ϕ, as given by equation (28-34), *is positive, the applied electromotive force leads the current by ϕ degrees, whereas, if ϕ is negative, the current leads the applied electromotive force.*

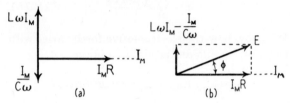

Fig. 28-11.—Electromotive-force triangle

For effective values of current and electromotive force, equation (28-33) becomes

$$E = I\sqrt{R^2 + \left(L\omega - \frac{1}{C\omega}\right)^2}.$$

By definition the impedance of a circuit, Z, is the ratio E/I, i.e.,

$$Z \equiv \frac{E}{I}. \quad (28\text{-}35)$$

For this case

$$Z = \sqrt{R^2 + \left(L\omega - \frac{1}{C\omega}\right)^2}$$

is the impedance of the circuit and is measured in ohms.

If we divide each voltage of Fig. 28-11 (b) by I_M, we obtain the same diagram, but each quantity

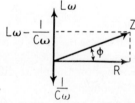

Fig. 28-12.—Impedance triangle

is changed in scale. This new diagram represents impedances and is called the *impedance triangle* (Fig. 28-12).

From the impedance and electromotive-force triangles we find the interesting result that a voltmeter across the resistor will indicate the IR drop; across the capacitor, the $I/C\omega$ drop; across the inductor, the $L\omega I$ drop; and across the entire circuit, the IZ drop, i.e., the impressed electromotive force. This is an important idea and must be kept constantly in mind when dealing with alternating-current circuits. What it means practically is that the arithmetic sum of the "drops" across a series path is usually greater, and sometimes much greater, than the drop across the whole path.

The analysis given here is of *general validity for a SERIES circuit*. If any one of the factors, such as R, L, or C, can be neglected, the expressions for Z and $\tan \phi$ are simplified. For example, if a circuit consists of R and L alone, then

$$Z = \sqrt{R^2 + (L\omega)^2}\,, \\ \tan \phi = \frac{L\omega}{R}; \left.\right\} \quad (28\text{-}36)$$

again, if C and R are present, then

$$Z = \sqrt{R^2 + \left(\frac{1}{C\omega}\right)^2}\,, \\ \tan \phi = \frac{-1/C\omega}{R}. \left.\right\} \quad (28\text{-}37)$$

28.10. Series Resonance

It is interesting to note that, for a given frequency, it is possible to have values of capacitance and inductance for which the impedances are equal but opposite in value, that is, for which $L\omega - 1/C\omega$ is equal to zero. The impedance Z, then, is a minimum; and the circuit is said to *resonate.*

If it is possible to vary the frequency, keeping L and C fixed, then the condition for resonance requires that

$$L\omega = \frac{1}{C\omega} \quad \text{or} \quad \omega^2 = \frac{1}{LC}.$$

Since $\omega = 2\pi f$, we obtain

$$f = \frac{1}{2\pi \sqrt{LC}}. \quad (28\text{-}38)$$

At resonance the current is found by the use of the simple form of Ohm's law:

$$I = \frac{E}{R} \text{(at resonance)}. \quad (28\text{-}39)$$

Figure 28-13 illustrates the behavior of a series

circuit such as that of Fig. 28-10 as the frequency is varied from 0 to 2,800 cycles/sec. The three sets of curves represent the same value of L (0.2 henry) and C (0.127 μf) but three different values of resistance. It is evident that the resonant peak current depends on the value of the resistance only. It is also clear that the addition of resistance to a series resonant circuit decreases the sharpness of the resonance peak, that is to say, the ratio of the current at the resonant frequency to the current at frequencies near resonance is sharply reduced as more resistance is inserted into the circuit. Another useful way of describing the sharp-

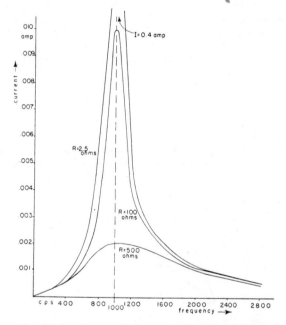

FIG. 28-13.—Resonance in a series circuit with various values of resistance ($L = 0.2$ henry; $C = 0.127$ μf; $E = 1$ volt).

ness of the resonance of the series circuit is to specify the width of the resonance curve at a point at which the current is half the value at resonance. For a resistance of 2.5 ohms, this so-called "band width" is 8 cycles/sec; for $R = 100$ ohms, 320 cycles/sec; and for $R = 500$ ohms, 1,500 cycles/sec.

The principle of resonance is widely used in electronics as a "tuned" circuit to select a particular frequency from a band of frequencies that may be impressed on a circuit. Also a resonant circuit may be used in lieu of a step-up transformer when a voltage greater than the impressed voltage is desired at some given frequency but with very little current drain (see Ex. 6).

EXAMPLE 4: Given the series circuit shown herewith, compute the voltmeter readings, V_1 and V_2, when $L_1 = 1/10\pi$ henry, $R_1 = 5$ ohms, $R_2 = 11$ ohms, $V_3 = 100$ volts, 60 cycles, and $L\omega = 1/10\pi \times 2\pi60 = 12$ ohms.

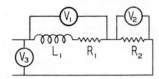

Solution: From equation (28-35),

$$I = \frac{100}{\sqrt{(5+11)^2 + 12^2}} = \frac{100}{\sqrt{400}} = 5 \text{ amp}.$$

In a series circuit the current is everywhere the same. Hence, from equation (28-35), using the impedance across the voltmeter, we have

$$5 = \frac{V_1}{\sqrt{5^2 + 12^2}} \quad \text{or} \quad V_1 = 5 \times 13 = 65 \text{ volts},$$

$$5 = \frac{V_2}{\sqrt{11^2 + 0}} \quad \text{or} \quad V_2 = 5 \times 11 = 55 \text{ volts}.$$

EXAMPLE 5: From the previous example determine the phase difference between V and I in different parts of the circuit, and hence show how 65 volts can combine with 55 volts to give 100 volts.

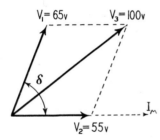

Solution:

$$\tan \delta_1 = \frac{L_1\omega}{R_1} = \tfrac{12}{5}; \quad \cos \delta_1 = \frac{5}{\sqrt{5^2 + 12^2}} = \tfrac{5}{13}. \quad *$$

Here δ_1 is the phase angle between the voltage V_1 and the current, with the current lagging, and $\tan \delta_2 = 0/R_2 = 0$, i.e., the voltage V_2 is in phase with the current.

Remember that δ_1 was the lag of the current behind the potential difference and hence is the lead of the potential difference over the current.

> * If $\tan \theta = a/b$, then $\sin \theta = a/\sqrt{a^2 + b^2}$ and $\cos \theta = b/\sqrt{a^2 + b^2}$. This may be verified by noting that
>
> $$\sin^2 \theta + \cos^2 \theta = 1 = \frac{a^2}{a^2 + b^2} + \frac{b^2}{a^2 + b^2}$$
>
> and that
>
> $$\tan \theta = \frac{\sin \theta}{\cos \theta} = \frac{a/\sqrt{a^2 + b^2}}{b/\sqrt{a^2 + b^2}} = \frac{a}{b}.$$

Using the cosine formula, we get

$$V_3 = \sqrt{V_1^2 + V_2^2 + 2V_1V_2 \cos \delta_1}$$

or

$$V_3 = \sqrt{55^2 + 65^2 + 2(55)(65)(\tfrac{5}{13})}$$
$$= \sqrt{10,000} = 100 \text{ volts}.$$

EXAMPLE 6: The resonance curves of Fig. 28-13 were plotted for the following data: $E = 1$ volt, $R = 2.5$ ohms, $L = 0.2$ henry, $C = 0.127$ μf. Compute (*a*) the frequency for resonance, (*b*) the current at resonance, and (*c*) the voltage across the capacitor at resonance.

Solution:

$$f = \frac{1}{2\pi \sqrt{LC}} = \frac{10^3}{2 \times 3.14 \sqrt{0.0254}} \quad (a)$$
$$= 10^3 \text{ cycle/sec},$$

$$I = \frac{E}{Z} = \frac{1}{2.5} = 0.4 \text{ amp}, \quad (b)$$

$$V_c = \frac{I}{C\omega} = \frac{0.4}{0.127 \times 10^{-6} \times 2\pi \times 10^3} \quad (c)$$
$$= 500 \text{ volts}.$$

This represents a 500-volt, 1,000 cycles/sec voltage source for very small current drain applications.

EXAMPLE 7: Given the following series circuit, compute the readings of V_1 and V_2 and check their sum against V_3, where $L = 1/20\pi$ henry, $C = 1/240\pi$ f, $R = 3$ ohms, and $V_3 = 100$ volts, 60 cycles.

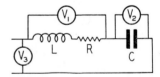

Solution:

$$L\omega = \frac{1}{20\pi} \times 2\pi60 = 6 \text{ ohms (inductor reactance)}$$

$$\frac{1}{\omega C} = \frac{1}{2\pi60 \times 1/240\pi}$$
$$= 2 \text{ ohms (capacitor reactance)};$$

therefore,

$$I = \frac{100}{\sqrt{3^2 + (6-2)^2}} = 20 \text{ amp (resultant current)}$$

and

$$20 = \frac{V_1}{\sqrt{3^2 + 6^2}} \quad \text{or} \quad V_1 = 20\sqrt{45} = 60\sqrt{5}$$
$$= 134.16 \text{ volts}.$$

Also,

$$20 = \frac{V_2}{\sqrt{0 + (0-2)^2}} \quad \text{or} \quad V_2 = 20 \times 2 = 40 \text{ volts}.$$

It follows that

$$\tan \delta_1 = \frac{6-0}{3} = \frac{2}{1},$$

where δ_1 is the phase angle between the voltage V_1 and the current, with the current lagging, and that

$$\cos \delta_1 = \frac{1}{\sqrt{5}} \quad \text{and} \quad \sin \delta_1 = \frac{2}{\sqrt{5}}.$$

Also,

$$\tan \delta_2 = \frac{0-2}{0} = -\infty.$$

Therefore, $\delta_2 = -90°$, the phase angle between the voltage V_2 and the current, with the current leading.

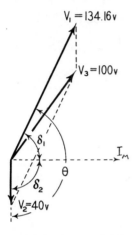

Applying the cosine formula, we obtain

$$V_3 = \sqrt{V_1^2 + V_2^2 + 2 V_1 V_2 \cos \theta}.$$

Since $\cos \theta = \cos (90 + \delta_1) = -\sin \delta_1$, it follows that

$$V_3 =$$

$$\sqrt{40^2 + (60\sqrt{5})^2 + 2 \times 40 \times 60\sqrt{5}\,(-2/\sqrt{5})}$$

$$= \sqrt{10,000} = 100 \text{ volts}.$$

EXAMPLE 8: For what frequency of alternating current would the greatest current exist in Example 7?

Solution: The resistance is assumed independent of the frequency, but $L\omega - 1/\omega C$ evidently depends upon ω. The greatest current will exist when $L\omega - 1/\omega C = 0$. Hence, in this example,

$$\frac{1}{20\pi} \times \omega - \frac{1}{\omega} \times \frac{1}{1/240\pi} = 0,$$

$$\frac{\omega}{20\pi} - \frac{240\pi}{\omega} = 0,$$

$$\omega^2 = 4,800\pi^2,$$

$$\omega = \sqrt{4,800}\,\pi \text{ rad/sec},$$

and

$$f = \frac{\omega}{2\pi} = \frac{1}{2}\sqrt{4,800} = 20\sqrt{3} \text{ cycles/sec}.$$

Then the current would be

$$I = \frac{100}{\sqrt{3^2 + (2\sqrt{3} - 2\sqrt{3})^2}} = 33.3 \text{ amp}.$$

28.11. Power and Power Factor

In order to calculate the power dissipated in the circuits we have discussed, we must multiply the impressed electromotive force by the corresponding current i set up in the circuit. If the electromotive force varies sinusoidally in accordance with the relation

$$e = E_M \sin \omega t,$$

with $\omega = 2\pi f$, and the current i is likewise varying sinusoidally, with the same frequency, f, but differing in phase by an angle ϕ, so that

$$i = I_M \sin (\omega t + \phi),$$

then the instantaneous power, p, is

$$p = I_M E_M \sin \omega t \sin (\omega t + \phi).$$

On expanding $\sin (\omega t + \phi)$, we get

$$p = I_M E_M [\sin^2 \omega t \cos \phi + \tfrac{1}{2} \sin (2\omega t) \sin \phi].$$

If we average this expression over one cycle, the $\overline{\sin^2}$ term will be equal to $\frac{1}{2}$, while the $\overline{\sin (2\omega t)}$ is zero. Hence,

$$P = \frac{I_M E_M \cos \phi}{2}, \qquad (28\text{-}40)$$

and, since $I = I_M/\sqrt{2}$ and $E = E_M/\sqrt{2}$, as read by alternating-current meters,

$$P = IE \cos \phi. \qquad (28\text{-}41)$$

From this we see that the product of the ammeter and the voltmeter readings will not give the true average power except at resonance, when $\phi = 0$ and $\cos \phi = 1$, i.e., when the current is in phase with the voltage. The angle ϕ is, of course, the phase angle between applied electromotive force and current. The number $\cos \phi$, by definition, is the *power factor*. The product IE is called the *volt-amperes*. Electrical machinery for alternating current is usually rated in kilovolt-amperes (kva).

The power factor may be determined in the laboratory by measuring, with a suitable alternating-current wattmeter (chap. 29), the true power dis- **393** sipated and dividing this by the product of the voltmeter and the ammeter readings. Thus

$$P = EI \cos \phi \; ,$$

or

$$\cos \phi = \frac{P}{EI} . \qquad (28\text{-}42)$$

28.12. The Alternating-Current Transformer

We saw in chapter 20 the necessity of transmitting electrical energy at high potential differences if the resistance of the transmission line is at all large. It is difficult to construct generators which generate high potential differences, and it is very dangerous to use electrical circuits in which a high potential difference is maintained. By the use of alternating current it is rather simple to generate electrical energy at whatever potential difference is most convenient, and then to "step up" that potential difference to an amount which assures efficient transmission and, finally, to "step down" the potential difference to a safe amount for use. The apparatus which thus efficiently changes the potential difference of an alternating-current line is a TRANSFORMER.

The modern iron-core transformer consists essentially of an iron core wound with two separate coils, a primary of N_p turns and a secondary of N_s turns, which are carefully insulated from each other and from the core, as shown in Fig. 28-14. Let us first assume that a varying voltage

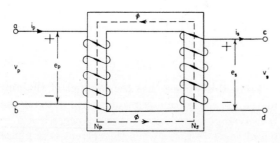

FIG. 28-14.—Simple transformer

of instantaneous value v_p and of constant amplitude and frequency is applied to the primary and that there is no load across the secondary. A small steady-state current i_p (but not a direct current) is established in the primary. The magnetomotive force, $N_p i_p$ (see § 24.7, eq. [24-22]), then establishes an alternating flux ϕ in the magnetic circuit,

as shown. This flux, in turn, induces a counter electromotive force in the primary equal to

$$e_p = N_p \frac{d\phi}{dt} .$$

By Lenz's law, the polarity of e_p with reference to v_p is that shown in Fig. 28-14. Thus, at some instant of time,

$$v_p = e_p + i_p r_p \; ,$$

that is to say, the counter electromotive force and potential drop through the resistance of the primary must balance. In the ideal transformer, which is closely approached by the modern practical iron-core transformer, the resistance of the primary is negligibly small, so that

$$v_p = e_p = N_p \frac{d\phi}{dt} .$$

The primary current adjusts itself so that the back electromotive force is nearly exactly the impressed electromotive force.

For the case of the ideal transformer, it is further assumed that the entire magnetic flux links all turns of both windings and that the permeability is so high that a negligibly small magnetomotive force will produce the necessary flux ϕ. Since the flux ϕ links the secondary, an electromotive force is induced, of value

$$e_s = N_s \frac{d\phi}{dt} ,$$

with the polarity as shown. Thus

$$\frac{e_p}{e_s} = \frac{N_p (d\phi/dt)}{N_s (d\phi/dt)} = \frac{N_p}{N_s} , \qquad (28\text{-}43)$$

and the ratio of the primary to the secondary potential can be controlled by the proper choice of the number of turns on the two coils. In Fig. 28-14, the windings are traced from a to b and from c to d, so that the core is encircled in the same direction by the windings. In this case the primary and secondary voltages are in the same phase, that is, at any instant, the a terminal is of the same relative polarity as the c terminal.

If a load is placed across the secondary, a current, i_s, will be established, as shown in Fig. 28-14, and therefore a magnetomotive force, $-N_s i_s$, in the magnetic circuit. The minus sign indicates that this magnetomotive force opposes that due to the primary current, $N_p i_p$. Since we assumed that, for an ideal transformer, there is no flux lost in the core and that a negligibly small magnetomotive force will produce the necessary flux ϕ, the algebraic sum of the magnetomotive forces is zero:

$$N_p i_p + N_s i_s = 0 \; ,$$

or

$$\frac{i_p}{i_s} = -\frac{N_s}{N_p} . \qquad (28\text{-}44)$$

By combining equations (28-43) and (28-44), we obtain

$$e_p i_p = -\, e_s i_s \; ;$$

that is, the instantaneous powers in the primary and secondary sides of the ideal transformer are equal, and, furthermore, when the secondary is delivering power to the load, the primary is absorbing an equivalent amount from the source generator.

It must be understood that these statements are only approximate. We have made no allowance for resistance of the coils or for hysteresis losses in the iron. However, in view of the fact that the efficiency of modern transformers may approach 99 per cent, it can be seen that the approximation is not a bad one. Nevertheless, the heat losses in very large transformers are so great that care must be taken to avoid overheating. By immersing the coils in circulating oil and mounting them in boxes with corrugated surfaces to increase radiation losses, their temperature may be kept down to safe levels. Overloaded transformers, however, occasionally burn out.

We inferred previously that connecting a load of impedance Z_s to the secondary of the transformer causes the source generator in the primary circuit to furnish current in the same manner that a load across the generator itself would draw current. However, the value of the current I_s through Z_s, when across the secondary of the transformer, is, in general, changed. In other words, the transformer placed between the generator and the load has effectively changed the impedance of the load as sensed by the generator. If V_s is the effective voltage of the secondary, Z_s the impedance of the load, and I_s the effective current through the load, then

$$Z_s = \frac{V_s}{I_s} ,$$

by definition of impedance. If I_p and V_p are the current and voltage of the generator with the transformer placed between it and the load Z_s, then the load across the generator that would absorb the same energy as Z_s across the secondary of the transformer is

$$Z_p = \frac{V_p}{I_p} ,$$

so that

$$\frac{Z_p}{Z_s} = \left(\frac{V_p}{V_s}\right)\left(\frac{I_s}{I_p}\right) ;$$

but

$$\frac{V_p}{V_s} = \frac{N_p}{N_s} ; \qquad \frac{I_s}{I_p} = \frac{N_p}{N_s} ,$$

and therefore

$$\frac{Z_p}{Z_s} = \left(\frac{N_p}{N_s}\right)^2 . \qquad (28\text{-}45)$$

We conclude from the result (eq. [28-45]) that the transformer may be used to match the impedance of a load to that of a generator for maximum power transfer to a load. This is a very useful property of transformers and is extensively employed in electronic circuits as an impedance-matching device.

Another common type of transformer that is used for voltage step-up or step-down is the autotransformer. This transformer does not use a secondary circuit that is electrically separated from the primary, as in the conventional transformer. The windings of the primary and secondary are around an iron core but are electrically connected, as shown in Fig. 28-15. In this figure (a)

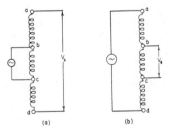

Fig. 28-15.—Circuit diagram of an autotransformer

is a step-up voltage transformer and (b) is a step-down transformer. The magnetic flux, ϕ, threads both the primary and the secondary, and the voltage ratio of V_s to V_p is proportional to N_s and N_p,

$$V_s = V_p \left(\frac{N_s}{N_p}\right) .$$

Thus in Fig. 28-15 (a), if the number of turns between a and d is 500 and between b and c is 250, then $V_s = 2V_p$. Similarly, if the connections are made as in Fig. 28-15 (b), then $V_s = \frac{1}{2}V_p$. For this case the load current I_s will divide at points b and c, and only a part will circulate through the secondary. For comparable ratings the I^2R losses of the autotransformer are thus less than those of the conventional type, where all the load current circulates through the secondary winding. One disadvantage of the autotransformer as a power transformer is that the entire primary voltage may be placed across the load if the secondary winding is accidentally opened.

EXAMPLE 9: The impedance of an alternating-current generator of value 10 ohms is to be matched to a load of impedance 490 ohms by means of a transformer. What ratio of turns should be used?

Solution:

$$Z_p = 10 \text{ ohms} ; \qquad Z_s = 490 \text{ ohms};$$

$$\frac{Z_p}{Z_s} = \left(\frac{N_p}{N_s}\right)^2 ; \qquad \frac{N_p}{N_s} = \frac{1}{7} .$$

28.13. Alternating-Current Motors

We have seen in § 27.11 that small ordinary direct-current motors may be used on alternating current. Besides these, there are several types of motor designed to run on alternating current only. The most simple of these to understand is the SYNCHRONOUS MOTOR. Suppose an alternating current were passed through a coil rotating in a uniform magnetic field (Fig. 27-1) and that the frequency of rotation of the coil were the same as the frequency of the alternating current. Then on each wire, as it passed under a pole of the magnet, there would be a thrust in the direction of rotation, and the coil would rotate as a motor. Such a motor is not "self-starting" and must be "separately excited" with direct current for the field magnets. Synchronous motors, particularly in large sizes, are in frequent use because of their excellent characteristics and efficient performance.

The most common type of alternating-current motor is the INDUCTION MOTOR. If we use two different alternating-current circuits having current of the same frequency but with 90° difference in phase and connect them to four coils, as indicated

in Fig. 28-16, then the changes in the two currents will result in changes in the polarity of the coils, as shown in Fig. 28-17. This results in a "rotating" magnetic field—even though the coils themselves do not rotate. If any sort of conductor is mounted

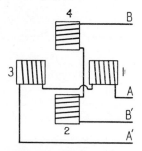

FIG. 28-16.—Field coils of a four-pole induction motor

on an axle and inserted into this rotating field, the moving field will induce electromotive force in the conductor; and, by having the resistance of the conductor low, large currents will be produced. These currents, reacting with the magnetic field, cause the conductor to rotate with a speed some-

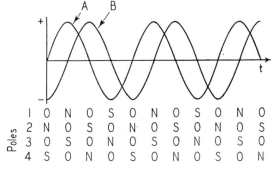

FIG. 28-17.—Illustrating the rotating magnetic field of the four-pole induction motor.

what less than the speed of rotation of the magnetic field.

The armature of an induction motor consists of a laminated core and heavy copper rods placed parallel to the axis of rotation and connected by

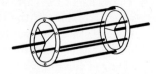

FIG. 28-18.—Copper conductors of a "squirrel-cage" armature of an induction motor.

heavy copper rings, as shown in Fig. 28-18. The iron laminations of the core are not drawn. Because of the shape of this rotor, it is called a "squirrel-cage" armature. In this motor there are no brushes or commutator or collector rings to wear out, and there is no chance for accidental sparking, with the resultant fire hazard. This motor is also self-starting. However, it requires a "two-phase" circuit and four line wires. A similar sort of motor could be run on "three-phase" circuits with either three or four line wires. Where induction motors are numerous, three-phase circuits are commonly used.

The induction motor will run on a single phase if it is once started, but it is not *self*-starting, as a two-phase motor is. There are several ways of overcoming this difficulty, particularly for small motors. We have seen that self-induction causes the current to lag behind the applied potential difference. A single-phase line may be split into two parts having different self-inductances. Then the two will not be in phase and, although not 90° apart, may still be sufficiently separated to cause the motor to start. In Fig. 28-19 the current in

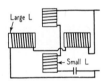

Fig. 28-19.—Split-phase induction motor

the horizontal coils will lag more than the current in the vertical coils. Such motors are sometimes called SPLIT-PHASE INDUCTION MOTORS.

Another method for starting single-phase induction motors is to start them as commutator motors. When they reach normal speed, a centripetal device built into them causes a weight, normally held close to the axle by a spring, to fly out. As the weight moves out, it lifts the brushes from the commutator, and then the motor continues to operate as an induction motor. Such motors give out a distinct "click" just after starting and just before they come to rest. This sound is caused by the motion of the weight and the brushes. Such motors are not free from sparking.

A more common type of single-phase motor is the *capacitor-start* induction motor. A capacitor is placed in series with the small inductance of Fig. 28-19. The presence of the capacitor introduces a large phase difference between the current in small inductance and that in large inductance, thus providing the necessary torque for starting. In a

properly designed motor the starting torque may be $3\frac{1}{2}$ times the full-load torque at all speeds from zero to 70 per cent of synchronous speed. At about 75 per cent synchronous speed a centripetal device disconnects the winding, *small L*, from the line (Fig. 28-19), and the motor runs as a single-phase induction motor.

28.14. Alternating-Current Parallel Circuits

In § 28.9 we considered at length the problem of a series circuit and investigated the conditions necessary for resonance. In this section we shall concern ourselves with an analysis of the *parallel circuit*, shown in Fig. 28-20 (*a*).

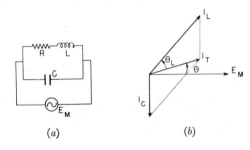

(*a*)   (*b*)

Fig. 28-20.—(*a*) Alternating-current parallel circuit; (*b*) vector representation of currents in the parallel circuit.

The effects of varying the frequency of the applied electromotive force across this circuit are widely different from the effects in a series circuit. As we shall presently see, as resonance is approached, the parallel circuit offers large impedance to the generator and the current becomes small, whereas in the series circuit the impedance becomes small and the current large as resonance is reached. Further, in the series circuit the *same current* passed through all the circuit elements, and the voltages across the elements were different. It is clear from the diagram that in the parallel circuit the same voltage is applied across each branch, but the current through each branch is different in general.

If E is the effective value of the applied electromotive force, then the current through the upper branch of the circuit is

$$I_L = \frac{E}{Z_L} = \frac{E}{\sqrt{R^2 + (L\omega)^2}}, \qquad (28\text{-}46)$$

397

where Z_L is the impedance of the branch. From our previous considerations we know that the current I_L lags behind the applied electromotive force by an angle less than 90°, since both resistance and inductance are present. This phase angle is given by

also

$$\left.\begin{array}{l} \tan\theta_L = \dfrac{L\omega}{R}; \\[2mm] \cos\theta_L = \dfrac{R}{\sqrt{R^2+(L\omega)^2}}; \\[2mm] \sin\theta_L = \dfrac{L\omega}{\sqrt{R^2+(L\omega)^2}}. \end{array}\right\} \quad (28\text{-}47)$$

In the lower branch the current, I_c, leads the applied electromotive force by 90°, since no resistance is present here. This current is given by

$$I_c = \frac{E}{Z_c} = E\omega C. \qquad (28\text{-}48)$$

The total current, I_T, in the network is obtained by vector addition of I_L and I_c, as shown in Fig. 28-20 (b). From this figure it is clear that the total current, I_T, and the applied electromotive force, E, are not in phase. The phase angle, θ, may be computed with the aid of this diagram,

$$\tan\theta = \frac{I_L\sin\theta_L - I_c}{I_L\cos\theta_L}. \qquad (28\text{-}49)$$

At resonance, $\theta = 0$, and the total current I_T is in phase with E. At resonance, then,

$$I_L\sin\theta_L = I_c. \qquad (28\text{-}50)$$

Substituting from above equations, we obtain, for resonance,

$$\frac{E}{\sqrt{R^2+(L\omega)^2}} \times \frac{L\omega}{\sqrt{R^2+(L\omega)^2}} = E\omega C,$$

or

$$\omega^2 = \frac{1}{LC} - \frac{R^2}{L^2}, \qquad \omega = \sqrt{\frac{1}{LC} - \frac{R^2}{L^2}},$$

$$f = \frac{1}{2\pi}\sqrt{\frac{1}{LC} - \frac{R^2}{L^2}}. \qquad (28\text{-}51)$$

Equation (28-51) represents the resonant frequency.

The equivalent impedance of this circuit is the ratio E/I_T, which, for the case of resonance, is

$$Z = \frac{E}{I_T} = \frac{E}{I_L\cos\theta_L} = \frac{R^2+(L\omega)^2}{R}$$

or

$$Z = R + \frac{(L\omega)^2}{R}. \qquad (28\text{-}52)$$

If R is small compared to L, the frequency at resonance is $1/2\pi\sqrt{LC}$, the same value found for a series circuit. However, a comparison of equations (28-35) and (28-52) reveals an important difference in these two circuits. As R approaches zero in the series circuit, the impedance approaches zero, and the current infinitely large values. In the parallel circuit, however, the impedance approaches infinitely large values and the in-phase current goes to zero.

Figure 28-21 illustrates the behavior of a parallel or "antiresonant" circuit with the same values of

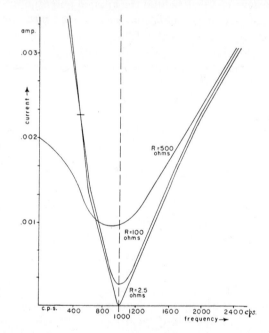

Fig. 28-21.—Antiresonance curves for different values of resistance ($L = 0.2$ henry; $C = 0.125$ μf; $E = 1$ volt).

L, C, and R as those used for the series circuit of Fig. 28-13. It is evident that here, too, the sharpness or selectivity of the antiresonant curves is markedly decreased with increasing resistance in the circuit. Also the resonant frequency is dependent on the resistance, decreasing with increasing resistance. It is possible for the circuit to lose its resonant characteristics for a sufficiently large resistance. According to equation (28-51), this resistance is given by $R = \sqrt{L/C}$ ohms.

EXAMPLE 10: Given the parallel circuit of Fig. 28-20 (a) with $L = 0.2$ henry, $C = 0.127$ μf, $R = 2.5$ ohms, and $E = 1$ volt, compute (*a*) the resonant frequency; (*b*) the current through the generator at resonance; (*c*) the resistance for which resonance is lost.

Solution:

$(a)\ f = \frac{1}{2\pi}\sqrt{\frac{10^6}{0.0254} - \frac{6.25}{0.04}} = 1{,}000$ cycles/sec ;

$(b)\ I = \frac{E}{Z} = \frac{1}{R + (L\omega)^2/R} =$

$\frac{1}{2.5 + (0.2 \times 2\pi \times 10^3)^2/2.5} \cong 1.6 \times 10^{-6}$ amp ;

$(c)\ R = \sqrt{\frac{L}{C}} = \sqrt{\frac{0.2 \times 10^6}{0.127}} = 1.26 \times 10^3$ ohms .

Both series and parallel circuits are widely used in electronics, where the vacuum tube is the source of power. When it is necessary to match low-impedance-power sources, the series circuit is used; for high-impedance-power sources, the parallel circuit is employed.

28.15. Résumé

The following is a brief listing of the definitions, principles, and theories, given in this chapter, which you should know:

The characteristics of alternating voltage and current; maximum and instantaneous values, frequency, and phase

Definition of effective voltage and current

Voltage and current relationships in a circuit containing resistance or inductance or capacitance alone and when these are connected in series

The electromotive-force and impedance triangles

The concept of resonance in a series-connected circuit

The transformer theory

The alternating-current induction motor

The theory of parallel resonance circuits

Queries

1. What reasons can you give for choosing the heating effect of a current to define 1 amp? Can one use chemical effects? Why?

2. If the current did not vary as a sine curve, would the formulas for effective current, voltage, etc., be modified? Explain in qualitative terms.

3. If, in Fig. 28-10, the two plates of the parallel-plate capacitor C were brought into electrical contact, would it be permissible to place $C = 0$ in equation (28-33)? Explain.

4. How much energy is expended in a circuit containing only a capacitance? In Fig. 28-9 what becomes of the energy represented by the positive area? In the purely inductive circuit of Fig. 28-6 what becomes of the energy represented by the positive area?

5. In the circuit in Fig. 28-10 why is there a net dissipation of energy? Where does this energy go?

6. What limits the value of the current in the primary of the transformer of Fig. 28-14?

7. How does the transformer adjust itself automatically to the load in the secondary?

8. In the induction motor of Fig. 28-17 will the motor rotate as fast as the rotating magnetic field? Explain.

9. Why is inductance sometimes called "electrical inertia"?

10. Why is capacitance sometimes compared to the inverse of a mechanical elastic constant?

11. Using the units, show that $L\omega$ and $1/C\omega$ are equivalent to ohms.

Practice Problems

1. Two generators in series give rise to maximum electromotive forces of 160 volts, each of 60 cycles. If the two are out of phase by 90°, with the second lagging behind the first, what is the resultant electromotive force? What is the phase angle between the resultant and the maximum electromotive force of generator No. 1? Ans.: 226 volts.

2. Find the resultant maximum electromotive force of two generators connected in series if one gives rise to an effective electromotive force of 100 volts and the other to 50 volts, the smaller electromotive force lagging 60° behind the other.

3. A choke coil of 5 henrys inductance and 2 ohms resistance is connected across 110-volt, 60-cycle mains. What current passes through the coil? How much power is dissipated? Ans.: 0.0583 amp; 0.00683 watt.

4. The choke coil of problem 3 is replaced by a pure **399**

resistance of such a value that the current remains the same. Compute the value of this resistance and the power dissipated in the resistance.

5. A capacitor of 0.0001-μf capacity is placed across a 2-volt, 1,000-cycle oscillator. What current passes through the circuit? Ans.: 1.26×10^{-6} amp, effective value.

6. If in an inductive coil of negligible resistance 15 amp pass when connected across a 220-volt, 40-cycle line, compute the reactance and the inductance of the coil.

7. What capacitance connected in series with a 0.001-henry inductance will produce resonance at a frequency of 60 cycles per second? Ans.: 7,040 μf.

8. A sinusoidal current alternates 60 times per second. If the current starts at zero when the time is zero, compute the value of the current $\frac{1}{16}$, $\frac{1}{8}$, and $\frac{1}{4}$ of a period later. The maximum value of the current is 15 amp. What is the effective value of the current?

9. A series circuit consists of a coil of 0.001 millihenry and a capacitor of 6 $\mu\mu$f. At what frequency will this circuit resonate? Ans.: 6.48×10^{7} cycles/sec.

10. An electromotive force of 10^{-4} volt, 10^{6} cycles/sec, is impressed across a series circuit consisting of a resistance of 2 ohms, an inductance of 0.2 millihenry, and a variable capacitor. What must be the capacitance of the capacitor for resonance? What current passes around the circuit at resonance?

11. A 50-volt, 1,000-cycle generator is placed in a circuit containing a 100-ohm resistance, a 40-millihenry inductance, and a capacitor of 0.12-μf capacitance. Compute the current in the series circuit and determine the phase angle between the applied electromotive force and the current. How much power is dissipated? Ans.: 0.0463 amp; 0.219 watt.

12. If, in problem 11, the inductance is removed, what it the value of the current and the phase angle between the impressed electromotive force and the resulting current?

13. In problem 11 if the capacitor is removed, what is the value of the current and the phase angle? Ans.: 0.185 amp; $\phi = 68°20'$.

14. Given the series circuit shown in the accompanying figure, where $R_1 = 5$ ohms, $L = 1/2\pi$ henry, $C = 1/360\pi$ farad, and $R_2 = 3$ ohms, what do the voltmeters and ammeters read?

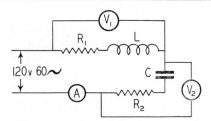

What is the power factor of the circuit? Determine the phase angle between V_1 and the current, V_2 and the current, and also the phase angle between the applied electromotive force and the current.

15. A capacitor having a capacitance of 50 μf has 500 volts at 60 cycles/sec impressed upon it. What is the current? Ans.: 9.43 amp.

16. A series circuit consists of an inductance of 0.2 henry and a resistance of 10 ohms. What is the power factor? What value of capacity must be inserted in the circuit to produce a power factor of 0.85? The frequency is 60 cycles/sec.

17. If the frequency of the generator of problem 15 is 25 cycles/sec, calculate the current.

18. A generator whose internal resistance is 1 ohm furnishes an electromotive force of 10 volts at a frequency of $6,000\pi$ cycles/sec. Design a series resonant circuit so that a voltage of 1,000 volts may be tapped across the capacitor. *Hint:* There is more than one circuit possible.

19. Three identical inductances in series each of 0.1 henry and 5 ohms have a 110-volt, 60-cycle/sec voltage impressed across them. What is the current through the inductance? Ans.: 0.65 amp.

20. In the parallel circuit of Fig. 28-20 (a), $L = 0.2$ henry, $C = 0.127$ μf, and $R = 1,000$ ohms. Calculate the resonant frequency and the current through the generator for $E = 10$ volts.

21. A 220-volt, 60-cycles/sec generator furnishes power to a load with a power factor of 0.8. If the load absorbs 7.5 kw of power, what current circulates in the load? Ans.: 42.6 amp.

22. A circuit of very high impedance, $Z = 200,000$ ohms, is to be coupled to a low-impedance load of 100 ohms through a transformer with 1,000 turns on the primary. How many turns must there be on the secondary?

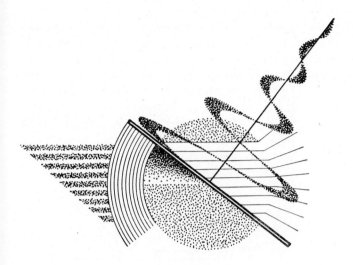

CHAPTER 29

Electrical Measuring Instruments
for Varying Currents

29.1. Alternating-Current Ammeters and Voltmeters; Dynamometers

Alternating-current ammeters and voltmeters, as used in electrical measurements, are calibrated to indicate effective values (cf. § 28.4). The DY-NAMOMETER-type ammeter, illustrated in Fig. 29-1, consists of a set of fixed coils, *1* and *2*, in

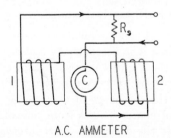

A.C. AMMETER

FIG. 29-1.—An alternating-current ammeter

series with a movable coil, *C*. A pointer attached to *C* is adjusted to read zero when the plane of *C* is parallel to the field set up by coils *1* and *2*. When a current is sent through the system, a torque, *L*, acts on the coil *C*, whose value is proportional to the field and the current in the coil, that is, $L \sim BI$. Since, however, the same current gives rise to *B*, itself proportional to *I* (if there is no iron present), it follows that

$$L \sim I^2 .$$

When the current reverses in direction, the current through *C* and also the field, *B*, reverse in direction, leaving the direction of the torque unchanged. The restoring torque is supplied by the torsion in the suspension. The angle θ through which the coil turns is then proportional to I^2, provided that the material of the suspension follows Hooke's law. Hence the deflections of the pointer are proportional to the square of the current through the instrument. This is a decided disadvantage, since it leads to a nonuniform scale, as illustrated in Fig. 29-2. The range of the instru-

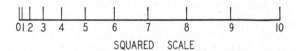

SQUARED SCALE

FIG. 29-2.—A squared scale

ment is determined by the value of the shunt, R_s (Fig. 29-1). By inserting a very high resistance, *R*, in series with the coil, the instrument becomes an alternating-current voltmeter of the dynamometer type (Fig. 29-3).

In the design of alternating-current instruments the self-inductance, as well as the pure resistance, must be considered. We need to distinguish between the alternating current of commercial frequencies (about 60 cycles/sec) and that of "radio"

401

frequencies, which may go up to millions of cycles per second. The instruments we have described can be designed to give reasonably accurate readings on all commercial frequencies, but they cannot be used at radio frequencies.

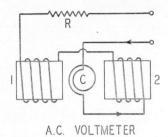

A.C. VOLTMETER

FIG. 29-3.—An alternating-current voltmeter

29.2. Hot-Wire Instruments

The frequency of alternation does not appear in the relation between average power and effective current that was used in equation (28-11). Therefore, this relation,

$$P = RI^2_{\text{eff}} ,$$

may be used as a basis for measuring instruments. These are of two types. In one, the current is passed through a single, straight, fine wire; and the change in length of the wire, owing to its expansion from the heat generated, is observed by a series of multiplying levers moving a pointer over a scale calibrated to indicate the current in the wire.

A second and more refined type of hot-wire instrument has one terminal of a thermocouple welded to the center of the fine heater wire or, in some instruments, held at a fixed short distance from the heater wire. The thermocouple is connected to an ordinary direct-current D'Arsonval galvanometer with a scale calibrated to indicate the current in the heater wire. Hot-wire instruments have scales divided roughly proportional to the square of the current.

The types of alternating-current meters we have described should work on direct current. However, a more reliable measurement is made if one reverses the terminals of the instrument and takes the average of the two readings, which frequently differ by a few per cent.

29.3. Wattmeters

The moving-coil alternating-current meters we have described give readings proportional to the

product of the currents in two coils—fixed-field coil and moving-armature coil. By connecting one of these coils as an ammeter, in series with the line, and connecting the other as a voltmeter, connected across the line, the resulting reading is proportional to the product of the two. The product of the current and the potential difference is the power; hence such an arrangement measures the power consumption of the device across which the potential leads are connected (Fig. 29-4).

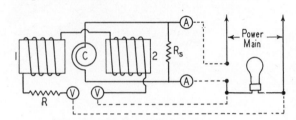

A.C. WATTMETER MEASURING POWER OF A LAMP

FIG. 29-4.—An alternating-current wattmeter for measuring power of a lamp.

Since there is some power used in the instrument, the manufacturer has calibrated the instrument for a particular set of connections. Such instruments will give more accurate readings and there will be less strain on the insulation if they are always connected exactly according to the maker's instructions. They operate on either alternating or direct current.

29.4. The Ballistic Galvanometer

In electrical measurements, occasions frequently arise when it is necessary to measure the total charge of electricity that passes around a circuit due to a current of very short duration. For example, in the discharge of a capacitor a definite quantity of charge passes from one plate of the capacitor to the other in a short interval of time. In the case of a coil that is quickly turned in a magnetic field, a definite charge passes around the circuit, in accordance with equation (23.16). If we could measure this accurately, it would be possible to determine magnetic flux and flux densities with some degree of precision.

An instrument designed to measure quantity of charge passing through it, in terms of the amplitude of swing of a coil due to a momentary current, is called a BALLISTIC GALVANOMETER. This instrument is a type of D'Arsonval galvanometer whose period of oscillation is made as large as possible, consistent with sensitivity. This is done either by loading the coil of a D'Arsonval galvanometer, to increase its moment of inertia, or by making the coil very wide, as shown in Fig. 29-5. If the discharge is extremely short, an ordinary D'Arsonval galvanometer can be used ballistically with but small error.

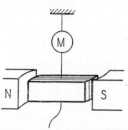

Fig. 29-5.—Diagram of ballistic galvanometer

We wish to determine how such an instrument can be used to measure charge. In the discussion that follows, we assume that the time of discharge is negligibly small, compared to the period of oscillation of the galvanometer.

In a time Δt, a current of I amp passes through the coil. The *impulsive torque* that acts on the coil is, then, $L \Delta t$, where L is the torque on a coil of N turns an area A, carrying current I in a magnetic field, B (see eq. [21-6]); i.e.,

$$L = BINA ,$$

and

$$L \Delta t = I \Delta t B N A = qB N A , \qquad (29\text{-}1)$$

since the charge q is equal to $I \Delta t$. Mechanically, this impulsive torque is measured by taking the difference between the final and the initial angular momenta. Since the coil is assumed to be at rest initially, it follows that

$$L \Delta t = I_M \omega ,$$

where I_M is the moment of inertia of the coil about the suspension as an axis and ω is the angular velocity of the coil. Therefore,

$$qB N A = I_M \omega . \qquad (29\text{-}2)$$

The rotating coil possesses kinetic energy of amount $\frac{1}{2} I_M \omega^2$ and is brought to rest by doing work in twisting the suspension through an angle θ_0 rad. The work done is $\frac{1}{2} L_0 \theta_0^2$, where L_0 is the moment of torsion of the suspension. By the law of conservation of energy,

$$\tfrac{1}{2} I_M \omega^2 = \tfrac{1}{2} L_0 \theta_0^2 ,$$

or

$$\omega^2 = \frac{L_0}{I_M} \theta_0^2 .$$

By equation (29-2),

$$\omega^2 = \frac{q^2 B^2 N^2 A^2}{I_M^2} .$$

Therefore,

$$\frac{q^2 B^2 N^2 A^2}{I_M^2} = \frac{L_0 \theta_0^2}{I_M} ,$$

or

$$q^2 = \frac{L_0^2}{B^2 N^2 A^2} \left(\frac{I_M}{L_0} \right) \theta_0^2 . \qquad (29\text{-}3)$$

The half-period of oscillation of the suspended coil, treated as a torsional pendulum with no damping, is

$$T_{1/2} = \pi \sqrt{\frac{I_M}{L_0}} .$$

Substituting for I_M/L_0 in equation (29-3), we obtain, for q,

$$q = \frac{L_0 T_{1/2}}{\pi B N A} \theta_0 .$$

In equation (21-10), § 21.4, we called L_0/BNA the current sensitivity and K the number of amperes required to produce a deflection of 1 rad. Therefore,

$$q = \frac{K T_{1/2}}{\pi} \theta_0 . \qquad (29\text{-}4)$$

We see from equation (29-4) that the charge q in coulombs is directly proportional to the angle θ_0 through which the coil turns.

If we set

$$Q_0 = \frac{K T_{1/2}}{\pi} ,$$

then

$$q = Q_0 \theta_0 , \qquad (29\text{-}5)$$

where we define Q_0 as the number of coulombs that will deflect the coil of the ballistic galvanometer through 1 rad (no damping). From Fig. 21-13, $\theta_0 = d/2D$, where d is taken as the deflection in millimeters on a scale D mm from the mirror. Substituting in equation (29-5), we obtain

$$q = \frac{Q_0 d}{2 D}$$

or

$$q = q_0 d . \qquad (29\text{-}6)$$

Here q_0 is the "charge sensitivity" of the ballistic galvanometer, which is numerically equal to the **403** number of coulombs that will cause a deflection of 1 mm on a scale 1 m away.

The d of equation (29-6) assumes no damping. Actually, there is damping present in all galvanometers, owing to air friction, eddy currents, etc. To correct for these effects, we must multiply *the first observed swing* by a damping factor, $\sqrt{\rho}$, where ρ is the ratio of two successive swings of the coil. It is found experimentally that if a number of successive swings, $d_1, d_2, d_3, \ldots, d_n$, are observed, the ratios,

$$\frac{d_1}{d_2} = \frac{d_2}{d_3} = \frac{d_3}{d_4} = \frac{d_{n-1}}{d_n} = \rho,$$

constant to a good order of approximation. This yields

$$\frac{d_1}{d_n} = \rho^{n-1} \qquad \text{or} \qquad d_n = d_1 \rho^{1-n}.$$

Since ρ is the damping for the interval between two maxima, for an interval between a zero and a maximum the damping would be $\sqrt{\rho}$. If d_1 is the first observed swing, then equation (29-6) may be written as

$$q = q_0 d_1 \sqrt{\rho}. \qquad (29-7)$$

29.5. Use of the Ballistic Galvanometer

We may measure the capacitance of a capacitor by the use of equation (29-7). If a capacitor of C farads capacitance is charged to a potential of V volts, the charge of electricity stored on the plates is

$$Q = CV.$$

The charge Q may be measured by discharging the capacitor through a ballistic galvanometer:

$$CV = q_0 d_1 \sqrt{\rho},$$

or

$$C = \frac{q_0 d_1 \sqrt{\rho}}{V}. \qquad (29-8)$$

If q_0 were measured in coulombs per millimeter, and V in volts, equation (29-8) would give C in farads.

A second use for equation (29-7) is found in the determination of the horizontal component of the earth's magnetic field, B_h. If a coil of N turns and area A m² is arranged with its plane normal to B_h and then is rotated through 180° about an axis, OO' (Fig. 29-6), parallel to the vertical component,

a charge will pass through the ballistic galvanometer. The value of this charge is given by

$$Q = \frac{N(\phi_2 - \phi_1)}{R}, \qquad (23-16)$$

where Q is in coulombs; ϕ_2 and ϕ_1 are in webers; R is in ohms; and N is a numeric. The change in flux is equal to $2B_h A$, so that

$$B_h = \frac{QR}{2AN} \text{ webers/m}^2.$$

Using equation (29-7), we obtain

$$B_h = \frac{q_0 d_1 R \sqrt{\rho}}{2AN} \text{ webers/m}^2. \qquad (29-9)$$

The vertical component, B_V, may be obtained by rotating the coil about an axis parallel to B_h. Such

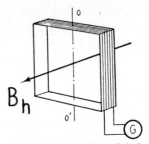

Fig. 29-6.—Diagram of earth inductor

an instrument as we have described is known as an EARTH INDUCTOR.

Another instrument for measuring magnetic flux—for example, in the Rowland ring of Fig. 24-3 —is the Grassot *flux-meter*. This is fundamentally a D'Arsonval galvanometer (see § 21.4), in which the mechanical damping and the restoring torque are deliberately made small. The flux-meter is generally connected to a small coil, called a *search-coil*, of N turns. In practice, this coil, of area A, is moved from a position where the flux density is B and the total flux linkage through the coil is NBA to a position where the linkage is zero. Because of the induced electromotive force and resulting current, a torque acts on the coil, rotating the moving element through θ degrees. It is shown in more advanced texts that this deflection is independent of the resistance of the circuit of the search-coil and of the time required to move the search-coil but depends only on the change in flux linkages through the search-coil. The scale, therefore, is calibrated to read weber-turns.

The flux density can be computed from a knowledge of the geometry of the search unit.

Since there are no restoring torques on the moving element, there is no definite zero position to the scale, and the deflection θ may fall anywhere on the scale. The change in flux linkages are thus equal to $k\Delta\theta$, where k is a calibrating constant of the meter and $\Delta\theta$ is the difference reading in weber-turns. In some instruments, however, provision is made by an auxiliary circuit to place the pointer at some convenient position.

29.6. Measurement of High Resistance; Charge and Discharge of a Capacitor

A method for determining the value of high resistance (of the order of 10 megohms) consists in

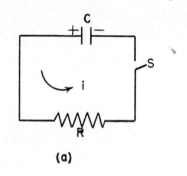

(a)

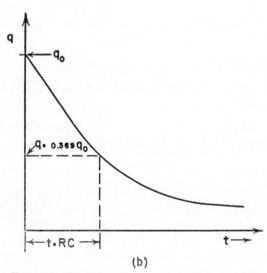

(b)

Fig. 29-7.—Discharge of a capacitor through a high resistance. (a) Basic circuit of C and R; (b) plot of $q = q_0 e^{-t/RC}$.

measuring the loss of charge of a capacitor shunted by the high resistance. The charge remaining on the capacitor after successive time intervals is measured by a calibrated ballistic galvanometer.

Let us assume that the capacitor of known value C and shown in Fig. 29.7 (a) has an initial charge

of q_0 with the switch S open. At any time t after the switch is closed, the algebraic sum of the iR drop and potential across the capacitor is zero:

$$iR + \frac{q}{C} = 0 ; \qquad (29\text{-}10)$$

and, since $i = dq/dt$, we may write equation (29-10) as

$$R\frac{dq}{dt} + \frac{q}{C} = 0$$

or

$$\frac{dq}{q} = -\frac{dt}{RC} ;$$

and, by integration,

$$\log_e \frac{q}{q_0} = -\frac{t}{RC}$$

or

$$q = q_0 e^{-t/RC} \text{ coulombs} , \qquad (29\text{-}11)$$

where q_0 is the initial charge for $t = 0$, and q is the charge remaining on the capacitor after a time of t sec. A plot of equation (29-11) is shown in Fig. 29-7 (b).

If we place $t = RC$ in equation (29-11), we obtain

$$q = \frac{q_0}{e} .$$

The product RC, called the *time constant*, thus represents the time required for the capacitor to discharge to $1/e(1/2.71 = 0.369)$ of its initial value. For $C = 1$ μf and $R = 10^7$ ohms, the time constant $RC = 10^{-6} \times 10^7 = 10$ sec. In 10 sec the capacitor has lost $(1 - 0.369)$, or 63 per cent of its charge. The student should compare this analysis with the growth of a current in an inductive circuit (§ 23.11).

Referring back to equation (29-11), we may rewrite this equation, solving for the unknown R,

$$R = \frac{t}{C \log_e (q_0/q)} \text{ ohms} . \qquad (29\text{-}12)$$

If t is in seconds, C in farads, and q in coulombs, then the units of R are ohms.

The circuit arrangement for measuring R is shown in Fig. 29-8. A known capacitor is shunted by the unknown resistance R. A two-position switch, S, provides for charging the capacitor by the battery, E. After charging the capacitor, the switch is quickly snapped to the ballistic galvanometer, and the initial charge q_0 is measured. A series of observations are then made at accurate **405** time intervals, measured after charging the capacitor each time. The value of R may then be computed from equation (29-12). For these measurements it is important that the capacitor be free from dielectric absorption and possess a very high internal resistance compared to R, for otherwise it is necessary to measure the resistance of the capacitor, which is in parallel with R. For practical reasons a time constant of about 10 sec is convenient.

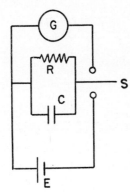

Fig. 29-8.—Experimental arrangement for the measurement of a high resistance.

The student may show that if a battery of electromotive force E volts is inserted into the circuit of Fig. 29-7 (*a*), then, with the capacitor initially uncharged, the rate of charge through the resistance R is given by

$$q = CE\,(1 - e^{-t/RC}) \text{ coulombs} , \quad (29\text{-}13)$$

and the charging current obtained by differentiating equation (29-13) is

$$i = \frac{E}{R}\,e^{-t/RC} \text{ amp} . \qquad (29\text{-}14)$$

The voltage v on the capacitor is given by

$$v = E\,(1 - e^{-t/RC}) \text{ volts} . \qquad (29\text{-}15)$$

A plot of equation (29-15) is given in Fig. 29-9.

29.7. Measurement of Wave Forms; the Cathode-Ray Oscilloscope

Perhaps the most versatile instrument in the electrical measurements laboratory is the cathode-ray oscilloscope. It is used to study slowly varying currents and voltages and transients too fast for mechanical devices; it is indispensable for the analysis of complex wave forms; it may be used

to study phase shifts in alternating-current circuits. When calibrated, it is useful as an alternating-current ammeter or voltmeter. The cathode-ray tube is also used to compare frequencies up to 10^8 cycles/sec.

The cathode-ray tube is shown diagrammatically in Fig. 29-10. Electrons emitted by the hot cathode, C, pass through the hole in the grid, G,

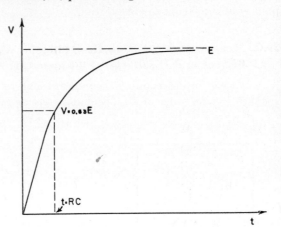

Fig. 29-9.—Graph of the rise of voltage across a capacitor; $v = E(1 - e^{-t/RC})$.

the number of electrons passing being controlled by the value of the negative potential on G. The electrons that emerge from G are accelerated by the positive potential of the first-focusing anode, A_1, and a number of them pass through a small hole, which thereby serves to confine the electrons to a small pencil beam making a small angle with the beam axis.

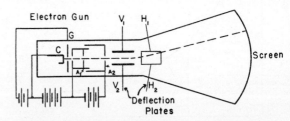

Fig. 29-10.—Diagram of a cathode-ray oscilloscope, showing the electron gun, deflection plates, and fluorescent screen.

The pencil of electrons is further accelerated by the positive potential on the second anode, A_2, and is then focused on the screen of the tube. This assemblage of cathode, grid, and focusing anodes constitutes the *electron gun*, which produces the high-velocity pencil of electrons. The brightness of the spot is controlled by the grid potential, G.

Near the end of the electron gun and positioned

around the electron beam is a set of plane electrodes, V_1, V_2, and H_1, H_2. If the vertical deflecting plates, V_1 and V_2, are charged, the electric field produced between the plates deflects the electrons upward or downward, depending upon the direction of the electric field between the plates. Similarly, the horizontal deflecting plates, H_1 and H_2, when charged, can deflect the electron beam to the left or right as we look down the tube. Since the electrons emerge from the electron gun with velocities in excess of 10^9 cm/sec, the time that the beam requires to traverse the tube is of the order of hundredths of a microsecond.

When used for the study of repetitive wave forms and transients, the horizontal plates H_1 and H_2 are connected to a source of potential that causes the electrons to be deflected linearly with time, moving the spot at a uniform rate across the screen and then snapping the spot quickly back to the starting point. The saw-tooth or sweeping voltage of Fig. 29-11 is used for this purpose. The

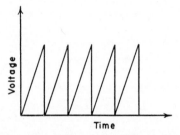

Fig. 29-11.—Saw-tooth or sweep voltage used for a linear time base.

method of generating and controlling the frequency of this sweeping voltage is described in (§ 30.11). Across the vertical plate are placed the voltages that are to be studied. For example, if a 60-cycle/sec alternating potential is across the vertical plates and the frequency of sweep properly adjusted, a sine curve is traced on the screen by the electron beam.

Frequency may be measured by the cathode-ray tube by noting the Lissajous figures (see § 8.16) obtained when a known frequency is applied to one set of the plates and the unknown frequency to the other. By referring to § 8.16, the student may determine the shapes of the Lissajous figures for frequency ratios of 2:1 and 3:2.

Several fluorescent materials are available for the screen of the cathode-ray tube, depending on the persistence of the trace desired. A high-persistence screen made of zinc sulphide is used for the

study of single transient phenomena. Such a screen may retain an image from a few seconds to several minutes. A very short-persistence screen with an image retention of only 25 μsec may be made of calcium tungstate.

29.8. Dimensions of Electrical and Magnetic Quantities

In § 2.11 we pointed out the usefulness of dimensional analysis as an aid in checking mechanical equations and formulas. The three physical dimensions used were those associated with the basic concepts of mass [M], length [L], and time [T]. The dimensions of all other mechanical quantities, such as velocity, force, work, etc., could be described in terms of [M], [L], and [T], as shown in Table 29-1.

TABLE 29-1

DIMENSIONS AND UNITS OF MECHANICAL QUANTITIES

Quantity	Dimensions	Units
1. Velocity, v = Distance/Time	[L] [T^{-1}]	M/sec
2. Force, f = Mass × acceleration	[M] [L] [T^{-2}]	Newtons
3. Work, w = Force × distance	[M] [L^2] [T^{-2}]	Joules

In describing thermal phenomena it was necessary to introduce the concept of temperature whose dimension is represented by [θ]. The dimensions of other thermal quantities could then be described in terms of [M], [L], [T], and [θ] (see Table 29-2).

TABLE 29-2

DIMENSIONS AND UNITS OF THERMAL QUANTITIES

Quantity	Dimensions	Units
4. Entropy, S = Heat/Temperature	[M] [L^2] [T^{-2}] [θ^{-1}]	Joules/degree
5. Stefan's constant, σ	[M] [T^{-3}] [θ^{-4}]	Watts/m^2-degree^4
6. Boltzmann's constant, k	[M] [L^2] [T^{-2}] [θ^{-1}]	Joules/degree

All the electrical and magnetic phenomena that we have described are due to electric charges, either at rest or in motion. Electric charge appears to be the fundamental concept in nature, perhaps more basic than mass itself. For this reason we shall introduce charge as a fifth basic quantity not **407** expressible in terms of mechanical or heat concepts. Its dimension is represented by [Q]. All physical quantities can be expressed in terms of these five fundamental quantities—mass, length, time, temperature, and electric charge—and therefore their dimensions in terms of [M], [L], [T], [θ], and [Q]. We shall list the more basic electric and magnetic quantities, with their dimensions and units as used in this text, in Table 29-3.

TABLE 29-3

DIMENSIONS AND UNITS OF ELECTRICAL AND
MAGNETIC QUANTITIES

Quantity	Dimension	Units
7. Charge, Q	[Q]	Coulombs
8. Current $I = (Q/t)$	[Q] [T^{-1}]	Amps
9. Electric field intensity, $\mathcal{E} = $ Force / Charge	[M] [L] [T^{-2}] [Q^{-1}]	Volts/m
10. Electric flux, $N = \mathcal{E} \times$ area	[M] [L^3] [T^{-2}] [Q^{-1}]	Volts-m
11. Potential or emf, $V = \mathcal{E} \times$ distance	[M] [L^2] [T^{-2}] [Q^{-1}]	Volts
12. Resistance, $R = V/I$ . . .	[M] [L^2] [T^{-1}] [Q^{-2}]	Ohms
13. Capacitance, $C = Q/V$.	[M^{-1}] [L^{-2}] [T^2] [Q^2]	Farads
14. Magnetic flux, $\phi = V \times$ time	[M] [L^2] [T^{-1}] [Q^{-1}]	Webers
15. Magnetic flux density, $B = \phi/A$	[M] [T^{-1}] [Q^{-1}]	Webers/m^2
16. Magnetomotive force = NI	[Q] [T^{-1}]	Amps-turns
17. Inductance = ϕ/I	[M] [L^2] [Q^{-2}]	Henrys
18. Permittivity of space, $\epsilon_0 = q_1 q_2 / f r^2$	[M^{-1}] [L^{-3}] [T^2] [Q^2]	Farads/m
19. Permeability of space, μ_0	[M] [L] [Q^{-2}]	Henrys/m
20. Magnetic pole, $m = f/H = f \mu_0 / B$	[M] [L^2] [T^{-1}] [Q^{-1}]	Webers
21. K, μ_r	Dimensionless	. . . .
22. $\mu_0 \epsilon_0$	[L^{-2}] [T^2]	Sec^2/m^2

We shall illustrate the use and consistency of these dimensions by the examples given in Table 29-4.

TABLE 29-4

CONSISTENCY OF DIMENSIONS

Formula	Dimensions	Units
23. $f = BLI$	= [M] [T^{-1}] [Q^{-1}] [L] [Q] [T^{-1}] = [M] [L] [T^{-2}]	Newtons
24. Energy = $Q^2/2C$	= [Q^2] [M] [L^2] [T^{-2}] [Q^{-2}] = [M] [L^2] [T^{-2}]	Joules
25. Power = RI^2 . . .	= [M] [L^2] [T^{-1}] [Q^2] [T^{-2}] [Q^{-2}] = [M] [L^2] [T^{-3}]	Watts

In § 21.12 we derived an experimental measure for the permeability of free space μ_0 and found it to be numerically equal to $4\pi \times 10^{-7}$. The numerical value of ϵ_0, the permittivity of free space, must also be determined by experiment. The method employed is to measure the capacitance of a capacitor of simple geometry—for example, the parallel-plate capacitor with a vacuum as dielectric. From equation (17-34),

$$C = \frac{A \epsilon_0}{d} \text{ farads} . \qquad (17\text{-}34)$$

The area and distance between plates can be directly measured and the capacitance can be found by means of equation (29-8). From measurements of this kind, ϵ_0 is found to have a value of

$$\epsilon_0 = \frac{1}{4\pi (2.9978)^2 \times 10^9} \text{ farads/m} , \quad (29\text{-}16)$$

where, for later convenience, we have separated a factor 4π.

From the dimensional quantity 22 (Table 29-3) we observe that the product, $\mu_0 \epsilon_0$, has the dimension of $1/(\text{velocity})^2$. Using the numerical values of μ_0 and ϵ_0 found experimentally, we have

$$\mu_0 \epsilon_0 = \frac{4\pi \times 10^{-7}}{4\pi (2.9978)^2 \times 10^9}$$

$$= \frac{1}{(2.9978 \times 10^8)^2 \text{ m}^2/\text{sec}^2} ; \qquad (29\text{-}17)$$

that is,

$$c^2 = \frac{1}{\mu_0 \epsilon_0} , \qquad (29\text{-}18)$$

where

$$c = 2.9978 \times 10^8 \text{ m/sec} .$$

The velocity of light, as directly measured, is

$$c = 2.99790 \times 10^8 \text{ m/sec} .$$

This excellent agreement is no mere coincidence but is a consequence of the fact that light waves are, in reality, electromagnetic waves. We shall take $c = 3 \times 10^8$ m/sec in all our formulas.

29.9. Résumé

The following is a brief listing of the definitions, principles, and theories, given in this chapter, which you should know:

The "squared scale" of the alternating-current ammeter and voltmeter; the dynamometer and hot-wire types

The theory of the alternating-current wattmeter

The theory of the ballistic galvanometer for measuring charge, flux density of the earth's magnetic field, and high resistance, using a capacitor

The operation of the cathode-ray oscilloscope

The dimensions of various physical quantities

Queries

1. In the alternating-current ammeter of Fig. 29-1 a squared scale is used. How would this scale be affected if a capacitor were placed in parallel with the shunt resistor R_s?
2. How would the readings of the alternating-current voltmeter in Fig. 29-3 be changed if a capacitor were placed across the input?
3. Should a ballistic galvanometer have a relatively long or short period, a relatively large or small moment of inertia? Give a brief explanation.
4. Explain why a cathode-ray oscilloscope is capable of following fast electrical pulses?
5. Explain why a wattmeter of the type shown in Fig. 29-4 will operate on either alternating or direct current.
6. When the capacitor of Fig. 29-7 (a) is discharged, is there any current between its plates, and, if so, does it decay in the same manner as the current through the resistor R?

Practice Problems

1. A capacitor of 0.1 μf is charged to a potential of 10 volts and then discharged through a ballistic galvanometer whose resistance is 50 ohms. If the first maximum swing is 10 cm on a scale 1 m away, compute the charge sensitivity of the galvanometer, assuming that the damping factor, ρ, equals 1.2. Ans.: 0.0913×10^{-6} coulomb/cm.
2. The ballistic galvanometer of problem 1 is connected to the earth inductor of Fig. 29-6. The coil has a cross-sectional area of 170 cm^2 and has 200 turns of wire, whose resistance is 120 ohms. The coil is arranged to cut the horizontal component of the earth's field ($B_H = 2.2 \times 10^{-5}$ weber/m^2). Calculate the deflection that results when the coil is turned through 180°.
3. Calculate the time constant of an RC circuit with $C = 1$ μf, $R = 10^3$ ohms; $C = 5$ $\mu\mu$f, and $R = 10^6$ ohms. Ans.: 1 millisec; 5 μsec.
4. In the experiment for measuring high resistance shown in Fig. 29-8 the following data were taken: $C = 10^{-7}$ f, the initial charge 1 μcoulomb. The charges after 1, 2, 3, 4, and 5 sec are 0.368, 0.136 0.05, 0.018, and 0.0067 μcoulomb, respectively. Determine R by plotting time against $\log_e (q_0/q)$. What kind of curve should one expect? How can one obtain RC and hence R?
5. Calculate the discharge currents for $t = 10^{-5}$ sec, 1 sec, and 10 sec if the initial charge on a capacitor is 10^{-4} coulomb and the shunting resistor has the value $R = 10^6$ ohms. Take $C = 1$ μf. *Hint:* Differentiate equation (29-11). Ans.: 10^{-4} amp; 3.67×10^{-5} amp; 4.5×10^{-9} amp.

6. By means of graph paper construct the Lissajous figure that results if the ratio of frequencies on the horizontal and vertical plates of a cathode-ray tube is 3:1. Assume that the voltages are of equal amplitude and in phase.
7. In the electron gun of a cathode-ray tube an electron falls through a constant potential gradient of 2×10^4 volts/m and remains in the field for 1.97×10^{-4} sec. With what velocity does the electron leave the gun? If the distance from the gun to the screen is 35 cm, how long does it take the electron to strike the screen? Ans.: 7×10^6 m/sec; 0.05 μsec.
8. By means of dimensional analysis, check the validity of equations (17-34), (18-10), (21-30), and (20-14).
9. The quantity $\sqrt{\mu_0/\epsilon_0}$ is called the "impedance of free space." Compute its value and show by dimensional analysis that the units are ohms. Ans.: 376.7 ohms.
10. What are the dimensions of reluctance?
11. Derive an expression for the deflection of electrons in a cathode-ray beam that have been accelerated by a potential of V_1 volts and travel between two parallel plates separated by a distance d and of length l, with a voltage V_2 applied between the plates. The plates are parallel to the direction of the beam. Neglect end effect of this simple capacitor. Ans.: Deflection $= V_2 l^2/4dV_1$, m.
12. Show by dimensional analysis that the product of the electric-field intensity and magnetic intensity, $\mathscr{E}H$, represents power density, watts per square meter.

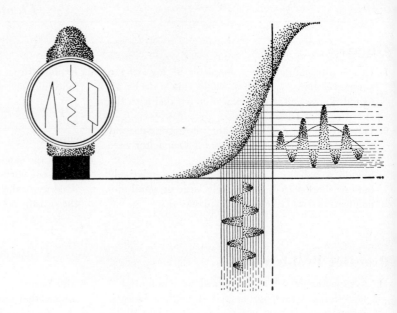

CHAPTER **30**

Electronics

30.1. Introduction

With the advent of radio and television, nearly everyone is acquainted with the name "electronics," even though he may know little or nothing about the science of it. Today this science is so extensive that it has become a separate branch of engineering. In a broad sense, electronics may be said to be concerned with the emission of electrons, the interaction of electrons with fields of force, i.e., with electric and magnetic fields, the flow of electrons in various kinds of circuits, and the production and interaction of electromagnetic radiation with electrons. An electronic circuit is generally composed of some or all of the following: electronic tubes, resistances, capacitances, inductances, transformers, and some form of electrical power. Different combinations of these go into making up the large variety of electronic circuits now in use. In all of these the electron plays a fundamental role.

There are several ways in which electrons can be liberated from metals. These are by thermal processes, known as *thermionic emission;* by electromagnetic radiation of suitable frequency, known as *photoelectric emission;* by bombardment of metals with electrons or positive ions, known as *secondary emission;* and by the application of very strong electric fields, thereby pulling electrons

from the metal, called *field emission*. We shall first describe thermionic emission, since this is of fundamental importance in all radio tubes.

30.2. The Edison or Thermionic Effect

In the latter part of the last century Thomas Edison was concerned with making a usable electric lamp. The filaments of the early lamps were

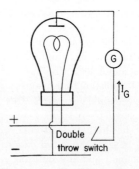

FIG. 30-1.—The Edison effect

made of carbon and caused considerable darkening of the envelope by the evaporation of the hot carbon. In the course of his experiments, Edison in 1883 placed a second electrode in the tube and connected it to a source of direct current (d-c), as shown in Fig. 30-1. The glass envelope was evacuated as highly as possible. Edison found that when the filament was heated, a current I_G flowed in the galvanometer G if the second electrode or plate was connected to the positive side of the voltage source, but there was no current if the connection was made to the negative side. There was no current at all in the galvanometer for either connection if the filament was not heated.

Since every electrical circuit in which there is current must be a closed one, it follows that the heated filament emits some charged particles, whose motion from the filament to the plate constitutes the current observed in the galvanometer. These particles must carry a negative charge, since the current is observed only when the plate is connected to the positive side of the voltage.

Though this discovery was of fundamental importance, it nevertheless did not aid Edison in reducing the evaporation of the heated filament, and so he did not investigate it further. However, Edison furnished a British engineer, Preece, with several of his modified bulbs. The latter with another engineer, Fleming, saw the possibility of using these bulbs as "valves" for rectifying alternating currents and for many other useful purposes.

About 1897, J. J. Thomson had shown that all atoms contain negative charges, which became known as *electrons*. These electrons have masses of about one two-thousandth of the mass of the lightest atom, hydrogen (see chap. 38). It was shown that the negatively charged particles emitted from the hot filament in the Edison bulb were electrons.

Experiments were then carried out to determine the relationship between the number of electrons emitted per unit time and the temperature and nature of the filament. These experiments will now be briefly discussed.

30.3. Richardson-Dushman Equation for Thermionic Emission

An apparatus for investigating thermionic emission is shown in Fig. 30-2. In this there is a filament f in the form of a thin, metallic wire surrounded by three concentric metallic cylinders. These are placed in a glass envelope, which is then highly evacuated. Only the central cylinder is used for collecting current, the other two being used to reduce end effects. The cylinders are connected to a variable source of positive voltage, and

the current, I_p, reaching the central cylinder is measured by the milliammeter, M. An ammeter, A, is used for measuring the filament current, I_f.

No current, I_p, between the filament and the central cylinder or plate, P, is observed until the temperature of the filament is quite high, as shown by its incandescence. As the filament current and temperature are further increased, the electronic

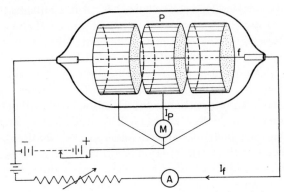

FIG. 30-2.—Apparatus for the study of thermionic emission

current, I_p, increases very rapidly, as shown in Fig. 30-3. If there is a large potential difference, V, between the filament and plate, which is larger than V_3, then the solid curve is obtained. For this situation, all the electrons emitted by the filament are collected at the plate and registered as

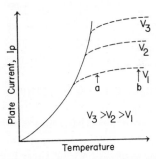

FIG. 30-3.—Plate current as a function of filament temperature for various plate voltages.

the plate current, I_p. The dashed lines, which show an approximately constant plate current, although the temperature of the filament is increasing, represent what is called *space-charge saturation*. This will be discussed in the next section. The solid curve represents the true variation of the electron emission with temperature.

This experimental curve and its theoretical interpretation were first given by O. W. Richardson in 1903. A later modification was made by Richardson in 1914 and another in 1923 by S. **411**

Dushman on the basis of quantum mechanics. For the variation of the electron current density, J, with the temperature, T, of the filament they gave the following relationship:

$$J = AT^2 e^{-b/T} , \qquad (30\text{-}1)$$

where J is the electron current density in amperes per square meter; T is the temperature of the filament in degrees absolute, $°$ K; A is a universal constant; b is a constant depending upon the nature of the material of the filament; and e is the natural base of logarithms ($e = 2.718 \ldots$).

The quantity b is related to the so-called *work function*, ϕ_0, that is, the energy required to remove an electron from the metal. It is low for the alkali metals in the first group of the periodic table and has the smallest value for cesium. We shall encounter the work function again in the discussion of the photoelectric effect (chap. 38). Values of b for different materials are given in Table 30-1.

TABLE 30-1

VALUES OF THE EMISSION CONSTANTS

Element	A (Amp/M²-Deg²)	b (° K)	ϕ_0 (Volts)
Carbon............	60.2×10^4	54,500	4.7
Cesium............	16.2×10^4	21,000	1.8
Molybdenum........	60.2×10^4	49,900	4.3
Tantalum..........	60.2×10^4	47,600	4.1
Thorium...........	60.2×10^4	39,400	3.4
Tungsten..........	60.2×10^4	52,400	4.5
Thoriated tungsten...	3×10^4	30,500	2.6

From equation (30-1) it is seen that the quantity b is of prime importance in giving the variation of the electron emission with temperature. It is for this reason that tungsten filaments are used only in thermionic tubes for which there are high plate voltages. For the ordinary radio tubes, oxide-coated electron emitters are used. These operate at much lower temperatures than the tungsten for the same electron emission.

In practice, the oxide-coated emitters are placed on the outside of a thin cylinder through which an insulated heating filament passes. This is of great convenience in electronic circuits, for then the electron emitter, called the *cathode*, is electrically separated from the heating current or filament circuit.

Let us now return to the Richardson-Dushman equation (30-1). According to Dushman's analy-

sis, the quantity A should be 60.2×10^4 amp/m²-deg². K. Table 30-1 shows that this value is found for most pure metals, while for cesium or thoriated tungsten the value is considerably lower. The exact reason for this discrepancy between theory and experiment is not fully understood. It is connected with the difficult problem of the theory of the solid state of matter.

However, the general features of the variation of electron emission with temperature are well represented by equation (30-1). From this and the solid curve of Fig. 30-3, it is seen that the electron emission increases rapidly with temperature. We may think of this variation as being due to the increase in kinetic energy of the electrons with temperature, so that more electrons can escape through the surface of the heated filament at high temperatures than at low temperatures. It now remains for us to discuss the dashed portions in Fig. 30-3, which was stated as being due to space charge.

30.4. Space-Charge Saturation

Let us consider the region between the heated filament and the cylindrical plate shown in Fig. 30-2. The filament is emitting a copious supply of electrons, and, unless these are rapidly pulled over to the plate, they will form an electron cloud about the filament. This cloud of electrons is a region of negative charge about the filament and is commonly called a *space charge*. If the cylindrical plate has a positive potential relative to the filament, then electrons are drawn from the periphery of the cloud to the plate. The negatively charged cloud exerts a retarding force on the electrons emitted from the filament.

Now consider the region ab in Fig. 30-3, for which the electron current, I_p, is practically constant and independent of the temperature of the filament. Though the temperature of the filament is increased, the volume density of electrons in the cloud remains constant, and the electron current to the plate remains constant. Suppose, now, that the plate-filament voltage, V_p, is raised from some value V_1 to V_2, as shown in Fig. 30-3. This rise in voltage increases the electric field between the plate and the filament, raising the volume density of electrons in the space around the filament and increasing the current, I_p. Again a flat region is found at which there is equilibrium between the number of electrons drawn to the plate and the number which get into the cloud from the filament. In the flat portions, such as *ab* in Fig. 30-3, the current is said to be *space-charge-limited*.

As is seen, the current I_p increases with the plate voltage V_p. In Fig. 30-4, the variation of I_p with V_p is shown by the solid curve, for which the filament temperature is higher than the value T_3. The plate current I_p is space-charge-limited, and the variation of I_p with V_p has been calculated

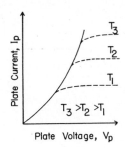

FIG. 30-4.—Space-charge saturation and temperature saturation.

independently by Child and Langmuir for different geometrical conditions of filament and plate. The Child-Langmuir equation for space-charge currents is

$$I_p = C V_p^{3/2} , \qquad (30\text{-}2)$$

where C is a constant whose value depends on the geometry of the electrodes. Note that this equation is quite different from Ohm's law, in which the current is proportional to the first power of the potential difference.

In the derivations of equation (30-2), both Child and Langmuir assumed that the electrons from the filament emerged with zero velocity and that the electric field at the cathode is zero. Later calculations were made by other investigators for which these assumptions were dropped. These calculations showed that, while the $\frac{3}{2}$-power law was valid, the form of the geometrical constant C was changed and also the minimum of potential between the filament and plate occurred at a short distance from the filament.

The dashed horizontal lines in Fig. 30-4 show that at a given temperature, say T_1, the plate current, I_p, is independent of the voltage, V_p. This horizontal portion is readily accounted for by assuming that in this region the plate is receiving all the electrons emitted at the constant temperature, T_1. It is impossible to obtain more plate cur-

rent than that emitted by the filament, so that there is a natural limit set to the current. This limitation of the plate current is called *emission limitation* or *temperature saturation*.

In summary, we may say that the Richardson-Dushman equation (30-1) predicts the number of electrons emitted per second from a metal at some temperature $T°\,\mathrm{K}$, while the Child-Langmuir equation (30-2) gives the numbers which actually reach the plate every second.

30.5. The Two-Element Vacuum Tube: Rectification

A device consisting of two electrodes—the filament or cathode and the plate or anode—in an evacuated envelope, such as we described in Fig. 30-2, is called a *high-vacuum diode*. The principle of operation of a diode is briefly this: With the plate charged positively, electronic currents may pass only from the hot cathode to the plate. The other alternative—that of an electron current from plate to cathode—is ruled out, since the plate is always maintained at a temperature so low that no thermionic emission from the plate is possible. The energy that is delivered to the tube by the source of plate potential is partly expended in accelerating the electrons from cathode to anode. The high kinetic energy of the electrons on striking the plate is then converted into heat that must be dissipated by the plate.

In addition to zero conduction when the anode is negative, note also that the current is not linearly related to the applied potential of the diode, as is evident from Fig. 30-4. The diode, therefore, is a *nonlinear* circuit element, in contrast to a pure resistance.

An important and common application of the diode is in the rectification of alternating-current (a-c) voltage, i.e., in converting an a-c voltage into a pulsating d-c voltage. The frequency of the a-c voltage may be low (60 cycles/sec), as in the case of power supplies, or it may reach radio frequencies.

The way this is done is illustrated in Fig. 30-5 (*a*), which shows a diode used as a half-wave rectifier. In this figure T represents a transformer with two windings—one for heating the filament of the diode and one for supplying the voltage e to be rectified. The current i_p is the instantaneous value in the circuit, while V_p is the plate potential **413** equal to the transformer voltage minus the iR drop in the load resistor, R, i.e., $V_p = e - i_p R$. The resulting current through the resistor is shown in Fig. 30-5 (c). The bottom halves of the curves are cut off because of zero conduction when the anode is negative.

As shown in Fig. 30-6 (a), two diodes may be connected to obtain full-wave rectification of the applied transformer voltage. For half a cycle the a-c voltage is such that the anode of tube *1* is

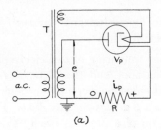

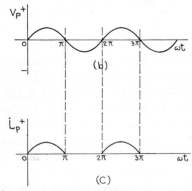

FIG. 30-5.—(a) Simple diode circuit for half-wave rectification; (b) potential of the plate as a function of time; (c) current in the tube and resistor R.

positive and that of tube *2* negative. Tube *1* conducts, and tube *2* does not. On the second half of the cycle the polarities are reversed, and tube *2* conducts. The current i_p is shown in Fig. 30-6 (b).

In many applications it is desired to obtain a reasonably steady direct current rather than the pulsating current shown in Fig. 30-6 (b). This can be accomplished by the use of *electrical filters*. Two of the simplest filters are shown in Fig. 30-7 (a), (c). In (a) a capacitor is shunted across the load resistor. Its capacitance is so chosen that the time constant for charging during conduction by the diode is very small compared to the time constant for discharge through the load resistor. The student will recall from § 29.6 that the time constant

414 is the product of the capacitance and the cor-

responding resistance. Thus, while the diode is conducting, the capacitor is very quickly charged to the maximum potential of the transformer voltage. During the nonconducting time, the capacitor

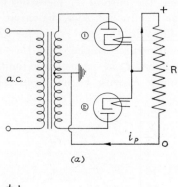

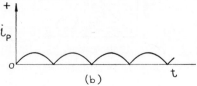

FIG. 30-6.—(a) Full wave rectifier; (b) current through resistor R.

discharges through the load resistance. If, however, the time constant of the load-resistance circuit, RC, is large compared to the period of

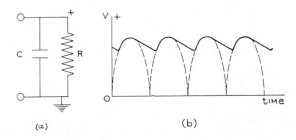

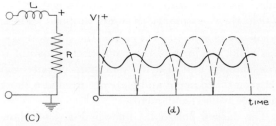

FIG. 30-7.—Electrical filters: (a) a shunt capacitor; (b) the load current with (a); (c) a series inductor; (d) the load current with (c).

oscillation of the applied voltage, a nearly steady voltage will be maintained, as shown in Fig. 30-7 (b).

An effective electrical filter can be constructed by combining an inductance in series with the load resistor and a capacitor in parallel, as shown in Fig. 30-8. The inductances tend to flatten out the current peaks because of the induced electromotive forces in the inductance (Fig. 30-7 [c], [d]), while the capacitor serves to maintain a constant

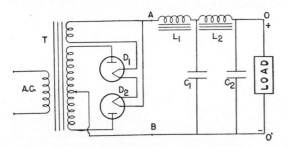

Fig. 30-8.—A rectifying and filtering circuit

potential across the resistor. The combination can produce a d-c voltage across the load with a very small ripple or fluctuation.

30.6. The Triode; Amplification Factor

In 1906, Lee Deforest modified the diode by inserting a third electrode, or *grid*, between the cathode or filament and the plate. Thus was formed the *triode* or three-electrode tube—a device destined to change the history of communication through radio, etc. The grid usually is in the form of an open helix of wire, the spacing between the turns being several times the diameter of the wire. This large spacing allows electrons to pass freely from the cathode to the plate.

In general, tubes are operated so that there is a plentiful supply of electrons from the cathode, that is, the tubes are space-charge-limited and not temperature-limited. The third electrode or grid of the triode is close to the cathode and consequently is in the region of space charge. Any voltage applied to the grid considerably changes the current to the plate, I_p. Deforest immediately found this effect and saw that the triode could be used for amplifying currents.

The grid is usually operated at a negative potential with respect to the cathode, and consequently very few electrons are collected on the grid. In general, then, the electron current to the grid, I_g, is negligibly small compared to the plate current, I_p. It is, of course, assumed that the tube is highly evacuated.

The plate current, I_p, depends on the grid-cathode voltage, V_g, and to a much lesser extent on the plate-cathode voltage, V_p, as shown in Fig. 30-9. Suppose the grid voltage, V_g, is changed by a small amount, then the plate current, I_p, is changed, even though the plate voltage remains constant. The plate current can now be brought back to its original value by changing the plate voltage. For example, suppose that a change of -0.1 volt on the grid decreased the plate current by a given amount and that bringing the current back to its original value required an increase in the plate voltage of 2.5 volts. The grid-voltage change is then twenty-five times as effective as an equal change in plate voltage. This tube would be

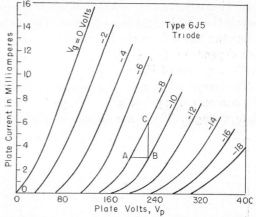

Fig. 30-9.—Plate-current–plate-voltage characteristics for 6J5 triode.

said to have an *amplification factor*, μ, of 25. The amplification factor, μ, of a triode is defined as the ratio of the change in plate voltage, ΔV_p, to the change in grid voltage, ΔV_g, so that the plate current, I_p, remains constant. Notice that if ΔV_g is positive, then ΔV_p must be negative for I_p to be constant. In symbols, μ is defined as

$$\mu = -\frac{\Delta V_p}{\Delta V_g}\bigg|_{I_p \text{ const}}. \qquad (30\text{-}3)$$

Since tubes are normally operated at space-charge saturation of current, the plate current varies with the factor $(V_g + V_p/\mu)$ in the same manner as the plate current varies with V_p in the diode (eq. [30-2]); thus the plate current may be written as

$$I_p = C\left(V_g + \frac{V_p}{\mu}\right)^{3/2}, \qquad (30\text{-}4)$$

where C is a constant depending on the geometry of the electrodes.

415

30.7. Characteristic Curves for Triodes

In order to learn how the plate current varies with changes of grid and plate voltages, it is convenient to plot curves relating these electrode voltages and currents. The most important relationships among these factors are (1) plate current and plate voltage with constant grid voltage and (2) plate current and grid voltage with constant plate voltage. Fig. 30-9 shows a family of plate-current–plate-voltage characteristic curves of a triode. In obtaining these curves the filament current is held constant at normal value, the grid voltage is held constant at values shown on the curves, and the plate voltage is varied, giving the plate-current values. Note that these curves are all of approximately the same shape and follow approximately the Child-Langmuir equation. These characteristics follow from the fact that the plate current depends on $(V_g + V_p/\mu)$ and not upon any particular value of V_g and V_p.

The second set of curves generally used is that of plate current against grid voltage for various plate voltages. These are shown in Fig. 30-10.

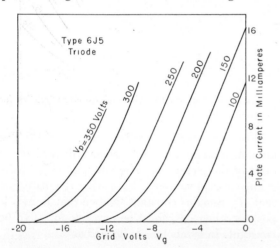

FIG. 30-10.—Plate-current–grid-voltage characteristics for 6J5 triode.

Note that the principal effect of changing the plate voltage is to displace the curves without appreciably changing their shape. A circuit whereby these sets of curves may be experimentally determined is shown in Fig. 30-11.

30.8. Tube Parameters

Three important tube parameters, obtainable from either set of curves, are useful in making quantitative evaluations of tube performance. These parameters are the amplification factor, μ; the internal plate resistance, r_p; and the mutual conductance or transconductance, g_m.

The *amplification factor*, μ, has already been defined as the ratio of the plate-voltage change to the grid-voltage change when the plate current is maintained constant. If ΔV_g and ΔV_p are the

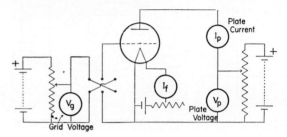

FIG. 30-11.—Circuit for measurement of characteristic curves.

voltage increments that keep the plate current constant, then

$$\text{Amplification factor} = \mu = -\frac{\Delta V_p}{\Delta V_g}\bigg|_{I_p \text{ const}}.$$

In Fig. 30-9, at points A and B, the plate currents are the same. Here ΔV_g is -2 volts, and ΔV_p is approximately 31.6 volts. Thus $\mu = 31.6/2 = 15.8$.

The *plate resistance*, r_p, of a vacuum tube is given by the ratio of the plate-voltage change to the plate-current change, the grid voltage being held constant:

$$\text{Plate resistance} = r_p = \frac{\Delta V_p}{\Delta i_p}\bigg|_{V_g \text{ const}}. \quad (30\text{-}5)$$

From this definition, the plate resistance is equivalent to the reciprocal of the slope of the plate-current–plate-voltage characteristics shown in Fig. 30-9. Since the slopes along the curve change, it is clear that, in general, the value of the plate resistance will depend on the point of the curve chosen for operation. From the figure, the tube resistance at a point along the straight portion of the curve (C) is 10,000 ohms.

The *mutual conductance*, g_m, is defined as the ratio of the plate-current change to the grid-voltage change, when the plate voltage is held constant:

$$\text{Mutual conductance} = g_m = \frac{\Delta i_p}{\Delta V_g}\bigg|_{V_p \text{ const}}. \quad (30\text{-}6)$$

Mutual conductance is a measure of the effectiveness of the grid in controlling the plate current.

Since in many problems a low plate resistance and high amplification are desirable, the value of g_m is a rough measure of the merit of the particular tube. Since mutual conductance has the dimensions of the reciprocal of resistance, it is measured in micromhos, typical values being 500–5,000 micromhos. For the tube of Fig. 30-10, g_m is 1,550 micromhos.

By combining equations (30-3), (30-5), and (30-6) it can be seen that the tube parameters are related by the following equation:

$$\mu = r_p g_m . \qquad (30\text{-}7)$$

30.9. Vacuum-Tube Amplifier

The triode is most frequently used in industry and in the laboratory as an amplifier of small alternating-current voltages. By maintaining the grid at a negative voltage, no appreciable electron current is drawn by the grid, so that the power dissipated by the grid is very small. Also, since the grid is much more effective than the plate in controlling plate current, the output voltage may be made to vary with many times the amplitude of the input voltage. Let us see how such a tube may be used as an amplifier.

A circuit of an amplifier, such as might be used in a telephone circuit, is shown in Fig. 30-12. Here

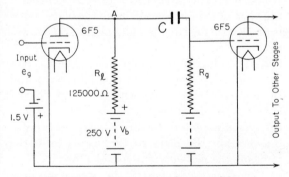

FIG. 30-12.—A two-stage voltage amplifier

we have drawn two stages of amplification, showing the manner in which the output voltage, e_0, appearing across the plate resistor, R_l, is introduced onto the grid of the second tube. The function of the capacitor, C, is to prevent the positive potential on the plate of the first tube from being applied to the grid of the second tube. We shall analyze only the first tube and the relation between the input voltage, e_g, and the output voltage, e_0, appearing across R_l. The tube we have

chosen is the 6F5, a tube with a high amplification factor. In Fig. 30-13 we have drawn the plate-current–plate-voltage curves for this tube.

Consider, now, that an alternating-current input signal, e_g, having a maximum value of 0.5 volt is applied to the grid (Fig. 30-12). When no current is passing through the tube, the entire 250 volts are across the tube, and the potential of A is +250 volts, giving point Y in Fig. 30-13. On the other hand, when a current of 2 milliamperes passes through the tube, the entire drop in potential of 250 volts is across the 125,000-ohm re-

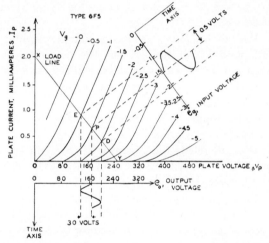

FIG. 30-13.—Graphical representation of output voltage against input voltage.

sistance, and hence A is at zero potential. This gives point X in Fig. 30-13. For intermediate values of the plate current, the plate voltage will have values from 0 to 250. The straight line XY of Fig. 30-13 gives the possible values of I_p and V_p under the conditions that have been imposed on the tube. For this reason the line XY is called the *operating* or *load line*. The variation of I_p and V_p along this line is controlled by the grid voltage. Let us bias the grid at -1.5 volts, i.e., make $V_g = -1.5$ volts with respect to the cathode (Fig. 30-12). For this bias the plate current and plate voltage are given by the coordinates of the point P of Fig. 30-13: $I_p = 0.6$ milliamperes, and $V_p = 175$ volts. The point P ($V_g = -1.5$ volts) on the load line is called the *quiescent point*.

We now impress a sinusoidally varying voltage, e_g, on the grid, whose peak value we shall assume to be 0.5 volt. Immediately the instantaneous operating point, P, of the tube will vary along the   *417* line XY between the points E and D. This variation will cause the plate voltage, V_p, at the point A to vary also. The plate-voltage variation, e_0, is impressed on the capacitor C and then onto the grid of the next tube. Note that e_g and e_0 are in opposite phase, that is, when e_g has its maximum value, e_0 has its minimum value.

The variation of e_g and e_0 with time is shown by the sinelike curves of Fig. 30-13. We observe that the shapes of the two are sensibly the same, but the voltage scales are quite different. Thus the peak value of e_0, as read off the graph, is 30 volts. *We have produced a voltage amplification of sixty times the input voltage without destroying the shape of the wave.*

We can, of course, produce a voltage amplification with a transformer, but the power output must always be somewhat less than the power input. However, with the vacuum tube we are able to add power and produce an output much greater than the input. The extra energy comes from the batteries of the circuit.

It is possible in the laboratory to design *current amplifiers* that will measure currents of the order of 10^{-18} amperes—less than 10 electrons/sec.

While this form of analysis can be used to determine the output voltage for a given input, it can become tedious and complicated for some circuits. A mathematical linear circuit analysis is then used, which assumes that the amplification factor, μ, is a constant. This form of analysis is found in any textbook on electronics. The gain, K, of a single stage of the amplifier shown in Fig. 30-12 is given by

$$K = \frac{\text{Output voltage } e_0}{\text{Input voltage } e_g} = \frac{\mu R_l}{r_p + R_l}, \quad (30\text{-}8)$$

which, for the particular values given in the circuit, for $\mu = 100$, and $r_p = 66{,}000$ ohms, is

$$K = 100 \frac{(125{,}000)}{125{,}000 + 66{,}000} = 65.$$

30.10. Tetrodes and Pentodes

After the successful introduction of a third electrode, several attempts were made to introduce still more electrodes. However, it was not until 1928 that the fourth electrode was introduced usefully, to form the *tetrode*. The second grid, called the *screen grid*, consisted of fine wires placed between the control grid and the plate. This screen grid is usually maintained at a fixed positive voltage somewhat less than the plate voltage.

The purpose of this screen grid was to overcome one of the inherent disadvantages of the triode. As we have already seen, not only does the plate of a triode collect the electrons, but the plate voltage also controls to some extent the plate current. This latter statement also means that the electric field from the plate reaches to the cathode, so that there is a relatively large capacitance between the plate and the cathode. The introduction of the positively charged screen grid reduces the electric field and capacitance between the plate and cathode by a factor of the order of one-thousandth. This then means that the plate collects only electrons and thus the plate current, I_p, is almost independent of the plate voltage. While the screen grid collects some electrons because of its positive charge, nevertheless this current is small compared to that going to the plate, since the physical dimensions of the screen grid are small compared to those of the plate.

There is, however, a serious disadvantage to a tetrode, inasmuch as when electrons strike a metal surface, electrons, commonly called *secondary electrons*, are emitted from the metal. This phenomenon is called *secondary emission* and is used to great advantage in some special tubes called *multipliers*. The secondary electrons emitted at the plate of the tetrode will leave the plate if there is an electric field at the plate pulling them away. Such an electric field would exist if the screen potential was higher or more positive than the plate potential. Since the screen potential is fixed while the plate potential varies with the input signal, it is quite possible for the plate potential to go below the screen potential. In such a case, electrons are drawn from the plate to the screen grid, thus reducing the plate current considerably.

Almost immediately after the tetrode was introduced, the pentode followed. The introduction of a third or *suppressor grid* for suppressing this secondary electron emission was readily accomplished. This suppressor grid again consists of fine wires or is a grid network placed between the screen grid and the plate, as shown in Fig. 30-14 (*a*) and (*b*). The suppressor grid is usually connected to the cathode, so that it is kept at a fixed low potential. Notice that the voltages in a tube are usually measured with respect to the cathode. The field at the plate is then always in such a direction as to drive back into the plate any electrons emitted by the plate. Thus the plate current is almost entirely independent of the plate voltage, as shown in Fig. 30-15.

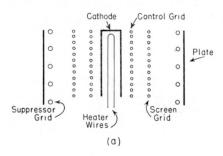

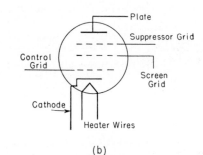

Fig. 30-14.—Schematic diagrams of a pentode

The pentode, then, has the following desirable characteristics: a very low plate-control-grid capacitance, a plate-grid-control mutual conductance which is of the same order of magnitude as

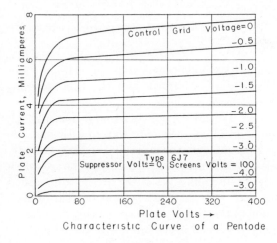

Fig. 30-15.—Characteristic curves for a pentode

the triode, and an amplification factor and plate resistance ten to a hundred times that of the triode. The pentode has vastly increased circuit-design possibilities. It is used extensively in ampli-

fiers, as a power-output tube and, because the plate current is nearly independent of the plate voltage, as a source of constant current.

30.11. Gas Tubes

If a small trace of gas, of the order of a fraction of a millimeter pressure, is purposely introduced into a two- or three-electrode tube, the characteristics of these tubes are considerably changed. The change is largely due to the fact that if electrons have sufficient energy, they are capable of ejecting an electron from one of the atoms of the gas. Since an atom is electrically neutral, the removal of an electron results in a positively charged entity called a *positive ion*. The process of removing an electron from an atom is called *ionization* and is discussed in chapter 37.

The presence of the negative and positive charges in the gas leaves a region which is largely electrically neutral or field-free, called the *plasma*. This results in a fairly large variety of gas tubes which have different functions. Two of these will be briefly discussed.

The first of these is the glow tube or a cold-cathode, gas-discharge tube which operates in the glow-discharge region. The important characteristic of this type of tube is that the voltage drop across it is largely independent of the current through the tube up to about 30 milliamperes. Depending on the type of gas and material of the electrodes, the fairly constant voltage drop can be varied to suitable amounts. Commercial glow lamps are available which have normal operating potentials of 75, 90, 105, and 150 volts. These tubes are used for regulating the voltage in a particular part of a circuit to those values given above. The tubes are generally referred to as "voltage-regulating tubes" or VR tubes—VR-75, VR-90, etc. The gases commonly used in the tubes are neon, helium, or argon, though some have a mixture of these gases.

A second type of gas-discharge tube in common use is called a *thyratron*. Before describing this, we must briefly discuss the hot-cathode gas diode. If a few drops of liquid mercury are placed within the vacuum diode of § 30.5, mercury-vapor pressure of an amount dependent on the ambient temperature will be present. This vapor changes the operating characteristics of the diode in the following way: With the application of the plate **419** potential to the diode, the plate current starts to increase in the same way as in a vacuum diode, but at some critical potential there is a sudden increase in plate current, with little or no increase in plate voltage. The current-voltage characteristic is shown in Fig. 30-16, where a comparison is made with the I_p-V_p characteristics of the high-vacuum diode.

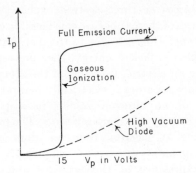

FIG. 30-16.—Conduction in a hot-cathode diode

This abrupt increase in plate current occurs when the plate voltage reaches the ionization potential of the gas (§ 37.7). The breakdown can also be detected by the presence of a luminous discharge in the tube. A discharge of the type we have described, characterized by a low cathode-to-plate potential drop and a high plate current, is called an *arc* and the tube an *arc-discharge diode*.

Thyratrons.—A tube possessing rather remarkable operating properties may be constructed by inserting a grid in the arc-discharge diode so as to provide an almost complete electrostatic shield be-

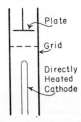

FIG. 30-17.—Electrodes of a thyratron

tween the cathode and the plate. By controlling the potential of the grid relative to the cathode, it is possible to control the initiation of the arc. The grid of this tube is usually a metal cylinder that incloses both the plate and the cathode and is provided with several holes between cathode and anode. A schematic cross-section of such a tube is shown in Fig. 30-17. Triodes of this type are known as *thyratrons*.

The action of the grid of the thyratron *before* breakdown of the gas occurs is similar to that of the vacuum triode. In both tubes the grid shields the plate from the cathode and controls the electron current from cathode to plate. In the case of the thyratron the shielding is so nearly complete that only a small negative potential suffices to counteract the field produced at the cathode by the high plate potential. For example, in one make of tube a negative potential of the grid of −6 volts will prevent discharge, even though the plate potential with respect to the cathode is 500 volts (see Fig. 30-18). As the grid is made more positive than

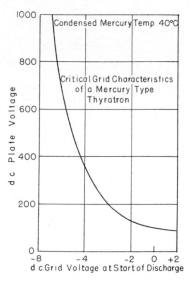

FIG. 30-18.—Characteristic curves for a mercury-type thyratron.

some critical value, ionization by collision takes place, an arc is formed, and the tube acts like the arc-discharge diode.

Once the arc is present, the grid completely loses control, and the only way to extinguish the arc is by removing or lowering the plate potential below a value necessary to maintain the arc (of the order of 15 volts). When the arc is extinguished, the grid once more gains control. This trigger action of the grid is the basis of many engineering applications and is perhaps one of the most useful control devices known. We shall describe one such application—a "saw-tooth" voltage generator.

The ability of the pentode to supply a constant current nearly independent of plate voltage, when combined with the trigger action of the thyratron, may be used to design a saw-tooth voltage generator, widely used in the sweep circuit of the cathode-ray tube (§ 29.7). The principle of operation is briefly this: If the capacitor, C_1, of Fig. 30-19 (*a*) is charged at a constant rate and then abruptly discharged, the voltage developed across the capacitor would appear as shown in Fig. 30-19 (*b*). Since the voltage across C_1 is given by $e_{c_1} = It/C_1$, where I is the constant current delivered to C_1, e_{c_1} would increase linearly with time, as shown.

In the circuit of Fig. 30-19 (*a*), assume that C_1 across the thyratron is initially uncharged. Also the control grid of the thyratron is adjusted to some critical value, e_{cg}, so that conduction will occur only if the voltage from plate to cathode is e_{cp}. With a fixed plate, grid, and

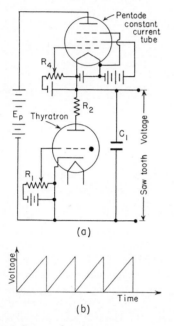

(a)

(b)

FIG. 30-19.—Saw-tooth voltage generator: (*a*) circuit; (*b*) voltage-time curve.

screen voltage on the pentode, a constant plate current, I, will be delivered to the capacitor, C_1, charging it uniformly until the voltage across it reaches the critical value e_{cp}. At this instant the thyratron fires and quickly discharges the capacitor, until the voltage across the capacitor falls to such a value that the thyratron stops conducting. The capacitor will then start to charge again, and the entire cycle will repeat itself, giving rise to the voltage wave form of Fig. 30-19 (*b*). Increasing the charging current, I, by adjusting the grid bias on the pentode (R_4), for example, will increase the rate of repetition. Changing the bias on the thyratron (R_1) will change the amplitude of the saw-tooth voltage, as well as the frequency. The resistor, R_2, limits the peak current delivered by the thyratron and thus determines, essentially, the minimum time for discharging C_1 and hence the steepness of the voltage drop of Fig. 30-19.

30.12. Photoelectric Cells

The fundamental laws of photoelectric emission are discussed in a later chapter (chap. 37). For our present purposes it is sufficient to state the well-established properties of the photoelectric effect:

a) Electrons are emitted from surfaces when exposed to radiation of sufficiently short wave length. For low-work-function surfaces, this radiation may be in the visible or near-visible part of the spectrum.

b) The number of electrons emitted depends upon the intensity of the radiation striking the surface.

c) The maximum velocity of the photoelectrons is independent of the intensity of the radiation but dependent upon the frequency of the radiation.

In this section we shall apply these principles to the design of the photoelectric cell—a device of great value in all kinds of physical measurements, control, and detection. There are three general classes of photocells: the *photoemissive cell*, in which radiant energy ejects electrons from a surface; the *photovoltaic* or barrier-layer cell, in which radiant energy striking the cell causes the generation of an electromotive force; and the *photoconductive* cell, in which the resistance of the cell changes with the intensity of radiation striking the cell. We shall briefly discuss the first two types.

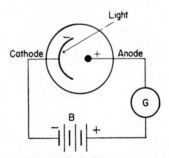

FIG. 30-20.—The photoemissive or photoelectric cell

The high-vacuum emissive type consists essentially of a sensitive cathode surface of large area and a collecting electrode contained in an evacuated glass envelope. The cathode consists of a semicylindrical metallic plate on which a photosensitive substance, generally an alkali, has been evaporated. The anode is a straight wire, practically coaxial with the cathode.

As illustrated in Fig. 30-20, light striking the sensitive cathode liberates electrons, which are pulled over to the anode by the electrostatic field. **421**

The volt-ampere characteristic curves of a typical high-vacuum photocell (PJ22) are shown in Fig. 30-21. The current present at zero accelerating potential is due essentially to the initial velocities of the electrons. Curves are drawn for various light intensities (see chap. 35).

The photovoltaic cell, of which the selenium type is typical, consists of a base of iron, on which is spread a thin layer of selenium, which, in turn, is covered with a semitransparent layer of silver.

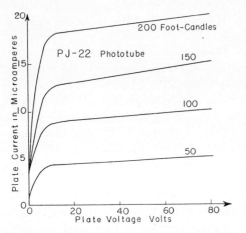

FIG. 30-21.—Plate-current–plate-voltage characteristics of a PJ22 photoelectric cell.

A copper ring in contact with the base acts as one electrode, and a copper ring in contact with the silver layer is the other electrode. A complete electrical circuit is made through a galvanometer, G, or microammeter, as shown in Fig. 30-22 (a). As long as the cell is illuminated, a current is present in the external circuit.

Under the influence of light the cell generates an electromotive force due to the difference in electron densities between the selenium and the iron base. The current in the external circuit is nearly proportional to the light intensity if the resistance of the external circuit is very small. This is illustrated in Fig. 30-22 (b).

As a simple application of photoelectric cells, we shall use the phototube in the circuit of Fig. 30-23. The object of the circuit is to turn on the lamp by having light from an auxiliary source fall on the phototube, keep the lamp on by allowing some of the radiant flux of the lamp to fall onto the cell, and then turn the light off by momentarily interposing a screen between the lamp and the cell.

As can be seen from this circuit, as long as no light falls on the phototube, no voltage appears across the

resistor, R, so that the grid bias, E_{cg}, is sufficiently negative to prevent the thyratron from firing on the positive half-cycle of the applied alternating-current voltage. If light from some source momentarily strikes the cell, a voltage of such a value appears across R that the thyratron conducts and the lamp lights. If the lamp

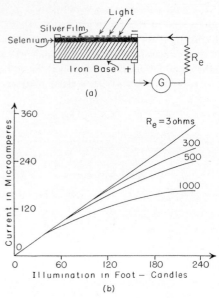

FIG. 30-22.—Photovoltaic cell: (a) electrical circuit of the cell; (b) operational characteristics of the cell.

and cell are so arranged that part of the flux from the lamp falls onto the cell, the lamp remains burning. If the light, however, is momentarily interrupted, the grid voltage on the thyratron again becomes sufficiently negative so that, on the negative cycle of the applied alternating current, the grid gains control, and the lamp remains extinguished.

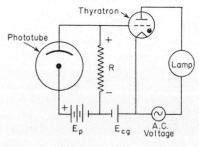

FIG. 30-23.—Typical application of a photoelectric cell

30.13. Production of Electromagnetic Radiation

In 1856, Maxwell showed theoretically that electromagnetic radiation, which is propagated with the velocity of light (3×10^8 m/sec), should be emitted from an oscillating electric circuit. He assumed the existence of a displacement current, $\epsilon dE/dt$, which, like any conduction current, has associated with it a magnetic field. Faraday had shown that a changing magnetic field dB/dt gives rise to a current and an electric field (see § 23.6). Thus Maxwell's theory showed essentially that a changing electric field gives rise to a changing magnetic field and a changing magnetic field to a changing electric field. This electromagnetic wave, consisting of changing magnetic and electric fields at right angles, is propagated with the velocity of light.

The existence of these electromagnetic waves was first experimentally shown by Henry in 1842 but, because of his lack of publication, the credit is given to Heinrich Hertz, who demonstrated this about 1888. Oscillations were set up in a capacitor-inductor circuit when a spark from an induction coil discharged the capacitor. Some short distance away, the electromagnetic radiation from the high-frequency oscillations of the capacitor-inductor circuit was picked up. The receiver was a single turn of wire with a small gap in it, and the electrical oscillations caused a spark to pass across the gap.

The mechanism of radiation of electromagnetic waves is suggested by the diagram of Fig. 30-24,

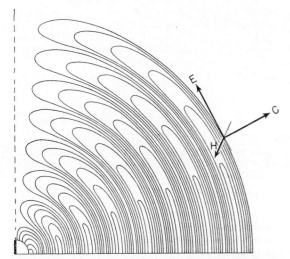

Fig. 30-24.—Electromagnetic radiation from a section of a dipole antenna.

showing the electric field distribution of a radiating dipole antenna. The figure should be rotated about the antenna as an axis in order to obtain the proper spatial distribution of the field; also the lower half, symmetrical with the upper half, has not been drawn. It is suggested that, because of the very high frequency of voltage change along the antenna, electric lines of force become detached, form closed loops, and travel out into space with the speed of light. This diagram is a copy of the one first published by Hertz in 1889.

The accompanying magnetic field is an integral part of the electromagnetic wave. In Fig. 30-24 the relationship between E, H, and the velocity of propagation, C, is indicated at one point on the wave. As is clear from Fig. 30-24, no energy is radiated from the ends of the antenna, so that the loops of lines of force never reach the vertical. The most intense radiation is broadside, as indicated by the crowding of the lines of force.

The oscillations producing the electromagnetic radiation in the Hertzian oscillator were quickly damped, and, though modifications were made in the circuit, it was only with some difficulty that dot-and-dashed code messages could be transmitted with this form of apparatus. In this connection, Marconi did pioneer work, for at the turn of the century he transmitted a short message across the Atlantic. However, it was not until the advent of the triode and other multigrid tubes that sustained oscillations could be set up so that speech and music could be transmitted.

In order to get sustained oscillators in a circuit, some of the output electrical energy is fed back into the input portion of the electrical circuit. An illustration of a feedback circuit, originally introduced by Armstrong,* for producing oscillations

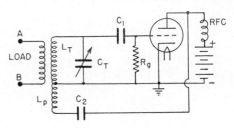

Fig. 30-25.—An *L-C* oscillator

is shown in Fig. 30-25. Energy is fed by the connection from the plate to the grid by means of the coupling between the inductances L_p and L_T. When properly set up, oscillations take place in the capacitor-inductor circuit, C_T-L_T, called the

* An interesting qualitative account of Armstrong's work is given by L. P. Lessing, in the *Scientific American*, April, 1954, p. 64.

tank circuit, and these oscillations are maintained by the feedback from the tube. The frequency of oscillations is very nearly that of the natural frequency of the tank circuit, i.e.,

$$f = \frac{1}{2\pi \sqrt{L_T C_T}}. \tag{30-9}$$

By adjusting C_T, the frequency range may be altered. Oscillators may be designed for audio frequencies up to 20,000 cycles/sec and for radio frequencies up to millions of cycles per second, i.e., megacycles per second.

EXAMPLE 1: Find the natural frequency of a tank circuit consisting of an inductance $L_T = 0.1$ millihenry and a capacitance $C_T = 1$ $\mu\mu$f.

Solution:

$$L_T = 10^{-4} \text{ henry} , \qquad C_T = 10^{-12} \text{ farad} .$$

From equation (30-9),

$$f = \frac{1}{2\pi \sqrt{10^{-16}}} = \frac{10^8}{2\pi} \text{ cycles/sec}$$
$$= 15.9 \text{ megacycles/sec} .$$

30.14. Radio Transmitting and Receiving

In order that electromagnetic waves may be used for voice communication, it is necessary to impress an audio-signal upon the wave. One method commonly employed is *amplitude modulation* (altering the amplitude) of the high-frequency oscillations by combining them with low-frequency oscillations, such as are produced by a microphone or telephone.

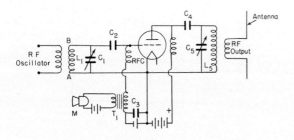

FIG. 30-26.—A simple transmitting circuit

A very simple transmitter that serves to illustrate the principle of amplitude modulation is shown in Fig. 30-26. The RF oscillator produces a continuous voltage fluctuation across the coil AB, as indicated in Fig. 30-27 (a). The RF voltage is then fed to the grid of the triode. A microphone

circuit, M, is also inductively coupled to the grid circuit of this triode, as shown, so that any voltage fluctuation in the microphone circuit (Fig. 30-27 [b]) will give rise to corresponding voltage fluctuations in the grid circuit. These audio-signals are combined with the RF signals in the grid circuit, thereby changing the amplitude or modulating the RF wave. The resulting voltage fluctuation, fed to the antenna system and radiated out into space as

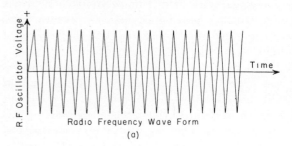

Radio Frequency Wave Form
(a)

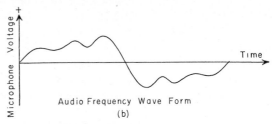

Audio Frequency Wave Form
(b)

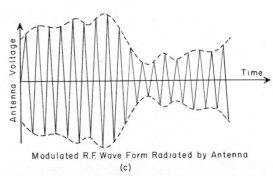

Modulated R.F. Wave Form Radiated by Antenna
(c)

FIG. 30-27.—The production of an amplitude-modulated wave.

an electromagnetic wave, is shown in Fig. 30-27 (c). This method of modulation is known as *grid modulation*.

The principle of reception of radio waves may be demonstrated by means of Figs. 30-28 and 30-29. A block diagram of a simple receiver is shown in Fig. 30-28. The receiving antenna intercepts a portion of the electromagnetic waves. As these waves move by, the electric field induces a RF voltage in the antenna system, which is then fed by the transmission line to the resonant circuit

AB, and thence to the RF amplifier. The resulting amplified RF voltage that appears across *EF* is shown in Fig. 30-29 (*a*). This wave is an exact replica of the transmitted wave of Fig. 30-27 (*c*). The amplified RF signal is next applied to the rectifier or detector tube (here a diode), wherein

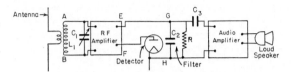

FIG. 30-28.—A simple radio receiver

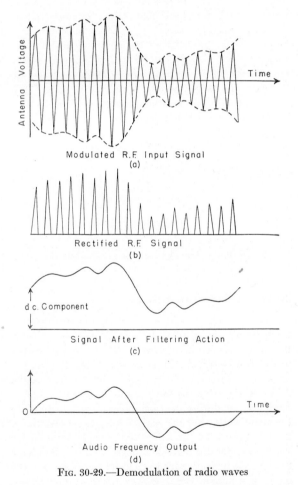

FIG. 30-29.—Demodulation of radio waves

the lower half of the RF signal is eliminated. The voltage appearing across *GH* is shown in Fig. 30-29 (*b*), the voltage across the resistor, *R*, after passing through the filter is that of Fig. 30-29 (*c*). The capacitor, C_3, prevents the direct-current component of the signal from reaching the audio-amplifier. The final audio-frequency output is

drawn in Fig. 30-29 (*d*). This is a replica of the audio-voltage generated by the microphone.

The student wishing to study in greater detail the refinements of modern transmitting and receiving circuits should consult the many excellent texts in this field. In transmitting speech or music by means of amplitude-modulated wave, the radio wave needs a band of frequencies centered about the carrier frequency.

Suppose f_c is the frequency of the carrier wave and f_v that of a frequency in the music, then the amplitude-modulated wave may be readily shown to contain three frequencies, namely, f_c, $(f_c + f_v)$, $(f_c - f_v)$. The frequencies $(f_c + f_v)$ and $(f_c - f_v)$ are called the *upper* and *lower* side bands. This phenomenon is analogous to that of beats in sound (chap. 32).

There is another manner of transmitting audio-signals by means of radio wave, which is known as *frequency modulation*. In this method the amplitude of the carrier wave remains constant, but its frequency is modulated by the impressed audio-frequencies. The wave shape of such a frequency-modulated wave has some resemblance to a concertina when it is being played. Frequency-modulation transmission obviously has a band width similar to that of the amplitude-modulated wave.

30.15. Microelectromagnetic Waves; Application to Radar

One of the most remarkable recent achievements in electronics is the development of intense sources of microelectromagnetic waves of wave lengths 10 cm, 3 cm, and shorter and with peak power of radiated energy of several million watts.

These sources of microwaves have been effectively employed with associated circuits in the design of systems for the detection and accurate location of airplanes, ships, clouds, land forms, and, as a matter of fact, any object capable of scattering electromagnetic waves. Such systems have been designated as *radar* (RAdio Detection And Ranging) *systems*. Radar techniques have been refined to a degree that permits not only detection of objects but also recognition of certain details of the objects that are scanned. This recognition of details is made possible because of the widely different scattering and absorption coefficients of materials for electromagnetic waves.

The accurate angular location of an object or **425**

target by radar was made possible through the generation of very high-frequency waves and the subsequent design of highly directional antenna systems. The determination of range is based on the experimental fact that electromagnetic waves travel through space with a constant speed of 299,792 km/sec. Therefore, a measure of the total elapsed time from the instant a wave leaves the antenna, strikes the target, and returns to the antenna is also a measure of the range of the target.

It is beyond the scope of this text to discuss the details of the electronic circuits and special tubes used in radar. Instead, a brief description will be given of a typical pulse-modulated radar system, using the block diagrams of Fig. 30-30. It should

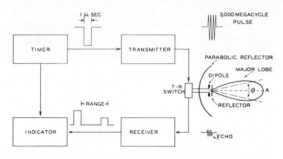

FIG. 30-30.—Block diagram of a radar system

be borne in mind that each block in the diagram contains many components necessary to produce the effects illustrated by the diagram.

The radar system that we shall describe is one that radiates energy in short, intense pulses whose time duration is about one-millionth of a second, i.e., 1 microsecond or 1 μsec. The transmitter is turned off after the pulse leaves the antenna and remains dormant until the echo is picked up by the antenna and passed to the indicator. Systems employing this technique are known as *pulse-modulated radar systems*.

The block diagram of Fig. 30-30 illustrates the five fundamental functional components of a radar system—the timer, transmitter, antenna system, receiver, and indicator. We shall briefly describe the function of each component.

30.16. Timer of a Radar System

Basically, the function of the timer is to supply synchronizing voltage pulses for the many parts of the radar. In one system the timer consists of a stabilized sine-wave generator and appropriate electronic circuits

that modify the sinusoidal voltage fluctuations into rectangular wave voltage pulses of high amplitude (about 12,000 volts) and of the proper width. The ideal rectangular voltage pulse is shown between the timer and the transmitter blocks of Fig. 30-30.

The time between pulses is known as the *pulse repetition time* and must be sufficient to permit an echo to return from a target within the workable range of the radar set before the next pulse is radiated. This range depends, among other factors, on the peak power radiated by the antenna and the sensitivity of the receiver.

With an 800-cycle/sec generator, for example, the pulse repetition time is 1,250 μsec. Since electromagnetic waves travel approximately 300,000 km/sec or 0.3 km/μsec, the maximum range is restricted to $\frac{1}{2}$(0.3 km/μsec)(1,250 μsec) = 187.5 km. The minimum range, set by the width of the pulse (of duration assumed to be 1 μsec), is $\frac{1}{2}$ (0.3 km/μsec) (1 μsec) = 0.15 km. Since a pulse of energy must travel from the antenna to the target and back again, the factor $\frac{1}{2}$ is introduced into the foregoing calculation. The relation between pulse width and pulse repetition time is shown in Fig. 30-31.

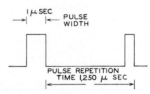

FIG. 30-31.—Pulse width and pulse repetition times

30.17. Radar Transmitters

The very large-amplitude, properly timed pulse from the timer is applied to an RF oscillator. It is the function of the oscillator to generate RF oscillations at the proper time intervals. These pulsed RF oscillations are shown to the right of the transmitter block of Fig. 30-30. As is evident from the foregoing discussion, the oscillator is on only during the time interval corresponding to the pulse width. For the case we have been describing, the oscillator is on for 1 μsec and remains dormant for 1,249 μsec. For this reason a properly designed oscillator can be greatly overloaded to yield RF peak power of the order of a million watts. However, the average power over the 1,250-μsec interval is a few hundred watts.

Another important constant of an RF oscillator is the choice of the carrier frequency. The frequency chosen for the carrier is determined by such considerations as the degree of directivity desired by the radar set and the feasibility of designing superhigh-frequency oscillators and receivers. In general, the shorter the wave length, the smaller the antenna system for a given degree of directivity. Carrier frequencies of the order of 3,000 megacycles/sec and higher are in use.

As we have indicated previously (§ 30.13), the frequency of oscillation is determined by the values of inductance and capacitance of the tank circuit of an oscillator. By making proper use of the distributed capacitance and inductance of the radio tube and its circuit elements, frequencies as high as 100 megacycles/sec can be attained with conventional-type tubes. To generate frequencies of 3,000 megacycles/sec or higher, corresponding to 10-cm waves or shorter, special oscillator tubes with very small distributed inductance and capacitance are required. The magnetron, shown diagrammatically in Fig. 30-32 (a), is one such tube.

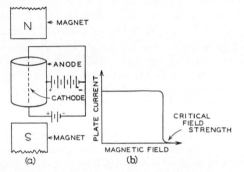

FIG. 30-32.—(a) A schematic diagram of a magnetron; (b) the plate current as a function of the magnetic field.

The magnetron is essentially a diode placed in a magnetic field in such a way that the straight cathode is parallel to the magnetic field (Fig. 30-32 [a]). The magnetic field may be supplied, for example, by permanent magnets of the Alnico type. With no magnetic field present, an electron leaving the heated cathode travels directly to the cylindrical anode that surrounds the cathode. As the field is increased, the electron no longer

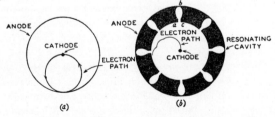

FIG. 30-33.—(a) Electron paths in a magnetron; (b) resonating cavities of a magnetron.

travels straight to the anode but in a curved path. As the field is increased, the curvature of the electron path is increased until, at a certain critical value of the magnetic field, the curvature is so great that the electron just misses the anode and returns to the cathode in a circular path, as shown in Fig. 30-33 (a). For greater magnetic field strengths the electron returns to the

cathode in a smaller circular path. It follows that the plate current for a given plate potential and filament temperature is constant and independent of the magnetic field until the critical field strength is attained, at which value the plate current drops abruptly to zero, as shown in Fig. 30-32 (b).

The magnetron operated at the critical magnetic field strength will produce oscillations of very high frequencies, owing to the currents induced by the fast-moving electronic charges. The frequency of oscillation is determined essentially by the time required for an electron to travel from cathode to plate and back again. In practice the magnetron is set in oscillation at the instant the 12,000-volt rectangular voltage pulse is placed across the tube.

The conventional magnetron has been modified by constructing the anode in the form of segments and connecting each pair of segments with resonating cavities, as shown in Fig. 30-33 (b). These cavities, because of the distributed inductance along *abc* and the capacitance between *a* and *c*, form a resonating tank circuit and give rise to microwaves of frequencies of the order of 3,000 megacycles/sec and higher. Owing to the combined effects of the magnetic fields and the electrostatic fields between the segments, the electron trajectory is unlike the loops of the conventional magnetron but consists, rather, of curves with many cusps, as shown in Fig. 30-33 (b). Plate 30-1 shows some interesting radar pictures.

30.18. Radar Antenna and Receiving Systems

The RF pulses from the transmitter are directed by suitable transmission lines and wave guides to an antenna that radiates the energy in the form of a highly directional beam or lobe, as shown in Fig. 30-30. In this figure the major lobe represents a polar diagram of the radiated signal strength in a vertical plane. A radius vector drawn from the dipole to any part of the curve is a measure of the signal strength in the direction of the radius vector. Radii vectors, drawn from the dipole to points on the major lobe corresponding to signal strengths 70 per cent of the maximum, determine an angle θ, defined as the width of the beam. Beam widths as small as $\frac{1}{2}°$ have been attained in both the vertical and the horizontal planes.

In some radar systems one antenna is used to radiate RF pulses and another to receive the echoes. It is possible to combine both functions in a single antenna. However, if a single antenna is used, provision must be made to prevent the high-amplitude RF pulse from entering the receiving circuit during transmission and destroying the sensitive detecting systems. A receiver protective device known as the "transmit-receive switch" or T-R switch is inserted between the transmitter and the receiver, as shown in Fig. 30-30. This T-R switch is electronic in character and is capable of acting within a few microseconds, since the receiver must be in position to receive an echo at the instant that the RF pulse leaves the antenna.

A practical form of an antenna to radiate microwaves of 10 cm or less is the parabolic reflector with the radiating dipole located at the focal point. An auxiliary reflector is placed about a quarter of a wave length in front of the dipole, in order to reflect nearly all the energy from the dipole back to the parabola, from which it is, in turn, reflected in a very narrow beam. The reflection of RF energy by the parabolic reflector or "dish," as it is commonly called, is analogous to the reflection of light by a parabolic mirror.

The receiver is required to amplify the very weak echoes from the antenna system, serve as a detector of the RF frequencies, and, in addition, amplify the detected pulse envelope sufficiently to be useful in operating the indicator. The amplified pulse envelope is shown in Fig. 30-30 between the receiver and the indicator blocks. As in the case of the transmitter, special tubes and circuits are required to amplify these high-frequency signals.

30.19. Radar Indicators

The indicator used in radar systems is the cathode-ray oscilloscope. The data received by the radar system may be presented by the cathode-ray tube in a variety of ways, depending on the use that will be made of the data. We shall briefly discuss two rather useful modes of data presentation: (1) the cathode-ray tube or oscilloscope may indicate only the amplitudes and time separation of the transmitted pulse and echo. This method of presenting data is called an "A scan presenta-

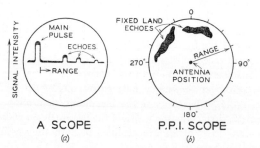

FIG. 30-34.—Radar traces in an oscilloscope

tion" and the cathode-ray tube an "A scope"; (2) a cathode-ray tube may present data in the form of a polar diagram, with range as distance from the center of the tube outward and azimuth angle along the periphery of the tube. Such a cathode-ray oscilloscope is called a "plan-position indicator" or simply a "P.P.I. scope" (see Fig. 30-34 [a], [b]).

An oscilloscope, employed as an A scope, is the conventional oscilloscope described in § 29.7. A linear sweep is applied to the horizontal deflecting plates, establishing a time base and thus the range. The output from the receiver is applied to the vertical plates, giving rise to the images shown in Fig. 30-34 (a). By rotating the antenna, both in elevation and in azimuth, until the signal amplitude of the echo is a maximum, the angular position of the reflecting target may be ascertained.

To obtain plan-position scanning, a cathode-ray tube employing electromagnetic deflection of the electron beam is generally used. Magnetic field coils mounted outside the tube are activated by currents controlled by the azimuth-position indicator of the antenna. An electron leaving the cathode of the oscilloscope and traveling toward the screen may be deflected to any desired position on the screen. If the antenna rotates continuously in azimuth, properly phased currents in the deflecting coils give rise to a rotating magnetic field in synchronization with the antenna rotation.

The sweep of this type of tube is adjusted to move the electron beam from the center of the tube radially outward. A negative potential is placed between the cathode and the anode of the electron run of the tube of such value that only a few electrons reach the screen. As soon as an echo is received, however, this amplified signal is applied between the cathode and the anode of the gun with such polarity as enormously to increase the intensity of the electron beam and hence produce a bright spot on the long-time persistence screen at the correct position. Thus, by proper synchronization, a polar map is described on the screen, with range plotted against postion in azimuth for a given elevation angle. Plate 30-1 contains actual photographs of cathode-ray screens showing various types of presentation.

The ability to generate microwaves of great frequency stability, large peak powers up to 10 megawatts for 30-cm waves, and covering a frequency range of 1,000–75,000 megacycles/sec (30 cm–4 mm) has opened up a new and important area for engineering applications and physical research. We shall touch upon a few of these applications.

Microwaves are presently employed for surveillance of an area to detect and guide aircraft; for navigational purposes on ships at sea; for the control of antiaircraft guns; for the control of guided missiles. Microwaves of 3-cm wave length are effective for the detection of storm areas and, at wave lengths of about 1 cm, are used to measure the top and bottom of clouds. Microwave relay stations, consisting of a series of microwave transmitters and receivers in the 7-cm band, have recently been constructed across the country for telephonic communications and for the transmission of television signals.

In the laboratory, microwaves are being successfully employed in the study of atomic and molecular struc-

1

2

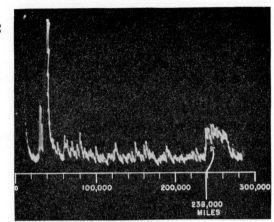

0-megacycle/sec search antenna used for the first radar
:tion of the moon (January 10, 1946). This antenna is
sentative of a broadside dipole array (see text).

"A scope" presentation of a radar echo from the moon,
showing the transmitted pulse and received echo.

3

4

3,000-megacycle/sec dipole radiator
its parabolic reflector, 6 ft in di-
ter.

A 10-cm wave-length cavity magnetron, capable of delivering approximately 100 kw
of RF peak power. The small size of the magnetron is indicated by the inch scale
(see text).

5

6

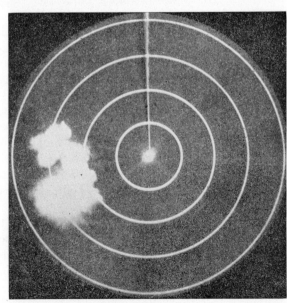

.I. scope presentation, showing a con-
leaving New York Harbor.

P.P.I. presentation of the precipitation pattern of a typical summer thunderstorm. The concentric circles are 5-mile markers.

ographs through the courtesy of Signal Corps Engineering Laboratory

tures. The linear accelerator designed to produce particles of energies of the order of 1,000 mev for nuclear research is based on the use of a series of microwave transmitters. Charged particles traversing a long, straight tube are given small accelerations by applying potentials at the proper time to a series of electrodes in the tube. Synchronized microwave generators supply the requisite voltages at the proper time intervals. One such accelerator designed to produce the 1,000-mev particles is expected to be over 200 ft long.

30.20. Television

A number of practical systems of television have been proposed that differ essentially in the techniques employed for scanning a given scene. Two techniques that have been especially successful are the "Image Dissector," developed by P. T. Farnsworth, and the "Iconoscope," of V. Zworykin. We shall briefly describe the iconoscope, a television pickup device illustrated schematically in Fig. 30-35.

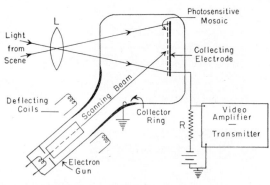

Fig. 30-35.—Television pickup system

The picture to be televised is focused on a photosensitive mosaic that contains an enormous number of tiny photocells electrically insulated from one another. The sensitive screen may consist of a thin sheet of mica, on one side of which is a mosaic of photosensitive globules of cesiated silver, whose diameters are of the order of 10^{-4} cm or less. The reverse side of the mica has a thin metal coating that serves as an output or collecting electrode. Each sensitive element forms, in effect, one side of a capacitor, the other being the collecting electrode.

When the optical image is focused upon the mosaic photosensitive surface, the elements lose electrons photoelectrically, the number of electrons emitted being dependent upon the light intensity on the element. These electrons are drawn to the collector ring on the inner surface of the tube, each element therefore becoming positively charged with respect to the back plate. The brightly illuminated portions are more positive than the dark portions. The image is then scanned by directing a focused cathode-ray beam back and forth across the mosaic, by means of magnetic-deflection methods (see Fig. 30-35). The electron beam then neutralizes each element of the mosaic, causing a current to pass through the resistance, R, proportional to the positive charge on the given element. The resulting voltage pulse or video signal developed across the resistance is finally amplified and impressed on a broadcast wave.

At the receiving end, the radio waves are intercepted by the receiving antenna, the video signal is properly amplified and is then applied to the deflecting plates of a cathode-ray tube similar in construction to the one described in § 29.7. By suitable synchronization the electron beam of the correct intensity is made to scan the screen of the cathode-ray tube. The result is a replica of the original image.

30.21. Electron Microscope

In § 21.14 we discussed the effects of electric and magnetic fields on the paths of moving electrons. It was pointed out there that, by proper design of the fields, it was possible to focus an electron beam. These fundamental principles have been brought together in the design of the *electron microscope*. In one instrument electrostatic fields of suitable design are used to focus the electron beam in very much the same manner as a lens focuses a beam of light (§ 34.17). In another type of instrument, magnetic fields are employed for this purpose. These fields are known as *electron lenses*.

The reason for the great interest in electron microscopes lies in the greatly increased resolving power (§ 35.18) possible with this instrument. As is pointed out in § 35.18, the smallest object that may be distinguished in an optical microscope is, because of diffraction effects, limited by the wave length of light employed. It has not been practicable to use light of wave length lower than about 3,000 angstroms (1 angstrom, 1 A = 10^{-10} m).

It was suggested by De Broglie in 1924 that moving particles, e.g., electrons, should exhibit properties of waves. He suggested that a particle of momentum, mv, has associated with it a wave length, λ, given by

$$\lambda = \frac{h}{mv}, \qquad (30\text{-}10)$$

where h is Planck's constant and has a value of about 6.6×10^{-27} erg-sec (§ 37.4). In 1927, Davisson and Germer and, independently, G. Thomson, experimenting with the diffration of electrons by crystals and thin foils, showed that electrons acted as waves with a wave length given by equation (30-10). Electrons with energies of 54 ev have, for example, a wave length, λ, of about 1.67×10^{-10} m or 1.67 A; for 100-ev electrons λ is about 1.22 A. Thus electrons may travel in straight lines with diffraction effects nearly negligible. By the use of 10^5-ev electrons, particles separated by distances of the order of 10^{-6} cm can be resolved and photographed. It thus becomes possible with the electron microscope to study the structure of the larger organic molecules, such as tobacco mosaic virus, of bacteria, and of any other objects for which very high magnification is necessary.

A schematic sketch of an electron microscope using magnetic lenses is shown in Fig. 30-36 (*a*). A corresponding optical system is drawn in Fig. 30-36 (*b*). The magnetic gaps on the coils are of such design as to produce the proper focusing action. (It is not intended that Fig. 30-36 [*a*] should show actual design details.) Electrons, accelerated by the source, proceed through the magnetic field condensing lens, where they are focused upon an object, *O*, placed on a very thin layer of cellulose. The electrons diverging from the object are focused by the magnetic objective lens forming the first image I_1, magnified some 100 times. The magnetic projecting lens forms the final image I_2, magnified again some 200 times, on a fluorescent screen or a photographic plate. A total magnification of some 20,000 diameters can be utilized, since the resolving power is large.

The mechanism for the production of an image by the electron microscope is basically different from that of the optical microscope. An image is formed because an appreciable fraction of the electrons, as they pass through the specimen, is scattered rather than refracted. The scattering process results from the interaction of the electronic charge with the electrostatic field of the atoms and molecules of the specimen. Thus more electrons are scattered from the denser parts of the specimen than from the less dense parts, thereby producing a variation in the blackening of the photographic film. The picture, therefore, is more like an X-ray picture than an optical one, the essential difference being that electrons can be readily focused, X-rays cannot.

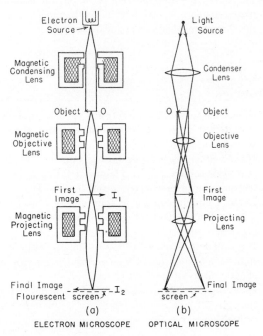

Fig. 30-36.—Schematic diagrams of the formation of images by electron and optical microscopes.

Although the modern electron microscope has a magnification and resolving power large enough to make individual virus particles visible, there is a distinct disadvantage to the instrument in its present form. It is necessary to introduce the specimens into a high vacuum to avoid scattering of the electron beam by gas particles. This procedure quickly dries out living organisms. Also, the intense electron beams heat and destroy the living microörganisms. Some techniques have been developed to minimize these defects by placing the object between very thin membranes. Much research in technique is necessary before the full capabilities of the electron microscope are exploited.

30.22. Résumé

The following is a brief listing of the definitions, principles, and theories, given in this chapter, which you should know:

The general principles of electron emission from hot bodies and the importance of the work function

The meaning of space charge and temperature saturation in electronic tubes

The principle of rectification and the advantages of electrical filters for reducing ripple

The importance of the third electrode or grid in an electronic tube

Definitions of the amplification factor, μ; the dynamic plate resistance, r_p; the mutual conductance, g_m; and the theoretical relationship of these quantities

Method of measuring the foregoing quantities

The determination of the voltage amplification of a circuit by graphic analysis

The principal characteristics of the tetrode and the pentode

The characteristics of the cold- and hot-cathode gas tubes

The use of a pentode and thyratron in a linear voltage or "saw-tooth" circuit

The basic theory of the production of electromagnetic radiation

The principles of a simple radio transmitter and receiver

Some of the characteristics of radar, of television, and of the electron microscope

Queries

1. Give some reasons why electronic vacuum tubes are operated at space-charge saturation rather than at temperature saturation.

2. If the voltage V_p between the cathode and the plate of a diode is increased by a factor of 4, then by what factor is the space-charge current increased?

3. Draw a circuit employing four diodes which produces full-wave and full transformer voltage rectification.

4. If a capacitor is placed across the load resistor R in Fig. 30-6 (a), is the voltage across the resistor increased, decreased, or not changed?

5. How does the plate resistance r_p definition differ from that given by Ohm's law and why is this not used rather than r_p?

6. For a pentode operating with normal voltages on the electrodes give the direction of the electric field from (a) the suppressor grid to the plate; (b) the suppressor grid to the screen grid; (c) the screen grid to the control grid; and (d) the control grid to the cathode.

7. The voltage-time curves for the saw-tooth generator Fig. 30-19 (b), are drawn as straight lines. Explain how far this is correct and state the factors on which the degree of correctness depends; see equation (29-11).

8. In the tank circuit of Fig. 30-25 oscillations are set up. State the types of energy involved in the oscillations and compare with an oscillating simple pendulum.

Practice Problems

1. Calculate the emission current density, J, in amperes per square centimeter for tungsten at a temperature of $2,600°$ K. (Take $e^{-20} = 2.06 \times 10^{-9}$.) Ans.: 0.84.

2. Calculate the emission current density, J, in amperes per square centimeter for thorium at a temperature of $2,600°$ K. (Take $e^{-15} = 3.06 \times 10^{-7}$.)

3. Calculate the emission current density, J, in amperes per square centimeter for thoriated tungsten at $1,500°$ K. Ans.: 0.014.

4. Repeat question 3 for a temperature of $2,600°$ K. (Take $e^{-11.7} = 8.3 \times 10^{-6}$.) Ans.: 168.

5. Show that the ratio of emitted current densities of thoriated tungsten and tungsten at $2,200°$ K is about $0.05\ e^{10}$ or about 1,100.

6. Show that, by differentiating equation (30-1),

$$\frac{dJ}{J} = \frac{2\,dT}{T} + \frac{b}{T}\frac{dT}{T}.$$

7. If there is a 1 per cent change in temperature of a tungsten filament operating at $2,600°$ K, show that the percentage change in the emission current is 2 per cent due to the T^2 term and about 20 per cent due to the exponential term, giving a total change of 22 per cent.

8. Compute the saturation current for a thoriated tungsten filament 2 cm long and 0.5 mm in diameter when its temperature is $1,800°$ K.

9. A diode consists of a central filament with a

cylindrical collecting electrode. Show that if the voltage between plate and filament is doubled, the space-charge current is increased by a factor of about 2.8.

10. Prove the relationship given in equation (30-7);
$$\mu = r_p g_m.$$

11. A two-electrode vacuum tube has 1,000 volts applied between the filament and the plate. What is the approximate speed of the electrons when they reach the plate? The mass and charge of an electron are 9×10^{-31} kg and 1.6×10^{-19} coulomb, respectively. Ans.: 1.9×10^7 m/sec.

12. If in problem 11 all the energy of the electrons goes into heating the plate, find the number of calories produced in the plate per minute.

13. Using equation (30-8), compute the voltage amplification of an amplifier which uses a 6J5 tube with characteristics given in Fig. 30-9. The amplifier uses a 50,000-ohm load resistor. Ans.: 13.

14. From Fig. 30-13 measure the amplification and plate resistance for the 6F5 tube and show that these are about 100 and 66,000 ohms, respectively.

15. A three-electrode tube has the following characteristics: a change of 10 volts in the plate voltage gives rise to a change in plate current of 5 milliamperes. A change in the grid voltage of 0.4 volt restores the plate current to its previous value. Find the approximate values of μ, r_p, and g_m for the tube. Ans.: 40; 2,000 ohms; 0.02 mho.

16. Show that electrons accelerated through a potential difference of 100 volts have a wave length of about 1.2×10^{-8} cm.

17. Show that if the accelerating voltage applied to electrons is doubled, the corresponding wave lengths of the electron waves are in the ratio of about 1 to 0.707.

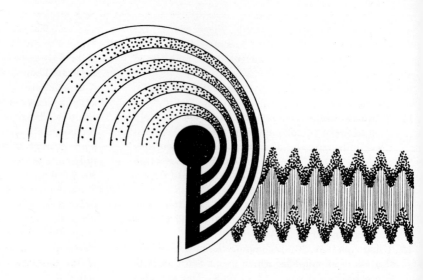

CHAPTER 31

Wave Motion

31.1. Introduction

In approaching the subject of wave motions we are entering upon the discussion of a great realm of physical phenomena. Most, if not all, of the knowledge we possess on the material world has come to us through such evidence as can be gleaned from our two major kinds of sense impressions: sight and hearing. A race of creatures totally blind and totally deaf would seem to have little chance of ever acquiring even such a fragmentary structure of knowledge about the environing world as we possess.

It is important, first, to note that both these major senses of ours respond to events that occur *at a distance* from the observer. Through them we have developed ideas of space, or distance. From the observation of other objects in motion in localities more or less remote from us and from certain psychological impressions of "sooner, now, and later," ideas of time, as well, are firmly imbedded in our thinking. In this dual framework, sometimes combined, for purposes of mathematical convenience, into a single four-dimensional "world" of "space-time," we formulate whatever precise descriptions we are able to obtain of the "events" of physical phenomena.

The information reaching our end organs of sense impression, in these two most useful senses, comes from distant places, where the phenomena are occurring. It does not come instantaneously but requires a finite lapse of time. Thus we never either hear or see anything happen at the moment it occurs, but only at some later moment. In the case of sound the mechanical disturbances that constitute *any* source of sound set up in the material medium (air, water, or solid material that surrounds the source) a series of *related* mechanical disturbances in that medium. These disturbances travel outward through the medium from the source for definite and well-known physical reasons. SOUND WAVES, which consist of regular and systematic motions of the molecules of the transmitting medium (air, water, steel), superimposed on the thermal motions of these molecules, travel away from a sound source and produce in our ears other mechanical effects, affecting sensory nerves and interpreted by our brains as sound.

It is not so simple in the case of vision. Mechanical, electrical, or thermal phenomena involve disturbances of the molecules, atoms, and electronic structures of objects; and associated with these disturbances is the rapid transport of energy over great distances. Mysterious as is the mode of transmission of energy through empty space, we can, nevertheless, experiment upon the effects it produces whenever this radiant energy is received and *absorbed*.

When we do this, we discover that it has certain characteristics that we call "wavelike." We can identify some of the same sorts of qualities—make some of the same kinds of descriptions—about it that we do in the well-defined, even photographable, material phenomena that we call "sound waves." Thus we talk of "light waves," "radio waves," etc. In all these cases, however, *the important property of the wave is its ability to transfer energy.* Other characteristic properties of waves are given in later sections.

The last three decades of research and investigation, especially into phenomena of the more minute orders of magnitude, have brought to light another astonishing fact. Wavelike attributes are to be found associated not merely with these aspects of energy which we call "radiant." Even those somewhat more localized entities like electrons, protons, neutrons, and other "particles" of matter, when in motion, possess wavelike characteristics. We have much reason to believe that even larger aggregates of matter, likewise, may have similar wavelike attributes. At least, it is possible by means of "wave equations" to describe a large portion of the content of modern physical theory.

These, in brief, are the reasons why it is most important for the novitiate in the field of physical science to have a thorough understanding of what is meant by "wave motion." Its most simplified form is that with which we begin, according to our custom—the idealized and almost purely geometrical form.

Most of us are familiar with the waves in a rope, in a spring, or on a large body of water. In each case the medium has inertia and exerts a restoring force on any portion of it when it is displaced from its equilibrium position. The inertia and the restoring forces are distributed uniformly throughout the medium. In the discussion of the simple harmonic motion of a spring (§ 8.9) we assumed that the restoring force and the inertia were entirely separate, namely, in the spring and in the mass hung on the end. For such a case there is no wave motion set up. If the spring had been heavy enough that the inertia resided entirely in it, then waves would have been produced in it.

31.2. Some Fundamental Definitions in Wave Motion

In order to understand some of the fundamental aspects of wave motion, let us consider a simple sine wave traveling along a rope. An instantaneous

picture of the rope at some time t is shown in Fig. 31-1. At this time the rope has maximum displacements at B, C, and D. Points such as B and D are said to be in the same phase as are the points O and E. Point C is 180° out of phase with B or D. The distance of the points B, C, and D from the undisplaced position OX of the rope is called the *amplitude* of the wave. Thus the amplitude A of

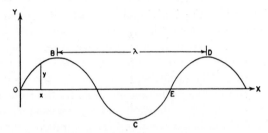

Fig. 31-1.—Transverse waves in a rope

the wave is the maximum height of a crest or the depth of a trough. Different positions x along the rope have different displacements y. The distance between two successive positions on the rope having the same displacement and the same phase is called a *wave length* of the wave. From this it follows that the wave length of the wave is the distance between two successive crests or two successive troughs. In Fig. 31-1 the wave length, λ, is equal to OE or BD.

Now a short time later or at time $t + \Delta t$ the wave has moved along the rope, and if the wave is moving to the right, it has the position indicated by the dashed lines in Fig. 31-2. The crests and

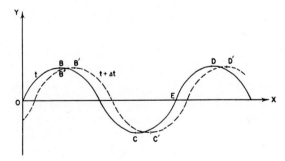

Fig. 31-2.—Transverse waves at time t and time $t + \Delta t$

troughs have moved to the right, while the individual elements of the rope have moved vertically up or down, that is, transverse to the direction of the wave. Such a wave is called a *transverse wave,* in contrast to longitudinal waves, in which the individual elements move back and forth in the **435** same direction as the wave. These we shall discuss later.

Any individual element of the rope undergoes simple harmonic motion about the undisplaced position OX. Thus in the short time Δt in Fig. 31-2 the element of rope at B has moved downward to B'', while the crest has moved to B'. It is the geometrical configuration of crests and troughs which constitutes the wave and travels to the right while the individual elements of the rope move up and down. The time for the portion of the rope at B to move from that of a crest to a trough and back to a crest is called the *period*, T, of the wave. It is the same definition used for the period of a simple harmonic motion, namely, the time T for a complete vibration. The *frequency*, f, of these vibrations is the number of complete vibrations per second so that, as in equation (8-2), we have

$$T = \frac{1}{f} \quad \text{or} \quad f = \frac{1}{T}. \qquad (31\text{-}1)$$

In the wave on the rope, every particle of the rope is undergoing simple harmonic motion, and it is such that during the time the element of rope of B undergoes one complete vibration, from crest to trough and back to crest, the crest at B has traveled from B to D, a distance of one wave length. A similar statement applies to any element of the rope, whatever its displacement, so that the wave moves to the right a distance of one wave length, λ, in a time equal to the period T. Thus the velocity, v, of the wave is the wave length divided by the period, T, or

$$v = \frac{\lambda}{T} = f\lambda. \qquad (31\text{-}2)$$

This velocity is called the *phase velocity* and is the only velocity involved in a simple harmonic wave. In contrast to this is a group velocity, which is important when several wave lengths and phase velocities travel through a medium. For the present we shall limit the discussion to simple harmonic waves, such as are shown in Fig. 31-2.

31.3. Analytical Expression for a Progressive Wave

In a simple harmonic wave such as that shown in Fig. 31-3, the displacement y depends on the position x of the particular element of the rope chosen and the time t which is being considered. In

other words, the displacement y is a function of the position x and time t, or

$$v = F(x, t).$$

For the rope in Fig. 31-2 taken at time $t = 0$, the function is a simple sine curve, and the equation of the curve is given by a displacement y at any position x, as

$$y = A \sin \frac{2\pi x}{\lambda}, \qquad (31\text{-}3)$$

where A is the amplitude of the sine curve at such points as B and C in Fig. 31-2. As can be seen from

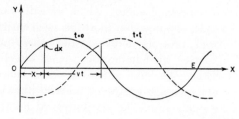

Fig. 31-3.—Displacement y at time zero and time t

Fig. 31-3, the value of y is zero at $x = 0$, and is also zero when $x = \lambda$, the distance OE.

Now at some time t later, the wave has traveled to the right a distance vt, as is shown by the curve with dashed lines. If the curve representing the wave at time t is moved backward through the distance $-vt$, then it will coincide with that at time zero. Thus the equation of the curve at time t is

$$y = A \sin \frac{2\pi}{\lambda} (x - vt). \qquad (31\text{-}4)$$

In a time equal to a period T, the wave has traveled a distance λ, so that there is the same displacement y at some value x and time t as there is at $(x + \lambda)$ and $(t + T)$, so that

$$v = A \sin \frac{2\pi}{\lambda} [(x + \lambda) - v(t + T)]$$

$$= A \sin \frac{2\pi}{\lambda} (x - vt),$$

since $\lambda = vT$.

A wave traveling to the left along the negative x-axis has the equation

$$v = A \sin \frac{2\pi}{\lambda} (x + vt). \qquad (31\text{-}5)$$

In both the waves represented by equations (31-4) and (31-5) it is assumed that the displacement y is zero at $x = 0$ and $t = 0$. This, of course, need not be the case, and the general expression for a wave traveling to the right is

$$y = A \sin\left[\frac{2\pi}{\lambda}(x - vt) + a\right], \quad (31\text{-}6)$$

where a is called the *phase angle*. If this phase angle is $90°$ or $\pi/2$ rad, then the displacement at $x = 0$ and $t = 0$ is $y = A$.

For a given value of x, that is, a given position on the rope, the displacement y can be written as

$$y = A \sin\left(\frac{2\pi vt}{\lambda} + \beta\right). \quad (31\text{-}7)$$

This is similar to equation (8-11) for simple harmonic motion, for which the phase angle β in equation (31-7) is equal to $\pi/2$ rad. Thus any particular element of the rope undergoes simple harmonic motion about its equilibrium position.

31.4. Velocity of a Wave in a Rope

The velocity of a wave in a uniform rope can easily be shown experimentally to depend only on the tension or force, F, exerted on the rope and the mass per unit length, μ, of the rope. By dimentional analysis we can easily find the velocity of the wave in a rope if we assume that the velocity depends only on the tension F and the mass per unit length μ. In terms of mass [M], length [L], and time [T], the dimensions of F and μ are

$$F: \ [\text{MLT}^{-2}],$$

$$\mu: \ [\text{ML}^{-1}].$$

The only manner in which these dimensions can be combined to yield a velocity v having dimensions $[\text{LT}^{-1}]$ is by taking the square root of F/μ. Thus the dimensions of F/μ are

$$[\text{MLT}^{-2}\,\text{M}^{-1}\text{L}] = [\text{L}^2\text{T}^{-2}],$$

and the velocity v is given by

$$v = \sqrt{[\text{L}^2\text{T}^{-2}]} = \sqrt{\frac{F}{\mu}}. \quad (31\text{-}8)$$

Dimensional analysis does not introduce any dimensionless quantities into the expression for velocity, so that equation (31-8) may or may not be the correct expression for the velocity. The most that we can say is that the velocity is equal to a constant times $\sqrt{F/\mu}$. The value of the con-

stant is determined by experiment or by a physical analysis of the problem. These methods show that the constant is equal to unity, and equation (31-8) is the correct expression for the velocity of a wave in a rope. The units which must be used in equation (31-8) are the force F, which must be in newtons, dynes, or lbf, with the corresponding units for the mass per unit length μ in kg/m, gm/cm, or slugs/ft. For these three sets of units the velocity is in m/sec, cm/sec, or ft/sec.

Suppose a rope has a value of μ of 2 lb/ft (which is equivalent to $\frac{2}{32}$ slugs/ft), and the tension on the rope is 4 lbf; then the velocity of a progressive transverse wave in the rope is

$$v = \sqrt{\tfrac{4}{2}\ \text{lb-ft/lb} \times 32\ \text{ft/sec}^2}$$

$$= 8\ \text{ft/sec}.$$

31.5. The Differential Equation for and Velocity of a Transverse Wave in a Rope

This section definitely presupposes a knowledge of the calculus and may be omitted without loss of continuity in this chapter. Suppose the rope of mass per unit length μ is subjected to a uniform tension F. Consider a small section of the rope in Fig. 31-3 at position x and having a length dx. Such a section, shown in Fig. 31-4, has a mass μdx, and the slopes at the two ends are

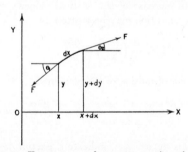

Fig. 31-4.—Forces on an elementary section of a rope

given by the angles θ_1 and θ_2. It is assumed that these angles are sufficiently small that their cosines may be taken as unity and their sines may be set equal to their tangents. From the expressions for the cosine or sine we have

$$\cos\theta = 1 - \frac{\theta^2}{2}\cdots,$$

$$\sin\theta = \theta - \frac{\theta^3}{3}\cdots,$$

so that we are assuming that θ^2 and higher powers of θ are negligibly small compared to unity. Resolving the **437** force F at the two ends into their horizontal and vertical components, we have

$$F_x = F \cos \theta_2 - F \cos \theta_1 \,,$$

$$F_y = F \sin \theta_2 - F \sin \theta_1 \,.$$

Thus, since we are assuming $\cos \theta_2 = \cos \theta_1 = 1$, it follows that the resultant horizontal force on the element is zero. To the same degree of approximation we may set

$$\sin \theta_2 = \tan \theta_2 \quad \text{and} \quad \sin \theta_1 = \tan \theta_1 \,;$$

so that the resultant vertical force in the upward direction is

$$F_y = F \tan \theta_2 - F \tan \theta_1 \,.$$

Now $\tan \theta_1$ is the slope of the rope at the points x, y, so that

$$\tan \theta_1 = \frac{\partial y}{\partial t} \,,$$

and $\tan \theta_2$ is the slope at $(x + dx)$, $(y + dy)$, and this is approximately equal to the slope at (x, y) plus the rate of increase of the slope times the element of length. Thus we have

$$\tan \theta_2 = \frac{\partial y}{\partial x} + \frac{\partial}{\partial x} \left(\frac{\partial y}{\partial x} \right) dx = \frac{\partial y}{\partial x} + \frac{\partial^2 y}{\partial x^2} \, dx \,.$$

Hence

$$F_y = F \tan \theta_2 - F \tan \theta_1 = F \frac{\partial^2 y}{\partial x^2} \, dx \,. \quad (31\text{-}9)$$

From Newton's Second Law the vertical acceleration of this element is $F_y / \mu dx$. This vertical acceleration is the time rate of increase of the vertical velocity $\partial y / \partial t$, so that the acceleration may be written as

$$\frac{\partial}{\partial t} \left(\frac{\partial y}{\partial t} \right) = \frac{\partial^2 y}{\partial t^2} \,.$$

The equation of motion of the element is

$$\frac{\partial^2 y}{\partial t^2} = \frac{F}{\mu \, dx} \frac{\partial^2 y}{\partial x^2} \, dx = \frac{F}{\mu} \frac{\partial^2 y}{\partial x^2} \,. \quad (31\text{-}10)$$

The symbols $\partial y / \partial x$, etc., indicate a partial derivative. Since the displacement y depends on, or is a function of, both x and t, the quantity $\partial y / \partial x$ means the rate of change of y with x when t is constant; similarly, $\partial y / \partial t$ means the rate of change of y with t when x is constant.

By substitution, it can be shown that the differential equation (31-10) is satisfied by any function of $(x - vt)$ or $(x + vt)$ where $v = \sqrt{F/\mu}$. Let us consider the sine function, since we have used this previously, that is, we shall substitute

$$y = A \sin k \, (x - vt) \quad (31\text{-}11)$$

in the differential equation and show that it is satisfied if $v = \sqrt{F/\mu}$. In this equation A and k are constants, though we know from our earlier discussion that A is

the amplitude of the wave and k is $2\pi/\lambda$, where λ is the wave length. By differentiating equation (31-11), we have

$$\frac{\partial y}{\partial t} = - \, vk A \cos k \, (x - vt) \,, \quad (31\text{-}12)$$

$$\frac{\partial^2 y}{\partial t^2} = \frac{\partial}{\partial t} [- \, vk A \cos k \, (x - vt)] \quad (31\text{-}13)$$
$$= - \, v^2 k^2 A \sin k \, (x - vt) \,,$$

$$\frac{\partial y}{\partial x} = k A \cos k \, (x - vt) \,, \quad (31\text{-}14)$$

$$\frac{\partial^2 y}{\partial x^2} = - \, k^2 A \sin k \, (x - vt) \,. \quad (31\text{-}15)$$

Substituting in equation (31-10) gives

$$- \, v^2 k^2 A \sin k \, (x - vt)$$
$$= - \sqrt{\frac{F}{\mu}} \, k^2 A \sin k \, (x - vt)$$

Hence

$$v^2 = \frac{F}{\mu}$$

or

$$v = \sqrt{\frac{F}{\mu}} \,.$$

Notice that in carrying our the differentiation with respect to t, all the other quantities, including x, are considered constant, whereas in finding $\partial y / \partial x$ the time t is considered constant. Thus we have shown that

$$y = A \sin k \, (x - vt)$$

is a solution of the differential equation

$$\frac{\partial^2 y}{\partial t^2} = \frac{F}{\mu} \frac{\partial^2 y}{\partial x^2} \,.$$

Actually, we could have taken any function of $(x - vt)$ or of $(x + vt)$ and have shown that it satisfied the differential equation with $v = \sqrt{F/\mu}$. Were we to try a solution of the type

$$y = A \sin k \, (x^2 - vt) \,,$$

we should find that it would not satisfy the differential equation or that this is not a solution of the differential equation.

The differential equation (31-10) is called the *wave equation* in one dimension, that is, along the X-axis in the YZ plane. For a drumhead the wave equation would be similar to equation (31-10) but would contain another term involving $\partial^2 y / \partial z^2$. The wave equation for a wave emanating from a point in space and moving uniformly in all directions would involve three terms.

As stated previously, the quantity k in equations (31-11)–(31-15) is given by

$$k = \frac{2\pi}{\lambda} \qquad (31\text{-}16)$$

and varies with the wave length of the wave.

31.6. Longitudinal Waves

Sound is propagated through the air or any material medium by longitudinal waves. For such waves the displacement of the medium is along the direction of propagation. In order to understand the mode of propagation of longitudinal waves, let us consider a series of equidistant planes in a continuous medium (Fig. 31-5). For

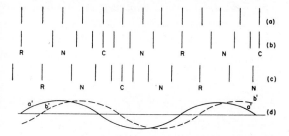

FIG. 31-5.—A longitudinal wave in a medium: (a) undisplaced positions; (b) displaced positions at some time t; (c) displaced positions at $t + \Delta t$; (d) horizontal displacements shown as vertical ones.

convenience, we shall consider a plane wave traveling along the X-axis. Some form of periodic disturbance is acting on the left side of the medium, giving rise to the displacements shown in (b). This is an instantaneous picture of a longitudinal wave in the medium, showing the positions of the compressions at C, rarefactions at R, and regions of normal density at N. The distance between two successive rarefactions is a wave length, λ. We used these positions for convenience, for a wave length is the distance between any two successive positions having the same phase.

These planes, as part of the medium, are in motion and vibrate with simple harmonic motion about their undisplaced or equilibrium positions. In diagram (c) the displacements are shown for a short time later (about one-sixth of a period) than in (b). The longitudinal wave has traveled to the right about one-sixth of a wave length. In a time equal to a period of vibration T, the wave has traveled a distance of one wave length, λ, so that

the velocity of these waves is λ/T, as given in equation (31-2).

For convenience of visualization, diagram (d) is shown in which displacements to the right in the longitudinal waves are placed vertically upward and corresponding displacements to the left are placed vertically downward. The horizontal displacements in (b) are shown as vertical displacements in a' and similarly for c and b'.

31.7. The Velocity of Longitudinal Waves in Various Media

The velocity of longitudinal waves depends on the properties of the medium in which they are traveling. In general, these properties are the modulus of elasticity and the density of the medium. It is the elasticity which provides restoring forces on the displaced regions; it is the mass per unit volume or density which provides the inertia of the system. By means of dimensional analysis, we can readily show that the velocity of a longitudinal wave in a medium is given by the square root of the modulus of elasticity, E, divided by the density, ρ.

A modulus of elasticity E is expressed in terms of force divided by an area and has the dimensions of

$$[MLT^{-2}L^{-2}] = [ML^{-1}T^{-2}] \,.$$

The dimensions of a density ρ are

$$[ML^{-3}] \,.$$

By division of these quantities we see that the dimensions of

$$\frac{E}{\rho} \quad \text{are} \quad \frac{[ML^{-1}T^{-2}]}{[ML^{-3}]} = [L^2T^{-2}] \,.$$

Hence it follows that the velocity, v, of these waves is given by

$$v = D\sqrt{\frac{E}{\rho}}, \qquad (31\text{-}17)$$

where D is a dimensionless constant and E is some modulus or a combination of moduli of elasticity as yet unspecified.

The particular modulus of elasticity used depends on the nature of the wave and the medium. Though the actual expressions for the velocities of the different kinds of waves are derived by physical analysis beyond the scope of this book, nevertheless we shall see that they all conform to **439** that given by equation (31-17). We shall now present a few results.

a) Velocity of longitudinal waves in the body of a gas or liquid but not along the surface.—In this case the modulus of elasticity E is the bulk modulus K, while the dimensionless constant D is unity. Hence the velocity of longitudinal waves in a gas or liquid is given by

$$v = \sqrt{\frac{K}{\rho}}. \qquad (31\text{-}18)$$

For water, $K = 2 \times 10^9$ newtons/m² and $\rho = 10^3$ kg/m³, so that the velocity of longitudinal waves or of sound in water is

$$v = \sqrt{\frac{2 \times 10^9}{10^3}} = \sqrt{2 \times 10^6} = 1{,}414 \text{ m/sec}.$$

b) Velocity of longitudinal waves in a gas.—In chapter 13, § 13.18, it is shown that the bulk modulus for isothermal processes in a gas is equal to the pressure of the gas, while for adiabatic processes the bulk modulus is equal to γp, where γ is the ratio of the specific heats of the gas at constant pressure and constant volume.

If the temperature of a gas remains constant when a longitudinal wave passes through it, then the velocity v of sound in the gas is

$$v = \sqrt{\frac{p}{\rho}}. \qquad (31\text{-}19)$$

For air at 0° C this is equal to about 280 m/sec, whereas the measured velocity of sound for this case is about 332 m/sec.

Newton was aware of this discrepancy between the theoretical and the experimental values, but he failed to make the correction in an acceptable manner. It was Laplace in 1816 who pointed out the error in the assumption that the temperature of a gas remains constant when a longitudinal wave passes through it. Consider a sound wave having a frequency of 400 vibrations/sec passing through air. The time between a compression and the succeeding rarefaction is then 1/800 sec. In a compression the gas is heated, while in a rarefaction it is cooled, and this goes on at a very rapid rate. The gas does not have time to lose to or gain heat from the surroundings between the successive heatings and coolings, so that when a longitudinal wave passes through a gas, the process is essential-

ly adiabatic. In equation (13-71) the bulk modulus for a gas under adiabatic conditions is

$$K = \gamma p \,,$$

where γ is the ratio of the specific heat at constant pressure to that at constant volume and has a value of about 1.40 for air. Thus the velocity of longitudinal waves or of sound in a gas is

$$v = \sqrt{\frac{\gamma p}{\rho}}. \qquad (31\text{-}20)$$

Substituting, in the values for air, the velocity of sound in air from equation (31-20) at 0° C, we find about 332 m/sec, which is in excellent agreement with the measured value. In the next chapter we shall discuss the effect of changes of pressure and temperature on the velocity of sound.

c) Velocity of longitudinal waves in a solid rod.—The modulus of elasticity, which applies in the case of a solid rod, is Young's modulus Y, so that, from equation (31-17), the velocity of sound in a rod is

$$v = \sqrt{\frac{Y}{\rho}}. \qquad (31\text{-}21)$$

For oak wood Young's modulus is about 1.4×10^{10} newtons/m², and its density is about 750 kg/m³, so that the velocity of sound in a rod of oak wood is

$$v = \sqrt{\frac{1.4 \times 10^{10}}{750}} = \sqrt{1.87 \times 10^7} = 4{,}320 \text{ m/sec}.$$

This is somewhat over ten times the velocity of sound in air and about three times that in water.

d) Velocity of waves in an unlimited solid.—In an unlimited solid such as is present for earthquake waves in the body of the earth, the velocity of longitudinal waves is

$$v = \sqrt{\frac{K + 4n/3}{\rho}}, \qquad (31\text{-}22)$$

where K is the bulk modulus and n is the rigidity modulus of the medium. These latter may be contrasted with the velocity of transverse waves such as occur in unlimited solids like the earth. The modulus of elasticity applicable to the transverse wave is the modulus of rigidity n, and the velocity is given by

$$v = \sqrt{\frac{n}{\rho}}. \qquad (31\text{-}23)$$

In an earthquake there is a readjustment of the rocks at some relatively small distance below the earth's surface which creates both longitudinal and transverse waves. These waves are then detected by seismographs at various places on the earth. The first wave to arrive is the longitudinal one, called the *primary wave*, then follows the transverse wave, called the *secondary wave*. Both these velocities vary with the depth below the surface, and it is by analysis of these waves that some idea is obtained of the elastic constants at various depths of the earth. There is some question as to whether the transverse wave can travel through the center of the earth, for a transverse wave cannot exist in a liquid. The analysis of the seismographic records is not easy, and thus the question of whether an interior zone of the earth is solid or liquid is still debatable.

31.8. Waves on the Surface of Liquids

Anyone who has been in a storm at sea or on a large lake is vividly aware of the energy transmitted by the waves on the surface of the water. In general, these are caused by winds through friction with the surface of the water. Pouring oil on the water may lessen this friction and the size of waves, so that the old adage of pouring oil on troubled waters may have some physical basis. The velocity of surface waves on a liquid depends on the physical conditions—whether the liquid is deep compared with the wave length or not. Before discussing these differences, we shall first take up the subject of the nature of the displacement in these waves.

It is obvious to any swimmer that the displacements which constitute waves on the surface of water are both horizontal and vertical. The waves are, at one and the same time, both longitudinal and transverse. The crest carries one upward and forward; in the trough one sinks down and is pulled back. Furthermore, these two components are not in phase. When one's upward velocity has ceased, one's forward velocity is greatest. At the lowest part of the trough, one has ceased to drop, but the backward pull is greatest there. As to the form of the wave, one knows that the crest is a relatively narrow uplifted region and that the troughs are broad and much more extensive.

Detailed analysis shows that both the vertical and the horizontal components of the motion are simple harmonic motions of equal amplitude and that the phase difference is $\pi/2$ rad. The sum of two such motions, as discussed in § 8.16, is a uniform circular motion. This, then, is the motion of the individual particles of water as the wave passes over them. Particles at the surface execute the largest displacements, i.e., move in circles of the largest radius. The amplitude A of the motion, that is, the radius of this circle, decreases rapidly with depth; if it is A at the surface, then it is $0.04 A$ at a depth of $\lambda/2$ and $0.002 A$ at a depth λ. Submarines, therefore, do not have to travel very far below the surface to be in a region quite free from tempestuous conditions that may be taking place above.

The foregoing facts give us a device for drawing the wave form of surface waves. Figure 31-6 shows

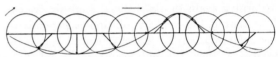

Fig. 31-6.—A wave surface in a liquid in which individual particles move in circles.

a series of particles, each constrained to move in a circle. All the particles, however, have definite phase relations between them as we pass from one circle to the next. Note the narrow crest and broad trough. Also the particles in the crest have their maximum backward velocity.

31.9. Velocity of Surface Waves on Water

For waves on the surface of water in which the depth is very much larger than the wave length, the controlling factors on the waves are gravity and their wave length. By dimensional analysis we can readily show that $\sqrt{g\lambda}$ has the dimensions of a velocity. A detailed analysis of the deep-water waves shows their velocity to be

$$v = \sqrt{\frac{g\lambda}{2\pi}}. \qquad (31\text{-}24)$$

EXAMPLE 1: (*a*) What is the velocity of a wave having a wave length of 3 ft at the surface of very deep water?

Solution:

$$v = \sqrt{\frac{32 \times 3}{2\pi}} \text{ ft/sec}^2 \times \text{ft} = 3.9 \text{ ft/sec}.$$

b) How many waves per mile are there in an ocean ground swell which travels 50 mi/hr (73.4 ft/sec)?

Solution:

$$\lambda = \frac{2\pi v^2}{g} = \frac{2\pi (73.4)^2}{32} \text{ ft}^2/\text{sec}^2 \times \text{sec}^2/\text{ft}$$

$$= 1,060 \text{ ft} .$$

Hence the number of waves per mile is

$$\frac{5,280}{1,060} = 5 \text{ (approximately)} .$$

When the waves on the surface of deep water have a very small wave length, $\lambda < 1$ cm, the forces that propel them are no longer due to their weight. These waves of very small wave length are called *ripples*, and, in contrast to larger waves, the liquid surface of ripples is very much more curved. We saw in chapter 10 that the curvature of a liquid surface depends on the surface tension. It is the surface-tension forces that are responsible for the propagation of ripples.

Turning again to dimensional analysis, we can show that if the velocity of ripples depends on the surface tension T', the wave length λ, and the density ρ, then the only combination of these three quantities which will have the dimensions of velocity is $\sqrt{T'/\lambda\rho}$.

The surface tension, T', has the dimensions of force/distance or $[MT^{-2}]$; the wave length, λ, has the dimensions of $[L]$; and the density ρ has the dimension $[ML^{-3}]$; so that the dimensions of

$$\sqrt{\frac{T'}{\lambda\rho}} \quad \text{are} \quad \sqrt{\frac{[MT^{-2}]}{[LML^{-3}]}} = [LT^{-1}] .$$

Hence the velocity of ripples, as given by dimensional analysis, is

$$v = D\sqrt{\frac{T'}{\lambda\rho}} = \text{Constant} \times \sqrt{\frac{\text{surface tension}}{\text{wave length} \times \text{density}}} ,$$

where D is a dimensionless constant which, for this case, can be shown to be $\sqrt{2\pi}$. Hence the velocity of ripples is given by

$$v = \sqrt{\frac{2\pi T'}{\lambda\rho}} . \tag{31-25}$$

So exact is the relation for small ripples that it forms the basis of an accurate method for measuring surface tension. These ripples can be produced by the electrostatic forces between a periodically charged wire placed near the surface of the water.

EXAMPLE 2: Ripples having a wave length of 0.320 cm are excited electrically on the surface of water by the electrostatic forces from an alternating current of 60 cycles/sec. One ripple is produced at each half-cycle or every $\frac{1}{120}$ sec. From these data find the surface tension of water.

Solution:

$$v = f\lambda ,$$

where f is the frequency of the ripples or 120 cycles/sec. Hence

$$(\text{Surface tension}) \; T^1 = \frac{f^2\lambda^3\rho}{2\pi}$$

$$= \frac{120^2 \times 0.320^3 \times 1}{2\pi} \text{ cm}^3\text{-gm}/\text{sec}^2\text{-cm}^3$$

$$= 75 \text{ gm-cm}/\text{sec}^2\text{-cm} = 75 \text{ dynes}/\text{cm} .$$

By combining equations (31-24) and (31-25), we obtain the equation for deep-water waves of any size. For waves intermediate between the two extremes, both terms will be of importance, and the velocity is given by

$$v = \sqrt{\frac{g\lambda}{2\pi} + \frac{2\pi T'}{\lambda\rho}} . \tag{31-26}$$

If the student is familiar with the calculus, he can find the wave length having the minimum velocity in water by setting $dv/d\lambda = 0$. This gives

$$\lambda_{\min} = \sqrt{\frac{4\pi^2 T'}{g\rho}} , \tag{31-27}$$

which is 1.73 cm for water, with a minimum velocity of 23.4 cm/sec.

These surface-water waves present a new situation to us, inasmuch as their velocity depends on their wave length. This gives rise to the phenomenon of dispersion, which we shall discuss more explicitly in optics. The waves having different wave lengths are "dispersed" because they travel with different velocities.

So far, our discussion of water waves has been limited to deep water, in which the depth of the water is much greater than the wave length. We shall now briefly discuss the situation in which the wave length of the waves is larger than the depth. When waves advance into water where the depth of the water is considerably less than the wave length, the circular motion of the individual water particles becomes modified by the presence of the unyielding bottom. The horizontal or longitudinal component is unaffected, while the vertical or transverse component decreases, as shown in Figure 31-7. The particles then move in ellipses of greater and greater eccentricity as the bottom is approached, and immediately adjacent to the bottom the water has no vertical motion whatsoever. At the bottom the water simply oscillates to a lesser degree in this same manner, and thus ripple patterns are built up on the sand. These are familiar to anyone who has ever waded along the sandy shore of any body of water large enough to have a surf.

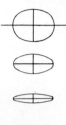

Fig. 31-7.—Elliptical orbits of water particles at different depths.

This modification of the motion of the particles profoundly affects the velocity of propagation of the surface wave, which, in shallow water of depth d, is

$$v = \sqrt{gd} . \qquad (31\text{-}28)$$

We may verify that this equation is dimensionally correct, though its derivation is obtained by analytical methods. The velocity of surface waves in which the wave length is much larger than the depth varies as the square root of the depth. It is this dependence of the velocity on depth that causes waves to come in to a shelving shore line parallel to the shore line. Since the frequency of the waves remains constant, the wave length must de-

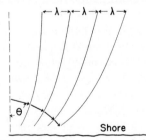

Fig. 31-8.—Waves approaching a shelving shore

crease with the velocity as the wave moves into shallower water. If the wave is inclined so that its direction of travel makes an angle, θ, with the normal to a shelving beach, this angle decreases as the wave proceeds, since the shoreward end is most retarded (Fig. 31-8). This is the phenomenon

of refraction and occurs where the velocity of the wave changes as it moves through the medium or from one medium to another in which the velocities are different. This latter phenomenon will be discussed in optics.

31.10. Velocity of a Group of Waves

In a great many cases of waves moving in a medium, there is not a single wave with a definite wave length but rather a group of waves with slightly varying wave lengths. If there is a different velocity corresponding to each of the different wave lengths, there is interference between the different waves and a so-called *group velocity* of the waves is set up (Fig. 31-9). Suppose v is the

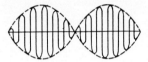

Fig. 31-9.—Group velocity shown in dotted curve produced by the interference of the individual waves shown in the full-curve pattern.

velocity of a wave of wave length λ and suppose $v + dv$ is the velocity of a wave of wave length $\lambda + d\lambda$; then it may be shown that, by interference between the waves, the velocity u of the group is given by

$$u = v - \lambda \frac{dv}{d\lambda}. \qquad (31\text{-}29)$$

The velocity v is the phase velocity of the wave of wave length λ. The quantity $dv/d\lambda$ is the rate of change of the velocity of the waves with their wave length. When the velocity of waves in a given medium is different for different wave lengths, then there is dispersion within the medium. If the longer waves have the larger velocity, that is, $dv/d\lambda$ is positive, then the longer waves outstrip the shorter ones, and the group velocity is smaller than the phase velocities of the waves. For this case the individual waves pass through the group toward the front, disappearing at the front as new ones appear. This is shown in photograph of waves produced by a ship when turning (Fig. 31-10).

This phenomenon of group velocity plays an important role in optics, where it was found to explain some apparent contradictions between the measured velocities of light in two media with the **443** measured refractive index. This latter should be equal to the ratio of the velocities, but it was found to be equal to the ratio of the group velocities in the media and not the ratio of phase velocities. Probably a much more common example of group velocities is that of a V-shaped bow wave produced by a fast-moving motor boat. In a more erudite field of physics there is the so-called Čerenkov radiation given off by electrons which are moving in a medium such as water, Lucite, etc., with a

FIG. 31-10.—Group waves from bow and stern of a fast-moving ship when turning.

speed greater than the speed of light in the medium. This radiation, which is visible, is satisfactorily explained on the basis of a group velocity of waves associated with the moving electrons.

31.11. Some Properties of Waves

We are all familiar with the fact that waves transmit energy and that they can be reflected and refracted. These three properties, however, are not distinguishing properties of waves, because moving particles can likewise show them. The two phenomena which waves exhibit but moving particles do not are those of *interference* and *diffraction*. At this point we shall give a qualitative discussion of these two phenomena, leaving the more quantitative one to the sections on sound and light.

If two waves with the same amplitude, velocity, frequency, and wave length are traveling in opposite directions in a medium, then there are places

at which a crest of one wave falls onto the trough of the other, and a position of no displacement, or a *node*, is formed, while at other places a crest of one falls at the same position as a crest of the other, and the resulting displacement is double that of either wave. This is called a *standing wave* and is of importance in the discussion of vibrating strings.

Suppose two sources produce waves on the surface of water with the same frequency and amplitude and also that they are in the same phase, that is, when the one source is producing a crest, the other source is also producing a crest. In Fig. 31-11

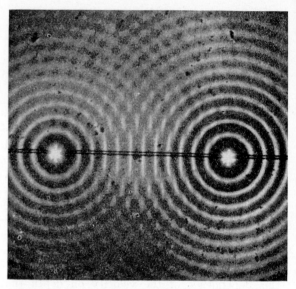

FIG. 31-11.—Interference between two sources photographed by stroboscopic illumination.

these are shown in a photograph taken with stroboscopic illumination using a rotating sector disk. If the full circles represent the crests and the dotted circles the troughs, then at the instant of time considered in Fig. 31-12 the interference pattern is set up. The points shown as small circles have maximum displacement or are loops, while the crosses show the places in which the one wave is producing a crest while the other is producing a trough. At these latter points the resultant displacement of the water is zero; these are nodes of motion. This interference pattern, produced by the superposition of two similar waves on the medium, will show lines of loci of loops and nodes such as aa' and bb'. Figure 31-13 shows a photograph of the loops and nodes. If S and S' represented two sounds in air of the same frequency phase and amplitude, then a person at b' would receive sound from either source alone but not when both sounds were emitted simultaneously, since at b' there was destructive interference. On the other hand, a person at a' would receive sound from both sources, the waves having double the amplitude of either source

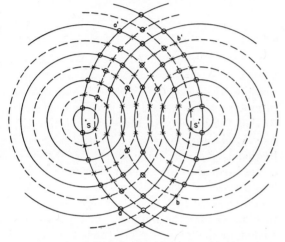

Fig. 31-12.—Interference between two sources

alone. At a' there is constructive interference of the waves producing the *loops* or the *antinodes* of motion. This phenomenon of interference is of importance for both sound and light and will be discussed further in those sections.

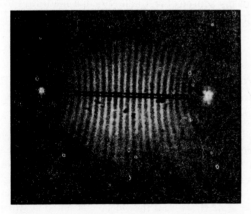

Fig. 31-13.—Photograph of nodes and loops from two sources.

Another phenomenon which is exhibited by waves but not by particles is that of diffraction. In a very qualitative matter we can say that diffraction involves the bending of waves around obstacles or through openings which are small compared to the wave length. Waves of different wave length are diffracted or bent by different amounts. Audible sound waves have wave lengths of a few feet, while visible light has wave lengths of the order of 10^{-5} cm, so that we can hear much more what is going on in a room than see what is going on by looking through the keyhole. Similarly, it is possible to hear people talking around a corner when it is not possible to see them. Diffraction plays an important role in the study of optics, and we shall discuss this phenomenon further in the chapter on light.

31.12. Wave Lengths Associated with Moving Particles

It may seem contradictory to speak of the wave length of moving particles after stating previously that the criterion of waves was diffraction and that particles did not exhibit it. This, however, was considering particles whose motion was governed by Newton's laws, in which the energy of the particles resided in them. But this is not a complete description of the motion of particles, for it was shown by Davisson and Germer in this country and G. P. Thomson in England about 1927 that subatomic particles—electrons—having a mass of about 10^{-27} gm could show the phenomenon of diffraction. From these diffraction experiments the wave length of the waves associated with the moving electrons was calculated as given in § 30.21 by De Broglie as

$$\lambda = \frac{h}{m\,v},$$

where h is a constant called Planck's constant after its discoverer and has a value of 6.62×10^{-27} erg-sec, m is the mass, and v the speed of the electron. However, for all macroscopic matter in motion, the diffraction effects and the wave lengths are negligibly small, so that there is no error in using Newton's laws for everyday objects; this is not completely the case for subatomic matter.

31.13. The Doppler Effect

The Doppler effect is concerned with the change in frequency of the waves received by an observer for the situation in which there is relative motion between the source and the observer. It can be readily observed for sound waves, though it is equally valid for all kinds of waves. The calculation of the frequency of a note received by an observer in motion from a source of sound in motion was first made by the Dutch physicist, Christian Doppler (1803–52), about 1842. At that **445** time the theory was checked by trained musicians, who used a trumpeter to provide a note of definite frequency and a railroad engine and flat car to provide the motion for the musicians.

Suppose a source of waves, such as that provided by a note of definite frequency f_0 and wave length λ_0 is moving with a velocity v_s toward a stationary observer. Because of the motion of the source, the apparent wave length of the waves is decreased, since in a sense the source is catching up with the front of the waves. This decrease in wave length is equivalent to an increase in frequency, so the observer hears a note having a frequency higher than f_0. Similarly, if the source is moving away from the stationary observer, he will receive a note of frequency lower than f_0.

Consider, next, the observer in motion, with the source at rest. If the observer is moving away from the source, then fewer waves will pass him per second than if he were at rest. That is, the moving observer will receive a note of frequency lower than f_0. Similarly, if the observer is moving toward the source, he will receive more notes per second and thus hear a note of frequency higher than f_0. We shall now proceed to analyze this phenomenon for a source and observer both in motion along the same straight line, with the medium transmitting the waves also moving along the straight line.

Suppose the source, emitting waves of frequency f_0, is moving with a velocity v_s toward an observer moving with a velocity v_0. The medium is also moving with a velocity v_m. All these velocities are assumed to be along the positive x-axis, as shown in Fig. 31-14. The velocity of the waves in

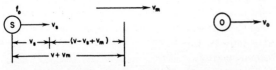

FIG. 31-14.—The Doppler effect. Source, observer, and medium all in motion along the positive x-axis.

the medium is v, and, since this is a property of the medium, it is not influenced by any motion of the source. Since this velocity v is relative to the medium and the medium is moving with a velocity v_m relative to the earth, it follows that the velocity of the waves in Fig. 31-14 relative to the earth is the sum of the two velocities, that is, $v + v_m$. We could consider this motion of the medium as a horizontal wind carrying the sound waves along with it.

In 1 sec the front of the waves has moved a distance $v + v_m$ to the right, and during this time the source has moved a distance v_s. Since the source has a frequency f_0, it is emitting f_0 waves per second, so that these f_0 waves are contained in a distance of $v + v_m - v_s$. The wave length λ of these waves, then, is

$$\lambda = \frac{v + v_m - v_s}{f_0}. \qquad (31\text{-}30)$$

Let us now consider the observer. Waves having a velocity $v + v_m$ relative to the earth are approaching him. However, the observer is assumed to be in motion with a velocity v_0 relative to the earth, so that the velocity of the waves relative to the moving observer in Fig. 31-14 is $v + v_m - v_0$. The frequency f of the note which the observer receives is equal to the number of waves which pass him per second. These waves have a wave length λ and travel with a velocity $v + v_m - v_0$ relative to him, so that the number which pass him per second, or the frequency f of the note received by the observer, is

$$f = \frac{v + v_m - v_0}{\lambda}. \qquad (31\text{-}31)$$

By substituting for λ from equation (31-30), we obtain the frequency f of the note received by a moving observer when the note is emitted from a moving source of frequency f_0 in a medium moving with a velocity v_m,

$$f = f_0 \frac{v + v_m - v_0}{v + v_m - v_s}. \qquad (31\text{-}32)$$

When using equation (31-32), it is necessary to be careful of the signs of the various velocities. An example will make this point clearer.

EXAMPLE 3: A source of sound emitting a note of frequency 400 vibrations/sec is moving toward the west with a velocity of 20 ft/sec. Directly to the east of the source is an observer, who is moving to the east with a velocity of 30 ft/sec. There is a wind blowing from the east at 45 ft/sec, and the velocity of sound in still air is 1,200 ft/sec. The problem is to find the frequency of the note received by the observer.

If velocities toward the east are taken as positive, as in Fig. 31-14, then it follows that

$$v_s = -20 \text{ ft/sec}; \qquad v_m = -45 \text{ ft/sec};$$

$$v_0 = 30 \text{ ft/sec}$$

$$f_0 = 400/\text{sec} \cdot \qquad v = 1{,}200 \text{ ft/sec}$$

From equation (31-32) the frequency received by the observer is

$$f = 400 \times \frac{1,200 - 45 - 30}{1,200 - 45 + 20}$$

$$= 400 \times \frac{1,125}{1,175} = 383 \text{ vibrations/sec}.$$

If the velocities are not along the line joining the source and observer, then the components of the velocities in that direction must be used.

31.14. Doppler Effect in the Case of Light

In the case of light or of other forms of radiant energy we have no knowledge of any medium transmitting the wave phenomena. Furthermore, the velocity of light is so great that any terrestrial sources, even if they have projectile velocities, exhibit such minute wave-length and frequency changes that the latter require most refined apparatus for their detection. Astronomical sources with higher velocities show more definite Doppler changes in wave length; and in the case of a relatively small number of very distant galactic nebulas a change in wave length that is large enough to constitute a real change in color of the radiation is observed. These large changes in wave length in nebulas are always such as to increase the wave length, i.e., always denote recession of the nebulas. In the case of nebulas it is even more astonishing to discover that their velocities seem to increase linearly with their distance from the solar system. Their distances are of the order of tens to scores of thousands of "light-years."* The obvious inference from this fact is, of course, that it is because of their swifter motion that the more distant nebulas have become the more remote, i.e., that at some earlier epoch there must have been something analogous to a cosmic explosion which blew out into space a gigantic assemblage of material formerly concentrated in one location. From this theory are derived the ideas of an "expanding universe" current today. There is great uncertainty about this, however, in view of our slender knowl-

edge as to the nature of radiation and of what may happen to it in traversing such prodigious distances. These wave-length changes in the light of very distant nebulas may not be due to Doppler effect at all! About 1950 Meinel measured the speed of the hydrogen atoms in the aurora from the Doppler shifts.

In the laboratory, on the other hand, small as is the largest Doppler effect which could conceivably be produced with respect to light, the equations we have given have been confirmed by experiments.

Radiating atoms in gases electrically excited to luminescence also show the Doppler effect. When only the random motion of atoms due to the random distribution of their speeds about an average is involved, a certain widening of the radiation over a region both greater and less than its intrinsic wave length results. Michelson was the first to verify that the range of wave-length spread under these circumstances was in the proper correlation with the temperature and the molecular weight of the gas demanded by the combination of optical theory with that of the kinetic theory of gases.

When a unidirectional stream of radiating ions can be produced, as in the case of positive rays, then there appears both the wave length of the light from those atoms not partaking of this motion and the wave length of the light from the moving atoms, shifted in the appropriate direction. Under such circumstances the confirmation of the Doppler equations in this field is very precise.

31.15. Résumé

The following is a brief listing of the definitions, principles, and theories, given in this chapter, which you should know:

The fundamental definitions for a wave motion—amplitude, period, frequency, velocity, and phase for transverse and longitudinal waves
The equation of a simple harmonic wave moving to the left or to the right; the velocity of a wave in a rope
The velocity of longitudinal waves or of sound waves in air and other media
Properties of waves
Criteria for waves, interference and diffraction
Wave length of moving particles
Doppler effect for waves

---

* One light-year
= Distance traveled by light in one year
= 186,000 mi/sec × 86,400 sec/day × 365 day/yr
= 186,000 mi/sec × 31.536 × 10^6 sec/yr
= 5.8657 × 10^{12} miles, i.e., nearly 6,000,000,000,000 miles.

Queries

1. What experimental proof can you give that energy is associated with a wave?

2. The depth of water in a tank diminishes from 5 to 1 cm. How will the speed of long waves be affected in this tank?

3. A cork floats on the surface of a pond. What will be the motion of the cork as a long wave passes beneath it?

4. Does interference of two waves involve a loss of energy? Explain.

5. An automobile carrying a siren is traveling in still air at 55 ft/sec. The frequency of the note emitted by the siren is 550 cycles/sec, and the speed of sound in the air is 1,100 ft/sec, so that the wave length of the emitted note is 2 ft. State whether the wave length of the sound waves (a) in front of, (b) behind, the automobile is greater than, less than, or equal to 2 ft.

6. In question 5 is the frequency of the sound heard by a fixed observer who is (a) in front of, (b) behind, the automobile greater than, smaller than, or equal to 550 cycles/sec?

7. How would the answers in questions 5 and 6 be modified if a wind was blowing in the direction of motion of the automobile?

8. Does one ever have to deal with a moving medium in the case of the Doppler effect in light?

Practice Problems

1. The equation of a simple harmonic wave is $y = 0.12 \sin \pi(x - 4t)$, where x and y are in meters and t is in seconds. (a) In which direction is the wave progressing? (b) What are the amplitude, wave length, and period of the wave? Ans.: (a) In the positive x direction; (b) 0.12 m, 2 m, 0.5 sec.

2. Write the equation for a progressive wave moving along the positive x-axis with the following characteristics: amplitude, 0.5 cm; period, 2 sec; wave length, 50 cm.

3. Write the equation for a progressive wave moving along the negative x-axis and having amplitude, 0.02 m; frequency, 550/sec; velocity, 330 m/sec. Ans.: $y = 0.02 \sin \pi/0.3(x + 330t)$.

4. The equation of a progressive wave is

$$y = 0.2 \sin 2\pi\left(\frac{x}{0.3} - \frac{t}{0.001}\right),$$

where x and y are in meters and t is in seconds. What are the wave length, amplitude, frequency, and velocity of the wave?

5. Find the velocity of a transverse wave in a rope having a length of 2 m and mass 0.04 kg under a tension of 50 kgf. Ans.: 156.5 m/sec.

6. Find the velocity of a transverse wave in a rope having a length of 2 yd and a weight of 0.02 lbf under a tension of 3 lbf.

7. Calculate the velocity of sound in air at 0° C, if the density of air at 0° C is 1.29 kg/m³ and at atmospheric pressure of 1.013×10^5 newtons/m². Ans.: 333 m/sec.

8. Calculate the velocity of sound in air at 27° C and atmospheric pressure.

9. Calculate the velocity of sound in air at 0° C and a pressure 1.2×10^5 newtons/m². Ans.: 333 m/sec.

10. Prove that the velocity of sound in a gas is proportional to the absolute temperature of the gas and is independent of the pressure of the gas. Assume that the gas obeys the ideal gas law.

11. Given the bulk modulus of rock as 2×10^{10} newtons/m² and the modulus of rigidity of this rock as 2×10^{10} newtons/m². If the density of this rock is 5×10^3 kg/m³, find the velocity of longitudinal waves and of transverse waves in this rock. Ans.: 3,055 m/sec; 2,000 m/sec.

12. Find the velocity of a deep-water wave 1 km in wave length.

13. Find the velocity of a deep-water wave 1 m in wave length. Ans.: 1.25 m/sec.

14. Find the velocity of a deep-water wave 1 cm in wave length.

15. Find the velocity of a deep-water wave 1 mm in wave length. Ans.: 0.69 m/sec.

16. Ripples are excited electrically on the surface of the water by the electrostatic forces from an alternating current of 60 cycles/sec. If the wave length of the ripples is 0.3 cm, find the surface tension of the liquid.

17. Find the velocity of propagation of a wave in shallow water of depth 0.2 m. Ans.: 1.4 m/sec.

18. Show that the group velocity of deep-water waves is half their phase velocity.

19. Find the wave lengths of the De Broglie waves for electrons of mass 9.1×10^{-28} gm, moving with speeds of (a) 10^6 cm/sec; (b) 10^8 cm/sec. Ans.: (a) 7.28×10^{-6} cm; (b) 7.28×10^{-8} cm.

20. A siren having a frequency of 1,000 cycles/sec is attached to an automobile moving with a speed of 60 mi/hr. If the automobile is approaching a stationary observer, what frequency does the observer

hear? Assume the velocity of sound to be 1,100 ft/sec.

21. If the automobile carrying the siren in problem 20 is at rest while the observer is approaching the siren with a speed of 60 mi/hr, find the frequency of the note received by the observer. Ans.: 1,080/sec.

22. If both the siren source and the observer are moving with a speed of 60 mi/hr and approaching each other, find the frequency heard by the observer.

23. Given the velocity of sound as 350 m/sec. A whistle with a frequency of 1,000 vibrations/sec is carried from a fixed observer toward a reflecting cliff with a velocity of 10 m/sec. What is the difference in the frequencies of the two sounds heard by the observer? Ans.: 57.3 vibrations/sec.

24. The radial velocity (relative velocity in the line of sight) of a certain star with respect to the earth is 30 km/sec toward the earth. The red radiation of the hydrogen gas in this star has a "true" wave length (to an observer moving with the star) of $6,563 \times 10^{-10}$ m. If the velocity of light is 3×10^5 km/sec, find the wave length of this radiation as measured by a terrestrial observer.

25. The "red shift" of a radiation from a distant nebula in which the light is known to have a wave length of $4,340 \times 10^{-10}$ m ($H\gamma$) is such as to make this radiation in the nebula appear to have a wave length of $6,562 \times 10^{-10}$ m. What is the velocity of the nebula in the line of sight with respect to the earth? Is it approaching or receding from the earth? Ans.: 1.54×10^5 km/sec; receding.

26. What difference, if any, would result from using the Doppler equations for a moving observer for light and for sound waves instead of those for a moving source?

27. A wire 2 m long has a mass of 0.01 kg and is under a tension of 1,000 newtons. Find the velocity of transverse waves in the wire. Ans.: 447 m/sec.

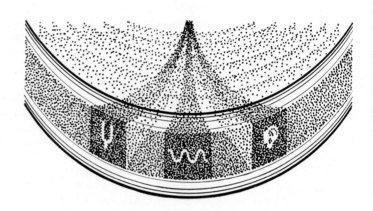

CHAPTER 32

Sound

32.1. Introduction

Important as is the sense of hearing and the human organs of sound production, the physical aspects of the subject of sound, like those of mechanics, attract little attention from present-day investigators and form a relatively small part of a year's curriculum in the study of physics. The subject of mechanics, on the other hand, of basic m portance in all machines, especially the research machines of the physicist, involves the alphabet, as it were, of the language in which a large part of both thermal and electric phenomena must be described and interpreted. Indeed, it is in large measure due to the completeness of our knowledge of mechanics that so much progress has been made subsequently in both these sister-subjects. In contrast to this, the study of sound, although providing many beautiful illustrations of mechanical principles, has not, in general, been of vital importance in the development of work that lies much nearer to the borders of the unknown—in atomic and nuclear physics. Recent studies of the properties of sound waves having very short wave length and large energy, called *supersonic* or *ultrasonic* waves, give promise of results of far-reaching importance. These waves are discussed at greater length in § 32.18.

32.2. Sound Waves

Sound waves are longitudinal vibrations of material media—solid, liquid, or gaseous—which are transmitted therein with a velocity determined precisely by the readily measurable mechanical properties of the medium itself. Sound waves in general find their origin in some sort of mechanical vibration, caused by impacts, friction, and other forms of energy dissipation in mechanical arrangements, either natural or artificial. In extended isotropic media the wave fronts are spherically expanding surfaces, subject to all the vicissitudes that any train of waves of any other sort might encounter. Reflection and refraction, dispersion, diffraction, and interference—all are exhibited by sound waves. In limited sections of material media, such as rods or columns of solids, liquids, or gases, sound waves may be largely confined to progress in but one direction and may be treated as one-dimensional affairs. Vibration forms of strings and rods and air columns thus form the simpler analytical types of problems and serve to illustrate principles equally valid with respect to the two-dimensional cases of vibrating plates, membranes, and diaphragms. The study of the three-dimensional wave, indeed, is usually reduced to the study of two-dimensional plane sections of the same. Visualization of wave fronts and rays already utilized in the case of surface-water waves are possible also for the two-dimensional case in sound.

32.3. Velocity of Sound in Air; Effects of Pressure and Temperature

We have already seen in equation (31-20) that the velocity of sound in a gas is

$$v = \sqrt{\frac{\gamma P}{\rho}}, \qquad (32\text{-}1)$$

where γ is the ratio of the specific heat of the gas at constant pressure to that at constant volume and ρ is the density. For air at $0°$ C we found the velocity to be about 332 m/sec.

Let us consider the effect of increase of atmospheric pressure on the velocity of sound, with the temperature remaining constant. If problem 10 of the last chapter has been worked, then this material has already been covered. From Boyle's law, the ratio P/ρ is constant for a gas at constant temperature, for if the pressure of a fixed mass of the gas is doubled, the volume is halved or the density is doubled. Thus, according to equation (32-1), we should expect the velocity of sound to be independent of changes in pressure. While this is confirmed for small pressure changes such as occur in the atmosphere, nevertheless for large pressures the velocity is increased. The velocity of sound in air at 100-atm pressure has been measured to be about 351 m/sec at $0°$ C. This does not invalidate equation (32-1) for the velocity but merely indicates that Boyle's law is not valid for air at 100-atm pressure.

To the extent that air may be considered a perfect gas, we can write the general gas law as

$$\frac{P}{\rho} = \frac{RT}{M}, \qquad (32\text{-}2)$$

where R is the gas constant having a value of 8.31×10^7 ergs/mole and M is the molecular weight of the gas, which has a mean value of about 29 for air.

Substituting for P/ρ from the gas law in equation (32-1), we have the velocity of sound in a gas of molecular weight M at temperature $T°$ absolute,

$$v_t = \sqrt{\frac{\gamma RT}{M}}. \qquad (32\text{-}3)$$

For a given gas, the values of γ, R, and M are constant, so that, to the extent that the gas obeys the general gas law, the velocity of sound increases as the square root of the absolute temperature. If v_0 is the velocity of sound in the gas at $0°$ C, then the velocity at $t°$ C can be given as

$$v_t = v_0 \sqrt{\frac{273+t}{273}} = v_0 \sqrt{1+0.00366t} \quad (32\text{-}4)$$

If t is not too large, then, very approximately,

$$\left. \begin{array}{l} v_t = v_0\,(1+0.00183t) \\ = 332 + 0.6t \text{ m/sec}. \end{array} \right\} \quad (32\text{-}5)$$

Thus the velocity of sound in a gas increases by about 0.6 m/sec for every degree centigrade rise in temperature. This result was confirmed in 1890 in the Arctic by Greely for a temperature range of $-10°$ to $-45°$ C.

A simple transformation gives the relation in British units,

$$v_t = 1,054 + 1.1t \text{ ft/sec}.$$

where t is in degrees Fahrenheit. This velocity of 720 mi/hr at which sound is transmitted at about $32°$ F is now exceeded by the faster planes. Sound waves produced in the air by sources that have a speed greater than that of sound itself in no way resemble ordinary sound waves. So important are the changes in the forces which take place when an object moves in air with the velocity of sound that a new unit called the *Mach number* has been introduced. This dimensionless Mach number is the ratio of the speed of the object to the speed of sound. Hence, for supersonic speed, the Mach number is greater than 1, while for the usual subsonic speeds it is less than 1.

32.4. Effect of Humidity and Other Factors

The effect of humidity in the atmosphere is to reduce the latter's density below that which dry air would have at the same temperature and to alter slightly the ratio of specific heats; hence the speed of sound increases slightly as the humidity increases.

The velocity of sound also depends, to a great extent, upon the intensity of the sound *if the intensities are very great*, as in the sounds from explosions. The propagation of the energy in an explosion is initially in the form of a shock wave, which has a speed greater than that of normal sound waves at the same temperature and pressure. The more violent the explosion, the more intense the accompanying shock wave and the greater the velocity of its propagation. Ultimately, the shock wave becomes dissipated and becomes a normal sound wave. The results of some experiments on the velocity of sound are given in Table 32-1.

Somewhat puzzling, in the light of this table, are the observations that were made on the most gigantic sound wave of recorded history. This was the sound of the explosion of Krakatao, a volcanic island not far from Java in the Indian Ocean, on August 27, 1883. This explosion, which caused a water wave 50 feet high and great destruction along near-by coasts, resulted also in a sound wave that was clearly heard 3,000 miles away and affected barometers all over the world. From a subsequent study of the records of many of the self-recording barometers, it was found that the opposite fronts of this wave, which went in opposite directions

TABLE 32-1

VELOCITY OF EXPLOSIVE WAVES*

Powder Charge (Gm)	Velocity (M/Sec)	Powder Charge (Gm)	Velocity (M/Sec)
0.24	336	17.40	931
3.80	500	45.60	1,268

* From D. C. Miller, *Sound Waves and Their Uses* (New York: Macmillan Co., 1937).

around the earth, met 18 hours later at a point directly opposite the island and produced a maximum disturbance there. Again and again this great wave swept through the atmosphere, leaving its record on the instruments for more than 5 days. From these records the velocity was calculated to be only 700 mi/hr, as compared with the 720-mi/hr velocity of low-intensity sounds through the air at 0° C.

32.5. Variations of Pressure in a Sound Wave

In a sound wave in air there are compressions and rarefactions at which the pressure goes above and below the normal or atmospheric pressure. We have seen in Fig. 31-5 that a compression occurs where there is no displacement at the center but that the displacements of the air molecules on both sides of the center are toward this central position. Similarly, at a rarefaction there is no displacement at the center, but the displacements on both sides of the center are away from the center. As the wave passes through the air, successive compressions and rarefactions occur, with corresponding pressure variations.

Suppose a sound wave of amplitude A, velocity v, and wave length λ is traveling through the air so that

the displacement y from the mean position at time t and position x is given by equation (31-4), namely,

$$y = A \sin \frac{2\pi}{\lambda} \left(x - vt \right). \qquad (32\text{-}6)$$

In such a wave the pressure variations p, above or below normal atmospheric pressure in air of density ρ, is shown in more advanced texts to be

$$p = -\frac{2\pi A v^2 \rho}{\lambda} \cos \frac{2\pi}{\lambda} (x - vt) \qquad (32\text{-}7)$$

$$= -P \cos \frac{2\pi}{\lambda} (x - vt) , \qquad (32\text{-}8)$$

where P is the maximum pressure variation or the pressure amplitude obtained by setting the cosine term equal to unity, that is,

$$P = \frac{2\pi A v^2 \rho}{\lambda}. \qquad (32\text{-}9)$$

The amplitude or maximum displacement of the particles in the air through which the wave is passing is given in terms of the maximum pressure variation in the wave as

$$A = \frac{\lambda P}{2\pi \rho v^2}. \qquad (32\text{-}10)$$

From Fig. 32-1, giving the audibility range, we see that the maximum pressure variation in a sound wave which the ear can tolerate without pain is of the order of 25 newtons/m² or 250 dynes/cm².

Let us calculate the maximum displacement, A, of the particles, that is, the amplitude of the sound wave, for a sound having a frequency of 500 cycles/sec and wave length of 68 cm at 20° C in air.

From equation (32-10) the value of A is given by

$$A = \frac{\lambda P}{2\pi \rho v^2}$$

$$= \frac{68 \times 250}{2\pi \times 0.0012 \times (3.42 \times 10^4)^2}$$

$$= 0.00193 \text{ cm} .$$

In this calculation the density of air was taken as $0°00120$ gm/cm³ at 20° C and the velocity of sound as 3.42×10^4 cm/sec at 20° C. This small displacement of the loudest sound tolerated is about one hundred times the mean free path of a gas molecule of oxygen or nitrogen at atmospheric pressure. The faintest sound heard, of frequency 500 cycles/sec, is associated with a maximum pressure variation of about 0.0002 dynes/cm², which corresponds to a maximum displacement of the particles in the wave of about 2×10^{-9} cm. This is much less than the mean free path of the gas molecules, which is about 10^{-5} cm at atmospheric pressure. Of course, the motions of the gas molecules produced by a sound wave traversing the gas are superimposed on the random molecular motions. Since atmospheric pressure is about 10^6 dynes/cm^2, it means that the human ear is sensitive to pressure variations in a sound wave as small as about 2×10^{-10} atm. The loudest sound the ear can tolerate has pressure variations of about 0.004 atm.

32.6. Intensities of Sound Waves

By the "intensity" of a wave is meant the amount of energy which the wave transmits each second through a unit of area perpendicular to the direction of propagation of the wave. The usual units for measuring intensity are ergs per square centimeter per second, joules per square meter per second (equivalent to watts per square meter) or a hybrid unit frequently used in acoustic circles of watts per square centimeter.

In order to calculate the intensity of a sound wave, we shall first find the average kinetic energy of the vibrating particles per unit volume of the medium. To do this, we find the rate of change with time of the displacement of the vibrating particles. The displacement y of the particles in a wave is given by equation (31-11) as

$$y = A \sin \frac{2\pi}{\lambda}(x - vt) , \qquad (32\text{-}11)$$

and the velocity of these particles at any time t is

$$\frac{dy}{dt} = -\frac{2\pi A v}{\lambda} \cos \frac{2\pi}{\lambda}(x - vt) . \ (32\text{-}12)$$

The kinetic energy per unit volume of these particles is

$$\tfrac{1}{2} \rho \left(\frac{dy}{dt}\right)^2 = \frac{1}{2} \frac{\rho 4\pi^2 v^2 A^2}{\lambda^2} \cos^2 \frac{2\pi}{\lambda}(x - vt) , \ (32\text{-}13)$$

where ρ is mass per unit volume or density of the medium.

In order to find the average value of this kinetic energy over a complete vibration, we must obtain the average value of the cosine^2 term over a complete cycle. This can be done by the calculus or by recognizing that the average value of a cosine^2 α is the same as that of sine^2 α over a complete cycle. Now, for any angle α, the sum $\cos^2 \alpha + \sin^2 \alpha = 1$, so that the average value of a cosine^2 α or sine^2 α over a complete cycle is $\frac{1}{2}$. Thus the average

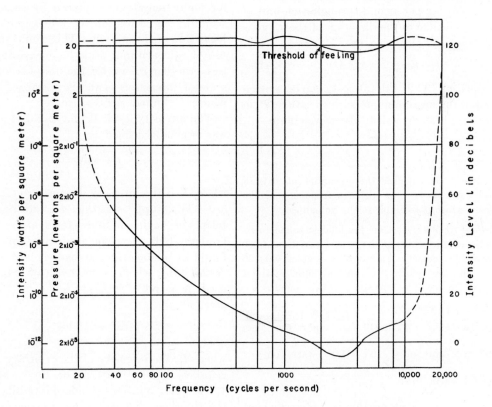

Fig. 32-1.—Audiogram, showing limits of hearing for different frequencies

value of the kinetic energy, $\overline{KE}$ per unit volume, over a complete cycle, is obtained from equation (32-13) by setting the cosine² term equal to $\frac{1}{2}$, hence

$$\overline{KE} = \frac{\rho \pi^2 v^2 A^2}{\lambda^2}. \qquad (32\text{-}14)$$

In § 8.10 it was shown that, for any particle executing simple harmonic motion without damping, the average value of the kinetic energy of the particle over a complete cycle was equal to the average value of the potential energy over a complete cycle. Since the particles in a wave are undergoing simple harmonic motion, it follows that the average value of the total energy $\bar{E}$, kinetic plus potential, of the particles in a wave over a complete cycle is double the average of the KE given in equation (32-14), or the average energy per unit volume is

$$\left. \begin{aligned} \bar{E} &= \frac{2\pi^2 \rho v^2 A^2}{\lambda^2} \\ &= 2\pi^2 \rho f^2 A^2, \end{aligned} \right\} \qquad (32\text{-}15)$$

since $v = f\lambda$.

This energy of the wave is transmitted with the velocity v of the wave, so that the average amount of energy flowing per second through each unit of area perpendicular to the direction of the wave is the intensity I of the wave, or

$$I = \bar{E}v = 2\pi^2 \rho f^2 A^2 v. \qquad (32\text{-}16)$$

From equation (32-9) this intensity can be written in terms of the maximum pressure variation, P, in the wave, so that, by substitution, we obtain

$$I = \frac{P^2}{2\rho v}. \qquad (32\text{-}17)$$

EXAMPLE 1: To calculate the maximum intensity of a sound wave at 20° C in air at a frequency of 500 cycles/sec, which is tolerable to the human ear.

Solution: At 500 cycles/sec the loudest sound tolerable to the ear has a maximum pressure variation of 250 dynes/cm². The density of air at 20° C is about 0.00120 gm/cm³, and the velocity of sound at 20° C is about 3.42×10^4 cm/sec. The intensity I of this sound is

$$I = \frac{(250)^2}{2 \times 0.0012 \times 3.42 \times 10^4} \left(\frac{\text{gm-cm}}{\text{sec}^2\text{-cm}^2}\right)^2 \frac{\text{cm}^3}{\text{gm}} \times \frac{\text{sec}}{\text{cm}}$$

$$= 761 \text{ ergs/cm}^2\text{-sec} = 7.61 \times 10^{-5} \text{ watts/cm}^2$$

$$= 0.761 \text{ watts/m}^2.$$

454 An intensity as great as this is, of course, not used in ordinary practice. The power generated in the form of sound in an ordinary conversation is about 10 microwatts (10^{-5} watts). A violin generates about 50 microwatts, while an organ pipe can generate as high as 3,000 microwatts. The human ear can respond to intensities in a range from about 1 watt/m² to the faintest one heard, of about 10^{-6} microwatts/m² or 10^{-12} watts/m². Thus the human ear responds to intensities varying by a factor of 10^{12}. Its frequency range is from about 20 to 16,000 cycles/sec, though the upper limit decreases with the age of the individual. For average phonographs and radios the range of frequencies produced as sound is much less than that of the human ear, of the order of 40–7,000 cycles/sec.

32.7. Intensity Levels in Decibels

While intensity and loudness are associated with each other, it is not correct to say that if sound A has double the intensity of sound B, then sound A will be twice as loud as B. Intensity is a physical quantity, while loudness is a physiological one. Weber and Fechner have shown that, for many sensations produced by some form of stimuli, the sensation is proportional to the logarithm of the stimulus. This would suggest the use of a logarithmic scale for measuring intensities. Another reason for the use of a logarithmic scale of intensities is the large range, 10^{12}, over which the human ear is sensitive.

The intensity level, L, of a sound of intensity I was originally defined as

$$L = \log_{10} \frac{I}{I_0}, \qquad (32\text{-}18)$$

where I_0 is the intensity of some arbitrary standard. The unit of L was the *bel*, a unit proposed by telephone engineers and named after Alexander Graham Bell, of telephone fame. This unit was found to be too large, so another unit, called the *decibel* (db), which is one-tenth of a bel, is now used. A sound of intensity has a level of l (db), given as

$$l \text{ (db)} = 10 \log \frac{I}{I_0}, \qquad (32\text{-}19)$$

where I_0, the intensity of the arbitrary standard, is taken as 10^{-12} watt/m² or 10^{-16} watt/cm² at 1,000 cycles/sec. This standard intensity, I_0, is on the threshold of hearing. From equation (32-19) a sound of intensity of 10^{-12} watt/m² would have a sound level of zero decibels, since the logarithm of unity is zero. Since there is a factor of 10^{12} in the sounds tolerated, it follows that the threshold of pain from sound is 12 bels or 120 db. Table 32-2 gives the approximate level of various sounds. From the table it follows that the intensity of the sound I in a noisy factory is given by equation (32-19) as

$$90 = 10 \log_{10} \frac{I}{I_0}$$

or

$$I = I_0 10^9$$
$$= 10^{-12} \times 10^9 = 10^{-3} \text{ watt/m}^2$$
$$= 10^{-7} \text{ watt/cm}^2 .$$

The threshold of hearing, as well as the threshold of pain or the threshold of feeling from sounds,

TABLE 32-2

LEVELS OF VARIOUS SOUNDS

Source of Sound	Sound Level (db)	Source of Sound	Sound Level (db)
Threshold of hearing .	0	Noisy factory	90
Quiet home	40	Very loud thunder . .	110
Ordinary conversation	60	Threshold of pain . . .	120

varies considerably with the frequency of the sound. These limits have been measured by Dr. Harvey Fletcher, of the Bell Telephone Laboratories, and are usually represented in diagrammatic form, commonly called an *audiogram*, as shown in Fig. 32-1.

32.8. Production of Standing Waves

Up to the present the discussion of sound as wave motion has been entirely theoretical, with no methods of measurement of the various quantities. The velocity of sound in air can be measured directly, for example, by timing an echo from a surface at a known distance. In this a sound of relatively short duration is produced, and the reflection of this sound by a distant surface, called an *echo*, is heard. If the distance to the reflecting surface is known and the time from the production of the sound to the receiving of its echo is known, then it is easy to obtain the velocity of the sound. More elaborate methods, using electrically produced sounds recorded at some distance by electrical means, can be used to measure accurately the velocity of sound.

If the sound has a single known frequency f and its velocity v is also known, then, from equation (31-2), $v = f\lambda$, the wave length of the sound wave is obtained by

$$\lambda = \frac{v}{f}.$$

Most of the laboratory methods of measuring the wave length of sound waves involves the phenomenon of interference. In general, this consists of a wave producing a standing wave by interference with its own reflected wave. There are then two waves of the same frequency, velocity, and wave length traveling in opposite directions through the medium. In Fig. 32-2 are shown two

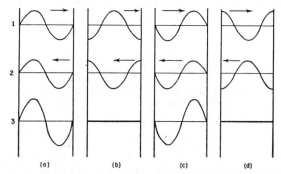

FIG. 32-2.—Standing waves, shown on lower line 3, produced by interference of two similar progressive waves 1 and 2 traveling to the right and left, respectively.

transverse waves of the same velocity, wave length, and amplitude, 1 traveling to the right and 2 to the left. The four wave forms a, b, c, and d, show the wave shapes at successive intervals of a quarter of a period or a displacement of a quarter of a wave length. Since both the waves are acting on the medium, then, by the principle of superposition, the resultant displacement of the medium is the sum of the two individual displacements. The curves in 3, Fig. 32-2, show the resultant displacement. In diagram 3a the two waves are in phase, so that the amplitude of this displacement is twice that of either wave. In 3b the waves are 180° out of phase, so there is no displacement of the medium, while in 3c the two progressive waves are in phase and the amplitude is double, and in 3d the waves are again 180° out of phase, so there is no displacement. These four positions are shown in Fig. 32-3, from which we see that there are three positions, marked N, called *nodes*, in which **455** there is no displacement of the medium. Halfway between the nodes the medium can undergo maximum displacement or any displacement between zero and double that of either wave. These positions are called *loops*, *L*, or sometimes *antinodes*. Such a configuration as is shown in Fig. 32-3 is called a *standing wave*, inasmuch as the two progressive waves interfere so as to produce a stationary or nonprogressive wave pattern. From the figure it is seen that the distance between two successive nodes or loops is half a wave length,

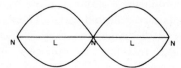

FIG. 32-3.—Standing-wave pattern

$\lambda/2$, while between a node and the next loop it is a quarter of a wave length, $\lambda/4$. In order to produce a standing wave, not only is it necessary to have the two similar waves moving in opposite directions, but their relative phases must be correct, as shown in Fig. 32-2. This is most easily attained by producing the second wave by reflection of the first. We must then discuss the conditions of reflection of waves. Finally, we shall show how the wave length of these standing waves can be measured.

32.9. Reflection of Waves

First, we shall consider the reflection at a fixed end, such as a rigid wall for a sound wave, or at a fixed point for a transverse wave on a rope. At the fixed end there can be no resultant displacement of the particles in the sound wave or of the rope in the transverse wave. At the fixed end the displacement and the velocity of the reflected wave are both reversed relative to the incident wave. A fixed end must be a node of displacement. In a sound wave a node of displacement occurs at a loop of pressure variation, as shown in Fig. 31-5, for there it is seen that condensations and rarefactions occur at positions of zero displacement. At a closed end a compression in a sound wave is reflected as a compression, and a rarefaction is reflected as a rarefaction. From a region of normal pressure to a compression in a progressive wave the displacement of the particles goes from a maximum to zero, and, since the fixed end reverses

the displacement, it follows that a compression incident on a fixed end is reflected as a compression.

The reflection of waves at a free end may be shown by a long rope hanging vertically, with the lower end free or, for sound, by sending a wave motion down a pipe whose diameter is several times smaller than the wave length of the waves. In both cases there is a reflected wave such that at a free end there is a displacement loop or antinode and, for the resultant sound wave, a pressure node. Since the end of the pipe is open to the atmosphere, one would find the pressure there remaining constant or becoming a pressure node. Suppose a compression moves up to an open end of the pipe, then the air moves out so as to reflect a rarefaction down the pipe. The resultant of these two forms a region of normal pressure at the end. Thus at an open end a compression is reflected as a rarefaction, and, similarly, a rarefaction is reflected as a compression. Actually, the node of pressure does not occur exactly at the free end but about six-tenths of the radius beyond the opening.

32.10. Standing Waves on a String Fixed at Both Ends

Consider a string fixed at both ends and plucked or bowed, as in a violin, so as to set up waves along the string. Each end of the string must be a displacement node so that if the length of the string is l, then the wave length, λ_1, of this standing wave is

$$\lambda_1 = 2l .$$

This vibration is called the *fundamental mode of vibration*. The velocity of a progressive wave in the string is given by equation (31-8) as

$$v = \sqrt{\frac{F}{\mu}},$$

where F is the tension and μ is the mass per unit length of the string. The frequency of the fundamental note, f_1, is

$$f_1 = \frac{v}{\lambda_1} = \frac{1}{2l} \sqrt{\frac{F}{\mu}} . \qquad (32\text{-}20)$$

The only condition which the string must fulfil, so far, is that the two ends must be nodes of displacement. In general, then, there can be many different wave lengths; the next one having a node at the center with a wave length, λ_2, equal to the length l of the string. Thus the string can have a standing wave of frequency given by

$$f_2 = \frac{v}{\lambda_2} = \frac{1}{l} \sqrt{\frac{F}{\mu}}. \qquad (32\text{-}21)$$

Similarly, there can be standing waves set up with wave lengths λ_3, λ_4, λ_5, given by

$$\lambda_3 = \frac{2l}{3}; \qquad \lambda_4 = \frac{2l}{4}; \qquad \lambda_5 = \frac{2l}{5}, \qquad (32\text{-}22)$$

with the corresponding frequencies,

$$f_3 = \frac{3}{2l} \sqrt{\frac{F}{\mu}}; \quad f_4 = \frac{4}{2l} \sqrt{\frac{F}{\mu}}; \quad f_5 = \frac{5}{2l} \sqrt{\frac{F}{\mu}}. \qquad (32\text{-}23)$$

The lowest frequency f_1 is that of the fundamental, while the standing waves of frequencies f_2, f_3, f_4, etc., are called *overtones*. Thus, from the foregoing, the overtones have frequencies $2f_1$, $3f_1$, $4f_1$, $5f_1$, etc. To the reader who is familiar with music, if the fundamental frequency f_1 is that of C, then the overtones are the upper partials: C′, G′, C″, E″, G″, etc., where C′ is the octave of C, and C″ is the octave of C′, etc. The amplitudes of the overtones depend on how the string is set in motion, but, in general, they decrease with the frequency of the overtones, as shown in Fig. 32-4.

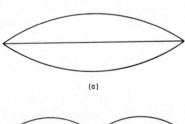

(a)

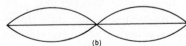

(b)

Fig. 32-4.—Vibration of a string fixed at both ends: (a) fundamental, (b) first overtone.

32.11. Analysis of Sounds

If a string is plucked at the mid-point, then the point would be a loop of displacement, so that the second, fourth, sixth, etc., overtones of frequencies $2f_1$, $4f_1$, $6f_1$, etc., would not be present. Suppose the center point of a string of length l is pulled up a distance h and then let go. The displacement y at any point x on the string

at time t can be shown, by a so-called "Fourier analysis," to be given by

$$y = \frac{8h}{\pi^2} \left(\sin \frac{\pi x}{l} \cos \frac{\pi vt}{l} - \frac{1}{9} \sin \frac{3\pi x}{l} \right.$$
$$\times \cos \frac{3\pi vt}{l} + \frac{1}{25} \sin \frac{5\pi x}{l} \cos \frac{5\pi vt}{l} \qquad (32\text{-}24)$$
$$\left. - \frac{1}{49} \sin \frac{7\pi x}{l} \cos \frac{7\pi x}{l} + \ldots \right),$$

where v is the velocity of the wave in the string.

The period T_1 of the fundamental from equation (32-24) is given by

$$\frac{\pi v T_1}{l} = 2\pi \qquad \text{or} \qquad T_1 = \frac{2l}{v},$$

and its frequency, f_1, is

$$f_1 = \frac{1}{T_1} = \frac{v}{2l}.$$

Thus we see that the even overtone frequencies $2f_1$, $4f_1$, $6f_1$, etc., are absent and that the amplitudes of the odd overtones decrease in the ratio of the square of the odd numbers, namely, $1/9$, $1/25$, $1/49$, etc. In Fig. 32-5 is

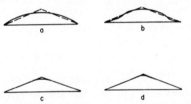

Fig. 32-5.—The fundamental: (a) alone; (b) with first overtone; (c) with two overtones; (d) with three overtones for a string plucked at mid-point.

shown the initial displacement of the string and how this is approximated by four terms in equation (32-24) for the condition $l = 5h$.

In 1807 Fourier showed that any periodic function can be expressed as the sum of a number of sine and cosine functions with appropriate amplitudes. By this Fourier analysis it is possible to analyze a complex wave shape into its harmonic components. This mode of analysis is of great value in almost every branch of physics. Fourier's pioneer work was in the analysis of the propagation of temperature waves in various substances.

As another example of the use of Fourier analysis, let us consider the actual wave shape of the note produced by a violin under some given condition. The resultant wave shape of the tone produced can be recorded by mechanical means, as in the phonodeik* used by Miller, or by electrical means, using a microphone,

* See D. C. Miller, *The Science of Musical Sounds* (New York: Macmillan Co., 1916).

amplifier, and oscilloscope. This composite wave form can then be analyzed into its component waves by a machine known as a "harmonic analyzer." If the equation of the composite wave form is known, a mathematical analysis can be carried out; otherwise one goes through a process of intelligent guessing with a harmonic analyzer. For the violin note, see Fig. 32-6 (a) and (b), where y represents the pressure in the composite wave and θ is proportional to the time. Thus the upper curve in Fig. 32-6 is the sum of the six lower

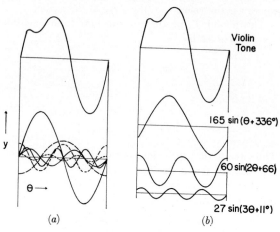

(a)   (b)

Fig. 32-6.—Analysis of note of violin string: (a) three sines (*full lines*) and three cosines (*dashed lines*); (b) same tone represented by three sine terms and three phase constants.

curves: $y = 151 \sin \theta + 24 \sin 2\theta + 27 \sin 3\theta - 67 \cos \theta + 55 \cos 2\theta + 5 \cos 3\theta$. By simple trigonometric transformations these six terms with sines and cosines can be combined into three sine terms with appropriate phase angles and amplitudes, as shown in Fig. 32-6 (b). This form of analysis is most useful to the acoustical investigator, for it tells him that the tone in question is the sum of three simple harmonic motions of frequencies one, two, and three times that of a fundamental, compounded with various amplitudes and phases.

32.12. Quality of a Tone

Analytically, we expect that a musical tone may be represented by a sum of a number of simple tones and that these simple tones have frequencies that are multiples of that of a fundamental. This expectation is verified physically by the use of phonodeiks or oscilloscopes and, to some extent, even by the trained unaided ear. Strike vigorously the next to the lowest C on the piano, place the foot on the sustaining pedal, and listen carefully as the sound dies away. You may be able to hear the next highest C, G above that, then C, E, G, B♭. . . . These faint tones of quite definite mul-

tiple frequencies are called the *overtones* or the *partial* tones of the fundamental that was struck. It is the presence of the overtones that allows us to distinguish a flute's high C from that of a violin or other musical instrument producing the same fundamental tone of high C. The distinction of musical sounds which enables us to identify their source is referred to as the *quality* of the note. Thus we can say that the quality of a musical tone is determined by the relative amplitudes and frequencies of its partial tones or overtones.

32.13. Influence of Phase

In the statement concerning the quality of a musical tone, no mention is made of the relative phases of the partial tones. The relative phases of the component partial tones might be altered at will, resulting in changes in the resultant wave shape, but the resultant sensory impression would remain the same—a fact discovered by the German scientist, Helmholtz. The waves illustrated in Fig. 32-7, (a) and (b), would give rise to identical

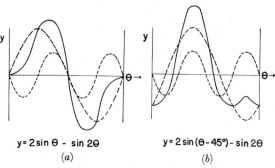

$y = 2 \sin \theta - \sin 2\theta$ $y = 2 \sin (\theta - 45°) - \sin 2\theta$

(a)   (b)

Fig. 32-7.—Effect of change of phase of partials on wave shapes.

sounds, because the amplitudes of the partial tones are in the same proportions, although the phase and wave shapes are quite different. The lack of discrimination of phase differences materially simplifies the description of musical sounds but, at the same time, remains a fact that must be incorporated into any theory of auditory response of the inner ear—a subject which we shall not attempt to discuss.

32.14. Stationary Waves in Pipes— Resonance

Suppose a column of air in a pipe closed at one end is set into vibration. Since the upper open end must be a displacement loop and the closed end a node, it follows that the length l of the pipe can be a quarter of a wave length, $\lambda/4$ of a stationary longitudinal wave. Similarly, the length l can be three-quarters of a shorter wave, λ_2, or five-quarters of a still shorter wave length, λ_3, or

$$l = \frac{\lambda_1}{4} = \frac{3\lambda_2}{4} = \frac{5\lambda_3}{4}. \quad (32\text{-}25)$$

The frequencies of the waves or notes shown in Fig. 32-8 are

$$f_1 = \frac{v}{\lambda_1} = \frac{v}{4l}; \quad f_2 = \frac{v}{\lambda_2} = \frac{3v}{4l}; \quad f_3 = \frac{v}{\lambda_3} = \frac{5v}{4l}, \quad (32\text{-}26)$$

where v is the velocity of sound in the air of the tube. The ratios of the frequencies of the notes emitted are $1:3:5$, etc., or the odd overtones or

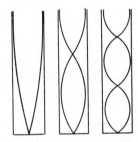

Fig. 32-8.—Stationary waves in a closed pipe, showing displacements.

harmonics. Now these notes can be excited by blowing air over a sharp edge, as in an organ pipe or in different ways for the various wind instruments.

An open organ pipe shown in Fig. 32-9 has a loop of displacement or a pressure node at both

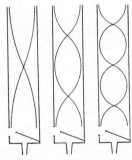

Fig. 32-9.—Displacements in waves in an open organ pipe

ends. If the length of the pipe is l, then the wave lengths of the stationary waves are

$$\lambda_1 = 2l; \quad \lambda_2 = l; \quad \lambda_3 = \frac{2l}{3}, \quad (32\text{-}27)$$

and the corresponding frequencies are

$$f_1 = \frac{v}{2l}; \quad f_2 = \frac{2v}{2l}; \quad f_3 = \frac{3v}{2l}. \quad (32\text{-}28)$$

Thus an open pipe emits notes having frequency ratios of $1:2:3:4$, etc., or all the harmonics, and is therefore a note of different quality from that emitted by an organ pipe with a closed end.

Practically all musical instruments depend on the phenomenon of resonance, which was discussed for the forced vibrations of a spring in § 8.17. In blowing the air past the sharp edge in an organ pipe, a large number of vibrations with different frequencies are set up, and the ones heard are those to which the pipe resonates, namely, the fundamental and overtones given above. The phenomena of resonance can be well illustrated by holding a vibrating tuning fork over a column of air whose length is adjustable.

Suppose a tuning fork of frequency f is held over an adjustable air column, as shown in Fig. 32-10.

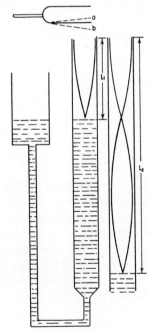

Fig. 32-10.—Resonance between a fork and an adjustable air column.

A tuning fork is one of the few musical devices which emit a single-frequency note, that is, a fundamental with no overtones. If the fork has a frequency f, then the shortest length, l_1, to which the fork will resonate is, from equation (32-26),

$$l_1 = \frac{v}{4f}, \quad (32\text{-}29) \quad \textbf{459}$$

since from Fig. 32-8 the shortest length of pipe emitting a frequency f has a wave length of $4l$. The resonating length, l_1, can be found experimentally to within a few millimeters. The resonance point corresponds to the sound of maximum intensity, for at resonance the fork and air column have the same frequencies. The frequency of the driving mechanism, the fork, is the same as the natural frequency of the driven mechanism, the air column.

If the length of the air column is increased by lowering the water level in the apparatus of Fig. 32-10, then another length, l_2, will be found at which resonance takes place. From the discussion at the beginning of this section it follows that the length l_2 corresponds to three-quarters of a wave length or $l_2 = 3l_1$. Similarly, a length $l_3 = 5l_1$ will also produce resonance. In all these cases there is a displacement loop at the upper end of the pipe and a displacement node at the closed end.

There is another instructive way to regard this resonance. Suppose the tuning fork of frequency f is producing a compression, that is, the lower prong of the fork in Fig. 32-10 is moving from a to b. This compression travels down the air column with the velocity of sound, v, and is reflected at the lower end as a compression. The time t_1 for this compression to travel down the tube and back a total length $2l$ is

$$t_1 = \frac{2l}{v}.\qquad(32\text{-}30)$$

Now when this compression reaches the open end, it is reflected as a rarefaction, and if at the same time the fork is in the position to produce a rarefaction, then there will be a reinforcement or resonance between the fork and the air column. The time between the fork producing a compression, moving a to b, then a rarefaction, moving b to a, is half its period, T. Thus there is resonance when

$$\frac{T}{2} = t_1 = \frac{2l}{v}\qquad(32\text{-}31)$$

or

$$T = \frac{4l}{v}.\qquad(32\text{-}32)$$

Since the period T is the inverse of the frequency f, the resonance occurs when

$$\frac{1}{f} = \frac{4l}{v}\qquad\text{or}\qquad f = \frac{v}{4l},$$

as given in equation (32-26). If the frequency of the fork is known, by measuring l_1, l_2, etc., the velocity of the sound is readily determined.

32.15. Velocity of Sound in Rods and Gases—Kundt's Tube

In the apparatus invented by Kundt, a horizontal rod is clamped rigidly at its center point and has a disk on one end which fits loosely into the end of a horizontal glass tube, as shown in Fig. 32-11. The other end of the tube is closed with a

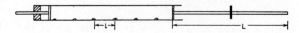

Fig. 32-11.—Kundt's apparatus

plug, which can be moved. Some dry powder (lycopodium or cork) is spread lightly over the bottom of the tube. The horizontal rod is now set into longitudinal vibration by pulling a cloth or chamois over it from the center outward. This causes the light disk on the end of the rod to be set into longitudinal vibration with the frequency of the note emitted. The horizontal vibrations cause the gas in the tube to vibrate, and, by adjusting the position of the plug at the far end, standing waves can be set up in the gas. The presence of these standing waves is clearly shown by the powder which collects at the displacement nodes.

If l is the mean distance between the centers of the nodes, then the wave length λ_g of the note in the gas is

$$\lambda_g = 2l\ ,$$

and similarly for the wave length λ_R in the rod,

$$\lambda_R = 2L\ .$$

Thus if V_R is the velocity of longitudinal waves in the rod and v_g is the velocity in the gas, then

$$V_R = f\lambda_R\qquad\text{and}\qquad v_g = f\lambda_g\ ,$$

where f is the frequency of the longitudinal wave and is the same in the gas as in the rod. Hence

$$v_g = \frac{V_R \lambda_g}{\lambda_R} = V_R \frac{l}{L}.\qquad(32\text{-}33)$$

From equation (31-21) the velocity of longitudinal waves in a solid rod having a density ρ and a value of Young's modulus of Y is

$$V_R = \sqrt{\frac{Y}{\rho}}.$$

Thus, if V_R is calculated, the velocity of sound in a gas can be obtained.

32.16. Analytical Expression for Standing Waves

As we have already discussed, standing waves are set up by two waves of the same wave length, velocity, and amplitude traveling in opposite directions through the medium. A wave of amplitude A, wave length λ, and velocity v traveling along the positive X direction has a displacement y_1 at distance x and time t given by equation (31-4) as

$$y_1 = A \sin \frac{2\pi}{\lambda} (x - vt) \ .$$

Suppose this wave is reflected at a fixed end; then, as seen in § 32.9, the velocity and displacement of the reflected wave are both reversed relative to the incident wave. The equation of the reflected wave then is

$$y_2 = - A \sin \frac{2\pi}{\lambda} (x + vt) \ .$$

The total displacement y at any time t and position x is

$$\left. \begin{aligned} y &= y_1 + y_2 \\ &= A \left[\sin \frac{2\pi}{\lambda} (x - vt) \right. \\ &\qquad \left. - \sin \frac{2\pi}{\lambda} (x + vt) \right] \\ &= - 2A \cos \frac{2\pi x}{\lambda} \sin \frac{2\pi vt}{\lambda} . \end{aligned} \right\} \quad (32\text{-}34)$$

From equation (32-34) we see that the displacement y of the medium is zero at all times for

$$\cos \frac{2\pi x}{\lambda} = 0 \quad \text{or} \quad \frac{2\pi x}{\lambda} = \frac{\pi}{2}, \ \frac{3\pi}{2}, \ \frac{5\pi}{2}, \ \text{etc.} ;$$

or, if the positions are x_1, x_2, x_3, etc.,

$$\frac{2\pi x_1}{\lambda} = \frac{\pi}{2} \quad \text{or} \quad x_1 = \frac{\lambda}{4}$$

and

$$x_2 = \frac{3\lambda}{4}; \quad x_3 = \frac{5\lambda}{4} .$$

These positions x_1, x_2, x_3, etc., mark the positions of the displacement nodes and are half a wave length apart, as already seen. The positions of the loops or antinodes occur at the positions where

$$\cos \frac{2\pi x}{\lambda} = 1$$

or

$$\frac{2\pi x}{\lambda} = \pi, \ 2\pi, \ 3\pi, \ \text{etc.},$$

corresponding to

$$x_1 = \frac{\lambda}{2}; \quad x_2 = \frac{3\lambda}{2}; \quad x_3 = \frac{5\lambda}{2} .$$

It is seen that the loops are halfway between the nodes, or there is a quarter of a wave length between a node and the next loop. This confirms mathematically what we had previously arrived at by physical reasoning.

32.17. Beats between Two Notes

If two tuning forks having frequencies of 256 and 258 cycles/sec, respectively, are sounded together, a note is heard whose amplitude or intensity rises and falls twice each second. The notes are said to have a beat frequency of 2 per second. If we wished to find out which fork had the highest frequency, the information about the beat frequency would not supply this. Suppose fork A has the frequency of 256 cycles/sec and fork B a frequency of 258 cycles/sec. On the tip of one of the forks, say A, we place a small piece of beeswax, thus increasing the inertia of the prong and lowering the frequency of the fork to, say, 254. When the modified fork A is sounded with B, there will be 4 beats per second. Of course, it might be possible to put enough wax on fork B so as to lower its frequency from 258 to 252 cycles/sec so that, when sounded with the unwaxed A fork, there would be 4 beats/sec. One should easily be able to see how this apparent difficulty can be overcome.

This beat frequency will now be analyzed for two frequencies of f_1 and f_2 cycles/sec. Suppose the amplitudes of the two notes are equal; then at some point in the medium, the displacements produced by the note can be written

$$v_1 = A \sin 2\pi f_1 t \ ,$$

$$y_2 = A \sin 2\pi f_2 t \ .$$

The resultant displacement, y, by the principle of superposition is

$$\left. \begin{aligned} v &= y_1 + y_2 = A \left(\sin 2\pi f_1 t + \sin 2\pi f_2 t \right) \\ &= 2A \sin 2\pi \left(\frac{f_1 + f_2}{2} \right) t \\ &\quad \times \cos 2\pi \frac{(f_1 - f_2)}{2} t . \end{aligned} \right\} \quad (32\text{-}35)$$

Let us set the frequency difference,

$$(f_1 - f_2) = 2\delta$$

461

and the average frequency,

$$\bar{f} = \frac{f_1 + f_2}{2}.$$

Then we can unite thus:

$$y = 2 A \cos 2\pi\delta t \sin 2\pi \bar{f} t . \quad (32\text{-}36)$$

The factor $2A \cos 2\pi\delta t$ can be regarded as the amplitude of the sine function. This amplitude changes very slowly, since the factor δ is very small compared with the factor $\bar{f}$ in the argument of the sine function. Suppose the two frequencies are 252 and 248 cycles/sec; then the mean frequency $\bar{f}$ is 250 cycles/sec, and the difference is 4 cycles/sec or $\delta = 2/\text{sec}$. From equation (32-36) we see that the displacement y is zero when

$$\cos 2\pi\delta t = 0 \quad \text{or} \quad 2\pi\delta t = \frac{\pi}{2}, \ \frac{3\pi}{2}, \ \frac{5\pi}{2}, \ \text{etc.},$$

or for

$$y = 0, \qquad t = \frac{1}{4\delta}, \ \frac{3}{4\delta}, \ \frac{5}{4\delta}, \ \text{etc.}$$

In terms of the numerical example the amplitude is zero at times of 0.125, 0.375, 0.625 sec, etc. It has its maximum value when $\cos 2\pi\delta t$ is equal to unity, that is,

$$\cos 2\pi\delta t = 1 \quad \text{or} \quad -1$$

or

$$2\pi\delta t = 0, \ \pi, \ 2\pi, \ 3\pi, \ \text{etc.},$$

or for

$$y = \text{Maximum}, \quad t = 0, \ \frac{1}{2\delta}, \ \frac{2}{2\delta}, \ \frac{3}{2\delta}, \ \text{etc.}$$

It is the rising and falling of the amplitude which gives rise to the beats, and the time between two maxima or between two zeros of amplitude is the period of the beats. Thus the beat period is

$$\frac{1}{2\delta} = \frac{1}{(f_1 - f_2)},$$

and the beat frequency is the inverse of the period, or

$$2\delta = (f_1 - f_2),$$

which in our example is 4/sec.

Notice that the displacement y in equation (32-36) is zero when $\sin 2\pi \bar{f} t$ is zero, that is, for our numerical example 500 times/sec. The maximum value of the amplitude is governed by the $\cos 2\pi\delta t$ term, and it is this which gives rise to the fluctuations in amplitude or to the beats. This phenomenon is a form of amplitude modulation which has a counterpart in radio receivers (see § 30.14).

32.18. Supersonic Waves

Certain crystals, among them quartz, Rochelle salt, and tourmaline, exhibit a phenomenon known as the PIEZOELECTRIC EFFECT. If a crystal of quartz is subjected to mechanical stresses, charges of electricity appear on the surface. Conversely, if the crystal is suitably electrified by coating two sides of a slab of the crystal with a metal and applying an alternating voltage to these sides, the crystal will be set into vigorous longitudinal vibrations, the frequency of which may be quite beyond the range of audibility. Frequencies as high as a million cycles per second have been obtained.

Since the mechanical vibrations are here longitudinal, equation (31-21) is applicable. In the case of quartz, ρ is 2.654 gm/cm³, Young's modulus may be taken as 8×10^{11} dynes/cm², and so the velocity of a progressive longitudinal wave is 5.5×10^5 cm/sec. If the boundaries of the crystal slab are free, the fundamental vibration has a node at the center of the plate, so that if l is the length of the crystal in millimeters, the fundamental wave length is $2l$, and the frequency f, given $f\lambda = v$, is

$$f = \frac{2,750 \times 10^3}{l} \text{ cycles/sec}.$$

The importance of these supersonic waves lies in the large amount of energy that may be transmitted. These waves have been used by Langevin and others in the detection of icebergs and in depth soundings. If these short waves are projected across the boundary of two liquids, a rather stable emulsion may form. With frequencies approaching 3×10^5 cycles/sec, heating effects in liquid become pronounced. Small animals have been killed by exposure to these supersonic waves. Pulsed supersonic waves have also been used for the detection of submarines.

32.19. Architectural Acoustics

We are all aware that speech or music can be heard much more clearly in some churches or concert halls than in others. The primary reason for this difference is the time which a sound persists after the source has been stopped. This is referred to as the *time of reverberation of the sound*. In a room with hard walls and little sound-absorbing material the reverberation time may be several seconds, so that successive sounds become jumbled up with one another. On the other hand, there may be much sound-absorbing material in a room, such as heavy rugs, draperies, etc., so that there is practically no reverberation and the room is said to be "dead."

The study of the properties of a room which

determine how effective the room is for speech or music was first undertaken by Professor W. C. Sabine, at Harvard University, about 1895. In a new lecture room at the university, audiences found it practically impossible to understand a speaker because of the excessive reverberation. Sabine was asked to study the problem and correct the difficulty. This he did so thoroughly that he established the science of architectural acoustics, and relatively little basic knowledge has been added since his time.

When a sound of definite frequency and duration is produced, the sound waves impinging on the floor, walls, or ceiling of a hall are partly reflected and partly absorbed. If there were no absorption—an impossible situation in practice—then the sound would persist forever, or the reverberation time would be infinite. On the other hand, if there was no reflection or complete absorption on striking a surface, then the time of reverberation would be zero. In practice the reverberation times of rooms lie between these limits. In order to have a quantitative measurement of reverberation times, some agreement must be reached on how much the intensity of the sound must be decreased during this reverberation time. This has been set at one-millionth of the original intensity. Thus if I_0 is the original intensity of a sound, then the time of reverberation, t_r, is the time after the sound ceases for its intensity I to become

$$I = 10^{-6} I_0 .$$

This means that the intensity level of the sound is lowered by 60 db during the reverberation time.

Now this reverberation time depends on the volume, V, of the room and the amount of absorption taking place. It also depends to a much smaller extent on the frequency of the sound or on whether speech or music is being produced. Some compromise has to be made if a hall is used for both speech and music. The best or most satisfactory reverberation time depends on the volume of the room, a small room requiring a smaller time than a larger one for clearly hearing the sounds. This is a psychological factor and must be agreed on by a number of people. Table 32-3 gives the best average values for the reverberation times and volume. These are only approximate, since they depend on the frequency of the sounds.

Having decided on the best reverberation time,

one must next measure the absorption coefficient of various substances. This is done in terms of square feet of surface and is measured for a pitch of 512 vibrations/sec under controlled conditions, usually with electronic devices to determine the intensities. The most perfect absorber is an open window, inasmuch as the sound passes out through

TABLE 32-3

VOLUME AND REVERBERATION TIMES

Volume (Ft³)	Time (Sec)	Volume (Ft³)	Time (Sec)
1,000	0.8	500,000	1.9
10,000	1.2	1,000,000	2.0
100,000	1.5		

the window and none returns. Table 32-4 gives the absorption coefficient, a, for a square foot of various materials.

TABLE 32-4

ABSORPTION COEFFICIENTS

MATERIAL IN UNITS OF a/FT²

Open window	1.00	Celotex	0.35
Hair felt, 1 in thick	0.58	Carpets	0.20
U.S.G. acoustic tiles		Brick wall, plaster,	
$\frac{3}{4}$ in thick	0.56	or glass	0.03
Heavy curtains	0.40	Marble	0.01

INDIVIDUAL OBJECTS IN UNITS OF TOTAL
ABSORPTION COEFFICIENT, A

Adult person	4.7
Seat cushions	1.00–2.00
Wooden seats, per seat	0.20

In order to obtain the total absorption, A, the number of square feet, a, of the different materials must be known and these multiplied by the corresponding absorption coefficients a, so that

$$A = a_1 a_1 + a_2 a_2 + a_3 a_3 + A_1 + A_2 + \ldots ,$$

where A_1, A_2, etc., are the absorption of the individual objects.

By much careful work Sabine found that the time of reverberation, t_r, could be satisfactorily expressed as

$$t_r = 0.05 \frac{V}{A} \text{ sec}, \qquad (32\text{-}37)$$

where V is the volume of the room in cubic feet and the coefficients a are per square foot. In the metric system, in which V is given in cubic meters and the coefficients a are per square meter, then

$$t_r = 0.16 \frac{V}{A} \text{ sec} .$$

The problem, then, for a given hall is to adjust the total absorption A so that the reverberation time is as close as possible to the best value for that volume of the hall given in Table 32-3. An illustrative example will show how this is done.

EXAMPLE 2: A concert hall has a volume 150,000 ft³ and is made up of the materials given in the table. The problem is to determine whether the hall is acoustically satisfactory.

From Table 32-3 a satisfactory reverberation time for this volume is about 1.7 sec. Thus, from equation (32-27) the total absorption, A, must be such that

$$A = \frac{0.05 \times 150,000}{1.7} = 4,400 .$$

The absorbing material was made up principally of the following:

		Units
Wood................	10,000 ft² at 0.06 =	600
Stone................	3,500 ft² at 0.02 =	70
Carpet..............	2,500 ft² at 0.20 =	50
Curtains	800 ft² at 0.40 =	320
Vents................	700 ft² at 0.75 =	520
Seats cushioned........	600 ft² at 1.50 =	900
Plaster..............	20,000 ft² at 0.03 =	600
Total...........................		3,060

Thus, if the hall were empty, the reverberation time would be somewhat too large. However, if the hall were two-thirds filled with 400 persons, the additional absorption would be

$$400 \times (4.7 - 1.5) = 1,280 .$$

This makes the total 4,340, which is sufficiently close to the desired amount of 4,400 as to be acceptable. The value of 1.5 for the cushioned seats has already been included and therefore must be subtracted, since the 400 people cover up the 400 cushioned seats. If less audience is expected, then it might be desirable to increase the absorbing material in the hall.

There are other acoustical effects besides reverberation which must be considered in constructing a concert hall. One of the chief ones is the reflection by curved surfaces, especially in the ceiling, so that a large intensity is concentrated at some spot or over small areas. If this is serious, then the smooth curve of the ceiling may have to be changed or the surface corrugated, so that the sound is reflected in different directions. In most large halls there are several loud-speakers at different places, so that the intensity can be kept fairly uniform over the whole of the hall.

32.20. Résumé

The following is a brief listing of the definitions, principles, and theories, given in this chapter, which you should know:

The effects of temperature and pressure on the velocity of sound in gases

The variation of the pressure in a sound wave; limits of audibility of the human ear

The expression for the intensity in a sound wave

Intensity levels in sound waves; the bel and the decibel

The production of standing waves

The reflection of waves at fixed and free ends

Stationary waves in a cord and in pipes; the relative frequencies of the possible vibrations

The theory of beats between notes of nearly the same frequency

The theory of architectural acoustics

Queries

1. When a vibrating tuning fork is brought near a closed organ pipe of the proper length, the intensity of the sound from the pipe becomes much greater. Is the pipe a mechanism for creating energy?

2. Why is it better to have a foghorn on top of a high structure than near the ground?

3. One source of sound propagates 10^7 times as much energy per unit time as another does. How many decibels higher is it?

4. Occasionally the report of an explosion is inaudible a few miles away, although it may be heard plainly at greater distances. What possible explanation can you give?

5. An observer is at rest, relative to the air. What would the observer hear if a source of frequency f moved away with the speed of sound?

6. If one vigorously strikes a note in the middle of a piano keyboard and then, on lifting the finger, immediately thereafter depresses the sustaining pedal, the same tone may be heard persisting. Explain this.

7. Can one get a response from a piano by singing strongly on a certain pitch while the sustaining (or "loud") pedal is held down? Why?

8. The passage of an electric spark through the air or that of a bolt of lightning is pratically instantaneous. Why, then, does thunder roll?

9. Is the intensity of a wave always given by the product of the energy per unit volume and the velocity of the wave?

Practice Problems

1. What is the ratio of the specific heats of a gas if the velocity of sound in that gas is 224 m/sec and the density is 0.0025 gm/cm³, all under standard conditions? Ans.: 1.24.

2. The velocity of sound in chlorine at 0° C and 76 cm pressure of mercury is 206 m/sec. Assuming that chlorine is a perfect gas, how much does the velocity of sound in it change for each degree rise in temperature?

3. The velocity of sound in a certain gas is 206 m/sec. The density of the gas is 3.214 gm/liter under standard conditions, and $c_p = 0.1125$ cal/gm °C. Compute c_v. Ans.: 0.084 cal/gm ° C.

4. How many beats per second would be heard by the observer in problem 23 of chapter 31.

5. Two tuning forks, A and B, give 5 beats per second when sounded together. Fork A is known to have a frequency of 300 cycles/sec. When a small piece of wax is added to fork B, the number of beats with fork A becomes 3 per sec. What is the frequency of fork B without wax? Ans.: 305 cycles/sec.

6. Two tuning forks, A and B, give a beat every 5 sec. Fork A has a frequency of 200 cycles/sec. If a small piece of wax is put on fork B, there is one beat every 2 sec. From these data is it possible to determine the frequency of fork B, and, if so, what is it? Give reasons for your answer.

7. The fundamental frequency of an open organ pipe 100 cm long is 180 vibrations/sec. What is the velocity of sound in the pipe? What is the frequency of the second possible overtone of that open pipe? Make a diagram showing the loops and nodes in this problem. Ans.: 360 m/sec; 360 vibrations/sec.

8. A closed organ pipe has a length of 2 m. If the velocity of sound is 360 m/sec, what are the frequencies of the fundamental and first harmonic emitted by the pipe?

9. A tuning fork of frequency 350 vibrations/sec is held over an adjustable column of air, as in Fig. 32-10. If the velocity of sound is 350 m/sec, find the lengths of the air column to which the fork resonates. Ans.: 0.25 m; 0.75 m; 1.25 m; etc.

10. Two identical tuning forks give forth a note of 500 vibrations/sec in frequency. A listener moves slowly at a rate of 10 cm/sec along the straight line between the two forks. Will he hear beats? Why, and of what frequency? He moves at the same rate along the perpendicular bisector of the line joining the forks. Again, will he hear beats? Why, and of what frequency?

11. If a quartz crystal is 10 mm long, compute the frequency of the supersonic waves set up. Ans.: 2.75×10^5/sec.

12. Two pipes, each of 256 vibrations/sec, are mounted at opposite ends of a rod 3 m long and capable of rotation about a vertical axis. Where the rod is making 1 revolution in 3 sec, does an observer in the plane of the rod's revolution hear beats? If so, describe them. What does an observer located in the line of the axis hear? Does the interference pattern set up by such a device exhibit any nodal surfaces of revolution? Has it any nodal planes or nodal lines? Assume that the observer is at a distance large in comparison with 3 m and take the velocity of sound to be 332 m/sec.

13. An open organ pipe has a length of 2 m. What length of closed pipe would give the same fundamental frequency as the open? If the velocity of sound is 350 m/sec, what is this fundamental frequency? Give the frequencies of the overtones for the closed and open pipes. Ans.: 1 m; 87.5 vibrations/sec. Closed: 262.5, 437.5 vibrations/sec. Open: 175, 262.5 vibrations/sec.

14. In a Kundt dust-tube experiment, a rod of length 1 m, clamped at its mid-point, is set into longitudinal oscillation. The distance between the dust heaps in the air is 10 cm. If the velocity of sound is 350 m/sec, find (a) the frequency of the note emitted by the rod and (b) the velocity of sound in the rod.

15. An aluminum rod of density 2.7 gm/cm³ and length 80 cm is clamped at its center point, and one end, with a disk attached, is placed in a horizontal glass tube to form a Kundt dust tube. When the rod is set into oscillation, the dust heaps in the air of the tube are 5.5 cm apart. If the velocity of sound in the air is 348 m/sec, find (a) the frequency of the note emitted by the aluminum rod; (b) the velocity of sound in the rod; (c) Young's modulus for the rod. Ans.: (a) 3,160 vibrations/sec; (b) 5.05×10^3 m/sec; (c) 6.9×10^{10} newtons/m².

16. A wire 1 m long has a mass of 0.010 kg and is under a tension of 10 kgf. If the wire is rigidly held at both ends and is set into vibration, find the frequencies of the fundamental and first two overtones.

17. If the intensity in a sound wave is doubled, find how many times the pressure amplitude is increased. If the pressure amplitude in a sound wave is increased fourfold, by how many times is the intensity of the wave increased? Ans.: 1.4; 16.

18. A sound wave in air has a maximum pressure variation of 10 dynes/cm². Find the intensity of this sound wave, assuming the density of air to be 0.00125 gm/cm³ and the velocity of sound to be 345 m/sec.

19. Find the intensity level in decibels of a sound wave **465**

whose intensity is 10^{-10} watt/cm² relative to the arbitrary reference intensity level of 10^{-16} watt/cm². Ans.: 60 db.

20. Find the ratio of intensities of two sounds whose intensity levels differ by (*a*) 20 db; (*b*) 10 db; (*c*) 5 db.

21. A hall has a volume of 10,000 ft³. What is the total absorption which the room must have to have a satisfactory reverberation time.? Ans.: 4,170.

22. A room has a floor of 20 by 30 ft and a height of 10 feet. If the average absorption coefficient *a* for all the surfaces of this room is 0.08, find the reverberation time for this room.

23. Suppose that, in order to reduce the reverberation time of the room in problem 22, the room has the ceiling covered with Celotex and the floor covered with carpets. Find the reverberation time. Ans.: 0.73 sec.

Musical Sounds and the Science of Music*

33.1. Description of Musical Sounds

Music, since the advent of the phonograph and the radio, is an important part of every person's experience, and great numbers of persons today are highly sophisticated with respect to it. It is a very complex phenomenon, however. Its physical basis is none too well understood by musicians, who are primarily interested in it as an art. It has been somewhat neglected by physicists, to whom detailed study yields insufficient reward, since little knowledge fundamental in character and bearing on other physical problems is to be unearthed. Thus both groups lose much of interest and understanding in a field having great aesthetic appeal to all classes of people.

Musical phenomena involve matters that pertain especially to the source, to the carrier waves, and to the receiving and interpreting mechanism. We shall discuss somewhat further the essential characteristics of the waves—pitch, intensity, and quality. However, much of this chapter will interest principally the student who already has some knowledge of, or interest in, music.

33.2. Pitch

Pitch, measured by frequency, is the single characteristic that, more than any other, dis-

* The material in this chapter has been furnished by Dr. Franklin Miller, Jr., of Kenyon College, and the authors' acknowledgments are hereby gratefully recorded.

tinguishes MUSICAL SOUNDS from that heterogeneous mixture of frequencies we call NOISE. A musical TONE is defined by its frequency. "Middle C" is about 262 cycles/sec; tones that have greater (or lesser) frequencies are said to be "higher" (or "lower") in pitch, although there is no good physical or psychological reason for associating such a spatial magnitude with frequency. Greek musicians, indeed, used the opposite convention. As we have previously noted, audible musical sounds range in pitch from about 16 to 16,000 cycles/sec, with great variation as to the limits heard by different individuals' ears.

33.3. Intensity

INTENSITY, or loudness, we have seen (§ 32.6), is measured by the amount of energy carried across a unit area in 1 sec or, in practical units, by decibels. Expressions showing that the intensity of a wave of simple harmonic motion depends upon the squares of both amplitude and frequency of vibrations have also been derived. The simultaneous production of several sounds of different frequencies, we have noted (§ 32.17), results in a sound of periodically changing amplitude and hence intensity—the phenomenon of BEATS. If the beat frequency is small, the individual pulsations are registered as such by the ear; if the beat frequency is sufficiently high that it lies well in the audible range, a musical tone can be distinguished. This BEAT TONE is physically present, for the definition of a musical tone implies only the existence of a pressure which has a periodically fluctuating amplitude. DIFFERENCE TONES, as beat tones are sometimes called, are usually of quite low intensity but are of great importance in the study of harmony and the aesthetics of musical perception.

33.4. Quality

As we have noted previously, not all tones of the same frequency and loudness have the same psychological effect upon the ear. A flute's high C can easily be distinguished from the same frequency produced by a violin, and the difference is essentially due to the differences in the relative intensity of the harmonics in the two tones, and these, in turn, to the different methods by which the sounds are produced. We shall now develop a quantitative method for describing this obvious physical attribute, which is known as the QUALITY of a tone.

If, by any sort of mechanism that enables us to follow accurately the variations of pressure with the time, we look at the graph of the resulting WAVE FORM, we seldom discover one that is a pure sine (or cosine) wave. In Fig. 33-1 we illustrate several types of waves characteristic of different instruments. Pressure is the ordinate, and time the abscissa. Only a gently bowed tuning fork or a gently blown open organ pipe or flute gives approximately SIMPLE TONES, in the sinusoidal sense. Indeed, the ear can distinguish only with difficulty any difference between the three examples just cited. A fundamental principle of acoustics is that the sensation of a "simple" tone is given only by a sinusoidal wave and that the method of production of such a simple tone is immaterial. This is psychological justification for considering the simple harmonic motion to be of importance in the analysis of a complex tone.

33.5. Musical Instruments

A musical instrument is a machine for imparting energy to the air in the form of pulsations. Table 33-1 gives a list of several musical instruments and their characteristics. Such machines, in general, consist of two parts: a GENERATOR, which, by its vibrations, supplies complex musical tones,

and a RESONATOR, which amplifies certain of the incoming tones up to an audible intensity. The resulting tone may be analyzed by Fourier's method, provided that the partial tones have integral frequency ratios (are HARMONIC). If the partial tones are INHARMONIC, analysis must be

TABLE 33-1
PARTIAL TONES OF INSTRUMENTS
HARMONIC

Instrument	Generator	Resonator	Partial Tones in Order of Intensity
Tuning fork	Fork	Sometimes hollow wooden box	1
Flute	Air stream at blow-hole	Cylindrical air column	1, 2 (4, 3)
Violin	One of four strings	Belly, bridge, back (wood); also partly inclosed volume of air	E string: 1, 3 (2, 4) A string: 1, 5, 2, 4, 3 (6, 7, 8) D string: 1, 2, 3, 5, 4, 6 (7, 8, 9, 11, 12) G string: 4, 5, 3, 2, 1, 6 (7, 8, 9, 11, 12, 13, 14)
Clarinet, saxophone	Single reed	Cylindrical air column	8, 9, 10, 3, 1, 7, 11, 12 (5, 4)
Oboe	Double reed	Conical air column	5, 4, 6 (10, 37, 11, 8, 2, 1)
Horn	Lips of player	Conical air column	All partials up to the thirtieth; relative intensities depend greatly upon manner of playing instrument
Voice	Stretched vocal cords	Air column in throat and mouth	Soprano voice intoning *a* as in father, 488 cycles/sec; 2, 1 (3, 4, 5, 6, 7) Bass voice intoning *a* as in father, 92 cycles/sec; 8, 7, 9, 11 (12, 4, 5, 1, 2, 3, 13, 14, 15)
Piano	String	Wood sounding board	Low pitch: weak fundamental; partials up to the forty-second Medium pitch: 1–10 about equal High pitch: 2, 1, . . .
Organ pipes	Air stream *or* Reed	Cylindrical air column	1 (2, 3) 1, 2, 3, 5 (4, 6) . .

INHARMONIC

Instrument	Generator	Resonator	Theoretical Frequencies of Overtones Expressed as Multiples of the Lowest
Circular drum	Drum beat	Volume of air	1.000, 1.594, 2.136, 2.296 . . .
Xylophone	Bar of wood, free at both ends	Cylindrical metal tubes sometimes	1.000, 2.76, 5.43, . . .
Thin bells	Metal plate	. . . .	1.000, 2.928, 5.423, 8.771,* . . .

* See Rayleigh, *Theory of Sound*, with historical introduction by R. B. Lindsay (2d rev. ed.; New York: Dover Publications, 1945), I, 387, 390 ff.

made by trial and error, until a series of sinusoidal terms can be found which represents the complex wave form with the minimum number of terms. The frequencies of the terms thus determined are those of the partial tones. Thus, in theory at least, a vibrating circular drumhead has partial tones with frequencies in the ratios of 1.000, 1.594, 2.136, 2.296, 2.653, 2.918, ... These theoretical fre-

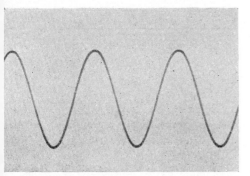

Tuning fork, $n = 256$ cycles/sec

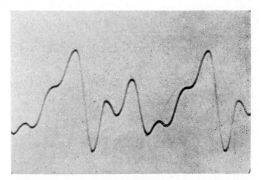

Voice, "Ah," $n = 165 \pm$ cycles/sec

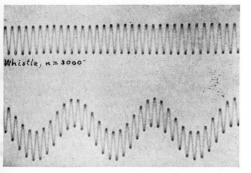

Fork, $n = 256$, and whistle, $n = 3,000$ cycles/sec

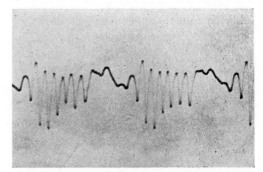

Clarinet, C_3, $n = 230$ cycles/sec

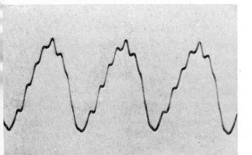

Violin, D_3, $n = 290$ cycles/sec

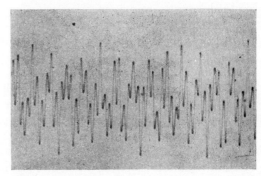

Bell, $n = 110 \pm$ cycles/sec

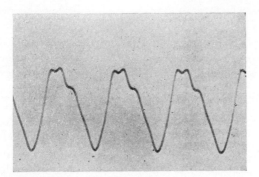

Flute, G_3, $n = 387$ cycles/sec

The speaking voice saying —

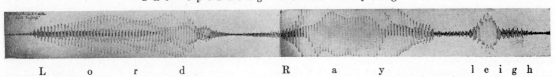

L   o   r   d   R   a   y   l e i g h

FIG. 33-1.—Wave forms of sounds of pressure and time, for various instruments as recorded by D. C. Miller's phonodeik

quencies are capable of exact calculation,* if one makes use of irrational functions invented by Bessel; and the experimentally observed frequencies fit the theory with remarkable accuracy.† For reasons to be discussed later, such inharmonic sounds are less pleasing than sounds whose overtones are in simple integral ratios; hence most musical instruments attempt to avoid such inharmonic partials.

33.6. Synthesis of Musical Tones

We have seen that a complex tone may be broken down into a number of simple tones, each a simple harmonic motion with a different period. The reverse is also true: a complex tone may be synthesized out of a number of simple tones, provided that the amplitudes and frequencies of the partial tones are the same as exist in the original complex tone. Machines exist for doing this mechanically, so that, as verification of the analysis by Fourier series, the various simple sinusoidal curves are put together and their sum compared with the original wave form. The electric organ is based upon this principle. A number of sinusoidal pulsations are produced in a loudspeaker by electrical and mechanical means. The frequencies of the pulsations are integral multiples of a fundamental, and the relative intensities are adjusted by the player. In practice, only eight partial tones are mixed; although many of the higher tones are omitted, the resulting sound can be made to imitate the organ pipes, flute, trumpet, and clarinet. The percussion effect of a struck bell may be imitated on the piano, which is also a percussion instrument; it is necessary only that the frequencies and intensities of the overtones found in the bell tone be artificially introduced (see Fig. 33-2). The chord may be transposed higher and

---

* See Rayleigh, *Theory of Sound*, with historical introduction by R. B. Lindsay (2d rev. ed.; New York: Dover Publications, 1945), I, 331.

† The frequencies of overtones for any simple resonator can be found by the solution of a differential equation which is characteristic of the resonator and which must satisfy boundary conditions imposed by the physical nature of the object. With respect to the organ pipes discussed in chap. 32, the boundary conditions were that at a closed end the air particles remain stationary, etc. If the generator is free to move in more than one dimension, the boundary conditions are such that the overtones permitted by the differential equation are not simple integral multiples of the lowest one.

---

lower to imitate bells of different pitches; however, the lower half of the piano is not useful, because the overtones of the piano strings themselves, being harmonic, conflict with the desired tones of the bell, which are inharmonic. In general, synthesis of musical tones can be made only if *simple* tones are used as building blocks; tuning forks

FIG. 33-2.—The clang of bells may be imitated on the piano. Two different bell tones are illustrated. The sustaining pedal may be used if desired.

were used by Helmholtz and stopped organ pipes by D. C. Miller; electrically produced simple harmonic motions are used in some electric organs.

33.7. Melody and Scales

We will now consider some of the physical and psychological factors which are related to musical art as a whole. The most obvious fact of music is MELODY—a succession of tones of different pitches. At first glance it would seem that there was an infinite array of frequencies from which to choose the frequencies which form a melody; the fact is that the most pleasing melodies are constructed from a finite array of frequencies, known as the MUSICAL SCALE. As an experimental fact, a melody sounds best if the ratio of the frequency of any tone, divided by the frequency of the tone immediately preceding, is a simple rational fraction. Thus a tone of frequency 513 following a tone of frequency 300 sounds much worse than a tone of frequency 500 following the 300-cycle tone.

Making use of this fact, the reason for which is obscure, we can construct a scale. The most simple ratio is 2:1, designated as an OCTAVE; and the scale must divide the octave into a number of smaller intervals with simple ratios. An important psychological fact is that intervals which seem identical differ in the same *ratio*. Thus two tones of frequencies 600 and 300, respectively, are separated by an octave; two tones of frequencies 700 and 350 are separated by the *same* interval, the octave, since $\frac{600}{300} = \frac{700}{350} = \frac{2}{1}$. Intervals will therefore be compared by taking ratios rather than by taking differences of frequencies; and, when we *add* intervals, we *multiply* the ratios together.

Thus two octaves is an interval of $\frac{2}{1} \times \frac{2}{1} = \frac{4}{1}$. With these rules in mind, we can define a MUSICAL SCALE as "a division of the octave into intervals suitable for musical purposes."*

Such a scale is that of PTOLEMY,† also called the "just" or "true" scale (Fig. 33-3). If we designate

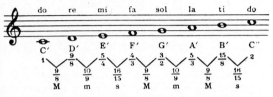

FIG. 33-3.—The just scale of Ptolemy

the frequency of the first note of the scale as "1," then the last note of the scale must be the octave, or "2." On the piano keyboard the sequence of white notes from C′ to C″ is approximately in these ratios; the absolute frequencies of these notes cover a range of from 261.6 to 523.2 cycles/sec.

It will be noted that the various allowed frequencies of the Ptolemaic scale bear fairly simple ratios to the first, or TONIC, tone of arbitrary frequency 1; the various allowed frequencies also bear simple ratios *to one another* in many instances. Thus $\frac{15}{8} \div \frac{5}{4} = \frac{3}{2}$; another simple ratio is $\frac{5}{3} \div \frac{4}{3} = \frac{5}{4}$; etc. In particular, the ratio of *each frequency to that of its immediate predecessor* in the series is either $\frac{9}{8}$, $\frac{10}{9}$, or $\frac{16}{15}$.

33.8. Intervals

When two tones are sounded either simultaneously or consecutively, they are said to constitute an INTERVAL. Depending upon the frequency ratios, the simple intervals have been given names, as follows: The intervals between successive tones of a Ptolemaic scale, then, are either the MAJOR TONE, M (9:8), the MINOR TONE, m (10:9), or the SEMITONE, s (16:15). When we say that the octave is the sum of three major tones, two minor tones, and two semitones, we are merely stating the arithmetical identity that

$$\frac{9}{8} \times \frac{10}{9} \times \frac{16}{15} \times \frac{9}{8} \times \frac{10}{9} \times \frac{9}{8} \times \frac{16}{15} = \frac{2}{1} ,$$

* John Redfield, *Music: A Science and Art* (New York: Alfred Knopf, 1928), p. 68.

† Flourished about A.D. 130.

or, from Table 33-2,

$$M + m + s + M + m + M + s = \text{Octave} ,$$

since we *add* intervals by *multiplication* of frequency ratios.

TABLE 33-2

MUSICAL INTERVALS

Name of Interval	Frequency Ratio	Name of Interval	Frequency Ratio
Unison	1:1	Major sixth	5:3
Octave	2:1	Minor sixth	8:5
Fifth	3:2	Major tone (M)	9:8
Fourth	4:3	Minor tone (m)	10:9
Major third	5:4	Semitone (s)	16:15
Minor third	6:5		

33.9. Diatonic Scales and Modes

Only a very few acute ears are able to distinguish between the major tone and the minor tone.‡ The Ptolemaic, or DIATONIC, scale is therefore considered to be made up of five WHOLE TONES (w) and two SEMITONES (s). The particular arrangement of these seven intervals, which we know as the "C scale" on the piano, is $MmsMmMs$, or approximately $wwswwws$; we shall consider later the reasons underlying the survival of this arrangement, for there were, in all, seven arrangements or MODES commonly employed by the

TABLE 33-3

MODES

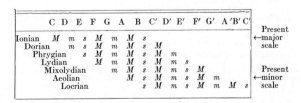

Greeks for their melodies. These modes were later adopted by the early Christian church with some alterations; they are known as the "ecclesiastical modes." The sequence $MmsMmMs$ that we have been discussing was the IONIAN MODE, and other modes are tabulated in Table 33-3. All can be played on the piano, starting upon the proper white note and extending to a white note one

‡ Comparing these intervals by division, as always, we find that $\frac{9}{8} \div \frac{10}{9} = \frac{81}{80} = 1$, correct to about 1 per cent.

octave higher; more will be said later concerning these ecclesiastical modes and the reasons why the Ionian and Aeolian modes have survived as our present-day major and minor scales, respectively.

33.10. Consonance and Dissonance

We have so far followed the historical development of music in that we have avoided mention of the possibility of sounding two or more tones simultaneously. In modern music, since about 1300, it has been recognized that if two frequencies are in simple ratios (e.g., 3:2), they seem pleasing when sounded simultaneously as well as when sounded consecutively. This leads us to define HARMONIC CONSONANCE and MELODIC CONSONANCE as the temporal juxtaposition or overlapping of tones whose frequencies bear simple ratios. All other combinations, whether harmonic (simultaneous) or melodic (consecutive), are DISSONANCES.

At the outset it should be stressed that the words "consonant" and "dissonant" have meaning only in relation to the mental state of the listener. What are "simple" ratios to us, in view of our childhood musical environment, were quite horrible to Pythagoras and Ptolemy, who admitted only the octave (2:1), the fifth (3:2), and the fourth (4:3) to the list of harmonic consonances. At present, all the intervals of Table 33-2 except the semitone are regarded as exceedingly good harmonic consonances. Acceptance of a melodic consonance has always taken place earlier than acceptance of the same interval as a harmonic consonance; the Greeks used the major and minor third melodically, and present-day composers use such melodic intervals as 32:15 with telling effect.

Today we consider the major triad F'–A'–C'', for instance, ♫, as the most perfect consonance. This group of several notes, or CHORD, contains within itself a fifth (F'–C''), a major third (F'–A'), and a minor third (A–C''). The reader can verify that the frequencies of the triad C'–F'–A', ♫, are in the ratio of 3:4:5 and that the triad F'–A'–C'' has frequencies in the ratio of 4:5:6. Another triad, of course, is A–C'–F'. These are INVERSIONS of the same chord, that of F major, since one can be derived from the other by a change of an octave (3:4:5 is equivalent to 6:4:5, since 3 and 6 differ by a factor of 2, or an octave). It is obvious, upon a little consideration, that the major triad, 4:5:6, and its inversions represent the combination of three tones *within an octave* having the simplest frequency ratios; hence the consonance is good. The point to be made is that the Greeks, who did not admit

the consonance of the third (4:5 and 5:6), had to do without the major triad.

They were incapable of forming a theory of HARMONY, since such a theory and practice deal with relations between various chords and since the essential chords themselves were lacking from early music.* Musical history is inextricably bound up in the struggle which

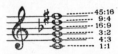

FIG. 33-4.—The augmented eleventh as used by Gershwin in the *Rhapsody in Blue*. Among the ratios included in this chord are: 15:8 (G'–F#''); 16:9 (C'–Bb'); 9:4 (C'–D''); 45:16 (C'–F#''); and 405:256 (Bb'–F#'').

took place (A.D. 1300–1600) concerning the admission of the major and minor thirds to the list of consonances; since 1600, music has developed along harmonic lines rather than melodic lines, with the result that today the chord of the augmented eleventh is commonplace in jazz. Dissonances of Stravinsky, Sibelius, and other modern composers are much more complex. Or are they actually consonances to those of other backgrounds? Musical environment cannot be made retroactive, and the present generation may never be able to decide this.

33.11. The Physical Basis of Consonance

The fact that musical tones generally have a series of overtones with frequencies two, three, four, etc., times that of a fundamental gives a clue to the phenomenon of harmonic consonance. Briefly, dissonance occurs between two tones sounded together if a beat frequency between 10 and 50 per second is formed by the tones themselves *or by any pair of partial tones* contained in the original complex tones. This hypothesis is borne out by the fact that marked dissonance does not occur with stopped pipes of the organ, whose tones are very nearly pure and contain no overtones. Notes separated by an octave and a semitone, which are quite dissonant on the piano, do not sound bad at all on such pipes, presumably because of the lack, in these purer tones, of overtones which might produce beats. The figures 10–50/sec are only approximate and depend upon the individual and the pitch; the dissonance is most acute near the middle of the keyboard if beats of about 30/sec are produced. This enables us to predict an interesting effect, namely, the decrease of the "most dissonant interval"

* Greek music consisted of melody alone, sometimes duplicated, or "magadized," at the interval of an octave. Probably a melody could, on occasion, be "organized" also at the interval of a fifth, but this imposed restrictions upon the melodies which seemed insuperable. The expedient of admitting other intervals to the list of consonances seems not to have been suggested until about A.D. 900.

as the pitch is increased. Thus piano tones of frequencies B and C′ (495 and 528) are separated by a semitone and produce a very painful dissonance between the fundamentals, giving a beat frequency of 33/sec. Four octaves lower, the frequencies are $2^{-4} = \frac{1}{16}$ of their former value; the difference between 31 and 33 cycles/sec is now only 2 beats/sec (this could also have been found directly by dividing the former beat frequency by 16); and if it were not for overtones, which are prominent in the low strings of the piano, the semitone B″–C′ would not be particularly dissonant. The most dissonant interval measured upward from B‴ = 31 cycles/sec would be that between B″ (31) and about F′ (44), or 13 beats/sec. These effects are, to some extent, masked by beats produced by overtones, but it is a well-known fact that even the major triad sounds dissonant when played at a low pitch:

E′ — 49.5 } 16.5
G — 41.25 } beats
C′ — 33

Although the frequencies are in the simple ratio of 4:5:6, there are beats produced that lie in the dissonant range of 10–40 beats/sec.

A consonance is therefore to be interpreted psychologically as a lack of dissonance; this is the reverse of our analytical definition and goes to show that our reasoning on this subject has wandered a long way from the physiological and psychological experiments upon which our knowledge must rest; for, after all, is not music, in the last analysis, a phenomenon of the inner ear and the brain? It would have been desirable, but hardly feasible, to consider musical sounds directly from the biological viewpoint; unfortunately, the present knowledge of the mechanism of the auditory receptor process is in a rudimentary state, compared with our knowledge of the photoreceptor process, for example. It is, of course, of interest to see what can be gleaned from the consensus among physiologists, but limits of space make this out of place in this volume.

The remainder of this chapter will be taken up with matters which will be of importance to one who seeks an understanding of the nature of music and the physical considerations which have influenced and are still influencing the course of musical history.

33.12. Relative Merits of Various Scales

Table 33-4 gives an analysis of the Ptolemaic scale which was illustrated in Fig. 33-3. Extending the scale into the next higher octave where necessary, we can determine the frequency ratios of all the 29 intervals larger than a whole tone and less than a seventh in the scale. Of the 29 possible intervals, 23 are perfect and can be formed as simple ratios. The four "Doric" inter-

vals are fairly close to other intervals that *are* perfect, and only the tritones are neither simple ratios nor close to other intervals that are so expressible. The superiority of the just scale is thus apparent—a relatively large number of consonances can be formed by using notes belonging to the sequence.*

TABLE 33-4

CONSONANCES AND DISSONANCES OF
THE PTOLEMAIC SCALE

	$\frac{9}{8}$	$\frac{5}{4}$	$\frac{4}{3}$	$\frac{3}{2}$	$\frac{5}{3}$	$\frac{15}{8}$	2								
C	D	E	F	G	A	B	C′	D′	E′	F′	G′	A′	B′	C″, etc.
M	*m*	*s*	*M*	*m*	*M*	*s*	*M*	*m*	*s*	*M*	*m*	*M*	*s*, etc.

Decimal Ratio	Interval	Composition of Interval	Frequency Ratio	Tones Used (in Ptolemaic Scale Based on C as Tonic)
2.0000....	Octave	3M+2m+2s	2:1	(CC′)
1.6875....	Doric major sixth	3M+ m+ s	27:16	(FD′)
1.6667....	Major sixth	2M+2m+ s	5:3	(CA) (DB) (GE′)
1.6000....	Minor sixth	2M+ m+2s	8:5	(EC′) (AF′) (BG′)
1.5000....	Fifth	2M+ m+ s	3:2	(CG) (EB) (FC′) (GD′) (AE′)
1.4815....	Doric fifth	M+2m+ s	40:27	(DA)
1.4250...⎫	Tritone	⎰2M+ m	45:32	(BF)
1.4222...⎭		⎱ M+ m+2s	64:45	(FB)
1.3500....	Doric fourth	2M + s	27:20	(AD′)
1.3333....	Fourth	M+ m+ s	4:3	(CF) (DG) (EA) (GC′) (BE′)
1.2500....	Major third	M+ m	5:4	(CE) (FA) (GB) (BD′)
1.2000....	Minor third	M+ s	6:5	(EG) (AC′) (BD′)
1.1852....	Doric minor third	m+ s	32:27	(DF)

33.13. Temperament and Intonation

We have described the modes and the diatonic scale from which they may be constructed, and we have inferred that the melodies are diatonic, or nearly so. Yet it is a fact that the strict diatonic scale is almost unknown in modern music. The piano keyboard contains twelve, not seven, intervals to the octave, and five black keys have been added to produce these additional tones. Why was this done, and what complications have arisen as a consequence of this action?

The addition of the black notes (ACCIDENTALS) was a confusing process (1200–1600). The strict modal melodies suffered from a certain monotony; harmony was frowned upon except at intervals of the fourth and fifth; and so occasionally it was found advantageous to MODULATE from one key to

* To form similar scales composed of seven intervals, we must solve the Diophantine equation $(\frac{9}{8})^p \times (\frac{10}{9})^q \times (\frac{16}{15})^r = 2$, where p, q, and r are integers. The only solution is $p = 3$, $q = 2$, $r = 2$; and by trial it is found that the particular arrangement $MmsMmMs$ of Ptolemy yields the maximum number of consonances and near-consonances; Redfield (*op. cit.*, pp. 191 ff.) proposes an arrangement $mMsMmMs$, which yields almost as many consonances, near-consonances, and dissonances as the 23, 4, and 2 of the Ptolemaic scale.

another. Singing in the Ionian mode with C as tonic, or first tone, a few notes might be sung in the Ionian mode based upon F. Now in the Ionian F, the fourth above F should be of frequency $\frac{4}{3} \times \frac{4}{3} = \frac{16}{9}$ (using Table 33-2). This is much flatter than B ($\frac{15}{8}$); and, in fact, the required note lies about midway between A and B. Thus B flat (B♭) was added in order to be able to write in several modes during the progress of a single composition. Another reason for adding the B♭ was to avoid the dissonance of the tritone, which occurred if a strict parallel movement of voices was to be maintained at a distance of a fifth and if the lower voice chose to sing B (BF′ = tritone). The note B was not always flatted, however, and our present knowledge of musical practice of the fourteenth and fifteenth centuries is incomplete because the ALTERED NOTES were rarely marked.* The performer used his judgment, much as a modern jazz performer supplements the printed notes with his own musical experience.

The whole difficulty with the new music was this: In spite of a certain modulatory facility, too many altered notes were required, and often it was desirable to alter a note in several directions at once! To illustrate, let us compare a just scale based upon E′ with another which has C′ as its third tone. Such a comparison is of the utmost importance, as a very pleasing method of modulation would be to let E′ (whose frequency let us fix at $2 \times \frac{5}{4} = \frac{5}{2}$) be regarded as the third tone of a C′ scale. Then, by letting the C′ so determined be the third tone of a new scale, we find that a new tone is needed, lying between G and A. Call this note A♭, it being altered from the A which is the natural third tone below C′. Similarly, in the E′ scale we find that the third tone above E′ will also lie between G′ and A′; and we will call this new tone G♯′, since the unaltered third tone is G. Lowering this G♯′ by an octave, let us compare with the A♭ already determined. Remembering that the interval of a major third is 5:4, the calculation looks like this:

$$\text{E}' = \tfrac{5}{2} \text{ (assumed)} ; \quad \text{hence } \text{C}' = \tfrac{5}{2} \times \tfrac{4}{5} = 2 ;$$

but, if C′ = 2, then

$$\text{A}\flat = 2 \times \tfrac{4}{5} = \tfrac{8}{5} = 1.6000 .$$

* To do so would violate the letter of the canonical laws for ecclesiastical singing. Hence early harmony which involved altered notes was called *musica ficta*.

On the other hand,

$$\text{E}' = \tfrac{5}{2} \text{ (assumed) and } \text{G}\sharp = \frac{\tfrac{5}{2} \times \tfrac{5}{4}}{2} = \tfrac{25}{16} = 1.5625 .$$

Hence, we need *two* tones between G and A!

We have chosen an extreme case, but the difficulty is a fundamental one. A very simple manifestation of this difficulty is the obvious fact that in the Ptolemaic scale based on C the interval (DA) is not a true fifth (3:2) but a Doric fifth, 40:27.

Even placing two accidental notes between every white or natural note will not help us here; and, besides, modern music has developed to its present harmonic state upon the assumption that each whole tone is divided into two semitones in one, and only one, way.

We have two choices: refrain from modulation into remote keys or make concessions in the tuning of the existing scales. The former was done by the early church musicians, and the latter method of adjustment of frequencies we know as TEMPERAMENT.

EQUAL TEMPERAMENT is discussed in many places, not always correctly, however. The procedure is to divide the octave ($\frac{2}{1}$) into twelve *equal* semitones of value x. To determine x, we must have $(x)^{12} = 2$, whence $x = 2^{1/12} = 1.059463$. ... Thus all the intervals of a fifth are alike, being seven semitones, or $2^{7/12}$; G♯ and A♭ are identical and are actually midway between G and A. If C = 1, then G♯ = A♭ = $(2^{1/12})^8 = 1.587401$, since there are eight semitones from C to A♭, a major sixth. This is so close to the 1.600000 ratio demanded by the Ptolemaic scale (Tables 33-4) that only the exceptionally acute ear can tell that the tempered minor sixth is slightly flat.

The tempered fifth turns out even better: $(2^{1/12})^7 = 1.498307$; few musicians can detect the difference of 17 parts in 15,000 between the tempered fifth and the true fifth, whose ratio is 1.5000000. Other intervals are tabulated in Table 33-5.

With a compromise of the equal temperament, G♯ becomes identical with A♭, and it is just this identity which allows the rich harmonic modulations that we associate with the music of the last three centuries. In general terms, the necessity for a temperament of some sort is attributable to the fact that a number is incommensurable with its logarithm unless the number is some power or root of the base. For our problem the base is 2;

and, since any integral root of 2 is irrational, the octave cannot be divided into *equal* parts except by *irrational* ratios. These ratios are therefore dissonant by our fundamental hypothesis of *simple* ratios. The reason that the twelve-semitone division happens to work so well is that $2^{7/12}$ is approximately equal to $\frac{3}{2}$, to within the threshold of pitch-difference perception; and it is to this finite threshold—a physiological phenomenon—that we owe the tremendous development of harmony since 1700, a development still taking place.

TABLE 33-5

TEMPERED VERSUS TRUE INTERVALS

Tone	True Scale	Tempered Scale	Percentage of Difference	Interval from C
C.........	1.00000	1.00000	0	Unison
C♯, D♭.....	....	1.05946	....	Semitone
D.........	1.12500	1.12246	....	....
D♯, E♭.....	1.20000	1.18921	−0.9	Minor third
E.........	1.25000	1.25992	+0.8	Major third
F.........	1.33333	1.33484	+0.1	Fourth
F♯, G♭.....	....	$1.414214 = \sqrt{2}$	....	Tritone
G.........	1.50000	1.49831	−0.1	Fifth
G♯, A♭.....	1.60000	1.58740	−0.8	Minor sixth
A.........	1.66667	1.68179	+0.9	Major sixth
A♯, B♭.....	....	1.78180	....	....
B.........	1.87500	1.88775	....	....
C'.........	2.00000	2.00000	0	Octave

It is often stated that, while temperament is necessary for a keyed instrument, such as piano, organ, or flute, players of the bowed instruments need not concern themselves with temperament, since they can play any tone, however irrational. It is true that, as far as possible, they will play the just scale, in solo passages written in one or two closely related modes; and in this respect the keyed instrument is inferior. But the inconsistencies of the just scale are numerical; it is impossible for a violinist or singer to be at all times in tune even with himself,[*] and there are frequent occasions[†] when a string quartet—its members free to play all possible rational and irrational intervals—must resort to a tempered scale if dissonance is to be avoided. The note G♯ is actually different from A♭, and the change, when

[*] See *American Physics Teacher*, II (1934), 81, for an interesting discussion of temperament.

[†] E.g., in the development section of the first movement of the Brahms quartet, *Op. 67, Quartet in B♭*.

necessary, is known as an "ENHARMONIC modulation." If one instrument changes while the others do not, dissonance will result; if all change at once, there will be a distinct melodic discontinuity; many quartets prefer the use of the tempered scale upon such occasions. Opera singers might also well profit by a study of the limitations of their just scale.

33.14. Difference Tones

In conclusion, we shall touch briefly upon some of the other scientific aspects of music. We have seen that two tones will always produce a third tone, whose frequency is the difference of their frequencies. When rapid enough, this DIFFERENCE TONE is integrated by the ear and has the subjective characteristics of an ordinary tone—pitch, loudness, and perhaps quality. Difference tones, illustrated in Fig. 33-5, themselves are faint, since they

FIG. 33-5.—Difference tones. Note that the extra tones supply a harmony.

are merely relatively slow variations in intensity of a sound that is already varying quite rapidly but can be detected fairly easily when diapason notes are sounded loudly on the organ. (Sometimes it is possible to detect such tones on the piano.[‡]) Undoubtedly, there are difference tones due to beating of partial tones present in the original tones; but the partials are of low intensity to begin with, and the difference-tone process has a very low efficiency, so that only difference tones produced by the fundamentals need be considered.

We can thus gain an insight into the reason why the minor triad (CE♭G) seems vaguely unsatisfying and inconclusive, while the major triad (CEG) seems complete in itself. (These notes are shown in Fig. 33-6.) One of the difference tones formed by elements of the minor triad is at a dissonant interval (A♭–G); this is not true of the major triad. Composers of today are beginning to take notice of this phenomenon, which undoubtedly has something to do with the tonal effect of music.

33.15. Instrumentation

A composer's choice of instruments for a given composition will depend upon the effect he wishes to produce

[‡] The integration of the difference tone into a note with a definite pitch depends upon the fact that the ear, as an amplifier mechanism, has a response that is not exactly linear. This is due to the asymmetry of the tympanum, perhaps.

and upon the instruments which have, up to that date, been invented. A string quartet (two violins, viola, and cello) gives a very well-blended tone, because the various instruments involved produce tones which have roughly similar sequences of partial tones. Also, in this arrangement, frequencies can be produced to cover the entire range from 65 to about 2,000 cycles/sec, with no gaps—a fact of great importance in considerations of melody. Other blended combinations can today be formed which have the same characteristics as enumerated for the string quartet—one such is the single-reed combination of two B♭ clarinets, a basset horn, and a

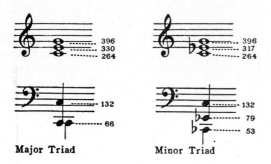

Major Triad   **Minor Triad**

Fig. 33-6.—Difference tones produced by the major triad and by the minor triad.

bass clarinet. The double-reed combination of oboe, English horn, heckelphone, and bassoon would be another combination, with a blended, though distinctive, tone color. Brass ensembles were used with great effect by Wagner, but composers of today seem to lag behind the scientific instrument-makers, who have provided other combinations that are rarely used. By making use of a table similar to Table 33-1, groups of instruments could be assembled which would be of great use to the composer; impetus might thereby be given to the development of new instruments and the improvement of old, a problem for the physicist and the engineer. The string quartet was favored by the classical composers partly because the other instruments had not as yet been perfected—nor are they, even now, perfected.

Considerations of tone quality should help in blending instruments of different families. The combination of piano and string quartet to form a quintet is, in some respects, unfortunate, because the piano is essentially a percussion instrument. Replace the piano by a clarinet, and we have a combination, rich in overtones, which has been sadly neglected by composers—the clarinet quintet.* The partial tones of the clarinet are sufficiently like those of the strings that homogeneity of tone is preserved when needed; on the other hand, the clarinet

* Brahms's *Opus 115* and Mozart's K. 581 rank among the masterpieces of these composers; the only other notable clarinet quintets are those of Reger (1873–1916) and Bax (1883–1953).

tone is sufficiently different to enable its melodic line to be discerned with ease in an intricate developmental section or in a quiet melodic section. Similar analysis should be applied by composers to other possible combinations.

33.16. Counterpoint

Before the introduction of the tempered scale the development of music was chiefly along melodic lines. The consonance of the third was discovered by accident, when *two* melodies simultaneously performed happened to be separated by this interval. The great polyphonic† music of Palestrina was equaled by that of Bach, who also popularized the equal-tempered scale, which led to the employment of a wider circle of keys.‡

Now, if we are to discern several melodies at once, instrumentation plays a great role. The organ is a very unsuitable instrument for the purpose, since the same set of pipes must often be used for several melodies, or VOICES of a fugue, and it is difficult, even on modern organs, to give emphasis to any one voice to the exclusion of the others. The piano has the defect of poor survival value (a tone dies away in from 2 to 15 sec) but does have "accent value." The string quartet was used by Beethoven for some great fugues (cf. especially *Op. 131* and *Op. 133*), but the very unity of tone color mentioned above detracts from the usefulness of the string quartet for contrapuntal music.

Perhaps the solution is to use a totally different type of instrument for each voice, and there will be less danger of putting the individual notes of the separate melodies together in a vertical array to form a chord. The conflict between the contrapuntists and the harmonists has been likened to that between the locksmiths and the burglars. The contrapuntists concoct melodies which, when viewed vertically, seem perfectly safe from possible interpretation as harmony. With passing time, the ear becomes accustomed to the dissonances and calls them harmony, so that the contrapuntists must resort to new and more complexly dissonant melodies to escape the rising tide of harmonic consonance. A very interesting terzetto of Holst (1874–1935) is scored for three totally different instruments—flute, oboe, and viola—and, to make sure that the listener does not combine the parts into a harmony, the flute plays in the key of A major, the oboe in the key of A flat, and the viola in the key of C. Harmony is thrown overboard here (Holst hopes), and the listener is

† "Many voices."

‡ The *Well Tempered Clavichord* of J. S. Bach (1685–1750) is a series of forty-eight preludes and fugues, two in each major and minor key. These compositions could be played without cacophony only with a tempered scale. "The Forty-eight" constitute a document of the utmost importance with respect to harmony as well as counterpoint. They influenced musical development greatly.

"free" to enjoy the separate melodies in their rhythmic and melodic interrelations.

33.17. Résumé

The following is a brief listing of the definitions, principles, and theories, given in this chapter, which you should know:

The definitions of pitch, intensity or loudness, and quality of a sound in terms of the characteristics of waves

The analysis and the synthesis of sounds

Musical scales—the diatonic and the equal temperament scales

Various problems in music and their physical counterpart

Practice Problems

1. Compute the frequencies of the overtones of the thin bell of Table 33-1, assuming the fundamental to be 264 cycles/sec (middle C). Determine the corresponding notes on the piano (approximately), using Table 33-5, extended into other octaves where necessary.

2. Play C on a well-tuned piano, then sing (unaccompanied) upward "do, re, mi" to what seems the most consonant value of mi, or E. Then strike E on the piano. Which will be higher, you or the piano? Which is wrong, you or the piano? (This experiment does not require a particularly good singing voice; if in doubt of the result, try singing upward "la, ti, do" in a similar experiment, starting on C and ending on E♭. The result should be in the opposite direction this time, showing that the effect is actually present.)

3. A bugle call of do, fa, la, do′ starts with the note do of frequency 300 vibrations/sec. Give the frequencies of the other notes.

4. Could a piano be tuned to play the diatonic scale in *any* one key? If so, why is this not done?

5. Explain why in the tempered scale F♯ or G♭ is $\sqrt{2}$ of C is 1.000.

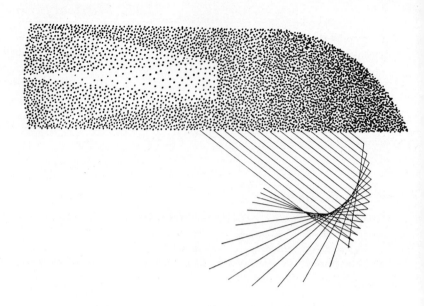

CHAPTER 34

Light: Geometrical Optics

34.1. Introduction

Since the major portion of our knowledge of our immediate environment comes to us through the sense of sight, the study of light is of great importance. With respect to our more remote environment—the outer universe, of which our world is so insignificant a part—the beams of radiation we receive from it comprise the only roads that our inquiring intelligence can travel. Equally true is it that we would have barely guessed the existence of that inner universe of molecules and atoms, electrons, and nuclei, had it not been for the telltale story of radiation coming through the interstices of matter—a story which we decode by the aid of precise relationships of wave lengths as sorted out into a spectrum.

It must be remembered that *visible light* is but a fraction of that mysterious essence we call "radiant energy"—one octave, as it were, out of sixty or more. To all these other fifty-nine-odd octaves our physical senses are totally blind; and yet, indirectly, we have explored this vast range, by methods that overlap in more than one area of physical experimentation. Thus we are furnished with valuable checks and much certainty in many respects. In spite of all this progress in certain directions, there are, on the other hand, many aspects of radiant energy that are still baffling to us. Only by absorbing it and thus transmuting its energy into some other form, are we able to recognize its existence. Even when absorbed and reduced to experimentation, light and all other aspects of radiation have what appear to be curious and, at first sight, quite contradictory properties. The wave properties of visible light are its most conspicuous ones. The longer the waves, as we proceed beyond the visible red, the more conspicuous do the wave aspects become. As we go in the other direction, beyond the ultraviolet to X-rays, these wave aspects become not only less conspicuous but increasingly difficult to find, let alone to measure. Conspicuous in these short ranges of wave length are the CORPUSCULAR PROPERTIES of radiation, bullet-like attributes seemingly quite irreconcilable with wave phenomena. Accidents of discovery and careful search, however, have revealed that visible light possesses these corpuscular properties also, no less real and arresting because of lesser magnitude.

For these reasons, among others, the study of light and allied phenomena is today, as it has always been, one of the most fascinating aspects of physics. As formerly, in other aspects of this great science, we shall begin with the simplest and most certain things.

34.2. Rectilinear Propagation of Light

In discussing the properties of light, we shall make the division between those which depend essentially on the property that light travels in straight lines in a homogeneous medium—that is, the geometrical properties—and those which depend essentially on the wave nature of light—interference and diffraction. These latter are discussed in the next chapter under the subject of physical optics. In the present chapter we shall consider some of the consequences of rectilinear propagation, the measurement of the finite velocity of light, the laws of reflection and refraction, and the properties of mirrors and lenses.

One hardly needs to cite the evidence that light travels in straight lines in a homogeneous medium. Indeed, we often call a line "straight" when, by sighting points that lie in it, we find them "lining up" one behind another. The production of shadows in general and the myriad little images of a partly eclipsed sun through leaves at the time of an eclipse are other examples.

If the source of light cannot be considered a point relative to the object casting the shadow, then there is a region from which light is entirely excluded, called *umbra*, and another region in which the light is only partially obscured, called the *penumbra*. In Fig. 34-1 is shown an exagger-

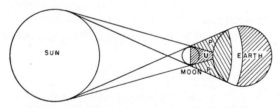

FIG. 34-1.—Partial and total eclipse of the sun by the moon in the penumbra portion, *p*, and the umbra portion, *u*.

ated drawing of a total eclipse of the sun in the portion of the earth covered by the umbra, *u*, and the partial eclipse in a portion of the earth covered by the penumbra, *p*. As can be seen, we can assume that light travels in straight lines and that the shadow is a geometrical projection. However, it must also be admitted that, even with the most minute sources or the finest of pinholes, shadow edges under magnification are not *perfectly* sharp. To this extent, and to this extent only, is there a minute failure of light to travel in geometrically straight lines—a fact that we shall ignore for the moment and discuss in the next chapter.

The rectilinear propagation of light leads immediately to the concept of a *ray of light as the rectilinear path in a homogeneous medium along which light is propagated*. If, then, we have a luminous object and choose one point on the object and draw a straight line from the point in the direction of the propagation of the light, we shall have drawn a representation of a ray of light.

It is clear that from a point source of light an infinite number of rays of light may be drawn. This collection of rays, or cone of light, from a point source is called a *pencil of rays*. Two rays from such a pencil suffice to locate the origin of the pencil.

If the luminous object is an extended object—the filament of an incandescent lamp, for example—we may, from each point of the luminous object draw a pencil of rays. The collection of pencils of rays from all the points on the object is called a *beam of light*.

34.3. The Velocity of Light

Another one of the most definitely known and, indeed, one of the most precisely measured quantities in the entire field of physical phenomena is the velocity with which light (and all other forms of radiant energy) travel. This constant, furthermore, is one of the most fundamentally important constants in all physical theory. The history of the search for its value is naturally, therefore, as old as science itself. Empedocles was the first to suggest that light probably requires a finite time for its passage between two points. Galileo first proposed a method of attempting to measure it.

Galileo's proposal was to station, at as great a distance from each other as possible, two men with lanterns which could be flashed on and off. One of them, A, uncovered his light, so that the other, B, could see it. In turn, B uncovered his the instant he saw the flash from A, and A measured the time between uncovering his own and seeing the flash from B. The experiment failed, of course, since the reaction times of the two individuals were far greater and also had far greater variation than the time (10^{-5} sec) required for the light to traverse the few kilometers between the two observers.

Römer, a Danish astronomer, in 1675 made the first measurement over an astronomical, instead of a terrestrial, distance. He noted that the eclipses **479** of the first satellite of Jupiter occurred at slightly shorter intervals as the earth was approaching Jupiter ($E_1 \rightarrow E_2$ [Fig. 34-2]) than when it was receding ($E_2 \rightarrow E_1$). Since the time between eclipses, averaged over an entire year, was very constant (in spite of a 6 months' total gain of 16 min and 26 sec, followed by a 6 months' loss of the same amount), Römer correctly interpreted the gain or loss as being the time necessary for the light signals

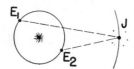

FIG. 34-2.—Römer's method for the determination of the velocity of light, using a satellite of Jupiter.

of the eclipse to traverse the diameter of the earth's orbit. From the then available average diameter of the earth's orbit, 189,000,000 mi, and the time of 986 sec, he calculated the speed of light to be 192,000 mi/sec.

A second determination, by quite a different method, came in 1728 from another astronomer, Bradley. He pointed out that the position of a star observed in a direction at right angles to the earth's orbital motion is displaced from its true position by an angle of 20.44 seconds of arc. This is called the angle of ABERRATION and results from the fact that, while the light is traveling down the telescope tube, the tube is carried, by the motion of the earth, a distance not entirely negligible (see Fig. 34-3). Here it will be noted that

$$\tan a = \frac{v}{c},$$

where v is the velocity of the earth in its orbit and c is the velocity of light.

FIG. 34-3.—The velocity of light determined from its aberration.

If D is the diameter of the earth's orbit and s is the number of seconds in a year, then

$$v = \frac{\pi D}{s} \quad \text{and} \quad c = \frac{\pi D}{s \tan a}.$$

Since we know the value of c today with far greater precision than D and, indeed, quite as accurately as a can be measured, we now use this relation, solved for D instead of c, as one *method of determining the distance of the sun* from the earth, i.e., $D/2$.

The first *laboratory* method of measuring the speed of light over terrestrial distances was that of Fizeau (1849). He used a large, many-toothed wheel, spun rapidly in front of a brilliant source. The flash of light emerging from between two teeth went to a distant mirror, was returned, and was focused at the wheel's periphery. If the wheel were at rest, this return beam was visible between two teeth other than the pair from which it emerged. When the wheel was spun with an appropriate speed, then during the time of transit of the light a tooth would have moved up into the viewing aperture, and the return beam would no longer be visible. At double this speed, the beam would again become visible, since now the next aperture would have moved over in front of the viewing point. It is obvious that the method of Fizeau was, indeed, a highly mechanized adaptation of the method proposed by Galileo. Cornu, who improved the details of Fizeau's arrangement, in 1872 obtained a value which, when corrected, was 299,950 km/sec (in a vacuum).

34.4. The Rotating-Mirror Method of Foucault

Arago, in 1842, proposed an entirely new method, that of the rotating mirror, for measuring this important constant. Foucault, however, was the first to use it, in 1850. It was destined to become the modern precision scheme. As seen from Fig. 34-4, a source, S, is focused by a lens, L, on a dis-

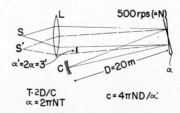

FIG. 34-4.—Foucault's rotating mirror

tant concave mirror, C, after reflection from a rotating plane mirror.

Since the center of curvature of C is at the axis of the plane mirror, the beam falling on C is sent exactly back upon itself. On its return, the plane mirror has turned through a slight angle, a, and thus the reflected image of S is displaced slightly to S', through an angle a' $(2a)$. With as great a distance $(D = 20 \text{ m})$ as Foucault could use and still see the returning image, a' amounted to but $0°2'40''$. This small angle, measurable to an accuracy of about 0.6 per cent, limited the precision of the experiment to the same extent. His value for the velocity, c, was 2.98×10^{10} cm/sec.

34.5. The Michelson Determinations

Simon Newcomb, while he was professor at the United States Naval Academy, repeated Foucault's early method with somewhat increased precision. One of his midshipmen students, A. A. Michelson, assigned to demonstrate the method, ingeniously modified the arrangement of the apparatus. By this means he was able, with relatively simple apparatus on the grounds at Annapolis, to get a more accurate value than that previously obtained. This was the first experiment which Michelson performed, and after he retired as a professor at the University of Chicago, where he was the first American to receive the Nobel Prize, he turned again to the measurement of the velocity of light. He then increased the distance from the fixed to the rotating mirror to about 20 mi. In 1927, in his seventy-fifth year, Michelson published the results of this determination, giving the velocity of light as

$$c = 299,796 \pm 4 \text{ km/sec}.$$

At the time of his death in 1931, Michelson was engaged in another attempt in which a steel tube was used, 3 ft in diameter and 1 mi long, evacuated to pressures ranging from 0.5 to 5.5 mm of mercury. By multiple reflections, the effective length of the tube was increased from 8 to 10 mi. A schematic diagram of the apparatus taken from the published paper of Michelson, Pease, and Pearson* is shown in Fig. 34-5. Light from an arc A, after passing through the slit C, impinges on the rotating mirror D, after which it makes several

* *Astrophysical Journal*, Vol. LXXXII (1935).

reflections and strikes mirror G, from which it returns on an almost identical path. During the time the light has been traversing the path, the mirror has been rotating, so that the returning beam striking the mirror is reflected in a different direction from that at which it entered. With the amount of deflection in the observing eyepiece, the length of the optical paths, and the speed of

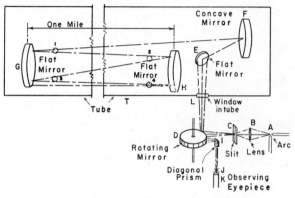

Fig. 34-5.—Schematic diagram of apparatus used for the measurement of the velocity of light by Michelson.

rotation of the mirror known, the velocity of light can be calculated. Actually, several corrections have to be applied if the desired accuracy is to be obtained. As an example of the extreme care taken, the time of the rotating mirror was measured stroboscopically through successive steps by the use of a tuning fork synchronized with the rotating mirror, a free-swinging pendulum, a chronometer, and radio signals from Arlington. There were nearly 3,000 determinations of the velocity, the simple mean of which was

$$c = 299,744 \text{ km/sec},$$

with an average deviation of 11 km/sec from the mean.

Along with these more or less direct methods of measuring the velocity of light there have been others which used electrical means, also very high-frequency or microwave radar techniques (chap. 30). This latter method gives very accurate results. The best weighted average from various methods gives the velocity of light in a vacuum (or in air) as

$$c = 299,793.0 \pm 0.3 \text{ km/sec}$$

$$= (2.99793 \pm 0.0000003) \times 10^{10} \text{ cm/sec}.$$

As we shall see, there is very little difference in the velocity of light in air or in a vacuum, but the velocity in glass or water is very different from that in air. For the British system of units,

$$c = 186,000 \text{ mi/sec (approximately)} .$$

34.6. The Electromagnetic Spectrum

Experiments by Young in England and Fresnel in France, as early as 1830, had demonstrated that light shows interference and diffraction. Thus light has wave characteristics, and with each color may be associated a frequency, f, and a wave length, λ. These are related to the velocity of propagation c by equation (31-2),

$$c = f\lambda .$$

The wave lengths to which the human eye is sensitive are very small and are usually given in units of angstroms, where

$$1 \text{ angstrom} (1 \text{ A}) = 10^{-8} \text{ cm} = 10^{-10} \text{ m} .$$

Another unit commonly used is the micron, which is one-millionth of a meter:

$$1 \text{ micron} (1 \text{ } \mu) = 10^{-6} \text{ m} = 10^{-4} \text{ cm} ,$$

$$1 \text{ millimicron} (1 \text{ m}\mu) = 10^{-9} \text{ m} = 10^{-7} \text{ cm} .$$

The human eye is sensitive to a range of colors from red to violet, whose wave lengths in free space are approximately

$$\lambda \text{ (red)} = 7000 \text{ A} = 0.7 \mu = 700 \text{ m}\mu$$

$$= 7 \times 10^{-7} \text{ m} = 7 \times 10^{-5} \text{ cm} ,$$

$$\lambda \text{ (violet)} = 3800 \text{ A} = 0.38 \text{ } \mu = 380 \text{ m}\mu$$

$$= 3.8 \times 10^{-7} \text{ m} = 3.8 \times 10^{-5} \text{ cm} .$$

There is some variation in this range of colors for different individuals.

As will be discussed later under the subject of refractive index, the wave length of light or any electromagnetic radiation is different in different media. Thus the wave length of visible light in water is about 0.75 of its wave length in air. However, the frequency, f, of the light in water and in air is the same. For violet light of wave length 3500 A in air, its frequency f is given approximately as

$$f = \frac{c}{\lambda} = \frac{3 \times 10^8}{3.5 \times 10^{-7}} \text{ m/sec-m}$$

$$= 8.6 \times 10^{14} \text{ cycles/sec} ,$$

where the velocity of light in empty space is taken as 3×10^8 m/sec rather than the more precise value given above.

The electromagnetic spectrum consists of a whole gamut of radiations, ranging in wave length in free space from the shortest, about 10^{-14} m, to the longest, about 10^6 m. These are not absolute limits, since future work may extend them in both directions. These electromagnetic radiations are characterized by the common velocity c in free space. To different portions of this spectrum are given different names, which usually indicate their mode of production. As seen in the chart (Fig. 34-6) of the electromagnetic spectrum, portions

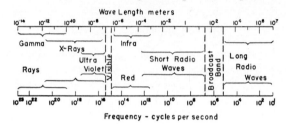

Fig. 34-6.—The electromagnetic spectrum

with different names overlap, as, for example, it is possible to produce X-rays having the same wave length as gamma, γ, rays.

The name *electromagnetic spectrum* came from the work of Maxwell, who about 1873 showed, § 30.13, by assuming a displacement current $\epsilon_0(d\mathcal{E}/dt)$ in free space which gave rise to a magnetic field, that an oscillating electrical circuit should produce oscillating electric and magnetic fields whose velocity of propagation in free space is c. Thus from electrical measurements the velocity c can be measured. In the last chapter many of the modes of production of this electromagnetic spectrum will be discussed.

34.7. Fundamental Law of Reflection of Light

Light entering the eye may come directly from an incandescent source or more commonly by reflection or by scattering from some body. The page of this book scatters or diffusely reflects the light incident on it, and some of this enters the eye. On the other hand, a mirror made from highly polished metal or a smooth surface of glass coated with silver reflects the light in a definite manner, usually called *regular* (or *specular*) *reflec-*

tion. A mirror has a reflecting surface which is smooth, that is, its irregularities are small compared to the wave length of visible light of the order of 10^{-5} cm. Diffuse surfaces are relatively rough, so that any light incident on them is scattered more or less uniformly in all directions. Here we are concerned with regular reflection.

Suppose a ray of light from a source, S, is incident on a mirror (Fig. 34-7). The angle of inci-

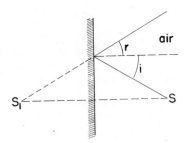

Fig. 34-7.—Reflection of a ray of light

dence, i, of this ray is the angle between the ray and the normal to the mirror at the point of inci dence. This ray is then reflected so that the angle between the reflected ray and the normal, called the *angle of reflection*, r, is equal to the angle of incidence. Thus we have, as an experimental conclusion, the law of regular reflection, which may be stated as follows:

When a ray of light is reflected from a surface, the angle of incidence is equal to the angle of reflection, and, further, the incident ray, the reflected ray, and the normal all lie in a plane.

In Fig. 34-7 the reflected ray appears to come from the point S_1, called the *image of the source S* in the mirror. From similar triangles it is readily proved that the image is as far behind the mirror as the object is in front. Actually, the eye receives a narrow cone of light from every point on the object which is appropriately reflected in the mirror. Two rays from the ends of the object OO' are shown in Fig. 34-8, with the reflected rays appearing to come from the image II'. Every ray is reflected in the mirror in such a way that its angle of reflection is equal to its angle of incidence, so that points in the image are as far behind the mirror as are the corresponding points on the object. The image II' is called a *virtual* image, inasmuch as the reflected rays do not actually pass through II' but only appear to do so. A virtual image is never really formed but only appears to

be formed. To the eye the rays of light appear to be coming from II', and thus the observer sees the image at II'. If the eye is moved, then the image still appears at II', but the rays of light reflected in the mirror are different from those shown in Fig. 34-8.

There is a reversal of the image in the mirror, for, if one looks directly at the object with the projection P to the front, then O' is on the ob-

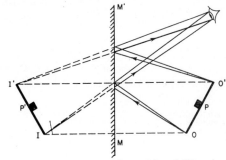

Fig. 34-8.—Light reflected from object OO' in mirror MM' to the eye.

server's left and O on the observer's right. However, in the mirror the reflection of the point O appears at I, which is on the left, and I', corresponding to I, is on the right. This is more vividly shown in Fig. 34-9, where, if the reader looks at the image in the mirror, he will see that it is correct. One can draw the necessary diagram and

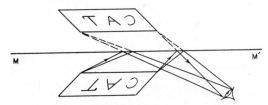

Fig. 34-9.—Reversal of an image in reflection

prove for one's self that, for the case of two plane mirrors set at right angles to each other, no reversal takes place for beams reflected from both mirrors. This can be generalized into three dimensions when three mirrors are set up mutually perpendicular, thus forming the apex of a tetrahedron. Such a triple mirror will return any beam from any direction exactly back in the direction from which it came. This principle is widely used in reflectors for illuminating roadside signs along motor highways at night. Small reflector buttons are generally used. The back reflecting surface of these consists of a series of little pits, each one of which is roughly a tetrahedral apex, with three perpendicular plane surfaces. Such buttons shine brilliantly when illuminated by the headlights of approaching vehicles; their intensity does not change appreciably as the angle of illumination varies, and the advertiser is enabled to emblazon his message at night without any monthly electric bill to pay.

It is of interest to investigate graphically the great number of images formed by a single pair of mirrors. In Fig. 34-10 we illustrate the case where

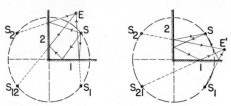

FIG. 34-10.—Images formed by reflection in two mirrors at right angles.

the angle between them is 90°. The eye, E, sees the source, S, reflected in one mirror at S_1 and in the other at S_2, and also a third image—the *doubly* reflected one—in position S_{12}, *as he looks into mirror 2*. This image S_{12} is a virtual image of a virtual image. If the eye is shifted to position E', one still sees the same three images but now gets the double-reflected one as S_{21}, since one now looks into mirror *1* and a quite different path is involved from that which gave S_{12}.

It is readily shown that an object and its three images lie on one and the same circle.

The number of images multiplies rapidly when the angle between the two mirrors becomes acute (see Fig. 34-11). It can be shown that if the mirrors

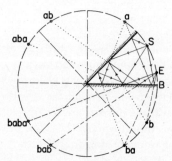

FIG. 34-11.—Images formed by two mirrors at 45°

make an angle of θ with one another, there are images to the number of

$$\frac{360}{\theta} - 1$$

and that these all fall on a circle on which the source also lies. This is seen to have been true for the right-angled case, where $\frac{360}{90} - 1 = 3$, and in Fig. 34-11, where $\frac{360}{45} - 1 = 7$.

When a source lies between two parallel mirrors, there is an indefinitely large number of images, all on *one straight line*. The construction for this case is almost self-evident.

34.8. The Aplanatic Condition

As we shall see, in almost no cases where a mirror or a lens forms a real image, does a perfectly sharp or undistorted image result. Therefore, it is of interest to draw attention to two simple geometrical situations, which, though both are not reducible to practical devices, at least in theory give perfect images. Technically speaking, these conditions are called APLANATIC. Consider an ellipsoid of revolution of which any plane section is an ellipse (Fig. 34-12). It is well known that an ellipse

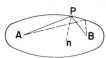

FIG. 34-12.—The aplanatic condition with an ellipsoid

may be drawn by a pencil confined to a path determined by two pegs, A and B, and a loose string between them. The string (whose length is greater than the peg separation) is held taut as the pencil, P, moves. This means that the total distance, ABP, is constant for every point on the resulting curve.

It can be proved by the geometry of this situation[*] that if n is the normal to the ellipse at P, the angles APn and BPn are equal, wherever P is situated. This means that if A is a source of light and the ellipse is a mirror surface, *every* ray emanating in any direction from A as a source will pass through B, which thus becomes a real image point for an object at A. This case is occasionally used for sound in "whispering galleries"—rooms of ellipsoidal cross-section.

As the two foci of the ellipse draw together and assume identical positions, the ellipse becomes a circle (the ellipsoid of which the ellipse is a section becomes a sphere). Here every ray emerging from the center along a radius is reflected back along

[*] See E. Edser, *Light for Students* (New York: Macmillan Co., 1931), pp. 42–43.

itself to the center. Object and image assume identical positions at the center and aplanatically.

On the other hand, as the two foci of the ellipse move apart (Fig. 34-13), one to an indefinitely great distance from the other, the ellipse becomes a parabola (the ellipsoid—a paraboloid of revolution). Now rays from the infinitely distant focus are parallel; and all pass precisely, i.e., aplanatically, through the other focus of the parabola.

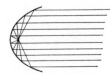

FIG. 34-13.—The aplanatic condition with a paraboloid surface.

This case is eminently practicable. One finds paraboloids of revolution used as reflectors in all high-grade motor headlights and searchlights, even though scattering lenses may be used in front of the mirror to render the parallel rays again divergent. The great mirrors of astronomical telescopes are always made of parabolic section and are thus aplanatic in character. In Fig. 34-14 the

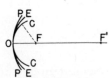

FIG. 34-14.—For small apertures, ellipsoid, *EE*, and paraboloid, *PP*, surfaces approach that of a spherical surface, *CC*.

line *F'FO* is called the OPTICAL AXIS of the mirrors, which may be considered to be small portions of a paraboloid *PP*, an ellipsoid *EE*, or a sphere *CC*.

Finally, one interesting approximation may be pointed out. One can see that if a relatively small section of surface is used, centered on its optical axis, *a spherical surface approximates* aplanatic conditions for either diverging, parallel, or converging rays. Therefore, for practical or simple laboratory conditions, spherical surfaces are used for small mirrors.

34.9. Equation Relating Object and Image Distances and Focal Length for Concave and Convex Mirrors

Let us consider a ray of light parallel and close to the optical axis incident on a concave mirror

(Fig. 34-15) and also on a convex mirror (Fig. 34-16). The centers of curvature of these spherical mirrors are at *C*, and the angles of incidence and reflection are *i* and *r*, respectively. In the case of the concave mirror the incident ray is reflected so as to cross the optical axis at *F* in front of the mirror, whereas for the convex mirror the light

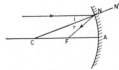

FIG. 34-15.—Parallel light incident on a concave mirror

only appears to, but does not actually, pass through point *F*. The point *F* in each case is called the *focus* of the mirror.

For a spherical surface, the normal *NN'* at any point *N* is obtained by joining the point to the center of curvature—in other words, any radius is normal to a spherical surface at the point of con-

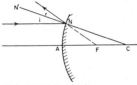

FIG. 34-16.—Parallel light incident on a convex mirror

tact. Since the angle of incidence *i* is equal to the angle of reflection *r*, it can readily be proved that if we assume the point *N* to be sufficiently close to *A* that *FN* is equal to *FA*, then

$$CF = FA = \frac{r}{2},$$

where *r* is the radius of curvature of the spherical surface. The distance *FA* is called the focal length, *f*, of the spherical mirrors. Thus for rays sufficiently close to the optical axis we have

$$f = \frac{r}{2}. \qquad (34\text{-}1)$$

Point *F*, the focus, may be regarded as the image of an object very far away or at infinity. We now turn our attention to the case of an object *O* on the axis and close to a concave mirror (Fig. 34-17). The image of this object is formed on the optical axis at *I*. Since the exterior angles of triangles are equal to the sum of the opposite interior angles, it follows from Fig. 34-17 that

$$\gamma = a + i$$

and
$$\beta = \gamma + r ;$$

whence, by subtraction and substituting $i = r$, we have
$$2\gamma = a + \beta .$$

Also, since the angles are all very small, they may be written as

$$2\frac{NA}{CA} = \frac{NA}{OA} + \frac{NA}{IA} \left.\begin{array}{c} \\ \\ \end{array}\right\}$$
or
$$\frac{2}{r} = \frac{1}{p} + \frac{1}{q}, \qquad (34\text{-}2)$$

where $p = OA$ is the object distance and $q = IA$ is the image distance. Since, from equation (34-1),

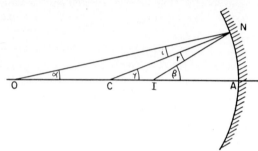

FIG. 34-17.—Object and image positions for concave mirror

$r = f/2$, the relationship between the focal length and the object and image distances is

$$\frac{1}{f} = \frac{1}{p} + \frac{1}{q}. \qquad (34\text{-}3)$$

Suppose we now apply this same analysis to an object O placed on the axis in front of a convex mirror, as shown in Fig. 34-18. From this figure

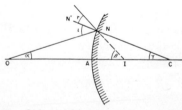

FIG. 34-18.—Object and image positions for convex mirror

we see that
$$i = a + \gamma ,$$
$$i + r = \beta + a ;$$
so that, since
$$i = r$$
$$2a + 2\gamma = \beta + a$$
or
$$2\gamma = \beta - a$$

486

and

$$2\frac{NA}{AC} = \frac{NA}{AI} - \frac{NA}{AO} \left.\begin{array}{c} \\ \\ \end{array}\right\}$$
or
$$\frac{2}{r} = \frac{1}{f} = \frac{1}{q} - \frac{1}{p}. \qquad (34\text{-}4)$$

We see that for the convex mirror the center of curvature is behind the mirror and the image I is virtual. In order to make one equation apply to both convex or concave mirrors, we shall use the following convention of signs:

a) *The radius of curvature and focal length of a concave mirror are positive and of a convex mirror are negative.*

b) *Distances measured to real objects and real images are positive, whereas to virtual objects or images they are negative.*

With this convention we see that in equation (34-4) for a convex mirror r, f, and q are negative, so that we must write this equation as

$$-\frac{2}{r} = -\frac{1}{f} = -\frac{1}{q} - \frac{1}{p}$$
or
$$\frac{1}{f} = \frac{1}{p} + \frac{1}{q}.$$

This is now equation (34-3) and is applicable to both concave and convex mirrors if the sign convention given here is adopted.

34.10. Relative Size of Image and Object

In order to find the position and size of an object OO' of finite size placed perpendicular to the axis, the following construction is useful:

a) From the tip O' (Figs. 34-19 and 34-20) of the object draw a ray parallel to the optical axis.

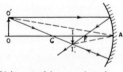

FIG. 34-19.—Object and image sizes for concave mirror

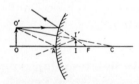

FIG. 34-20.—Object and image sizes for convex mirror

This ray, on reflection, passes through, or appears to pass through, the focus F.

b) A ray from the tip O' through the center of curvature C is reflected back along its own path.

The intersection of these two rays gives the tip of the image as shown in Figs. 34-19 and 34-20. From Fig. 34-19 it is seen that the image II' is *real, inverted,* and *diminished.* It is real because the reflected rays actually pass through II', or the image could be seen on a screen placed at II'. By drawing figures for different object positions, it can be seen that when the object is placed anywhere between the center of curvature and infinity from the concave mirror, the image is real and inverted and approaches the size of the object as the object approaches the center of curvature. When the object is placed between the center of curvature and the focus of the concave mirror, the image is real, inverted, and magnified. If placed between the mirror and the focus, the image is virtual, erect, and magnified.

In the case of a convex mirror (Fig. 34-20) the image is always virtual, erect, and diminished, no matter where the object is placed. The criterion of real or virtual is readily checked from equation (34-3); for the real image, q is positive; for the virtual image, q is negative.

The relative sizes of image and object can be most readily obtained by joining O' to A so that, from the law of reflection, the angle $O'AO$ is equal to the angle IAI' in Figs. 34-18 and 34-19. Thus the triangles $O'AO$ and IAI' are similar, so that

$$\frac{II'}{OO'} = \frac{AI}{AO} = \frac{q}{p}, \qquad (34\text{-}5)$$

or the ratio of the image size to object size is equal to the ratio of image distance to object distance. It is good practice always to draw a diagram roughly to scale of the object and the image formation.

For a plane mirror, which we may regard as a spherical mirror with an infinite radius of curvature or infinite focal length, we have, from equation (34-3),

$$\frac{1}{p} + \frac{1}{q} = 0; \qquad \frac{1}{p} = -\frac{1}{q}; \qquad \text{or} \qquad p = -q,$$

so that the image is virtual, erect, and of the same size as the object.

EXAMPLE 1: An object 2 cm high is placed 50 cm from a concave mirror of focal length 15 cm. Find the position, nature, and size of the image.

Solution: $p = 50$ cm; $f = 15$ cm, and, from equation (34-3), we have

$$\frac{1}{p} + \frac{1}{q} = \frac{1}{f}$$

or

$$\frac{1}{50} + \frac{1}{q} = \frac{1}{15}; \qquad \frac{1}{q} = \frac{1}{15} - \frac{1}{50} = \frac{10 - 3}{150} = \frac{7}{150}.$$

Hence

$$q = \frac{150}{7} = 21.4 \text{ cm}.$$

From equation (34-5),

$$\frac{\text{Image size}}{\text{Object size}} = \frac{II'}{OO'} = \frac{q}{p}$$

or

$$II' = \frac{2 \times 150}{7 \times 50} = \frac{6}{7} = 0.86 \text{ cm}.$$

The image is real, inverted, and diminished.

EXAMPLE 2: An object 2 cm high is placed 10 cm from a concave mirror of focal length 15 cm. Find the position, nature, and size of the image.

Solution: $p = 10$ cm; $f = 15$ cm; and

$$\frac{1}{10} + \frac{1}{q} = \frac{1}{15},$$

$$\frac{1}{q} = \frac{1}{15} - \frac{1}{10} = \frac{2 - 3}{30} = -\frac{1}{30}; \qquad q = -30 \text{ cm}.$$

Also

$$II' = \frac{q}{p} OO' = \frac{30}{10} \times 2 = 6 \text{ cm}.$$

The image is virtual, erect, and magnified.

EXAMPLE 3: An object 2 cm high is placed 20 cm from a convex mirror of focal length 15 cm. Find the position, nature, and size of the image.

Solution: $p = 20$ cm; $f = -15$ cm; and

$$\frac{1}{20} + \frac{1}{q} = -\frac{1}{15},$$

$$\frac{1}{q} = -\frac{1}{15} - \frac{1}{20} = -\frac{7}{60},$$

$$q = -\frac{60}{7} = -8.6 \text{ cm}.$$

Also

$$II' = \frac{q}{p} OO' = \frac{60 \times 2}{7 \times 20} = \frac{6}{7} = 0.86 \text{ cm}.$$

The image is virtual, erect, and diminished. Notice that the image of a real object in a convex mirror is always virtual and erect.

34.11. How To Look at an Image by Eye

It is often convenient to inspect images by means of the unaided eye. This is perfectly simple in the case of virtual images, since they are, in most cases, sufficiently far behind the mirror to be clearly visible to normal eyes at *the distance for distinct vision*, which is about 10 in, or 25 cm. It must be remembered, of course, that the normal eye cannot see distinctly anything which is placed nearer than this to it.

This fact must also be borne in mind when looking for real images. Since they constitute regions of diverging rays, just like any actual source of light, they can always be seen by unaided vision if the eye is placed (1) so that it will intercept these diverging pencils, i.e., *facing the mirror*, and (2) at a sufficient distance from the position of the image so that distinct vision of that point is possible (Fig. 34-21). *The eye must never be placed close*

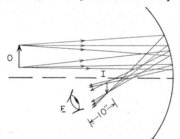

Fig. 34-21.—Viewing a real image

to the position of the image itself. From this location one sees the surface of the mirror uniformly illuminated, i.e., one's eye catches all the pencils at the point of convergence and, indeed, may be injured from the resulting intensity, especially if the source of light is bright and the mirror has a rather large aperture. It is often convenient to place a ground glass or piece of paper at the positions of the image, which will then be projected upon the glass or paper. We use the word "convenient" advisedly. The person who has acquired a little skill does not find this aid necessary—his eye suffices.

34.12. Spherical Aberration

It is a fact of common observation and it has already been noted that one cannot get sharp and clear images by means of spherical concave mirrors, especially when they are of rather large aperture. Why this is so may be seen by the construction of even a crude drawing of such a mirror receiving a beam of rays from a distant source of light, i.e., a parallel beam (Fig. 34-22). Using the simple law of reflection for each ray, it is at once apparent that the rays falling near the outer portions of the mirror cross one another and "focus" at points *near the mirror and far from its axis.* Central rays, on the other hand, cross one another at points much nearer the axis and much farther away from the surface of the mirror. In other

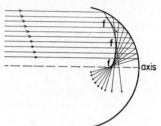

Fig. 34-22.—Spherical aberration

words, the mirror does not bring to any one point, f, all the rays entering it. It is not aplanatic except for the very limited bundle of rays close to its axis. For *all* rays there is, rather, a "focal line" ($f, f, f,$ of Fig. 34-22), along which the entire beam is concentrated. This is called the CAUSTIC CURVE of a spherical mirror from the fact that, if sunlight is used with a large mirror, a great amount of heat is generated along this line. The phenomenon, in general, goes by the name of SPHERICAL ABERRATION.

It must be remembered, of course, that in the plane drawings which we must use for illustrations this region is shown as a line. In case of an actual mirror, the phenomena take place along a surface in space. Since all our drawings have radial symmetry about an axis, these surfaces could be generated in space by rotating these drawings about their central or OPTICAL AXIS.

34.13. Refraction of Light

When a ray of light passes obliquely from air to water, air to glass, etc., the ray changes direction at the interface, that is, the ray is bent or refracted. This refraction of a ray of light depends on the two media and to a much smaller extent on the color or the frequency of the light. By actual measurement, Foucault and Michelson showed that the velocity of light is smaller in glass or water than in air. As we shall show later, the ratio of the wave lengths of a given color in air and water, for example, is the same as the ratio of their velocities in these media.

As shown in Fig. 34-23, an oblique ray is always bent or refracted toward the normal in proceeding from a medium of high velocity to one of low velocity. If NN' is the normal to a glass surface, then, for the rays shown, the angle i is the angle of incidence and r is the angle of refraction. As early as 1621, Willebrord Snell showed that the ratio of $\sin i$ to $\sin r$ is a constant for a given pair of media and color of light. This constant is

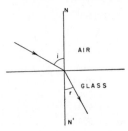

FIG. 34-23.—Refraction of a ray of light in proceeding from air to glass.

called the *refractive index*. Thus in Fig. 34-23 the refractive index from air to glass, $_a\mu_g$, is given by

$$_a\mu_g = \frac{\sin i}{\sin r}. \qquad (34\text{-}6)$$

Equation (34-6) is the analytical expression of Snell's law for the two media of air and glass. Furthermore, the incident ray, the refracted ray, and the normal all lie in the same plane.

Rays of light proceeding from water into air are shown in Fig. 34-24. The ray numbered 6 strikes the water surface perpendicularly and is not refracted, since the angle of incidence is zero. Suppose the ray numbered 8 has an angle incidence, $i = C$, such that its refracted ray just grazes the surface or has an angle of refraction of 90°; then, from equation (34-6),

$$_w\mu_a = \frac{\sin C}{\sin 90} = \sin C; \qquad (34\text{-}7)$$

or, regarding the ray proceeding from the air to the water with an angle of incidence of 90° and an angle of refraction of C, we have

$$_a\mu_w = \frac{\sin 90}{\sin C} = \frac{1}{\sin C}. \qquad (34\text{-}8)$$

The angle C is called the *critical angle*.

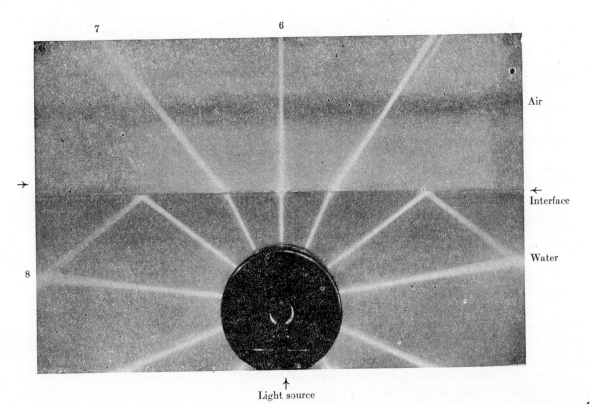

FIG. 34-24.—Refraction of rays of light from water to air

At the critical angle some of the light is internally reflected, as shown, and when the angle of incidence in the water is greater than this critical angle C, no refracted ray appears in the air, and the ray of light is reflected entirely internally. The refractive index of a ray proceeding from air to water, $_a\mu_w$, is about 1.33, so that the value of the critical angle C is given by equation (34-8) as

$$\frac{1}{\sin C} = 1.33 ; \qquad \sin C = 0.75$$

or

$$C = 48°35' .$$

If you should ever look upward when submerged in water, you would have a most interesting view. You could see everything above the surface and also anything on the bottom whose ray made an angle with the surface greater than 48°35'.

The refractive index of a flint glass is $_a\mu_g = 1.587$, so that its critical angle is about 39°21'. Since this is less than 45°, it is thus possible to reflect a beam of light critically in a 45°–90° flint-glass prism, as shown in Fig. 34-25 (a), and to

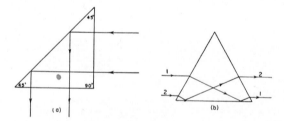

Fig. 34-25.—(a) Reflection of light through 90°; (b) reversal of two rays of means of a 60° prism.

reverse two rays, as shown in (b). All the light within the prism is reflected, and, unlike mirrors, prisms need no silvering and never tarnish. When the light enters the prism, some of it is reflected by the glass surface. This reflected light can be almost eliminated by coating the glass surface with a thin layer of transparent material whose refractive index is smaller than that of the glass. The thickness of the layer must be a quarter of a wave length so that no reflected light is produced by interference (see chap. 35).

In Table 34-1 are given the refractive indices of various solids, liquids, and gases for yellow light, whose wave length in air is 5890 A. The values of the refractive index are with respect to a vacuum or free space. It is seen that the refractive index

of the gases is almost that of free space. For example, dry air at atmospheric pressure and 15° C has a refractive index with respect to a vacuum, $_v\mu_a$, of

$$(_v\mu_a - 1) \times 10^7 = 2{,}765 ,$$

or

$$_v\mu_a = 1 + 2{,}765 \times 10^{-7} = 1.0002765 .$$

TABLE 34-1

REFRACTIVE INDEX

SOLIDS

(μ at $\lambda = 5893 \times 10^{-8}$ cm)

Crown glass*	1.5171	Fluorite	1.4339
Light flint glass	1.5803	Calcite	1.6584
Dense flint glass	1.6555	Nitroso-dimethyl	
Barium glass	1.5681	aniline	
Barium-crown glass	1.5741	At λ 5840 A	1.815
Quartz	1.5442	At λ 5000 A	2.114
Rock salt	1.5443	Diamond	2.423
Sylvite	1.4904	Ice at $-8°$ C	1.31

LIQUIDS

Nitrogen at $-190°$ C	1.205	Carbon disulphide at	
Oxygen at $-181°$ C	1.221	20° C	1.643
Carbon dioxide at		Carbon tetrachloride	
15° C	1.195	at 20° C	1.461
Alcohol at 20° C	1.329	Water at 0° C	1.3338
Benzene at 20° C	1.502	Water at 20° C	1.3330
		Water at 40° C	1.3307
		Water at 80° C	1.3230

GASES

$[(_v\mu_a - 1) \times 10^7$ at $\lambda = 5890$ A]

Dry air at 15° C, 76 cm Hg	2765
Hydrogen at 0° C, 76 cm Hg	1320
Water vapor at 0° C, 76 cm Hg	2500

* Percentage composition (major components) of a few of the glasses mentioned here:

Type of Glass	SiO$_2$	Na$_2$O	PbO	BaO
Crown	67	12	0	10.6
Light flint	54	1	35.2	0
Dense flint	39	3	49	0
Barium-flint	54	1.7	16.7	14.0
Barium-crown	48	2	0	29.5

From equations (34-7) and (34-8) we see that

$$_a\mu_w = \frac{1}{_w\mu_a} .$$

Now, as will be seen later, the refractive index air to water is equal to the ratio of the velocities in air and in water. Thus

$$_a\mu_w = \frac{v_a}{v_w}, \qquad (34\text{-}9)$$

where v_a is the velocity of propagation of light in air and v_w that in water. From this we can readily deduce refractive indices with respect to two media from their values with respect to other media. Suppose we are given the values of $_a\mu_g$ and $_a\mu_w$; then we can deduce the value of $_w\mu_g$:

$$_w\mu_g = \frac{v_w}{v_g} = \frac{v_w}{v_a}\frac{v_a}{v_g} = {_w\mu_a} \times {_a\mu_g} = \frac{_a\mu_g}{_a\mu_w}. \quad (34\text{-}10)$$

Choosing from the table $_a\mu_g = 1.517$ for crown glass and $_a\mu_w = 1.333$ for water at $20°$ C, we have

$$_w\mu_g = \frac{1.517}{1.333} = 1.138.$$

34.14. Measurement of Refractive Index

A prism is an exceedingly useful optical device for measuring the refractive index of the material of which it is composed. Suppose a ray of light is incident on one of the faces, as shown in Fig. 34-26; then this ray is deviated through some angle

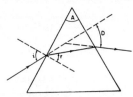

FIG. 34-26.—Deviation of a ray of light by a prism

D. This angle of deviation D can be calculated in terms of the angle of incidence i, the angle A of the prism, and the refractive index of the material of the prism. This gives no simple result. However, if the prism is rotated about an axis perpendicular to the plane of the paper, it is found that the angle of deviation D passes through a minimum value. In practice it is relatively easy to set the prism and the incident ray so that this deviation is a minimum. When this is done, it is found that the rays proceed symmetrically through the prism, as shown in Fig. 34-27. The angle of incidence i at

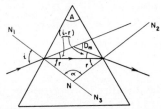

FIG. 34-27.—Minimum deviation of a ray of light

which the ray strikes the prism is equal to the angle i of emergence of the ray of light into the

air. Similarly, the angles which the ray makes with the normals in the prism are both equal to r. From Fig. 34-27 it is seen that the angle of minimum deviation, D_m, is

$$D_m = 2(i - r). \quad (34\text{-}11)$$

Since N_1N and N_2N are normals to the surfaces, then angles

$$A + a = 180.$$

But

$$a + 2r = 180.$$

Hence

$$A = 2r \quad \text{or} \quad r = \frac{A}{2}.$$

From equation (34-11),

$$\left.\begin{array}{l} i = \dfrac{D_m}{2} + r \\[2mm] \quad = \dfrac{D_m + A}{2}. \end{array}\right\} \quad (34\text{-}12)$$

The refractive index of the material of which the prism is made, say glass, is given by

$$_a\mu_g = \frac{\sin i}{\sin r} = \frac{\sin (A + D_m)/2}{\sin A/2}. \quad (34\text{-}13)$$

Since the angles A and D_m can be measured quite accurately with a spectrometer, it follows that the refractive index of any suitable material made into the form of a prism can be obtained.

Before proceeding with refraction at a curved surface, we shall illustrate another method of measuring the refractive index of some material. Suppose a ray of light EB proceeds from the point E on the bottom of a slab of a parallel-sided piece of glass. On refraction at the glass-air surface, it enters the eye, which then sees a virtual image of E at the point C perpendicularly above E. Actually,

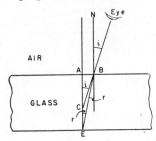

FIG. 34-28.—Refractive index from apparent depth

the eye sees a small cone of rays which intersect at C. From Fig. 34-28 the refractive index from air to glass, $_a\mu_g$, is

$$\mu_g = \frac{\sin i}{\sin r},$$

where the line BN is normal to the surface. Since AE is perpendicular to the glass faces, it follows that angle i is equal to angle ACB and angle r to angle AEB as marked in the figure. Hence

$$_a\mu_g = \frac{\sin i}{\sin r} = \frac{AB}{BC}\frac{BE}{AB} = \frac{BE}{BC}.$$

Now suppose the angles are made smaller so that the eye views the point E perpendicularly; then B approaches A and BC becomes equal to AC and BE to AE. This is equivalent to setting the sines of these angles equal to their tangents, which is permissible if the angles are small. Hence, for perpendicular incidence,

$$_a\mu_g = \frac{AE}{AC} = \frac{\text{Real depth}}{\text{Apparent depth}}. \quad (34\text{-}14)$$

The apparent depth can be readily obtained with measuring microscope. First the microscope is focused on the point E without the glass present; then the glass is replaced, and the microscope is again focused on E. Since the point E now appears at C, the microscope must be raised the distance EC. Thus from the measurement EC and the thickness of the glass slab, we can obtain its refractive index.

34.15. Refraction at a Curved Surface

In general, refraction at curved surfaces is very complex, hence we shall take the simple case of a ray of light entering glass at a spherical surface at a small angle, as shown in Fig. 34-29. From an

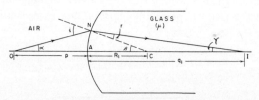

Fig. 34-29.—Refraction into a slab of curved glass

object O a ray of light is incident on the spherical surface, radius R_1, at an angle i and is refracted into the glass at the angle r. This refracted ray crosses the axis at I in the glass. From Snell's law we have

$$_a\mu_g = \mu = \frac{\sin i}{\sin r}. \quad (34\text{-}15)$$

The incident ray ON makes an angle a with the

axis, which is assumed to be sufficiently small that it may be set equal to its tangent, or

$$a = \frac{AN}{p}.$$

Similarly for the refracted ray, at an angle γ, we have

$$\gamma = \frac{AN}{q_1}.$$

The angle β subtended by the radius CN is given by

$$\beta = \frac{AN}{R_1}.$$

From Fig. 34-29 we have

$$i = a + \beta = \frac{AN}{p} + \frac{AN}{R_1}, \quad (34\text{-}16)$$

$$\beta = r + \gamma \quad \text{or} \quad r = \beta - \gamma = \frac{AN}{R_1} - \frac{AN}{q_1}. \quad (34\text{-}17)$$

To the degree of approximation we are assuming here, equation (34-15) may be written

$$\mu = \frac{\sin i}{\sin r} = \frac{i}{r},$$

where we are using μ for the refractive index of glass relative to air, $_a\mu_g$. Hence, substituting in equations (34-16) and (34-17) and canceling AN, we have

$$\frac{1}{p} + \frac{1}{R_1} = \frac{\mu}{R_1} - \frac{\mu}{q_1}$$

or

$$\frac{1}{p} + \frac{\mu}{q_1} = \frac{\mu - 1}{R_1}. \quad (34\text{-}18)$$

34.16. The Aplanatic Condition and the Oil-Immersion Lens

Suppose the object is moved a great distance away from the spherical surface of the glass so that parallel rays of light fall on the surface. Then the image of these rays is formed at the point called the *focus*, a term similarly applied in the case of mirrors. Thus, when

$$p_1 = \infty, \qquad q_1 = f,$$

where f is the distance of the focus from the point A on the optical axis (Fig. 34-29). Substituting in equation (34-18), we have

$$\frac{\mu}{f} = \frac{\mu - 1}{R}.$$

Hence

$$f = \frac{R\mu}{\mu - 1}, \qquad (34\text{-}19)$$

so that a real image of an object at infinity is formed within the glass at a distance $\mu/\mu - 1$ of the radius from A.

Unless the aperture is small, that is, the angles a, β, and γ in Fig. 34-29 are small, the refracted rays are not all brought to a focus at the same point. Parallel rays at some distance from the axis are brought to a different focus than parallel rays close to the optical axis, or, in other words, there is spherical aberration present.

A simple case, in which there is no spherical aberration, i.e., in which a perfect image is produced for a beam of any aperture, can be produced in a glass sphere. The refraction is then *aplanatic*.

Suppose a source of light O is produced inside a sphere of glass at a distance R/μ from the center C of the glass sphere of radius R (Fig. 34-30). For

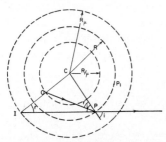

Fig. 34-30.—The aplanatic condition in a spherical medium

a particular source, O, on the inner surface the virtual image I appears at the same point, distant $R\mu$ from the center, no matter what angle it is viewed from. This may be proved in the following manner. From the construction,

$$\frac{IC}{CP} = \frac{R\mu}{R} = \mu = \frac{\sin i}{\sin r} = \text{Constant}.$$

Thus the ray of light OP from the source O emerges from the sphere along the line IP appearing to come from the virtual image I.

Suppose another ray OP_1 is chosen. Then, by a similar construction, the emergent ray would proceed along the direction IP_1 so that the image of O would again appear to come from the virtual image at I.

While it would be difficult, if not impossible, to place a light source inside a glass sphere, nevertheless the equivalent of this is done in an oil-immersion microscope objective (§ 34.30). Es-

sentially, a portion of the glass sphere is ground away to leave a flat surface through the point O. The object to be viewed in the microscope is placed at O in an oil whose refractive index is the same as that of this glass (Fig. 34-31). Thus the

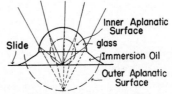

Fig. 34-31.—The oil-immersion microscope objective

object is placed on the inner aplanatic surface of radius R/μ. All rays radiating out from the object through the entire hemisphere travel from a virtual image lying below the object O on the outer aplanatic surface of radius μR.

This is of importance in the resolving power of the objective lens of a microscope, that is, how close two objects can be and still be seen as two objects, or how fine the detail is that can be seen in a miscroscope. In a microscope this ability to distinguish detail clearly is measured by a quantity called the *numerical aperture*, NA. Without oil immersion, the practical limit for the numerical aperture of a microscope is about 0.95, whereas with an oil-immersion objective this can be increased to about 1.60. Thus we find that all high-power microscopes employ an oil-immersion objective.

34.17. Refraction by Thin Lenses

In this book we shall limit ourselves to the analysis of thin lenses. This omits a great deal of optics, for thick lenses or complicated lens systems are fairly common but require a much more extensive treatment than can be given here. By a "thin lens" is meant one in which the thickness of the lens itself can be neglected so that object and image distances can be measured from the center of the lens. In this first analysis we shall relate the object and image distances to the radii of curvature of the two faces of the lens and to the refractive index of the material of which the lens is made. To do this, we shall consider the ray in Fig. 34-29, which was refracted into the glass, now being refracted into the air, as it would be in a lens. The angle which this refracted ray in the glass made with the optical axis in Fig. 34-29 was **493**

γ and is so marked in Fig. 34-32. The radius of curvature of the glass surface is R_2. A final image is formed in air at I, making an angle ϕ with the axis. From Fig. 34-32 we see that angle

$$i = \theta + \phi$$

and angle

$$r = \theta + \gamma .$$

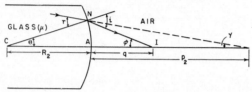

FIG. 34-32.—Refraction from glass into air at curved surface.

Also for small angles i and r, we again have

$$i = \mu r .$$

By substitution,

$$\theta + \phi \doteq \mu (\theta + \gamma) .$$

Hence

$$\frac{AN}{R_2} + \frac{AN}{q} = \mu \left(\frac{AN}{R_2} + \frac{AN}{p_2} \right)$$

and

$$\frac{1}{q} - \frac{\mu}{p_2} = \frac{\mu - 1}{R_2} . \qquad (34\text{-}20)$$

Now let us combine Figs. 34-29 and 34-32 so as to form a thin, double convex lens, as shown in Fig. 34-33. By so doing, we combine equations

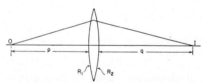

FIG. 34-33.—Object and image distances for a thin lens having radii R_1 and R_2.

(34-18) and (34-20), using the image distance q_1 in Fig. 34-29 as the object distance p_2 in Fig. 34-23. We thus arrive at the result, by adding equations (34-19) and (34-20):

$$\frac{1}{p} + \frac{1}{q} = (\mu - 1) \left(\frac{1}{R_1} + \frac{1}{R_2} \right) . \qquad (34\text{-}21)$$

This is often known as the *lens-maker's equation.*

If the object O is placed at a great distance from the lens so that we make $p = \infty$ and $1/p = 0$, then any rays from the object are parallel to the axis and, after refraction through the lens, form an image at the focus. The distance from the focus

to the lens is called the *focal length, f.* Hence we may rewrite equation (34-21) as

$$\frac{1}{p} + \frac{1}{q} = \frac{1}{f} = (\mu - 1) \left(\frac{1}{R_1} + \frac{1}{R_2} \right) . \qquad (34\text{-}22)$$

In Fig. 34-34 (a) and (b) are shown parallel rays of light falling on a double convex lens and a

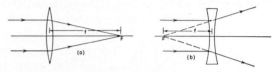

FIG. 34-34.—Focal length of (a) convex lens and (b) concave lens.

double concave lens, respectively. The convex lens converges the rays and produces a real image at the focus F. A concave lens diverges the rays and produces a virtual image at the focus F. In keeping with our sign convention for mirrors, we shall assign a positive sign to real object or image distances and a negative sign to virtual object or image distances.

Thus any converging lens, which is thicker at the center than at the edges, has a positive focal length, and any diverging lens, which is thinner at the center than at the edges, has a negative focal length. The radii of curvature R_1 and R_2 of the double convex lens (Fig. 34-34[a]) are considered to be both positive, while for the double concave lens (Fig. 34-34[b]), the radii R_1 and R_2 are both

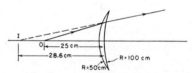

FIG. 34-35.—Image I formed by concavo-convex lens

negative. If one has a concavo-convex or a convexo-concave lens, then one of the radii is positive and the other negative. It is a relatively simple matter to determine which is positive and which is negative from a consideration of the focal length, as the following example will show.

EXAMPLE 4: A concavo-convex lens, made of glass of refractive index 1.5, has radii of curvature of 50 and 100 cm. If an object is placed at 25 cm from the lens, Fig. 34-35, find the position and nature of the image.

Solution: The lens is a convex or converging lens and hence has a positive focal length. In order to obtain a positive focal length from the right-hand side of equation (34-22), we must have the radius of curvature o-

50 cm positive and that of 100 cm negative. Thus, from equation (34-22), we have

$$\frac{1}{f} = (\mu - 1)\left(\frac{1}{R_1} + \frac{1}{R_2}\right)$$

$$= (1.5 - 1)\left(\frac{1}{50} - \frac{1}{100}\right) = \frac{1}{200}.$$

Hence $f = 200$ cm. Also

$$\frac{1}{p} + \frac{1}{q} = \frac{1}{f},$$

$$\frac{1}{25} + \frac{1}{q} = \frac{1}{200},$$

$$\frac{1}{q} = \frac{-7}{200},$$

$$q = -28.6 \text{ cm}.$$

The image is therefore virtual and is on the same side of the lens as the object, as shown in Fig. 34-35. Notice that the ray shown is made to converge toward the axis but that, since the focal length is so large, the ray never crosses the axis.

34.18. Approximate Geometrical Treatment of Thin Lenses

As the preceding paragraphs have indicated, refraction at curved surfaces is more complicated analytically than in the case of mirrors, because of the entry of the refractive index into the equations. Methods of treating wave fronts instead of rays, which will be discussed in the next chapter, are more adaptable to the visualization of both mirror and lens problems. However, a relatively simple geometrical method exists which enables one to trace rays with relative ease through any kind of thin lens or simple combinations of thin lenses. The two rays from an object which are most useful in determining the position and size of an image produced by a thin lens are:

a) A ray from the top of the object parallel to the optical axis of the lens. This ray, after refraction, crosses the axis at the focus a distance of the focal length f from the lens.

b) A ray from the top of the object through the center of the lens. To the degree of approximation which we are using, this ray passes through the lens without any change in direction or lateral deviation. Near the center of the lens, on the optical axis, tangent planes to the two lens surfaces are parallel, and, since the lens is thin, the refracted ray is a continuation of the incident ray.

This method is shown for converging and diverging lenses in Fig. 34-36 (a) and (b). The convex lens forms a real, inverted, and magnified image II' of the object OO', whereas the concave lens forms a virtual, upright, and diminished image of the object. It can easily be verified that a concave lens always forms a virtual image of a real object, no matter where it is placed, and a convex lens

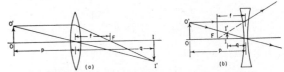

Fig. 34-36.—(a) Real, inverted image formed by convex lens; (b) virtual, upright image formed by concave lens.

forms a virtual image of a real object when it is placed closer to the lens than the focal length. From both the diagrams it can be seen from similar triangles that

$$\frac{\text{Size of image } II'}{\text{Size of object } OO'} = \frac{\text{Image distance } q}{\text{Object distance } p}$$

or

$$\left.\frac{II'}{OO'} = \frac{q}{p}.\right\} \quad (34\text{-}23)$$

Also the equation relating image and object distance with the focal length,

$$\frac{1}{p} + \frac{1}{q} = \frac{1}{f},$$

can be readily proved by the use of the diagrams.

It should be noted that the two rays used for locating the image are only two of a very large number which come from the object to the lens and go to the image.

To locate the image of a single lens, the use of the two rays given above is always sufficient. When there is a train of lenses, however, the continuation of the diverging rays *1* and *2*, as in Fig. 34-37, may not have a direction which will result

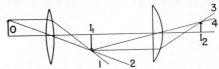

Fig. 34-37.—Location of an image with two lenses

in their encountering the next lens at all. Other rays from this first image, *3* and *4*, may, of course, be drawn parallel to the axis and to the center of the second lens, and the construction repeated; but then the course of *one and the same ray* through the entire system will not be followed. **495**

To rectify this, we recall the fact not only that rays parallel to the optical axis intersect in the focal point but that bundles or rays in any direction, *if they are parallel* to one another, will meet in the plane perpendicular to the optical axis which contains the focal point (Fig. 34-38). This plane is the FOCAL PLANE.

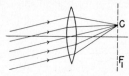

FIG. 34-38.—The focal plane of a convex lens

Hence a ray from an object or an image in any direction projected forward and a ray parallel to it through the lens center, and hence undeviated, will meet in the point C in the focal plane where the central ray pierces it. Thus the *refracted* non-central ray's direction is determined (Fig. 34-39 [a]).

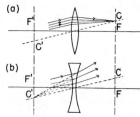

FIG. 34-39.—Focal planes for convex and concave lenses

If the lens is a negative lens, the focal plane on the side of the incident beam must be used. The central ray is then projected back to this, and the point C' is the point from which the refracted ray appears to have come; i.e., the image here is virtual (Fig. 34-39 [b]).

EXAMPLE 5: In applying this construction, let us choose, for our example, a system of three lenses, the last of which is that of the observing eye (always a part of any complete system). We shall also specify *numerical focal lengths* and distances between lenses and to the object (Fig. 34-40). Lenses L_1 and L_2 have focal lengths

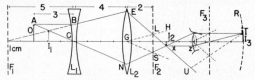

FIG. 34-40.—Location of an image, using two lenses and the eye.

of -5 and $+2$, respectively, and are placed 4 units apart, with the object 3 units from the diverging lens. **496** Lines AB and AC are drawn as formerly, but AB then

diverges, as if from the virtual focus of L_1. The ray BE, extended backward, intersects AC and locates the virtual image, I_1, of O.

Through the second lens, GH is drawn parallel to BE and through the center of lens L_2. The dotted line, F_2, is the focal plane of L_2, and GH intersects it in L, indicating the direction of BE after refraction. Similarly, GS is parallel to CN, which thus is indicated as being refracted in the direction NS. The intersection of NS with the other ray, EL, is I_2, the second image of A, now real.

Ray NS, extended, enters the pupil of the eye, and a repetition of the construction directs it to the retina, R. Our other ray, EL, extended, misses the pupil of the eye altogether! If this lens were large enough, the construction would indicate a direction, UT, for this ray. It must not be forgotten that we have been tracing *but two* of an infinite multitude of rays emerging from any image. Some *other* ray from I_2 *will* strike the pupil. Take any of these, say XZ. The same construction again brings this ray to the retina, R, where the crossing of rays at T again indicates the third image, I_3.

Solution: Let us compute the locations and sizes of images I_1 and I_2:

$$\frac{1}{3} + \frac{1}{q_1} = \frac{1}{-5}; \quad q_1 = -\frac{15}{8} \text{ (a virtual image)}.$$

Now

$$p_2 = 4 + \frac{15}{8} = \frac{47}{8},$$

where p_2 is taken as positive, since the rays are diverging when they reach L_2, just as they would if I_1 were an actual object and there was no lens L_1. Therefore,

$$\frac{8}{47} + \frac{1}{q_2} = \frac{1}{2}; \quad q_2 = +\frac{94}{31} \text{ (a real image)}.$$

To determine the sizes of images, we observe that

$$\frac{O}{I_1} = \frac{3}{\frac{15}{8}} = \frac{24}{15}; \qquad I_1 = \frac{15}{24}O;$$

also that

$$\frac{I_2}{I_1} = \frac{I_°}{\frac{15}{24}O} = \frac{q_2}{p_2} = \frac{\frac{94}{31}}{\frac{47}{8}}.$$

Therefore,

$$\frac{I_2}{O} = \frac{10}{31} \qquad \text{or} \qquad I_2 = \frac{10}{31}O.$$

34.19. Combined Effects of Two or More Thin Lenses in Contact

If two thin lenses of focal lengths f_1 and f_2 are in contact, it is possible to replace the combination by a single lens whose focal length F is given by

$$\frac{1}{F} = \frac{1}{f_1} + \frac{1}{f_2}. \qquad (34\text{-}24)$$

We may justify this formula as follows: Consider a real object at a distance p from the first lens. Then, for this lens,

$$\frac{1}{p}+\frac{1}{q_1}=\frac{1}{f_1},$$

where q_1 is the image distance of the first lens. The light from this first lens converging toward the image is intercepted by the second lens. The object distance for this second lens is the image distance of the first lens, but with the sign changed. Thus

$$-\frac{1}{q_1}+\frac{1}{q}=\frac{1}{f_2},$$

where q is the final image distance. Since the lenses are assumed to be thin and in contact, all distances (q_1, q, p, etc.) are measured to the center of the lens system. Adding the two equations, we obtain

$$\frac{1}{p}+\frac{1}{q}=\frac{1}{F}=\frac{1}{f_1}+\frac{1}{f_2}.$$

The result is most nearly correct when the combined thickness of the two lenses is small compared to the focal length of either lens. If a convex and a concave lens of the same numerical focal length are placed in contact, they behave optically like a piece of plane glass. The focal lengths are equal and opposite, so the combined focal length, F, is theoretically infinite.

As we have seen, the shorter the focal length of lenses, the more convergence or divergence they produce in a beam of light. This reciprocal of the focal length is called by optometrists the *power P* of the lens and is measured in *diopters*, when the focal length is expressed in meters. Thus a converging lens of focal length of 2 m has a power of 0.5 diopter, whereas a diverging or concave lens of focal length -2 m has a power of -0.5 diopter. The total power of a lens can be expressed as the sum of the powers of the two lens surfaces. Consider the lens shown in Fig. 34-33, with radii of curvature R_1 and R_2. The total power of this lens, from equation (34-22), is

$$P=\frac{1}{f}=(\mu-1)\left(\frac{1}{R_1}+\frac{1}{R_2}\right).$$

Now this lens can be considered as two plano-convex lenses having their plane surfaces in contact. The power P_1 of the surface of radius R_1 is

$$P_1=\frac{1}{f_1}=(\mu-1)\left(\frac{1}{R_1}+\frac{1}{\infty}\right)=\frac{\mu-1}{R_1}.$$

Similarly, the power of the second plano-convex lens radius R_2 is

$$P_2=\frac{\mu-1}{R_2}.$$

Hence the total power of the double convex lens is given as the sum of the individual powers of the two plano-convex lenses, or

$$P=P_1+P_2. \qquad (34\text{-}25)$$

Suppose the optometrist has to make spectacle lenses of power P. He usually has lenses available from the manufacturers, and from these he selects a suitable one having a power P_1 for one surface and then grinds the other surface to a power P_2 such that $P = P_1 + P_2$. Actually, the focal length is only one aspect in the correction of sight by glasses, as may be seen from § 34.25.

34.20. Dispersion

The phenomenon of dispersion was discovered by Newton in 1666. Using a triangular prism of his own manufacture, placed in a beam of sunlight entering a darkened chamber through a circular aperture, he obtained a colored and elongated spot of light about five times as long as it was wide. This elongation puzzled him, and he made various hypotheses to account for it—none very satisfactory. Upon using two prisms, however, and directing various portions of the elongated beam from the first prism through a second aperture upon the second prism, he observed that the colors (blue) which were the more refracted by the first prism were likewise more refracted by the second and, conversely, that red colors which were less refracted by the first prism were likewise less refracted by the second, since the beam from the second aperture and that from the second prism were fixed in direction and position throughout. Newton correctly interpreted this experiment as proof that *white light* was composed of different colors which were differently refrangible and that, in his own words, "to the same Degree of Refrangibility ever belongs the same colour," and vice versa.

The same observation may be more simply made if one looks through a prism at the deviated virtual image of an incandescent lamp filament. The latter is seen to be not white, like its reflection in a mirror, but colored. If a *linear* filament is under observation and the edges of the prism are held parallel to the direction of the filament, *no part of the latter is white*, but all is colored into a brilliant band with red on one side, that of least deviation, and violet on the other

These observations are conclusive proof that the amount of refraction of a beam of light *depends upon the color* of the light and also that white light is really a *mixture of a number of different colors.*

Figure 34-41 shows in a diagrammatic manner the difference in the refraction of different colors

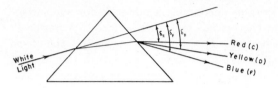

FIG. 34-41.—Dispersion of white light by a prism

produced when white light falls on a prism. The white light has been dispersed or spread out by the prism into a spectrum. This series of colors or spectrum is that seen in a rainbow, since it is refraction in water droplets that produces the rainbow. If one wishes to produce a clear spectrum with a prism in the manner shown in Fig. 34-41, then parallel beams of light must be used, so that, in practice, lenses would be used for focusing the light. The angular dispersion produced by the prism would be measured by the difference in deviations of the red and blue colors, i.e., $\delta_F - \delta_C$.

This terminology, calling the red light C, yellow D, etc., was first devised by Fraunhofer. By a careful examination of the spectrum of the sun or of daylight he found certain dark lines crossing the otherwise continuous spectrum. A dark line in the red portion of the spectrum was called a C line, and its wave length has been measured to be 6563 A. These dark lines are due to absorption and will be explained in the chapter on spectra. In Table 34-2 the refractive index for the C, D, F, and H lines are given for various substances.

Since the deviation of the red light by a prism or a lens is less than that for the blue light, it follows that the focal length of the lens for red light is longer than that for blue. Thus we see that a single lens must show chromatic or color aberration. We shall first investigate how this chromatic aberration can be partially corrected for the simpler case of thin prisms.

34.21. A Compound Achromatic Prism System

The deviation, δ, produced by a thin prism of small apex angle, A, can be obtained from equation (34-13),

$$\mu = \frac{\sin (A + \delta)/2}{\sin A/2}.$$

TABLE 34-2

VARIATION OF THE INDEX OF REFRACTION
WITH WAVE LENGTH

Color	Solar Spectrum Line	$\lambda \times 10^8$ (Cm)	Light Crown Glass	Medium Flint Glass	Borosilicate Crown lass*	Dense Flint Glass	Carbon Disulphide
Red.....	C	6563	1.5146	1.6224	1.5219	1.6500	1.6182
Yellow...	D	5893	1.5171	1.6272†	1.5243	1.6555	1.6276†
Blue....	F	4861	1.5233	1.6385	1.5297	1.6691	1.6523
Violet...	H	3969	1.5325	1.6625	1.6592	1.6940	1.6994

* Composition of borosilicate crown glass: SiO_2, K_2O, B_2O_3, BAO, and Na_2O.

† Note that flint glass and carbon disulphide have about the same index for D; also that flint has a higher index for C but smaller indices for F and H than has carbon disulphide.

If A and δ are small, then, approximately,

$$\mu = \frac{(A + \delta)/2}{A/2} = \frac{A + \delta}{A}$$

and

$$\delta = A(\mu - 1). \qquad (34-26)$$

The deviation of the red C line is denoted by δ_C, and the corresponding refractive index by μ_C, thus

$$\delta_C = A(\mu_C - 1).$$

Likewise for the blue F line we may write

$$\delta_F = A(\mu_F - 1). \qquad (34-27)$$

Hence the dispersion or angular separation, $_C\Delta_F$, between the C and F lines is given by

$$_C\Delta_F = \delta_F - \delta_C = A(\mu_F - \mu_C). \qquad (34-28)$$

From the table of the values of the refractive index at different wave lengths it follows that

For crown glass: $\mu_F - \mu_C = 0.0087$.

For dense flint glass: $\mu_F - \mu_C = 0.0191$.

Thus there arises the possibility of adjusting the angle A of a prism of crown glass and one of flint glass so that the dispersions are equal and opposite. This would produce an achromatic pair of prisms, which would still produce some deviation of the light.

Suppose we refer to the crown-glass prism by a prime and to the flint-glass prism by a double prime. The problem is to construct an achromatic pair of prisms with a crown-glass prism of angle A' and one of flint of angle A''. Both A' and A'' are considered small angles. The two prisms must be placed so that their dispersions are opposite, as

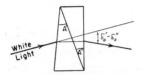

Fig. 34-42.—An achromatic compound prism

shown in Fig. 34-42. If the dispersions are to be equal and opposite, then, from equation (34-28),

$$\delta'_F - \delta'_C = \delta''_F - \delta''_C ,$$

or

$$A'(\mu'_F - \mu'_C) = A''(\mu''_F - \mu''_C) . \quad (34\text{-}29)$$

Because the two dispersions are equal and opposite, it does not follow that the compound prism will have no deviation. Let us determine the deviation of the yellow D line which is intermediate between the C and F deviations. For the crown-glass prism the deviation of the D line, from equation (34-27), is

$$\delta'_D = A'(\mu'_D - 1) ,$$

and of the flint-glass prism it is

$$\delta''_D = A''(\mu''_D - 1) .$$

Thus the total deviation of the compound prism in Fig. 34-42 is

$$\delta'_D - \delta''_D = A'(\mu'_D - 1) - A''_D(\mu'_D - 1) ; \quad (34\text{-}30)$$

and, as we can see from an actual example, this quantity is not zero when the total dispersion is zero.

EXAMPLE 6: Given a medium flint-glass prism with $A'' = 5°$, the problem is to find the angle A' of a light crown-glass prism so that the two prisms form an achromatic pair, and also to find the mean deviation produced by the pair.

Solution: From the table of refractive indices we have for achromatism, from equation (34-29),

$$A'(1.5233 - 1.5146) = 5°(1.6385 - 1.6224) ,$$

or the angle A' of the crown-glass prism is

$$A' = \frac{5 \times 0.0161}{0.0087} = 9°.25 .$$

The total deviation of the D line by this compound achromatic pair, from equation (34-30), is

$$\delta'_D - \delta''_D = 9°.25 \times 0.5171 - 5° \times 0.6272$$

$$= 1°.65 .$$

Such a compound prism could then deviate a beam of white light without any colors being separated. Though the achromatism is not perfect, it is quite good, and such combinations are used in optical instruments. The principal use comes in making achromatic lenses for cameras, thus overcoming to a large extent the chromatic aberration.

34.22. A Direct-Vision Compound-Prism System

It is often convenient to have a compound prism such that one can look directly through it at a source and see a spectrum of the source. Such a direct-vision prism produces dispersion without deviation of the mean color. Let us design two prisms which do not bend the yellow light at all (Fig. 34-43) and then examine the resulting

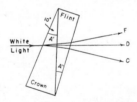

Fig. 34-43.—A direct-vision compound prism

dispersion. As a numerical example, consider a 10° light crown prism. Then, for no deviation of the D line, we have

$$\delta'_D = \delta''_D ,$$

or from equation (34-27),

$$A'(\mu'_D - 1) = A''(\mu''_D - 1) .$$

Substituting values from the table and setting $A' = 10°$, we have

$$10° \times 0.5171 = A'' \times 0.6272 ,$$

so that

$$A'' = 8°.24 .$$

The dispersion of the C and F lines in the crown-glass prism is, from equation (34-28),

$$_C\Delta'_F = \delta'_F - \delta'_C = A'(\mu'_F - \mu'_C)$$

$$= 10°(1.5233 - 1.5146) = 0°.087 , \quad \textbf{499}$$

and the dispersion in the medium flint prism is

$$\delta''_F - \delta''_C = 8°.24\,(1.6385 - 1.6224) = 0°.133\ .$$

Since the prisms are reversed (Fig. 34-43), their dispersions are in opposite directions, so the resultant dispersion is

$$_c\Delta''_F - {}_c\Delta'_F = 0°.133 - 0°.087 = 0°.046\ .$$

We have here used thin prisms to simplify the mathematical treatment. Thin prisms are not very useful, so thicker ones are used in practice. The adjoining surfaces of the prisms are cemented together, so that total reflection at the common surfaces is avoided.

34.23. Chromatic Aberration in Lenses

Consider two simple lenses (Fig. 34-44 [a] and [b])—one positive and convergent, the other negative and divergent. White light is not all focused at one point by the former, and the position of the focus of the other likewise depends upon color, but in the opposite sense.

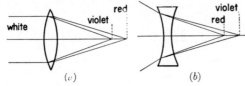

FIG. 34-44.—Chromatic aberration in convex and concave lenses.

Since the size of an image depends upon the focal length of the lens, we see that the positive lens forms images of different sizes in different colors at various distances along its optical axis. Such confusion of colors makes the images formed by such single lenses hard to see sharply and gives them a very unnatural appearance—colored borders to white objects, etc., characteristic of images in cheap telescopes, spyglasses, microscopes, and magnifiers. The reversal in the sequence of colored images in positive and negative lenses suggests the manner of attempting a solution. If a positive lens be combined with a negative lens of different material and less curvature, a lens of total positive character may be produced, as in Fig. 34-45, where, with proper selection of materials of different dispersive power at different regions, *at least two* entirely different colors may be brought accurately to the same focal point.

In discussing the achromatic prisms, we saw that the angle of the crown-glass prism was about twice that of the flint-glass prism. Then, if the crown-glass lens is plano-convex, the flint-glass lens will be convexo-concave. The focal length, f, of the two lenses in contact, from equation (34-24) is

$$\frac{1}{f} = \frac{1}{f'} + \frac{1}{f''}\ .$$

From equation (34-22) we have

$$\frac{1}{f} = (\mu - 1)\left(\frac{1}{R_1} + \frac{1}{R_2}\right).$$

For convenience, let $(1/R'_1 + 1/R'_2) = \alpha$ for the crown-glass lens, and for the flint-glass lens let $(1/R''_1 + 1/R''_2) = \beta$.

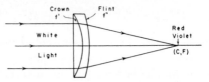

FIG. 34-45.—An achromatic lens combination

Since the focal length for the C line must be the same as that of the F line, for the combination we have

$$(\mu'_F - 1)\,\alpha + (\mu''_F - 1)\,\beta = (\mu'_C - 1)\,\alpha + (\mu''_C - 1)\,\beta$$

or

$$\frac{\alpha}{\beta} = \frac{\mu''_F - \mu''_C}{\mu'_C - \mu'_F}\ . \tag{34-31}$$

The focal length f'_D of the crown-glass lens for the intermediate D line is

$$\frac{1}{f_D} = (\mu'_D - 1)\,\alpha\ ,$$

and for the flint-glass lens,

$$\frac{1}{f''_D} = (\mu''_D - 1)\,\beta\ ,$$

so that the focal length of the combination for the D line is

$$\frac{1}{f_D} = (\mu'_D - 1)\,\alpha + (\mu''_D - 1)\,\beta\ . \tag{34-32}$$

EXAMPLE 7: Let us calculate the radii of curvature of the plano-convex crown-glass lens and the concave flint lens shown in Fig. 34-45 which form an achromatic doublet of focal length 50 cm.

Solution: Using the values of the refractive indices from Table 34-2 in equation (34-31), we have

$$\frac{a}{\beta} = \frac{1.6385 - 1.6224}{1.5146 - 1.5233} = \frac{0.0161}{-0.0087} = \frac{-161}{87}.$$

Substituting in equation (34-32), we obtain

$$\tfrac{1}{50} = 0.5171a + 0.6272\beta$$

or

$$2 = 51.7a + 62.7\beta.$$

From our conventions of signs, the focal length and radii of curvature of the convex crown-glass lens are positive, so that a is positive and β is negative. Solving the foregoing equations for a and β gives

$$a = 0.1121, \qquad \beta = -0.0606.$$

For the crown glass, $R_1' = \infty$ and R_2' has the same numerical value, but opposite sign, of the radius R_1'' of the common surface with the flint-glass lens. Hence

$$\frac{1}{R_2'} = 0.1121, \qquad \text{or} \qquad R_2' = 8.93 \text{ cm},$$

$$\beta = \left(\frac{1}{R_1''} + \frac{1}{R_2''}\right) = -0.0606,$$

$$\frac{1}{R_2''} = -0.0606 + 0.1121 = 0.0515,$$

$$R_2'' = 19.4 \text{ cm}.$$

This achromatic doublet therefore consists of a plano-convex crown-glass lens with radii of curvature ∞ and 8.93 cm, and a convexo-concave lens with radii of curvature -8.93 and 19.4 cm. The focal length of the crown-glass convex lens for the D line or yellow color is

$$\frac{1}{f_D} = (\mu_D' - 1)\left(\frac{1}{R_1} + \frac{1}{R_2}\right) = 0.5171 \times 0.1121$$

$$= 0.0580,$$

$$f_D' = 17.25 \text{ cm};$$

and for the convexo-concave flint-glass lens it is

$$\frac{1}{f_D''} = (\mu_D'' - 1)\left(\frac{1}{R_2'} + \frac{1}{R_2''}\right) = -0.6272 \times 0.0606$$

$$= 0.0380,$$

$$f_D'' = -26.3 \text{ cm}.$$

As a check you should show that the focal length of the combination of the two lenses for the yellow, red, and blue light is 50 cm.

Chromatic aberration can also be reduced by having two lenses of the same kind of glass if the lenses are separated by a distance of half the sum of the focal lengths. This method is usually used in the eyepieces of optical instruments.

34.24. Other Types of Aberration

As any camera enthusiast knows, the price of the camera depends largely on the lens system, and if he is to use a large opening on the lens and have a clear picture, then a complicated lens system must be used. A single lens would give considerable distortion, especially at points removed from the axis. Our simple first-order theory is valid only for rays near the axis. When the rays are at some distance from the axis, then serious image defects or aberrations arise. It is not possible to give anything more than a brief mention of these without any real account of how they are corrected. The following aberrations are concerned with light of a single wave length.

a) Spherical aberration.—In this, rays from a point object on the axis do not form a point image; or rays parallel to the axis are not brought to a single focus, as shown in Fig. 34-46 (*a*) and (*b*). This effect can be mini-

Fig. 34-46.—Spherical aberration for (*a*) point on axis; (*b*) rays parallel to axis.

mized but not eliminated by dividing the deviation equally between the glass surfaces, that is, by using a plano-convex lens with the convex surface facing the incident light. Of course, it is reduced if the outer rays *2* and *3* are eliminated by a suitable stop.

b) Coma.—This has to do with the comet-like appearance of the image of points not on the axis of the lens. The magnification is different for different zones of the lens and is shown in Fig. 34-47 for an infinitely dis-

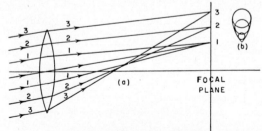

Fig. 34-47.—Production of coma

tant point slightly off the axis. If the magnification for the outer rays is larger than that for the central rays, the **501** coma is said to be positive and is of the type shown in Fig. 34-47 (*b*). This aberration can be minimized in a manner similar to that used for spherical aberration, namely, by using a plano-convex lens with the convex side to the incident light or by eliminating the marginal rays with a stop.

c) *Astigmatism.*—This has to do with the failure of a lens to bring rays from a point object at a considerable distance from the axis to a point image. The name comes from the Greek word *stigma*, meaning a "point." A horizontal line image, called the *primary image*, is formed at I_1, where the rays in the vertical plane *CD* intersect (Fig. 34-48). A secondary image, a vertical line at I_2, is

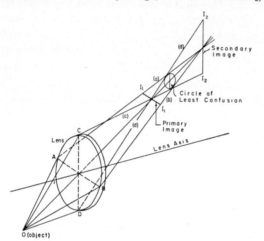

FIG. 34-48.—Astigmatic images produced by a lens

formed where the rays in the horizontal plane *AB* intersect. Almost midway between these line images the rays form a circle called the *circle of least confusion*.

The amount of astigmatism measured by the distance between I_1 and I_2 is approximately proportional to the focal length and is not appreciably changed by changing the shape of the lens surfaces. Related to this astigmatism is the curvature of the field in which all points in a plane object are considered. In this case the location of the primary images is on a surface I_1 (Fig. 34-49) and

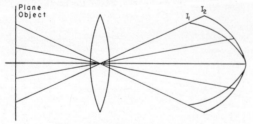

FIG. 34-49.—Curvature of field

on a surface I_2 of the secondary images. For a thin lens, these are paraboloids of revolution and are not flat surfaces. This form of aberration cannot be eliminated for a single thin lens, although it can be minimized by

suitable stops placed in front of the lens at appropriate distances. This is frequently done in cheap cameras to reduce the curvature of the field. A system of lenses with suitably located stops is also used to reduce these defects. Such lenses are called *anastigmats*. All the foregoing aberrations are concerned with the failure of a lens to bring a point object to a point image. We shall now consider a final aberration which is not concerned with sharpness of image.

d) *Distortion.*—This image defect is produced by a variation in the magnification with distance from the optical axis and does not arise from any failure in the focusing of the image. In Fig. 34-50 (*a*) is shown the

FIG. 34-50.—(*a*) Object; (*b*) pincushion distortion; (*c*) barrel distortion.

object placed perpendicular to the lens axis so that the center of the wire mesh lies on the optical axis. If the magnification increases with distance from the axis, then the pincushion distortion shown at (*b*) is produced. If the magnification decreases with distance from the axis, then the image shows the barrel distortion, as in (*c*). This kind of distortion is considerably influenced by the position of the stop, whether it is in front of or behind the lens. It is a difficult form of distortion to eliminate, but lens systems with suitable stops are made which are almost free of distortion for such critical work as aerial mapping.

34.25. The Human Eye

We conclude this chapter on geometrical optics by appending, in summarized form, brief descriptions of the human eye, together with the more common optical instruments.

Certain aberrations of the human eye can be corrected by appropriate lenses. In normal (emmetropic) eyes (Fig. 34-51 [*a*]) the curvature of the lens can be altered by tension on the muscles, *M*. Thus the eye, which, when relaxed, brings parallel rays to focus on the retina, can "accommodate" and shorten its focal length, thus becoming essentially myopic, so that nearer objects will focus on the retina. The radii of curvature of the lens surfaces are given in Table 34-3. In most individuals the gradual failure of the power to accommodate, which begins in youth, does not become apparent until about the age of fifty, when "reading glasses" need to be used. These, of course, are positive lenses. This is called the "elderly" condition, PRESBYOPIA.

EXAMPLE 8: Suppose a person suffering from presbyopia can plainly see objects at 75 cm and beyond. The problem is to find the focal length of the lens which moves this near point to 25 cm.

Solution: $p = 25$ cm, $q = -75$ cm,

$$\frac{1}{f} = \frac{1}{p} + \frac{1}{q} = \frac{1}{25} - \frac{1}{75} = \frac{2}{75},$$

$$f = 37.5 \text{ cm}.$$

Thus the glasses must consist of convex lenses of focal length 37.5 cm.

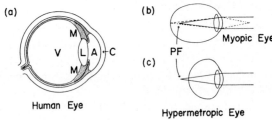

FIG. 34-51.—(a) Human eye; (b) myopic eye; (c) hypermetropic eye.

(a) *Human eye.*—
 $C =$ Cornea (transparent modification of skin),
 $\mu_A = 1.330$ (water with 0.77 per cent glucose),
 $\mu_L = $ (1.36 at edges, 1.42 in nucleus), transparent interlacing fibers hexagonally packed,
 $\mu_V = 1.337$ jelly,
 $M =$ Muscles for changing focal length of L.

(b) *Myopic eye.*—Eyeball too long. Parallel light focuses in front of retina, very near objects *on* it; hence the term *near-sighted*. A diverging negative lens corrects this condition.

(c) *Hypermetropic eye.*—Eyeball too short. Parallel light focuses behind the retina. So-called "*farsighted*" eye," but incorrectly named, since only *converging* light can focus on retina. Corrected by positive-lens spectacles.

Eyes originally myopic, normally wearing diverging spectacles, can, of course, be adjusted to some extent for reading by removing spectacles as the presbyopic condition increases. Bifocal glasses, for myopics, have the upper lens negative for distance; the lower, positive for close reading. An intermediate lens is sometimes included (in trifocal glasses) for middle distance. Hypermetropic bifocals are made of two positive lenses, the lower one of greater power.

Astigmatism, a common defect in human eyes, is not to be confused with the astigmatism which is one of the aberrations of spherical lenses or mirrors. Ocular astigmatism is due to a certain amount of cylindrical curvature of the eye's cornea or lens. Along some transverse axis of the lens its curvature may be greater than in other directions. This results in a shorter focal length in the plane containing this axis. The person with this defect, on viewing a series of lines diverging radially from a point, will see those in the plane of his astigmatic cylindrical curvature less sharply, since these focus in front of the retina. Negative cylindrical curvature, properly oriented in eyeglasses, corrects this condition.

TABLE 34-3

OPTICAL CONSTANTS OF THE UNACCOMMODATED EYE

Refractive indices:

Aqueous humor (A)	1.330
Vitreous humor (V)	1.337
Cystalline lens (in nucleus)	1.42

Radii of curvature of:

Front surface of cornea (C)	7.83 mm
Front surface of lens (L)	10.0 mm
Back surface of lens (L)	6.0 mm

Distance from the cornea of:

Front surface of lens (L)	3.6 mm
Back surface of lens (L)	7.2 mm
Fovea	24.0 mm

Astigmatism often occurs in combination with myopia or hypermetropia, and expert diagnosis and measurement by a qualified oculist should always be relied on for the proper optical prescription for lenses.

34.26. Simple Optical Instruments

The first intelligent use of optical instruments essentially began when Galileo made his first telescope and began using it on celestial objects. The *magnifying power* of any optical instrument is defined as the ratio of the size of the image formed on the retina of the eye, when the instrument is in front of the eye, to the size of the retinal image for the unaided eye. These relative sizes may be expressed in terms of the angles subtended by the images on the retina at a point on the optical axis of the eye at the crystalline lens. The size of the retinal image obviously depends upon the distance of the object from the eye. Since, however, the object cannot be placed closer than 25 cm from the normal eye and still obtain a distinct image, there is a limit to the size of the retinal image that can be formed. Let us see how a simple converging lens, the *magnifier*, will enable us to increase the size of the retinal image by permitting the eye to get quite near the object.

In Fig. 34-52 an object of height OO' is placed at a distance p from a convex lens of focal length

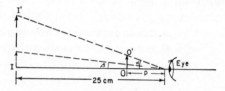

FIG. 34-52.—Simple magnifying lens

f so that a virtual upright image, II', is produced at the near point of 25 cm from the eye. The object distance, p, is given by

$$\frac{1}{p} - \frac{1}{25} = \frac{1}{f},$$

$$\frac{1}{p} = \frac{25 + f}{25f} \quad \text{and} \quad p = \frac{25f}{25 + f}.$$

The angle a subtended at the eye by the object at a distance p is approximately

$$a = \frac{OO'}{p} = OO' \left(\frac{25 + f}{25f} \right).$$

Without the magnifying lens, the shortest distance at which the object can be clearly seen is 25 cm, so that the largest angle, β, which the object itself can subtend at the eye is approximately

$$\beta = \frac{OO'}{25}.$$

The *angular magnification M* produced by this lens is the ratio of angles subtended at the eye by the image and object, respectively. For the angular magnification with the image at 25 cm,

$$M_{25} = \frac{a}{\beta} = \frac{25 + f}{f} = \frac{25}{f} + 1, \quad (34\text{-}33)$$

where f must be given in centimeters, the same unit as the 25 cm.

By placing the object OO' at the focus f of the lens, the image II' is at infinity, or the rays forming the image are parallel, so that $p = f$,

$$a = \frac{OO'}{f}, \qquad \beta = \frac{OO'}{25}.$$

Hence

$$M_\infty = \frac{25}{f}. \qquad (34\text{-}34)$$

Though it might appear that the angular magnification could be made as large as desirable, it is found in practice that, with a single lens, the various aberrations limit this magnification to two or three. If in equation (34-33) the magnification is 3, usually denoted 3×, then

$$3 = \frac{25}{f} + 1 \quad \text{or} \quad f = 12.5 \text{ cm}.$$

For higher magnification, compound lenses are used, as is discussed in § 34.29.

34.27. Galilean Telescope and the Opera Glass

Historically the first telescope that we have definite evidence about was constructed by Galileo about 1609. In this instrument (Fig. 34-53) the objective lens, L_1, brings the light from a very distant object subtending an angle β to a focus at F, a distance f_1 from L_1. The lens, L_1, produces an image of height h at F, where, approximately,

$$\beta = \frac{h}{f_1}.$$

Interposed between L_1 and F is a concave eyepiece lens, L_2, at a distance of its focal length f_2 from F, as shown in Fig. 34-53. In other words,

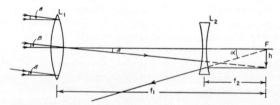

FIG. 34-53.—The Galilean telescope or opera glass in diagrammatic form.

the distance between the lenses is the magnitude of the difference $f_1 - f_2$ of the focal lengths. Now consider the image of height h at F as the object for the final image produced by L_2. A ray parallel to the axis through the tip of the arrow will, after refraction through the lens L_2, appear to come from F. The angle a subtended by this ray is given by

$$a = \frac{h}{f_2}.$$

Hence the angular magnification, M, for object and image at relatively large distances is

$$M = \frac{a}{\beta} = \frac{f_1}{f_2}. \qquad (34\text{-}35)$$

The chief advantages of this type of telescope are (1) simple construction, (2) short length, and (3) the production of an erect image without the use of a special erecting eyepiece. Its disadvantages are (1) the aberrations are large, (2) crosshairs cannot be used, and (3) the magnifying power is limited. For these reasons the instrument is generally confined to viewing terrestrial objects, as in an opera glass.

34.28. The Astronomical Telescope

The astronomical telescope consists essentially of an object lens, L_1, of long focal length (Fig. 34-54) and an eyepiece lens, L_2. The object lens

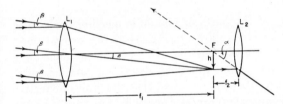

Fig. 34-54.—The astronomical telescope in diagrammatic form.

produces a real image of height h at its focus F, a distance of its focal length f_1 from L_1. The object, which is at a large distance compared to the focal length f_1, subtends an angle β, where, approximately,

$$\beta = \frac{h}{f_1}.$$

A convex eyepiece lens, L_2, is placed at a distance of its focal length f_2 from the real image, of height h. This eyepiece lens produces a virtual image at a large distance. The angular height of the final image is

$$a = \frac{h}{f_2}.$$

Thus the angular magnification of this telescope for very distant objects is

$$M = \frac{a}{\beta} = \frac{f_1}{f_2}.$$

This form of telescope is ordinarily used for viewing astronomical objects, since cross-hairs can be placed at F for accurately locating the relative positions of celestial objects. For large magnifications the simple thin lenses shown in the diagram are replaced by compound lenses or lens systems. The largest telescope of this type is the Yerkes telescope of the University of Chicago, in which $f_1 = 60$ ft and $f_2 = 1$ in, giving a magnification of 720. This magnification cannot be greatly increased in practice, because a limitation is imposed by diffraction of the light, which limits the clarity of the image or the resolving power of the telescope. This depends on the diameter of the objective lens, L_1. This objective lens must be corrected for chromatic aberration and as well as possible for the other aberrations.

Newton, whose limited experiments led him to believe that the chromatic aberration could not be corrected, decided to construct a reflecting telescope in 1688. A concave mirror, originally having a spherical, now a paraboloidal, surface, is used in place of the objective lens to bring the light to a focus in front of the mirror. In order not to cut out the light on the mirror, it is necessary to use some means to bring the light away from the optical axis of the mirror. Newton used a small plane mirror just in front of the focus of the mirror and reflected the light out to where it could be viewed.

The reflecting telescope produces a very sharp image in the center, but the oblique aberrations, such as astigmatism and coma, are considerable a short distance from the axis. For this reason, all but the largest telescopes are of the refracting type. A limit is set to the size of these, in that it is extremely difficult to obtain optical glass in large sizes which is homogeneous and free from striae. The largest refractor in use is the Yerkes telescope, which has an objective lens 40 inches in diameter. The largest reflecting telescope is that on Mount Palomar in southern California, which has a mirror about 200 inches in diameter. In order to decrease temperature effects on the mirror, it is made of fused quartz, and the surface, after being correctly shaped, is silvered. The mounting of such a huge mirror is a major engineering task. It should be noted that while the light-gathering power of a telescope increases approximately as the square of the diameter of the objective lens or mirror, nevertheless the use of sensitive photographic plates, to replace the eye, has increased the usefulness of telescopes many hundred fold. 505

34.29. Eyepieces

In order that much of the spherical and chromatic aberration present in the simple magnifier may be eliminated, optical instruments known as *eyepieces* or *oculars* have been designed. There are two types in common use, known as the Huygens eyepiece and the Ramsden eyepiece. In the Huygens eyepiece (Fig. 34-55) there are

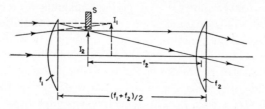

FIG. 34-55.—The Huygens eyepiece

two plano-convex lenses of crown glass and of focal lengths f_1 and f_2. They are separated by a distance equal to half the sum of the focal lengths, and the convex surfaces are placed toward the incident light. This is the condition for minimum lateral chromatic aberration. Light from an objective on the left is forming an image at I_1. The first lens of the eyepiece of focal length f_1 produces a real image at I_2, which is at a distance of f_2 from the eye lens. The light entering the eye is therefore parallel. A field stop is usually placed at the position of the image I_2 and cross-hairs or a scale put on this stop. Some changes in this fundamental design have been used to improve the curvature of the field and the longitudinal achromatic aberration.

In the Ramsden eyepiece there are two plano-convex lenses of crown glass of the same focal length. The distance between the two lenses is about two-thirds their focal length (Fig. 34-56). An image is formed by an ob-

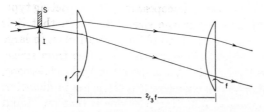

FIG. 34-56.—The Ramsden eyepiece

jective at I and a field stop, S, is placed at this position. On this is mounted any cross-hairs or scale. This eyepiece is used as a magnifier on any real object placed in the position I. The Ramsden eyepiece is relatively free from distortion and chromatic aberration, and the field is much flatter than for the Huygens eyepiece.

34.30. The Microscope

To obtain large magnification of small objects, the microscope is used. This consists essentially of an objective lens system (shown in Fig. 34-57 [a] as a simple lens, L_1) that produces a greatly enlarged image, I, of an object, O, and an eyepiece to view this real image. In Fig. 34-57 (a), only the angular relationships between O and I and an eyepiece are shown.

In practice, the objective lens L_1 is a highly perfected compound-lens system. A schematic draw-

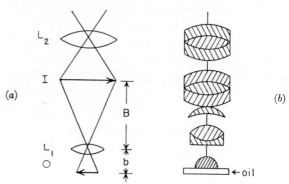

FIG. 34-57.—The microscope

ing of the parts of one oil-immersion microscope objective is shown in Fig. 34-57 (b), several times actual size. This objective of ten components is corrected for spherical aberration and for chromatic aberration so well that the limit of effective magnifying power is essentially determined by the diffraction effects of light. The eyepiece used to view the enlarged image is usually a Ramsden or Huygenian type lens system.

From Fig. 34-57 (a) the approximate magnifying power of the microscope is found: that of the objective lens is the ratio of B to b. Since, however, the object is placed very close to the principal focus of L_1, we may replace b by the focal length F_1 of L_1, so that

$$M_1 = \frac{B}{F_1}.$$

The eyepiece used to view the real image, I, has additional magnifying power approximately equal to $25/f_2$, where f_2 is the focal length of the eyepiece and the factor 25 represents the distance of distinct vision. The total magnifying power, then, is

$$M_0 = \frac{25B}{F_1 f_2}. \tag{34-36}$$

The distance B is practically the length of the microscope tube; thus large tube lengths and short focal lengths of objective and eyepiece are indicated for high power.

In many of the more expensive microscopes there are three objectives provided with different magnifying power. A typical list could be a low power with magnification of about $10\times$ and focal length of 16 mm; a medium power with magnification about $44\times$ and focal length of 4 mm; and a high-power, oil-immersion objective with magnification about $97\times$ and focal length of 1.9 mm.

34.31. The Photographic Camera

The essential features of a camera are a lens and a shutter, set into a light-tight box, in the back of which may be inserted a photographic film or plate. The lens forms an image on the photographic plate, which is permanently recorded on development. The purpose of the shutter is to allow light to pass through the lens and affect the photographic plate. The amount of light reaching the photographic plate depends on the area of the lens and the length of time it is open. A continuously adjustable diaphragm is used for varying the area or diameter of the lens. This is marked in f-values, such as 4.5, 6.3, etc. The focal ratio or f-value is defined as the ratio of the focal length f to the diameter of the lens opening d, or

$$f\text{-value} = \frac{f}{d}. \qquad (34\text{-}37)$$

Thus a lens with a focal length of 12 cm and a lens opening or linear aperture of 3 cm has an f-value of 4, or the lens is set at $f/4$. Since the amount of light entering the camera varies as the square of the lens diameter, the f-values are commonly given so that a move from one of the f-values to the next changes the amount of light entering the camera by a factor of 2. Thus the ratio of the amount of light entering a camera with $f/6.3$ and $f/4.5$ is

$$\frac{(4.5)^2}{(6.3)^2} = \frac{20.2}{39.6} \approx \tfrac{1}{2}.$$

Since no camera lens system is completely free of aberrations, it is always wise to minimize these by using as large an f-value—that is, as small a lens opening—as practical.

In many cameras the lens is movable relative to the photographic plate so as to focus objects at different distances on the plate. If a fixed-focus camera is used, than a relatively small lens opening or large f-value, $f/11$ or $f/16$, is used. A pinhole camera focuses objects at all distances or has an infinite depth of focus.

34.32. Résumé

The following is a brief listing of the definitions, principles, and theories, given in this chapter, which you should know:

The velocity of light and its measurement

The electromagnetic spectrum and the range of the visible radiation

The law of reflection of light

The relationship of the object and image distances and focal length for concave and convex mirrors; real and virtual images

The relationship of the size of the image to that of the object

The law of refraction of light

The refractive index of a substance and its measurement

Refraction at a curved surface

The achievement of the aplanatic condition with a spherical lens

The relationship of the object and image distance to the refractive index and the radii of curvature of a thin lens

The power of lenses in diopters

The dispersion of light by a prism

Chromatic aberration and its correction with prisms and lenses

The human eye, some of its defects and their corrections

The meaning of linear and angular magnification

A brief explanation of telescopes, microscopes, and cameras

Queries

1. Why is it necessary to include the statement "The reflected ray, incident ray, and normal all lie in a given plane" in the law of reflection?
2. Why does a pool of clear water appear shallower than it really is?
3. Why does the apparent depth of water in a tank change with the position of the observer?
4. Why does it seem to a person about half-immersed in a pool which is everywhere of equal depth that he is standing in a portion that is deeper than anywhere else?
5. What explanation can you give for the phenomenon known as "mirage"?
6. Can a convex mirror show spherical aberration?

7. How does the eye see distinctly objects at different distances?

8. What kind of image would a cylindrical lens give of a point source?

9. Why are near objects indistinct to a person under water?

10. How could you determine the position of the center of curvature of a concave mirror with a pin?

11. How could you prove that a beam of light is a parallel beam?

12. Can you give any physical reasons why the visual brightness of an object of finite size cannot be increased by the use of a lens?

13. Does a concave mirror show chromatic aberration?

14. If a camera is focused for distant objects, must the distance between the photographic plate and lens be decreased or increased to focus on relatively near objects?

15. If p is the object distance, q the image distance, and f the focal length for either a mirror or a lens, then show that if $p + q$ is plotted against p, the graph has a minimum at $p = 2f$ and $4f$.

Practice Problems

Drawings, where necessary, must be made with a ruler; freehand drawings cannot be sufficiently exact.

1. An object 3 cm high is placed 15 cm in front of a concave mirror of focal length 10 cm. Locate by drawing and computation the position and size of the image. Ans.: 30 cm; 6 cm.

2. An object 3 cm high is placed 15 cm in front of a convex mirror of focal length 10 cm. Locate by drawing and computation the position and size of the image.

3. A spherical mirror has a focal length of 5 cm. Locate by drawing and by computation the images formed of objects, 3, 8, 10, and 12 cm from the mirror. Ans.: $-\frac{15}{2}$ cm; 13.3 cm; 10 cm; 8.6 cm.

4. A mirror has a focal length of -5 cm. Locate by drawing and by computation the images formed of objects 3, 8, 10, and 12 cm from the mirror.

5. An object 2 cm high is placed 20 cm in front of a convex lens of focal length 10 cm. Locate by drawing and computation the position and size of the image. Ans.: 20 cm; 2 cm.

6. An object 2 cm high is placed 20 cm in front of a concave lens of focal length 10 cm. Locate by drawing and computation the position and size of the image.

7. A lens has a focal length of 5 cm. Locate by drawing and by computation the images formed of objects 3, 8, 10, and 12 cm from the lens. Ans.: -7.5 cm; 13.3 cm; 10 cm; 8.6 cm.

8. A lens has a focal length of -5 cm. Locate by drawing and by computation the images formed of objects 3, 8, 10, and 12 cm from the lens.

9. $F = 4$ cm. From the illustration herewith, determine, by a drawing, where the ray will cut the principal axis.

10. A thin converging lens of 24 cm focal length is placed in contact with a diverging lens of 12 cm focal length. What is the focal length of the combination?

11. What single lens is equivalent to a converging lens of focal length 5 cm in contact with a diverging lens of focal length 10 cm? Ans.: Converging lens, focal length 10 cm.

12. Given two lenses, as indicated in the accompanying figure, locate, by drawing and by computation, all images formed. Trace rays of light from the object to an eye which can see the final image. *Given:* $p_1 = 3$, $F_1 = 5$, $F_2 = -2\frac{5}{3}$, and $D = 5$.

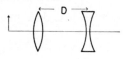

13. A beam of light is converging toward a point 20 cm in front of a lens. After going through the lens, the light converges 30 cm in front of the lens. Compute the focal length of the lens. Ans.: 60 cm.

14. A large piece of glass has a spherical cap of radius 8 cm. A point source of light is placed 100 cm from the cap. If the index of refraction of the glass is 1.6, locate the image of the point source.

15. What is the velocity of light in water? If the wave length of red light in air is 6000 A, what will it be in water? What is the frequency of the light in air and water? Ans.: 2.25×10^{10} cm/sec; 4500 A; 5×10^{14} vibrations/sec.

16. The refractive index of flint glass is 1.65 and of water is 1.33, both relative to air. What is the refractive index of water relative to flint glass, $_w\mu_g$? Find the critical angle for light going from the flint glass to water.

17. A small object is placed at the bottom of a tank of water which is 20 cm deep. What is the apparent depth of this object? Ans.: 15 cm.

18. If a block of flint glass 5 cm thick is placed over the object in the water in question 17 so that the water is 15 cm above the glass block, find the apparent depth of the object below the top of the water.

19. A 60° prism has an angle of minimum deviation of 38°. Find the refractive index of the prism. Ans.: 1.51.

20. A plano-convex converging lens is made of glass of refractive index 1.5 and has a focal length of 24 cm. Find the radius of curvature of the convex face.

21. A plano-concave diverging lens is made of glass of refractive index 1.6 and has a focal length of 20 cm. Find the radius of curvature of the concave surface. Ans.: 12 cm.

22. Give the power of the lenses in questions 20 and 21. If these lenses are placed together, what is the total power and focal length of the combination?

23. A magnifying glass of focal length 5 cm is used to view print so that the image is at 25 cm. What is the magnification produced? Draw a diagram of the rays. Ans.: 6.

24. An astronomical telescope (Fig. 34-54) is used to view an object placed at 100 m. The eyepiece is adjusted for distinct vision. If the focal length of the object glass is 50 cm and that of the eyepiece 5 cm, find the magnifying power of the system. What will be the magnifying power of the telescope when used to view very distant objects? (Solve this problem by first principles.)

25. Two stars are observed through a telescope. One star lies on the principal axis. Light from the other one makes an angle $\tan^{-1} \frac{1}{10}$ with the principal axis. *Given:* $F_1 = 5$ cm, $F_2 = 1.25$ cm, D (distance between lenses) $= 6.2$ cm. Locate images by computation and by drawing. Trace rays of light through the telescope to an eye which can see the final image. Ans.: -30 cm from F_2 to final image.

26. Given two lenses, as indicated in the figure herewith, trace rays of light from the object to an eye which can see the final image, locate the images, and compute the positions of the images.

27. A thin crown-glass prism has an angle of 10°. What size medium flint-glass prism should be put with it so that the two will be achromatic? (See Table 34-2 for necessary data.) Show by a diagram how the light will be bent by the achromatic prism, and compute the angle of deviation. Ans.: $A_{\text{flint}} = 5°40$ for red, yellow, and blue colors.

28. A thin, medium, flint-glass prism has an angle of 10°. What size crown-glass prism should be put with it so that the two together will pass yellow light (D line) without deviation? Compute the dispersion of the compound prism and indicate the direction of the dispersion. Ans.: $A_D^C = 12°15$; $\Delta = 0°055$.

29. A compound microscope has, as objective and eyepiece, thin lenses of focal lengths 1 cm and 3 cm, respectively. An object is placed 1.2 cm from the objective. If the virtual image produced by the eyepiece is 25 cm from the eye, compute the magnifying power of the microscope and the separation of the lenses, from fundamental principles. Ans.: 46.6; 8.68 cm.

30. A point source of light is submerged in a tank of carbon tetrachloride. At what angle, measured from the normal, does critical reflection take place?

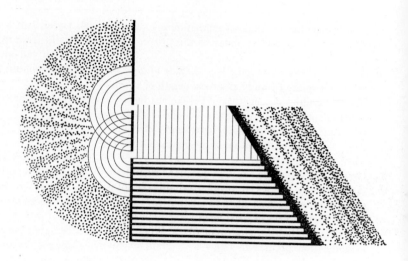

CHAPTER 35

Physical Optics

35.1. Introduction

In the study of light in the preceding chapter we considered primarily the geometrical behavior of the "ray" of light. Since rectilinear propagation is one of the characteristics of all forms of radiant energy as well as of that single octave that comprises visible light, we most naturally, perhaps, think of the rectilinear tracks or rays in visualizing the fundamental phenomena of reflection, refraction, dispersion, etc. In earlier chapters, however, we laid considerable emphasis upon the fact that *wave phenomena* were of the utmost importance in consideration of transmission of radiation. It is true that the possession of wave characteristics offers a far more comprehensive specification for light and all other forms of radiant energy than the straight-line propagation of this energy. Indeed, if one examines the facts with more acuteness, one realizes that rectilinear propagation is actually not *strictly* compatible with wave phenomena at all. Newton himself recognized this. We have continually hedged our statements about rectilinear propagation with qualifications implying that we were neglecting small departures from it. Indeed, Newton's adherence to a corpuscular theory of light was not a matter of prejudice. He recognized that many optical effects could well be interpreted on the basis of wave theories. He knew, however, that wave motions *must*, by their very wave nature,

bend around corners of obstacles and must deviate markedly into the region of a shadow from a path that was geometrically rectilinear. As far as he knew, small sources of light cast sharp shadows of obstacles, and no experiments upon it had ever revealed departures from the strictest of straight paths. Therefore, if light was lacking in the crucial test of undulatory character, Newton was compelled to conclude that it presumably was not of undulatory nature.

Now, of course, Newton himself had discovered the famous colored bands seen in white light between plates of glass separated by thin films of air and in the films of soap bubbles. He, indeed, never thought of these cases, clear to us as examples of interference phenomena, as in any way revealing undulatory characteristics.

PHYSICAL OPTICS deals primarily with light from the point of view of its physical character—therefore, with LIGHT WAVES. It must be emphasized, however, when we seem to speak so confidently as knowing something about the *physical character* of light, that even in this twentieth century we know relatively little about the nature of this form of energy. We do know that certain corpuscular characteristics—of a kind, of course, undreamed of by Newton—are as definite and as universal as are the wave lengths. Indeed, so intimately associated with the latter are these discrete packets of radiant energy, which we call by the name of

PHOTONS, that they are actually *measured* in terms of frequencies or of reciprocal wave lengths.

In this chapter, we shall look at optical matters exclusively from that other partial perspective, let us say, that visualizes *only waves*.

35.2. Huygens' Construction Again

Wave interpretations, as well as those based on rays, are most easily cast into geometrical forms. Indeed, this was first recognized by the Dutch physicist, Christian Huygens, who was born fourteen years before Newton and who did much to pave the way in mechanics for his illustrious British contemporary. Huygens always championed the undulatory point of view but lacked any experimental proof of its correctness. He gave to science, however, a construction for watching waves of light proceed, so to speak—a construction that to this day serves as a basis for many optical calculations, one that solves all cases which can be solved by ray geometry and not a few cases that cannot. Most important of all is the fact that, for one's study of this field at the more elementary level, Huygens' method is far more picturesque and fruitful, since it keeps wave characteristics, unquestionably existent, always to the fore.

In chapter 31 on wave motion we saw that a wave front is the locus of all points in the same phase. For example, a stone thrown into a pond generates waves which travel in circles from the center of the disturbance. These circles are the wave fronts. Huygens' construction is concerned with the method of obtaining the wave front at some later time from that present at an earlier time. He imagined every point on a wave front to be a center of a disturbance, sending out wavelets such that the new wave front is the envelope of all the wavelets. This is illustrated in Fig. 35-1, in

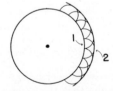

FIG. 35-1.—Illustration of Huygens' construction

which the circle *1* is the initial wave front, on which each point is emitting wavelets so that at a later time the new wave front *2* is tangent to the

wavelets. As given here, this construction may appear highly artificial and of little use, but we shall see that this is not the case. First we shall apply Huygens' construction to the cases of reflection and refraction and later to interference and diffraction.

35.3. Reflection by Huygens' Construction

Let us consider a plane wave, having wave fronts CC', AA', etc., incident on a plane mirror MM' (Fig. 35-2). These wave fronts may be con-

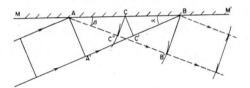

FIG. 35-2.—Reflection by Huygens' construction

sidered to be a wave length apart, and all points on these wave fronts are in phase. For clarity, we shall show only two secondary wavelets, though we should consider every point on a wave front as giving rise to wavelets. Consider the wave front AA'. At A a wavelet is sent out, and this progresses a distance AC'' while the incident wave front progresses a distance $A'C'$, where $A'C' = AC''$. The new wave fronts consist of the incident one, CC', and the reflected one, CC''. As the wave from A' moves to B, the wavelet generated at A has moved to B', so that $A'B = AB'$. The wave front BB' of the reflected wave is obtained by drawing a tangent to the wavelet at B' so that the angle $BB'A$ is a right angle.

From the construction we see that the triangles ABA' and ABB' are congruent, so that the angle α is equal to the angle β. Now α and β are, respectively, 90° minus the angles of incidence i and reflection r for the incident ray $A'B$ and the reflected ray AB'. Thus

$$90 - i = 90 - r$$

and

$$i = r ,$$

which is the fundamental law of reflection.

The virtual image in Fig. 35-3, where secondary waves are omitted for simplification, is the point **511** from which the reflected wave front appears to proceed. This point can be located by the direction of the rays, which in isotropic media are always perpendicular to the wave-front surfaces, or by applying the construction in the reverse direction.

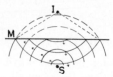

Fig. 35-3.—Source S and image I of waves in a plane mirror

The change in curvature imposed on a wave front by its reflection from a curved mirror surface is readily constructed (Fig. 35-4 [a] and [b]).

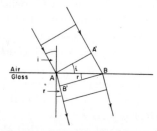

(a)   (b)

Fig. 35-4.—Source S and image I for convex and concave mirrors.

35.4. Refraction by Huygens' Construction

Let us consider a plane wave incident obliquely on a sheet of glass (Fig. 35-5). As experimentally

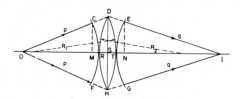

Fig. 35-5.—Huygens' construction for refraction of light

shown first by Foucault, the velocity of light in air, v_a, is greater than that in glass, v_g. The ratio of these velocities is equal to the refractive index, $_a\mu_g$, of the glass relative to the air.

Suppose AA' is the plane wave front of the incident wave; then, while the wave front at A' is traveling to B, the wavelet emitted at A is traveling in the glass from A to B'. Thus

$$\frac{A'B}{AB'} = \frac{v_a t}{v_g t} = \frac{v_a}{v_g} = {_a\mu_g} , \qquad (35\text{-}1)$$

where the wave front BB' has been drawn tangent
512 to the wavelet at B'. If i is the angle of incidence

and r is the angle of refraction, from Fig. 35-5 it is seen that

> angle $A'AB = i$ and   angle $ABB' = r$.

Also

$$\frac{A'B}{AB} = \sin i \qquad \text{and} \qquad \frac{AB'}{AB} = \sin r ,$$

so that, by division and equation (35-1), we have

$$\frac{A'B}{AB'} = \frac{\sin i}{\sin r} = \frac{v_a}{v_g} = {_a\mu_g} . \qquad (35\text{-}2)$$

This is Snell's law for refraction. To complete the diagram according to Huygens' construction, an infinite number of wavelets should have been shown progressing from the air to the glass. These are omitted for the purpose of clarity.

35.5. The Lens Equation, Using Huygens' Construction

At the expense of repetition, we shall show how the lens-maker's equation (eq. [34-22]) can be derived by means of Huygens' construction. Suppose a point source of light O is placed a distance p from a double convex lens of radii R_1 and R_2 (Fig. 35-6). From O there are

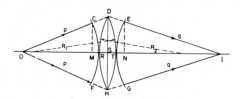

Fig. 35-6.—The lens equation and wave front

emitted waves whose wave fronts are spheres with O as center. A portion of one of these wave fronts is CRF, which is tangent to the lens at R. The radius of this wave front is p, the object distance. During the time this wave front travels in air from C to D to E, it travels in the glass at the middle a distance RT. All the wavelets from the wave front ETG then converge on the image I, where q is the radius of curvature of the wave front ETG. For the image formation at I the necessary optical condition is

$$CD + DE = {_a\mu_g} (RT) . \qquad (35\text{-}3)$$

Now, as in our previous development of the lens equation, we shall assume that the incident and refracted rays make small angles with the optical axis, which in this case amounts to assuming that the distance ($CD + DE$) is equal to CE, so that equation (35-3) can be written

$$CE = {_a\mu_g} (RT) .$$

By dropping perpendiculars from C, D, and E to the optic axis of the lens, we see that

$$CE = MR + RS + ST + TN ,$$

where MR is the sagittal distance of the arc CRF of radius p; RS is the sagittal distance of the arc DRH of the surface of the lens of radius R_2 and similarly for the sagittal distances ST and TN. Also we have

$$RT = RS + ST .$$

The approximation that CE is equal to CDE is equivalent to stating that the perpendicular distances CM, DS, and EN are equal. If these distances are called y, then from a proposition in geometry we have, for the distance MR,

$$y^2 = (2p - MR) MR \text{ or, approximately, } MR = \frac{y^2}{2p} .$$

If this is applied to equation (35.3), we have

$$\frac{y^2}{2p} + \frac{y^2}{2R_1} + \frac{y^2}{2R_2} + \frac{y^2}{2q} = {}_a\mu_g \left(\frac{y^2}{2R_1} + \frac{y^2}{2R_2} \right)$$

or

$$\frac{1}{p} + \frac{1}{q} = {}_a\mu_g \left(\frac{1}{R_1} + \frac{1}{R_2} \right) - \left(\frac{1}{R_1} + \frac{1}{R_2} \right) .$$

Thus the lens-maker's equation can then be written as

$$\frac{1}{p} + \frac{1}{q} = \frac{1}{f} = ({}_a\mu_g - 1) \left(\frac{1}{R_1} + \frac{1}{R_2} \right) .$$

This agrees with equation (34·22), derived by the analysis of the rays of light. Note that, for the biconvex lens, R_1 and R_2 are considered positive quantities, whereas for a biconcave lens these radii are both negative.

35.6. Interference

We have already discussed interference in the case of sound waves, as, for example, the production of standing waves. This will now be amplified in order to explain some optical phenomena. Interference involves the principle of superposition, which states that the resultant displacement of a wave at any point in a medium is the sum of the instantaneous displacements at the point produced by the two or more waves transversing the medium. For light or any other form of electromagnetic wave the displacements would be the value of the electric or magnetic field present in the waves at the point considered.

In order to produce interference, the two or more waves must have the same velocity, frequency, and wave length, and their relative phases must remain constant. In practically all the cases of interference there are only two sets of waves, and, if these are to produce complete destructive interference, as at a node in a standing wave, then the amplitudes of the waves must be equal. If the waves are transverse, the vibrations must be in the same plane, or the waves must be polarized in the same plane for complete destructive interference. We shall see how these conditions are met in practice as the discussion proceeds.

First let us consider the interference from two sources B and C, as shown in Fig. 35-7. In order

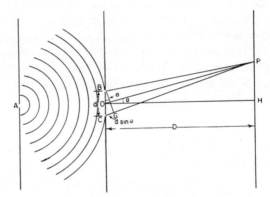

FIG. 35-7.—Interference from two sources

for the phases of the waves emitted from B and C to be the same, it is usual for them to come from a common source, such as A. If two separate light sources were used, no interference pattern would be observed along PH, for the radiating atoms of the sources emit light with random phase relations and two such sources could never produce interference. In Fig. 35-7, A, B, and C are slits with a source placed to the left of slit A.

If a monochromatic source of light is placed at A and the distances are suitable, then alternate light and dark slit images are seen on the screen PH. If white light is used, the images are colored. Such an experiment was first performed about 1800 by Thomas Young, and he correctly interpreted the light and dark images as being due to interference and thus established that light is propagated as a wave motion.

Suppose the slits B and C are 1 mm or so apart and the distance D from the slits to the screen HP is 1 m or more, then the light and dark images or interference fringes on the screen are a few millimeters apart. Let P be any point on the screen where OH is the perpendicular distance D from the center of the two slits of width d. The line OP makes a small angle θ with the perpendicular OH. **513**

Draw lines from P to B and P to C and drop a perpendicular BG from B onto CP. With the distances adopted in this system, the angle θ is very small, so that the lines BP and GP can be considered equal.

Consider, now, waves of wave length λ emitted from slits B and C. The path difference, δ, of these waves arriving at P is

$$\delta = CP - BP = CG .$$

If this path difference, δ, is 0, λ, 2λ, . . . , $n\lambda$, then, since the waves start out in phase, they will arrive in phase at P. That is, if a crest is arriving at P from source B, then there is also a crest arriving at P from source C, and there will be constructive interference; hence the resultant amplitude is the sum of the amplitudes of the two individual waves. From Fig. 35-7 it is seen that

$$CG = d \sin \theta .$$

Therefore, the condition for constructive interference at P is

$$d \sin \theta = n\lambda , \qquad (35\text{-}4)$$

where n is any integer, 0, 1, 2, etc. The value of $n = 0$ corresponds to the point H on the perpendicular bisector of BC. At this point waves of any wave length will be in phase, because the distances BH and CH are equal in length.

As the point P moves away from H, there will first appear a point at which the difference in distances, $d \sin \theta$, will be equal to half a wave length, $\lambda/2$. There will then be destructive interference such that when a crest is arriving from one source, there is a trough from the other. The general condition for destructive interference is

$$d \sin \theta = \frac{\lambda}{2}, \ \frac{3\lambda}{2}, \ldots, (2n+1)\frac{\lambda}{2}, \ (35\text{-}5)$$

where n is any integer, 0, 1, 2, 3, etc. From this discussion it follows that there will be a central maximum at H, followed on each side by a place of destructive interference, corresponding to a path difference of $\lambda/2$, then a place of constructive interference of path difference λ. The positions of destructive and constructive interference alternate as the position P moves outward from the center H.

Now what Young observed in his experiment was a central bright maximum followed by dark and bright bands of light. These corresponded

exactly with what the wave theory predicted, so that he concluded that light is propagated as a wave motion, and from the positions of the bright and dark fringes he was able to measure the wave length of the light used. This will be shown in the following example.

EXAMPLE 1: Two slits placed 0.5 mm apart are illuminated with monochromatic light so that interference fringes are observed on a screen 1 m distant. The center of the first bright fringe is found to be 1.16 mm from the central maximum. The problem is to find the wave length of the light.

Solution: From Fig. 35-7 it is seen that the triangles POH and CBG are similar; hence

$$\frac{PH}{D} = \frac{CG}{d} .$$

For the first bright fringe, equation (35-4) gives, for $n = 1$,

$$CG = d \sin \theta = \lambda$$

or

$$\frac{PH}{D} = \frac{\lambda}{d} .$$

Now $PH = 1.16$ mm, $d = 0.5$ mm, $D = 1$ m, so the wave length λ of the light in centimeters is

$$\lambda = \frac{d \times PH}{D} = \frac{0.05 \times 0.116}{100} = 5.8 \times 10^{-5} \text{ cm}$$

$$= 5800 \times 10^{-8} \text{ cm} = 5800 \text{A} .$$

The light in this example would be yellow.

The brightness of the fringes decreases from the central maximum out, so that the distinction between the light and dark fringes becomes less and less. An interpretation of this phenomenon can be given in terms of the Huygens wavelets. These wavelets are issuing from the slits B and C, and it is assumed that the amplitude of the wavelets decreases from the center outward, so that at 180° or backward the amplitude is zero. Thus we should expect the central maximum at H to be the brightest, the first bright fringe to be brighter than the second, and this, in turn, brighter than the third, and so on. This is in agreement with the experiments.

Fresnel achieved the same result as that of Young by two very brilliant experiments which produced much more intense illumination of their optical patterns. In the first, he used two plane mirrors set at a very slight angle to each other

(Fig. 35-8 [*a*]), each of which reflected, at nearly grazing incidence, the light emanating from a single slit, set accurately parallel to the line of intersection of the mirrors. Interference fringes appeared across the overlapping areas of the two beams. Thus he showed that light from the two virtual images of a single source maintained its wave aspects unimpaired. In his second experi-

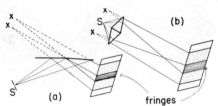

Fig. 35-8.—Fresnel's mirrors and biprism

ment (Fig. 35-8 [*b*]), he used two prisms of very small angle, set base to base, producing thereby two virtual images by refraction, and obtained identical results even more brilliantly illuminated.

35.7. Newton's Rings

Isaac Newton gave his adherence to a corpuscular theory of light partly because of his failure, as he thought, to find any experimental evidence of interference phenomena in optics. He failed to interpret as interference the beautiful colored rings he discovered when light was reflected from the interfaces of a convex surface resting on a plane glass surface; and yet this is exactly what the phenomenon demonstrated. The colored surface of oil on a wet pavement is another example of interference. Let us now consider Newton's rings, produced by a beam of monochromatic light of wave length λ incident perpendicularly on a

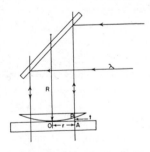

Fig. 35-9.—Newton's rings

plano-convex surface of radius R resting with its convex surface on a plane sheet of glass (Fig. 35-9). Suppose a parallel beam of light from the

right is incident on a semireflecting mirror set at 45°, so that a beam is incident perpendicularly on the glass surface. At the glass surface some of the light is reflected and some transmitted. What we are interested in is the light that is partially reflected and transmitted in the region between the convex and the plane glass regions.

Now Newton's rings are centered about the point of contact O of the convex and plane surfaces. These may be viewed in reflected light by looking from above or by transmitted light by looking from beneath. Near the center O there is a small region in which the separation, t, between the convex and plane surfaces is a small fraction of a wave length and can be considered to be zero. The reflected beam consists of light reflected at points, such as P, interfering with light reflected from the plane glass surface, as at A. The path difference in air of these two rays is $2PA$ or $2t$. If we are looking at the reflected beam close to the center O, where the path difference is practically zero, we might expect the two beams to interfere constructively and produce a bright spot at the center. Now this is not observed but, instead, a black or dark spot, indicating destructive interference. On the other hand, the transmitted beam shows a bright central spot. This apparent discrepancy is satisfactorily explained in terms of a change in phase of half a wave length when light is reflected in proceeding from a region of low refractive index to one of higher refractive index. In the electromagnetic theory this consists in a reversal of the electric vector, that is, a change of $\lambda/2$ between the incident and the reflected rays. No such change takes place when light is reflected at a glass-air interface, that is, when the light is proceeding from a region of high refractive index to one in which it is lower.

Thus for the interference pattern observed in reflected light there is a phase change of $\lambda/2$ at A and none at P in Fig. 35-9. The *condition for constructive interference or brightness* at some point P *in reflected light* is

$$2t + \frac{\lambda}{2} = \lambda, \ 2\lambda, \ 3\lambda, \ \text{etc.},$$

or

$$t = \frac{\lambda}{4}, \frac{3\lambda}{4}, \ \ldots, \ (n + \tfrac{1}{2})\frac{\lambda}{2}, \ (35\text{-}6)$$

where n is any integer, 0, 1, 2, etc.

To relate this distance t to the radius r of the **515** interference ring, we have, from the geometry relating a length of a chord to its tangent,

$$(2R - t)t = r^2.$$

In this experiment t is small compared to R, the radius of the convex surface, so that, very approximately,

$$2Rt = r^2 \quad \text{or} \quad t = \frac{r^2}{2R}. \tag{35-7}$$

Hence the radius r of the nth bright ring seen in reflected light is given by

$$\frac{r^2}{2R} = (n + \tfrac{1}{2})\frac{\lambda}{2}. \tag{35-8}$$

If r is measured with a microscope and R with a spherometer, then the wave length λ can be calculated. The results are in agreement with those obtained by other methods. If monochromatic light is used, the bright rings are of the color of the light used and the dark rings are black, or there is no light. On the other hand, if white light is used, there are a number of wave lengths, each undergoing interference, and a gradual merging of one color into another. For separations t of several wave lengths the pattern becomes blurred, for it is possible, for example, for t to be equal to three long wave lengths and four shorter ones. At such a point two waves of different wave lengths or colors are producing interference, so that the resultant does not have a definite position.

Interference fringes are used extensively in industry for testing for optical flatness. The glass plate labeled "Unknown" in Fig. 35-10 is first

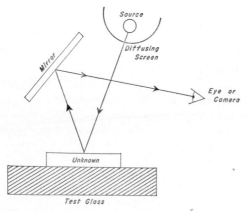

Fig. 35-10.—Apparatus for testing optical flatness

fine-ground to mechanical standards of flatness and then placed on an optically flat glass surface.
516 Diffuse light from a mercury lamp strikes the

interface of the glass plates, and the fringe system produced is reflected to the mirror and then to the eye or camera. If the two surfaces of the interface are plane surfaces separated by a wedge of air, the interference fringes will exhibit no curvature, as shown in photograph 6 of Fig. 35-11. On the other

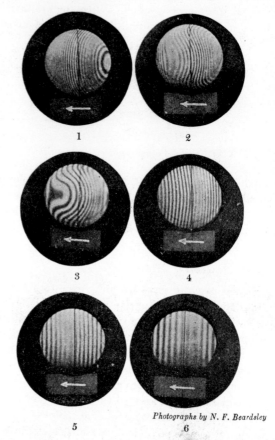

Photographs by N. F. Beardsley

Fig. 35-11.—Photographs of fringes between an optical flat and a glass plate in process of being made optically flat.

hand, if at the interface the unknown has portions of its surface concave ("valley") or convex ("hill"), the fringe system will be curved, as shown by photographs *1–5*. It is not possible to determine whether a hill or a valley is present by studying the photographs alone. Additional information is necessary. If the eye (Fig. 35-10) is moved downward, so that the angle of reflection of the light from the interface is increased, the fringe system will move. If the fringe system moves *away* from its center of curvature, *a hill* is present at the interface; if the fringe system moves *toward* the center of a curvature, *a valley* is present.

In the series of photographs (Fig. 35-11) the white arrow indicates the direction of movement of the fringe system as the angle of reflection of light from the interface is increased. Thus, in *1*, the fringes, or "contour lines," surround a hill on the right of the surface. By counting the number of fringes that cross the vertical black line between one edge and the center of the glass surface, it is possible to calculate the height of the hill. In *1*, the height of the hill is 5 fringes, or 5 half wave lengths of the blue light of the source. In *2*, this hill has been overcorrected by polishing and made into a valley about 1.7 fringes deep. The fringes are reversed in curvature in the center, but the edges are still high. In *3*, the valley has been widened but, of necessity, deepened at the same time to a maximum of $3\frac{1}{2}$ fringes. Now, however, the edges are lower. In *4*, the entire surface is of nearly uniform concave curvature, the valley being $1\frac{1}{2}$ fringes deep. In *5*, the valley has been reduced to $\frac{1}{3}$ of a fringe, and in *6*, the finished surface is plane to ± 0.03 of 1 fringe. One fringe is roughly equivalent to $\frac{1}{100,000}$ of an inch, so that the surface is plane to within 3×10^{-7} in.

35.8. Michelson's Interferometer

A. A. Michelson, in 1882, published an account of his INTERFERENTIAL REFRACTOMETER, later abbreviated into the name INTERFEROMETER, which was to serve as the model for most of the later forms of similarly named devices destined to establish a new precision of measurement in a host of physical problems where these tiny waves of light might be used as measuring rods. Instead of being limited to narrow pencils of light and small apertures, the Michelson arrangement uses broad illuminated surfaces and parallel light if desired (see Fig. 35-12). An extended wave front, *W*,

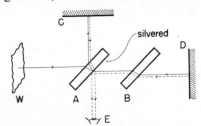

FIG. 35-12.—Diagram of Michelson interferometer

which may be plane, convergent, divergent, or even diffusely scattered in direction, impinges on a "dividing surface," a plane-parallel block of glass *half-silvered* on the back. After refraction, half the beam is reflected up to the plane mirror, *C*.

The other half of the beam, transmitted through *A*, passes through a "compensator," *B*—a block cut from the same piece as *A* and hence identical in thickness and set parallel to *A*. The beam then proceeds to the plane mirror, *D*.

After reflection from the mirrors, *C* and *D*, the beams return, via the dotted lines, as nearly parallel to their original directions as may be desired, by proper tilting of *C* and *D*. The first beam, refracted through *A*, now in part proceeds to the eye below. The second, refracted back through *B*, is in part reflected down to *E* by the first surface of *A* that it encounters, now the half-silvered one.

Now, if the distances of *C* and *D* are properly set with respect to the dividing back surface of *A*, both beams may be made to traverse *exactly* the same optical path not only in the air between the mirrors *but also in the glass of* A *and* B. (Each beam will be seen to have traversed the same distance through glass three times.)

If either mirror *C* or *D* is movable parallel to the direction of the incident light upon it, by the use of a finely made screw and suitable guides it is seen that we can arbitrarily impose *any* path difference *desired* from precisely zero upward through any number of wave lengths. Thus the entire surface of the mirrors, wherein one looks at two superimposed virtual images of *one and the same incident wave front*, i.e., that split by the back of *A*, may be made alternately light and dark; or, if the mirrors *C* and *D* are slightly inclined, we have the results of Fresnel's and Young's experiments, but of any brilliancy desired. Phenomena identical with Newton's rings may also be reproduced.

Space does not permit treatment of the numerous applications of interference made by Michelson and his followers to precision measurements in astronomy, physics, and other physical science. We can but list a few of them.

1. International standards of length have been calibrated in terms of light waves.

2. When the "domains" or regions containing about 10^{15} atoms that give iron its magnetic properties are lined up in an external magnetic field, the length of the specimen of iron is altered by an extremely small amount, but is clearly recorded by means of an interferometer, one of whose mirrors is attached to the bar. The phenomenon is called MAGNETOSTRICTION.

3. By interference methods one can get a degree of "resolving power"* far greater than that of any tele-

* The technical aspects of resolving power are treated in § 35.18.

scope. A few of the nearer stars thus become not too remote for their angular dimensions to be determined, and so the actual diameters of stars are brought within the range of human measurement.

4. By interference methods, measurements with instruments may be made with precision almost a thousand fold greater than by any other means.

5. The Michelson-Morley experiment for determining the motion of the earth relative to the so-called "luminiferous ether." It was to measure this effect, as a matter of fact, that Michelson invented the interferometer. The experiment is conceded to have given a completely zero result by the vast majority of workers. It was on the assumption of this zero result that Einstein was led to enunciate his famous theory of relativity.

Thus interference methods have opened up great areas for future exploration in theoretical, as well as in experimental, science.

For a brief and more technical account and a complete bibliography of Michelson's work in optics, reference is made to his *Studies in Optics.**

The following illustrations show two applications of the interferometer.

Measurement of the thickness of a thin transparent film.—A very thin film of collodion was placed in the path of light of a Michelson interferometer in such a way that about half the beam of light had to pass through the film before striking one of the parallel plates, say *H* of Fig. 35-15. Since the index of refraction of collodion, $\mu = 1.18$, is different from that of air, the beam of light through the film is slightly retarded, and the parallel fringe system is shifted. The shift of the upper half of the fringe system to the right, relative to the lower half, is shown in Fig. 35-13.

Light through film

Light through air

Fig. 35-13.—Shift of interference fringes in passing through a collodion film.

Since a displacement of a complete fringe means a retardation of one wave length, λ, and since a film of thickness, d, and refractive index, μ, produces a retardation, $2d(\mu - 1)$, because the light passes twice through the film, it follows that a fractional displace-

* Chicago: University of Chicago Press, 1927.

ment, x, of a wave length is related to the film thickness by

$$x\lambda = 2d(\mu - 1) \qquad \text{or} \qquad d = \frac{x\lambda}{2(\mu - 1)}.$$

The displacement shown in Fig. 35-13 is about $\frac{1}{5}$ fringe, so that $d = 3 \times 10^{-5}$ cm. The wave length of the light used was 5461 A.

Measurement of small angles.—In Figs. 35-14 and 35-15 the mirrors, M_1 and M_2, of a Michelson inter-

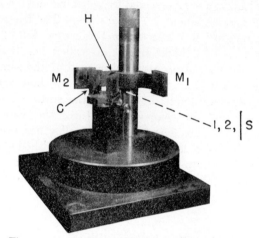

Fig. 35-14.—Photograph of a Michelson interferometer for testing the twist in a steel post.

ferometer are rigidly connected to a heavy steel post and base, as shown. The parallel plates, *C* and *H*, are attached to the heavy base alone. Twisting the post, relative to the base, shortens one path, *1*, (Fig. 35-15), and lengthens the other, *2*, causing the fringe to shift. These fringes may best be seen by sighting through the

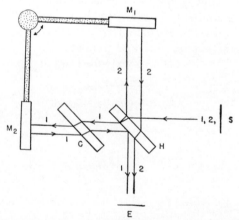

Fig. 35-15.—Diagram of Michelson interferometer for twisting of a steel post.

dividing surface, *H*. This optical arrangement is very sensitive; a slight twist of the post by the fingers suffices to produce a shift of many fringes.

35.9. Rectilinear Propagation and Its Failure

The wave-front treatment brings to view, out of an application of Huygens' construction, the fact that propagation of light, like that of any other kind of waves, is not strictly rectilinear when it passes by the edges of obstacles or when it passes through a narrow opening between the opposed edges of two obstacles, i.e., through a slit. Under such circumstances we encounter DIFFRACTION phenomena, the existence of which the ray treatment of propagation would never lead us to suspect. Huygens' construction, discussed in § 35.2, and here illustrated in Fig. 35-16, is applied

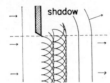

FIG. 35-16.—Bending of wavelets into geometrical shadow

to a plane wave passing the edge of a screen. The envelope of the secondary waves does not cease abruptly at the geometrical boundary of its shadow but hooks around into the shaded region, swiftly diminishing in intensity, of course. This accords with the facts, which show not only that, in reality, there is a gradual falling-off in intensity

within the geometrical shadow but that bands of *unequal illumination* result in the *illuminated region close by*, from the interference by the secondary waves that originate from points immediately nearest to the obstacle on the incident wave (Fig. 35-17). With a very small distant point

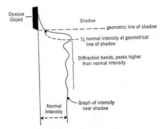

FIG. 35-17.—Intensities in a diffraction pattern at a straight edge.

of light and a sharply defined obstacle, like a razor blade, these DIFFRACTION BANDS, near the edge of a shadow, may be seen and readily photographed (Fig. 35-18). Only the small scale of this phenomenon in light and the need for a source of extremely small angular dimensions, when viewed from the obstacle, prevented Newton from finding it.

The curving of the wave fronts into the shadow is clearly visible in waves on water and other liquid surfaces. Diffraction is a universal wave

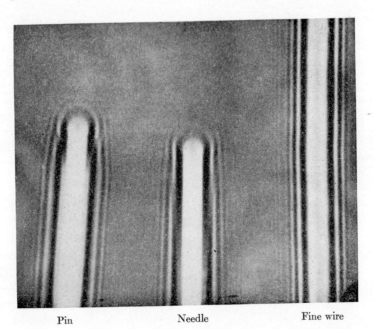

Pin   Needle   Fine wire

FIG. 35-18.—Enlargement of a photographic negative obtained by placing the object in front of the plate and illuminating it with a point source of light several meters distant.

phenomenon; indeed, like interference (which it is, from the secondary waves), it is a crucial test of wave characteristics.

35.10. Diffraction at a Straight Edge; Fresnel Zones

The French physicist Fresnel attributed the diffraction phenomena to the mutual interference of the Huygens secondary wavelets which diverge from a wave front. Diffraction occurs when a portion of a wave front is removed by an obstacle. Each portion of a primary wave is considered as the source of secondary waves, and Fresnel expressed the resultant amplitude of these wavelets at any point by means of two integrals within limits determined by the particular problem. The problem of diffraction can thus be solved by the principle of Huygens combined with the principle of interference. We shall now qualitatively apply these principles to the diffraction of a plane wave at a straight edge (Fig. 35-19).

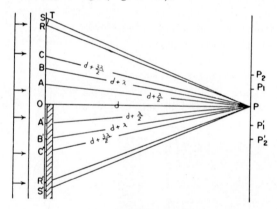

Fig. 35-19.—Fresnel zones for diffraction at a straight edge

Let us consider the wavelets from the plane wave front SS' at some point P, distant d from O, the edge of the obstacle. For the present we shall apply Fresnel's method to the whole wave front, SS', considering the obstacle removed. Fresnel's method consists in breaking up the wave front into a series of zones, OA, AB, BC, . . . , RS, etc., such that

$$PA - PO = PA' - PO = \frac{\lambda}{2},$$

$$PB - PA = PB' - PA' = \frac{\lambda}{2},$$

$$. \quad . \quad . \quad . \quad . \quad . \quad . \quad . \quad . \quad . \quad ,$$

$$PS - PR = PS' - PR' = \frac{\lambda}{2},$$

or the distance from the edges of the zones increases by $\lambda/2$ for each adjacent zone as measured from P.

Consider any zone RS, and from R project a perpendicular RT on PS; then, from the construction, $ST = \lambda/2$ and the triangles RST and POS are similar, so that

$$\frac{RS}{ST} = \frac{PS}{OS} \quad \text{or} \quad \frac{RS}{PS} = \frac{\lambda}{2\,(OS)}.$$

Neglecting the effect of obliquity, the amplitude of the vibration contributed by the zone RS is proportional to RS/PS, which is equal to $\lambda/2$ (OS). Hence, if the wave front is divided into Fresnel zones, or half-period elements of uniformly increasing phases, their effects at the point P decrease inversely as the distance of the zone from the pole O of the wave front. If obliquity is taken into account, this decrease in amplitude at P is still more rapid. Thus if a_1 is the amplitude at P due to the wavelets in the first half-zone OA, a_2 that due to the second half-zone AB, a_3 that due to the third half-zone, etc., then the total amplitude a at P due to the two halves of the wave front can be written as

$$\left.\begin{aligned} a &= 2\,(a_1 - a_2 + a_3 - a_4 + \ldots) \\ &= 2\left\{\frac{a_1}{2} + \left(\frac{a_1}{2} - a_2 + \frac{a_3}{2}\right)\right. \\ &\qquad \left. + \left(\frac{a_3}{2} - a_4 + \frac{a_5}{2}\right) \ldots\right\}. \end{aligned}\right\} \quad (35\text{-}9)$$

(The amplitude a_2 is opposite to that of a_1 at P, since the zones OA and AB are $\frac{1}{2}$ wave length difference in distance from P.) The three terms in the parentheses add to practically zero, since a_1 is larger than a_2 and a_2 larger than a_3. Thus, since the amplitudes decrease rapidly with the distance from the pole O, it follows that the resultant amplitude a of the whole wave front at P is

$$a = a_1. \quad (35\text{-}10)$$

That is, the effect due to the whole wave is equivalent to that from the half of the first or central zone OA whose contribution to the amplitude is a_1. In other words, if one eliminates the whole wave front, leaving only the first half-zone OA, the effect at P is the same as that from the whole wave front. If the even half-zones having amplitudes a_2, a_4, etc., at P are blocked out, the resultant amplitude at P would be

$$a = 2\,(a_1 + a_3 + a_5 + \ldots),$$

which is many times the resultant amplitude from the whole wave. Such partially obscured plates have been made and are called *zone plates*, and from what has been said they focus parallel light just as a lens does.

Let us now return to the problem of the plane wave and the straight-edge obstruction (Fig. 35-19). Consider, first, the point P at the geometrical edge of the shadow. Only the wavelets from the upper portion, OS, contribute, and their resultant amplitude, from equation (35-10), is $a_1/2$. Next consider the point P_1', within the region of the geometrical shadow by a distance OA', the width of the first half-zone. To obtain the resultant amplitude at P_1', we should construct another diagram similar to that in Fig. 35-19 with its center now at P_1'. This is equivalent to moving the straight edge up to A in Fig. 35-19, so that, instead of drawing a series of new diagrams, we shall use the one already drawn and consider the straight edge to be moved. With the straight edge at A, the resultant amplitude a_1' is that due to the second, third, etc., half-zones in the upper region, so that

$$
\begin{aligned}
_1' &= a_2 - a_3 + a_4 - a_5 + \ldots - a_n \\
&= \frac{a_2}{2} + \left(\frac{a_2}{2} - a_3 + \frac{a_4}{2}\right) \\
&\quad + \left(\frac{a_4}{2} - a_5 + \frac{a_6}{2}\right) + \ldots \\
&= \frac{a_2}{2}.
\end{aligned} \qquad (35\text{-}11)
$$

Similarly, if the screen is moved up to B so as to cover the first two half-zones, the amplitude a_2' at P_2', two half-zones below the geometrical shadow, is

$$
_2' = \frac{a_3}{2}.
$$

Since the intensities are proportional to the square of the amplitudes and since the amplitudes a_1, $a_2, \ldots$, decrease relatively rapidly, it follows that the light intensity drops off rapidly within the geometrical shadow. The light within the geometrical shadow shows no fluctuations in intensity but decreases rapidly with the distance into the shadow. This is not the case for positions above the geometrical shadow.

In order to find the intensity at P_1, the first half-zone distance, OA, above the edge of the shadow, consider the straight edge moved down from O to A'. Then the whole of the upper half of the wave $(a_1/2)$ and the first lower half-zone a_1 contribute to the amplitude. The resultant amplitude a_1 at P_1 is

$$
a_1 = \frac{a_1}{2} + a_1 = \frac{3a_1}{2}.
$$

Since intensities are proportional to the square of the amplitude of a wave, it follows that the ratio of intensities at P' and P (the geometrical edge of the shadow) is

$$
\left(\frac{a_1}{a}\right)^2 = \left(\frac{3a_1}{2}\right)^2 \left(\frac{2}{a_1}\right)^2 = 9. \qquad (35\text{-}12)
$$

At point P_2, two half-zones above P_1, the amplitude a_2 is that due to the upper half $(a_1/2)$ plus the first two lower half-zones $(a_1 - a_2)$. Hence

$$
a_2 = \frac{a_1}{2} + a_1 - a_2 = \frac{3a_1}{2} - a_2.
$$

The amplitude a_2 is smaller than a_1, so that the amplitude at P_2 is larger than that at the geometrical edge of the shadow but is less than that at P_1, where only one half-zone is exposed. Proceeding in this manner, it follows that, as one moves away from P, the edge of the geometrical shadow exposing half-zones of smaller but alternating amplitudes, the intensity of the light shows fluctuations which decrease in magnitude, the farther one moves from P. This is shown in Fig. 35-17.

35.11. Fresnel and Fraunhofer Diffraction

There are two general classes into which the study of diffraction phenomena may be divided. In the first, called the "Fresnel type of diffraction," the source of light is at a finite distance from the diffracting obstacle, and the resulting diffraction pattern is studied on a screen near the diffracting obstacle. The diffraction phenomena illustrated in Figs. 35-16 and 35-17 are of this type.

In the second class, called "Fraunhofer diffraction," the oncoming plane wave front has been collimated by a lens. Another lens is placed beyond the diffracting obstacle, and the resulting diffraction pattern is examined on a screen placed in the focal plane of the lens.

We shall discuss in the next two sections the Fraunhofer diffraction patterns due to a single and double slit.

35.12. Fraunhofer Diffraction by a Single Slit

In Fig. 35-20 the plane wave front of mono-chromatic light, W_1, previously collimated by a lens system, not shown, is diffracted by the slit opening, AC. Consider the bundle of parallel rays diffracted through the angle θ, as shown in the

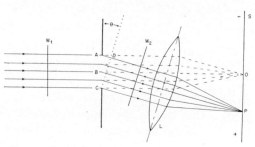

FIG. 35-20.—Fraunhofer diffraction by a single slit

figure. The corresponding diffracted wave front is W_2. The problem to be solved is the calculation of the intensity distribution of the light along the screen, S.

The parallel bundle of rays diffracted through the angle θ will be brought to a focus, P, in the focal plane of the lens, L. The line CD, perpendicular to the diffracted rays, represents the diffracted wave front. If the angle θ is such that the distance AD is one wave length, then the parallel bundles of rays crossing at P will annul each other. The reason for this is clear if we realize that a wavelet leaving from B, midway between A and C, and one leaving A have a path difference of $\lambda/2$. When these two wavelets reach P, they annul each other's contribution to the intensity. In a similar way we may pair off all wavelets between A and B with wavelets between B and C, each pair having a path difference of $\lambda/2$. By an extension of this argument, we see that if AD is equal to $n\lambda$, we obtain annulment of the parallel rays when they cross in the focal plane of the lens.

If we represent the slit-width AC by the letter b, then the path difference AD is given by

$$AD = b \sin \theta \; ;$$

and, if $AD = n\lambda$, where $n = 1, 2, \ldots$, the condition for *dark fringes* is

$$b \sin \theta = n\lambda \; . \qquad (35\text{-}13)$$

The rays of the bundle corresponding to $\theta = 0$ and shown by dotted lines in Fig. 35-20 have zero path difference and hence, when brought to a focus at O, give rise to a bright image of the slit. For angles between $\theta = 0$ and $\theta = \sin^{-1}(\lambda/b)$, the intensity of the light decreases from a maximum to zero. The intensity pattern due to the single slit and viewed on the screen S is shown (but not to scale) in Fig. 35-21.

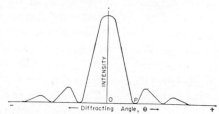

FIG. 35-21.—Intensities in a diffraction pattern from a single slit.

The central maximum is called the *zero order of diffraction*. The bright fringe between $n = 1$ and $n = 2$, as given by equation (35-13), is the first order of diffraction. It will be noticed that the intensity of the central maximum is much greater than the maxima on either side of it.

EXAMPLE 2: Light of wave length $\lambda = 4 \times 10^{-5}$ cm strikes a slit of width $\frac{1}{4}$ mm in the manner shown in Fig. 35-20. Calculate the angular spread of the light in the central maximum.

Solution: The half-width of the central image is given by

$$\sin \theta_1 = \frac{4 \times 10^{-5}}{\frac{1}{4} \times 10^{-1}} = 16 \times 10^{-4} \; ,$$

$$\theta_1 = 5'.5 \; ,$$

$$2\,\theta_1 = 11' \text{ (angular width of central maximum)} \; .$$

If the slit-width is $\frac{1}{400}$ mm, then

$$\sin \theta_2 = \frac{4 \times 10^{-4}}{\frac{1}{400}} = 16 \times 10^{-2} \; ,$$

$$\theta_2 = 9^\circ.2 \; ,$$

$$2\,\theta_2 = 18^\circ.4 \text{ (angular width of central maximum)} \; .$$

We see from these calculations that when the aperture is many times the wave length of light, the angular half-width, θ_1, of the central image is small. The other orders of diffraction would hardly be seen, and hence quite a sharp image of the aperture is formed by the lens. On the other hand, when the aperture is very small, θ increases, so that the central band spreads over quite a large area of the screen.

35.13. Fraunhofer Diffraction from Two Slits

Let us next consider the nature of the Fraunhofer diffraction that results when a plane wave front is interrupted by two slits of equal aperture separated by an opaque portion equal to twice the width of either aperture. The optical arrangement is shown in Fig. 35-22.

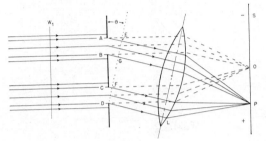

FIG. 35-22.—Fraunhofer diffraction by two slits

The discussion of the previous section is applicable to *each slit separately;* i.e., a diffraction pattern such as is shown in Fig. 35-21 results from the parallel bundle of rays emerging from each slit. The lens, L, intercepts these two bundles of rays and brings them to a focus in the focal plane of the lens, as shown. We have, then, in addition to the diffraction by each slit, interference phenomena produced by the two bundles of rays.

The condition that the point P shall be a region of darkness is that the path difference between rays leaving AB and CD shall be $\lambda/2$. The reader will observe that the path difference between rays leaving D and B is the same as the path difference of rays leaving C and A. If, therefore, BG is equal to $\lambda/2$ or *any odd number of half wave lengths,* the point P will lie in a region of darkness; i.e.,

$$BG = (2n + 1)\frac{\lambda}{2},$$

where n is equal to 0, 1, 2, 3, ... We may rewrite this equation in a more convenient form if we let $BC = b$ and $CD = a$ and note that $BG = (a + b) \sin \theta$; then

$$(a + b) \sin \theta = (2n + 1)\frac{\lambda}{2} \qquad (35\text{-}14)$$

(condition for a dark fringe).

The condition that P shall be in a bright band

is that BG shall equal an even number of half wave lengths. Thus

$$(a + b) \sin \theta = 2n \frac{\lambda}{2} \qquad (35\text{-}15)$$

(condition for a bright fringe).

When $n = 0$, the path difference is zero, and we obtain the central bright fringe. For $n = 1$, we obtain the second bright fringe, etc.

The distribution of intensity for a double slit with $b = 2a$ is shown in Fig. 35-23. The dotted

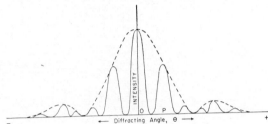

FIG. 35-23.—Intensity pattern from two slits

envelope is the diffraction pattern of a single slit. The lines within the envelope are the interference fringes.

Photographs, taken with monochromatic light, of Fraunhofer diffraction patterns produced by a number of equidistant and identical slit apertures are shown in Fig. 35-24, together with the corresponding intensity graphs. The intensity pattern of the single slit, showing clearly the central maximum and an additional one on each side of this central maximum, is portrayed in Fig. 35-24. In the sequence of photographs the number of slits is increased, the spacing between slits being such that the opaque portion is three times the width of the aperture. This ratio of slit-width to distance between slits determines the number of *principal maxima* touching the enveloping single-slit diffraction pattern (*broken curves*). For the slit arrangement shown here, there are three principal maxima—the first, second, and third to the right and left of the central maximum, respectively.

As the number of slits is increased, we observe from the photographs and intensity graphs that (1) *subsidiary maxima* begin to appear, their intensities being very small compared to the principal maxima; (2) these subsidiary maxima progressively decrease in intensity; (3) the principal maxima become *narrower* until finally, for many thousands of narrow slits (diffraction grating), the principal maxima become very sharp lines—spectral lines.

523

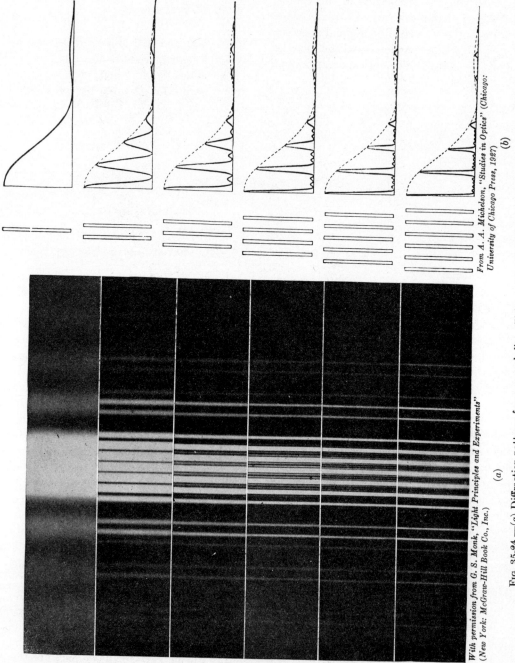

With permission from G. S. Monk, "Light Principles and Experiments" (New York: McGraw-Hill Book Co., Inc.)

(a)

From A. A. Michelson, "Studies in Optics" (Chicago: University of Chicago Press, 1927)

(b)

Fig. 35-24.—(a) Diffraction patterns from several slits; (b) intensities in diffraction patterns from several slits

If, instead of monochromatic light, the incident light contained many wave lengths, the diffraction patterns of these slits would consist of a superposition of a number of differently spaced patterns of the type illustrated here. Each wave length would have its principal maxima. The collection of _first principal maxima_ of the different wave lengths is called the _first-order spectrum_ of the incident light; the collection of second principal maxima, the _second-order spectrum_, and so on. As we have seen, the greater the number of slits, the sharper are the maxima. Thus, if two wave lengths of nearly the same value are in the incident light, the ability to resolve them as separate wave lengths is greatly increased by the use of many slits (§ 35.16).

35.14. Spectra Produced by Diffraction

We have seen previously that, if heterochromatic or white light is diffracted by a pair of slits, the diffraction pattern is colored, since the maxima and minima of different wave lengths fall at different positions in each order of the diffracted beams. Since with but two slits the regions of maxima and minima are rather broad, there is much overlapping of colors, and the spectrum so produced is so impure that it really should hardly be called a spectrum at all.

If large numbers of slits or other types of diffracting elements are used, the purity of the color at any given region may be made very great. Such arrangements are called DIFFRACTION GRATINGS. Their production has been, for decades, one of the most difficult and important techniques in the field of experimental optics. To discuss the ruling engines that have been developed to produce such gratings will take us too far afield.

Diffraction gratings are of two types: _transmission_ and _reflection_ gratings. It is the latter type that is usually produced by the ruling engines. The former type may be prepared from the latter by making a thin cast from one of the latter by some transparent medium. Both types consist of spaces of clear, transparent (or reflecting) surfaces, equivalent to slits, alternating with opaque (or inclined) surfaces, which prevent direct transmission (or reflection) of light and are equivalent to opaque material in which slits may be cut.

Ideal gratings would have indefinitely narrow slits, absolutely uniformly spaced at a constant distance—say _d_—apart. In actual gratings this constancy of spacing is excellent, this being the major task of the ruling engine. The slit spaces must obviously be of appreciable width—say _b_—to transmit or reflect any light at all, and such widths of slits must be quite as carefully constant as the distances between them are. We can here discuss only the theory of rather ideal gratings, but we will indicate how practical limitations modify the diffraction patterns resulting.

35.15. Theory of Plane Diffraction Gratings

Let Fig. 35-25 represent an enlarged view of a small portion of a transmission grating on which a parallel beam of monochromatic light of wave length λ is incident in a normal direction. Suppose

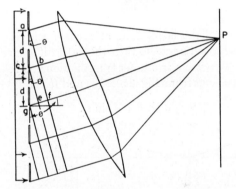

FIG. 35-25.—Diffraction by a plane transmission grating

d is the grating space, that is, the length of the bar and the slit. These slits must be narrow enough that the diffracted beam from one of them can interfere with the diffracted beams of others. Consider the diffracted wave fronts at an angle θ with the grating. These are represented in the figure as abf and ce. Between each grating space there is a path difference to this wave front of $cb = ge = ef = d \sin \theta$.

These wave fronts are then brought to a focus at P by the convex lens. If the point is to be a bright diffracted image of the light of wave length λ, then the necessary condition is that the path difference $d \sin \theta$ between the successive wave fronts must be an integral number of wave lengths or:

For brightness: $\qquad d \sin \theta = n\lambda \,,$ $\qquad$ (35-16) $\quad$ **525** where n is any integer, 1, 2, . . . , etc., called the *order of the spectrum*. When n is equal to zero, then the angle θ is zero, which corresponds to the un-diffracted central image.

If the angle θ corresponds to a dark spot, so that there is destructive interference in the wave front, the path difference between the successive wave fronts must be an odd number of half wave lengths or:

For darkness: $\quad d \sin \theta = (2n+1)\dfrac{\lambda}{2}.\quad$ (35-17)

Thus there will be successive dark and bright diffracted images as the angle θ is increased from zero upward. Following the central image is a dark spot at angle θ_1, given by

$$d \sin \theta_1 = \frac{\lambda}{2};$$

then a bright spot at angle θ_2, given by

$$d \sin \theta_2 = \lambda \; ;$$

and so on.

If the light is incident at angle i and is diffracted at an angle θ, then the path difference of successive wave fronts for transmission and reflection gratings from Fig. 35-26 (a) and (b) is

$$d \sin \theta + d \sin \imath \; .$$

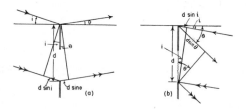

FIG. 35-26.—(a) A transmission grating; (b) a reflection grating.

If the incident and diffracted beams are on the opposite sides of the normal, then the path difference is

$$d \sin \theta - d \sin \imath \; .$$

Thus, for a bright diffracted image in a direction θ, the condition is

$$d \sin \theta \pm d \sin \imath = n\lambda$$

or

$$\sin \theta = \frac{n\lambda}{d} \mp \sin \imath . \quad (35\text{-}18)$$

We have here assumed a parallel incident beam and a parallel diffracted beam, so that lenses must be used. Frequently the reflection gratings are ruled on concave metal surfaces so that the focusing can be done without the use of lenses.

35.16. Dispersion and Resolving Power of Gratings

By the DISPERSION or *dispersive power*, D, of a grating we mean the rate of change of angular position of the diffracted light in any order with the change of wave length of the light, i.e., $D = \Delta\theta/\Delta\lambda$. Since

$$n\lambda = d \sin \theta_n , \quad (35\text{-}19)$$

for small changes (by differentiation),

$$n\Delta\lambda = d \cos \theta \Delta \theta , \quad (35\text{-}20)$$

so that the dispersion becomes

$$D = \frac{\Delta \theta}{\Delta \lambda} = \frac{n}{d \cos \theta}. \quad (35\text{-}21)$$

Thus the openness of the spectrum increases with the order and also as the grating space is smaller. The effect of cosine θ complicates the relation. For $\lambda = 6000 \times 10^{-8}$ cm and $d = 0.0002$ cm, $\cos \theta_3 = 0.436$. The dispersion at this region in the third order is 6.9 times that of the first order.

Another feature of equation (35-21) is of great importance. If we tilt the grating with reference to the incident light so that the diffracted light comes off as close to the normal to the grating as possible, cosine θ is unity or very near it. Thus the dispersion is constant, n/d, and we have so-called NORMAL SPECTRA, where measurements of distances between diffracted images are directly proportional to the wave lengths of the light responsible for those images. This is a reason for the use of gratings, rather than of prisms, for quantitative measures of spectra.

By the *limit of resolution of two different wave lengths* in a diffracted spectrum we mean: *How close in wave length* may two different radiations be, so that their diffraction patterns may be sufficiently separated in the spectrum to be recognized as distinct? Fig. 35-27 illustrates the problem diagrammatically.

Grating AG of m apertures and of space d diffracts light of wave length λ into the nth order at point p. The maximum of the pattern at p with its adjacent minima is shown. Here p is a *maximum corresponding to* $mn\lambda$, the difference in path.

We will assume that the incident light also contains a wave length, $\lambda + \Delta\lambda$, which is slightly different but just resolvable from the given one; the maximum of its diffraction pattern must fall at the point q, *exactly at the location of the minimum of the pattern* due to λ. The maxima and minima of light, $\lambda + \Delta\lambda$, are shown by dotted lines. *Here* q *is a maximum corresponding to* $mn(\lambda + \Delta\lambda)$.

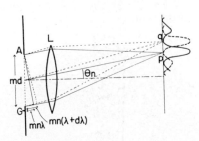

FIG. 35-27.—Resolving power of a grating

But now an entirely independent condition also determines the location of q as a minimum for pattern λ. If the difference in path to p ($mn\lambda$) is increased by *one wave length* (across the entire aperture of the grating), then light from every slit below the center will arrive in phase exactly opposite to light from the corresponding slits above the center and *cancel*, so that the place of arrival, q, will be totally dark. This, of course, is what creates the first minimum q on either side of p. *Here* q *is a minimum corresponding to* $mn\lambda + \lambda$.

Equating the last two equivalent and independent quantities,

$$mn(\lambda + \Delta\lambda) = mn\lambda + \lambda, \quad (35\text{-}22)$$

we obtain, for the limit of resolution,

$$\frac{\Delta\lambda}{\lambda} = \frac{1}{mn} \quad (35\text{-}23)$$

and the resolving power

$$R = \frac{\lambda}{\Delta\lambda} = mn. \quad (35\text{-}24)$$

Thus, for a grating to be able to separate the finest details of the light falling upon it, the product of the total number of apertures by the order of diffraction used must be as large as possible.

Usually this result is achieved by making the grating space very small, i.e., ruling its apertures as close together as possible and carrying this ruling over the widest possible area and then using as high an order (fourth to sixth) as possible. For

example, if we can rule 10,000 lines per centimeter for a distance of 25 cm, there would be a total, m, apertures of a quarter-million. If an order, n, as high as the sixth could be used, the resolving power—theoretically, at least—would be a million and a half.

Joseph Fraunhofer was the first to use gratings in the measurement of wave lengths. By modern standards these were very coarse—about 20 open spaces to the millimeter. The first fine diffraction gratings in the United States were ruled by Professor Henry A. Rowland at Johns Hopkins University in the early nineties of the past century. He ruled fine lines on glass and metallic mirror surfaces—up to as many as 15,000–20,000 per inch. In 1898 A. A. Michelson, at the then new University of Chicago, built his first experimental ruling engine. From then until the present time, there has been steady improvement and slow development of these instruments.

35.17. The Nature of Optical Images

As we have seen, diffraction patterns like those we have been describing cannot be rendered visible unless the envelopes of the interfering secondary waves are made convergent and the illumination (or lack of it) is focused. The insertion of a positive lens (or mirror), as shown in Figs. 35-20, 35-22, and 35-28, accomplishes

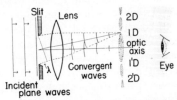

FIG. 35-28.—Diffraction pattern showing dark regions, D

this. Since we have been discussing a one-dimensional pattern, from slits whose length lies perpendicular to the paper, the lens shown might be supposed to be cylindrical rather than spherical. If, however, we limit the length of the slit to include only that actually in, or very close to, and including the plane of the page and *use a narrow slit*, then we are discussing the results of diffraction for a small radiant POINT SOURCE, and our diagrams represent sections of the intensity distribution of such a front surface in any plane that contains the optical axis. In other words, rotating (Fig. 35-28) about the optical axis produces the actual result—now two-dimensional. Seen by an eye directed toward the source and along the optical axis of radial symmetry, the diffraction pattern takes the form of a series of concentric circles of illumination, alternating with circular regions of darkness. In printing this, we have applied *ink* where light is concentrated, thus producing a negative impression.

This is shown in an attempt at perspective in Fig. 35-29 and is drawn on a larger scale for three different

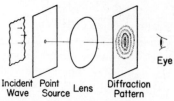

FIG. 35-29.—Diffraction pattern from point source with a lens.

sizes of circular apertures b in Fig. 35-30, where the plane of the paper is now perpendicular to the optical axis.

We note from equation (35-13) (since angles are usually small) that the angular scale of the pattern, $\theta = \sin \theta$, varies *inversely* as the size of the aperture, b

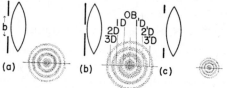

FIG. 35-30.—Diffraction patterns with different-sized openings, showing dark regions, D.

A very small aperture (Fig. 35-30 [*b*]) gives a large-scale pattern; a *very* large aperture (Fig. 35-30 [*c*]), a *very* small one. Of course, since all lenses and mirrors are themselves of finite size, *this itself limits the aperture*. Some lenses, notably in cameras, are provided with diaphragms so that their apertures may be changed at will, for reasons that we cannot discuss here.

MOST IMPORTANT is it to realize that *all optical images formed by lenses or mirrors are built up of a series of diffraction patterns coming from each point in the object.* No lenses or mirrors of finite aperture can ever produce perfect images, i.e., point-to-point correspondence, for this most fundamental reason. Figure 35-31 shows a photograph of this diffraction pattern.

The best illustrations are the cases of celestial objects—stars whose distances are so remote that they constitute for us practically the ideal POINT SOURCE of light. When we look at a star or any other very small source through a telescope, we look at the diffraction pattern only. Through a telescope with a pinhole diaphragm we should see, were it not too faint, a large pattern that would *not in any way suggest the star's appearance.* This pattern, of course, can be readily photographed. As the aperture or available lens size of the telescope increases in dimensions, the diffraction pattern

shrinks in extent and gains in intensity, so that, even with moderate sizes, the details of the pattern are entirely lost to sight, and we see an image which *appears to be a point of light*, like the star. It appears brighter, of course, since the lens gathers more light to our eye than the eye's small pupil would normally receive. Actually, this image is *not* a point, or even a tiny circle of light, but a series of circular diffraction FRINGES around a central smaller area generally illuminated all over, but brightest at the center.

Photograph with permission from G. S. Monk, "Light Principles and Experiments" (New York: McGraw-Hill Book Co., Inc.)

FIG. 35-31.—The circular diffraction pattern from a point source formed by a lens.

Contrary to common belief, *what we can see and separate as distinctly different objects* in telescopes, microscopes, field glasses, or, indeed, by our unaided vision depends entirely on the *size of the diffraction patterns* with which we have to deal and *not at all on the magnifying power** of the instrument.

Such information is vital for the intelligent use of optical tools and will be given in the next section under its technical name, "limit of resolution."

35.18. Limit of Resolution and Resolving Power

Suppose we are observing two sources, S_1 and S_2 (Fig. 35-32), with telescope (or microscope) and that their angular separation is a. The telescope (or microscope) lens is L, whose aperture (indi-

---

* This statement is true only on the usually legitimate assumption that the magnifying power of the telescope is sufficient to enlarge what it resolves beyond the limit of resolution of the eye. Otherwise the eye is unable to resolve what the telescope has already resolved.

cated by wide slit or diaphragm, S) is of width b. Central rays from S_1 are solid lines; from S_2, dotted; the optic axis is dash-dotted. The details of diffraction are omitted, but the intensity graphs are plotted as formerly. The intensity graph of S_1, the central maximum with but two adjacent maxima and four adjacent minima, is plotted as

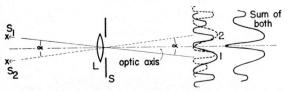

FIG. 35-32.—Diffraction patterns from two sources, S_1 and S_2.

solid lines; that from S_2, by dotted lines. The angular separation of the two sources is here sufficient to throw the maximum from S_2 somewhat beyond the first minimum on one side of S_1, so that the resultant intensity curve (*heavy solid*) has a sharp valley in it. This is visible as a fine dark band across the broad illumination of the central band from both sources. Just as soon as this band is wide enough to be visible, it is possible to distinguish that the light comes from *two* sources and not from one. Experience of the observer is a large factor in detecting the onset of this condition; so a more exact criterion must be made for the angular limit *below which* two sources may *not be* RESOLVED or separated.

This is done by stating arbitrarily that when the angular separation of the two sources, a, is such that the central maximum from one *falls exactly on the first minimum of the other*, the LIMIT OF RESOLUTION of the optical system has been attained. We recall that the first minimum from either source comes at an angle of diffraction θ_1, given by equation (35-13) as

$$\sin \theta_1 = \frac{\lambda}{b}. \qquad (35\text{-}25)$$

These angles are always very small, and no appreciable error results from writing

$$\theta_1 = \frac{\lambda}{b}.$$

Moreover, here

$$a \equiv \theta_1,$$

because the definition states that when a becomes identical with θ_1, the criterion is fulfilled for *defin-*

ing the limit of resolution. Hence the angular limit of resolution is

$$a = \frac{\lambda}{b}. \qquad * \qquad (35\text{-}26)$$

The RESOLVING POWER of an aperture is defined as the reciprocal expression,

$$R \equiv \frac{1}{a} = \frac{b}{\lambda}, \qquad (35\text{-}27)$$

that is, equals the number of "light waves contained in the aperture."

If white light, containing many wave lengths, is used, it is clear that a sort of average value for the wave length must be taken. This is usually chosen as the value of the yellow-green, to which the eye is most sensitive, i.e., $\lambda = 5550 \times 10^{-8}$ cm.

It is further to be noted that even the most experienced eyes cannot separate two sources as close together as demanded by the theoretical limit of resolution defined. Actually, observable limits are always above this value. Thus theoretical resolving powers are always somewhat higher than the values realizable in practice.

35.19. Photometry

Another important aspect of light is its intensity or brightness. Since light is a form of energy—radiant energy—*visible* light, as the name implies, is radiant energy to which the retina of the human eye is sensitive. We find experimentally that only a very small fraction of the radiant energy of an incandescent body is visible to the eye; that is to say, the eye is sensitive to but a small range of the wave lengths emitted by the lamp. If, therefore, we wish to compare the "intensities" of two sources of light, a rather serious practical problem arises. The physicist could readily determine the relative intensities of two sources by measuring the total number of watts or ergs per second emitted by the sources, using a thermopile. However, this result would be of little practical value, for we want to know the relative intensities of the sources as *emitters of visible light*. The human eye, therefore, must be the judge. The science of measuring and comparing light quantities is called *photometry*. We shall, in the following sections,

---

* Considerations quite beyond our scope show that for circular apertures, instead of slits, a factor 0.61 is also involved; i.e., $a = 0.61\lambda/b$.

introduce the reader to the vocabulary of this science.

Radiant flux is the flow of electromagnetic energy and can be measured with a thermopile or some other suitable temperature-measuring device. It is usually measured in watts or in ergs per second. Now equal amounts of monochromatic radiant flux of different wave lengths or colors do not evoke the same sensation of brightness in the human eye. For a large number of people, it has been found that the eye is most sensitive to the green color of wave length 5550 A. The relative luminosities for other colors are shown in Fig. 35-33. At about 6100 A, the relative luminosity is

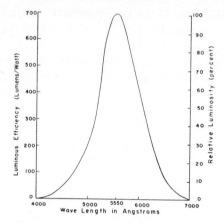

Fig. 35-33.—Relative luminosity curve

only 50 per cent. That is, if equal fluxes of radiant energy measured in watts are viewed for the two colors of wave lengths 5550 and 6100 A, then the latter appears only half as bright as the former. From this relative luminosity curve, it appears that the average human eye is sensitive to radiant energy between about 4000 and 7000 A. The limits of this visible spectrum are somewhat indefinite. It has been found that, with subdued lighting, the eye is more sensitive to shorter wave lengths than with normal lighting conditions. With the subdued lighting, the wave length of maximum luminosity moves from 5550 to close to 5000 A. We shall assume normal lighting conditions or that Fig. 35-33 is the correct luminosity curve.

The unit of *luminous flux* is called the *lumen*. At the peak of the relative luminosity curve 5550 A, it is found that *1 watt of radiant flux of the monochromatic radiation corresponds to 685 lumens of luminous flux.* Thus, for the normal eye, 1 lumen is equivalent to 1/685 watt (0.00146 watt)

of monochromatic light of green color of wave length 5550 A. If we choose some other color, for example, 6100 A, at which the luminous efficiency is only 50 per cent, then it would require 0.00292 (2 × 1/685) watt of radiant flux to produce 1 lumen of luminous flux of this color. Table 35-1

TABLE 35-1

RELATIVE RATINGS OF SOME COMMON SOURCES OF LIGHT

Source	Lumens per Watt	Total Flux in Watts ÷ Input in Watts (Per Cent)
Open-flame gas burner.............	0.22	0.35
Kerosene lamp....................	0.26	0.04
Acetylene lamp...................	0.67	0.11
Carbon-filament lamp.............	2.6	0.41
Nernst porcelain-filament lamp......	4.8	0.76
Tungsten vacuum lamp............	8.0	1.3
Carbon arc lamp, using 9.6 amp.....	11.8	1.9
Mazda C gas-filled, 500-watt lamp, 7 watts/candle..................	15.0	2.4
Open yellow-flame carbon arc, 10 amp d-c..........................	44.7	7.1

contains the lumen per watt ratings of different sources emitting light of many wave lengths. It is to be remembered that some of these wave lengths have no appreciable visual effect.

If the source emitting luminous flux is a *point source*, then the *luminous intensity* of the point source in a given direction is defined as the number of lumens per unit solid angle radiated in that direction. By *unit solid angle* we mean the angle subtended at the center of a sphere of radius r by a portion of its surface area equal to r^2. If the solid angle is $\Delta\omega$ steradians and if a luminous flux of amount ΔF lumens is emitted by a point source within the solid angle, then the luminous intensity, I, by our definition, is

$$I = \frac{\Delta F}{\Delta\omega} \text{ lumens/steradians or candles}. \quad (35\text{-}28)$$

The unit of luminous intensity, I, is *1 lumen per steradian*, also called *1 candle*. Since the total solid angle subtended by a point source is 4π steradians, a point source of luminous intensity of 1 candle emitting uniformly in all directions radiates 4π lumens (Fig. 35-34).

Instead of determining a laboratory standard of flux from which, in turn, we may determine the unit of luminous intensity, it proves more practical to provide a standard of luminous intensity, the candle.

The official ruling in this country, as given by Congress in 1949, is: "The unit of intensity of light shall be the candle, which is one-sixtieth of the intensity of one square centimeter of a perfect radiator, known as a 'black-body' when operated at the temperature of freezing platinum." The next section of the bill gives: "The unit of flux of

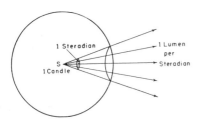

Fig. 35-34.—Uniform point source of intensity 1 candle emits 1 lumen per steradian or a total of 4π lumens.

light shall be the lumen, which is the flux in a unit of solid angle from a source of which the intensity is one candle." Now the freezing point of platinum is $2,042°$ K $(1,769°$ C), and it is by theoretical calculation on the radiation emitted by a perfect radiator at this temperature that the value of 685 lumens/watt for the maximum of the luminosity curve at 5550 A is obtained. For a point source of luminous intensity 1 candle, the luminous flux in unit solid angle is 1 lumen. Table 35-2 gives the

TABLE 35-2

CANDLE RATINGS OF COMMON SOURCES

Ordinary gas jet..........	10–18
Welsbach mantle.........	45
Electric lamp............	25
Electric arc.............	1,000–2,000

approximate candle ratings of some common sources treated as point sources.

Another quantity frequently used in practice is *illumination*, E. The luminous flux falling on a unit area of a given surface, the flux being normal to the area, is a measure of the illumination of the surface. The units are lumens per unit area—lumens per square meter or lumens per square foot, depending upon the unit of area chosen. The illumination produced by a point source is given by the well-known inverse-square law,

$$E = \frac{I}{r^2} \cos\theta ,\qquad (35\text{-}29)$$

where E is the illumination in lumens per unit area, I the luminous intensity in candles, r the distance of the surface from the point source, and θ the angle between the direction of the flux and the normal, N, to the surface (Fig. 35-35).

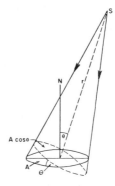

Fig. 35-35.—Point source S of intensity I candles or I lumens per steradian gives illumination of E lumens per unit area A at angle θ, where $E = I \cos\theta/r^2$.

In engineering practice, illumination is measured in a unit called the *foot-candle*, which is the illumination produced when the light from a point source of 1 candle falls normally on a surface at a distance of 1 ft. The foot-candle is thus numerically equal to 1 lumen/ft^2.

In the mks system the unit of illumination is lumens per square meter (lumens/m^2), and this unit is also called the *lux*. Since 1 ft^2 = 0.093 m^2, it follows that an illumination of 1 lumen/ft^2 or 1 foot-candle is approximately equal to 10 lumens/m^2 or 10 lux. A few typical values of the illumination in various places are given in Table 35-3.

TABLE 35-3

THE ILLUMINATION OF SOME TYPICAL PLACES

MODE OR PLACE OF ILLUMINATION	APPROXIMATE ILLUMINATION	
	Lumens/ft^2 (Foot-Candles)	Lumens/M^2 (Lux)
Bright sunlight.................	10,000	100,000
Classrooms, daylight (recommended).....................	12	120
Classrooms, artificial light (recommended).....................	10	100
Moonlight (approximate).........	0.03	0.3

Since the concept of luminous intensity of a source is strictly applicable only to point sources or to sources sufficiently small to be approximated by a point source, another quantity must be used to specify luminous intensity of unit area of an extended surface, which may be self-luminous or not. This quantity is called *brightness* or *luminance*. By definition, the brightness or *luminance in a given direction* is the luminous intensity of unit area of the surface. The units of brightness are, therefore, candles per square centimeter. The difference between the concepts of brightness and illumination must be kept clear. Thus this page is nearly uniformly illuminated; but the printed letter, because it reflects less of the incident light, is *less bright* than the white paper. The brightness of a surface will be numerically equal to the illumination on the surface only if the surface reflects all the light that falls on it. Table 35-4 lists

TABLE 35-4

BRIGHTNESS OR LUMINANCE IN CANDLES
PER SQUARE CENTIMETER

Sun's disk..................	165,000
Crater of an arc.............	15,000
Electric lamp filament........	500
Moon's disk.................	0.5
Clear blue sky..............	0.4
Paper for reading...........	0.002
Freezing platinum (2,042° K)...	60

the approximate brightness or luminance values of some common objects; some are self-luminous, others are not.

In general, the brightness of a surface depends upon the direction from which it is viewed. There are numerous substances that scatter light in such a way that the brightness is the same for every angle of observation. Plaster of Paris, magnesium oxide, and freshly fallen snow are good examples of these diffusing surfaces. For such surfaces another unit of brightness has been used, the *lambert*. By definition, the lambert is the brightness of a perfectly diffusing surface emitting or scattering 1 lumen per square centimeter. If a diffusing surface reflects all the light that is incident upon it, its brightness in lamberts is numerically equal to its illumination. On the other hand, if less light is reflected than is incident, the brightness in lamberts is equal to the product of illumination and a reflecting coefficient.

EXAMPLE 3: How many lumens pass through a spherical area of 240 cm² having a radius of 20 cm when a point source of 100 candles is at the center?

Solution: The solid angle subtended by the area at the center is equal to $\frac{240}{400} = \frac{6}{10}$ steradians. The luminous flux is then $\frac{6}{10}100 = 60$ lumens.

We state without proof the following theorem, which is of use in optical instruments: *The visual brightness of an object of finite size cannot be increased above that of the object by the use of a lens.*

Since many of the terms used in this section are probably unfamiliar, we shall summarize them in Table 35-5.

TABLE 35-5

PHOTOMETRIC TERMS AND DEFINITIONS

Quantity	Symbol	Unit or Units	Definition or Comments
Source or luminous intensity.......	I	Lumens/steradian ≡ candles	Luminous flux per steradian or $\frac{1}{60}$ of intensity from 1 cm² of platinum at 2,042° K
Luminous flux....	F	Lumen	0.00146 watt of radiant flux of 5550 A wave length or flux in unit solid angle from one candle
Illumination......	E	Lumens/cm²; lumens/m² ≡ lux; or lumens/ft² ≡ foot-candles	Luminous flux incident per unit area
Brightness or luminance......	B	Candles/cm² ≡ lamberts; candles/m² ≡ meter-lamberts; candles/ft² ≡ foot-lamberts	Luminous intensity per unit area

35.20. Comparison of the Intensities of Two Sources

The BUNSEN GREASE-SPOT PHOTOMETER is an instrument for comparing the intensities of two similar sources. The photometer consists of a sheet of white paper, P (Fig. 35-36), which we assume

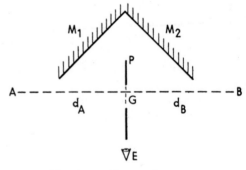

FIG. 35-36.—Bunsen photometer

reflects *all* the radiant energy that falls on it. A spot G is made translucent by dropping a little grease on it. The translucent spot thus transmits a considerable portion of the incident light and reflects the rest, thus appearing dark, when viewed by reflected light, with respect to the surrounding white paper. Two sources are mounted at A and B, as shown, and the distances d_A and d_B are adjusted until the spot disappears, as viewed from either side of the paper. The mirrors, M_1 and M_2, are placed so that the eye, E, can see both sides of the paper at the same time. If the spot cannot be distinguished from the surrounding paper, the amount of light coming from a unit area of spot and paper must be the same.

Let E_A and E_B represent the illumination on the two sides of the spot. If a represents the fraction of the light reflected by the spot, aE_A is the amount of light reflected by the grease spot. A quantity $E_B(1 - a)$ is transmitted to the left by the spot. The total amount of light toward the left, then, is $aE_A + E_B(1 - a)$; and, if the grease spot disappears, then this amount of light coming from the spot is equal to the amount E_A coming from the white paper, i.e.,

$$E_A = aE_A + E_B(1 - a)$$

or

$$E_A(1 - a) = E_B(1 - a);$$

and, since

$$(1 - a) \neq 0,$$

therefore,

$$E_A = E_B.$$

Moreover, if I_A and I_B are the luminous intensities of the two sources, then, by equation (35-29), $\theta = 0$, it follows that

$$E_A = \frac{I_A}{d_A^2}, \qquad E_B = \frac{I_B}{d_B^2},$$

and so

$$\frac{I_A}{I_B} = \frac{d_A^2}{d_B^2}. \qquad (35\text{-}30)$$

If one of the sources is taken as a standard, the intensity of the other may be computed by equation (35-30).

Since it is not an easy matter to determine when the grease spot disappears, one of the sources may be moved so as to alter the illumination by as much as 5 per cent before a noticeable change occurs in the appearance of the spot.

There are many forms of photometers in actual use, and these are nearly all similar to one originated by Lummer and Brodhum. In this, light from two sources, S_1 and S_2, falls on the opposite sides of a photometric surface, P (Fig. 35-37), which is made of plaster of Paris or some other white diffusing substance. The diffusely reflected light from P falls onto the mirrors, M (or 45°–90° prisms), and then onto the photometric cube, C. In this there are two right-angled prisms, the hypotenuse of a portion of one being ground away before it was cemented to the other. Where the two surfaces are together, the light passes straight through, as from S_2, whereas where they are not in contact, the light is totally reflected and the

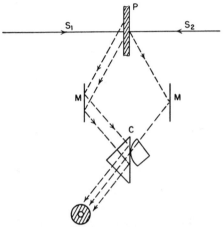

Fig. 35-37.—Lummer-Brodhum photometer

light from S_1 is seen. The field of view consists of an inner circle of light from S_2 surrounded by a ring of light from S_1. When the illuminations from the two sources are equal, then the dividing line between the inner and outer circles disappears. This setting can be made quite accurately, and the ratio of the luminous intensities of the two sources S_1 and S_2 is given by equation (35-30).

35.21. Color: Its Description and Measurement

The physicist may use the term "color" only in a casual and descriptive way. His technical specification of radiation invariably is confined to a statement of the wave lengths or wave numbers included in any specified range of radiation and the respective intensities of each. If the radiation involved is continuous in character, the physicist's specification is that of energy distribution as a function of the wave length, frequency, or wave number throughout the region involved. This region may or may not include that scant octave visible to the eye that we call "visible light."

For the artist, the paint and dye manufacturers, weavers, and laymen in general or for any persons whose business or pleasure is closely related to the use of colored objects, there is little necessity for acquaintance with even the rudiments of scientific knowledge concerning radiation. Yet visible colors and the necessity for adequate means of describing them constitute a large part of the content of the daily lives of these persons. Only in comparatively recent years have the foundations of an adequate empirical science of color description and measurement in terms of the subjective sensations been laid. That this new science contains within itself the possibility of satisfying an almost unrecognized need for a universal color language is known to but a relatively very small number of professionals.

We can order, specify, and describe a great many of the objects of daily use in terms of numbers—collar, shoe, glove, and clothing sizes—but colors have to be matched to samples or specified vaguely by such terms as "rose pink" or "Alice blue."

Since the development of the science of colorimetry, a numerical system capable of describing within the limits of visual perception the various aspects of hue, purity, and intensity (the latter being of lesser importance for relative comparisons)—indeed, a universal "color language"—is possible.

35.22. Color of Illuminant

The light reflected to the eye from a colored surface depends upon two factors: the color of the illuminating source of light and the color of the surface. We shall not discuss here the technical details of the development of sources of illumination which form close approximations to the color of full daylight: sky plus sunshine upon an object. Daylight color is rather naturally chosen as the standard color for the illumination of colored objects. This color can be closely approximated artificially by the use of gas-filled tungsten lamps operating at temperatures of about $2,850°$ K (obtainable in calibrated forms from the National Bureau of Standards) when used in conjunction with certain specified filters—quantitatively reproducible by means of definite solutions and qualitatively obtainable in colored glasses, e.g., the so-called "daylight lamps" of the incandescent types with bluish bulbs. These are largely superseded by fluorescent types.

35.23. Color of Reflecting Surface

Inherently and entirely independent of the nature of the illuminant are those qualities of, let us say, a reflecting surface which give it color. The color we see, however, is a function of both illuminant and surface.

The REFLECTION FACTORS of a surface are usually expressed as decimal fractions lying between 0 and 1 which represent the ratio of reflected to incident light, a ratio found to be independent of intensity. This ratio must be given for as many narrow bands of frequencies or wave-length regions as the precision of the specification demands. A WHITE surface is one which reflects equally all colors with a factor 1 for each. Freshly prepared magnesium oxide is almost such a surface. The graph of the spectral distribution of diffusely reflected white light from a yellow-green plant, for example, might look like Fig. 35-38. The

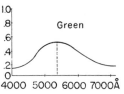

FIG. 35-38.—Light from a yellow-green plant

"dominant color" is given by the maximum ordinate. It is characteristic of reds, oranges, and yellows to reflect highly throughout the longer-wavelength region. The ENERGY reflected from any colored surface would be found by multiplying the energy of the illuminant in each wave length by the reflecting power of the surface at that wave length.

35.24. Tricolor Stimulus

It is a well-known empirical fact that any color sensation may be matched by the sensation produced by the mixture of any three colors (with certain exceptions noted later), taken in the proper amounts. Suppose one projects from three lanterns the three component colors, red, green, and violet, on a white screen, as shown in Fig. 35-39. In the regions where the colors overlap, the eye perceives that a new color is produced which is not analyzed into the components by the eye. By adjusting the intensities of the components, the color produced in the overlapping regions can be made to match a large range of colors, though not all possible colors. However, it is possible to take the unknown color and mix it with one or possibly two of the components and match this with the other component or components. In this sense any color can be matched with three other component colors. So long as the three components cover the visible spectrum, it is of no consequence what they are, though red, green, and blue components will match directly the widest range of colors. Thus we

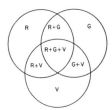

FIG. 35-39.—Matching colors with three component colors

may consider red, green, and blue to be primary colors. If one of the components has to be added to the unknown in order to produce a match with the other compounds, then the amount added to the unknown is considered a negative quantity. To avoid these negative values, the International Commission on Illumination (I.C.I.) has, on mathematical grounds, selected three spectral distributions for components, roughly shown in Fig. 35-40, and marked for identification B for

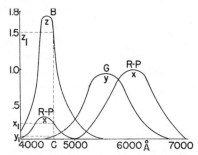

FIG. 35-40.—The I.C.I. component colors

blue, G for green, and R-P for red-purple. That these spectral distributions do not correspond to any real color is of no consequence, since, as we have seen, any three components are equivalent to any other three within certain limits. It would take us too far afield to discuss how the particular distributions shown in Fig. 35-40 were arrived at.

In terms of the I.C.I. components, the unit energy of a *pure* or *monochromatic* spectral color, C_1, can be matched by fractions X_1, Y_1, and Z_1 of the "unreal primaries," P, G, B, thus:

$$C_1 = X_1 P + Y_1 G + Z_1 B . \qquad (35\text{-}31)$$

Now a three-dimensional model would be required to represent all pure spectral colors, using the so-called "tristimulus coefficients" X, Y, and Z. In order to avoid this, a new series of *chromatic coefficients*, a, b, and c, are used. These are defined in the following manner:

$$a = \frac{X}{X + Y + Z},$$
$$b = \frac{Y}{X + Y + Z}, \qquad (35\text{-}32)$$
$$c = \frac{Z}{X + Y + Z}.$$

It is seen from equations (35-32) that $a + b + c = 1$, so that if any two of the chromatic coefficients are given, the third can immediately be found. It is usual to take the coefficients a and b and plot them on a two-dimensional graph (Fig. 35-41). The curved outer boundary represents the

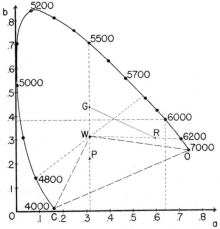

FIG. 35-41.—Color-mixture diagram of chromatic coefficients a and b.

locus of the pure spectral colors. As an example we shall show how the coefficients a, b, and c are obtained for the wave length 6000 A. The accurate values of the tristimulus coefficients X, Y, and Z, taken from curves similar to Fig. 35-40 for the wave length 6000 A, are

$$X = 1.0622 , \qquad Y = 0.6310 , \qquad Z = 0.0008 .$$

Hence
$$X + Y + Z = 1.6940 .$$

From equations (35-32),

$$a = \frac{1.0622}{1.6940} = 0.6270; \qquad b = \frac{0.6310}{1.6940} = 0.3725;$$

$$c = 1 - (a + b) = 1 - 0.9995 = 0.0005 .$$

The coordinates of the point 6000 A are then 0.6270 and 0.3725 and are so marked in Fig. 35-41. Thus the red color corresponding to a wave length of 6000 A can be obtained by taking the fractional intensities of the I.C.I. component colors given above. The other coefficients for different monochromatic colors are similarly calculated and shown on the boundary of the graph. The standard daylight source, which we shall call *white*, W (Fig. 35-41), has its own pair of chromatic coefficients a and b which can be found by a rather tedious calculation.* White then turns out to have the values $W = (0.3101, 0.3163)$, $c = 0.3736$, located in Fig. 35-41, as shown.

The tremendous usefulness of a color diagram such as that shown in Fig. 35-41 lies in the fact that all points within the curve represent real colors—approximately spectral purity as the points approach the boundary and pure white (daylight) as the point W is approached.

If two points G and R represent a certain "pale" green and "warmer" red, all mixtures of these two colors lie in the straight line RG. Furthermore, the colors G and R themselves represent dilutions by white of the spectral lines at which straight lines from W through G and R intersect the boundary, i.e., 5500 and 6200 A, respectively. The "purity" of G and R is measured by the ratio of the distance of either from the point W to its distance from the boundary curve—in Fig. 35-41, 31 per cent for G and 73 per cent for R.

The directions of straight lines drawn from white W through any given color to the boundary of the curve thus provide two alternative and more meaningful numbers for the designation of any particular line. These are the DOMINANT WAVE LENGTH, λ_0, which is the point of intersection of the line with the boundary in terms of the wave-length scale along the boundary, and PURITY, σ,

* This computation involves selecting at suitable small wave-length intervals the observed values of x, y, and z in the source, multiplying each by the relative energies E_x, E_y, and E_z in that wave length, and summing separately $\Sigma x E_x$, $\Sigma y E_y$, and $\Sigma z E_z$ when coordinates a and b are

$$a = \frac{\Sigma x E_x}{\Sigma x E_x + \Sigma y E_y + \Sigma z E_z},$$

$$b = \frac{\Sigma y E_y}{\Sigma x E_x + \Sigma y E_y + \Sigma z E_z},$$

$$c = 1 - (a + b).$$

which is the ratio in which the given color divides this line, i.e.,

$$\frac{\text{Distance, color to white}}{\text{Distance, white to boundary through color}}.$$

Thus the colors G and R of Fig. 35-41 may be represented either by their chromatic coefficients or by their dominant wave lengths and purity:

$$G = 0.31, 0.43 \quad \text{or} \quad \lambda_0 = 5500\,\text{A},$$
$$\sigma = 31 \text{ per cent},$$

$$R = 0.60, 0.28 \quad \text{or} \quad \lambda_0 = 6200\,\text{A},$$
$$\sigma = 73 \text{ per cent}.$$

Any spectral color is completely "pure" ($\sigma = \infty$) or entirely "saturated." White is completely "impure" or "unsaturated," i.e., $\sigma = 0$.

Complementary pure colors, that is, colors which add up to form white, on the boundary curve are at opposite ends of a straight line through the white point; e.g., 4800 A spectral blue and 5800 A spectral yellow are complementary. However, lines drawn from a large section of the boundary through W will not intersect the boundary a second time but will pass across the broken line, CO. The region within the triangle CWO contains all purple and magenta shades which cannot be composed of a single spectral color mixed with white light. Indeed, to get the purple, P, white has to be added to the "complementary color of 5500 A" in the ratio of $\frac{12}{18}$, i.e., 67 per cent.

Thus, by generalizing the procedure used with pure spectral colors to the region within the triangle CWO and labeling points on the line CO as saturated colors complementary to those directly opposite (through W) on the true wave-length scale (the chromatic diagram), their coefficients and the alternative dominant wave-length and purity numerology can be applied to all shades of purple and magenta also.

As an example of the use of numbers, instead of names, to describe commonly known colors, the selections in Table 35-6, from a table from Hardy's *Handbook of Colorimetry*, are interesting. A number indicating the brightness of a color is often included to complete its numerical description.

This possibility has been taken care of in the derivation of the tristimulus values plotted in Fig. 35-40. There the curve y was chosen (by suitable selection of the basic primary colors) so that it conformed precisely to the color-visibility curve for the normal human eye on a scale in which a hole in a black body is zero and the perfect white is 100.

Thus, to indicate, in addition to dominant wave length and purity of a color, its brightness as seen

TABLE 35-6

COLORS AND THEIR SPECIFICATIONS

$\Sigma y E_y$ (Per Cent)	Name	λ_0	σ
50........	Baby pink	6100	10
23........	Old rose	6080	20
7........	Maroon	6000	35
5........	Chocolate	5860	30
35........	Apple green	6580	40
32........	Turquoise	5000	30
3........	Navy blue	4750	20
4........	Royal blue	5600	50
33........	Orchid	5000	15
17........	Blood red	6000	65

by the eye, one has to list merely the component $\Sigma y E_y$ of the chromatic coefficents formed as in the footnote in this section, in a separate column alongside of λ_0 and σ. This has been done in Table 35-6, to the left of the name for each color.

35.25. Résumé

The following is a brief listing of the definitions, principles, and theories, given in this chapter, which you should know:

Huygens' principle and its use in establishing the laws of reflection and refraction of light

The interpretation of interference phenomena on the basis of Huygens' principle

Newton's rings and the measurement of the wave length of light of different colors

The theory of the Michelson interferometer

The phenomenon of diffraction at a straight edge and its interpretation by means of Fresnel zones

The diffraction pattern from a single slit, from two slits, and from *n* slits

The theory of the diffraction grating

The resolving power of a grating and of a single slit

Radiant and luminous flux; the lumen and the unit of luminous intensity

The measurement of luminous intensity by a photometer

Color and its specification; the meaning of tricolor stimulus; the chromatic coefficients; and the color-mixture diagram

Queries

1. Describe and explain the appearance of a small source of light when viewed with a single layer of handkerchief held in front of the eye. Describe how one can test by this method the fineness of weave of the cloth.
2. Have the colors of soap bubbles, of oil films on the surface of water, and of Newton's rings anything in common?
3. What pitch of sound waves would one expect to be diffracted by a picket fence whose pickets were 10 cm apart?
4. When one looks at one picket fence behind another as one drives along a highway, one sees a pattern of very coarse regions of light and dark moving along the two fences. Has this anything to do with interference?
5. How do colors of red, green, blue, and violet cloth, as seen by daylight, look under candlelight?

Practice Problems

1. One surface of a lens is convex, with a radius of curvature of 12 cm; the other surface is flat. Calculate the focal length of this lens if the index of refraction of the glass is 1.57. Ans.: 21 cm.
2. A lens made of glass whose index of refraction is 1.62 has one convex surface of a radius of curvature of 10 cm and the other a concave surface of a radius of curvature of −18 cm. Compute the focal length of the lens.
3. The separation *d* of the slits *B* and *C* in Fig. 35-7 is 0.2 mm, and the distance *D* to the viewing screen is 1 m. Find the distance on the screen between the central maximum and the first dark band for light of wave lengths 5000 and 6000 A. Ans.: 1.25 mm; 1.5 mm.
4. Find the distance between the first and second bright fringes on the screen in the interference experiment of problem 3 for both colors.
5. A very thin layer of water, $\mu = 1.33$, on the surface of an oil film shows interference colors. Assuming normal incidence of the light and a first-order reflection of blue light of wave length 4000 A, find the thickness of the water layer. Ans.: 7.5×10^{-6} cm.
6. A plano-convex lens of radius of curvature 50 cm is

placed with its convex surface on a plane glass plate as in Newton's ring experiment. The illumination is perpendicular on the plate and has a wave length of 6000 A. If the rings are viewed in reflected light, find (a) the separation t for the first ring and (b) the radius of this ring.

7. The Newton rings in question 6 are viewed in transmitted light. Find (a) the separation t for the first bright ring and (b) the radius of this ring. Ans.: 3×10^{-5} cm; 0.548 mm.

8. A glass plate whose index of refraction is 1.57 is placed in the path of one of the beams of the Michelson interferometer. Compute the thickness of the plate if a displacement of 390 fringes of sodium light is observed.

9. A Michelson interferometer is illuminated with monochromatic light of wave length $\lambda = 6438.47 \times 10^{-8}$ cm. Calculate the distance its mirror is moved if 12,000 fringes sweep across the field. Ans.: 0.39 cm.

10. Derive the lens-maker's equation for a biconcave lens in a manner similar to that used in § 35.5.

11. Using the argument given in § 35.10, show that the intensity at the geometrical line of the shadow produced by a straight edge is 0.25 of normal intensity of light (Fig. 35-17).

12. Show that the width of the Fresnel zones OA, AB, and BC (Fig. 35-19) are $\sqrt{\lambda d}$, $0.41\sqrt{\lambda d}$, and $0.32 \sqrt{\lambda d}$.

13. Assume that the amplitude of the wave produced by each Fresnel zone is proportional to the width of the zone. Show that the ratio of the intensity at P_2 (Fig. 35-19) to that at P is approximately 4.75.

14. Two plane-parallel glass plates, in contact at one end and separated at the other end by the diameter of a wire, form a wedge-shaped air film. If the air film is viewed by reflected monocromatic light of wave length 5890×10^{-8} cm, it is found to be crossed by 96 dark fringes. What is the diameter of the wire?

15. A grating with 230 lines/cm is placed 100 cm from a slit in a screen. Sodium light passes from the lamp through the slit and grating to the eye. How many centimeters from the slit will the first-order image appear? The tenth image? Here $\lambda = 5890$ A. Ans.: $d_1 = 1.36$ cm and $d_{10} = 13.7$ cm.

16. If white light were placed behind the slit of problem 15, how many centimeters wide would the first-order spectrum of the visible (7000–4000 A) light extend?

17. Light of wave length 5700×10^{-8} cm strikes a slit of width 1 mm in the manner shown in Fig. 35-20. Calculate the angular spread of the light in the central maximum. Ans.: 1.14×10^{-3} rad.

18. If the slit width of problem 17 is 10^{-2} mm, what is the angular spread in the central maximum?

19. The two yellow lines in the spectrum of mercury have $\lambda_1 = 5770$ A and $\lambda_2 = 5790$ A. What resolving power is needed to separate these lines? How many lines would be needed on a grating so that these lines would be just resolved in the second order? Ans.: $R = 289$; $n = 145$ lines/cm.

20. The two yellow lines in the spectrum of sodium have wave lengths of 5890 and 5896 A. What resolving power is needed to separate them? How many lines must a grating have to accomplish this in the third-order spectrum?

21. If the distance from a slit, 0.2 mm wide, to a screen is 2 m, how far from the center of the diffraction pattern will the first, second, and third lines appear? Use $\lambda = 6400$ A. The first, second, and third dark lines? Ans.: 0.96, 1.60, and 2.24; 0.64, 1.28, and 1.92.

22. The tristimulus coefficients X, Y, and Z for the wave length 5000 A are $X = 0.0049$, $Y = 0.3230$, and $Z = 0.2720$. Find the chromatic coefficients a, b, and c and show that these are 0.0082, 0.5384, and 0.4534, as given in Fig. 35-41.

23. Some radiant flux consists of a mixture of 1 watt of monochromatic light of wave length 5000 A and 4 watts of monochromatic light of wave length 6000 A. Find (a) the total radiant flux, (b) the total luminous flux, (c) the luminous efficiency of this light. (Use Fig. 35-33.) Ans.: (a) 5 watts; (b) about 1,900 lumens; (c) 380 lumens/watt.

24. How many lumens pass through a spherical area of 50 cm² having a radius of 10 cm when a point source of 50 candles is at the center?

25. A 100-candle lamp is suspended 5 ft above the center of a square table, 4 ft on a side. Find the illumination at the center of the table and at the corners. Ans.: 4 lumens/ft² or 4 foot-candles; 2.64 lumens/ft².

26. Using a Bunsen photometer, the illumination from a standard 32-candle source at 100 cm is equal to that of an unknown source at 120 cm. Find the candles in the unknown source.

27. What is the surface brightness in lamberts of a slab of plaster of Paris with an illumination of 10 lumens/cm², the reflecting coefficient being 0.4? Ans.: 4.

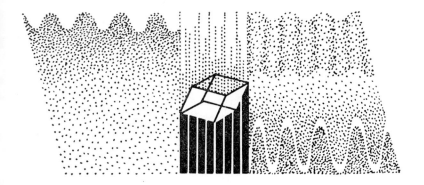

CHAPTER 36

Polarized Light

36.1. Introduction

In the previous chapter we discussed the phenomena of interference and diffraction which give us, perhaps, our best evidence that light has wave characteristics. These phenomena, however, do not tell us whether the undulatory motion is longitudinal or transverse. The reader will recall that sound waves, longitudinal in character, exhibit interference and diffraction effects and that water waves, which have both transverse and longitudinal characteristics, likewise display both diffraction and interference. There is, however, one phenomenon of light and other forms of radiant energy that shows conclusively that its wave properties are transverse. This phenomenon is known as POLARIZATION.

36.2. Polarization of Waves

The polarization of simple transverse waves may be clearly illustrated by means of transverse waves in a rope. The apparatus, shown in Fig. 36-1, consists of the rope, XY, and two identical

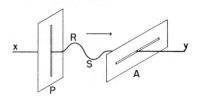

FIG. 36-1.—The polarization of transverse waves

boards, P and A, with slits cut in as shown. Let us assume at first that the two boards are not present and that an experimenter grasps the end of the rope at X and sets up waves by moving the end X

always perpendicular to the direction XY. A set of transverse waves will be set up, and the resulting wave forms may lie in any conceivable plane that contains the rope. This type of wave motion is said to be *unpolarized*. If the rope now passes through the slit of board P, as shown in the diagram, *only those vibrations of the oncoming unpolarized transverse wave that are in the plane of the paper* will pass through the slit, such as, for example, RS, whose vibrations are *parallel* to the slit. These waves are said to be PLANE-POLARIZED. The slit P is called the POLARIZER.

If the rope passes through another slit which is parallel to P, the plane-polarized wave will go through this slit undisturbed. If, however, this slit A is perpendicular to the slit P, as shown in the figure, the wave RS will not pass through. Indeed, scarcely any motion at all gets through this slit. The two slits are then said to be "crossed." The board A with its slit is called the ANALYZER of the plane-polarized wave. The plane that contains the wave form RS is called the PLANE OF VIBRATION; the plane perpendicular to it, the PLANE OF POLARIZATION.

The reader will observe, however, that a longitudinal wave which vibrates *in the direction of propagation of the wave* will pass through both slits, whatever their orientation. In other words, transverse waves do not exhibit the same symmetry about the line of propagation as do longitudinal waves.

Light coming from ordinary sources is unpolarized in the sense that the vibrations of the rope in the absence of slits are unpolarized. By interacting with matter, unpolarized light can be made plane- **539** polarized.* This interaction can result from reflection from surfaces, refraction through crystals, scattering by small particles, or selective absorption by certain substances. In the sections that follow we shall describe the different ways that light may become polarized and shall discuss the properties and practical applications of this polarized light.

36.3. Polarization of Reflected Light Reveals Transverse Character

Suppose we take a plane piece of ordinary glass (① of Fig. 36-2), having a refractive index $\mu =$

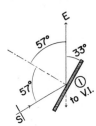

Fig. 36-2.—Light reflected from a mirror

1.55, and use it as a mirror to reflect the light from a candle flame or lamp bulb, S. We will, furthermore, make a special selection of the angle of incidence, as shown in Fig. 36-2. Setting the plane of the mirror at about 33° to the vertical and looking vertically down upon it, we will place the candle so that its virtual image also lies in the vertical direction. Thus we make the angle of incidence, measured from the normal to the mirror, 57°. We next take another piece of glass (② in Fig. 36-3)

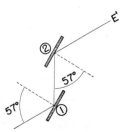

Fig. 36-3.—Reflection from two parallel mirrors

and arrange it in a mount, again at about 33° with the vertical. We place this mirror in place of our

* According to the electromagnetic theory of light, a polarized light wave consists of an electric vibration (displacement current) in one plane associated with a magnetic vibration in a plane at right angles to it. In the discussions of this chapter the light vibrations are to be identified with the electric vibrations.

eye (E of Fig. 36-2), as shown in Fig. 36-3, and look at the light source from E', the light now being reflected from both mirrors. So far, nothing unexpected has turned up: the reflections are entirely clear, bright, and distinct.

Next we rotate either (a) mirror ① and the light source with it or (b) mirror ② and the position of our eye (the latter is easier) through 90°, with the ray between mirror ① and mirror ② as an axis, into the position shown in Fig. 36-4. Our eye, E',

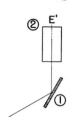

Fig. 36-4.—Mirror ② rotated through 90° gives no reflection.

has now come out of the plane of the paper and is in front of it, looking into mirror ②. *The image of the light source,* S, *has become quite faint!* If we had been precise in making the angles as indicated, 57°.2, *no trace of the light source* (if it is small in angle) *remains! One cannot form a virtual image of a virtual image with two glass mirrors in the positions shown in Fig. 36-4.*

To get at the bottom of this unique situation, let us look at the direction of the refracted ray in either mirror ① or mirror ②, a small section of the

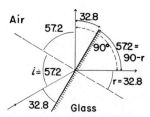

Fig. 36-5.—Relation of angles in polarization through reflection.

front surface of either being shown in Fig. 36-5 From Snell's law,

$$\mu = \frac{\sin 57°.2}{\sin r} = 1.55 \, ,$$

$$\sin r = \frac{0.8406}{1.55} = 0.542 \, ,$$

$$r = 32°.8 \, ,$$

$$90 - r = 57°.2 \, .$$

Our choice of 57°.2 for the angle of incidence on glass of index $\mu = 1.55$ resulted in the *refracted ray's taking a direction at right angles to the reflected ray*. This results in some sort of modification of the reflected light beam, so that it cannot be reflected again under similar circumstances when at the next reflection it encounters a mirror ② (Fig. 36-4), in which the PLANE OF INCIDENCE is at right angles to that of the first reflected ray from mirror ①. (The "plane of incidence" is that plane which contains the incident ray and the normal to the mirror surface.)

To understand how this can be, we have but to *assume* that the light contains no *longitudinal components*, i.e., that its vibrations are *exclusively transverse*. In Fig. 36-6 we have represented nat-

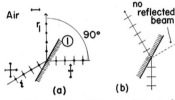

FIG. 36-6.—Vibrations in reflected and transmitted beams

ural light, which has vibrations in all directions perpendicular to that of its propagation, as consisting of two vibrations at right angles, ⊹ or ↕, since any direction can be resolved into its components along any arbitrary rectangular axes.

The incident beam in Fig. 36-6 (a) shows these two directions as a line (|) in the plane of the drawing and a dot (·) representing the line perpendicular to the page. On encountering the glass of mirror ①, the direction of the reflected ray is such as to be identical with that of one of the two components of the vibration in the light as it becomes refracted at this surface. If this vibration were transmitted, it would be longitudinal in character, i.e., have the direction of the wave's propagation. If it were not transmitted, the reflected ray would contain vibrations at right angles to the plane of the figure only, i.e., would be plane-polarized.

Upon the second reflection on ② (Fig. 36-6 [b]), with the plane of incidence now 90° from its position on ① (Fig. 36-6 [a]), the residual beam, r_1, now encounters the mirror so that, if it is reflected, its *only remaining vibration direction* lies in the direction of the reflected ray. Light thus reflected would consist *exclusively* of a longitudinal component. The complete *absence* of any light in this direction clearly establishes the *complete inability for transmission of any* LONGITUDINAL *components in radiation*. Thus the experiment establishes the transverse character of light.

If the angle of incidence, i, of Fig. 36-5 is such that the refracted and reflected rays are at right angles to each other, then it follows from the geometry of the figure that

$$\mu = \tan i . \qquad (36\text{-}1)$$

This equation is known as *Brewster's law*, which states that the angle of incidence for complete polarization is that angle whose tangent is equal to the index of refraction of the reflecting material.

36.4. The Partial Polarization of the Refracted Beam

The phenomena observed and analyzed above in the reflected beam present certain inevitable implications regarding the characteristics of the *refracted* rays. Fig. 36-7 diagrammatically illustrates

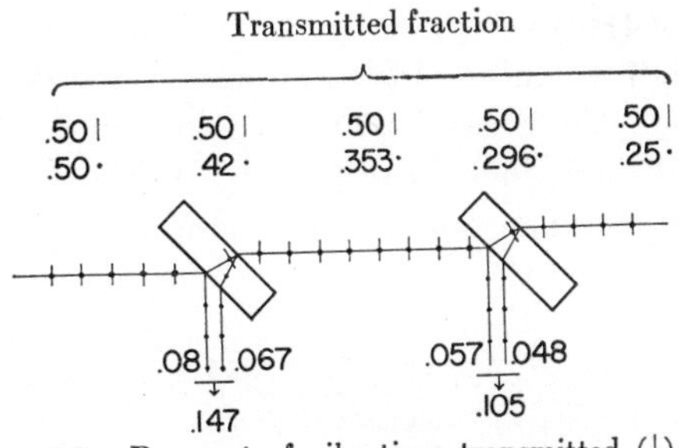

Transmitted fraction

.50		.50		.50		.50		.50	
.50 ·	.42 ·	.353·	.296·	.25 ·

.08 ┃.067   .057┃ .048
.147   .105

FIG. 36-7.—Per cent of vibrations transmitted (|) in the plane of incidence and (·) perpendicular to the plane of incidence.

the passage of a beam in sequence through several plates. Photometric measurements show that 16 per cent of the light that has vibrations perpendicular to the plane of incidence is reflected. Thus, if the incident beam has half its vibrations in the plane of (↑), and half perpendicular to (·), the plane of incidence, 8 per cent of the incident light comes plane-polarized from the front surface of the first mirror. From the back surface of the same mirror, 16 per cent of the remaining 42 per cent (0.067) is reflected out, leaving the *transmitted* beam in the ratio of 50 : 35 plane-polarized in the plane of incidence. The same argument, repeated on transmission at the second plate, leaves an excess of 2:1 in the intensity ratios of vibrations in and at right angles to the plane of incidence. **541**

Thus, although transmission never can completely reduce the refracted beam to pure plane polarization, ten to fifteen plates are sufficient to make it practically so. Absorption and scattering of light within the glass, of course, have been neglected. The figures are not exact but are given as illustrative only of the principles involved.

36.5. Double Refraction

It was discovered by the Danish physicist, Bartholinus (1625–98), that many transparent crystals transmit light with different velocities in different directions; and, since two directions suffice for the resolution of the component velocities, the phenomenon is called *double refraction*. Polarization associated with double refraction was noted by Huygens in 1690.

If a crystal of Iceland spar (calcite, $CaOCO_2$), which is rhomboidal in form, is laid over a black dot on a piece of white paper or over an illuminated pinhole, one observes, on looking down through the crystal, *two* dots or holes. The separation of the two is greater, the thicker the crystal through which one looks. However, one direction through the crystal may be found along which only a single image can be seen. Along this direction there is no double refraction, and this direction is called the OPTIC AXIS (*op* of Fig. 36-8).

FIG. 36-8.—Crystal of Iceland spar showing optic axis, *op*

If one rotates the crystal, one image remains stationary, and the other moves in a circle around it. Tracing the ray that emerges from the holes to give the stationary image, one finds that it is displaced just as it would be through a parallel-sided piece of glass, and it is called the ORDINARY RAY. The other ray, which rotates around the ordinary ray as the crystal is turned, is called the EXTRAORDINARY RAY.

Figure 36-9 shows diagrammatically how a ray, R, entering normally one face of such a crystal and having a direction not along its optic axis, *op*, divides into two components: RO, the ordinary, and RE, the extraordinary rays. Huygens, as we have already indicated, noted that these two rays were

plane-polarized at right angles to each other. The plane of vibrations of the extraordinary ray is parallel to that which contains the optic axis and the incident ray, R; of the ordinary ray, it is normal to this plane.

Furthermore, the extraordinary ray emerges from the crystal at a point in the line determined

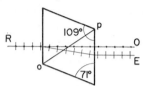

FIG. 36-9.—Ordinary (*O*) and extraordinary (*E*) rays in Iceland spar.

by the obtuse solid angle p and the point of emergence of the ordinary ray and on the other side of the latter from p.

If the obtuse-angled points of the crystal are ground and polished, as shown in Fig. 36-10, along

FIG. 36-10.—No difference in refraction for ordinary and extraordinary rays along optic axis *op*.

two planes normal to the optic axis, *op*, it can be experimentally demonstrated that a ray of light, RO, entering along, or parallel to, this axis is neither doubly refracted nor polarized.

That the velocity of the ordinary and extraordinary rays is different in the crystal may be seen by looking at a dot on a piece of white paper with the crystal in the position of Fig. 36-9. The ordinary image of the dot appears nearer the eye than the extraordinary image. This shows that the index of refraction is greater for the ordinary or that its velocity through the crystal is less than that of the extraordinary ray. Indeed, the two indices of refraction for calcite are $\mu_0 = 1.658$ and $\mu_E = 1.486$ for ordinary and extraordinary rays, respectively.*

36.6. The Nicol Prism

Plane-polarized light can be produced by using a calcite crystal in such a manner as to separate

* Calcite is an example of what is called a "negative" crystal, the ordinary ray having the *larger* index. A crystal in which the extraordinary ray has the larger index is called a "positive" crystal. Of the latter, ice is one example: $\mu_0 = 1.309$, $\mu_E = 1.313$. Quartz (SiO_2) is another: $\mu_0 = 1.544$, $\mu_E = 1.553$.

completely the ordinary from the extraordinary ray. This may be done by a special construction made with a calcite crystal, first achieved by a Scotch physicist, Nicol, in 1828. A calcite crystal (Fig. 36-11) has two opposite surfaces, polished

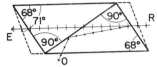

FIG. 36-11.—A Nicol prism

down so that the smaller angle between them and their adjacent sides is 68° instead of the natural 71°. The length of crystal is chosen so that it can be sawed diagonally in two in a plane *perpendicular* to these two new faces. The sawed surfaces are polished and replaced together, using Canada balsam as cement.

The angles are such that the extraordinary ray, *RE* (refracted away from the normal on normal entry [Fig. 36-11]), now travels *straight through the prism*. The ordinary ray, bent toward the normal in the denser medium, as usual takes an oblique direction, *RO*. Since the index of refraction for the balsam ($\mu = 1.530$) is intermediate between that of the two rays, the ordinary ray is *totally reflected out* at the balsam interface, and the extraordinary ray passes through.

By cutting calcite crystals four times longer than their width, it is possible to have Nicol prisms whose ends may be ground 90° to the sides.

36.7. Wave-Front Sections in the Principal Plane

Confining our attention next to that plane section of the crystal which contains the optic axis and is perpendicular to the upper and lower faces (this defines the PRINCIPAL PLANE), let us look at diagrammatic representations of the ordinary and extraordinary rays that travel in this plane up through the crystal (Fig. 36-12) and examine the wave fronts of the light passing through.

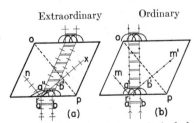

FIG. 36-12.—Wave fronts in a principal plane

First we will make the construction of Huygens; and we will further assume that, for the ray whose vibrations occur in the plane, i.e., extraordinary ray, the velocity is *greatest* when the vibrations are parallel to the optic axis, direction *rx*, and least when perpendicular to it, direction *rn*. The Huygens secondary wavelet is then not a circular segment but *the segment of an ellipse* (Fig. 36-13). The

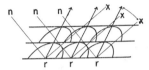

FIG. 36-13.—Huygens wavelets for extraordinary ray in principal plane.

envelope of the consecutive ellipses, i.e., the new wave front, moves upward *in a direction which is inclined to its normal*, i.e., in a direction which is the resultant of the vectors along *rx* and *rn*, representing the major and minor axes of these ellipses, as shown in Fig. 36-12 (*a*) for the extraordinary ray.

For the ordinary ray the components of its vibrations are perpendicular to the principal plane and to the optic axis. The vibrations thus have the same velocities in directions *rm* and *rm'*. The sections of wave fronts in this plane for this ray are *circles*, and the ray travels in the "ordinary" way at right angles to its wave front (Fig. 36-12 [*b*]).

In other directions and planes than those chosen, this simple analysis cannot be made. More rigorous methods, however, show that the surface representing a wave front through a doubly refracting crystal is an interesting dimpled surface.

36.8. Biaxial Crystals

In the preceding discussion of the passage of light through crystals we have assumed that there was but a *single optic axis*, i.e., a single direction through which light can pass without exhibiting the phenomena of double refraction. There are many such crystals—besides Iceland spar, or calcite ($CaOCO_2$); there are beryl ($3BeO \cdot Al_2O_3 \cdot 6SiO_2$) and tourmaline. However, Brewster discovered that there are many other crystals that contain *two optic axes* and hence are called "biaxial." The simple construction of Huygens is not valid for these crystals, but a theory first worked out by Fresnel made predictions which were subsequently verified and so provided us with a **543** satisfactory understanding of the refraction and the wave-surface form in these cases also. A discussion of biaxial crystals is out of place here, and the reader is referred to more advanced books, such as *Physical Optics* by R. W. Wood. Examples of biaxial crystals are gypsum ($CaO \cdot SO_3 \cdot 2H_2O$), turquoise ($CuO \cdot 3Al_2O_3 2P_2O_5 \cdot 9H_2O$), and topaz ($2AlOF \cdot SiO_2$).*

36.9. Crystals in Polarized Light

To examine any transparent material in polarized light, we need two plane-polarizing devices called the POLARIZER and the ANALYZER, i.e., two mirrors at appropriate angles, two piles of glass plates, two Nicol prisms, or two polaroids. Polaroids (§ 36.14) make the most effective analyzers and polarizers.

These are arranged so that the transparent object to be examined may be inserted between them. If the transmitting plane of the polarizer is at right angles to that of the analyzer, no light is transmitted: the field is dark.

If crystals, transparent celluloid, Cellophane, molded glass, and certain other transparent materials are placed between crossed polarizer and analyzer, light reappears through the material, and this light *is usually colored*. When the analyzer is rotated so that its transmitting plane is parallel to that of the polarizer and the general field is bright, the color of the transparent specimen changes to the complementary one.

To understand these color changes, the reader must recall some matters he learned in the chapter on simple harmonic motion.

Suppose a plane-polarized beam ($\rightarrow$) falls upon a doubly refracting crystal. If the crystal is so oriented that the directions of vibration of the two rays in it (at right angles to each other) make angles of 45° with the vibration of the incident ray ($\times_E^O$), the incident ray is split into two equal components (O, E) in these directions. Now one of these components, the extraordinary, travels faster than the other in the crystal. If the crystal is thin enough that, on passage through it, the faster-moving beam gets just one-half wave length ahead of the other, its phase is reversed ($\times_E^O$; and,

on emergence, these two beams unite into a plane-polarized one whose direction of polarization is at right angles to that of the incident beam ($\uparrow$).

Such a crystal is called a "half-wave plate." If placed between crossed polaroids or any other polarizer-analyzer combination that is crossed and hence transmits no light, the introduction of such a plate restores the light to full intensity. Conversely, if the polarizer-analyzer combination is set for full transmission, a half-wave plate inserted in the path extinguishes the light.

Suppose, however, one uses another crystal which is only half as thick as a half-wave plate, i.e., is a "quarter-wave plate"; the phases of the two rays are altered by but 90°; and the emergent beam consists of two equal vibrations at right angles ($\times_E^O$) and of phase difference $\pi/2$. The resultant is a circular vibration ($\bigcirc$), which because of its symmetry across any diameter in the analyzer, shows *no variation* of intensity as the analyzer is rotated. Such light is "circularly polarized."

Sections of a crystal intermediate in thickness between zero and quarter (quarter- and half-; half- and three-fourths-; etc.)-wave plates produce emergent components of intermediate phase difference ($\times_E^O$), and these combined into "elliptically polarized light" ($\bigcirc$). Rotation of the analyzer produces variations of intensity of such beams, but the *minimum is not zero*.

If any crystal is set between a polarizer and an analyzer, either crossed or parallel, with the plane of one or the other of its transmitted rays parallel to either the polarizer or the analyzer plane, one of the transmitted rays is quenched, and the foregoing resolutions cannot take place. The crystal in such positions shows no evidence of double refraction and in no way alters the character of light passing polarizer and analyzer. Consequently, when a doubly refracting crystal is rotated between crossed polaroids or Nicols, there are four positions in which all colors disappear. On opposite sides of any one of these colorless positions the colors are complementary.

36.10. The Production of Colors in Polarized Light

What is the source of these colored effects that are produced when crystals are placed between

---

* It must not be inferred from the foregoing that all crystal-line materials are doubly refracting. Isotropic monorefringent crystals are numerous, including opal, fluorite (CaF_2), sylvite (KCl), bauxite ($Al_2O_3 \cdot H_2O$), diamond, rock salt (NaCl).

polarizer and analyzer? These effects happen only when the crystals are of such dimensions in the direction of the light through them as to produce retardations of several, but not too many, wave lengths.

Since blue light is about half the wave length of red, a blue half-wave plate would be a red quarter-wave plate. In white light such a crystal between parallel polaroids would entirely *suppress the blue rays* (emergent ray plane-polarized at right angles to the incident ray and hence stopped in the analyzer that transmitted the original beam), but the *red light would come out circularly polarized* and hence unaffected by *any* position of the analyzer. The color, of course, is complementary to that of the blue light removed and, in hue and saturation, will vary with the orientation of the crystal, the polarizer, and the analyzer.

As crystals become quite thick, the retardation becomes so great that only for a narrower and narrower wave-length range is it exactly right for quenching. The range of color thus removed is less and less, and consequently what passes through is more and more perfectly white.

36.11. Rotation of the Plane of Polarization

If a crystal of quartz, cut with faces normal to the optic axis, is placed between crossed polaroids or Nicols, the light reappears, but it is again extinguished by further rotation of the analyzer. This rotation is an amount which is almost $22°/$ mm of thickness of the quartz plate if yellow (sodium) light is used. For blue light the same thickness of quartz rotates the plane of polarization by nearly $45°$. This phenomenon was discovered by Arago in 1811. The fact that solutions of certain substances, like sugar, will do the same thing was discovered by Biot in 1815.

Since the rotation produced by an optically active substance is proportional to the amount of the substance present, an accurate means is thus available for determining the amount of the substance—for example, sugar—present in an inactive solvent. The *specific rotation, s,* is defined as the rotation produced by a 10-cm column of liquid containing 1 gm of active substance for every cubic centimeter of solution; thus $s = 10\theta/dl$, where θ is the angle of rotation, d is the grams of active substance per cubic centimeter, and l is the length of the light path in centimeters.

Cane sugar occurs in two different forms of the same chemical constitution but of different molecular arrangements. One of these rotates the plane of polarization to the right with respect to an observer toward whom the beam of polarized light is traveling and is called "dextrose"; the other form, rotating the plane of polarization opposite to the left, is therefore called "levulose."

Similarly, two different crystal forms of quartz are known, one crystal being the mirror image of the other. These two forms rotate the plane of polarization of light passing parallel to their optic axes oppositely.

In 1845 Faraday found that a block of glass subjected to a strong magnetic field became optically active. Plane-polarized light sent through such a glass block in a direction parallel to the magnetic field has its plane of polarization rotated. The amount of rotation is found experimentally to be proportional to the field strength and to the distance the light travels through the medium. This "Faraday effect" has been observed in many solids, liquids, and gases.

An analogous effect, called the "Kerr electro-optic effect," is present whenever a substance is placed in a strong electric field. When placed in an electric field, many liquids, solids, and gases behave optically like a uniaxial crystal with the optic axis parallel to the field.

36.12. Double Refraction in Nonhomogeneous Media

There are transparent media other than crystalline which, because of a lack of homogeneity inherent in them or produced by external stresses, become doubly refracting. Glass and celluloid and some of the more modern plastics will show localized colored bands and spots when put under a stress and placed between crossed polaroids or Nicols. The modern machine industries and designing laboratories find this fact most useful in fashioning materials of such shape as to eliminate dangerous strains due to excess stress or for so altering the location of these strains in the material that fracture and failure are prevented.

Transparent plastic models are studied to guide in the casting and machining of the metal parts. Strains due to rapid motion may be investigated **545** by putting the plastic model in the required state of motion, then chilling it suitably, so that the strains "freeze in" and may be examined in detail with the model at rest later. This phenomenon, which is extensively used in engineering for stress analysis, is known as *photoelasticity*.

36.13. Polarization of Scattered Light

If a beam of light from the sun or a powerful lantern is sent through a transparent-sided trough filled with a carefully filtered solution in which a fine precipitate may be formed slowly (e.g., sodium thiosulphate and a small addition of sulphuric acid), the beam, when viewed transversely at some distance from one side and perpendicular to its direction, slowly becomes visible and of an intensely blue color when the precipitate begins to come out in extremely fine particles.

As the particles increase in size, the light scattered from the solution becomes whiter, and ultimately the solution becomes almost opaque. As this goes on, the light transmitted through the solution changes from white through yellow and orange to deep dark red. During the earlier stages of scattering, when the particles are small, the scattered light is strongly polarized in the plane that contains the line of the light beam and the line of view of the observer.

The blue color of the sky, due likewise to scattering of sunlight by atmospheric molecular aggregates, shows polarization in the same sense, i.e., in a plane containing the sun, the observer, and the point in the sky at which the observer is looking.

36.14. Polarization by Selective Absorption

Certain crystals, known as *dichroic*, of which tourmaline is an example, are doubly refracting and, in addition, strongly absorb the ordinary ray, transmitting the extraordinary ray. Thus a crystal of tourmaline 2 mm thick transmits practically only the extraordinary ray. Unfortunately, tourmaline absorbs the various wave lengths of white light differently, so that the plane-polarized light is colored.

Crystals such as herapathite (quinine iodosulphate) also absorb the ordinary ray completely and, in addition, are nearly free from selective color absorption. Herapathite crystals, as ordinarily available, are very tiny, and thence their use as polarizers is limited. However, E. H. Land has developed a method whereby these tiny crystals are deposited uniformly at about 10^{11} per square centimeter on a cellulose film. The crystals are oriented with their axes all parallel to one another, so that the film acts as one huge crystal. These polarizing films, known commercially as *Polaroid*, have the advantage of very large size, small cost, and a polarizing efficiency equal to that of a Nicol prism except at the extreme ends of the spectrum.

The commercial possibilities of Polaroid are far-reaching. By equipping the headlights and windshields of automobiles with Polaroid, it is possible to eliminate headlight glare. The reader is already familiar with Polaroid glasses for preventing the glare of sunlight reflected from sidewalks and pavements.

36.15. Résumé

The following is a brief listing of the definitions, principles, and theories, given in this chapter, which you should know:

The meaning of polarization of light

Various means of producing polarized light

The phenomenon of double refraction

The Nicol prism as a means of producing plane-polarized light

Interference with polarized light; plane, circular, and elliptical polarized light

The rotation of the plane of polarization by various means

Queries

1. How would you distinguish a beam of unpolarized light from plane-polarized light?
2. How would you distinguish unpolarized light from circularly polarized light? Would a quarter-wave plate help?
3. How must the axis of the Polaroid on the headlights on an auto be oriented with respect to the axis of the Polaroid on the windshield in order to eliminate glare?
4. If in the interference phenomena discussed in the last chapter the two interfering beams are polarized in the planes perpendicular to each other, the interference effects disappear, even though the two beams come from the same source. What conclusion can you draw?

Practice Problems

1. Water has an index of refraction of 1.33. At what angle must a beam of light strike the water surface in order that the reflected beam may be plane-polarized? Ans.: $53°.07$.

2. A 60° prism is made of Iceland spar and is cut so that the refracting edge is parallel to the optic axis. Draw the paths of the ordinary and extraordinary rays through the prism. Assume minimum deviation for the ordinary ray.

3. Determine the polarizing angles for sodium light of light crown glass, dense flint, rock salt, and diamond. (Use Table 34-1.) Ans.: $56°.58$; $58°.87$; $57°.07$; $67°.57$.

4. A glass tube 30 cm long contains 10 gm of cane sugar per 100 cm³ of water. If the specific rotation for cane sugar for sodium light is $66°.5$ cm³/cm-gm, compute the angle of rotation of the plane of polarization.

5. Calculate the thickness of a quartz half-wave plate for sodium light, if the indices of refraction are 1.5442 and 1.5533. Ans.: 0.00324 cm.

6. How much cane sugar in grams is contained in a 30-cm column whose volume is 50 cm³ if the plane polarization for sodium light is rotated through $39°.9$?

7. A thin slab of calcite is cut with the optic axis parallel to the plane of the slab. Calculate the minimum thickness required to produce destructive interference between crossed Nicols in sodium light. The extraordinary and ordinary refractive indices of calcite for sodium light are 1.486 and 1.658, respectively. Ans.: 3.4×10^{-4} cm.

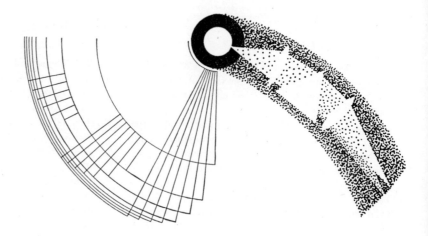

CHAPTER 37

Spectra; Quantum Theory; Solid-State Phenomena

37.1. Introduction

We have already described the instruments for producing a spectrum—a prism or a diffraction grating with a suitable lens system for focusing—but we have had no concern with the spectrum produced. In this chapter we shall be concerned mainly with the sources and their spectra. The interpretation of the spectrum of various sources has been of immense importance in physics, for it was largely by this means that an understanding of the outer structure of the atom was obtained. It is because of this that many universities and industrial research laboratories have large rooms in which are installed concave gratings by which spectra can be photographed under high dispersion. Whether a prism or a grating is used to produce the spectrum of a source depends largely on the particular problem being investigated, though for most purposes the concave grating is preferable. We shall now give an outline of the various types of spectra and their sources, together with a brief statement about the information which is obtained from an analysis of the spectra. Following this will be more detailed discussion of the problems.

37.2. Types of Spectra

There are four main types of spectra which are characterized as continuous, line, band, and ab-

sorption spectra. We shall consider that each of these is produced by a spectroscope having as its source an illuminated narrow slit. The spectral images are therefore the images of this slit, which may be either viewed or photographed. Each wave length, λ, is deviated through an angle θ, which depends on λ and for a grating is given by equation (35-19), $n\lambda = d \sin \theta$. Thus if only one wave length were present in the source, the spectrum would consist of a bright line, a slit image, corresponding to the color of the wave length λ, and no other colors would be present. As we shall see, the names of the various types of spectra are descriptive of what is produced when using an illuminated narrow slit as source.

a) Continuous spectrum.—This consists of a continuous band of color corresponding to a large continuous range of wave lengths extending from the ultraviolet through the visible (violet, blue, green, yellow, red) into the infrared. The sources which can produce this continuous spectrum are incandescent solids, liquids, or gases at very high pressures. The temperature of the source can be obtained from the analysis of the spectra. On the theoretical side the analysis of these spectra was of great importance, for it was from this that the quantum theory was first developed by Planck about 1900.

b) Bright-line spectra.—This spectrum consists of a series of separated bright lines corresponding to a number of distinct wave lengths. Gases at a fairly low pressure in an electrical discharge tube, such as a neon sign, or vapors in a flame are the sources which produce the bright-line spectra. The bright lines are characteristic of the type of atom, so that these spectra can be used to identify atoms. The analysis of the bright-line spectra of hydrogen gave us our first understanding of the structure of the atom. This theory is known as the Bohr theory and, although it has been modified by wave mechanics, from a pictorial point of view the Bohr theory is still very useful.

c) Band spectra.—In these spectra there is a series of bands which, under high dispersion, can be shown to be separate lines exceedingly close together. In general, they are much more complicated than the line spectra. Band spectra are produced by molecules, the source frequently being a small sample of the molecular compound placed in the crater of a carbon arc. The analysis of band spectra is very complex and has to be handled by the highly mathematical theory of wave mechanics. The spectra are characteristic of the molecules in the source and, as such, may be used to identify the molecular types.

d) Absorption spectra.—This consists of a continuous spectrum crossed by dark lines or dark bands. The source is therefore an incandescent solid or liquid in front of which is placed some absorber. The dark lines correspond to wave lengths which are the same as those which would be produced if the absorbing material were itself radiating. Thus if the absorber consists of atoms, there is line absorption, whereas if it consists of molecules, there is band absorption. The information obtained from an analysis of such spectra is the temperature of the incandescent source and the types of atoms or molecules in the absorber. Absorption spectra are characteristic of the sun and of 98 per cent of the stars.

37.3. Continuous Spectra

In the 1890's the German investigators Lummer, Pringsheim, and Kurlbaum, among others, made a very careful experimental study of the distribution of energy in the continuous spectra from a source at different temperatures. The source was known as a "black body" and consisted of a furnace kept at a carefully controlled temperature. A small hole in the furnace allowed radiation to

escape which was characteristic of a black body at that temperature. The radiation was then spread out into a spectrum, and a temperature-measuring device, such as a thermopile, was used to measure the energy in a small spread of wave lengths over the whole of the spectrum. The energy radiated per unit time per unit area and per unit range in wave length E_λ was then plotted against the corresponding wave lengths. The curves in Fig. 37-2 are typical of the furnace, while Fig. 37-1 shows the curve for a body at about the temperature of the sun.

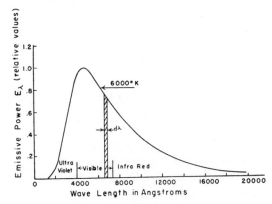

FIG. 37-1.—Black-body radiation curve for source at 6,000° K.

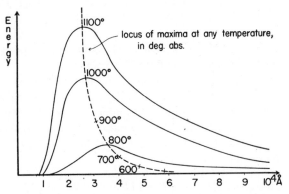

FIG. 37-2.—Energy radiated per second per unit area per unit wave-length interval for sources at different temperatures.

If the thermopile or other energy-measuring device covers a small range of wave lengths, its reading is proportional to $E_\lambda d\lambda$, shown in the shaded portion in Fig. 37-1. The ordinate E_λ, called the *emissive power*, is the energy radiated per second per unit area per unit wave length interval and in the mks system could be measured in watts per square meter per angstrom. It is the **549** amount of energy radiated per second per unit area by the source about some wave length, λ. In Fig. 37-1 are shown the approximate limits of the visible spectrum, and we see that accompanying this are ultraviolet and infrared radiations. For the sun, which is approximately at the temperature of 6,000° K, the latter two radiations—the ultraviolet and the infrared—are strongly absorbed by the earth's atmosphere. Thus the spectrum of the sun received at the earth would have absorption regions in it, so that only its main outline would appear like Fig. 37-1.

The sources producing the spectra shown in Fig. 37-2 are relatively low-temperature sources, and very little, if any, of the radiation would be visible. Most of it is in the infrared region. The total energy radiated in all wave lengths at any one of these temperatures is proportional to the total area inclosed by the curve, and we see that this increases considerably with temperature. This is a fact that we are all qualitatively familiar with, for the higher the temperature of a body, the more energy it radiates. The total area under the curve is the summation of all the small areas $E_\lambda d\lambda$; this can be written as

$$\text{Total area} = \Sigma E_\lambda d\lambda = \int_0^\infty E_\lambda d\lambda = E,$$

where E is the total energy in all wave lengths radiated per second per unit area of the radiating body.

As stated in § 15.5, Stefan showed experimentally that the energy E is proportional to the fourth power of the absolute or Kelvin temperature. Some years later Boltzmann derived this law from theoretical reasoning; hence the law relating the total energy radiated by a black body at temperature $T°$ K, known as the *Stefan-Boltzmann law* (eq. [15-7]), is

$$E = \sigma T^4 , \qquad (37\text{-}1)$$

where σ is the Stefan-Boltzmann constant. Its value is

$$\sigma = 5.6687 \times 10^{-5} \text{ erg/cm}^2 \text{ sec (°K)}$$
$$= 5.6687 \times 10^{-8} \text{ watt/m}^2 (°K)^4 .$$

If the radiating body is not a black body but has an emissivity of e, then, as given in chapter 15, the Stefan-Boltzmann law is written as (eq. [15-12])

$$E = e\sigma T^4 . \qquad (37\text{-}2)$$

For a black body which absorbs all radiation and reflects none, $e = 1$, whereas for a perfect reflector which reflects all the radiation incident on it and absorbs none, $e = 0$.

A second interesting point which can be drawn from Fig 37-2 is that the maximum of the curves moves to shorter wave lengths as the temperature increases, as shown by the dotted curve. If λ_m is used to indicate the wave length having the maximum energy or the peak of the curve, then Wien (1864–1928) showed theoretically that λ_m is inversely proportional to the absolute temperature of the black-body radiator. This result is given as *Wien's displacement law*,

$$\lambda_m T = \text{Constant} . \qquad (37\text{-}3)$$

If λ_m is measured in centimeters and T in degrees Kelvin, then

$$\lambda_m T = 0.2898 \text{ cm °K} ,$$

whereas, if λ_m is in angstrom units and T in ° K, then

$$\lambda_m T = 2.898 \times 10^7 \text{ A °K} .$$

It is this law which is used in determining the temperature of a black body. All that is necessary is to find the wave length of maximum energy, λ_m, in the continuous spectrum of the black body.

37.4. Planck's Radiation Law; Origin of Quantum Theory

The shape of the radiation curves given in Fig. 37-2 depends on the temperature of the black body or furnace and not at all on the type of atom or molecule composing the furnace. Thus in creating a theory for the production of a continuous spectrum, a simple type of oscillator could be assumed for the radiating atoms or molecules. Attempts were made by Rayleigh and Jeans in England and Wien in Germany by using classical theory, i.e., Newton's laws and wave-motion theory. With somewhat different modes of attack, Rayleigh arrived at an expression which agreed with experiments at long wave lengths but disagreed markedly at short wave lengths, whereas for Wien's expression the reverse was true. The graphs using these derivations are shown in Fig. 37-3, in which R stands for Rayleigh and W for Wien.

In 1900 Planck showed that the concepts of classical physics were not all applicable on an atomic scale. In particular, he showed that, in order to account satisfactorily for the experimental results, he had to assume that an atomic oscillator could *not* have all possible energy values but only those which were integral multiples of the frequency, f, of the oscillator. Thus an oscillator, vibrating with frequency f, could absorb or emit

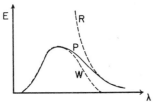

Fig. 37-3.—Graphs of equations for energy distribution given by Rayleigh (R), Wein (W), and Planck (P).

energy only of amounts E which were integral multiples of hf, where h is a universal constant, now known as Planck's constant or for oscillator energies, where

$$E = nhf \,, \qquad (37\text{-}4)$$

where n is any integer.

The dimensions of h are energy/frequency or energy $\times$ time, and h has a value of 6.625×10^{-34} joule-sec, or 6.625×10^{-27} erg-sec. Since the radiation emitted by the oscillators is propagated with the velocity of light, c, it is possible to give the distribution-of-energy expression in terms of frequency or wave length, since $c = \lambda f$.

The derivation of Planck's equation for the energy distribution is far beyond the scope of this book, so we shall just state it. The energy per unit area per second in the wave-length interval $d\lambda$, which we have previously given as E_λ, is given as

$$E_\lambda = \frac{2\pi c^2 h}{\lambda^5 \,(e^{hc/\lambda k T} - 1)} \,, \qquad (37\text{-}5)$$

where k = Boltzmann's constant = 1.3804×10^{-23} joule/° K = 1.3804×10^{-16} erg/° K; e = base of natural logarithms = $2.7183 \ldots$; T = temperature in degrees absolute of the black body emitting the radiation; and c = velocity of light in a vacuum = 2.998×10^8 m/sec = 2.998×10^{10} cm/sec. This equation gives the shape of the curves E_λ plotted against λ, which agrees exceedingly well with the experimental curves, as indicated by P in Fig. 37-3.

The important aspect of Planck's work, so far as we are concerned, is that the energy of the oscillators is quantized in packets or quanta of integral multiples of hf. In 1905 Einstein extended this concept of quantization to the radiation itself.

He assumed that the energy of the radiation was in quanta or packets, of amounts hf. These radiation energy quanta he called *photons*. Thus the energy E of the photons of electromagnetic radiation of frequency f is given by

$$E = hf \,. \qquad (37\text{-}6)$$

In bringing forward the concept of photons, Einstein was interested in explaining the emission of electricity from metals by the absorption of electromagnetic radiation—the so-called *photoelectric effect*. It was some years before Millikan gave experimental confirmation of Einstein's proposals. We shall discuss the photoelectric effect in § 37.11.

From what has been said up to the present, it appears that light or electromagnetic radiation has dual characteristics—those of waves and those of particles. In general, one can say that if the absorption or emission of radiation is involved, then the particle or photon characteristics show up, whereas during the transmission of radiation, as in experiments on interference or diffraction, the wave characteristics are the important ones. It appears that natural phenomena are not so simple that they can be explained in terms of waves or of particles. The much more comprehensive theory of wave mechanics brings these two aspects together, but this is beyond the scope of this book. We shall now discuss the phenomenon of line spectra and then show how the photon concept was used to explain the line spectra of hydrogen.

37.5. Line Spectra

It is an experimental fact that every element in the periodic table has associated with it a characteristic spectrum. By this we mean that if a gas, such as hydrogen, is electrically* excited until it emits radiation, this radiation, when allowed to fall on a dispersive instrument, such as a diffraction grating or a prism, will be spread out into a large number of lines, the relative positions of which are characteristic of hydrogen. The totality of lines emitted by atomic hydrogen is called its

---

* We should add "or thermally," since temperature alone is entirely adequate to cause a gas or vapor to emit its characteristic spectrum. With a gas as nearly "ideal" as hydrogen, the temperatures must be very high. Sodium vapor, on the other hand, emits its characteristic spectra at the relatively low temperature of an ordinary blue-flame Bunsen **burner.**

"line spectrum." Figure 37-4 is drawn from a photograph of a *small portion* of the spectrum of atomic hydrogen; many other lines, showing several similar groupings or patterns but lying in the

FIG. 37-4.—The two main series in hydrogen

infrared, are not shown. Figure 37-5 shows some of the lines of sodium; these are typical data obtained by the experimental spectroscopist.

Note the presence of two distinct, but similarly arranged, patterns, SERIES, of lines in that portion

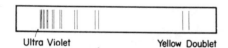

FIG. 37-5.—The principal series of sodium

of the spectrum of hydrogen shown in Fig. 37-4, the so-called "Balmer series" in the visible region and the "Lyman series" in the far ultraviolet. Both series converge to definite limits, 3646 A (1 A = 10^{-8} cm) in the case of the Balmer series and 911 A for the Lyman lines. The more prominent lines in the visible region are called H_α, H_β, and H_γ. Contrast the spectrum of hydrogen with that of sodium shown in Fig. 37-5. The PRINCIPAL SERIES of sodium consists of a number of doublets, starting with the familiar yellow doublet 5890–96 A and extending into the far ultraviolet. The separation of the doublets decreases as the limit is approached. There are other "series" of lines not shown in the case of sodium as well as in that of hydrogen.

We shall, in conformity with modern spectroscopic usage, identify the position of each line of the spectrum, not by its wave length, λ cm, but by the number of wave lengths in a centimeter, $1/\lambda$ cm^{-1}, called the "wave number." The reason for this choice is that the formulas that we shall derive take on a much simpler form thereby. Table 37-1 gives the wave lengths and wave numbers of the more important Balmer and Lyman lines.

The wave lengths or wave numbers do not appear to follow any simple equation, and many attempts have been made to find one.. In 1885 the German scientist Balmer found that the wave numbers of the hydrogen lines in the visible spectrum could be represented by an equation of the following type:

$$\frac{1}{\lambda} = R\left(\frac{1}{2^2} - \frac{1}{n^2}\right), \qquad (37\text{-}7)$$

where $n = 3, 4, 5, \ldots, \infty$. This was purely an empirical result, but nevertheless it led to important developments. In honor of this work, equation (37-7) is known as the "Balmer series for hydrogen."

Other series for hydrogen were found somewhat later, and the complete system for hydrogen, named for their discoverers, is:

Lyman series: $\dfrac{1}{\lambda} = R\left(\dfrac{1}{1^2} - \dfrac{1}{n^2}\right)$

$(n = 2, 3, 4, \ldots, \infty, \text{ultraviolet region})$,

Balmer series: $\dfrac{1}{\lambda} = R\left(\dfrac{1}{2^2} - \dfrac{1}{n^2}\right)$

$(n = 3, 4, 5, \ldots, \infty, \text{visible region})$,

Paschen series: $\dfrac{1}{\lambda} = R\left(\dfrac{1}{3^2} - \dfrac{1}{n^2}\right)$

$(n = 4, 5, 6, \ldots, \infty, \text{infrared region})$,

Brackett series: $\dfrac{1}{\lambda} = R\left(\dfrac{1}{4^2} - \dfrac{1}{n^2}\right)$

$(n = 5, 6, 7, \ldots, \infty, \text{infrared region})$,

Pfund series: $\dfrac{1}{\lambda} = R\left(\dfrac{1}{5^2} - \dfrac{1}{n^2}\right)$

$(n = 6, 7, 8, \ldots, \infty, \text{far infrared region})$. $\qquad (37\text{-}8)$

The quantity R in these equations is called the *Rydberg constant*, in honor of Rydberg, who did

TABLE 37-1

WAVE LENGTHS AND WAVE NUMBERS OF THE
BALMER AND LYMAN SERIES

BALMER SERIES		LYMAN SERIES	
H_α	6562.2 A......15,233 cm^{-1}	1215 A......	82,258 cm^{-1}
H_β	4861.3.........20,264	1026........	97,482
H_γ	4340.5.........23,032	972.........	102,823
	4101.7.........24,373	950.........	105,290
		939.........	106,630
	. .	.	.
	. .	.	.
	. .	.	.
	3658.6.........27,325		
Limit	3646.0.........27,419	*Limit* 911.........109,677	

much pioneer work in this field. From the experimental data, the Rydberg constant has the value

$$R = 109,677.6 \text{ cm}^{-1} = 1.096776 \times 10^7 \text{ m}^{-1}.$$

The accuracy with which the wave numbers or wave lengths are given by these relatively simple formulas is indeed remarkable. For example, consider the H_a line of the Balmer series, the wave number of which is given by

$$\frac{1}{\lambda} = 109,677.6 \left(\frac{1}{2^2} - \frac{1}{3^2} \right) \text{ cm}^{-1}$$

$$= 27,419.4 - 12,186.4$$

$$= 15,233.0 \text{ cm}^{-1},$$

which is the value derived from experiment, as given in Table 37-1.

The success with the spectra of hydrogen led to much activity in spectroscopy. It was found that a few other elements gave rise to hydrogen-like spectra—for example, sodium, potassium, singly ionized helium, doubly ionized lithium, etc. In Fig. 37-6 is shown the bright-line emission spectrum of singly ionized helium. This single electron structure, similar to hydrogen, gives simple series spectra which in structure resemble those of hydrogen. Thirteen members of this series are shown from the fifth to the eighteenth. Mercury in small amounts was present in the discharge tube, and some of its stronger lines appear. It is of interest to note that this spectrum was discovered first in the star ξ Puppis by Pickering and erroneously ascribed to hydrogen prior to the development of modern spectroscopic theory.

Series spectra are characteristic of all atomic sources; but, as the atoms increase in atomic number and hence in the complexity of their outer electronic structures, the simple and obvious regularities, so conspicuous in elements like hydrogen, lithium, etc., become entirely masked. The identification of series in the spectrum of an element like iron, shown here, becomes a complex technical problem. The wave lengths of iron, whose spectrum is readily produced by forming an electric arc between iron electrodes (Fig. 37-7), are frequently used as secondary standards for comparison purposes.

In all these cases attempts were made to find a series of lines whose wave lengths could be expressed as the difference of two terms, as given in equation (37-8). The result of considerable work led to the formulation of an empirical principle called the *Ritz combination principle*. This we shall now discuss, starting with the spectrum of hydrogen.

37.6. The Ritz Combination Principle

If we examine the wave numbers partially given in Table 37-1 for hydrogen, we discover some interesting interrelationships. As an example, consider the differences in the wave numbers in the Lyman series:

$$97,482 - 82,258 = 15,234 \text{ cm}^{-1}$$

(H_a in Balmer series),

$$102,823 - 82,258 = 20,265 \text{ cm}^{-1}$$

(H_β in Balmer series),

$$105,290 - 82,258 = 23,032 \text{ cm}^{-1}$$

(H_γ in Balmer series).

If this procedure is carried out more extensively, we find that, by making a plot of the wave number of the Lyman series, choosing as the initial point

2967.3 A ↓   ↓ He 2945.1 A   ↓ Hg 2655.1 A   Hg 2536.5 A ↓

5   6   7   8   9   10 11 12 14   18

Fig. 37-6.—The series spectrum of singly ionized helium; mercury was also present in the discharge tube

2647.6 A   ↓ 2631.3 A   ↓ 2617.6 A   ↓ 2599.4 A   ↓ 2591.6 A   ↓ 2582.3 A   2566.9 A ↓

Fig. 37-7.—The iron-arc spectrum

the series limit 109,677 cm⁻¹ of the Lyman series and plotting each wave number of the series as shown in Fig. 37-8, the wave numbers of the members of the other series are obtained by taking differences. Because the spectrum lines in Fig. 37-4 converge, the levels in Fig. 37-8 also converge.

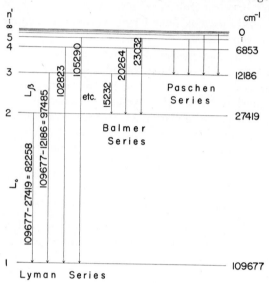

FIG. 37-8.—Term values for hydrogen

If we draw vertical lines (which we shall call *transitions*) from each level to the lowest for which $n' = 1$, we get, by taking the difference of the wave numbers of the two levels involved, the wave numbers of the Lyman series. Similarly, taking the difference between the next lowest $n' = 2$ and the higher levels gives the wave numbers of the Balmer lines. Thus the wave number of H_a is given by taking the difference between the levels $n' = 2$ and $n' = 3$, or

$$\frac{1}{\lambda}(H_a) = 27,419 - 12,186 = 15,233 \text{ cm}^{-1}.$$

Further, if we draw various transitions to $n' = 3$, 4, etc., we obtain the Paschen and other series located in the infrared.

What we have discovered is truly a fundamental law in spectroscopy; it is called the "Ritz combination principle," after its discoverer. Stated in words, it is:

To the spectrum of every atom can be ascribed a series of wave numbers, called TERM VALUES, *small in number, such that every observed spectral line can be obtained by differences of these term values.*

The Ritz principle implies the existence of different fixed levels of energy within every atom.

The horizontal lines of Fig. 37-8 represent these ENERGY LEVELS of the hydrogen atom. This concept of "energy levels" is of fundamental importance in spectroscopy, because it has been found that every line observed in a spectrum can be obtained by a particular transition from one level to another. Experience has shown, however, that the converse is not always true, i.e., not every possible transition gives rise to observed lines; but it is necessary to restrict the possible transitions by appropriate SELECTION RULES. The hydrogen spectrum which we have described is unique in its simplicity. The determination of the energy levels for the hydrogen atom is a relatively simple task. In more complex spectra—that of iron, mercury, etc.—the spectra are so complicated that the effort to obtain the energy levels or term values is quite formidable.

Let us summarize the information that the experimental spectroscopist can obtain from studying photographs of the spectra of the elements:

1. The position of the spectral lines, i.e., the wave length or preferably the wave number of each line

2. The relative intensity of the lines—e g., whether in hydrogen the H_a is more intense than the H_β

3. The degree of polarization of each line

In addition to these three primary observations, he notes:

4. These spectral lines are extremely sharp.

5. The atom giving rise to these sharp lines is extraordinarily stable; i e., no matter how many times the particular sample has been used in chemical reactions or subjected to extreme physical conditions, when placed in suitable conditions to radiate its light, it gives rise to sharp lines.

37.7. The Bohr Theory of the Hydrogen Atom

The Ritz combination principle was proposed in 1908, and there was much theoretical activity which attempted to account for the spectrum of the hydrogen on the basis of a model of the atom. Along with the research in spectroscopy, there had been some fundamental discoveries in other fields. Those discoveries which are pertinent to the present discussion will be mentioned here, leaving a more detailed account of them for later. In 1897 J. J. Thomson had discovered the *electron*, a negatively charged particle, and had found it to be a universal constituent of all atoms. The charge on an electron is -1.602×10^{-19} coulomb, and its mass is 9.107×10^{-28} gm. In 1895 Roentgen discovered X-rays, and immediately afterward the phenomenon of radioactivity was discovered by Becquerel. By bombarding a thin gold foil with the alpha particles from radium and counting the number scattered in different directions, Rutherford had concluded in 1911 that the positive portion of the atom is concentrated in a small nucleus at the center. The nucleus contains all the positive charge in the atom and practically all the mass, since the electrons are only about one two-thousandth of the mass of the lightest atom known. The Rutherford nuclear atom then consists of a central positive nucleus and electrons in orbits outside the atom. Atoms are of the order of 10^{-8} cm in size, while nuclei and electrons are of the order of 10^{-12} or 10^{-13} cm in size, so that any atom may be likened to be a miniature solar system. The number of electrons outside the nucleus is called the *atomic number* of the atom and is usually represented by Z.

Now ordinary hydrogen is the lightest atom known and has an atomic number of 1, so that it has one electron moving around the nucleus. The nucleus of the hydrogen atom is a very important entity and is called a *proton*. Later we shall see that the nuclei of all atoms are considered to contain protons. The electrical charge of a proton is equal and opposite to that of the electron, while the mass of a proton is 1,836.5 times the mass of an electron. Thus we may consider the hydrogen atom to consist of a central proton with a light electron revolving around it. Since the positive proton attracts the negative electron, the latter must revolve about the proton in a manner similar to the revolution of the planets about the sun under the gravitational force of attraction. There is, however, one big difficulty with the electron revolving about the nucleus and experiencing a centripetal acceleration. According to classical electrodynamics, an accelerated electric charge emits electromagnetic radiation. If this is the situation in the hydrogen atom, this atom should be continuously giving out energy, so that the electron would follow a spiral path into the nucleus. There would then be no hydrogen atom, which is obviously not so, for hydrogen atoms do exist in a stable form.

The foregoing roughly represents the physical situation as it must have appeared to a young Danish student, Niels Bohr, about 1913, who was then working in Rutherford's laboratory in Manchester, England. With some deep-sighted intuition Bohr discarded parts of classical electrodynamics, retaining some portions of it; and, by weaving this in with the then known quantum ideas, he was able to account for the spectrum lines of hydrogen with remarkable accuracy. In particular, Bohr made three hypotheses which were:

1. The electron in the hydrogen atom can revolve about the nucleus only in certain prescribed orbits in which the angular momentum, or moment of momentum, of the revolving electron is an integral multiple of $h/2\pi$.

2. These allowable orbits are stationary states, in the sense that the electron does not radiate energy when revolving in them about the nucleus.

3. Energy is radiated in the form of photons when the electron moves from an outer to an inner orbit.

We shall now calculate the frequency and wave length of the spectrum lines from hydrogen in a manner similar to that used by Bohr with circular orbits. Let us consider that the nucleus is so heavy that it remains stationary while the electron revolves in the circular stationary orbits shown in Fig. 37-9. Suppose m is the

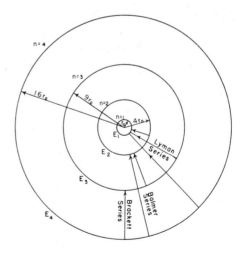

Fig. 37-9.—The Bohr atom for hydrogen

mass and $-e$ is the charge on the electron. The charge on the nucleus or proton, then, is e. If the electron is revolving about the nucleus in a stationary orbit of radius r and speed v, then the centripetal force is provided by the electrostatic attraction between the proton and the electron. Thus, from classical physics, we have

$$\frac{m v^2}{r} = \frac{e^2}{4\pi\epsilon_0 r^2}, \qquad (37\text{-}9)$$

where ϵ_0 is the permittivity of free space and has a value of 8.85×10^{-12} mks units. From assumption 1 above, we have, for the moment of momentum or angular momentum,

$$m v r = \frac{nh}{2\pi}, \qquad (37\text{-}10)$$

where n is an integer 1, 2, 3, etc. Dividing the two equations gives

$$v = \frac{e^2}{2\epsilon_0 n h}, \qquad (37\text{-}11)$$

and, by substitution,

$$r = \frac{nh \, 2\epsilon_0 n h}{2\pi m e^2} = \frac{n^2 h^2 \epsilon_0}{\pi m e^2}. \qquad (37\text{-}12)$$

The radius of the first Bohr orbit, r_0, is given by setting $n = 1$ in equation (37-12). Hence

$$r_0 = \frac{h^2 \epsilon_0}{\pi m e^2}. \qquad (37\text{-}13)$$

Substituting the following values in the mks system in equation (37-13),

$$h = 6.62 \times 10^{-34} \text{ joule-sec},$$

$$m = 9.11 \times 10^{-31} \text{ kg},$$

$$e = 1.60 \times 10^{-19} \text{ coulomb},$$

$$\epsilon_0 = 8.85 \times 10^{-12} \text{ (coulomb)}^2/\text{newton-m}^2,$$

the radius r_0 of the first Bohr orbit is

$$r_0 = 5.3 \times 10^{-11} m = 0.53 \text{ A}.$$

The presently accepted value is 0.52920 A. The radii of the other Bohr orbits are given in equation (37-12) by setting $n = 2, 3, 4$, etc. For the radius of the second orbit we have

$$r_2 = 4 r_0 = 2.1 \text{ A}.$$

Similarly,

$$r_3 = 9 r_0, \qquad r_4 = 16 r_0,$$

and, in general, for the nth orbit,

$$r_n = n^2 r_0. \qquad (37\text{-}14)$$

As was stated earlier, the size of atoms is of the order of 10^{-8} cm, and this is in agreement with the foregoing values.

If we are to find the frequency of the light emitted when the electron moves from one orbit to another, we must calculate the energy of the system when the elec-

tron is in a given orbit. This energy is partly kinetic and partly potential. The kinetic energy is

$$\text{KE} = \tfrac{1}{2} m v^2 = \frac{m e^4}{8\epsilon_0^2 n^2 h^2}. \qquad (37\text{-}15)$$

The potential energy of an electron, $-e$, at a distance r from a charge, $+e$, is the work done on the electron in bringing it from infinity to the distance r, or

$$\text{PE} = \frac{-e^2}{4\pi\epsilon_0 r}. \qquad (37\text{-}16)$$

It is negative because the force is one of attraction and the potential energy at an infinite distance is zero. Substituting for r, we have, for the potential energy,

$$\text{PE} = \frac{-m e^4}{4\epsilon_0^2 n^2 h^2}. \qquad (37\text{-}17)$$

As we see, the PE is twice the KE and of the opposite sign, so that the total energy E of the atom in orbit n is

$$E = \text{PE} + \text{KE} = \frac{-m e^4}{8\epsilon_0^2 n^2 h^2}. \qquad (37\text{-}18)$$

From Bohr's third assumption, the frequency f of the light radiated by the atom when it moves from outer orbit n_2 to inner orbit n_1, in which the corresponding energies are E_2 and E_1, is given by

$$E_2 - E_1 = h f_{12}, \qquad (37\text{-}19)$$

or, from equation (37-18),

$$f_{12} = \frac{m e^4}{8\epsilon_0^2 h^3} \left(\frac{1}{n_1^2} - \frac{1}{n_2^2} \right). \qquad (37\text{-}20)$$

Since $c = f\lambda$, we can write equation (37-20) in terms of the wave length of light emitted as

$$\frac{1}{\lambda_{12}} = \frac{m e^4}{8 c \epsilon_0^2 h^3} \left(\frac{1}{n_1^2} - \frac{1}{n_2^2} \right). \qquad (37\text{-}21)$$

This equation is similar in form to that given for the Balmer, Lyman, etc., series (eq. [37-8]). Actually, the equations are identical, for if we substitute the numerical values for the various quantities in equation (37-20), we have

$$\frac{m e^4}{8 c \epsilon_0^2 h^3} =$$

$$\frac{9.11 \times 10^{-31} \times (1.60)^4 \times 10^{-76}}{8 \times 3 \times 10^8 \times (8.85)^2 \times 10^{-24} \times (6.62)^3 \times 10^{-102}}$$

$$= 1.09 \times 10^7 \, m^{-1}.$$

This is the value given for the Rydberg constant, and when one considers the different sources which the numbers used above came from, the agreement is quite astonishing. Thus the term values given in Fig. 37-8 are the stationary states or energy levels of the Bohr atom shown in Fig. 37-9.

The energy of the atom is a minimum when the electron is in the innermost orbit, $n = 1$, and this is the configuration the atom would normally have. If the atom is subjected to an electric discharge, as it is in the neon-sign type of tube, then the atom absorbs some energy, and the electron moves from the inner to an outer orbit. From this outer orbit the electron falls to a lower one, giving out a photon, as shown in Fig. 37-9. If the atom receives sufficient energy, the electron may be completely removed from the atom, and the atom is said to be *ionized*. The number of volts to which the electron is subjected in order to ionize the atom is called the *ionization potential*. From equation (37-20) the energy for ionization is found by setting $n_1 = 1$ and $n_2 = \infty$ and multiplying by h to give units of hf, or

$$h f_\infty = \frac{m e^4}{8 \epsilon_0^2 h^2} = 2.18 \times 10^{-18} \text{ joule}.$$

Now, as given earlier, 1 electron-volt (1 ev) is the energy given to an electron in falling through a potential difference of 1 volt. From the relationship of

$$1 \text{ joule} = 1 \text{ volt} \times 1 \text{ coulomb}$$

we have

$$1 \text{ ev} = 1 \times 1.60 \times 10^{-19} \text{ joule} . \quad (37\text{-}22)$$

Hence the ionization potential of the hydrogen atom is

$$\frac{2.18 \times 10^{-18}}{1.60 \times 10^{-19}} = 13.6 \text{ volts}.$$

The accepted value for this is 13.595 volts.

37.8. Excitation Potentials

A brilliant confirmation of these ideas of discrete atomic energy levels was experimentally shown by Franck and Hertz. If a stream of electrons is shot into rarefied hydrogen gas at room temperature, these electrons are found to collide elastically and to lose no energy at low velocities of projection. As the velocities of projection are increased, however, at the instant the energies of the projected electrons reach 1.6×10^{-18} joule, inelastic collisions between elctrons and hydrogen atoms make their appearance. Simultaneously, the *first line only* of the Lyman series appears. The energy lost in collision has gone into excitation of the atoms, i.e., has been absorbed by them in lifting their

energy from the lowest to the next-lowest-but-one stationary state. Subsequently, the atom gives off this energy as radiation in the form of this single spectral line. Furthermore, if one further increases the velocity of the bombarding electrons, no new phenomena appear until an energy exactly enough to raise the atoms to the second state of energy above the lowest is reached. All this excess is now removed by inelastic collisions, and now both the first and second lines of the Lyman series and *the first line of the Balmer series* are found in the radiation from the gas. The latter line in the red is visible. This shows that atoms excited to this second stage may give off the excess energy in "one jump" (second Lyman line) or in "two jumps," first to the next-to-lowest level (first Balmer line) and then to the lowest level (first Lyman line).

37.9. Other Spectra

The success which the Bohr theory had with hydrogen and other one-electron systems, He^+ $(Z = 2)$, $Li^{++}(Z = 3)$, etc., was not repeated with atoms having two or more electrons. Refinements were made in the Bohr theory; elliptical orbits were substituted for the circular one, and relativity corrections were introduced. Despite this, it appeared that the theory had limited application. However, the energy levels in the atom corresponding to the term values in the spectra had real significance. There was some uneasiness about the apparently arbitrary manner in which the assumptions of the Bohr theory were made, particularly the first assumption concerning the angular momentum which had to be an integral multiple of $h/2\pi$. This arbitrariness was somewhat removed by the application of the discovery of De Broglie (§ 30.21) to the orbits of the hydrogen atom. In the section referred to, it was stated that De Broglie had associated a wave length λ to particle of mass m and speed v given in equation (30-10) as

$$\lambda = \frac{h}{m v}.$$

Thus the electron revolving in its orbit about the nucleus can be considered as a *matter or De Broglie wave*. In the allowed orbits these waves form standing waves in a manner similar to the standing waves on a rope tied at both ends. For the standing De Broglie waves the orbit and wave length must be such that there is an integral number of **557** wave lengths in the orbit. If the orbit has a radius r, then, for a circular orbit, the condition is

$$\frac{2\pi r}{\lambda} = n,$$

where n is any integer 1, 2, 3, ... Substituting h/mv for λ, we have

$$\frac{2\pi r m v}{h} = n,$$

or the angular momentum of the electron in its orbit mvr is

$$m\,vr = \frac{nh}{2\pi}.$$

This is the condition imposed on the angular momentum by Bohr, and it thus appears that it is somehow associated with the wave nature of the electron. Soon after this work of De Broglie, wave mechanics was introduced by Schrödinger in 1926. In this the motion of an electron in an atom is governed by laws other than the classical ones, its position and velocity being given in terms of a probability. This theory has been greatly extended by a large number of mathematical physicists— in particular, by Heisenberg, Bohr, Dirac, and others. This theory, which is highly mathematical, has been successful in interpreting the line and band spectra in a satisfactory manner. Being mathematical, it has no satisfactory pictorial representation, and for this we must use the energy levels of Bohr which are still retained, though they do not have the definite positions shown in Fig. 37-9. In the Bohr theory the integer n for the orbit is often called the *principal quantum number*. With the new theory there are four quantum numbers associated with an atom, the principal n, azimuthal l, magnetic m, and spin s. For each of these there are selection rules which pro-

hibit their taking on all possible values, so that it is obvious that the understanding of the extranuclear part of the atom has considerably advanced since Bohr's original work in 1913. We shall conclude this section with a brief statement on band or molecular spectra.

Molecules, in contrast to atoms, have, in general, very much more complex spectra than the latter. A molecule is considered to possess three forms of energy. These are the vibrational and rotational energies of the atoms composing the molecule as well as the electronic energy of the atoms themselves. Each of these energy states is quantized. The smallest energy differences are between the rotational states and the largest between the electronic levels of the atoms. The various lines in a single band arise from transitions between different rotational states of the molecule; different bands from vibrational states of the atoms within the molecule; and different *band systems* (only part of one system is shown in Fig. 37-10) from electronic transitions of the electrons of the molecule. It will be noted that, instead of "series limits," as in atomic spectra, molecular spectra have "band heads." Four of these band heads are marked with their appropriate wave lengths. The beautiful regularity of line arrangement within a band is most clearly seen in the first band (0–1) on the left. In the other bands, overlapping of the spectra mars this regularity.

Solids, liquids, and gases absorb light transmitted through them. Sometimes the absorption is over a large wave-length region, sometimes, especially in gases, with a high degree of selectivity for certain particular wave lengths. In Fig. 37-11 is a photograph of the absorption spectrum (molecular) of a triatomic molecule, SO_2, produced by

(0–1) 4216.0 A   (1–2) 4197.2 A   (2–3) 4181.0 A   (3–4) 4167.8 A

Fig. 37-10.—The band spectrum of the molecule CN

3100 A   3000 A   2900 A   2800 A   2700 A   2600 A

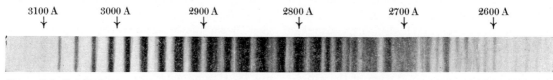

Photographs by Hans Beutler

Fig. 37-11.—Band-absorption spectrum of the molecule SO_2

the ultraviolet absorption of this visually colorless gas. Many band heads may be seen, but their complex structure is not resolved in this photograph.

37.10. X-rays

In 1895 Roentgen was experimenting with an electrical discharge tube (Fig. 37-12) at very low

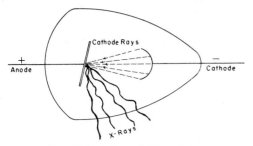

Fig. 37-12.—An early X-ray tube

pressure and was particularly interested in the fluorescence produced in a near-by zinc sulphide screen by radiation coming from the tube. He covered the discharge tube with black paper and still found the fluorescence. It was then discovered that the radiation coming from the tube could penetrate leather and other light materials. Not knowing the nature of the radiation, he called it the unknown X, so that this Roentgen radiation is now commonly known as X-rays.

When a fairly high voltage of the order of 20,000 volts is applied to a gas-discharge tube of the type used by Roentgen (pressure about 0.001 mm of Hg), there are cathode rays or electrons produced which move from the negative electrode, called the *cathode*, toward the positive electrode, the *anode*. When the electrons strike the anode, or target as it is sometimes called, then X-rays are produced. Modern X-ray tubes (Fig. 37-13) all have a high vacuum, and the electrons are produced at the cathode by a heated filament. Since the electrons give up most of their energy as heat and relatively little (less than 1 per cent at common voltages) as X-rays, it is usual to make the anode of a block of heavy metal. This block or target usually has cooling fins on the outside.

About 1912 Laue showed that X-rays could be diffracted, and he thus established the wave nature of this radiation. Measurements on the diffraction patterns show that the wave lengths are of the order of 1 A, or 10^{-8} cm. The spectrum from an X-ray tube consists of a continuous spectrum,

whose short-wave-length limit depends on the voltage applied, and a bright-line spectrum, which is characteristic of the atoms in the target. The frequency or wave length of the short-wave-length limit of the continuous spectrum can be readily found from energy considerations.

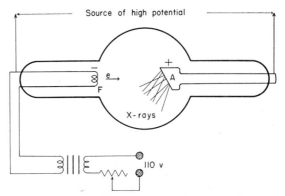

Fig. 37-13.—Diagram of a modern X-ray tube

Suppose there is a potential difference of V volts applied to an X-ray tube, then the energy of the electrons when they strike the target is eV. This energy may all go into producing a photon of frequency f_{max}, so that

$$h f_{max} = e V . \qquad (37-23)$$

The short-wave-length limit, λ_{min}, of the continuous spectrum is given by

$$\lambda_{min} = \frac{c}{f_{max}}$$

$$= \frac{h c}{e V} .$$

If a potential difference of 20,000 volts is applied to the tube, the minimum wave length of the continous spectrum is

$$\lambda_{min} = \frac{6.62 \times 10^{-34} \times 3 \times 10^{8}}{1.60 \times 10^{-19} \times 20,000}$$

$$= 6.21 \times 10^{-11} m = 0.621 A .$$

Most of the electrons striking the target give up energy, which goes into heat, before producing a photon, so that this continuous spectrum has a maximum at about double λ_{min} and then tails more or less indefinitely in the long wave lengths, as shown in Fig. 37-14. If the voltage applied to the X-ray tube is high enough to remove an electron from the inner orbits of the atom, then a bright-line or characteristic X-ray spectrum is also **559** produced. The wave lengths of these bright lines are usually measured by diffraction by means of a natural crystal, such as a sodium chloride, NaCl, as the diffraction grating. The grating space, d, of such crystals is of the order of 1 A, so that these are very suitable for use with X-rays.

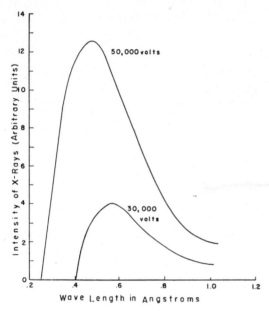

Fig. 37-14.—The continuous X-ray spectrum from a tungsten target.

In 1913 Moseley, working in Rutherford's laboratory, made a systematic study of the characteristic X-ray spectra of a number of elements. A typical picture of the photographically measured wave lengths from the elements arsenic ($Z = 33$) to rhodium ($Z = 45$) is shown in Fig. 37-15. This

figure shows that, as the radiating atom gets heavier, the atomic number increases, the characteristic wave lengths decrease, or their frequencies increase. A simple explanation of this characteristic line spectrum can be given in terms of the Bohr atom.

Moseley found, on studying the X-ray spectrum of many atoms, that the wave number, $1/\lambda$, of any X-ray line could be obtained by taking the difference in term values, as in the hydrogen spectrum. Thus he found that

$$\frac{1}{\lambda} = R\,(Z - s)^2 \left(\frac{1}{n_1^2} - \frac{1}{n_2^2} \right), \quad (37\text{-}24)$$

where R is the Rydberg number and s is a constant for a particular series, called the *screening constant*. The K-series X-ray lines correspond to $n_1 = 1$ and $n_2 = 2, 3, 4, \ldots$. The striking result found by Moseley was that equation (37-24) represents the K lines for all elements, provided that Z is given its appropriate ordinal number $1, 2, 3, \ldots$, 92 and $n_1 = 1$. This ordinal number was later known as the atomic number Z of the element and is numerically equal to the number of electrons outside the nucleus or the number of protons inside the nucleus. For hydrogen, $Z = 1$; for helium, $Z = 2$; and so on up to uranium, $Z = 92$, and now to the artificially produced elements as far as fermium (Fm), $Z = 100$.

If $n_1 = 1$ and $n_2 = 2$, a particular line of the K series, called the K_a line, is obtained. For the K_a lines of the different elements, equation (37-24) becomes

$$\frac{1}{\lambda} = \tfrac{3}{4} R\,(Z - s)^2. \quad (37\text{-}25)$$

A graph with $\sqrt{1/\lambda}$ as ordinate and Z as abscissa gives a straight line, and this is verified by experiment.

In order to interpret these results, let us consider the application of the Bohr theory to a heavy atom of atomic number Z. Such atoms have a nucleus with Z protons and Z electrons outside the atom. These Z electrons are arranged in definite energy states; the innermost one, corresponding to $n = 1$, is called the *K shell*, the next one, $n = 2$, the *L shell*, and so on for M, N, and O shells. The K shell can contain 2 electrons, the L shell 8 electrons, and so on. While there is only one energy state for the K electrons, the L electrons are in three subgroups, known as L_I, L_{II}, and

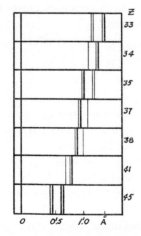

Fig. 37-15.—K series of X-ray spectra

L_{III}, and so on for the outer shells. These are correlated with the four quantum numbers previously mentioned. We shall ignore these divisions of the divisions of the shells and speak of the K, L, M, etc., shells. To a first approximation, these correspond to the orbits of the hydrogen atom shown in Fig. 37-9 with $n = 1, 2, 3$, etc.

In a heavy atom the innermost shells are all filled, so that, in order to produce the K-series X-rays, one of the K electrons must be removed. Thus the bombarding electron in the X-ray tube must have at least an energy of the binding energy of the electron in the K shell. With this vacancy in the K shell, we may picture an electron falling from the L shell, giving rise to the K_a line, so that

$$h f_{(Ka)} = E_K - E_L , \qquad (37\text{-}26)$$

where E_K and E_L are the respective energies in K and L shells. There is then a vacancy in the L shell, and an electron, falling from the M to the L shell, gives out the so-called L_a line.

In determining these energy levels, it is not correct in general to consider the electrostatic force to be between the charge Ze on the nucleus and the total charge Ze on the electrons but something less, because some of the electrons effectively screen the one considered, so that equation (37-24) has the factor $(Z - s)^2$ rather than Z^2, where s is the screening constant. For any particular series of lines such as the K lines, the quantity s is considered a constant. For the K series of molybdenum, $Z = 42$, the value of s is about 0.5.

37.11. The Photoelectric Effect

As early as 1887 it was found that electrical sparks jump much more readily between charged electrodes if the electrodes are illuminated by light, especially if the light contains some ultraviolet. This phenomenon is directly related to the photoelectric effect, as we shall now show. In order to do this, let us consider a clean zinc plate placed on a negatively charged electroscope, as shown in Fig. 37-16. If ultraviolet light is allowed to shine on the negatively charged zinc plate, it is found that the system is discharged. That the ultraviolet is needed is shown by the fact that the system does not discharge if a plate of ordinary glass, which absorbs ultraviolet light, is placed between the light source and the zinc plate. This experiment indicates that zinc emits negative

charges under the action of ultraviolet light but not with ordinary daylight. Many careful experiments by Millikan (1868–1953) and others showed that the negative particles emitted were electrons, often called *photoelectrons* to indicate their source. The experiments also showed that when a surface is illuminated with suitable monochromatic light

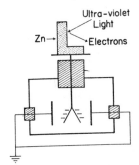

FIG. 37-16.—The photoelectric effect

of frequency f, (1) the number of photoelectrons emitted is proportional to the intensity of the light striking the surface and does not depend on the frequency f, and (2) the maximum velocity of the photoelectrons depends on the frequency f but is independent of the intensity of the light.

These experimental results are directly contrary to the classical idea of light, because, according to that, the velocity or kinetic energy of the photoelectrons should be proportional to the intensity of light incident in ergs per square centimeter per second or watts per square meter. In 1905 Einstein introduced the concept of a photon and gave a quantitative explanation of the photoelectric effect. Actually, the experiments were not done until later, and the incentive to do them came from Einstein's work. In Einstein's interpretation, which is a consequence of the principle of conservation of energy, each photon in the incident beam is assumed to give up *all* its energy, hf, to an electron. In order for this electron to get out of the metal, it has to expend some energy against the potential barrier of the metal. This energy is called the *work function* and is denoted by ϕ. The Einstein photoelectric equation can then be written as

$$\tfrac{1}{2} m v_m^2 = h f - \phi , \qquad (37\text{-}27)$$

where v_m is the maximum velocity of the ejected electrons and $v_m^2/2$ is their kinetic energy. Most of the photoelectrons give up more energy than ϕ in escaping from the metal surface, by undergoing **561** collisions within the metal. The value of ϕ, the work function, is least in the electropositive metals such as sodium, potassium, rubidium, and cesium. It is for this reason that these metals are generally used in photoelectric cells. From equation (37-27) it is seen that if ϕ is larger than hf, no photoelectrons would be emitted. This is the case with zinc and visible light. The photoelectric effect is an important means by which X-rays and γ-rays lose energy in passing through matter. It should be emphasized that the incident photon loses all its energy in the photoelectric effect. Notice that the short-wave-length limit of the continuous X-ray spectrum is an application of this equation, for in this case ϕ is negligible.

37.12. The Compton Effect

The Compton effect is another phenomenon in which X-rays lose energy on impact with electrons and whose interpretation requires the quantum theory. This effect was discovered by A. H. Compton in 1923. In this experiment a monochromatic beam of X-rays, the K_a line of molybdenum, was allowed to fall on a block of carbon, and the wave length of X-rays scattered by the carbon was measured with a crystal spectrometer. Compton found that, in addition to the original incident wave length, there was another companion line of longer wave length whose value depended on the angle between the incident and the scattered beams. Electrons called *recoil-electrons* were also produced. Carbon was used as the scatterer because it contains a number of relatively free electrons.

In order to explain this phenomenon, let us consider the impact of a photon of energy hf with a free electron initially at rest. In the impact between the photon and the electron there is conservation of energy and momentum, though the relativistic expressions must be used for these quantities in this case. A photon of frequency f is considered to have an energy of hf, momentum hf/c, and mass hf/c^2, where c is the velocity of electromagnetic radiation in a vacuum. After impact with a free electron there is produced a scattered photon of frequency f' (smaller than f) and a recoil electron, as shown in Fig. 37-17. The energy and momentum of the incident photon before impact are equal, respectively, to the energy and momentum of the scattered photon and recoil

electron after impact. Since momentum is a vector quantity, it follows that in Fig. 37-17 the components at right angles to the incident photon direction must be equal, or $hf' \sin \theta/c$ must equal the momentum of the recoil electron times $\sin \phi$.

In the Compton process the energy of the photon is degraded from hf to hf', so that there is always scattered radiation present. The photo-

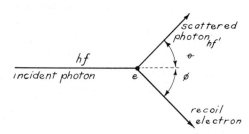

Fig. 37-17.—The Compton effect

electric effect and the Compton effect are the principal means by which radiation loses energy in its passage through matter. This is true until the energy of the incident photon is larger than 10^6 ev, which is the case for some gamma rays in radioactivity. This effect, in which a positive and negative electron are created, called *pair production*, will be discussed in the next chapter.

37.13. The Uncertainty Principle and the Principle of Complementarity

Up to the present we have been in the unsatisfactory position of having shown that electromagnetic radiation exhibits both wave characteristics (interference and diffraction) and particle or photon characteristics (Planck's black-body interpretation, the photoelectric and Compton effects). We have shown that there is a similar duality of wave and particle characteristics for electrons or any moving matter. This apparent dilemma has been solved, and we shall indicate how this has been done.

Let us consider the passage of light through a narrow slit of width Δx, as was done in § 35.12. On a screen, L distance from the slit, there is a diffraction pattern formed. The distance, d, of the first dark band from the central maximum is given by

$$\frac{\lambda}{\Delta x} = \frac{d}{L},$$

where λ is the wave length of the incident light.

Suppose we consider this phenomenon from the point of view of photons of frequency f and energy hf, where h is Planck's constant. The momentum, p, of these photons is given by

$$p = \frac{hf}{c} = \frac{h}{\lambda},$$

where c is the velocity of light and, from the wave equation, $c = f\lambda$. The photons do not all go directly through the slit, but some of them are given transverse momentum, as is shown by the fact that some of the light is diffracted from the central maximum. If p_x is the component of momentum of the diffracted photons going somewhere between the central maximum and the first minimum, then

$$\frac{p_x}{p} = \frac{d}{L} = \frac{\lambda}{\Delta x},$$

or

$$p_x = \frac{h\lambda}{\lambda \Delta x} = \frac{h}{\Delta x}.$$

For the photons which are diffracted beyond the first minimum, the component of transverse momentum, p_x, is larger than $h/\Delta x$. Notice that the narrower the width of the slit Δx, the larger is the value of the transverse momentum component. In other words, the closer we try to locate the position of a photon, by narrowing the slit, the larger becomes the uncertainty in the momentum of the photon. This is a particular example of what is known as *Heisenberg's principle of uncertainty or indeterminacy, which states that it is impossible to determine simultaneously, with absolute accuracy, the position and momentum of a photon.* This principle can be written as

$$\Delta p_x \Delta x \geqslant h .$$

This statement also applies to moving matter, such as electrons in motion, since they also show diffraction effects. In this case the waves are the De Broglie or matter waves. It is assumed that the square of the amplitude of these matter waves at any point is a measure of the probability of finding the moving particle at that point.

As we have seen, the allowed or Bohr orbits in the hydrogen atom are such that there is an integral number of these matter waves in the orbit, and outside such orbits the amplitude of the standing waves rapidly decreases. Thus the probability of finding the electron outside the allowed orbits is very small, for the amplitude of the standing waves is a maximum at the position of the allowed orbits and gradually tapers off to zero outside the allowed orbits. Thus we must think of the allowed orbits or energy levels in an atom as somewhat fuzzy or as having a range instead of the original narrow energy levels of Bohr.

We have seen previously that the uncertainty principle applies both to so-called "waves" and to so-called "particles." Natural phenomena show this duality of wave and particle characteristics, so that it is not possible to find simultaneously the position and the momentum of a particle or a photon with absolute accuracy. This also shows that a given phenomenon cannot be described both by the particle and by the wave theory. This has led Bohr to present the *principle of complementarity* for the two modes of describing nature. The wave aspect and the particle aspect of a given phenomenon can never be observed at one and the same time. If the wave theory provides the explanation, then the particle theory will not, and vice versa. For a description of nature, both theories are necessary and either theory alone is insufficient. The wave theory and the particle theory complement each other in the interpretation of natural phenomenon.

37.14. The Solid State and Crystal Types

In the discussion of properties of solids we have only mentioned the gross properties without attempting an interpretation. The interpretation or the theory of the solid state is quite complicated and still contains many unsolved problems. Although the scope of solid-state physics is quite broad and encompasses many important problems, such as plasticity, elasticity, ferromagnetic phenomena, electrical properties of matter, and optical properties of metals, to name a few of them, we shall confine our attention mainly to a brief study of some of the properties of the so-called "semiconductors."

For a broad division of solids we find some which have no sharp melting points, called *amorphous* materials, in contrast to a much larger group of *crystalline* materials with sharp melting points. From X-ray diffraction studies it is possible to locate the atomic groupings within substances. The amorphous substances, such as pitch or glass, have irregular atomic arrangements and are considered to be highly viscous liquids. In the crystal-line substances the atoms are in a regular array in space. It is this regular array of the atoms which permits one to use crystalline substances as diffraction gratings for X-rays. In what follows we shall use the words "solid state" to mean the "crystalline state."

In a crystal there is an orderly array of atoms or molecules built up in three dimensions. This orderly array implies an elementary lattice structure or unit cell from which the crystal can be constructed by repetition in three dimensions. One basis for classifying crystals is the nature of the forces holding the atoms within the crystal lattice. The three crystal types which concern us are called the *ionic*, *covalent*, and *metallic* types.

Salt crystals like sodium chloride (NaCl) or lithium fluoride (LiF) are examples of ionic crystals. The name *ionic crystal* is given because the binding forces keeping the crystal intact are electrostatic ones between the ions present, such as Na⁺ and Cl⁻ in the NaCl molecule. In Fig. 37-18 the cubic lattice structure of the NaCl

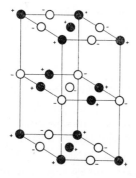

Fig. 37-18.—The cubic lattice structure of the salt crystal, sodium chloride (NaCl). The sodium ions are shown as + ● and the chlorine ions as − ○.

crystal is shown. While the electrical conductivity of ionic crystals is very low at low temperatures, nevertheless, if sufficient thermal energy is supplied to the crystal—that is, sufficiently high temperature is attained—the binding forces break down, the crystal melts, and this melted salt is a good conductor of electricity.

In the covalent crystals there is no transfer of charge as in the ionic ones but a sharing of electron-pairs between the atoms. Diamond, silicon, and germanium are examples of the covalent crystals. The atomic number and number of elec-

trons in the successive energy levels of these atoms are given in Table 37-2.

From the table it is seen that each of these atoms has four electrons in the outer level or each

TABLE 37-2

ELECTRONIC ARRANGEMENT OF
THREE ATOMS

Atom	Atomic No.	Electron Arrangement
C (diamond).........	6	2, 4
Si..................	14	2, 8, 4
Ge..................	32	2, 8, 18, 4

has four valence electrons or each is tetravalent. In the crystal each atom tends to form covalent or electron-pair bonds with four neighboring atoms, as shown in Fig. 37-19. For purposes of discussion

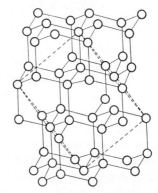

Fig. 37-19.—The arrangement of atoms in the germanium structure. Each germanium atom is surrounded by four others, symmetrically disposed at the corners of a tetrahedron. This is also the structure of the diamond crystal, with carbon atoms replacing germanium.

we shall represent the three-dimensional tetrahedron array of atoms (Fig. 37-19), in a two-dimensional array (Fig. 37-20). Each germanium atom is electrically neutral and is bound to its neighbors by sharing its four valence electrons, so that to each atom there are four valence bonds, each representing two shared electrons. Now these shared valence electrons can, according to quantum mechanics, only have a range of energies or lie in a band of energy and still remain attached to the atoms in the valence or sharing process.

In such a valence-bond structure every electron is tightly bound, there are no free electrons to partake in the conduction process, and the ideally perfect covalent crystal type would be an insulator. Conductivity can be produced in the perfect covalent crystal in a number of ways, all of which involve the breaking-down of the valence bonds. Energy in the form of high-energy particles or quanta of radiation or thermal energy can break the bonds. Conductivity in diamond has been used for detecting high-energy nuclear particles in the so-called "crystal counters." Another manner

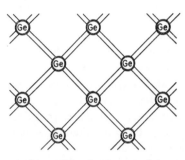

Fig. 37-20.—A two-dimensional array of germanium atoms schematically depicting the covalent or electron-pair bonds of the germanium crystal.

of speaking of the energy required to break the covalent bonds is that of giving the valence electrons sufficient energy to move them from the valence energy band to the conduction energy band. Both statements imply that an electron has been removed from the sharing arrangement to become free. The atom core from which this electron is taken has acquired a unit positive charge. Now a neighboring valence electron may, because of thermal vibrations, etc., become attracted to this positive core and neutralize it, leaving behind another positive core. Thus these positive cores, or *holes* as they are called, can migrate about the crystal in a manner similar to that of the free or conduction electron. Quantum mechanics has shown that there is a very high degree of symmetry in the behavior of electrons and the behavior of these holes. Basic to the theory of semiconductors is the theory that conduction can occur in two ways, one being conduction by electrons and the other called "deficit conduction" or conduction by holes. Before discussing the semiconductors, further, we shall briefly mention the third crystalline type, the metallic type.

In metals the positive ions are held together by a cloud of mobile free electrons not localized to any particular lattice site. This simple model of the metallic state is capable of explaining many char-

acteristics of metals—their high electrical and heat conductivities, optical properties, and many metallurgical phenomena.

37.15. Crystal Imperfections

In an ideal crystal the atoms or ions are arranged in a perfectly orderly array throughout the structure. Such perfect crystals are very rare indeed, and consideration must therefore be given to real crystals, which possess imperfections and flaws. These imperfections may take several forms: interruptions of perfect periodicity in the structure due to random thermal vibrations of the atoms or due to the absence of an atom from its proper lattice site (a vacancy); presence of extra atoms between lattice points (an interstitial atom); presence of impurity atoms (impurity centers); shifts of the entire atomic planes within the crystals (dislocations); etc. (see Fig. 37-21 [*a*], [*b*]).

It has been observed that the gross properties of crystals can be broadly divided into two classes: (*a*) structure-sensitive properties, such as elastic limit, tensile strength, ferromagnetic permeabilities, thermoelectric effects, conductivity of insulators and semiconductors, and (*b*) structure-insensitive properties, such as specific heat, density, and coefficient of expansion, to name a few. The structure-insensitive properties of a crystal can be explained in terms of the structure of an ideal crystal, whereas many of the structure-sensitive properties of crystals have been explained in terms of crystal imperfections of various types. It is interesting to observe that theoretical deduction of the stress necessary to rupture a crystal based on a perfect crystal model yields values about a thousand times greater than observed values. This discrepancy has been ascribed to the presence of minute flaws or faults within a real crystal. Many of the unusual electrical properties of the germanium crystal are due to the deliberate introduction of imperfections into the germanium lattice.

In our discussion of atomic structure we pointed out the importance of energy levels of individual atoms in explaining many observed properties of atoms—for example, the existence of characteristic line spectra. In that discussion we assumed that the individual atoms were sufficiently far apart that the orbits of their valence electrons did not interfere with one another. As a consequence, the allowed energy levels were discrete and separated, **565** as shown in Fig. 37-8. We wish to consider what modifications occur to these energy levels if a collection of atoms are brought together to form a crystal.

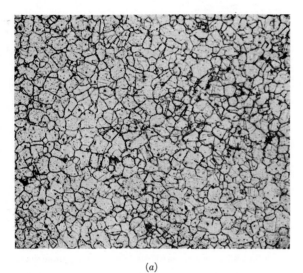

(a)

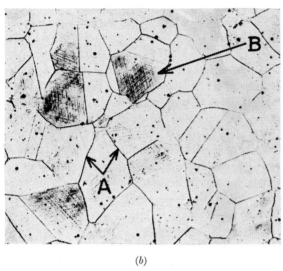

(b)

Fig. 37-21.—(a) A microphotograph (magnification 150×) of a stainless-steel rod, hot-rolled from a 2-in diameter to $\frac{7}{8}$ in and showing small polycrystalline aggregates. The surface was prepared by etching with aqua regia in glycerine. (b) A microphotograph (magnification 150×) of the same stainless-steel sample but annealed at 2,100° F and then quenched in water. Shown at A are the grain boundaries of a single crystal. The coarser grain size resulted from the high annealing temperature. The lines shown at B are due to imperfections within the single crystal.

Let us assume that N atoms, initially far apart, so that no interaction occurs between atoms, are brought together to form a crystalline solid. As

soon as the atoms are brought close together, interaction between atoms takes place, giving rise to the forces that keep the atoms together in the form of a crystal. Under this condition, we can no longer speak of the energy levels of the individual atoms, but, instead, we must speak of the energy levels of the system of atoms. Each discrete energy level of the isolated atoms spreads out into a band of allowed energy levels, the width of the bands depending on the strength of the coupling between atoms.

Each band contains as many allowed discrete levels as there are atoms in the solid. Since the density of atoms in a crystalline solid is of the order of 10^{22} atoms/cm³, each band contains a tremendous number of levels separated as very small energy differences, so that, for practical purposes, the band may be considered as a continuum of allowed energy levels.

Let us first consider the effect of thermal vibrations in the germanium lattice of Fig. 37-20. An electron is held in the covalent bond with an energy of about 0.7 ev. At absolute zero we assume that the electrons are filling the valency band and that none are in the conduction band. The conductivity of the germanium crystal at 0° K is zero. As the temperature is increased, some of the valency electrons gain the energy of 0.7 ev and become free or move from the valency band to the conduction band. The atomic core from which the valence electron leaves has a net unit positive charge and becomes a positive hole. Both the electrons in the conduction band and, to a lesser extent, the positive holes in the valence band can contribute to the electrical conductivity. This intrinsic conductivity tends to be dominant at high temperatures, while the conductivity due to impurity atoms is dominant at low temperatures.

It is by the deliberate introduction of suitable impurity atoms into the crystal lattice that the semiconductors have become important. (The presence of one boron atom in 10^5 silicon atoms has been found to increase the electrical conductivity of silicon by a factor of 10^3.) Let us consider the introduction of impurity atoms of valency 5 and also of valency 3 into the tetravalent covalent crystal of germanium. Phosphorus, P, atomic No. 15 and electron arrangement (2, 8, 5); arsenic, As, 33 (2, 8, 18, 5); and antimony, Sb, 51 (2, 8, 18, 18, 5) all have five valence electrons. When such atoms are introduced in very low concentration into the germanium melt, it is found that an impurity atom replaces a germanium atom in the crystal. Consider an arsenic atom in a germanium crystal (Fig. 37-22). The four covalent bonds of the germanium lattice require only four of the five valence electrons of the arsenic atom. At low tem-

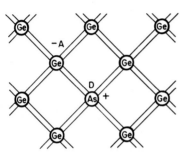

Fig. 37-22.—The introduction of arsenic (As) atoms in the Ge crystal introduces a positive hole at D and an electron at A.

perature the extra electron is weakly bound to the arsenic $+5$ core, but at room temperature the thermal energy is sufficient to break this bond and free this electron. The arsenic atom is called a *donor* atom because it donates an electron to the conduction band, and the electrical conductivity from this cause is by *negative electrons*, so that the material is said to be *n-type*.

Suppose a trivalent atom such as boron, B, having an electron arrangement of 2, 3, is introduced in very low concentration into a germanium crystal (Fig. 37-23). Boron has three valence elec-

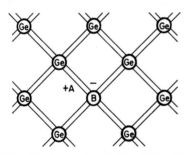

Fig. 37-23.—The germanium lattice (Ge) with boron (B) acceptor atom giving rise to p-type conduction by the hole.

trons and a core of charge $+3$. The three electrons are not enough to complete the four covalent bonds of germanium. At room temperature the thermal energies are sufficient for an electron from a neighboring valence bond to move in and fill up the incomplete band at the boron atom site. This leaves an electron deficiency or positive hole in the atom site from which the electron was removed.

The boron atom is known as an *acceptor* atom, and, as we see, there is an immobile negative core at the boron site and a mobile hole or positive carrier. Such a semiconductor in which the conduction is mainly by *holes* is called a *p-type* or *positive carrier* type semiconductor.

We may summarize the foregoing qualitative statements in the following manner:

a) In an intrinsic semiconductor the conduction is by both electrons and holes.

b) In an n-type semiconductor (donor atoms) the conduction is solely by electrons at low temperature and by electrons (major carrier) and holes (minor carrier) at high temperatures.

c) In p-type semiconductor (acceptor atoms) the conduction is solely by holes at low temperatures and by holes (major) and electrons (minor) at high temperatures.

Since the energy necessary for electron transitions of n- and p-type materials is supplied by the thermal vibrations of the lattice, it can be shown that the electrical conductivity, σ, at absolute temperature, T, is given by

$$\sigma = \sigma_0 e^{-u/kT} ,$$

where σ_0 depends on the impurity concentration or is the value of σ at absolute zero, u is the activation energy (energy difference of donor or acceptor and conduction bands), k is the Boltzmann constant, and e the exponential. The electron and hole concentrations in semiconductors of practical interest range from 10^{13} to $10^9/cm^3$, many orders of magnitude less than electron concentrations in metals.

37.16. A Semiconductor Rectifier; an n-p Junction

Let us consider the junction between n-type germanium and p-type germanium as shown in Fig. 37-24. If this junction is to have rectifying

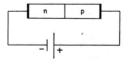

Fig. 37-24.—An n-p junction illustrating the direction of easy flow and low resistance through the junction.

properties, there must be a direction in which the resistance to current flow is high and in the opposite direction a low resistance. The direction of **567** easy flow, or the forward direction, is that for which the n-type is at negative potential and the p-type at positive potential.

When the junction is formed, there will be a migration of electrons from the n to the p side and of positive holes from the p to the n side. This sets up a potential barrier which prevents any further flow.

If a battery is now connected so that the n side is negative and the p side positive (Fig. 37-24), the potential barrier is lessened, and an electron current flows from n to p and a hole current from p to n.

When the battery voltage is reversed (Fig. 37-25), the current is very small and is due principal-

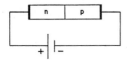

Fig. 37-25.—An n-p junction illustrating the direction of high resistance.

ly to the migration of the small number of positive holes in n and the corresponding small number of electrons in p. The voltage-current characteristic of an n-p junction rectifier is shown in Fig. 37-26,

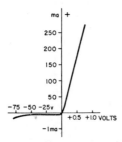

Fig. 37-26.—The voltage-current characteristics of an n-p junction. Note the change in scales for the forward and reverse voltage bias.

in which different scales are used to show up the reverse current (Fig. 37-24) as compared to that in the forward direction (Fig. 37-25).

37.17. The n-p-n Transistor

The name *transistor* was coined at the Bell Telephone Laboratories, where it was invented in 1948 by Bardeen and Brattain. It was so named because it is a resistor or semiconductor device which is able to amplify electrical signals when they are transferred through it from the input to the output

568

terminals. Already it is displacing the vacuum-tube amplifier in certain cases, though there is no fear that the usual vacuum tube will disappear. We shall give a brief account of one form of transistor called the *n-p-n junction transistor*. Such a transistor is made of a single-crystal germanium bar which, through the efforts of chemists and metallurgists, can be made as shown in Fig. 37-27,

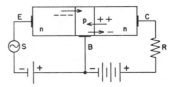

Fig. 37-27.—The n-p-n transistor showing the p portion much exaggerated in size, with emitter electrode *E*, base *B*, and collector *C*. *R* is the load resistor and *S* the applied signal.

with the p-type made a fraction of a millimeter or so in width. The transistor is less than 0.5 in long.

To this transistor three contacts are securely fastened in the emitter *E* portion, the base *B*, and the collector *C* portion. The electrode *C* is biased positively by a battery, which may range up to the order of 40 volts, while the emitter is biased negatively a small fraction of a volt. Because of the positive bias on the collector, there will be a small residual current of positive holes and electrons from the collector to the base. This current is largely independent of the collector voltage, as shown in Fig. 37-28 (*a*).

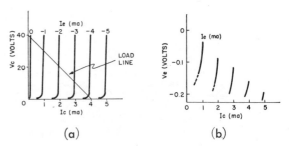

(a)   (b)

Fig. 37-28.—Static characteristic curves for a particular n-p-n junction transistor.

The small negative bias on the emitter electrode causes the electron-rich n-type emitter to produce an electron current in the base or p portion. Since the p-type region is very thin, practically all the electrons from the n-type collector region diffuse through the base region into the n-type region *C* and are collected there. Now the forward impedance of the emitter region is very low compared to the high reverse impedance of the collector region. Practically the same current flows in the E and C regions, while the E voltage is a fraction of a volt and the C voltage is up to 40 or so volts. This result is seen from the static characteristic curves (Fig. 37-28 [*a*], [*b*]) for one of these transistors. Here it is seen that the collector current, I_c, is practically equal to the emitter current, I_e, and that the impedance of the collector (V_c/I_c) is much larger than that of the emitter (V_e/I_e). Although the current gain can never exceed unity ($I_c < I_e$), the voltage and power gain can be very large because of the large ratio of impedances. This brief introduction gives only the barest glimpse into the phenomena associated with the solid state. On the practical side, amplifiers and oscillators are built with transistors so that their size becomes almost ridiculously small. A small signal having a range of a fraction of a volt produces current changes in the large-load resistor R of Fig. 37-27. If the collector battery has a voltage of 20 volts, then a load line for a load of 10,000 ohms can be drawn as shown in Fig. 37-28 (*a*). The output signal can then be readily obtained from the input signal and these characteristic curves, as was done in the electronics chapter.

Since the theory of the solid state is developing rapidly, it is to be expected that modifications in the transistor described will appear. Fundamental research and technological development are proceeding in this field at an almost unprecedented rate.

37.18. Résumé

The following is a brief listing of the definitions, principles, and theories, given in this chapter, which you should know:

The types of spectra and their importance

The continuous spectra from which the Stefan-Boltzmann and Wien laws are confirmed, together with Planck's quantum hypothesis

The series line spectra for hydrogen and the empirical equation for the wave numbers of the lines

The Ritz combination principle and the importance of term values

The Bohr theory for the line spectra of hydrogen; the assumptions and predictions of this theory

Ionization potentials

De Broglie's interpretation of the angular-momentum assumption of Bohr

X-rays—their production, analysis, and properties

The particle properties of radiation as shown by the photoelectric and Compton effects

The Heisenberg principle of uncertainty and the Bohr principle of complementarity

The experimental and theoretical distinctions between amorphous and crystalline materials

The distinction between ionic and covalent crystals

The comparison of metallic conduction with conduction in semiconductors by electrons and holes

The effects of very small quantities of impurities in semiconductor crystals

Donor atoms with n-type carriers contrasted with acceptor atoms and p-type carriers

Theory of rectification by a semiconductor

The theory of voltage amplification with an n-p-n semiconductor crystal

Queries

1. A simple method of observing spectra is to hold a small transmission grating in front of the eye. If the source were a Bunsen flame in which some sodium compound had been placed, how would the spectrum appear?

2. If the absolute temperature of a black body is doubled, by what factor is the total emitted radiation increased?

3. If the wave length of a monochromatic violet line is half that of a monochromatic red line, what is the ratio of the energy of a photon of violet light to one of red light?

4. In which case does the hydrogen atom have more energy: (*a*) when the electron is in the innermost orbit, (*b*) when the electron is in one of the outer orbits?

5. Explain why the De Broglie waves for a moving macroscopic object, such as a moving tennis ball, could never be detected.

6. Is it theoretically possible to have a hydrogen spectrum beyond the Pfund series (eq. [37-8]), starting with $n = 7$? If so, why is it not listed in § 37.5?

7. Explain qualitatively the essential differences between the photoelectric effect and the Compton effect.

8. If the zinc plate in Fig. 37-16 were positively charged, would it be discharged by ultraviolet rays? Explain.

9. Why would you expect the spectra of sodium, potassium, etc., or of atoms in the first column of **569**

the periodic table to be similar to that of hydrogen?

10. Would the spectra of the singly ionized atoms in the second column, such as Be$^+$, Mg$^+$, etc., be similar to that of hydrogen?

11. Does a piece of ice emit radiation, and, if so, what kind would it be?

12. Does the principle of uncertainty have any practical application in the motion of a billiard ball?

Practice Problems

1. Find the energy of a photon of red light of wave length 6000 A. Ans.: 3.31 × 10^{-19} joule.

2. A tungsten lamp filament has a surface temperature of 2,700° K. Find the wave length having maximum energy and the total energy emitted per square centimeter per second from the filament.

3. Wien's displacement law can be obtained from Planck's radiation law in the following manner. Find λ_m by differentiating equation (37-5), setting $dE_\lambda/d\lambda = 0$, and giving the transcendental equation

$$\frac{h\,c}{\lambda_m kT} = 5\,(1 - e^{-hc/\lambda_m k\,T}).$$

Show that this is satisfied for $hc/\lambda_m kT = 4.965$; and, by substitution, show that $\lambda_m T = 0.2897$ (cm ° K).

4. Using equation (37-8), calculate the wave lengths of the first four lines of the Balmer series. From these data find the wave length of the first line of the Paschen series.

5. Explain why the spectrum of helium should be similar to that of hydrogen. How would equation (37-20) be modified for an ionized helium atom. Ans.: Multiplied by 4 (give reasons for answer).

6. Find the energy in electron-volts of a photon of red light of wave length 6000 A; of violet light of wave length 2000 A; of X-rays of wave length 1 A; and of gamma rays of wave length 0.1 A.

7. The work function ϕ in equation (37-27) has a threshold wave length of 6000 A for a certain metal. (a) What is the longest-wave-length photon which could cause photoelectrons to be emitted from the surface? (b) What is the maximum energy of the photoelectrons emitted when light of wave length 4000 A shines on the surface? Ans.: 6000 A; 1.65 × 10^{-19} joule.

8. Light of wave length 3000 A falls on a surface from which are emitted photolectrons having the maximum energy of 2 ev. Find the work function ϕ for the surface in electron-volts.

9. Find the radii of the first three Bohr orbits for hydrogen and the energy of the atom when the electron is in these orbits. Ans.: 0.53 A; 2.1 A; 4.77 A; 2.18 × 10^{-18} joule; 0.54 × 10^{-18} joule; 0.24 × 10^{-18} joule.

10. Find the wave length of a De Broglie wave for an electron moving with speeds of 10^8 cm/sec and 10^9 cm/sec.

11. Find the voltage which must be applied to an X-ray tube to give a short-wave-length limit of 1 A. Ans.: 12,412 volts.

12. How much energy is required to raise the hydrogen atom from the ground state $n = 1$, to the excited state $n = 5$? What wave length is emitted if the atom returns directly to its ground state?

13. Show that for an element having an atomic number Z the expression for the energy given by equation (37-18) would be approximately increased by Z^2. Show that the wave length of the K_a line of molybdenum $Z = 42$, $s = 0.5$, has a wave length of about 0.71 A. (Use eq. [37-25].)

14. In § 37.8 it is stated that when an electron reaches an energy of about 1.6 × 10^{-18} joule in hydrogen gas, the first line of the Lyman series for hydrogen appears. Show that this should be so; use the data given in Table 37-1.

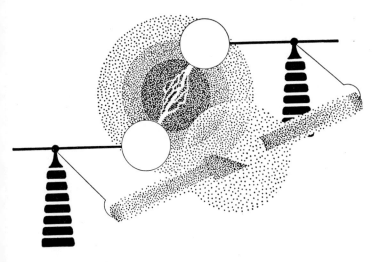

CHAPTER 38

The Electron, Isotopes, and Nuclear Processes

38.1. Cathode Rays

If the air in a discharge tube (Fig. 38-1) is gradually pumped out and a high voltage applied to the tube, then, as the pressure reaches a few centimeters of mercury, long, thin, reddish streamers pass down the tube. These expand as the pressure is further reduced until the whole inside of the tube is filled with a soft red glow. This illumina-

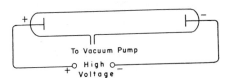

FIG. 38-1.—A gas-discharge tube

tion, from a theoretical point of view, is quite complicated. In practice it is employed extensively for illuminating signs.

A still further reduction of pressure causes this glow to begin to retreat from the negative electrode, and the tube takes on the appearance shown in Fig. 38-2 at a pressure of about 0.1 mm of mercury. The positive column is generally striated, the number, color, and width of the striations depending upon the pressure, the nature of the gas,

and the current density. If the pressure of the gas is reduced below 0.01 mm of mercury, the illumination almost completely disappears, the Crookes dark space having filled the tube. At still lower pressures, of the order of 10^{-4} mm of mercury, another phenomenon ensues which may be de-

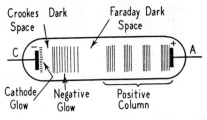

FIG. 38-2.—Discharge at a pressure of about 0.1 mm of mercury.

scribed as due to the presence of some sort of ray proceeding from the cathode and producing a fluorescence of yellowish-green color on the glass opposite the cathode. These rays are called *cathode rays*, since they appear to come in straight lines from the cathode. The cathode rays are made up of electrons which are mainly ejected from the cathode by positive ions. These electrons while traveling down the tube occasionally strike an atom of the gas, knocking out one or more electrons and leaving behind an atom minus one or more electrons, called a *positive ion*. In general, the electrons and positive ions pass down the discharge tube with few, if any, collisions with gas atoms, since the pressure of the gas is so low. It was with one of these cathode-ray tubes that Roentgen discovered X-rays. We shall now briefly describe the methods by which the mass and the charge of the particles making up the cathode rays were determined.

38.2. Measurement of e/m for Cathode Rays

In order to find the ratio of the charge, e, to the mass, m, of the particles making up the cathode rays, J. J. Thomson, about 1897, measured the deflection of these rays produced by known electric and magnetic fields. A diagrammatic sketch of the apparatus used by Thomson is shown in Fig. 38-3.

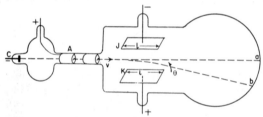

FIG. 38-3.—Thomson's apparatus for measuring e/m

A discharge tube shown at the left has a cathode, C, and an anode, A, with fine slits in it to define the cathode-ray beam. By connecting the two parallel plates J and K to a battery, the cathode rays can be deflected. There is also a magnet whose field is perpendicular to the plane of the paper in the region of the plates. This magnet, not shown in the figure, can also be used to deflect the cathode rays.

Suppose a constant potential difference of V volts is applied to the discharge tube. Then, if the pressure is sufficiently low, cathode rays will pass through the slit system in the anode A with a velocity v, given by

$$V e = \tfrac{1}{2} m v^2 . \qquad (38\text{-}1)$$

If the mass, m, of the particles or electrons making up the cathode rays is expressed in kilograms and their charge, e, in coulombs, then the velocity, v, will be given in meters per second. These cathode rays travel in a straight line and will strike at a if there is no electric or magnetic field in the inter-

vening space. The end of the tube at the right is coated with some fluorescent material such as zinc sulphide, which fluoresces when cathode rays strike it.

Suppose an electric field $\mathcal{E}$ is applied in an upward direction between the plates J and K by connecting them to a battery. It is found that the particles comprising the cathode rays are deflected downward,* showing that they have a negative charge. The vertical electrical field, $\mathcal{E}$, gives the cathode rays a vertical acceleration, a, which is given by

$$a = \frac{\mathcal{E} e}{m} . \qquad (38\text{-}2)$$

In the mks system $\mathcal{E}$ is measured in volts per meter, e in coulombs, and m in kilograms, so that the acceleration a is expressed in meters per second per second. If the length of the plates is l, then the time t for a cathode-ray particle of velocity v to pass through the plates is

$$t = \frac{l}{v} . \qquad (38\text{-}3)$$

The vertical component, v_y, of the velocity of the cathode-ray particles as they emerge from the plates is

$$v_y = a t = \frac{\mathcal{E} e l}{m v} . \qquad (38\text{-}4)$$

Thus the cathode-ray particles have two perpendicular components of velocity as they emerge from the plates, the horizontal one, v, and the downward vertical one, v_y. They must then emerge at some angle, θ, with the horizontal given by

$$\tan \theta = \frac{v_y}{v} = \frac{\mathcal{E} e l}{m v^2} . \qquad (38\text{-}5)$$

The cathode rays then proceed in a straight line and strike the fluorescent screen at some point b (Fig. 38-3).

Suppose a magnetic field is now introduced into the region between the plates at right angles to the plane of the paper so as to deflect the cathode

---

* Several investigators before Thomson had tried to deflect cathode rays with an electric field and had not succeeded. The failure was shown to be due to the presence of positive and negative ions between the plates J and K. These ions move relatively slowly and set up a space charge which effectively neutralizes the electric field. Thomson overcame this difficulty by keeping the pressure of gas between the plates so low that no positive or negative ions were formed by collision between the cathode rays and the atoms of the gas.

rays upward. The cathode rays are then acted on by a downward force, $\mathcal{E}'e$, due to the electric field $\mathcal{E}'$ and an upward force $B'ev$ due to the magnetic field B'. If these fields are adjusted so that the resultant vertical force is zero, then the cathode rays will move through the plates undeflected. Suppose $\mathcal{E}'$ and B' are the values of the electric and magnetic fields, respectively, so that the resultant force is zero. Then

$$e\mathcal{E}' = B'ev , \qquad (38\text{-}6)$$

where v is the horizontal velocity of the cathode rays. Hence

$$v = \frac{\mathcal{E}'}{B'} . \qquad (38\text{-}7)$$

In this manner the velocity of the cathode rays is determined. The units employed in equation (38-7) are $\mathcal{E}'$ in volts per meter and B' in webers per square meter, so that v is given in meters per second.

Suppose the electric field is removed, leaving only a magnetic field B present. The force Bev on the cathode rays is perpendicular to their velocity v. This is a centripetal force and causes the path of the cathode rays to be bent in the arc of a circle of radius R in the region of the magnetic field. Hence

$$Bev = \frac{mv^2}{R}$$

or

$$BR = \frac{mv}{e} . \qquad (38\text{-}8)$$

From the deflection of the cathode particles it is possible to determine the radius R so that, if B and R are known, the quantity mv/e can be determined. Similarly with an electric field alone, the quantity e/mv^2 can be determined from equation (38-5) plus various measurements. Thus if v is measured in a separate experiment, using equation (38-7), the ratio of the charge to the mass (e/m) of the cathode particles can be determined.

As far as Thomson could determine, the ratio e/m was independent of the kind of gas used in the discharge tube. He therefore concluded that the cathode particles were a universal constituent of all atoms. These particles forming the cathode rays became known as electrons, a name which had been proposed some time earlier, about 1891, by the Irish physicist, Johnstone Stoney.

Many experiments have been performed since Thomson's time to determine the ratio e/m of an electron. These all involve the use of electric and magnetic fields. The presently accepted value of the ratio of charge to mass of an electron is

$$\frac{e}{m} = 1.75890 \times 10^{11} \text{ coulombs/kg} .$$

The mass m here is the rest mass of an electron and, since it is assumed that the velocity of the cathode-ray electron is small compared to the velocity of light, it follows that the relativistic correction to the mass is negligibly small.

This value for e/m is 1,836.12 times as large as that for a hydrogen ion found in electrolysis. Now at the time Thomson found this value, or at least approximately this value, there was no direct way of knowing whether the charge or mass of the cathode particle was larger or smaller than that of the hydrogen ion. However, there were certain lines of evidence which led Thomson to consider that the mass of the cathode particle was much smaller than that of the hydrogen ion, although the charges were numerically equal. In particular, Lenard in Germany had shown that cathode rays would pass through thin films which were impervious to gases. In 1881 Helmholtz had written: "The most startling result of Faraday's Law [of electrolysis] is perhaps this: if we accept the hypothesis that the elementary substances are composed of atoms, we cannot avoid concluding that electricity also, positive as well as negative, is divided into definite elementary portions which behave like atoms of electricity." However great the indirect evidence for the existence of elementary particles of negative electricity of about one two-thousandth the mass of the hydrogen atom, it remained for other investigators to determine the charge e definitely by experiment. Some of these experiments will now be described.

38.3. The Charge on the Electron

In 1897 Townsend, a student in Thomson's laboratory, made an approximate measurement of the charge e by watching a charged cloud fall in a vessel. This was followed by the work of H. A. Wilson in 1903 at Cambridge and of R. A. Millikan, working from about 1909 to 1917 at the University of Chicago. Since Millikan's method is a refinement of the earlier methods, we shall briefly describe his experiment. This experiment, which is known as "the oil-drop experiment" consists essentially in following the motion of a tiny drop of oil in an electric field. The apparatus consists of a pair of parallel plates, PP (Fig. 38-4), inclosed in a glass container to avoid air currents. An atomizer, S, is used to produce tiny drops of oil. One of these finds its way through a small hole in

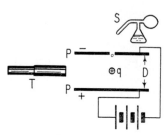

FIG. 38-4.—Millikan's oil-drop experiment

the upper plate and moves into the air between the plates. The atomizing action of the sprayer leaves the oil drops electrically charged, so that the one between the plates has a charge which may be positive or negative. First the motion of this charged oil drop is determined in the absence of an electric field and then later under known electric fields between the plates.

In the absence of an electric field the oil drop falls under gravity with a constant speed. The viscosity of the air exerts a resistance force on the oil drop which eventually becomes equal to the weight of the drop. (This is a situation similar to that of an open parachute falling through the air.) In the case of the oil drop the resistance force due to viscosity is given by equation (9-45) as

$$F = 6\pi\eta v_g R \ ,$$

where R is the radius of the drop, v_g is the constant downward speed under gravity, and η is the coefficient of viscosity.

The net weight of the oil drop, taking into account the buoyant force of the air, is

$$m g = \tfrac{4}{3}\pi R^3 \left(\sigma - \rho\right) g,$$

where σ is the density of the oil drop, ρ the density of air, and g the acceleration due to gravity.

For a constant terminal speed v_g, that is, no acceleration, we have

$$6\pi R\eta v_g = \tfrac{4}{3}\pi R^3 \left(\sigma - \rho\right) g$$

or

$$R^2 = \frac{9}{2}\frac{\eta v_g}{(\sigma - \rho) g}. \qquad (38\text{-}9)$$

Since all the quantities on the right-hand side of equation (38-9) either are known or can be measured, it follows that the radius R of the drop can be determined. The velocity v_g is measured by timing the oil drop over a known distance by means of the telescope, T, in Fig. 38-4.

Suppose that an electric field, $\mathcal{E}_1$, is established between the plates by the battery shown in the figure. If the electric field $\mathcal{E}_1$ is upward and there is a charge $+q_1$ on the oil drop, then it experiences an upward force of $\mathcal{E}_1 q_1$ and a resultant upward force of $\mathcal{E}_1 q_1 - mg$. The oil drop then moves upward with a constant speed v_1, given by

$$6\pi R\eta v_1 = \mathcal{E}_1 q_1 - m g \ . \qquad (38\text{-}10)$$

If v_1 is measured, the value of q_1 can be found from equation (38-10).

Millikan then changed the charge on the oil drop by ionizing the air between the plates with either X-rays or radioactive material. This produced positive and negative ions of air, some of which became attached to the oil drop, thus changing its charge from q_1 to some new value, q_2. The ions of air have a negligible weight compared to that of the oil drop, so that the mass m of the oil drop remains constant. With a charge q_2 on the oil drop, the electric field $\mathcal{E}$ must be changed to some value, $\mathcal{E}_2$, in order for the drop to move with some constant speed, v_2. This speed v_2 is given by an equation similar to equation (38-10), or

$$6\pi R\eta v_2 = \mathcal{E}_2 q_2 - m g \ .$$

The charge q_2 is then found in the same manner as that used for q_1. This procedure was repeated many times on the same oil drop, thus obtaining a large number of values for the different charges on the oil drop. It was found that these charges were always an integral multiple of a charge e, the charge on the electron.

Some refinements were applied to the theory given above which are described in an interesting book by Millikan.* In 1917 he gave the value of the charge on the electron as 1.59×10^{-19} coulomb. Later independent work in X-rays and radioactivity indicated that the electronic charge was higher than this value, so about 1928 there were redeterminations of the viscosity of air. This led to a slightly higher value for e, which was in

* R. A. Millikan, *Electrons (+ and −), Protons, Photons, Neutrons, Mesotrons, and Cosmic Rays* (2d ed.; Chicago: University of Chicago Press, 1947).

agreement with that found by other methods. The presently accepted value for the charge on the electron is

$$e = 1.60206 \times 10^{-19} \text{ coulomb} .$$

From the values of e/m and e for the electron the rest mass of the electron is found to be

$$m = 9.1083 \times 10^{-31} \text{ kg} .$$

In the discussion of electrolysis it was shown that 1 faraday or 96,521.9 coulombs liberates 1 gm equivalent of any substance. If N_0 is Avogadro's number or the number of atoms in a gram atomic weight and if each of the ions in electrolysis has a single electronic charge e, then

$$N_0 e = 96,521.9 \text{ coulombs/gm-atom} ,$$

and Avogadro's number is

$$N_0 = \frac{96,521.9}{1.60206 \times 10^{-19}}$$

$$= 6.02486 \times 10^{23} \text{ atoms/gm-atom} .$$

38.4. Positive Rays; Isotopes

Since atoms are electrically neutral, it was natural for the physicists to investigate the positive portion of the atom. There was much research along this line, and we shall largely limit ourselves to the experiments of J. J. Thomson. In 1907 he constructed the apparatus shown in Fig. 38-5

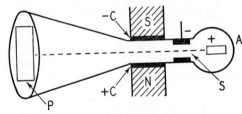

FIG. 38-5.—Positive-ray analysis tube

to determine the ratio of charge to mass of the positive portion of the atom.

In Fig. 38-5, A is a discharge tube at low pressure, and S is the cathode with a fine slit system. Some of the positive ions or positive rays move from the anode through the cathode into a magnetic field produced by a magnet NS and an electric field produced by connecting the plates CC to a battery. As shown in Fig. 38-5, these fields are both in the same direction, namely, upward. (Notice that in determining e/m for cathode rays the electric and magnetic fields were at right angles.) If a photographic plate, P, is placed perpendicular to the original direction of the rays, as shown in the figure, then the image formed on the plate is found to be a family of parabolas, as shown in Fig. 38-6.

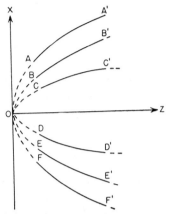

FIG. 38-6.—Positive-ray parabolas

In order to analyze the motion of the positive rays, consider the deflection z produced by the electric field alone. Suppose a positive ion of mass m and charge $+e$ is traveling with a horizontal velocity v such that it would strike the point O in the absence of any electric or magnetic fields. With the electric field alone, there would be a force $\mathcal{E}e$ in the Z direction and an acceleration,

$$a_z = \frac{\mathcal{E} e}{m} .$$

This vertical acceleration would act during the time the positive ray is passing between the plates, that is, for a time

$$t = \frac{l}{v},$$

where l is the length of the plates. The displacement in the Z direction between the plates would be

$$z_1 = \tfrac{1}{2} a_z l^2 = \frac{\mathcal{E} e}{2 m} \frac{l^2}{v^2},$$

and the displacement in the Z direction on the photographic plate is

$$z = k_1 \frac{\mathcal{E} e}{m v^2},$$

where k_1 is a constant depending on the dimensions of the apparatus.

Similarly, the magnetic field B alone exerts a

force Bev along the X direction and produces a displacement along the X direction of

$$x = k_2 \frac{B \, e \, v}{m \, v^2} = k_2 \frac{B \, e}{m \, v},$$

where k_2 is another constant depending on the dimensions of the apparatus.

If the electric field $\mathcal{E}$ and the magnetic field B are both acting, then the resulting displacement in the Z and X directions is obtained by eliminating v between the two foregoing equations. Hence

$$\left. \begin{aligned} z &= \frac{k_1}{k_2^2} \frac{m}{e} \frac{\mathcal{E}}{B^2} x^2 \\ &= k \frac{m}{e} x^2, \end{aligned} \right\} \qquad (38\text{-}11)$$

where k is a constant of the apparatus.

The positive ions do not all have the same velocity, for they are produced at different positions in the discharge tube. Thus, according to equation (38-11), which applies to one particular velocity, the positive ions of different velocities and the same m/e would be distributed along a parabola. Thus in Fig. 38-6 the parabolas AA', BB', and CC' correspond to different isotopes having different ratios of m/e. The corresponding lower halves of these parabolas, FF', EE', and DD', are obtained by reversing the magnetic field during the experiment.

In 1912 Thomson used neon as the gas in the discharge tube. Now neon has an atomic weight of about 20.2 relative to the oxygen atom of assumed atomic weight 16.0, and one might expect a single parabola from neon, corresponding to mass 20.2, but Thomson found two parabolas, corresponding to masses 20 and 22. This gave rise to much speculation and experimentation, and it was not until after a year or more that Thomson was thoroughly convinced that the two parabolas really belonged to neon. From the intensity of the two parabolas Thomson was able to estimate the relative abundance of atoms of mass 20 and 22 in neon gas. This he found to be roughly in the ratio of 9 to 1, so that the average mass came out to be 20.2, the atomic weight of neon.

From this experiment Thomson concluded that neon gas consists of atoms of two different masses, always found mixed in the same proportion so as to give the same atomic weight. The two atoms of neon are called *isotopes*, which means that both occupy the same place in the periodic table, namely, the place occupied by neon.

Isotopic atoms have the same chemical properties and approximately the same physical properties. It is now known that about three-fourths of the elements in the periodic table are composed of two or more stable isotopes. Oxygen has three, of masses 16, 17, and 18 and relative abundances of 99.76, 0.039 and 0.20 per cent, respectively. From more recent measurements it is found that neon has three isotopes, of masses 20, 21, and 22, with relative abundances of 90.5, 0.28, and 9.21 per cent, respectively. Tin has eleven isotopes, whose masses range from 112 to 114, while beryllium consists of only one isotope, of mass 9. The masses given here are the approximate masses of the atoms, for, as we shall see, very precise measurements show that the masses differ slightly from a whole number. This difference, though slight, is very important and will be shown to be related to the stability of the atom. The integer closest to the mass M of an isotope relative to oxygen of mass 16.0 is called the *mass number A* of the atom. The unit of measurement in isotope work is $\frac{1}{16}$ of the mass of the oxygen isotope of mass number 16, which is assumed to have a mass of exactly 16. This physical atomic mass unit, abbreviated amu, has an actual mass of

$$1 \text{ amu} = 1.65917 \times 10^{-27} \text{ kg}.$$

Notice that weight and mass have been used interchangeably, although strictly it is the mass which is involved, since gravity plays no part in any of this work.

As we shall see later, the nuclei of atoms are considered to be made up of protons and neutrons. The proton has a mass number of 1 and a charge of $+1$ electronic units, whereas the neutron has a mass number of 1 and a charge of zero. As we have already mentioned, the atomic number of an atom is equal to the number of protons in the nucleus or the number of electrons outside the nucleus. The mass number A of an atom, then, is the sum of the number of protons and neutrons in the nucleus. Thus we can summarize our knowledge of isotopes by stating that isotopic atoms have the same chemical properties, the same number of protons in the nucleus or electrons outside the nucleus, but a different number of neutrons in the nucleus.

38.5. The Accurate Measurement of Isotopic Masses

After Thomson's original work there were many investigators who constructed more accurate instruments for measuring isotopic masses. The one shown in Fig. 38-7 was designed by Dempster for

Fig. 38-7.—Photograph of Dempster's mass spectrograph

increased precision in mass analysis of heavy isotopes. A diagrammatic section of the apparatus is shown in Fig. 38-8. Ions, produced by a high-fre-

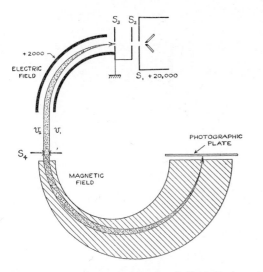

Fig. 38-8.—Diagrammatic sketch of Dempster's mass spectrograph.

quency oscillating spark between metal electrodes housed in the fin-cooled chamber, are accelerated between the slits S_1 and S_2. The electric and magnetic fields are so designed that ions of the same mass and charge but of different velocities are brought to a focus on the photographic plate, P. This perfect focusing is attained for only one ionic mass. For slightly heavier or lighter ions the focus becomes gradually less perfect as the distance from the true focus increases. This limitation is of minor importance, as the quality of the mass-spectrograph lines in Fig. 38-9 shows. Photograph 1 of Fig. 38-9 shows an analysis of the ions produced by electrodes of commercial cerium metal. In addition to the isotopes of cerium, of masses 136, 138, 148, and 150, there appear many impurities of the rare-earth metals. The next photograph shows the five isotopes of platinum which are doubly charged and a single line of rhodium from a singly charged ion. Since the mass spectrograph brings to a focus ions having the same charge to mass ratio, it follows that the line associated with Rh^+, mass 103, coincides with a mass of 206 and with a double $(++)$ charge. Photographs 3 and 4 show the six isotopes of palladium and the eight isotopes of cadmium. Photograph 5 was the first evidence of the isotope of mass number 235 of uranium which has become of considerable importance because the first atomic bomb was made of this material.

There is an interesting second-order effect produced in the highly resolved optical spectra of several isotopes of an element. In Fig. 38-10 is shown the line of mercury of wave length 2536.5 A, which, under high resolution, appears to be a very close combination of several different lines, five clearly separated. No less than six different isotopes of mercury, as indicated below the photograph, are responsible for these radiations. Lines due to odd-numbered and even-numbered isotopes may theoretically be differentiated. However, in the far ultraviolet, where these lines appear, the available resolving power is not usually sufficient to yield complete separation. In the case of the isotopes with odd mass numbers, 199 and 201, theory indicates that the spectral lines of these isotopes should be split—two components for 199 and three for 201. This is indicated in the legend.

As an example of the extreme accuracy which is attained in the determination of the masses of **577**

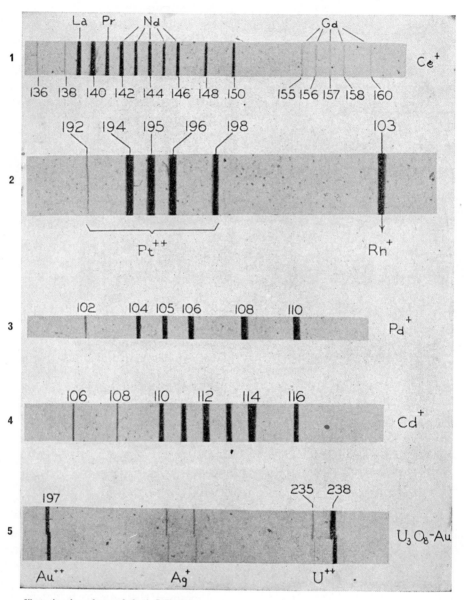

FIG. 38-9.—Mass spectrograms of various isotopes

isotopes, we shall give the masses in amu of the lighter isotopes:

Isotope	Mass (amu)
H¹	1.008142 ± 0.000003
H²	2.014735 ± .000006
C¹²	12.003847 ± 0.000016

38.6. Natural Radioactivity

In 1896 Henri Becquerel became interested in the fluorescence produced by some uranium compounds. These, he thought, had to be exposed to

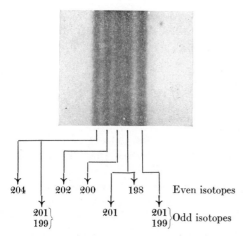

FIG. 38-10.—Isotopic effect in spectra of 2536.5 A line of mercury. (Photograph by H. Beutler.)

sunlight to produce any effect, until one cloudy day he developed some photographic plates which were wrapped in black paper but which had been close to a uranium compound. Much to his surprise, he found that the plates were blackened on development. Thus was ushered in our present atomic or nuclear age! Becquerel soon found that his uranium compound could discharge an electroscope; that is, the uranium compound emitted something which could ionize air. These two properties of uranium compounds—affecting a photographic plate and ionizing a gas—are the main ones still used for their detection.

In 1898 Pierre and Marie Curie put their efforts together and in a monumental piece of work produced two new compounds, polonium and radium, which were very much more active than the uranium compound. The name *radium* came from the Latin *radius* ("ray"), and this new phenomenon associated with uranium, radium, and polonium was called *radioactivity* by the Curies. It was relatively soon found that all the heavy elements at the end of the periodic table after bismuth are naturally radioactive. Some thirty years later the Curies' daughter Irene and her husband Joliot discovered that it is possible to make some isotopes of the stable elements radioactive. Today there are far more artificially produced radioactive isotopes than there are natural ones. Before proceeding further, we shall briefly describe the radiations emitted by radioactive substances.

38.7. Radiations from Radioactive Elements

In 1899 Rutherford was experimenting on the extent to which radioactive radiations could pass through thin sheets of aluminum. He found that two types could be definitely differentiated—one which was stopped or absorbed by a thin sheet, 0.002 cm, of aluminum he called "alpha" (α) rays or particles, and another which required a few millimeters of aluminum completely to absorb them he called "beta" (β) rays. The alpha and beta particles were also shown to be bent in opposite directions by a magnetic field, and from this it was concluded that the alpha particles were positively charged and that the beta particles were negatively charged. In 1900 a third type of radiation called "gamma" (γ) rays was found by Villard in France.

38.8. Alpha (α) Particles

The ratio of charge to mass, e/m, for alpha particles was found by magnetic deflection measurements to be about half that for a proton. This led to some controversy about the identification of the alpha particle. The problem was finally settled by Rutherford and Royds in 1909. They passed an electric discharge through a vessel in which alpha particles had been collected and found the spectrum of helium. This confirmed what they had thought, namely, that alpha particles are the nuclei of helium atoms and that after the alpha particles are slowed down, they capture two electrons to become normal helium atoms.

At this point we shall digress to present a nomenclature which is extensively used in nuclear physics. The helium isotope of mass number 4 and atomic number 2 is written as $_2He^4$. Actually, there is a repetition here, for an atom of atomic number 2 must be helium. The alpha particle is sometimes written as $_2\alpha^4$ or it may be written as $_2He^4$ or as He^4.

Alpha particles, then, have a charge of $+2$ electronic units and a mass of 4 units. They are ejected from radioactive substances with velocities of the order of one-twentieth the velocity of light. Because of their charge and mass, they produce intense ionization in a gas. In air at atmospheric pressure, alpha particles produce up to 60,000 ion-pairs per centimeter of path. As the alpha particles slow down, they produce more ionization per unit path length. Since the ionization is so intense, it follows that the alpha particle rapidly loses all its energy, so that its range in air is of the order of 2–10 cm, depending on its initial energy. Alpha particles are emitted from radioactive substances with definite energies ranging from 2 to 10 Mev, depending on the source. (1 Mev $= 10^6$ ev $= 1.6 \times 10^{-13}$ joule or 1.6×10^{-6} erg, as was given in eq. [37-22].)

38.9. Beta (β) Particles

By magnetic deflection the beta particles were shown to be electrons. Though the weight of the evidence is that electrons do not exist within the nucleus of the atom, it appears that the electron is created in the act of emission by the transformation of a neutron into a proton according to the following equation:

$$_0n^1 \rightarrow {}_1p^1 + {}_{-1}e^0 . \tag{38-12}$$

Depending on the energy conditions within the radioactive nucleus, the beta particles may be ejected with enormous velocities up to 0.998 of the velocity of light.

Now the strange fact is that beta particles appear with almost all velocities up to the maximum, that is, they have a continuous energy or velocity spectrum. From energy and other considerations it appears that the nucleus loses a constant amount of energy in the production of a beta particle which is equal to the maximum energy of the beta-particle spectrum. For those of lesser energy it was assumed by Pauli in 1927 that another particle is emitted, called a *neutrino*. This evasive particle has zero rest mass and hardly interacts with matter at all. Thus it is extremely

difficult to observe, and the experimental evidence for its existence is hardly as conclusive as we might wish. However, experiments have been performed which seem to be unequivocal, so that the neutrino is an established member of the relatively large family of particles associated with nuclear physics. From what has been said, it follows that the neutrinos must also have a continuous energy spectrum.

Beta particles produce ionization in gases which is of the order of one-hundredth of that due to alpha particles of comparable energy. Thus beta particles can traverse a meter or more of air at atmospheric pressure or a few millimeters of aluminum before their energy is lost. Fast-moving beta particles produce X-rays in materials, just as high-speed electrons do.

38.10. Gamma (γ) Rays

Another highly penetrating form of radiation, which accompanies many alpha and beta emissions, is that of gamma (γ) rays. In 1914 Rutherford and Andrade showed that the gamma rays could be diffracted, and they found the wave length to be of the order of one-hundredth that of X-rays. Thus gamma rays are electromagnetic radiations of very short wave length or of very high frequency.

Gamma rays, like X-rays, can penetrate very large thicknesses of materials, such as several centimeters of lead. Unlike alpha or beta particles, they do not have a definite range after which they no longer appear, but they are gradually attenuated, and their intensity is reduced in an exponential manner. In order to prove this attenuation law, consider a beam of X- or gamma rays of intensity I_0 incident perpendicularly on a slab of material, as shown in Fig. 38-11. Slits SS are used to confine the beam.

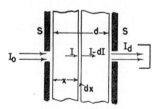

FIG. 38-11.—Absorption of X-rays or gamma rays

Let I be the intensity at a depth x, and suppose that in a small thickness dx the intensity is reduced by an amount $-dI$. It is found that dI is proportional to the intensity I and to the thickness dx, or

$$dI = -\mu I\, dx\,, \qquad (38\text{-}13)$$

where μ is a constant, called the *coefficient of linear absorption*. This equation can be written as

$$\frac{dI}{I} = -\mu\, dx$$

and can be integrated as

$$\int_{I_0}^{I_d} \frac{dI}{I} = -\mu \int_0^d dx\,.$$

The limits or boundary conditions in this integration are: at $x = 0$, the intensity is I_0, and at $x = d$ the intensity is I_d. The integration on the left-hand side yields a logarithm, so that we may write the foregoing equation as

$$\log_e I_d - \log_e I_0 = -\mu d\,, \qquad (38\text{-}14)$$

where e (2.71828 . . .) is the base of natural logarithms and

$$\log_e \frac{I_d}{I_0} = -\mu d\,,$$

$$\frac{I_d}{I_0} = e^{-\mu d}\,,$$

$$I_d = I_0 e^{-\mu d}\,. \qquad (38\text{-}15)$$

Thus the intensity I_d at a depth d is the initial intensity I_0 multiplied by the negative exponential $e^{-\mu d}$.

38.11. Positive and Negative Electrons from Gamma Rays

The absorption processes for X-rays or gamma rays which have been discussed so far are the photoelectric effect and the Compton effect. A third mode of absorption is that due to *pair production*, which becomes important when the energy, hf, of the X- or gamma ray is greater than about 1 Mev. In this the energy hf disappears, and there is created a pair of particles—a negative electron and a positive electron or positron. These are usually written as β^- and β^+ and have the same mass and equal but opposite charges. The positron β^+ has an exceedingly short life, of the order of 10^{-6} sec, since it unites with a negative electron to produce a gamma ray. In this process there is a transformation of energy in the form of electromagnetic radiation into that of the

masses of the electron and positron and their kinetic energy. Thus we have a direct application of Einstein's equation relating matter and energy, where m units of mass have an energy equivalent of E, given by

$$E = mc^2\,. \qquad (38\text{-}16)$$

If m is in grams and c in centimeters per second, then E is in ergs; likewise if m is in kilograms and c in meters per second, then E is in joules. Thus 1 gm of matter has an energy equivalent of 9×10^{20} ergs, and 1 kg of matter has an energy equivalent of 9×10^{16} joules. The velocity of light, c, is approximately 3×10^{10} cm/sec or 3×10^8 m/sec.

It is a relatively simple problem to show that a gamma-ray photon must have a minimum energy equivalent of about 10^6 ev in order to produce a positive and negative pair of electrons. In the mks system the sum of the masses, m_e, of the positive and negative electrons is approximately

$$2\,m_e = 2 \times 9.11 \times 10^{-31}\ \text{kg}\,,$$

and the energy equivalent, E, of this mass from equation (38-16) is

$$E = 2 \times 9.11 \times 10^{-31} \times 9 \times 10^{16}\ \text{kg-m}^2/\text{sec}^2$$

$$= 1.64 \times 10^{-13}\ \text{joule}\,.$$

From equation (37-12) we have

$$1\ \text{ev} = 1.60 \times 10^{-19}\ \text{joule}\,,$$

$$1\ \text{Mev} = 10^6\ \text{ev} = 1.60 \times 10^{-13}\ \text{joule}\,.$$

Thus the energy equivalent of the electron pair is

$$\frac{1.64}{1.60} = 1.02\ \text{Mev}\,.$$

In the annihilation of the positron it is first reduced to thermal speeds and then unites with an electron to produce two oppositely directed gamma rays, each of energy 0.51 Mev. The two gamma rays are necessary for conservation of momentum. Such monochromatic 0.51 Mev gamma rays are observed around an artificially radioactive source which is emitting positrons. Actually, positrons were theoretically predicted in 1928 by Dirac in England before they were first observed by C. D. Anderson at the California Institute of Technology in 1933, when he was experimenting with cosmic rays. Late in 1955 it was reported that a new particle, called the antiproton, p^-, had been discovered. This was found at the University of California using the bevatron and involved the transformation of several billion electron volts of energy into mass. The negative proton has a very short life and quickly decays into a meson.

38.12. The Law of Radioactive Disintegration

The two main characteristics of a radioactive substance are its half-life and the type of radiation emitted. By the *half-life* is meant the time in which half the original number of radioactive atoms decay and are transformed into some new kind of atom. For example, radium $_{88}Ra^{226}$ has a half-life of 1,620 years. If one had 1 gm of radium now, there would be 0.5 gm in 1,620 years, 0.25 gm in 3,240 years, and so on. Every 1,620 years the amount of radium would be decreased by a factor of 2. Among the hundreds of radioactive isotopes there are measured half-lives which range from fractions of a second to billions of years.

In the early history of radioactivity, attempts were made to change the half-life by excessive pressures and temperatures. These experiments were completely unsuccessful, and this provided strong evidence for the idea that radioactivity was associated with the nucleus and not with the electrons outside the nucleus, as is the case in chemical reactions.

While radioactivity is associated with individual atoms, it is never possible to say which particular atom will be the next one to disintegrate. It is a problem similar to that of the vital statistics in which insurance companies are interested. What one can say is that there is a certain probability, λ, that a radioactive atom will decay in some small time interval, dt. Suppose there are N radioactive atoms of the same kind present at some time t. Then in the interval of time dt, a number $-dN$ will disintegrate, where

$$-dN = \lambda N \, dt . \qquad (38\text{-}17)$$

The quantity $-dN/dt$ is the time rate of decrease of the radioactive atoms, and this is equal to the number, N, present and the decay constant, λ. Now equation (38-17) is very similar to equation (38-13) for the absorption of gamma rays. Thus we can integrate the two equations in the same manner. From equation (38-17) we have

$$\frac{dN}{N} = -\lambda dt$$

or

$$\int_{N_0}^{N} \frac{dN}{N} = -\lambda \int_{0}^{t} dt ,$$

where we are assuming that at time zero there are N_0 radioactive atoms present and that at time t there are N remaining. By integration,

$$\log_e \frac{N}{N_0} = -\lambda t ,$$

$$N = N_0 e^{-\lambda t} . \qquad (38\text{-}18)$$

Thus the original number of radioactive atoms decreases exponentially, as shown by equation (38-18).

38.13. The Half-life of Radioactive Substances

The half-life, τ, can be related to the disintegration constant, λ, by equation (38-18). After a time τ, the number of radioactive atoms N is half the original number, N_0, or

$$\tfrac{1}{2} = e^{-\lambda \tau}$$

or

$$e^{\lambda \tau} = 2 ,$$

$$\lambda \tau = \log_e 2 = 0.693 .$$

Hence

$$\lambda = \frac{0.693}{\tau} . \qquad (38\text{-}19)$$

For radium $\tau = 1,620$ years, so that its disintegration constant, λ, is

$$\lambda = \frac{0.693}{1,620} = 4.28 \times 10^{-4} \ \text{yr}^{-1} .$$

38.14. Radioactive Transformations

Let us consider the product produced when uranium I, $_{92}U^{238}$, emits an alpha particle according to the reaction

$$_{92}U^{238} \rightarrow {}_{2}\alpha^{4} + {}_{90}Th^{234} .$$

Since the alpha particle has the mass number 4 and atomic number 2, it follows that when the $_{92}U^{238}$ emits an alpha particle there is left behind a nucleus of mass number 234 and atomic number 90. The atom having atomic number 90 is thorium, Th, which is shown in the foregoing equation. Notice that in such equations the mass numbers and atomic numbers must balance on the left- and right-hand sides. The half-life of U^{238} is the huge number of 4.5×10^9 years. *Thus when a radioactive atom emits an alpha particle, there re-sults a new atom having an atomic number two units less and mass number four units less than the parent-atom.*

Now the Th^{234} atom, which was formerly called uranium X_1, is itself radioactive, emitting a β^--ray with a half-life of 24.1 days, so that the disintegration of Th^{234} produces a new atom called protactinium, Pa, according to the equation

$$_{90}Th^{234} \rightarrow _{-1}\beta^0 + _{91}Pa^{234} .$$

Thus *the emission of a β^--ray results in a new nucleus with an atomic number one unit larger than the parent but with the same mass number.* In the emission of gamma rays there is no change in atomic number or mass number. This represents the emission of energy only, so that the nucleus goes from a higher- to a lower-energy state. The foregoing two examples are from the natural radio-active substances, and we shall now give some from the artificial radioactive substances.

There are several methods of making stable isotopes into radioactive ones, but this discussion we shall leave until later. One of the first artificial radioactive substances discovered by the Joliots in 1934 was phosphorus of mass number 30, $_{15}P^{30}$. Now stable phosphorus is mono-isotopic and con-sists of a single kind of atom of mass number 31, $_{15}P^{31}$. The radioactive $_{15}P^{30}$ atom decays with posi-tron emission, $_{+1}\beta^0$, to produce silicon, $_{14}Si^{30}$:

$$_{15}P^{30} \rightarrow _{+1}\beta^0 + _{14}Si^{30} .$$

The half-life of this phosphorus isotope, $_{15}P^{30}$, is 2.5 min, and the positron has a maximum energy of 3.5 Mev.

There is another form of radioactivity that can take place, namely, orbital-electron capture, usually called *K capture*. In this process the nucleus captures one of the orbital electrons, and, since the K electrons are the closest to the nucleus, it is usually one of these which is captured. *In K capture, as in positron emission, the mass number is unchanged, but the atomic number is lowered by one unit.* The first evidence of the K-capture process was given by Alvarez at Berkeley in 1938 for radioactive gallium of mass number 67, $_{31}Ga^{67}$. The reaction may be written

$$_{31}Ga^{67} + _{-1}e^0 \rightarrow _{30}Zn^{67} ,$$

in which $_{-1}e^0$ stands for the K electron which is captured, thus producing a stable zinc atom.

The question naturally arises as to how one knows that this reaction has taken place. The answer to this is that when the $_{30}Zn^{67}$ atom is pro-duced, it has one electron missing in its K shell and would thus produce the characteristic X-radiation of the K series of zinc. It is the identifica-tion of the X-rays with the zinc atom that gave the clue to the K-capture process in the radio-active gallium. The intensity of the X-rays de-creases exponentially with the half-life of Ga^{67}, which is 78 hours.

Gamma rays may accompany any of the fore-going processes if, after the process, the nucleus is excited or still has some energy remaining above that of its ground state. The excess energy can be given out as gamma rays or by an interaction, be-tween the excited nucleus and an outer electron, in which the latter is ejected. This latter process is called *internal conversion* because, from an energy standpoint, the electron is ejected with an energy which is the difference between the energy of the gamma-ray photon and the binding energy of the electron. If E_a represents the binding energy of an electron in the K, L, and M shells and $mv^2/2$ is its kinetic energy, then

$$hf - E_a = \tfrac{1}{2}mv^2 ,$$

where hf is the energy of the gamma-ray photon. Notice the similarity between this equation and equation (37-27) for the photoelectric effect. Some of the de-excitation energy is usually given out as gamma rays and some as internal conversion electrons. As we can begin to appreciate, the nucleus is a complex structure, and there is much to be done before a complete theory can be given. Before discussing the nucleus further, we shall briefly take up the subject of the four radioactive series.

38.15. The Four Radioactive Series

There is no essential difference between arti-ficial and natural radioactive substances. In nature we find three radioactive series—the thorium, uranium-radium, and actinium series. In each of these the first member has an excep-tionally long half-life. There is a fourth series, which, though made artificially, takes its place with the other three. These four series are char-acterized by having mass numbers which are $4n$, $4n + 1$, $4n + 2$, and $4n + 3$, where n is an integer. **583**

The first and final members of these series are shown in Table 38-1.

TABLE 38-1

THE FOUR RADIOACTIVE SERIES

Name of Series	Isotope Having Largest Half-life (Years)		Final Stable Isotope	Type
Thorium	$_{90}\text{Th}^{232}$	1.39×10^{10}	$_{82}\text{Pb}^{208}$ (ThD)	$4n$
Neptunium	$_{93}\text{Np}^{237}$	2.20×10^{6}	$_{83}\text{Bi}^{209}$	$4n+1$
Uranium-radium	$_{92}\text{U}^{238}$	4.50×10^{9}	$_{82}\text{Pb}^{206}$ (RaG)	$4n+2$
Actinium	$_{92}\text{U}^{235}$	8.52×10^{8}	$_{82}\text{Pb}^{207}$ (AcD)	$4n+3$

The neptunium series does not occur in nature. Its members were discovered during the preparation of the atomic bomb. Presumably in the distant past radioactive substances were formed in the earth, but only those having the longest half-lives have survived to the present day. It is by means of these long-lived radioactive substances that the age of the earth has been measured and found to be of the order of 5×10^9 years.

Each of these series decays with alpha and beta (β^-) emissions, finally forming a stable isotope. For the natural series these are different isotopes of lead, though when they were discovered, they were known by the names given in parentheses in Table 38-1, largely to indicate their origin. One line of evidence presented for the existence of isotopes was that the measured atomic weights of ThD, RaG, and AcD were different, even though they were known to be lead. Another piece of evidence obtained about 1911 was that, although there were some forty recognizable radioactive substances, there were only twelve positions in the periodic table available. It was further work on these natural radioactive substances which led Soddy, an English chemist, in 1913 to propose the concept and name of isotopes.

Table 38-2 shows the transformations in the uranium-radium series from U^{238} to RaG or Pb^{206}. The older names are in the left-hand column, with the newer ones in parentheses. From this table it is seen that the emission of an alpha particle reduces the mass number by 4 and the atomic number by 2, whereas the emission of a beta particle (β^-) does not change the mass number but increases the atomic number by 1.

Before proceeding with any further discussion of radioactivity, we shall digress to give a brief discussion of various radiation-detecting devices.

TABLE 38-2

URANIUM-RADIUM SERIES

Substance	Atomic Weight	Atomic No.	Radiation	Half-life	Range of Air of α-Particles (Cm)	Velocity (Cm/Sec)
Uranium I (U^{238})	238	92	α	4.5×10^9 yr	2.73	1.401×10^9
Uranium X_1 (Th^{234})	234	90	β	24.5 days		
Uranium X_2 (Pa^{234})	234	91	β	1.14 min		
Uranium II (U^{234})	234	92	α	2×10^6 yr	3.28	1.495
Ionium (Th^{230})	230	90	α	7.6×10^4 yr	3.194	1.482
Radium (Ra^{226})	226	88	α	1,600 yr	3.389	1.511
Radon (Rn^{222})	222	86	α	3.83 days	4.122	1.613
Radium A (Po^{218})	218	84	α	3 min	4.722	1.688
Radium B (Pb^{214})	214	82	β	26.8 min		
Radium C (Bi^{214})	214	83	α, β	26.8 min		
Radium C' (Po^{214})	214	84	α	10^{-6} sec	6.971	1.922
Radium D (Pb^{210})	210	82	β	25 yr		
Radium E (Bi^{210})	210	83	β	5.0 days		
Radium F } Polonium } (Po^{210})	210	84	α	136 days	3.925	1.587
Radium G } Lead } (Pb^{206})	206	82	Stable			

38.16. The Ionization Chamber

We have already mentioned that Becquerel used photographic plates and an electroscope for measuring the intensity of the radioactive emissions. These are still used today, though in a somewhat different form. First we shall briefly describe the types of instrument which depend on the ionization of gases.

Consider the simple ionization chamber, consisting of two parallel plates connected to a battery and a current-measuring instrument, G, shown in Fig. 38-12. A constant radioactive source

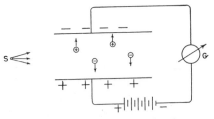

FIG. 38-12.—An ionization chamber

is placed at S. The alpha or beta particles from this source produce a constant amount of ionization in the air between the plates. At atmospheric pressure the electrons which are ejected in the process of ionization quickly join up with neutral atoms to form negative ions. In the absence of any electric field between the plates, these positive and negative ions recombine. However, when the two parallel plates are connected to a battery, an electric field is established between the plates, and the ions of opposite sign move in opposite directions. The motion of these ions constitutes an electric current, which is measured by the instrument G. As the voltage, V, and the electric field are increased between the plates, the ionization current changes in the manner shown in Fig. 38-13.

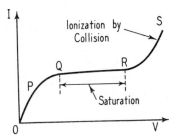

FIG. 38-13.—Increase of ionization current I with voltage V

In the region O to Q there is some recombination of the ions, but this decreases as the voltage is increased, since the larger electric field quickly sweeps the ions apart. A voltage is eventually reached at which all the ions produced are collected on the plates. The current in this region QR is sensibly constant and is known as the *saturation current*. When used in this manner, the

apparatus is called an *ionization chamber*. In this region the ionization current gives a measure of the intensity of the ionizing agent.

38.17. The Roentgen Unit, r

Ionization chambers are extensively used in radioactive measurements, especially for aiding in personnel protection. It is well known that ionization within the body, produced by radioactive substances, may be harmful. A unit called the *roentgen* has been established for the purpose of monitoring personnel. The roentgen or r unit is defined as "that quantity of X or γ radiation such that the associated corpuscular emission per 0.001293 gram of air (1 cm³ of dry air) produces, in air, ions carrying 1 esu (1/[3 × 10⁹] coulomb) of quantity of electricity of either sign." Thus the ionization-chamber current is proportional to the dosage rate expressed in r/sec or r/hr, etc. Since this definition is expressly formulated for X- and gamma radiation, it follows that it must not be used for alpha or beta particles. However, it is possible to find the ionizations of alpha or beta particles relative to those of X- or gamma rays, and these are then used. The common safety practice is not to allow anyone to be exposed to more than 0.3 r per week.

38.18. The Proportional Counter

Returning, now, to Fig. 38-13, we see that if the voltage is increased beyond the saturation limit at R, the ionization current increases, as shown by the region RS. In this region the electric field is great enough that some of the ions gain enough energy between collisions to produce further ions. This form of ionization is called *ionization by collision*. It has a cumulative effect, and, if the voltage is sufficiently increased, sparking will take place between the electrodes. This leads us into a discussion of two other types of instrument—the proportional and the Geiger counter.

These instruments usually take the form of a cylindrical metal tube, A, with a fine (1 mil) wire, B, down the axis (Fig. 38-14). These parts are inclosed in an airtight glass container filled with a gas, such as argon, at about 10 cm of mercury. A high voltage is connected between the wire and the cylinder, so as to make the wire positive.

Suppose some ions are formed in the gas in the counter by an ionizing particle. The electrons produced in the process of ionization do not, in general, unite with neutral atoms to form negative ions. There is little tendency for this to happen in a gas like argon and also if the gas pressure is relatively small. If the voltage applied to the counter is very low, there is no ionization by collision, and the counter acts as an ionization chamber. As the voltage is increased, new ions are formed by collision, and there is a multiplication of the number of ions. With a small central wire, the electric field is concentrated about the wire, and most of the ioni-

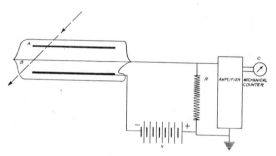

Fig. 38-14.—Diagram of Geiger counter and counting equipment.

zation by collision takes place near it. The ratio of the total number of ions produced to the number produced in the original ionizing event is called the *gas amplification factor*, A. In the low-voltage region, where there is no gas amplification and the counter acts as an ionization chamber, A is unity.

When ionization by collision does occur, there is a region of voltage in which the gas amplification is a constant which depends on the voltage. This region, called the *proportional region*, is very important in differentiating between various types of ionizing agents. For example, electrons, protons, and alpha particles produce much more intense ionization than do gamma rays. Thus if, as frequently happens, one wishes to measure the intensity of a beam of charged particles in the presence of gamma rays, it is relatively easy to do so with a proportional counter.

For each charged particle or photon passing through the counter, there are a certain number of ions produced. These are then multiplied 10^3 or 10^4 times by gas amplification, and the resultant charge is collected. The electrons quickly reach the central wire, and the heavier positive ions move slowly toward the outer cylinder. A voltage change or pulse is then produced on the central wire by each ionizing particle. This pulse is fed to an amplifier and counted by a mechanical counter. By biasing the grid of the first amplifier tube so that it accepts pulses only over a certain

minimum voltage, it is possible to count the larger pulses from the ionizing particles while rejecting the smaller ones from the gamma-ray photons. Thus the intensity of a beam of charged particles can be measured in the presence of gamma rays.

38.19. The Geiger Counter

Suppose, now, that the voltage between the wire and the cylinder is increased to the order of 1,000 volts. It is found that there is a certain minimum voltage called the *Geiger threshold*, *at which the output pulses are all the same size, irrespective of the amount of ionization produced in the original ionizing event*. The proportionality region has disappeared, and, though the pulse size increases with voltage, it is independent of the initial number of ions formed.

The mechanism for this process is quite complicated; essentially there is multiplication near the central wire, so that the discharge spreads over the whole length of the wire. Increase in voltage above the Geiger threshold increases the thickness of the discharge. The electrons are relatively rapidly collected on the central wire, while the heavy positive ions move slowly to the outer electrode. On reaching the outer electrode, these positive ions may cause electrons to be ejected from it, and these may start another discharge. One method of preventing this was to lower the voltage on the central wire by means of the voltage drop (IR) in a high resistance, R (Fig. 38-14). Electronic methods were also used. However, these limited the number of ionizing particles which could be counted in any time, for the time constant (RC) of the circuit was relatively long. In other words, there was a relatively long time, a millisecond or more, in which the counter was dead or inactive. A second ionizing event occurring within the dead time of the first is not counted, so it is important to have this dead time as small as possible.

A method which is now almost universally used in Geiger counters today is to put a small amount (about 1 cm of mercury pressure) of a polyatomic gas, such as ethyl alcohol, in the counter with the argon. This prevents secondary electrons being emitted, largely by suppressing the ultraviolet radiation given off by the positive ions when they are neutralized or in any way lose energy. In this manner the dead time is limited to approximately the time taken for the positive ions to reach the outer electrode, which, depending on the size of the counter, is of the order of 100–200 μsec.

The pulses from a Geiger counter are relatively large, since the gas amplification is a million or more. For a given ionizing source there is a region of the order of 200 or 300 volts above the Geiger threshold at which the number of recorded counts per unit time is independent of voltage. This is called the *Geiger plateau* and is important because it is not necessary to keep the voltage on the counter precisely constant. If the voltage is increased beyond the plateau region, the counter may go into continuous discharge and be ruined.

38.20. The Scintillation Counter

An early technique used in counting alpha particles was that of observing the bright flashes of light produced when they struck a zinc sulphide screen. This has now been modernized into the scintillation counter and can be used for detecting and measuring the energy of alpha, beta, or gamma radiation. For counting beta particles, a fairly large crystal of anthracene is used, while for gamma rays a crystal of sodium iodide is fairly efficient. The energy of the incident beta particles or gamma-ray photons is converted into a proportional amount of visible-light energy. This light is allowed to fall on a photomultiplier tube in which electrons are first ejected by the photoelectric effect. These electrons are then multiplied by a factor of the order of a million by allowing them to strike successive charged plates, thus producing secondary electrons. The final output pulse is proportional to the energy of the incoming radiation. These counters are rapidly displacing Geiger counters in research work, because they give not only information on the number of incoming particles but also their energies, whereas the Geiger counter gives only the number of ionizing particles entering the counter.

38.21. The Wilson Cloud Chamber and the Bubble Chamber

By means of a cloud-expansion chamber devised by C. T. R. Wilson in 1912, it is possible to make visible the paths of the radiations emitted by a radioactive particle. Wilson was interested in the fogs in the Scottish Highlands and started experiments on the artificial production of fogs. He

found that if water vapor confined in a closed region is suddenly expanded, it is possible to cool the vapor to a point where water will condense on any dust particles present. If the region is dust-free, the vapor cannot readily condense, and a supersaturated water vapor is produced. If ionization is produced in the chamber while it is in the supersaturated condition, then these ions can act as condensation nuclei so that water vapor condenses on them (Fig. 38-15). The ionization trail

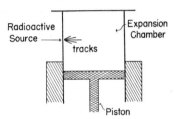

FIG. 38-15.—Diagram of a Wilson cloud chamber

becomes a fog trail of tiny water droplets condensed on the ions. By illuminating the chamber with a strong light, the condensed water droplets can be seen and photographed.

Some idea of the appearance of photographs of cloud-expansion tracks is given in the drawings of Fig. 38-16. Notice the straight, heavy path, con-

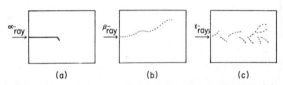

FIG. 38-16.—Drawings of cloud-chamber tracks: (a) alpha particle, (b) beta particle, (c) gamma ray.

sisting of many thousands of closely packed ions, produced by an alpha particle, as contrasted with the more tortuous path of a beta particle. The gamma ray, as it proceeds through the chamber, ejects electrons from the atoms it encounters by the photoelectric, Compton, or pair-production processes. These electrons, which frequently come off with high energy, are the ones which produce the tracks shown in (c).

By placing the cloud chamber in a magnetic field, the ion tracks can be bent into a circular track, from which one can calculate the momentum of the ionizing particle if its charge is known. The density of the water droplets along the track can be used to give an estimate of the charge and mass of the ionizing particle. Thus we find extensive use of the Wilson cloud chamber in various forms in nuclear and cosmic-ray research. It was by means of the cloud chamber that Anderson in California discovered the positron in cosmic rays.

In 1952 Glaser,* at the University of Michigan, made a so-called *bubble chamber* for the detection of ionizing particles. He argued that a superheated liquid, like the supersaturated vapor in the Wilson chamber, might be sufficiently unstable that it could be triggered by a small ionizing event. A superheated liquid is one above its boiling point and is very unstable, so that boiling may occur easily. He found that ionizing radiation could trigger the boiling and that the density of bubbles along an ionizing track is a measure of the ionizing power and energy of the incident particle. Since the density of a liquid is several hundred times the density of a gas, it follows that the number of ions produced in the bubble chamber is very much greater than in the Wilson chamber. A 6-in bubble chamber is the equivalent of a cloud chamber of about 140 ft. Ionizing events due to mesons, which are very rare or nonexistent in cloud chambers, are relatively commonplace in the bubble chamber. Thus a new research tool of great importance has been produced.

38.22. The Discovery of the Nuclear Atom

After Thomson's discovery of the electron as a universal constituent of all atoms, there was considerable conjecture as to how an atom was built up from electrons and positive charges. The problem was solved by Rutherford about 1911 from the results of an experiment in which alpha particles were shot into a thin gold foil. He measured the number of scintillations produced by the particles at various angles with the incident beam, using a zinc sulphide screen. It was found that most of the alpha particles went straight through the gold foil but that a few, about 1 in 8,000, were deflected or scattered through an angle of 90° in passing through the foil. In reporting on an experiment Rutherford, Geiger, and Marsden said: "If the high velocity (about 1.8×10^9 cm/sec) and mass of the alpha particle be taken into account it seems surprising that some alpha particles . . . can be turned within a layer of 6×10^{-5} cm of gold through an angle of 90° and even more." Sometime later Rutherford said of this result: "It

* An interesting account of the bubble chamber is given in the *Scientific American*, February, 1955.

was about as credible as if you had fired a 15-inch shell at a piece of tissue paper and it came back and hit you."

The results of these scattering experiments were quantitatively satisfied by assuming that there was a coulomb force of repulsion between the positively charged alpha particle and a positively charged small nucleus at the center of the atom. From these experiments it was possible to find the charge on the nucleus. After this work there was little room for doubt about the broad picture of an atom—namely, a central nucleus of very small size containing all the positive charge and practically all the mass, surrounded at some distance by the electrons. It was with this picture that Bohr started when analyzing the hydrogen atom and its spectrum.

It now appears from various lines of evidence that the radius of a nucleus varies with its mass number, A, approximately as given by

$$r = 1.5 \times 10^{-13} A^{1/3} \text{ cm} . \quad (38\text{-}20)$$

Thus the radius of the aluminum isotope of mass number $A = 27$ is

$$r = 4.5 \times 10^{-13} \text{ cm} .$$

This value, as compared with the radius of the atom, which is about 10^{-8} cm, shows that an atom has much "empty space."

As we stated earlier, the nuclei of atoms are composed of protons and neutrons. We shall now briefly give an account of the discovery of the neutron, though we shall not be able to give the evidence for its being a constituent of the nucleus. This is somewhat beyond the scope of this book.

38.23. The Discovery of the Neutron

The story of the discovery of the neutron is an interesting example of the scientific method. In 1930 Bothe and Becker noticed that when beryllium was bombarded with the alpha particles from polonium, a highly penetrating radiation was produced. It was thought that this radiation was composed of gamma rays of very short wave length. These experiments were repeated by the Joliots in 1932, who also showed that when the radiation was allowed to fall on a hydrogen-containing material such as paraffin, protons with a large amount of energy were ejected. By measuring the energy of the protons, it turned out that if the radiation were gamma rays, the energy of its photons should be about 52 Mev. This was much greater than the energy of any known photons, and the Joliots thought that they had discovered a new mode of interaction of radiation and matter.

The dilemma was solved by Chadwick in 1932, who saw that these and his own results were easily explained if one assumed that the radiation was really made up of a neutral particle of about the same mass as a proton. This particle was called a *neutron*. The nuclear reaction involved between the alpha particles from polonium and the beryllium nuclei can be represented by the following equation:

$$_4Be^9 + {_2}He^4 \rightarrow {_6}C^{12} + {_0}n^1 . \qquad (38\text{-}21)$$

This reaction is frequently written in shorthand notation:

$$Be^9 (\alpha, n) C^{12} . \qquad (38\text{-}22)$$

The isotope or nuclide on the left is the bombarded nucleus, and the first particle in parentheses is the bombarding particle, while the second is the emitted particle, leaving the resultant isotope or nuclide given on the right.

Since neutrons have no charge, they do not interact with electrons and hence leave no track in a cloud chamber. Having no charge, however, they are not repelled by the positive charge on a nucleus and consequently can enter a nucleus relatively easily. This frequently leaves the nucleus radioactive, and it is because of this radioactivity that the neutrons can be observed. This we shall discuss in a later section.

The neutrons emitted from the beryllium interaction have a continuous spectrum of energy with a maximum of the order of 12 Mev. They can be slowed down or reduced in energy by billiard-ball-like elastic collisions with light nuclei. For example, it requires about 18 collisions with hydrogen nuclei to reduce the energy of a 2-Mev neutron to the 0.025 ev of thermal energies.

Neutrons, photons, positrons, and mesons (see cosmic-ray section) have a temporary existence, unlike electrons and protons. If a neutron does not end its existence in a free state by being absorbed by a nucleus, then it decays with a half-life of about 20 min into a proton and an electron.

38.24. The First Artificial Transmutation

Rutherford had noticed as early as 1914 that when alpha particles were shot into nitrogen,

some high-speed, long-range particles, presumably protons, were given out. These were detected by a zinc sulphide scintillation screen. A few years later he conducted a systematic study of several gases to find out whether the long-range particles were emitted when bombarded with alpha particles. After much careful work he felt assured that they were emitted by nitrogen.

The number of these particles, which he considered to be protons, was exceedingly small. It was estimated that only 1 proton was ejected from a nitrogen nucleus for every 300,000 alpha particles incident on the gas. The result of this interaction of alpha particles with nitrogen nuclei was found by Rutherford in 1919 to be the production of protons (or hydrogen) and oxygen, according to the reaction

$$_7N^{14} + {_2}He^4 \rightarrow {_1}H^1 + {_8}O^{17} . \qquad (38\text{-}23)$$

Here, then, was the first artificial transmutation of nitrogen into oxygen and hydrogen.

This somewhat startling conjecture was confirmed by cloud-chamber photographs at a later date. A track taken from a photograph (Fig. 38-17) shows the incident alpha track and the

Fig. 38-17.—The collision of an alpha particle with a nitrogen nucleus.

long-range proton track, with the short one due to the heavy oxygen nucleus. Such photographs are exceedingly rare, for Blackett obtained only 8 photographs like Fig. 38-17 from a total of over 20,000 photographs taken in the cloud chamber.

38.25. The First Transmutation with Artificially Accelerated Particles

The next problem of great interest to nuclear physicists was that of using high voltages to accelerate charged particles, such as protons, in order to produce transmutations. Simple calculations of the coulomb repulsion between a positively charged nucleus and a positively charged proton showed that the energy of the incoming proton must be very large if it is to enter the nucleus. However, Gamow showed that, according to wave mechanics, there was a finite probability of a proton's entering a nucleus, even though, according to classical physics, the energy of the proton was not sufficient for it to overcome the coulomb potential barrier and enter the nucleus.

This theoretical prediction was verified by Cockroft and Walton in 1932. They constructed what was at that time a high-voltage source of about 300,000 volts and used this to accelerate protons—hydrogen ions—to bombard lithium. A zinc sulphide screen was used to observe any scintillations produced by the reaction. At 125,000 volts, scintillations were first observed, and these increased in number with the increase in voltage or energy of the incident protons. From the brightness of the scintillations and from the density of the tracks in later cloud-chamber pictures they concluded that alpha particles were being given off. The reaction may be written as

$$_3Li^7 + _1H^1 \rightarrow _2He^4 + _2He^4 \qquad (38\text{-}24)$$

or

$$Li^7 (p, a) He^4 \, .$$

The success of this experiment led to a whole new field of nuclear exploration so that today there are enormously large machines—cyclotrons, synchrocyclotrons, betatrons, etc. (see §§ 21.16; 23.15) —for producing particles with energies in the range of a hundred to several thousand million electron-volts. To begin to describe this work with any detail would be far beyond this book, so we shall merely indicate some of the typical nuclear reactions.

38.26. Types of Nuclear Reactions

By far the most common types of particles which are given very large energies for producing nuclear reactions are the nuclei of the first two elements. These include the three isotopes of hydrogen: protons or H^1 nuclei, deuterons or H^2 nuclei, and tritons or H^3 nuclei. Then there are helium nuclei or He^4, gamma rays, electrons, and neutrons. Of these, by far the most used for producing radioactive isotopes is the slow-neutron reaction, because very copious sources of neutrons are available in reactors or atomic piles. Using the symbols n for neutrons, p for protons, d for deu-

terons, and a for the helium nucleus, the following types of reactions have been observed in a number of nuclei: (p, n), (p, d), (p, a), (p, γ); (a, p), (a, n); (d, p), (d, a), (d, n), $(d, 2n)$; (a, p), (a, n); (n, p), (n, a), $(n, 2n)$, (n, γ); (γ, n). This is not an exhaustive list. Along with these, in which a different particle is emitted from that which is incident, are the so-called "scattering" reactions, in which the same particle is emitted as the incident particle. Among these are (n, n), (p, p), and (γ, γ) reactions. If the emitted particle has less energy than the incident particle, the reaction is said to be *inelastic*, in conformity with our use of this term in mechanics. For the nuclear reaction the absorption of the incident particle leads to an excited energy state of the initial nucleus, which then emits the same kind of particle as well as a gamma-ray photon. Thus the emitted particle has less energy than the incident because of the energy of the emitted gamma-ray photon. If the particles have the same energy, there is *elastic* scattering.

38.27. Spallation and Fission

If the energy of the bombarding particles from the high-energy machines gets into the hundreds of millions of electron-volts as has already happened, then the nuclei eject not one particle but a great many. This phenomenon is called *spallation*. As an example, in the University of California frequency-modulated cyclotron, arsenic, $_{33}As^{79}$, was bombarded with 400-Mev alpha particles, $_2a^4$. Among the debris were found $_{17}Cl^{35}$, many protons, neutrons, and alpha particles. As more of the high-energy machines get into operation, we can look forward to seeing many of these spallation reactions.

The nuclear reaction which has received by far the most publicity is the fission reaction. In its common form a neutron is absorbed by the U^{235} nucleus, which then splits or fissions into roughly two equal parts, called the *fission products*. These products are highly radioactive and range in mass number from about 70 to 160. This fission process takes place in many ways in different uranium atoms, so that there are many different kinds of fission products. The importance of this reaction lies in the enormous amount of energy released in it and the fact that, on the average, 2.5 neutrons are also emitted. There are approximately 200 Mev of energy released for each fissioning nucleus, which is about twenty million times that emitted per reacting atom in chemical explosive reactions.

Fission has been produced in many of the heavy elements by neutrons, protons, deuterons, alpha particles, and gamma rays. However, with slow neutrons having an energy of about 0.025 ev, the important fissioning nuclei are U^{235}, U^{233}, and Pu^{239}. This latter element, called "plutonium," was first made in quantity during the war in the large reactors.

As we have seen, ordinary uranium consists principally of two isotopes, U^{235} and U^{238}, the former of which has an abundance of only about 0.7 per cent or about 1 part in 140. The latter, U^{238}, which can undergo fission with fast but not with slow neutrons, can also absorb a neutron to form some of the new elements called the *transuranic elements*. Of the average 2.5 neutrons emitted in the fission of U^{235}, one may be absorbed in U^{238} to form $_{92}U^{239}$:

$$_{92}U^{238} + _0n^1 \rightarrow _{92}U^{239} + \gamma . \qquad (38\text{-}25)$$

This product, $_{92}U^{239}$, is radioactive, with a half-life of 23.5 min, and emits a β^--ray, to form neptunium:

$$_{92}U^{239} \rightarrow _{93}Np^{239} + _{-1}\beta^0 + \gamma . \qquad (38\text{-}26)$$

Neptunium, in turn, is radioactive, emitting a β^{-1} ray with a half-life of 2.3 days and forms plutonium, $_{94}Pu^{239}$:

$$_{93}Np^{239} \rightarrow _{94}Pu^{239} + _{-1}\beta^0 + \gamma . \qquad (38\text{-}27)$$

Plutonium, like U^{235}, undergoes fission with slow neutrons. The second atom bomb dropped in 1945 was made from plutonium. This isotope is radioactive, with a half-life of 24,300 years, emitting alpha particles and gamma rays. More will be said on this subject in the section on reactors.

38.28. Detection and Absorption of Neutrons

As we have already seen, neutrons do not ionize a gas directly: hence their detection and measurement has to be made from nuclear reactions. In general, these take two forms: one in which a substance, such as indium, is made radioactive, and the β^--particles or gamma rays emitted are measured by a Geiger counter; the other in which a substance containing boron is used in a Geiger counter. Boron absorbs neutrons and emits alpha particles, which produce ionization in a Geiger counter (§ 38.30).

Let us first consider the activation of indium by neutrons. The reaction which is used is

$$_{49}In^{115} + _0n^1 \rightarrow _{49}In^{116} \rightarrow _{50}Sn^{116} + _{-1}\beta^0 + _0\gamma^0 . \quad (38\text{-}28)$$

The half-life of radioactive In^{116} is 54.3 min, and the residual nucleus, Sn^{116}, is stable. Now the reason indium is frequently used is that its half-life is a convenient length of time, neither very short nor very long. Indium also has a relatively high coefficient of absorption for neutrons.

A collimated beam of neutrons is absorbed exponentially, just as was the case for gamma rays (eq. [38-15]). Thus we may write, for the neutrons,

$$I = I_0 e^{-\Sigma d} , \qquad (38\text{-}29)$$

where I_0 is the initial intensity, I the intensity at a thickness d, and Σ is called the *macroscopic absorption* coefficient. The units of Σ are cm^{-1} or m^{-1}, depending on whether d is measured in centimeters or meters, since the exponent Σd has no dimensions.

If we divide Σ by the number of atoms or nuclei per unit volume, we obtain the absorption, σ, per atom or per nucleus. The dimensions of σ are seen to be those of an area, square centimeters or square meters. It is the absorption per atom or nucleus, called the *microscopic absorption*, cross-section σ, which is usually quoted in the literature. During the war a unit for nuclear reaction cross-sections was introduced and was called the *barn*. The definition of the barn is

$$1 \text{ barn} = 10^{-24} \text{ cm} .$$

This is of the order of magnitude of the area of a nucleus.

38.29. Resonance Absorption

Some nuclei absorb neutrons very considerably, and for these σ would be large, while others have a small absorption and a correspondingly small σ. In practically every case known, σ depends on the energy of the neutrons. It should be mentioned that one can have a cross-section for scattering as well as for absorption; in fact, any reaction between neutrons and nuclei has a cross-section corresponding to the reaction. Another manner of looking at these cross-sections is that of considering them as a measure of the probability of the reaction's taking place.

As an illustration, consider the variation in the total cross-section (absorption and scattering) of indium with the energies of the neutrons, as shown in Fig. 38-18. This figure, which is plotted on log-log scale, shows the enormous variation in cross-section, principally due to absorption, with the energy of the neutrons. At very low energies the

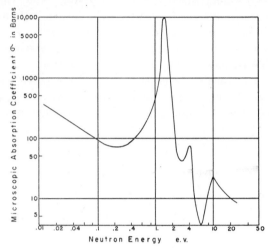

FIG. 38-18.—Cross-section σ of indium for neutrons of various energies.

absorption cross-section varies as the inverse of the velocity of the neutrons, i.e., σ is proportional to $1/v$. At the neutron energy of 1.44 ev, there is such an enormous increase in the absorption cross-section that σ is of the order of 20,000 barns. The indium nucleus is said to have a resonance level at the neutron energy of 1.44 ev. These resonance levels are of great importance in nuclear physics and are observed for almost every nucleus. For the absorption of thermal or low-energy neutrons, cadmium is frequently used. The thermal neutron cross-section for cadmium is very large, having a resonance at 0.18 ev of about 7,000 barns. In this neutron reaction Cd^{113} undergoes an (n, γ) reaction, to form stable Cd^{114}. A curve of the cross-section of cadmium is shown in Fig. 38-19. There is a large $1/v$ portion at low energies, followed by the relatively small resonance level at 0.18 ev.

38.30. Boron Counters for Neutrons

Another substance frequently used for absorption of neutrons is boron or one of the suitable boron compounds. Boron has an absorption cross-section which follows the $1/v$ law from very low neutron energies up to those of over 1,000 ev. At

very low neutron energies (0.015 ev) the boron nucleus has a value of σ of about 1,000 barns, while at 1,000 ev it is about 3.8 barns. When the isotope of boron B^{10} absorbs a neutron, it emits an alpha particle, according to the equation

$$_5B^{10} + _0n^1 \rightarrow _2He^4 + _3Li^7 + Q . \quad (38\text{-}30)$$

In this equation Q is called the *energy of the reaction.* If Q is positive, then energy is emitted and

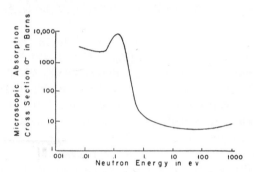

FIG. 38-19.—Absorption cross-section for cadmium

the reaction is *exothermic,* whereas if Q is negative, then energy must be supplied, to cause the reaction to take place; the reaction is said to be *endothermic.* For the boron reaction, Q is positive and has a value of about 2.5 Mev. This is divided between the alpha particle and the lithium nucleus. Both these charged particles can produce ionization in a gas. Thus boron may be used in the form of a gas as boron trifluoride, BF_3, or may be coated on the interior surface of an ionization chamber or Geiger counter for the detection and measurement of neutrons.

38.31. Energy Balance in a Nuclear Reaction

Let us take the simple reaction of a gamma ray disintegrating a deuterium $(_1H^2)$ nucleus into a neutron and a proton. The minimum energy of the gamma-ray photon which will cause this reaction has been found experimentally to be 2.22 Mev. We may write this reaction as

$$_0\gamma^0 + _1H^2 \rightarrow _1p^1 + _0n^1 + Q . \quad (38\text{-}31)$$

In this case, Q, the energy emitted, is negative, since the energy hf of the gamma-ray photon is necessary to make this reaction go. We see that when energy in the form of gamma rays is supplied to a deuterium or heavy hydrogen nucleus, it breaks this nucleus into its component parts of a proton and a neutron. The deuterium nucleus is stable, and from equation (38-31) we see that when a proton joins up with a neutron to form a deuterium nucleus, energy should be emitted. This comes out in the form of gamma rays and is observed when neutrons are in a hydrogen-containing medium. The cross-section for this reaction is relatively small.

Now the stability of the deuterium atom is considered to be due to the fact that the deuterium nucleus has a smaller mass than the sum of the masses of the proton and the neutron. From Einstein's law of the equivalence of mass and energy given in equation (38-16), it is considered that, in the union of a proton and a neutron to form the deuterium nucleus, the excess mass is emitted as energy. To see that this is the case, consider the masses of the various particles involved:

	amu
Mass of $_1H^1$ atom	$=1.00814$
Mass of neutron	$=1.00898$
Total mass	$=2.01712$
Mass of $_1H^2$ atom	$=2.01473$
Mass difference	$=0.00239$

By Einstein's relationship, equation (38-16), we have

$$E \text{ (ergs)} = m c^2 \text{ (gm-cm}^2/\text{sec}^2) .$$

Now a physical mass unit, 1 amu, represents a mass of 1.660×10^{-24} gm, so that the mass deficiency of 0.00239 amu has an energy equivalent, E, of

$$E = 2.39 \times 10^{-3} \times 1.66 \times 10^{-4} \times 8.99 \times 10^{20} \text{ erg}$$

$$= 3.56 \times 10^{-6} \text{ erg} .$$

From equation (37-22) we have

$$1 \text{ ev} = 1.60 \times 10^{-12} \text{ erg} .$$

Hence the energy equivalent, E, in electron-volt units of energy is

$$E = \frac{3.56 \times 10^{-6}}{1.60 \times 10^{-12}} = 2.22 \times 10^6 \text{ ev} .$$

This is equal to the value of the gamma-ray energy necessary to break up the deuterium nucleus into its component parts. A reaction of this kind in which gamma rays are involved is called *photodisintegration*. Actually, the experimental value of 2.22 Mev was used to measure the mass of the

neutron from the known values of the masses of $_1H^2$ and $_1H^1$. This is the reverse of the calculations given above.

For purposes of calculation it is convenient to have the energy equivalent in electron-volts of one physical atomic mass unit. From the foregoing data,

$$1 \text{ amu} = \frac{1.66 \times 10^{-24} \times 8.99 \times 10^{20}}{1.60 \times 10^{-12}} \text{ ev}$$

$$= 931 \times 10^6 \text{ ev} = 931 \text{ Mev} .$$

Thus the mass difference of 0.00239 amu of the deuterium nucleus is equivalent to $0.00239 \times 931 = 2.22$ Mev of energy, as given above.

38.32. Binding Energy of Nuclei

Suppose we have a nucleus of actual mass M, containing p protons of mass m_p and n neutrons of mass m_n. Then, in theory, we can take the nucleus apart into its constituent protons and neutrons which are referred to as *nucleons*. In every case we shall find that the mass of the nucleus is less than the sum of the masses of the nucleons. By "binding energy" we mean the energy which would have to be supplied to the nucleus to break it up into its component protons and neutrons. Thus the binding energy, B, may be written as

$$B = (p m_p + n m_n - M) c^2 ,$$

and the mean binding energy per nucleon, $\bar{B}$, for a nucleus of mass number A equal to $(p + n)$ is

$$\bar{B} = \frac{B}{A} = \frac{(p m_p + n m_n - M) c^2}{A} .$$

From the accurate values of the isotopic masses and the masses of the proton and neutron, it is possible to calculate the mean binding energy per particle for the various nuclei or nuclides. Such a curve is shown in Fig. 38-20, in which the mean binding energy is given in energy units of million electron-volts.

From this binding-energy curve it is seen that the most stable nuclides have mass numbers in the range from 50 to about 150. The stability is less at lower or higher mass numbers. Thus, if one could break up a heavy nuclide into two lighter and more stable ones, energy would be given out. This is essentially what occurs in the fission process of uranium or plutonium. A simple calculation allows **593** us to arrive at the value of 200 Mev for the fissioning of a uranium nucleus. From the curve, the value of $\bar{B}$ for nuclides of mass number 118 is about 8.5 Mev, and for mass number 236 it is about 7.6 Mev. There is then approximately 0.9 Mev per nucleon difference in the mean binding energy between the original uranium nucleus and the lighter fission products. This value multiplied

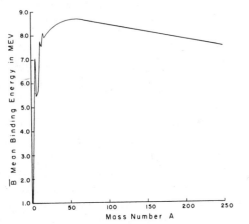

Fig. 38-20.—Mean binding energy per particle for nuclides of different mass numbers.

by 236, the number of nucleons involved, gives the value of

$$0.9 \times 236 = 212 \text{ Mev}$$

for the approximate energy released for each nucleus undergoing fission.

If, on the other hand, one could synthesize a stable nucleus from the lighter ones, then, on the basis of the curve, much more energy should be emitted. This is the basis for the fusion or so-called "hydrogen bomb." At this point it is appropriate to introduce a brief discussion on fission reactors and the peacetime use of atomic energy.

38.33. Thermal Reactors

The first nuclear reactor was built during World War II by the Manhattan Project at the University of Chicago. It was successfully operated at a few watts on December 2, 1942. The person who was largely responsible for its success was Enrico Fermi (1901–54).*

Let us consider a typical thermal reactor in which natural uranium is used as fuel. A reactor is

* An interesting account of Fermi's life is given in *Atoms in the Family* by Laura Fermi (Chicago: University of Chicago Press, 1954).

a device in which the fissioning process goes on at a controlled rate. In the fission process we have seen that, on the average, 2.5 neutrons are emitted. These are emitted with energies of the order of a few million electron-volts, whereas the U^{235} undergoes fission with very low-energy or thermal neutrons. These neutrons have about the same energy as molecules of a gas at ordinary temperatures and for this reason are called *thermal neutrons*. Thus it is necessary to slow down the fast fission neutrons if they are to produce more fissions and keep the reactor going. Let us consider what can happen to the fission neutrons when they are emitted. They can be absorbed in U^{238}, eventually to produce useful plutonium; they can be absorbed by the material used to slow down the neutrons, called the *moderating material;* they can be absorbed in the material used in the construction of the reactor; or they can leak out of the reactor. All these possibilities exist, and if the reactor is to keep going, there must be one thermal neutron available for fission to replace the one which originated the fast neutrons at the beginning of the cycle. In this case the multiplication factor k is unity, and the reactor maintains itself at a constant level of neutrons.

Suppose, however, that the multiplication factor k is very slightly greater than 1 or that there is more than one thermal neutron available for fission at the end of the cycle to replace the one which originated the cycle. Now it can be calculated that in a typical reactor the slowing-down time is of the order of 0.001 sec. If, for example, the multiplication factor is 1.005, then every thousandth of a second there are 0.005 excess neutrons available, so that after 1 sec the number of neutrons has increased by about 150 fold. This increase is exponential. It would therefore appear that in a matter of a few seconds the number of neutrons per unit volume in the reactor would be enormous and that control of a reactor would be virtually impossible.

38.34. Delayed Neutrons and Reactors

We have not told the whole story. Along with these average 2.5 prompt fission neutrons, which appear immediately on fission, there is also a very small number, 0.75 per cent, which appear as delayed neutrons. The source of the delayed neutrons is some of the fission products. Among the great variety of fission products produced, practically all of which are β^--emitters, there are some which are left in an energy state which leads to the emission of a neutron. As a typical example of a nuclear energy-level diagram, we give the decay scheme of Br^{87}, a fission product which emits delayed neutrons as well as β^--particles. From Fig. 38-21 it is seen that Br^{87} has a half-life of 55.6 sec

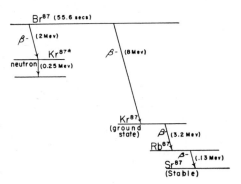

Fig. 38-21.—The decay scheme of Br^{87}, showing the delayed neutron emitter Kr^{87*}.

and has two modes of disintegration. The neutrons, which appear immediately after the krypton nucleus is left in the excited state, must appear with the half-life of the Br^{87}, namely, 55.6 sec. An excited state, such as Kr^{87}, is usually denoted by an asterisk, as shown in the diagram. The Br^{87} is called the *precursor* of this delayed neutron emitter. Actually, five groups of delayed neutron emitters are known.

The importance of these delayed neutrons lies in the fact that they make the control of reactors possible. For these delayed neutrons the weighted average time between two successive fissions or the time of a cycle is about 0.1 sec. Suppose the multiplication factor k of the reactor is, as before, 1.005. Then, of this, 0.0075 is due to the delayed neutrons and 0.9975 to the prompt ones. The running of the reactor is then controlled by the delayed neutrons, so that, after 1 sec, the neutron density would have increased by 1.05 instead of the former value of 150 times.

Thus there is time to push in some neutron-absorbing material in the form of control rods before the neutron density gets out of hand. (These control rods, made of cadmium or boron plated on steel rods, can be raised or lowered in the reactor.) The neutron density in the reactor is measured by ionization chambers and Geiger counters. If one wishes to increase the neutron density in the re-actor, that is, the number of fissions per second or the power produced, then the control rods are raised so as to increase the multiplication factor k. When the desired neutron or power level is reached, the multiplication factor k is set at exactly unity and the reactor continues to run at this level. To shut down the reactor, the control rods are pushed in, making k less than unity.

38.35. The Oak Ridge Thermal Reactor

As stated previously, the first reactor operated at a few watts, and this was soon replaced by a larger one at the Argonne Laboratory near Chicago. Later a large reactor was built and is still in operation at Oak Ridge, Tennessee. Today there are many reactors scattered around the world. However, the detailed structure of but few is available, so we shall limit ourselves to the one at Oak Ridge. In this the uranium is in the form of small cylindrical rods, 1.1 in in diameter and 4 in long, placed in aluminum jackets. These jackets prevent the fission products from spreading through the reactor and making it intensely radio-active. The uranium cylinders, or *slugs*, as they are called, are placed in slots in the graphite material used as the moderator for slowing down the neutrons. These graphite blocks make up a cube of approximately 24 ft on a side, pierced horizontally by 1,252 channels in which the canned uranium slugs lie. The channels are 1.75 in square, spaced on 8-in centers. Around the reactor proper is a thickness of a few feet of graphite which acts as a reflector of neutrons and thus prevents a large number from escaping. For personnel safety the reactor is inclosed in a 7-foot-thick concrete shield.

After the uranium slugs have been exposed for a sufficient time, they are pushed out of the channels and eventually go into a water-filled canal. These slugs are very radioactive and are much too dangerous to be handled by anything but remote-control equipment. After the radioactivity has decayed somewhat, the slugs are chemically treated to recover the plutonium and other valu-able materials such as may be required.

Figure 38-22 is a cutaway diagram of the Oak Ridge reactor. It is cooled by large cooling fans of 900 hp each. These exhaust the air at the rate of 120,000 ft^3/min and discharge it into the air through a stack 180 ft high. Knowing the inlet and outlet temperatures and the volume of air moving through it per unit time, it is possible to calculate the power developed in the reactor. This reactor operates at a maximum power of about 3,800 kw.

Other reactors have been built which use heavy water, deuterium oxide, as moderator; both the Canadian one at Chalk River, Ontario, and the

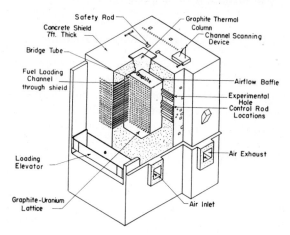

Safety Rod
Concrete Shield 7ft. Thick
Bridge Tube
Fuel Loading Channel through shield
Loading Elevator
Graphite-Uranium Lattice
Graphite Thermal Column
Channel Scanning Device
Airflow Baffle
Experimental Hole
Control Rod Locations
Air Exhaust
Air Inlet

Fig. 38-22.—Oak Ridge National Laboratory graphite reactor.

one at Argonne near Chicago use this liquid as moderator. There are several different types of reactors actually in use, including one in which the fissions take place at high neutron energies. A reactor is a complex structure and demands the skills of physicists, chemists, and engineers. In this work one first has to ask of any structural material what is its neutron absorption cross-section, for if this is too large, no matter how otherwise desirable, it cannot be used.

38.36. Breeder and Power Reactors

Another type of reactor which has been successfully built is the so-called *breeder*, in which the essentially nonfissionable material, such as U^{238}, is changed into some fissionable type, such as Pu^{239}. This is theoretically possible because of the average of 2.5 neutrons per thermal neutron fission in U^{235}, one of which is needed to keep the reaction going, leaving a total of 1.5 neutrons per fission to go into making fissionable material, such as plutonium, from U^{238}. This assumes, of course, that all neutrons are used in a profitable manner. The practical problem of reactor designers is to make sure that as large a proportion as possible of the

1.5 surplus neutrons goes into the manufacture of new fissionable material.

The enormous amount of heat energy produced in a reactor can be appreciated from the fact that 1 lb of fissionable material produces the energy equivalent of about $2\frac{1}{2}$ million tons of coal. It is a very difficult engineering problem to get the heat out of a reactor in a form that is useful for running turbines or electric generators. The fact that the launching of the nuclear-powered submarines "Nautilus" and "Sea Wolf" has taken place means that this problem is solved. One can expect stationary electric power plants to be built in which nuclear energy produces electrical energy at a competitive price with the present methods of production.

One should not leave this subject without mentioning another important aspect of reactors, namely, the making of radioactive substances for medical and other purposes. In preparing these, suitable isotopes are placed in the reactor, where, under bombardment by intense beams of neutrons, the desirable radioactive isotopes are produced. At Oak Ridge and other Atomic Energy Commission sites these radioactive isotopes are regularly produced for hospitals and other medical centers.

38.37. Stability of Nuclei

As we have seen, there are both stable and radioactive isotopes associated with every element in the periodic table. In Fig. 38-23 is shown a rough plot of the number of protons, Z, against the number of neutrons, N, for the known stable isotopes or nuclides. The straight line at an angle of 45° with the axes is the line on which there is the same number of protons as neutrons in the nuclei, i.e., the proton-neutron ratio is unity. From the figure it is seen that the proton-neutron ratio decreases, or the neutron-proton ratio increases, with atomic number. For example, at atomic number 80 there are about 120 neutrons in the stable nucleus, giving a ratio of neutrons to protons of 3 to 2. The stable nuclide of mercury, Hg^{200}, whose atomic number is 80 and whose mass number is 200, is such a nuclide.

From this curve of the stable nuclides we can draw certain interesting conclusions. Any nuclide whose position on the figure is to the right of the dotted stable region has too many neutrons relative to the number of protons for stability. Such a nuclide would be radioactive and would emit a β^--particle, thereby changing a neutron into a proton. Thus the nuclide of mercury, Hg^{205}, with a mass number of 205, is a β^--emitter with a half-life of 5.5 min. A nuclide on the left of the stable region contains too many protons relative to the number of neutrons for stability. This nuclide would be radioactive, changing a proton into a

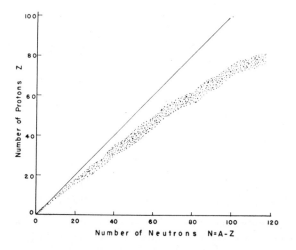

FIG. 38-23.—Plot of stable nuclides

neutron by either positron, β^+, emission, or by K-electron capture. For example, the mercury nucleus of mass number 192 emits β^+-particles and undergoes K capture, with a half-life of 5.0 hours. In a few cases in which the proton excess is large there are alpha particles emitted. Also, nuclei which are far from the stable region tend to disintegrate much more rapidly, or have shorter half-lives, than those which are closer to the stable region.

The foregoing information is empirical; in other words, it is not related to any theory concerning the structure of the nucleus. From our present knowledge of electrostatic forces we might expect all nuclei to be unstable, since protons in the free state repel one another. However, nuclei do exist, and from this we must infer that there are attractive forces in a nucleus. For reasons which are too abstruse to be given here, it is believed that when neutrons and protons are exceedingly close together, as they must be in a nucleus, there are attractive forces between neutron and neutron and neutron and proton. On the basis of certain assumptions it is possible to use these forces to set

up an equation for the binding energy of nuclei. Such an equation is found to be in reasonable agreement with the measured binding energies of nuclei.

The problem of the understanding of the nucleus is exceedingly difficult and as yet is far from solved. Besides the particles which we have already mentioned associated directly or indirectly with a nucleus—neutrons, protons, β^-, β^+, photons, and neutrons—there are also mesons. These have masses of about 200 electron masses and were first found in cosmic-ray research but relatively recently have been created in the laboratory by the high-energy machines. We shall now give a brief account of this cosmic radiation which comes from outside the earth.

38.38. Cosmic Radiation

In spite of all attempts to improve the insulation of electroscopes and to shield them from radioactive contamination in the surroundings, it has proved impossible to prevent a charged electroscope from slowly discharging. In other words, a constant slight ionization of the air around the electroscope persists. This has been known since the beginning of the century, when very serious attempts were made to shield electroscopes from radioactive substances. The origin of the residual radiation was not settled until 1912, when V. F. Hess took an ionization chamber up in a balloon and found that the radiation increased with altitude. This showed that the radiation had an extra-terrestrial origin, and it was therefore given the name of *cosmic radiation.*

Actually, the total amount of energy in the form of cosmic radiation reaching the earth is about equal to that produced by starlight. On the other hand, the energies of some individual cosmic-ray particles are enormous. Energies in the billion-electron-volt range are relatively common, while some particles in the upper atmosphere appear to have energies of the order of 10^{15} ev. At sea-level the cosmic radiation consists of a mixture of many forms—protons, neutrons, electrons (plus and minus), mesons, photons, and perhaps other nuclear species. Most of these arise from the interaction of the incoming primary radiation with nuclei of the atmosphere, and thus we find that the intensity and nature of the secondary radiation changes with altitude. The interest in cosmic rays **597** is twofold, namely, in the nature and origin of the primary radiation and in the mode of interaction of this with nuclei. It cannot be said that these problems are in any way settled, though much interesting and valuable work has been done.

38.39. Absorption of Cosmic Rays

One important piece of cosmic-ray research was the measurement of the intensity of the cosmic rays after passing through various thicknesses of absorber. In Fig. 38-24 is shown the intensity of

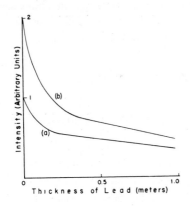

Fig. 38-24.—Change of intensity of cosmic rays with thickness of absorber (*a*) at sea-level, (*b*) at an altitude of 10,000 ft.

cosmic radiation after passing through lead at sea-level and at 10,000 ft. It is found that these curves are the same for all places on the earth between 45° and 90° geomagnetic latitude. This will be explained later.

Curve (*b*) at the high altitude shows a rapid decrease in intensity with absorber thickness at first and then a much lower decrease. This is explained by considering cosmic radiation as consisting essentially of two types of radiation: one, called the *soft component*, is readily absorbed, and the other, the *hard component*, is very penetrating. This soft component is largely absorbed by the atmosphere between 10,000 ft and sea-level. From this and other experiments it is inferred that the soft component is created from the hard component by nuclear processes in the atmosphere. Thus the soft component is composed of secondary radiation. At sea-level the hard component consists of mu-mesons (μ-mesons, described in § 38.41) and a small number of protons.

38.40. Latitude Effect on Cosmic Radiation

In the 1930's A. H. Compton, Clay, Millikan, and other workers undertook a world-wide survey of cosmic-ray intensities. Their results are partially summarized in Fig. 38-25. This shows the drop

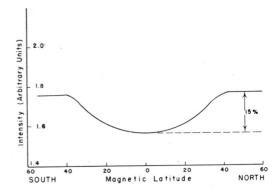

Fig. 38-25.—Cosmic-ray intensities at various magnetic latitudes at sea-level.

in cosmic-ray intensity of about 15 per cent from about 45° magnetic latitude to the magnetic equator at sea-level. These results apply to the hard component and show that the magnetic field of the earth has considerable effect on the intensity of the cosmic rays. The primary cosmic rays must be largely composed of charged particles, since these are deflected by a magnetic field. Because it is the magnetic field of the earth which is responsible for this deflection, the results are plotted with the magnetic latitude as abscissa. The magnetic poles are spots where a freely supported compass needle would point vertically downward, whereas the magnetic equator is a line along which such a needle would be horizontal. Though the intensity of the earth's magnetic field is very small at the earth's surface, it extends upward for large distances, so that at an elevation of 6,000 mi the intensity is about one-fifth that at the earth. The Mexican physicist, Vallarta, calculated the possible paths of incoming particles and showed that the experimental results were accounted for by assuming that the primary cosmic rays were largely high-energy protons.

These results have been confirmed by mounting Geiger counters above one another so as to form a cosmic-ray telescope and finding that the intensity is greater from the west than from the east. This east-west effect is again accounted for by the magnetic deflection of positively charged particles, that is, a preponderance of protons. There is evidence also from a few heavy, positively charged ions in the primary cosmic rays. These were found by sending photographic plates with sensitive nuclear emulsions to as high as 100,000 ft in free balloons.

38.41. Cosmic-Ray Particles

The complexity of the cosmic radiation in the atmosphere is very considerable, and in large measure this is associated with the huge energy of the incoming particles. These particles interact with nuclei in the atmosphere and produce almost every kind of subatomic particle. Two new kinds of particles have been discovered by cosmic-ray research—positive electrons or positrons and mesons. Both have now been created in the laboratory. In 1937 Anderson and Neddermeyer found a cloud-chamber track of a particle which, from the density of the water droplets and the curvature of the track due to a magnetic field, was interpreted to have a rest mass of about 216 electronic masses. This has been called a mu-meson (μ-meson). Some ten years later Powell in England found tracks in nuclear emulsion plates which were produced by a cosmic-ray particle having a rest mass of about 285 electronic masses. This has been called the pi-meson (π-meson).

This π-meson occurs at high altitudes, while the μ-meson is found at sea-level. It appears that the π-meson decays with a mean life of about 10^{-8} sec and produces, among other things, a μ-meson. The μ-meson is also radioactive and has a mean life of about 2×10^{-6} sec and produces a high-energy electron. Both types of meson can have positive or negative charges. They are apparently created by the primary cosmic ray in a nuclear explosion process.

The primary cosmic-ray protons can eject electrons from atoms with very high energies. These, in turn, can produce high-energy photons and negative and positive electron-pairs. In another form of cosmic-ray event, there is the interaction of single particles with nuclei. These may form a nuclear explosion, called a "star," in which protons, alpha particles, and slow mesons have been identified in cloud-chamber pictures. It is also probable that neutrons are produced in this cosmic-ray event, for they appear to increase with

altitude at about the same rate as the stars do. Then there are large bursts of ionization called *showers* (Fig. 38-26). In (*a*) a high-energy electron passes through a sheet of platinum and produces a shower, while in (*b*) a shower was started somewhere above the chamber and some of the radia-

Electron ↓

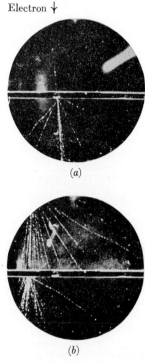

(*a*)

(*b*)

Fig. 38-26.—Cloud-chamber pictures of cosmic-ray showers.

tion was able to penetrate the platinum to produce other showers. For some of the showers photographed in cloud chambers, the total energy of the particles is of the order of a billion electron-volts. In the more energetic ones, mesons, protons, alpha particles, and other heavy nuclei are observed.

A meson of high energy (as determined by the ion density and slight curvature of the track in the magnetic field) is shown in Fig. 38-27 penetrating 9 cm of gold without much loss in energy. The Geiger counters in this experiment are so arranged that a cosmic ray must pass through the top counter, enter the first cloud chamber, and then pass through the second counter before the two cloud chambers operate. The cosmic ray then enters 9 cm of gold, and the second cloud chamber is present to observe whether the meson is absorbed and, if not, what loss of energy takes place **599** on penetrating the gold. From a very small beginning, cosmic-ray research has taken scientists all over the world, up high mountains, in airplanes, and into deep mines. The rewards in terms of understanding natural phenomena have, as is true of most basic research, been very great. There are still more exciting adventures ahead.

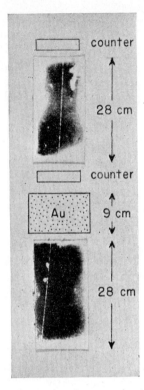

counter

28 cm

counter

Au   9 cm

28 cm

FIG. 38-27.—A meson track penetrating 9 cm of gold

38.42. Radioactivity from Fission

Prior to the discovery of fission, the amount of radioactive materials available was largely limited to the natural radioactive substances. One gram of radium, which gives 1 *curie* or 3.7×10^{10} disintegrations per second, was considered a huge amount. With the advent of reactors and the production of artificial radioactive substances and the radioactive fission products, the amount of radioactive material made today runs into many billions of curies. A very real problem with reactors is the safe disposal of the long-lived radioactive fission products. These have been placed in thick concrete containers and buried in the ground or sunk in the deep ocean. However, with the coming of a large number of reactors, as almost undoubtedly will occur for electrical power produc-

tion, the disposal of the radioactive waste will become more and more of a problem so that it has been seriously suggested that the limit to the number of reactors will be set by the disposal problem.

However serious this problem may be, it is dwarfed by that of the radioactivity produced in bombs. In this case the fission products are loosed into the world, there to be scattered about. The enormous energy released in a nuclear bomb creates a fire-ball with temperatures of the order of a few million degrees. This terrific heat creates a shock wave and causes the air to expand so that a mushroom-like column rises up to the stratosphere, there to fan out. If the bomb is detonated near the ground, pulverized material in various degrees of fineness is swept into the column and becomes radioactive mainly by becoming attached to the radioactive fission products. The material in the high cloud is then swept along by the winds and is at the same time falling down, the rate depending largely on the size of the particles. Thus the radioactive fission products come to earth in what is known as "fall-out" and create a potential danger. As the bombs have grown larger, so has the area of fall-out.

In the first nuclear bomb over Hiroshima in 1945 there were heavy casualties over about 7 mi²; in the tests of 1952 the area was said to be about 300 mi²; and in the Pacific tests of 1954 the lethal radioactive fall-out area was of the order of 7,000 mi². In this latter the diameter of the fire-ball column was about 4 mi, and within this there would be almost no chance of survival. The first nuclear bomb was stated to have an energy equivalent of 20,000 tons of TNT, while the 1954 bomb is estimated to be equivalent to about 20 million tons of TNT.

In order to give an idea of the danger of radioactive fall-out, we shall give the dosage in roentgens, r units, and their estimated over-all effects on human beings. What biological changes take place as the result of radiation is too far afield to discuss here, but suffice it to say that it runs the gamut from being fatal to various forms of radiation sickness. Table 38-3 gives the approximate result of radiation doses in r units when the radiation is delivered in less than 1 hour. If the radiation dose is delivered at a slower rate over several days, then the amount required is doubled.

It has been officially stated for the 1954 bomb that the fall-out covered an area of about 7,000 mi² and that, after 8 hr at a distance of 160 mi from the point of detonation, the radiation dosage was 500 r during the first 36 hr. If one assumes that the activity is uniform over the area, then the total amount of fall-out was about 10^{11} curies, a hundred thousand million curies.

The intensity of the radioactivity decreases with time according to an empirical relationship, $I = I_0 t^{-1.2}$, for all the fission products taken to-

TABLE 38-3

RADIATION EFFECTS

Dose (r)	Effect on Man
50–100	Small percentage have sickness
150–200	About 50 per cent have sickness
200–300	All become sick, with some fatalities
400–500	About 50 per cent fatalities
700	About 100 per cent fatalities

gether. If there is an initial intensity of 500 r/hr at 1 hour, then, the intensity after 1 day is 11 r/hr; after 1 week is 1 r/hr; after 1 month is 0.2 r/hr; and after 1 year is 0.01 r/hr. If one wishes to obtain the total radiation dose received in any time interval, then it is necessary to integrate the $t^{-1.2}$ expression over this interval. The preceding discussion applies to the gamma rays from the fission products, all the other radiations being readily absorbed. However, if there is any inhalation of the radioactive substances, then the beta-emitting substances can be very important.

If the fall-out area is well away from the primary blast and is more or less evenly spread over the ground, then adequate shelter can be provided fairly easily. A deep hole in the ground with a thin overhead cover or a basement provides quite good protection. It is a problem in absorption of gamma rays, for which one can assume that the exponential expression applies. Eighteen inches of packed soil or 11 in of concrete reduce the intensity by a factor of 50. Thus one might be able to avoid getting a large dose of radiation by an intelligent application of these principles.

At this point it is interesting to speculate on how the 1954 superbomb was made. From evidence on the fall-out the British physicist Rotblat has suggested that it is a fission-fusion-fission bomb. The bomb is surmised to be made up of a fissionable core, probably plutonium, surrounded by a fusionable shell of lithium deuteride and maybe some tritium, which is surrounded by a fissionable shell of ordinary uranium about 1 in thick.

When the innermost core undergoes fission, it produces a very high temperature and a number of neutrons. The high temperature causes a thermonuclear or fusion reaction in the lithium deuteride, producing many fast neutrons, which cause fission in the outer uranium shell. For a more detailed account of these problems it is recommended that the student read the articles in the *Bulletin of the Atomic Scientists* for 1955.

Even trusting that nuclear bombs will never be used as a means of warfare, there is still the problem of mutations produced by the increase in the background radiation as a result of the radioactivity from test bombs. This is really a long-term effect, but the geneticists of this and other countries are already concerned over the problem.

The advances in the sciences in the present century have been enormous, and the technological applications of these are everywhere around us. Science inquires impartially, and the fruits of this can be used for the betterment or the destruction of mankind. It is for each of us to say with Lincoln: "With malice toward none; with charity for all; with firmness in the right, as God gives us to see the right,—let us strive on to do all which may achieve and cherish a just and lasting peace among ourselves, and with all nations."

38.43. Résumé

The following is a brief listing of the definitions, principles, and theories, given in this chapter, which you should know:

The method for measuring the ratio of charge to mass of cathode rays in a discharge tube

The method for measuring the charge on an electron

The meaning of isotopes and the method of analysis of them

The principal properties of radioactivity

The absorption and ionization of alpha, beta, and gamma radiations

Positive and negative electrons produced by gamma rays and the reverse process

The law of radioactive decay

The meaning of half-life and its relation to the decay constant in radioactivity

The various types of radioactive transformations

The four radioactive series

Radiation-measuring instruments—the ionization chamber, the proportional counter, the Geiger counter, the Wilson cloud chamber, and the bubble chamber

The discoveries of the nuclear atom and the neutron

Transformations of nuclei by various means

The process of fission

The detection and absorption of neutrons

Energy balance in a nuclear atom and the stability of nuclei

Thermal fission reactors, the multiplication factor k of a reactor, and the importance of delayed neutrons in a reactor

Cosmic radiation; types of primary radiations, their absorption, and the production of secondary radiation

Fusion bombs and the problems of radioactive fall-out

Queries

1. Does the current in a gas (Fig. 38-13) differ essentially from that in a copper wire?

2. What evidence exists that leads us to believe that radioactivity is a nuclear phenomenon?

3. The age of rocks has been estimated by determining the amount of uranium and lead present in a rock sample. What is the justification for such an estimate?

4. In determining the mass of an atom, what distinct advantage is there in the mass spectrograph over standard chemical methods?

5. Why should the neutron be an effective projectile for penetrating the nucleus?

6. What evidence has been given that confirms the idea of energy-mass equivalence?

7. Why is it necessary for the moderating material in a reactor to be composed of atoms of small atomic weight and also for the material to have a small neutron absorption cross-section?

8. What reason would you give for not finding the neptunium radioactive series (Table 38-1) in nature, as is the case for the other series?

Practice Problems

1. An electron is accelerated through a potential difference of 30,000 volts. Find the kinetic energy of the electron in (a) ev, (b) joules. Ans.: (a) 30,000 ev; (b) 4.8×10^{-15} joule.

2. A proton whose mass is 1.67×10^{-27} kg is accelerated through a potential difference of 30,000 volts. Find the kinetic energy of the proton in (a) ev, (b) joules.

3. Find the speed of the proton in question 2. Ans.: 2.40×10^6 m/sec.

4. A gamma ray has a wave length of 0.1 A. Find the frequency of this gamma ray and the energy of the gamma-ray photon.

5. Find the momentum and mass of the gamma-ray photon whose wave length is 0.1 A. Ans.: 6.62×10^{-23} kg-m/sec; 2.21×10^{-31} kg.

6. An electron with a velocity of 2×10^5 m/sec is projected into a uniform magnetic field of 0.5 weber/m². If the electron moves at right angles to the field, find the radius of the circle which it describes.

7. A uniform beam of electrons is projected at right angles to a uniform electric field of 500 volts/m. If in a distance of 0.2 m the deflection is 0.02 m, find the velocity with which the electrons are projected into the field. Ans.: 9.38×10^6 m/sec.

8. The half-life of radium is about 1,600 years. If at some spot there is 1 gm of radium today, then find the amount of it that there would be in (a) 1,600 years, (b) 3,200 years, (c) 800 years, (d) 100 years.

9. A collimated beam of gamma rays falls on a sheet of lead whose thickness is 1.5 cm and whose absorption coefficient, μ, is 1.2 cm⁻¹. Find the ratio of final to initial intensities. Ans.: 0.165.

10. The half-value thickness of any material in respect to some specific radiation is that thickness which reduces the intensity of that radiation by a factor of 2. For the radiation of question 9 find the half-value thickness.

11. The thickness of an absorber is frequently given in terms of the areal density in grams per square centimeter. Show that the thickness of the absorber in question 9, which has a density of 11.35 gm/cm³, is 6.55 gm/cm².

12. A collimated beam of slow neutrons falls on a block of aluminum 2 cm thick, for which the macroscopic absorption cross-section Σ is 0.098 cm⁻¹. Show that the microscopic absorption cross-section σ is about 1.6 barns and find the absorption in the block. (Take the density of aluminum as 2.7 gm/cm³ and its atomic weight as 27.)

13. From equation (38-20) find the radius of a nucleus of mass number 8 and one of mass number 216. Ans.: 3×10^{-13} cm; 9×10^{-13} cm.

14. Find the number of disintegrations per second oc-

curring in 10^{-6} gm of uranium whose half-life is 4.5×10^9 years and whose atomic weight is approximately 238. *Note:* Use equation (38-17), making the time dt equal to 1 sec.

15. Show that there must be about 3.1×10^{10} fissions in U^{235} per second to produce 1 watt of power.

16. Thorium, $_{90}Th^{232}$, emits an alpha particle in a radioactive disintegration, to become mesothorium 1, which, in turn, emits a beta particle, to become mesothorium 2. Give the mass number and atomic number of mesothorium 1 and 2.

17. The first nuclear bomb had an energy equivalent of 20,000 tons of TNT, which is approximately equal to 2×10^{13} cal. Assuming that each fission act produces 200 Mev of energy, calculate how many atoms of U^{235} undergo fission and what the minimum weight of the bomb was. Ans.: 2.6×10^{24}; about 1 kg.

Periodic Table of the Elements

INERT GASES

IA													IVA	VA	VIA	VIIA	4.003 He 2

(Periodic table of the elements)

IA 1.0080 H 1																	
6.940 Li 3	IIA 9.013 Be 4	IIIA 10.82 B 5											12.010 C 6	14.008 N 7	16 O 8	19.00 F 9	20.183 Ne 10
22.997 Na 11	24.32 Mg 12	26.98 Al 13	IVB	VB	VIB	VIIB		VIII			IB	IIB	IIIB 28.09 Si 14	30.975 P 15	32.066 S 16	35.457 Cl 17	39.944 A 18
39.100 K 19	40.08 Ca 20	44.96 Sc 21	47.90 Ti 22	50.95 V 23	52.01 Cr 24	54.93 Mn 25	55.85 Fe 26	58.94 Co 27	58.69 Ni 28	63.54 Cu 29	65.38 Zn 30	69.72 Ga 31	72.60 Ge 32	74.91 As 33	78.96 Se 34	79.916 Br 35	83.80 Kr 36
85.48 Rb 37	87.63 Sr 38	88.92 Y 39	91.22 Zr 40	92.91 Nb 41	95.95 Mo 42	(99) Tc 43	101.7 Ru 44	102.91 Rh 45	106.7 Pd 46	107.880 Ag 47	112.41 Cd 48	114.76 In 49	118.70 Sn 50	121.76 Sb 51	127.61 Te 52	126.91 I 53	131.3 Xe 54
132.91 Cs 55	137.36 Ba 56	138.92 La 57	178.6 Hf 72	180.88 Ta 73	183.92 W 74	186.31 Re 75	190.2 Os 76	193.1 Ir 77	195.23 Pt 78	197.2 Au 79	200.61 Hg 80	204.39 Tl 81	207.21 Pb 82	209.00 Bi 83	210 Po 84	(210) At 85	222 Rn 86
(223) Fr 87	226.05 Ra 88	227 Ac 89															

Lanthanide Series

140.13 Ce 58	140.92 Pr 59	144.27 Nd 60	(145) Pm 61	150.43 Sm 62	152.0 Eu 63	156.9 Gd 64	159.2 Tb 65	162.46 Dy 66	164.94 Ho 67	167.2 Er 68	169.4 Tm 69	173.04 Yb 70	174.99 Lu 71

* Actinide Series

232.12 Th 90	231 Pa 91	238.07 U 92	(237) Np 93	(242) Pu 94	(243) Am 95	(243) Cm 96	(245) Bk 97	(246) Cf 98	99	100	101	102	103

* The elements of atomic numbers 99 and 100 have been named, respectively, Einsteinium, E, and Fermium, F (*Phys. Rev.*, 99, 1048 [1955]).

Some Important Physical Constants*

c, velocity of light in vacuum . . . . $299{,}793.0 \pm 0.3$ km/sec

G, gravitational constant . . . . $(6.670 \pm 0.005) \times 10^{-11}$ newton-m^2/kg^2

V_0, volume of an ideal gas ($0°$ C, P_0) (physical scale)† . . . . 22420.7 ± 0.6 cm^3/atm-mole

P_0, standard atmosphere . . . . $(1.013246 \pm 0.000004) \times 10^5$ newtons/m^2

R, gas constant (physical) . . . . 8.31696 ± 0.00034 joules/deg-mole

T_0, ice point (abs. scale) . . . . $273°.15 \pm 0°.01$ K

J_{15}, Joule's constant . . . . 4.1855 ± 0.0004 joules/cal

N_0, Avogadro's number (physical) . . . . $(6.02486 \pm 0.00016) \times 10^{23}$ gm-mole

k, Boltzmann's constant . . . . $(1.38044 \pm 0.00007) \times 10^{-23}$ joule/deg

h, Planck's constant . . . . $(6.62517 \pm 0.00023) \times 10^{-34}$ joule-sec

σ, Stefan-Boltzmann's constant . . . . $(5.6687 \pm 0.0010) \times 10^{-8}$ joule/m^2-sec-deg^4

F, faraday (physical) . . . . $96{,}521.9 \pm 1.1$ coulombs/gm-mole

e, electronic charge . . . . $(1.60206 \pm 0.00003) \times 10^{-19}$ coulomb

e/m_0, ratio of electronic charge to rest mass . . . . $(1.75890 \pm 0.00002) \times 10^{11}$ coulombs/kg

1 ev, 1 electron-volt . . . . $(1.60206 \pm 0.00003) \times 10^{-19}$ joule

n_0, Loschmidt's number (physical) . . . . $(2.68719 \pm 0.00010) \times 10^{19}$ cm^{-3}

H^+/m_0, ratio of proton mass to electron mass . . . . 1836.12 ± 0.02

m_0, electronic mass . . . . $(9.1083 \pm 0.0003) \times 10^{-31}$ kg

H^+, atomic-weight proton (phys. scale) (amu) . . . . 1.007593 ± 0.000003

H, atomic-weight hydrogen (phys. scale) (amu) . . . . 1.008142 ± 0.000003

n, neutron mass (phys. scale) (amu) . . . . 1.008982 ± 0.000003

amu, atomic mass unit (phys. scale) . . . . 931.141 ± 0.010 Mev

e, electron mass . . . . 0.510976 ± 0.000007 Mev

p, proton mass . . . . 938.211 ± 0.010 Mev

* Selected in part from Cohen, DuMond, Layton, and Rollett, *Rev. Mod. Phys.*, **27**, No. 4 (1955), 339. For a complete tabulation of conversion factors and physical constants the student is referred to the following books: *Handbook of Chemistry and Physics* (Chemical Rubber Pub. Co.); *International Critical Tables*; *Physical and Chemical Constants* (Kaye & Laby); and *Smithsonian Physical Tables* (Fowle).

† The physical scale of mass is smaller than that of the chemical mass used in atomic weights by about 0.0272 per cent.

n, neutron mass. . . . . .939.505 $\pm$ 0.010 Mev

ρ_{Hg}, density of mercury (0° C, P_0). . . . . (13.59504 $\pm$ 0.00005) $\times$ 10³ kg/m³

ρ_e, average density of the earth. . . . . (5.517 $\pm$ 0.004) $\times$ 10³ kg/m³

g_0, acceleration due to gravity (standard) . . . . .9.80665 m/sec²

g_{45}, acceleration due to gravity (45° lat.). . . . . .9.80616 m/sec²

Common Constants and Conversion Factors

$\pi = 3.1416\ldots$; $\pi^2 = 9.8696\ldots$; $1/\pi = 0.3183\ldots$
(Base of natural logarithms) $e = 2.7183\ldots$
$\log_e 10 = 2.3026\ldots$; $\log_e m = \ln m = 2.3026 \log_{10} m$
1 inch = 2.5400 cm;   1 meter = 39.37 in
1 angstrom unit = 1 A = 10^{-8} cm
1 micron unit = 1 μ = 10^{-6} m = 10^{-4} cm
1 mile (mi) = 1.609 . . . kilometers (km)
1 km = 0.6214 . . . mile
2π radians = 360°;   1 radian = 57°296 . . .
1 liter = 1,000.028 cm³
1 gallon = 3.785 . . . liters = 231 . . . in³

1 pound (lb) = 453.59 . . . gm
1 kilogram (kg) = 2.2046 . . . lb
1 slug = 32.2 . . . lb = 14.6 . . . kg
1 gm/cm³ = 62.43 . . . lb/ft³ = 8.345 . . . lb/gal
1 mi/min = 60 mph = 88 ft/sec
1 knot = 1 nautical mi/hr = 1.151 . . . mph
1 newton = 10^5 dynes = 0.224 . . . lbf
1 lbf = (4.45 . . .)10^5 dynes
1 ft-lbf = 1.3549 . . . joules
1 BTU = 252 cal = 778 ft-lbf (approx.)
1 horsepower (hp) = 550 ft-lbf/sec = 746 watts

Some Useful Formulas

$$e^x = 1 + x + \frac{x^2}{2} + \frac{x^3}{2 \times 3} + \ldots .$$

$$(1 \pm x)^n = 1 \pm nx + \frac{n(n-1)}{2} x^2 \pm \frac{n(n-1)(n-2)}{2 \times 3} x^3 + \ldots + x^n \qquad \text{if } x^2 < 1.$$

If $x \ll 1$: $\dfrac{1}{1+x} = 1 - x \ldots$; $\sqrt{1+x} = 1 + \dfrac{x}{2}\ldots$; $(1+x)^{1/3} = 1 + \dfrac{x}{3}\ldots$ etc.

Solution of $ax^2 + bx + c = 0$ is $x = \dfrac{-b \pm \sqrt{b^2 - 4ac}}{2a}$

If $n = b^x$,   then $x = \log_b n$.   Thus $100 = 10^2$ and $\log_{10} 100 = 2$.

CALCULUS

a) DIFFERENTIAL

$y = x^n$; $\dfrac{dy}{dx} = nx^{n-1}$;

$v = \sin \theta$; $\dfrac{dy}{d\theta} = \cos \theta$;

$y = \cos \theta$; $\dfrac{dy}{d\theta} = -\sin \theta$;

$v = \tan \theta$; $\dfrac{dy}{dx} = -\sec^2 \theta$;

$y = e^x$; $\dfrac{dy}{dx} = e^x$;

$y = e^{ax}$; $\dfrac{dy}{dx} = a e^{ax}$;

$v = uv$; $\dfrac{dy}{dx} = u\dfrac{dv}{dx} + v\dfrac{du}{dx}.$

b) INTEGRAL

$\int x^n dx = \dfrac{x^{n+1}}{n+1} + c$;

$\int \sin \theta d\theta = -\cos \theta + c$;

$\int \cos \theta d\theta = \sin \theta + c$;

$\int e^x dx = e^x + c$;

$\int e^{ax} dx = \dfrac{e^{ax}}{a} + c$;

$\int \dfrac{dx}{x} = \log_e x + c$;

$\int u\,dv = uv - \int v\,du + c,$

where c is the constant of integration

Angles

The number of radians in any given angle is

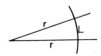

found by drawing a circle of arbitrary radius r and dividing the arc, L, intercepted by the unknown angle by the length of the radius:

$$\text{Radians} = \frac{L}{r}.$$

Similarly, a solid angle measured in STERADS can be defined in terms of the area intercepted by

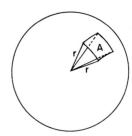

the unknown angle upon the surface of a sphere of radius r:

$$\text{Sterads} = \frac{A}{r^2}.$$

Since the total area of a sphere is $4\pi r^2$, there are 4π sterads in a complete sphere. This is analogous to the relation which stated that there are 2π rad in a complete circle.

Trigonometric Functions

In a right-angled triangle we define the sine, cosine, and tangent, of the angle θ by the following ratios:

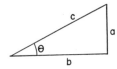

$$\sin \theta = \frac{a}{c} = \frac{\text{Opposite side}}{\text{Hypotenuse}}, \quad \sin \theta = \theta - \frac{\theta^3}{6} + \cdots ;$$

$$\cos \theta = \frac{b}{c} = \frac{\text{Adjacent side}}{\text{Hypotenuse}}, \quad \cos \theta = 1 - \frac{\theta^2}{2} + \cdots ;$$

$$\tan \theta = \frac{a}{b} = \frac{\text{Opposite side}}{\text{Adjacent side}}, \quad \tan \theta = \theta + \frac{\theta^3}{3} + \cdots .$$

$$\operatorname{cosec} \theta = 1/\sin \theta ; \ \sec \theta = 1/\cos \theta ; \ \cotan \theta = 1/\tan \theta$$

$$\sin^2 A + \cos^2 A = 1 ; \tag{1}$$

$$\sin (A \pm B) = \sin A \cos B \pm \cos A \sin B ; \tag{2}$$

$$\cos (A \pm B) = \cos A \cos B \mp \sin A \sin B . \tag{3}$$

If we consider that $A = B$, then equations (2) and (3), above, give us:

$$\sin 2A = 2 \sin A \cos A ; \tag{4}$$

$$\cos 2A = \cos^2 A - \sin^2 A . \tag{5}$$

From equations (2) and (3), above, it follows that:

$$\sin A + \sin B = 2 \sin \tfrac{1}{2}(A+B)\cos \tfrac{1}{2}(A-B); \tag{6}$$

$$\sin A - \sin B = 2 \cos \tfrac{1}{2}(A+B)\sin \tfrac{1}{2}(A-B); \tag{7}$$

$$\cos A + \cos B = 2 \cos \tfrac{1}{2}(A+B)\cos \tfrac{1}{2}(A-B); \tag{8}$$

$$\cos A - \cos B = -2 \sin \tfrac{1}{2}(A+B)\sin \tfrac{1}{2}(A-B). \tag{9}$$

RELATIONS OF THE FUNCTIONS

$$\sin A = \cos(90 - A) = \sin(180 - A); \tag{1}$$

$$\cos A = \sin(90 - A) = -\cos(180 - A); \tag{2}$$

$$\tan A = \cot(90 - A) = -\tan(180 - A); \tag{3}$$

$$\cot A = \tan(90 - A) = -\cot(180 - A). \tag{4}$$

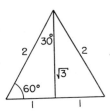

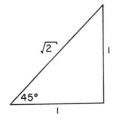

From the lengths of the sides of the triangles it follows that:

$$\sin 60° = \frac{\sqrt{3}}{2} = \frac{1.732}{2} \cdots = 0.866 \ldots = \cos 30° ;$$

$$\cos 60° = \tfrac{1}{2} = \sin 30° ;$$

$$\sin 45° = \frac{1}{\sqrt{2}} = 0.707 \ldots = \cos 45° ;$$

$$\tan 45° = 1 ; \qquad \tan 60° = \sqrt{3} = 1.732 \ldots .$$

The signs of the trigonometric functions in the four quadrants given in the figure are all positive.

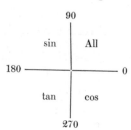

Thus

$$\sin 120° = \frac{\sqrt{3}}{2},$$

$$\cos 120° = -\tfrac{1}{2},$$

$$\tan 120° = -\sqrt{3}.$$

NATURAL TRIGONOMETRIC FUNCTIONS

Degrees	Radians	Sine	Cosine	Tangent	Degrees	Radians	Sine	Cosine	Tangent	Degrees	Radians	Sine	Cosine	Tangent
0....	0.000	0.000	1.000	0.000	31....	0.541	0.515	0.857	0.601	61...	1.065	0.875	0.485	1.804
1....	0.017	0.017	1.000	0.018	32....	0.556	0.530	0.848	0.625	62...	1.082	0.883	0.470	1.881
2....	0.035	0.035	0.999	0.035	33....	0.576	0.545	0.839	0.649	63...	1.100	0.891	0.454	1.963
3....	0.052	0.052	0.999	0.052	34....	0.593	0.559	0.829	0.675	64...	1.117	0.899	0.438	2.050
4....	0.070	0.070	0.998	0.070	35....	0.611	0.574	0.819	0.700	65...	1.135	0.906	0.423	2.145
5....	0.087	0.087	0.996	0.088										
6....	0.105	0.105	0.995	0.105	36....	0.628	0.588	0.809	0.727	66...	1.152	0.914	0.407	2.246
7....	0.122	0.122	0.993	0.123	37....	0.646	0.602	0.799	0.754	67...	1.169	0.921	0.391	2.356
8....	0.140	0.139	0.990	0.141	38....	0.663	0.616	0.788	0.781	68...	1.187	0.927	0.375	2.475
9....	0.157	0.156	0.988	0.158	39....	0.681	0.629	0.777	0.810	69...	1.204	0.934	0.358	2.605
10....	0.175	0.174	0.985	0.176	40....	0.698	0.643	0.766	0.839	70...	1.222	0.940	0.342	2.747
11....	0.192	0.191	0.982	0.194	41....	0.716	0.656	0.755	0.869	71...	1.239	0.946	0.326	2.904
12....	0.209	0.208	0.978	0.213	42....	0.733	0.669	0.743	0.900	72...	1.257	0.951	0.309	3.078
13....	0.227	0.225	0.974	0.231	43....	0.751	0.682	0.731	0.933	73...	1.274	0.956	0.292	3.271
14....	0.244	0.242	0.970	0.249	44....	0.768	0.695	0.719	0.966	74...	1.292	0.961	0.276	3.487
15....	0.262	0.259	0.966	0.268	45....	0.785	0.707	0.707	1.000	75...	1.309	0.966	0.259	3.732
16....	0.279	0.276	0.961	0.287	46....	0.803	0.719	0.695	1.036	76...	1.327	0.970	0.242	4.011
17....	0.297	0.292	0.956	0.306	47....	0.820	0.731	0.682	1.072	77...	1.344	0.974	0.225	4.331
18....	0.314	0.309	0.951	0.325	48....	0.838	0.743	0.669	1.111	78...	1.361	0.978	0.208	4.705
19....	0.332	0.326	0.946	0.344	49....	0.855	0.755	0.656	1.150	79...	1.379	0.982	0.191	5.145
20....	0.349	0.342	0.940	0.364	50....	0.873	0.766	0.643	1.192	80...	1.396	0.985	0.174	5.671
21....	0.367	0.358	0.934	0.384	51....	0.890	0.777	0.629	1.235	81...	1.414	0.988	0.156	6.314
22....	0.384	0.375	0.927	0.404	52....	0.908	0.788	0.616	1.280	82...	1.431	0.990	0.139	7.115
23....	0.401	0.391	0.921	0.425	53....	0.925	0.799	0.602	1.327	83...	1.449	0.993	0.122	8.144
24....	0.419	0.407	0.914	0.445	54....	0.943	0.809	0.588	1.376	84...	1.466	0.995	0.105	9.514
25....	0.436	0.423	0.906	0.466	55....	0.960	0.819	0.574	1.428	85...	1.484	0.996	0.087	11.43
26....	0.454	0.438	0.899	0.488	56....	0.977	0.829	0.559	1.483	86...	1.501	0.998	0.070	14.30
27....	0.471	0.454	0.891	0.510	57....	0.995	0.839	0.545	1.540	87...	1.518	0.999	0.052	19.08
28....	0.489	0.470	0.883	0.532	58....	1.012	0.848	0.530	1.600	88...	1.536	0.999	0.035	28.64
29....	0.506	0.485	0.875	0.554	59....	1.030	0.857	0.515	1.664	89...	1.553	1.000	0.018	57.29
30....	0.524	0.500	0.866	0.577	60....	1.047	0.866	0.500	1.732	90...	1.571	1.000	0.000	∞

GREEK ALPHABET

A, α	...........alpha	I, ι	...........iota	P, ρ	...........rho
B, β	...........beta	K, κ	...........kappa	Σ, σ	...........sigma
Γ, γ	...........gamma	Λ, λ	...........lambda	T, τ	...........tau
Δ, δ	...........delta	M, μ	...........mu	Υ, υ	...........upsilon
E, ε	...........epsilon	N, ν	...........nu	Φ, φ	...........phi
Z, ζ	...........zeta	Ξ, ξ	xi	X, χ	...........chi
H, η	...........eta	O, o	...........omicron	Ψ, ψ	...........psi
Θ, θ	...........theta	Π, π	...........pi	Ω, ω	...........omega

No.	0	1	2	3	4	5	6	7	8	9
10..........	00000	00432	00860	01284	01703	02119	02531	02938	03342	03743
11..........	04139	04532	04922	05308	05690	06070	06446	06819	07188	07555
12..........	07918	08279	08636	08991	09342	09691	10037	10380	10721	11059
13..........	11394	11727	12057	12385	12710	13033	13354	13672	13988	14301
14..........	14613	14922	15229	15533	15836	16137	16435	16732	17026	17319
15..........	17609	17898	18184	18469	18752	19033	19312	19590	19866	20140
16..........	20412	20683	20952	21219	21484	21748	22011	22272	22531	22789
17..........	23045	23300	23553	23805	24055	24304	24551	24797	25042	25285
18..........	25527	25768	26007	26245	26482	26717	26951	27184	27416	27646
19..........	27875	28103	28330	28556	28780	29003	29226	29447	29667	29885
20..........	30103	30320	30535	30730	30963	31175	31387	31597	31806	32015
21..........	32222	32428	32634	32838	33041	33244	33445	33646	33846	34044
22..........	34242	34439	34635	34830	35025	35218	35411	35603	35793	35984
23..........	36173	36361	36549	36736	36922	37107	37291	37475	37658	37840
24..........	38021	38202	38382	38561	38739	38916	39094	39270	39445	39620
25..........	39794	39967	40140	40312	40483	40654	40824	40993	41162	41330
26..........	41497	41664	41830	41996	42160	42325	42488	42651	42813	42975
27..........	43136	43297	43457	43616	43775	43933	44091	44248	44404	44560
28..........	44716	44871	45025	45179	45332	45484	45637	45788	45939	46090
29..........	46240	46389	46538	46687	46835	46982	47129	47276	47422	47567
30.......	47712	47857	48001	48144	48287	48430	48572	48714	48855	48996
31..........	49136	49276	49415	49554	49693	49831	49969	50106	50243	50379
32..........	50515	50651	50786	50920	51055	51188	51322	51455	51587	51720
33..........	51851	51983	52114	52244	52375	52504	52634	52763	52892	53020
34..........	53148	53275	53403	53529	53656	53782	53908	54033	54158	54283
35..........	54407	54531	54654	54777	54900	55023	55145	55267	55388	55509
36..........	55630	55751	55871	55991	56110	56229	56348	56467	56585	56703
37..........	56820	56937	57054	57171	57287	57403	57519	57634	57749	57864
38..........	57978	58092	58206	58320	58433	58546	58659	58771	58883	58995
39..........	59106	59218	59329	59439	59550	59660	59770	59879	59989	60097
40..........	60206	60314	60423	60531	60638	60746	60853	60959	61066	61172
41..........	61278	61384	61490	61595	61700	61805	61909	62014	62118	62221
42..........	62325	62428	62531	62634	62737	62839	62941	63043	63144	63246
43..........	63347	63448	63548	63649	63749	63849	63949	64048	64147	64246
44..........	64345	64444	64542	64640	64738	64836	64933	65031	65128	65225
45..........	65321	65418	65514	65610	65706	65801	65896	65992	66087	66181
46..........	66276	66370	66464	66558	66652	66745	66839	66932	67025	67117
47..........	67210	67302	67394	67486	67578	67669	67761	67852	67943	68034
48..........	68124	68215	68305	68395	68485	68574	68664	68753	68842	68931
49..........	69020	69108	69197	69285	69373	69461	69548	69636	69723	69810
50..........	69897	69984	70070	70157	70243	70329	70415	70501	70586	70672
51..........	70757	70842	70927	71012	71096	71181	71265	71349	71433	71517
52..........	71600	71684	71767	71850	71933	72016	72099	72181	72263	72346
53..........	72428	72509	72591	72673	72754	72835	72916	72997	73078	73159
54..........	73239	73320	73400	73480	73560	73640	73719	73799	73878	73957
55..........	74036	74115	74194	74273	74351	74429	74507	74586	74663	74741
56..........	74819	74896	74974	75051	75128	75205	75282	75358	75435	75511
57..........	75587	75664	75740	75815	75891	75967	76042	76118	76193	76268
58..........	76343	76418	76492	76567	76641	76716	76790	76864	76938	77012
59..........	77085	77159	77232	77305	77379	77452	77525	77597	77670	77743
60..........	77815	77887	77960	78032	78104	78176	78247	78319	78390	78462
61..........	78533	78604	78675	78746	78817	78888	78958	79029	79099	79169
62..........	79239	79309	79379	79449	79518	79588	79657	79727	79796	79865
63..........	79934	80003	80072	80140	80209	80277	80346	80414	80482	80550
64..........	80618	80686	80754	80821	80889	80956	81023	81090	81158	81224
65..........	81291	81358	81425	81491	81558	81624	81690	81757	81823	81889
66..........	81954	82020	82086	82151	82217	82282	82347	82413	82478	82543
67..........	82607	82672	82737	82802	82866	82930	82995	83059	83123	83187
68..........	83251	83315	83378	83442	83506	83569	83632	83696	83759	83822
69..........	83885	83948	84011	84073	84136	84198	84261	84323	84386	84448
70..........	84510	84572	84634	84696	84757	84819	84880	84942	85003	85065

No.	0	1	2	3	4	5	6	7	8	9
71	85126	85187	85248	85309	85370	85431	85491	85552	85612	85673
72	85733	85794	85854	85914	85974	86034	86094	86153	86213	86273
73	86332	86392	86451	86510	86570	86629	86688	86747	86806	86864
74	86923	86982	87040	87099	87157	87216	87274	87332	87390	87448
75	87506	87564	87622	87679	87737	87795	87852	87910	87967	88024
76	88081	88138	88195	88252	88309	88366	88423	88480	88536	88593
77	88649	88705	88762	88818	88874	88930	88986	89042	89098	89154
78	89209	89265	89321	89376	89432	89487	89542	89597	89653	89708
79	89763	89818	89873	89927	89982	90037	90091	90146	90200	90255
80	90309	90363	90417	90472	90526	90580	90634	90687	90741	90795
81	90849	90902	90956	91009	91062	91116	91169	91222	91275	91328
82	91381	91434	91487	91540	91593	91645	91698	91751	91803	91855
83	91908	91960	92012	92065	92117	92169	92221	92273	92324	92376
84	92428	92480	92531	92583	92634	92686	92737	92789	92830	92891
85	92942	92992	93044	93095	93146	93197	93247	93298	93349	93399
86	93450	93500	93551	93601	93651	93702	93752	93802	93852	93902
87	93952	94002	94052	94101	94151	94201	94250	94300	94349	94399
88	94448	94498	94547	94596	94645	94694	94743	94792	94841	94890
89	94939	94988	95036	95085	95134	95182	95231	95279	95328	95376
90	95424	95472	95521	95569	95617	95665	95713	95761	95809	95856
91	95904	95952	95999	96047	96095	96142	96190	96237	96284	96332
92	96379	96426	96473	96520	96567	96614	96661	96708	96755	96802
93	96848	96895	96942	96988	97035	97081	97128	97174	97220	97267
94	97313	97359	97405	97451	97497	97543	97589	97635	97681	97727
95	97772	97818	97864	97909	97955	98000	98046	98091	98137	98182
96	98227	98272	98318	98363	98408	98453	98498	98543	98588	98632
97	98677	98722	98767	98811	98856	98900	98945	98989	99034	99078
98	99123	99167	99211	99255	99300	99344	99388	99432	99476	99520
99	99564	99607	99651	99695	99739	99782	99826	99870	99913	99957

Index

611